£uK

PERGAMON INTE
of Science, Technology

The 1000-volume original p
industrial training ⸹

Publisher: Robert Maxwell, M.C.

Fast Breeder Reactors

Fast Breeder Reactors

ALAN E. WALTAR

Fast Reactor Safety Development
Westinghouse Hanford Company
Richland, Washington

ALBERT B. REYNOLDS

Professor of Nuclear Engineering
University of Virginia
Charlottesville, Virginia

PERGAMON PRESS

New York Oxford Toronto Sydney Paris Frankfurt

Pergamon Press Offices:

U.S.A.	Pergamon Press Inc., Maxwell House, Fairview Park, Elmsford, New York 10523, U.S.A.
U.K.	Pergamon Press Ltd., Headington Hill Hall, Oxford OX3 0BW, England
CANADA	Pergamon Press Canada Ltd., Suite 104, 150 Consumers Road, Willowdale, Ontario M2J 1P9, Canada
AUSTRALIA	Pergamon Press (Aust.) Pty. Ltd., P.O. Box 544, Potts Point, NSW 2011, Australia
FRANCE	Pergamon Press SARL, 24 rue des Ecoles, 75240 Paris, Cedex 05, France
FEDERAL REPUBLIC OF GERMANY	Pergamon Press GmbH, Hammerweg 6 6242 Kronberg/Taunus, Federal Republic of Germany

Library of Congress Cataloging in Publication Data

Waltar, A   E   1939-
   Fast breeder reactors.

   Includes index.
   1. Breeder reactors. 2. Fast reactors.
I. Reynolds, Albert Barnett, 1931-   joint
author. II. Title.
TK9203.B7W34 1980   621.48'34   80-25390
ISBN 0-08-025983-9
ISBN 0-08-025982-0 (pbk.)

Printed in the United States of America

We dedicate this book to two special groups of people:

First, to the scientists and engineers throughout the world who have contributed to the development of the Fast Breeder Reactor in order to assure the role of nuclear energy as a safe, economic, and inexhaustible source of energy for generations to come and,

Second, to our wives Anna and Helen who have continually provided us with the inspiration and support necessary for the writing of this book.

FOREWORD

It was as early as the 1940s that the special nature and the special role of Fast Breeder Reactors was recognized. While it is widely understood that nuclear energy is a qualitatively new dimension when compared with conventional energy sources, it is only through breeding that such a new dimension can become really operational. An easy way to point to a new dimension of nuclear energy is to consider the ratio of the energy content of 1 g of fissionable material to 1 g of carbon. It is close to 3×10^6, indeed a staggering number. But in normal reactors which essentially burn the fissile material, only about half a percent of the mined natural uranium can be used in this manner. And if eventually low grade ores are to be used when nuclear power is engaged at a globally significant scale, resources come into sight that might contain, for example, only 70 ppm of uranium. The effect is that this dilutes the factor 3×10^6 back to one:

$$3 \times 10^6 \times 5 \times 10^{-3} \times 7 \times 10^{-5} \cong 1.$$

Nuclear energy thereby becomes degenerated to something that our century knows—fossil fuels. One might allude to it by calling such uses of uranium "yellow coal." It is only through breeding that this can be avoided. Not half a percent but something like 60% of the mined natural uranium thereby becomes accessible and this in turn also prevents falling back to low grade uranium ores, at least for the next thousand years or so. Or in other words, when engaged on a truly significant scale, nuclear energy becomes nuclear energy only through the breeder.

Appreciating such relations requires a long range and strategic view. One must be open minded to the future. Developing the breeder is a long term task; it turns out to be outside of the field of normal market forces. The importance of the ability to supply energy virtually on an infinite basis is only visible when viewed from a global, long-term perspective. The world-wide international Fast Breeder community has always had that in mind and this was and is what makes it an enthusiastic community. But not all quarters in the nations of the world have this perspective. And this has sometimes made the Fast Breeder development so advanced that it became isolated. Indeed, it goes without question that the development of the Fast Breeder can only be pursued when put on the sound basis and experience of the first generation of nuclear power plants, that of burner reactors and most notably here, the Light Water Reactor. For this first generation of nuclear power plants, the issue was not so much the long-range strategic view, but the near term necessity to demonstrate a cheap, competitive, and reliable production of electricity. This turned out to be a more complex and time-consuming task than originally anticipated. So, sometimes not much capacity and attention has been left for the development of the Fast Breeder.

A second item becomes apparent when Fast Breeders are being developed: It then becomes necessary to close the nuclear fuel cycle, while in the case of the first generation of nuclear power plants this can in principle be delayed for a decade or two. In this respect as well, nuclear energy becomes nuclear energy through the breeder. The meticulousness which is required when a nuclear fuel cycle is to be operated is the price for the infinite supply of energy. The most general observation also applies in the case of the breeder: there is no such thing as a free lunch.

The next decade will force all of us to take a global and long range view of the energy problem. Considering that, today, a major share of the world's energy supply already comes from one single place on the globe, the Persian Gulf, then the global view is visibly forced upon all of us. There is no doubt that such a revelation of the dimensions of the energy problem will bring the breeder to its natural place. This might not happen at all places at the same time. The situation for Western Europe or Japan is much more precarious than that of the United States or the Soviet Union. But after the year 2000, breeding at a globally significant scale will be a worldwide necessity. And this requires demonstration and testing. Doing this during the 1980s is rather too late than too soon.

It is therefore very important that Alan E. Waltar and Albert B. Reynolds have accepted the task to write this book. It fills a long-standing gap as it collects and reviews otherwise wide-spread information about the Fast Breeder, thereby helping to identify and expand this body of knowledge. This will undoubtedly be of great help for teaching at universities. What must also be observed is the worldwide scope of the book. This is truly consistent with the wide-open international character of the Fast Breeder development work. So the book will also be used worldwide.

May it serve its purpose and may it fulfill its badly needed function.

Wolf Häfele

Deputy Director*
International Institute for Applied
Systems Analysis
Schloss Laxenburg, Austria

Vienna
September 1980

*Current Position:
Director
Kernforschungsanlage Jülich
Jülich, Federal Republic of Germany

PREFACE

The emerging importance of fast breeder reactors as an inexhaustible energy source for the generation of electricity has been elucidated by Professor Häfele in the Foreword. This book describes the major design features of fast breeder reactors and the methods used for their design and analysis.

The foremost objective of this book is to fulfill the need for a textbook on Fast Breeder Reactor (FBR) technology at the graduate level or the advanced undergraduate level. It is assumed that the reader has an introductory understanding of reactor theory, heat transfer, and fluid mechanics. The book is expected to be used most widely for a one-semester general course on fast breeder reactors, with the extent of material covered to vary according to the interest of the instructor. The book could also be used effectively for a two-quarter or a two-semester course. In addition, the book could serve as a text for a course on fast reactor safety since many topics other than those appearing in the safety chapters relate to FBR safety.

A second objective of this book is to serve as a background reference for engineers actively working in the FBR field. Research and development reports generally assume a basic knowledge in the area being reported, yet the paucity of introductory texts and reference books on FBR's often makes the acquisition of this knowledge difficult for the engineer entering the field. Elementary introductory analyses of many of the topics covered in this book have not been available in the open literature. Hence, the educationally oriented basic approach offered in this text should be useful to engineers in the FBR field.

Methodology in fast reactor design and analysis, together with physical descriptions of systems, is emphasized in this text more than numerical results. Analytical and design results continue to change with the ongoing evolution of FBR design whereas many design methods have remained fundamentally unchanged for a considerable time.

Appendix A is an exception to the above paragraph in that current reference material is provided. A considerable effort was expended herein to acquire and organize updated design data for the fast reactors built and designed at the time of writing.

The field of fast reactor technology is broad; hence, careful selection of topics was necessary to limit the scope of the text to those topics most appropriate for a graduate or advanced undergraduate course while, at the same time, including the concepts most important to FBR design. The analytical topics selected are ones which have received extensive treatment in fast reactor development and for which the methods are reasonably well established. A balance between the number of topics covered and the depth of treatment offered was carefully considered, where (except in the chapters on safety) a tendency toward in-depth treatment of fewer topics was followed rather than a curtailed treatment of more topics. It was necessary to limit severely the descriptions of components and heat transport and auxiliary systems, despite the great importance of these topics to FBR development. In the area of fast reactor safety, the phenomena requiring extensive analysis are so numerous that an adequate treatment of only a small fraction of them is possible in a general text. Therefore, in this area, emphasis was shifted from an in-depth treatment to a physical description of the important phenomena and design concepts being emphasized in FBR safety development.

Due to the principal role of this book as a textbook, no attempt was made to review all the literature on the subjects covered; thus the number of references is small relative to the published literature on each topic. More-over, the authors' primary experience with methods (especially computer programs) used in the United States has necessarily influenced the presenta-tion. It should be fully recognized, however, that FBR development is moving rapidly throughout the world and that the international FBR community is strongly influenced by developments in all countries.

The book is divided into five parts: I. Overview (Chapters 1–3), II. Neutronics (Chapters 4–7), III. Systems (Chapters 8–12), IV. Safety (Chapters 13–16), and V. Gas Cooled Fast Reactors (Chapters 17 and 18).

The first two chapters provide introductory material on breeding, nuclear fuel resources, breeder reactor programs, and a brief description of fast reactor design. A chapter on economic analysis of nuclear reactors has been included because of the importance of economics with regard to commer-cialization of the fast breeder reactor. This chapter is written by P. S. Owen and R. P. Omberg, of Westinghouse Hanford Company, who have been leaders in comparative economic analyses of reactor systems.

Chapters 4 through 7 treat the various topics of neutronics, including nuclear design methods, flux spectra and power distributions, cross sections, kinetics and reactivity effects, fuel cycle, breeding ratio and doubling time.

Chapters 8 through 12 cover topics in thermal-hydraulics and mechanical design, such as fuel pin stress analysis and thermal performance, coolant temperature analysis, hot-channel factors, heat-transport systems and components, and fast reactor materials.

A perspective on fast reactor safety is provided by the discussion of safety methodology and protected transients in Chapters 14 and 15. The next two chapters deal with phenomena in unprotected transients (i.e., transients for which the engineered shutdown control systems are assumed to fail) that have received extensive treatment in fast reactor safety research.

Due to the leading role of the Liquid Metal Fast Breeder Reactor (LMFBR) in breeder reactor development throughout the world, the LMFBR and sodium technology have been emphasized throughout most of the book. The Gas Cooled Fast Breeder Reactor (GCFR) is a potential alternative, however, and this concept is described in the last section of the book. GCFR design is described in Chapter 17. The final chapter treats GCFR safety and is written by A. F. Torri and B. Boyack of the General Atomic Company, the principal design organization for the GCFR.

Problems or review questions are provided at the end of each chapter. Some of the problems in the neutronics section require computer solutions (i.e., Problem 4-5 and later ones based on 4-5, and Problem 7-5), but they are limited to zero-dimensional geometry so that computer programming can be accomplished rapidly. Use of these computer problems in classes taught by both authors for several years indicates that they provide an effective way to gain an understanding of the neutron physics of fast reactors. Review questions are provided for the more descriptive, less analytical chapters.

Alan E. Waltar
Albert B. Reynolds

August 1981
Richland, Washington
Charlottesville, Virginia

ACKNOWLEDGMENTS

Numerous people have contributed heavily to the creation of this text. Most notable is H. Alter of the U.S. Department of Energy, who envisioned the need for such a book and provided constant support to bring it into being. D. E. Simpson and R. E. Peterson, of the Westinghouse Hanford Company, graciously provided the time required for the authors to write the book as well as a careful review of the entire manuscript. R. F. Pigeon of the U.S. Department of Energy and S. W. Berglin of the Westinghouse Hanford Company contributed particular expertise in making the necessary review and publishing arrangements.

Important technical contributions, coming at key junctures during the development of the text, were provided by W. W. Little, P. S. Owen, R. P. Omberg, D. R. Haffner, R. W. Hardie, C. W. Hunter, R. B. Baker, D. S. Dutt, R. J. Jackson, R. E. Masterson, E. H. Randklev, R. D. Leggett, J. E. Hanson, W. L. Bunch, A. Padilla, Jr., F. J. Martin, G. R. Armstrong, and F. S. Metzger of the Westinghouse Hanford Company; C. L. Cowan and P. Greebler of the General Electric Company; L. E. Strawbridge of the Westinghouse Advanced Reactors Division; R. Crosgrove of Rockwell International (Atomics International Division); D. R. Ferguson of the Argonne National Laboratory; W. E. Kastenberg of the University of California at Los Angeles; and B. Boyack and A. F. Torri of the General Atomic Company.

The final text has profited immeasurably by the skillful reviews and comments provided by a team comprising a wide variety of technical disciplines. A partial list of these individuals, to which the authors are particularly indebted, include J. Muraoka, H. G. Johnson, E. T. Weber, III, J. L. Straalsund, R. W. Powell, W. F. Sheely, K. R. Birney, J. M. Atwood, N. P. Wilburn, L. D. Muhlestein, R. E. Schenter, G. L. Fox, and W. T. Nutt of the Westinghouse Hanford Company; N. E. Todreas and J. T. Hawley of the Massachusetts Institute of Technology; J. Graham, P. W. Dickson, and R. O. Fox of the Westinghouse Advanced Reactors Division, J. F. Marchaterre of the Argonne National Laboratory; C. B. Baxi of the General

Atomic Company; E. L. Fuller of the Electric Power Research Institute; C. L. Wilson, S. M. Baker, G. D. Bouchey, J. Ford, and F. Kerze, Jr. of the U.S. Department of Energy; M. D. Weber of the American Nuclear Society; and R. A. Rydin of the University of Virginia.

Early encouragement for undertaking this project was offered by Professor K. Wirtz of Kernforschungszentrum Karlsruhe, Germany. Information for both Phénix and Super Phénix, which is used in several places in the text, was graciously provided by J. Costa of the Centre d'Études Nucléaires de Grenoble and M. Estavoyer of the Centre d'Études Nucléaires de Cadarache, France.

The organization, data collection and painstaking cross-checking of reactor data in Appendix A was skillfully carried out by L. D. Zeller of the Westinghouse Hanford Company. His efforts and the contributions of the FBR scientists and engineers listed at the front of Appendix A, who cooperated so constructively in supplying the data requested, are gratefully acknowledged.

An incalculable debt of gratitude is due the logistic team which molded the efforts of two harried and geographically separated authors into a manuscript of hopefully coherent value. Roland Parenteau provided extraordinary skill, patience, and dedication in editing the original version, and Linda Pfenning added a final professional touch by editing and imbedding the numerous reviewers' comments, preparing the index and list of symbols, and checking the references. The Westinghouse Hanford Company graphics team is to be commended for their splendid work in preparing all of the graphics included in the text. Typing and other logistic chores through the seemingly endless versions were cheerfully shared by Lois Martin, Carrie Locke, Helen Sullivan and Marsha Polk of the Westinghouse Hanford Company and by Vickie Shifflett of the University of Virginia. Lois deserves particular credit for maintaining at least a modicum of organization amidst the cross-fire of conflicting pressure points—especially during the concluding months of this effort.

In the final analysis, however, it was the willing sacrifice and loving support of two individuals, Anna Waltar and Helen Reynolds, who enabled two over-committed husbands to devote the time and energy necessary to allow this book to become a reality.

A.E.W.
A.B.R.

CONTENTS

─────────────── PART II   NEUTRONICS ───────────────

PART III   SYSTEMS

CHAPTER 8:   FUEL PIN AND ASSEMBLY DESIGN

———————————— PART IV SAFETY ————————————

CHAPTER 13: GENERAL SAFETY CONSIDERATIONS

CHAPTER 14: PROTECTED TRANSIENTS

PART I

OVERVIEW

The first part of this book is intended as an overview of the Fast Breeder Reactor field, from which the more detailed chapters to follow can be placed into perspective. Chapter 1 addresses the concept of breeding and the role which the fast breeder has grown to occupy in the energy structure of the industrialized nations. Chapter 2 provides a concise introduction to the basic design features which characterize a fast breeder reactor. Chapter 3 is included to provide the principal economic concepts which must be employed in assessing ultimate commercialization aspects.

CHAPTER 1

BREEDING AND THE ROLE OF FAST BREEDER REACTORS

1-1 INTRODUCTION

Energy, abundantly produced and wisely used, has always been needed for the advancement of civilization. Until the last few centuries, productivity was severely limited because only human and animal power were available as prime movers. By the early nineteenth century, wood burning, along with wind and water power, had considerably advanced the human capability to do work. Coal and then oil sequentially replaced wood, water, and wind as the world's primary energy sources.

The discovery of nuclear fission in the late 1930s provided hope that nuclear energy could relieve the strain on rapidly depleting fossil fuels and provide an abundant, clean, and relatively inexpensive new form of energy. Although the basic sources for nuclear energy—uranium and potentially thorium—were plentiful, they could not be used directly in the fission process. Early experiments indicated that specific isotopes of uranium and plutonium must be used to exploit this new energy source.

By the early 1940s it was known that the isotopes ^{233}U, ^{235}U, and ^{239}Pu would fission when bombarded by neutrons with energies in the low and intermediate to fast ranges, i.e., at thermal energies (< 1 eV) and at energies around 1 MeV. These isotopes are called *fissile* isotopes. It was also found that the abundant isotopes ^{238}U and ^{232}Th would fission only when struck by higher energy neutrons, of the order of 1 MeV or more. Fissile isotopes were found to be required to sustain a neutron chain reaction. But only a very small percentage (0.7%) of naturally occurring uranium is fissile; essentially all of the rest is ^{238}U. In order to derive full benefit from nuclear fission, it was soon recognized that a way must be found to utilize the remaining 99% of uranium's potential.

It was known that ^{238}U and ^{232}Th could *capture* neutrons at energies below the 1 MeV range and thereby convert ^{238}U into ^{239}Pu and ^{232}Th into ^{233}U. Therefore, ^{238}U and ^{232}Th were called *fertile* isotopes. If more fissile isotopes could be produced from fertile isotopes than were destroyed in the chain reaction, it would be possible to utilize the abundant fertile isotopes to make more fissile fuel. This process was shown to be possible. The name given to the process was *breeding*.

It was soon learned that the η value (where η for an isotope is the number of neutrons emitted in the fission process per neutron absorbed) for the fissile isotope ^{239}Pu is higher in a fast neutron spectrum than in a thermal spectrum. This higher number of emitted neutrons meant that more neutrons would be available for conversion of ^{238}U to ^{239}Pu. Hence, the idea developed that a breeder reactor operating on fast neutrons would utilize ^{238}U more efficiently than one operating on thermal neutrons. It was found possible to breed with thermal reactors, too, and the ^{232}Th-^{233}U cycle was determined to be the better cycle for thermal breeding. Although the ^{232}Th-^{233}U cycle could also be used in a fast breeder reactor, the ^{238}U-^{239}Pu cycle provided more efficient breeding; hence, this cycle was generally adopted for the fast breeder reactor. Another reason for giving preference to the ^{238}U-^{239}Pu cycle for the fast breeder was that it employed the same fuel cycle as the present commercial thermal reactors.

The first fast reactor was Clementine, built at Los Alamos in 1946. The next step was an experimental fast breeder reactor, EBR-I, designed by the Argonne National Laboratory. This reactor was built largely because of the enthusiasm of Enrico Fermi and Walter Zinn; and on December 20, 1951, EBR-I became the world's first nuclear plant of any type to generate electricity. The concept of the fast breeder reactor was demonstrated and gave impetus to the prospect of a long-term reliance on nuclear fuel as a new energy source.

Approximately thirty years after EBR-I, four *liquid metal* (sodium) cooled *fast breeder reactors* (LMFBR's) in the 250 to 600 MWe power range began producing electricity (and desalinizing water) in three European countries, and construction on a 1200 MWe LMFBR was nearing completion. By the end of the 1980s, it is expected that LMFBR's in the 1200 to 1600 MWe range will be under construction in five countries. Eight nations on three continents will be operating LMFBR power reactors or fast test reactors before 1990. For many countries with few indigenous energy resources, the dream which began in the middle of the twentieth century of an economical, inexhaustible, and practically independent energy source appears likely to become a reality early in the twenty-first century.

1-2 INTERNATIONAL FOCUS ON THE BREEDER

International interest in fast breeder reactor development is illustrated in Fig. 1-1. This figure provides an overview of the major fast reactor projects

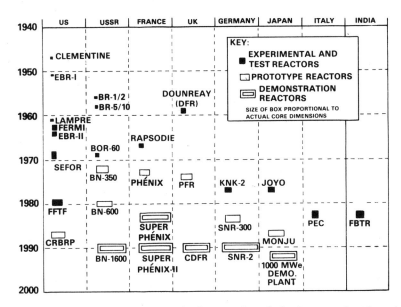

FIGURE 1-1. International fast reactor development through the demonstration plant phase.

that have been built or are in the construction or planning stages. In addition to those labeled, France has announced plans to begin construction on two 1500 MWe plants after Super Phénix begins operating. The projects are tabulated chronologically (from the date of first criticality) and by national origin. The illustration also shows the basic purpose of each project, i.e., whether it is primarily to gather experimental data or to demonstrate an integrated power production capability. The size of each box on the figure is scaled according to the dimensions of each core.

Tables 1-1 and 1-2 summarize the key features of the reactor systems shown in Fig. 1-1. Table 1-1 includes fast reactors built for experimental purposes or for fuels and materials testing, although electrical generating systems were included in some of these plants, as indicated in the table. Table 1-2 shows the fast reactor prototype and demonstration power plants. *Prototype plants* are intermediate-size plants of a particular type, generally in the 250 to 350 MWe range, built to provide data and experience to scale up to commercial size plants. *Demonstration plants* are full commercial size plants built to demonstrate the capability and reliability needed to build and operate commercial plants.

Two interesting observations are worth emphasizing from Tables 1-1 and 1-2. First, all of the fast reactors built and planned are based on the U-Pu fuel cycle. This is indicative of the advanced state of development of U-Pu fuel relative to Th-U fuel. The second observation is that all of the fast reactor plants built or planned use a liquid metal coolant; none uses a gas

TABLE 1-1 Fast Experimental and Test Reactors

Reactor	Country	Date Critical	Thermal Rating (MW)	Electrical Rating (MW)	Core Size (l)	Fuel	Coolant
CLEMENTINE	US	1946	0.025	–	2.5	Pu Metal	Hg
EBR-I	US	1951	1.2	0.2	6	U Metal	NaK
BR-1/2	USSR	1956	0.1	–	1.7	Pu Metal	Hg
BR-5/10	USSR	1958	5/10	–	17	$PuO_2, UC/PuO_2$	Na
Dounreay (DFR)	UK	1959	60	15	120	U Metal	NaK
LAMPRE	US	1961	1	–	3.2	Liquid Pu	Na
Fermi (EFFBR)	US	1963	200	65	400	U Metal	Na
EBR-II	US	1963	62	20	73	U Metal	Na
Rapsodie	France	1967	40	–	42	UO_2-PuO_2	Na
SEFOR	US*	1969	20	–	566	UO_2-PuO_2	Na
BOR-60	USSR	1969	60	12	60	UO_2	Na
KNK-2	Germany	1977	58	21	320	UO_2	Na
JOYO	Japan	1977	100	–	300	UO_2-PuO_2	Na
FFTF	US	1980	400	–	1040	UO_2-PuO_2	Na
FBTR	India	~1983	50	15	55	UO_2-$PuO_2^\dagger$	Na
PEC	Italy	~1985	118	–	325	UO_2-PuO_2	Na

*With participation by Germany and Euratom
†ThO_2 blanket

6

TABLE 1-2   Prototype and Demonstration Fast Breeder Reactors

Reactor	Country	Date Critical	Electrical Rating (MW)	Thermal Rating (MW)	Core Size (l)	Fuel	Coolant	Coolant Configuration
BN-350	USSR	1972	150*	1000	1900	UO_2	Na	Loop
PHÉNIX	France	1973	250	568	1300	UO_2-PuO_2	Na	Pool
PFR	UK	1974	250	600	1500	UO_2-PuO_2	Na	Pool
BN-600	USSR	1980	600	1470	2500	UO_2	Na	Pool
SUPER PHÉNIX	France‡	1983	1200	3000	10500	UO_2-PuO_2	Na	Pool
SNR-300	Germany†	1984	327	770	2300	UO_2-PuO_2	Na	Loop
MONJU	Japan	1987	300	714	2300	UO_2-PuO_2	Na	Loop
CRBRP	US	~1988	375	975	2900	UO_2-PuO_2	Na	Loop
SUPER PHÉNIX II	France‡	~1990	1500	3700	–	UO_2-PuO_2	Na	Pool
CDFR	UK	~1990	1320	3230	6660	UO_2-PuO_2	Na	Pool
SNR-2	Germany	~1990	1300	3420	12000	UO_2-PuO_2	Na	Loop
BN-1600	USSR	~1990	1600	4200	8800	UO_2-PuO_2	Na	Pool
Demonstration	Japan	~1990	1000	2400	8000	UO_2-PuO_2	Na	Loop

*Plus 5000 Mg desalted water/hr
†With participation by Belgium and the Netherlands
‡With participation by Italy, Germany, Belgium, and the Netherlands

coolant, even though gas cooling for fast breeder reactors is still being considered. (See Chapter 17.)

Although Tables 1-1 and 1-2 contain some basic data for the major fast reactor projects, the interested reader may want more detailed information, particularly as design considerations are introduced in later chapters. Appendix A contains more detailed information on these reactors. Further discussion of FBR development in each country is provided in the last section of this chapter.

1-3 BASIC PHYSICS OF BREEDING

A. CONVERSION CHAINS

In order to achieve breeding, a fertile isotope (^{238}U, ^{240}Pu, ^{232}Th, ^{234}U) must be converted via neutron capture (n, γ) into a fissile isotope (^{239}Pu, ^{241}Pu, ^{233}U, ^{235}U). The two important conversion chains are shown in Fig. 1-2. A more complete presentation of these chains is given in Chapter 7. All of the fertile and fissile isotopes appearing in Fig. 1-2 are long lived alpha emitters ($T_{1/2} \geqslant 10^4$ y) so that, with the exception of ^{241}Pu, they can be considered stable with regard to mass balances. Plutonium-241 is also a beta emitter with a half-life short enough to require consideration in fuel cycle calculations. Ignored in Fig. 1-2 are fission reactions (n, f) by all isotopes and neutron capture reactions by the short lived beta emitters.

B. BREEDING RATIO AND REQUIREMENTS FOR BREEDING

The degree of conversion that occurs in a reactor is denoted by the general term *conversion ratio*, CR, which is defined as

$$CR = \frac{\text{fissile material produced}}{\text{fissile material destroyed}} = \frac{FP}{FD}. \qquad (1\text{-}1)$$

Generally, the fissile material produced, FP, and fissile material destroyed, FD, during a fuel cycle—i.e., between periodic refueling—are used in the calculation.

A reactor is called a *breeder* if the conversion ratio is greater than unity. In this case, the conversion ratio is called a *breeding ratio*, BR; hence

$$BR = \frac{\text{fissile material produced}}{\text{fissile material destroyed}} = \frac{FP}{FD} > 1. \qquad (1\text{-}2)$$

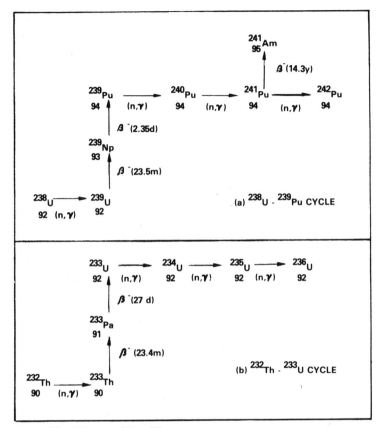

FIGURE 1-2. ^{238}U-^{239}Pu and ^{232}Th-^{233}U conversion chains.

A reactor for which the conversion ratio is less than unity is called a *converter*. Present commercial thermal reactors are converters. In a breeder reactor it is possible for the in-core conversion ratio to be less than one, while the breeding ratio for the entire reactor—core plus *blanket**—is greater than one.[†]

Another useful term is the *breeding gain*, G, which is defined as

$$G = BR - 1. \tag{1-3}$$

Using the notation FBOC for the fissile inventory in the reactor at the beginning of a cycle (i.e., directly after refueling) and FEOC at the end of the cycle (i.e., when the reactor is shut down for refueling), the breeding

---

[*]The blanket is the region of the reactor containing fertile fuel.

[†]The in-core conversion ratio of a breeder reactor has often been called the internal breeding ratio, despite the fact that it is generally less than one.

gain can be written as

$$G = \frac{\text{FEOC} - \text{FBOC}}{\text{FD}} \qquad (1\text{-}4a)$$

$$= \frac{\text{FG}}{\text{FD}}, \qquad (1\text{-}4b)$$

where FG is the fissile material gained per cycle. Clearly the breeding gain must be greater than zero in order for the reactor to be a breeder.

It is possible for a nuclear reactor to breed over a broad neutron energy spectrum, but adequate breeding ratios can be achieved for a given energy spectrum only by carefully selecting the appropriate fissile isotopes for that spectrum. The reason for this becomes evident from a review of a few key properties of basic reactor physics. This review also shows that a high breeding gain can be obtained only with a fast neutron spectrum, although a low breeding gain characteristic of some thermal breeder designs is not necessarily an overriding disadvantage (as we shall see later).

Recall from the fission process that

ν = number of neutrons produced per fission,

η = number of neutrons produced per neutron absorbed,

α = capture-to-fission ratio $\left(\sigma_c / \sigma_f \right)$.

These parameters are related by:

$$\eta = \frac{\nu \sigma_f}{\sigma_f + \sigma_c} = \frac{\nu}{1 + \sigma_c / \sigma_f} = \frac{\nu}{1 + \alpha}. \qquad (1\text{-}5)$$

The parameters ν and α are measured quantities, while η is a derived quantity.

For each of the primary fissile isotopes, ν is fairly constant for neutron energies up to about 1 MeV (about 2.9 for ^{239}Pu and about 2.5 for ^{233}U and ^{235}U) and slowly rises at higher energy. On the other hand, α varies considerably with energy and between isotopes. For ^{239}Pu and ^{235}U, α rises sharply in the intermediate energy range between 1 eV and 10 keV and then drops again at high energy; for ^{233}U, α never rises appreciably. This behavior of ν and α leads to the variations of η with energy shown in Fig. 1-3.

In order to appreciate the significance of Fig. 1-3, it is useful to establish the minimum criterion for breeding. Consider a simple neutron balance for which the basis is one neutron absorbed by a fissile nucleus (which is equivalent to the destruction of 1 fissile nucleus). In order to breed, the next

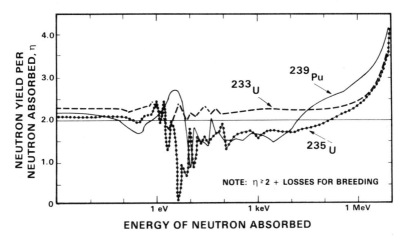

FIGURE 1-3. Neutrons produced per absorption vs energy for fissile isotopes.

generation of neutrons must, at a minimum, replace that 1 destroyed fissile nucleus.

The number of neutrons produced by this absorbed neutron is η. Let us examine the fate of these η neutrons:

1 neutron must be absorbed in a fissile isotope in order to continue the chain reaction;

L neutrons are lost unproductively, by parasitic absorption or by leakage from the reactor. For our present purpose, absorption by any material other than fissile or fertile materials is parasitic.

Hence, from the neutron balance, $[\eta-(1+L)]$ is the number of neutrons captured by the fertile material.

Since this value represents the number of new fissile nuclei produced, and since the basis of our neutron balance is the destruction of one fissile nucleus,

$$\eta-(1+L) \text{ must be } \geqslant 1 \qquad (1\text{-}6a)$$

in order to replace the 1 destroyed fissile nucleus, i.e., in order for breeding to occur. This relation defines a minimum value for η for breeding.*

Rewriting this relation gives

$$\eta \geqslant 2+L. \qquad (1\text{-}6b)$$

---

*It should be noted that the expressions are purposely simplified in order to elucidate the basic concept. In reality, $[\eta-(1+L)]$ can be slightly smaller than unity for the breeding condition because of the fast fission effect in ^{238}U.

Since the loss term is always greater than zero,

$$\eta > 2 \qquad (1\text{-}7)$$

becomes a further simplified minimum criterion for breeding.

The quantity $[\eta - (1 + L)]$ in Eq. (1-6a) is the ratio of fissile nuclei produced to fissile nuclei destroyed for this simplified model; hence, from Eq. (1-2), it is equal to the breeding ratio, i.e.,

$$BR = \eta - (1 + L). \qquad (1\text{-}8)$$

Therefore, a high value of η results in a high breeding ratio. This expression for breeding ratio is useful as a conceptual guide, but we will find in Chapter 7 that it is not very useful for actual computations.

Now we can re-examine Fig. 1-3 in light of our criterion that η must be greater than 2.0 in order to breed. In a thermal reactor most of the fissile absorptions occur in the 0.01 to 1 eV range. In an LMFBR using mixed oxide (UO_2-PuO_2) fuel, about 90% of the fissile absorptions occur above 10 keV. Examination of Fig. 1-3 reveals that ^{239}Pu is the best choice for breeding in a fast reactor, with ^{233}U a possibility and ^{235}U doubtful. The figure also shows that ^{233}U is the only realistic candidate for breeding in a thermal reactor.

These observations are reinforced by Table 1-3, which compares η for the three fissile isotopes averaged over typical light water reactor (LWR) and LMFBR neutron spectra. If relative loss events characterized by L in Eq. (1-8) can be made about the same for a fast and a thermal breeder reactor, the results of Fig. 1-3 and Table 1-3 show that larger breeding ratios can be achieved with fast reactors. Moreover, even higher breeding ratios can be obtained from fast reactors that use carbide fuel (UC-PuC) or metal fuel, partly because the average neutron energy for these fuels is higher and partly because these fuels have greater fissile and fertile material densities.

TABLE 1-3 Value for η Averaged Over Fast and Thermal Spectra[1]

	^{239}Pu	^{235}U	^{233}U
Average over an LWR spectrum ($\sim$0.025 eV)	2.04	2.06	2.26
Average over a typical oxide-fueled LMFBR spectrum	2.45	2.10	2.31

C. DOUBLING TIME

An obvious economic incentive for obtaining a high breeding ratio is to maximize the production of excess fissile material for its sale value. Another perhaps more important incentive can be illustrated by introducing a parameter called the *doubling time*. Although there are several ways to define the doubling time (cf. Chapter 7-7 for more detail), the simplest concept is the *reactor doubling time*, RDT. This is the time required for a particular breeder reactor to produce enough fissile material in excess of its own fissile inventory to fuel an identical reactor. Hence, it is the time necessary to double the initial load of fissile fuel.

A constantly debated aspect of fast breeder reactors involves the rate at which they will be needed in a nation's or region's power system and when commercialization should begin. At issue is the rate the breeder economy should grow to meet projected electrical growth and to replace older fossil and light water reactor plants. The introduction date for the commercial fast breeder reactor varies between countries, depending on many factors, including indigenous energy resources, energy growth rates, the supply of low-cost natural uranium, and the desire to become energy-independent. Indeed, this is a classical problem in systems engineering. Doubling times in the range of 10 to 15 years were generally considered as reasonable goals for breeder reactors during the 1970s, and meeting such a requirement appears to be possible for fast reactors. More recent electrical growth projections suggest that doubling times in the range of 15 to 20 years are probably adequate.

The reactor doubling time can be expressed quite simply in terms of the initial fissile inventory in a reactor, M_0 (kg), and the fissile material gained during one year, $\dot{M}_g$ (kg/y), where $\dot{M}_g$ is a time-averaged difference between the fissile inventory at the beginning of a year and the fissile inventory at the end of the year, i.e.,

$$\text{RDT} = M_0 / \dot{M}_g. \tag{1-9}$$

For example, if $\dot{M}_g$ were $0.1 M_0$, and every year this $0.1 M_0$ were set aside (since only M_0 is needed in the reactor), then after 10 years there would be $2 M_0$ of fissile material—M_0 still in the reactor and M_0 set aside.

Although Eq. (1-9) is simple, an accurate calculation of $\dot{M}_g$ is not. Computer techniques for this type of calculation are the subject of Chapter 7. Nevertheless, it is instructive to consider an approximation for $\dot{M}_g$ because this will correctly illustrate what parameters influence its value, and hence influence doubling time. The quantity $\dot{M}_g$ can be expressed in terms of the breeding gain, G, rated power in megawatts, P, fraction of time at rated power, f, and α as

$$\dot{M}_g = G \cdot (\text{fissile mass destroyed/year})$$

$$\cong G \cdot (1 + \alpha) \cdot (\text{fissile mass fissioned/year})$$

$$\dot{M}_g \cong G(1+\alpha)\left[\frac{\begin{array}{c}(P\times10^6)(2.93\times10^{10}\text{ fissions/W}\cdot\text{s})\\ \times(3.15\times10^7\text{ s/y})(f)(239\text{ kg/kg-mol})\end{array}}{(6.02\times10^{26}\text{ atoms/kg-mol})}\right]$$

$$\cong \frac{GPf(1+\alpha)}{2.7}. \tag{1-10}$$

Hence,

$$\text{RDT}\cong\frac{2.7M_0}{GPf(1+\alpha)}. \tag{1-11}$$

The doubling time in Eq. (1-11) is in years. It is proportional to the *fissile specific inventory**, M_0/P, and inversely proportional to the breeding gain, G. Allowances for time spent by the fuel in the fuel cycle outside the reactor, fuel cycle losses, fissions in fertile material, and variations during the burnup cycle need to be added to make Eq. (1-11) more precise, but the sensitivity of the doubling time to the breeding gain and fissile specific inventory is readily apparent. For example, an increase in breeding ratio from 1.2 to 1.4 results in a factor-of-two reduction in doubling time. The fissile specific inventory for an LMFBR with oxide fuel is in the range of 1 to 2 kg/MWth.

Equation (1-11) can also be used to estimate doubling times for thermal breeders. Two thermal breeder designs have received considerable development—the *light water seed-and-blanket breeder* and the *molten salt breeder*. Both utilize the ^{232}Th-^{233}U cycle. The light water breeder (Shippingport, Pennsylvania) has a calculated breeding ratio barely in excess of unity; hence, its doubling time is very long. The molten salt breeder has a breeding ratio higher than the light water breeder but still low relative to fast breeder reactors. However, since the fissile specific inventory of thermal reactors is generally significantly lower than for fast breeder reactors, the ratio of M_0/P to G, and hence the doubling time of the molten salt breeder, may be comparable to that of a fast breeder.

1-4 BREEDER STRATEGY ANALYSIS AND URANIUM RESOURCES

A. THE BREEDER AS AN INEXHAUSTIBLE ENERGY SOURCE

Fuel cycle studies indicate that approximately 60 to 80 times more energy can be obtained from a given quantity of uranium in a fast breeder reactor

---

*The inverse of this term, P/M_0, is also often quoted. This ratio is called fissile specific power.

than in a light water reactor. While the ratio of ^{238}U to ^{235}U is 140, LWR's convert some ^{238}U to plutonium; this changes the ratio of uranium utilization in an FBR to that in an LWR to a value considerably below 140. Moreover, detailed fuel cycle analyses must account for losses of heavy metals throughout the fuel cycle, and the net result is about 60 to 80 times more efficient fuel utilization for the FBR. Improved LWR designs and fuel cycles are being considered that may reduce this ratio slightly below the 60-80 range.

Since only a small fraction of the uranium is utilized in an LWR, the power cost is sensitive to the cost of the natural uranium, usually quoted as cost per kilogram or pound of the uranium oxide compound U_3O_8. Hence, low grade ores (i.e., of low uranium concentration) become economically unattractive for LWR use. In contrast, due to the FBR's factor-of-60 advantage in fuel utilization, the power cost for an FBR is insensitive to U_3O_8 costs. Therefore, low grade ores unacceptable for LWR use and uranium from seawater could be used in FBR's.

Increased utilization of uranium, together with insensitivity to uranium cost, means that uranium used in FBR's represents an inexhaustible energy source, at least to the extent that many thousands of years satisfies the term inexhaustible. In the same way, thorium is another inexhaustible fuel for use in FBR's, though, as we observed in Section 1-3, it is less efficient for breeding than uranium. In summary, then, fission energy through the breeder reactor joins solar energy (together with renewable sources derived from the sun), fusion energy, and geothermal energy as the four inexhaustible energy sources available for development to meet the world's long-term future energy needs.

Even though low grade uranium ores may be economical for FBR's (as explained above), in practice further mining of uranium beyond that needed for LWR's may never be needed. For example, there is enough uranium-238 currently stored as tails* from the U.S. gaseous diffusion separation plants to satisfy all U.S. electrical energy requirements for several hundred years. Moreover, this stockpile will grow as long as ^{235}U is needed for LWR's.

B. URANIUM RESOURCES AND BREEDER TIMING

While these long-term considerations are rather apparent, the near-term considerations that govern the timing for the introduction of the commercial FBR and the strategies for the transition to the FBR or to an FBR-LWR mix are more complex. Some of the issues involved were raised briefly in Section 1-3 in the discussion of doubling time.

---

*Tails represents the depleted ^{238}U which remains after completion of the enrichment process.

The remainder of this section is addressed to the influence of uranium *resources* on the timing of the commercialization of the FBR. An analysis of the influence of U_3O_8 *price* on LWR vs FBR power costs will be deferred to Chapter 3. Instead, a more simplistic view that postulates the existence of a specific limited quantity of U_3O_8 that is "economically attractive" for LWR's will be adopted in the present chapter. This assumption will allow us to introduce the methodology that illustrates the need for FBR commercialization. This methodology will then be expanded further when the bases for power costs are provided in Chapter 3.

In addition to uranium resources, the timing of FBR commercialization depends on electricity growth rates and the fraction of that growth to be supplied by nuclear energy. A schematic scenario for FBR introduction is given in Fig. 1-4. As economical uranium resources are depleted, LWR construction is replaced by FBR's. At the end of existing LWR plant lifetimes, LWR's are replaced by FBR's; hence the drop in the LWR curve after time t_4.

The upturn in the LWR curve illustrates another interesting potential strategy. In the period just after introduction of the FBR, the fissile material bred in existing FBR's will be used to start up new FBR's in order to increase the FBR fraction of the total electrical generating capacity. However, capital costs for LWR's will presumably always be lower than for FBR's. Therefore, it may eventually become more economical to renew the

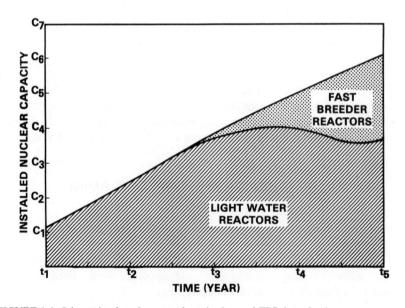

FIGURE 1-4. Schematic of nuclear growth projection and FBR introduction.

construction of LWR's to utilize the excess fissile material bred in FBR's and to hold the FBR fraction of total electrical capacity at a constant level. The upturn in the LWR curve illustrates this strategy. Moreover, the most economic fissile material for use in these LWR's may be ^{233}U bred from thorium FBR blankets.

Limitation on LWR Capacity Imposed by Uranium Resources

A starting point for determining the number of LWR's that can be fueled, and hence constructed, prior to FBR commercialization is the U_3O_8 require- ment for the operating lifetime of the LWR. Light water reactors can operate on one of three fuel cycles: (1) the once-through cycle which involves no recycle of fuel, (2) the recycle of uranium only following reprocessing (since the ^{235}U content of spent fuel is higher than that of natural uranium), and (3) the recycle of both uranium and plutonium. The U_3O_8 requirement also depends on the ^{235}U content in the tails of the isotope separation process—the lower the ^{235}U fraction in the tails, the lower the U_3O_8 requirement. Approximate values of U_3O_8 requirements for an LWR based on a 30-year plant lifetime (a frequently used value for planning and cost analysis) operating at 60% load factor (i.e., 18 full-power years) with 0.2% tails are listed in Table 1-4 for the three fuel cycles.

The results in Table 1-4 apply to LWR designs and enrichment tech- niques of the 1970s. Future improvements in enrichment technology to lower the ^{235}U tails fraction, along with LWR's designed for high fuel burnup and/or higher conversion ratios, may lower U_3O_8 requirements below these values.

TABLE 1-4   Approximate Uranium (U_3O_8) Requirements for a 1000-MWe LWR

| Based on 30-year plant lifetime, 60% load factor, 0.2% tails | |
Fuel Cycle	Mg (or tonnes) U_3O_8
No Recycle	5000
U Recycle only	4000
U and Pu Recycle	3000

These requirements can be compared with estimates of economically recoverable U_3O_8 resources to assess limitations on LWR capacity. To illustrate the relevant concepts, we will first consider potential LWR capac- ity that can be supplied by U.S. resources. These uranium resources are

TABLE 1-5   Estimated U.S. Uranium (U_3O_8) Resources[2]
(January 1980 U.S. Dept. of Energy Estimates)

Estimated Production Cost		Resources (Gg, or 1000 tonnes, U_3O_8)			
$/kg U_3O_8	$/lb U_3O_8	Reserves	Probable	Possible	Speculative
$33	$15	204	377	191	68
$33 to 66	$15 to 30	382	536	422	205
$66 to 110	$30 to 50	264	454	449	227
$110 to 220	$50 to 100	167	–	–	–
Total ⩽ $110	⩽ $50	850	1367	1062	500

estimated annually by the U.S. Department of Energy. The results for 1980 are shown in Table 1-5.

Values are reported for resources with "forward costs," or production costs,* less than $110/kg U_3O_8($50/lb). It is argued for a simplistic method of assessing resource availability that this cost represents a limit to economically attractive fuel for LWR's, which is the approach presented in this chapter (only to be modified in Chapter 3 because one knows that such an absolute cutoff never really exists). More uranium is available in shale, but its low concentration makes it expensive to recover.

The resources are listed in four categories—Reserves, Probable, Possible, and Speculative—depending on the level of confidence of the resource estimate. Reserves refers to the resources for which existence has been reasonably well established. The total resource value to be used for prudent planning purposes is not well established; as is frequently the case, there is room for reasonable disagreement among planners. If economically attractive U.S. resources turn out to be limited to those in the Reserve and Probable categories, the number of 1000-MWe LWR's that can be supported by this supply on the once-through cycle is slightly more than 400.

Estimates of world uranium resources (not including Soviet Bloc countries) with production costs below $110/kg U_3O_8 are given in Table 1-6[3]. The column labeled "Reasonably Assured" is most nearly comparable to the U.S. DOE's category of Reserves; "Estimated Additional" is comparable to DOE's "Probable." The world resources given in Table 1-6 would support only 1000 LWR's operating on the once-through cycle.

We next turn to methods for using the information in Table 1-4 by starting with a projection of installed capacity, as illustrated schematically

---

*Forward costs are considerably lower than the actual price to a utility. Forward costs include only current production costs. They do not include, for example, costs for mine construction, return on investment, and future exploration; and they do not account for market price fluctuations due to supply and demand.

TABLE 1-6 World Uranium Resources
(Not Including Soviet Bloc Countries)*

For U_3O_8 with Production Costs $<\$110/kg$		
	Resources (Gg U_3O_8)	
	Reasonably Assured	Estimated Additional
United States	835	1367
South Africa	410	85
Sweden	355	4
Australia	350	58
Canada	215	775
Niger	189	63
France	61	52
Argentina	49	—
India	35	28
Algeria	32	59
Other	129	108
World Total	2660	2599

*Values from a Ford Foundation Study[3] which obtained the results from U.S. DOE 1978 estimates and a 1977 joint report by the OECD Nuclear Energy Agency and the International Atomic Energy Agency.

in Fig. 1-5. Low, medium, and high projections are frequently given. The fraction of nuclear capacity must then be specified to obtain a low, medium, and high projection for nuclear (one such curve being shown previously in Fig. 1-4).

A nuclear capacity curve can be combined with the data of Table 1-4 to obtain a curve of U_3O_8 committed for the lifetime operation of all LWR's for the low, medium, and high growth rate options, as shown schematically for one of the growth rate options in Fig. 1-6. Two curves are shown; the curve labeled "Improvements" corresponds to potential fuel cycles using

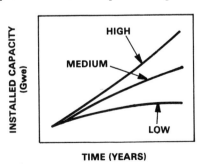

TIME (YEARS)

FIGURE 1-5. Projection of installed capacity.

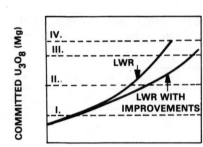

FIGURE 1-6. Schedule of U_3O_8 commitments. Dashed lines refer to available resources as categorized in Table 1-5. I includes reserves only. II includes reserves plus probable. III includes all resources through possible. IV includes all resources through speculative.

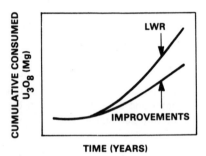

FIGURE 1-7. U_3O_8 consumed.

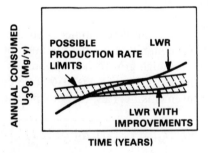

FIGURE 1-8. U_3O_8 annual consumption rate.

lower enrichment tails and/or LWR's with higher fuel burnup. Also sketched on the figure are the resources as categorized in Table 1-5. This figure shows how LWR deployment can be limited by resource availability, depending on which category is used as a basis for planning.

Next the cumulative consumption of U_3O_8 can also be projected, as in Fig. 1-7. Finally, the U_3O_8 consumed annually by all LWR's in the system can be calculated, as shown in Fig. 1-8. Also illustrated schematically on

Fig. 1-8 is an upper limit to U_3O_8 production rate. This rate is difficult to project, but its value is important because uranium production instead of the total resources of U_3O_8 may become the limiting factor on the number of LWR's in the system.

The curves in Figs. 1-5 through 1-8 can be obtained from the equations given below, together with the data of Table 1-4. Consider first the following definitions:

$\tau=$ plant lifetime (y)
$u=$ U_3O_8 commitment over the lifetime of a 1-GWe LWR plant (Mg/GWe)
$U=$ U_3O_8 commitment over the lifetime of all LWR plants (Mg)
$r=$ annual U_3O_8 requirement for a 1-GWe LWR plant (Mg/GWe·y)
$R=$ annual U_3O_8 requirement for all LWR plants (Mg/y)
$W=$ U_3O_8 consumed by all LWR plants (Mg)
$C=$ electrical capacity (GWe)
$N=$ number of 1-GWe LWR plants operating and decommissioned
$K=$ linear electrical growth rate (GWe/y)
$\lambda=$ exponential electrical growth rate constant (y^{-1})
$t=$ time (y), or date.

Subscripts:
$T=$ total (non-nuclear plus nuclear)
$N=$ nuclear
$0=$ initial value.

Electrical capacity can be approximated by either of two simple models. These include (a) constant linear growth rate, K, or (b) exponential rate, with the latter having been used more frequently in the past.

For (a), the total electrical capacity is governed by

$$\frac{dC_T}{dt} = K_T, \qquad (1\text{-}12)$$

which leads to

$$C_T(t) = C_{T0} + K_T(t - t_0), \qquad (1\text{-}13)$$

where C_{T0} is the capacity at the starting time, t_0. If K_N is the nuclear growth rate, then

$$C_N(t) = C_{N0} + K_N(t - t_0). \qquad (1\text{-}14)$$

Until nuclear plants begin to be decommissioned, the number of 1-GWe nuclear plants, $N(t)$, is equal numerically to $C_N(t)$. To simplify the nuclear accounting, we choose t_0 for the remainder of this analysis such that $C_{N0} = N_0 = 0$ at $t = t_0$. Then, between time t_0 and $t_0 + \tau$,

$$N(t) = K_N(t - t_0). \tag{1-15}$$

Between $t_0 + \tau$ and $t_0 + 2\tau$, nuclear plants that have operated for a lifetime are being decommissioned. If each decommissioned unit is replaced with a new nuclear unit, the rate at which new nuclear plants are introduced between $t_0 + \tau$ and $t_0 + 2\tau$ is

$$\frac{dN}{dt} = 2K_N \ (\text{for } \tau < t - t_0 < 2\tau). \tag{1-16}$$

The solution to this equation is (for $N_0 = 0$ at t_0)

$$N(t) = K_N\tau + 2K_N\big[t - (t_0 + \tau)\big], \tag{1-17}$$

which includes both operating and decommissioned nuclear plants.

For the exponential growth rate (b), the total electrical capacity growth rate is

$$\frac{dC_T}{dt} = \lambda_T C_T, \tag{1-18}$$

which gives

$$C_T(t) = C_{T0}e^{\lambda_T(t - t_0)}. \tag{1-19}$$

For the nuclear contribution,

$$C_N(t) = C_{N0}e^{\lambda_N(t - t_0)}. \tag{1-20}$$

The total U_3O_8 commitment, $U(t)$, is simply

$$U(t) = uN(t), \tag{1-21}$$

where the value of u is the appropriate value from Table 1-4.

The annual U_3O_8 requirement, r, for a 1-GWe LWR is

$$r = u/\tau. \tag{1-22}$$

The annual U_3O_8 requirement for all LWR's is

$$R = rC_N(t). \tag{1-23}$$

The cumulative U_3O_8 consumed, $W(t)$, is the integral

$$W(t) = \int_{t_0}^t R(t)\, dt. \tag{1-24}$$

For a constant linear growth rate (and with $C_{N0} = 0$ at t_0),

$$W(t) = r \int_{t_0}^t K(t - t_0)\, dt = rK \frac{(t - t_0)^2}{2}. \tag{1-25}$$

For an exponential growth rate (for which we can no longer assume that $C_{N0} = 0$ at t_0),

$$W(t) = r \int_{t_0}^t C_{N0} e^{\lambda_N (t - t_0)}\, dt \tag{1-26}$$

$$= \frac{rC_{n0}}{\lambda_N} e^{\lambda_N (t - t_0)}.$$

Modification of Analysis After FBR Introduction

The effect of FBR commercialization will be to reduce and eventually eliminate the construction of LWR's that use enriched ^{235}U fuel. This will reduce and eventually eliminate annual consumption of U_3O_8 for LWR's. The curves corresponding to Figs. 1-6 and 1-8 would be modified as shown in Figs. 1-9 and 1-10.

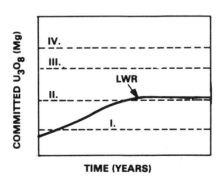

FIGURE 1-9. Modification of U_3O_8 commitments due to FBR introduction.

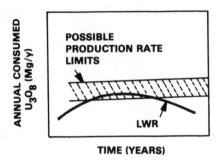

FIGURE 1-10. Modification of annual consumed U_3O_8 due to FBR introduction.

1-5 FAST BREEDER REACTOR PROGRAMS

The following sections give a brief account of the evolution of fast breeder reactor programs in each of the countries in which fast reactors have been or are being built. Although the narrative is brief, it will provide some background for the rationale and primary goals as perceived by those investigators who pioneered what is rapidly becoming a mature technology.

A. UNITED STATES

As early as the mid-1940s, the Los Alamos Scientific Laboratory (LASL) directed some attention to constructing a fast reactor to demonstrate the use of plutonium as a fuel for power production. The reactor, which became known as Clementine, had a 2.5 liter core fueled with plutonium metal and cooled by mercury. Construction started in mid-1946 and initial criticality was attained in November 1946, prior to completing the entire facility. Construction was later finished and full power (25 kWth) was attained in March 1949. The reactor was decommissioned in 1952.

In parallel to the Los Alamos effort, the Argonne National Laboratory (ANL) proceeded with the design of an Experimental Breeder Reactor, EBR-I, to be built at Idaho Falls. The principal purpose of this endeavor was to demonstrate that a fast spectrum reactor could be built as an integrated part of a central station power concept. Construction began in 1949 and criticality was attained in August 1951. This led directly to the historic moment on December 20, 1951 when electricity from the fission process was first introduced to the world. The electrical generating capacity was 200 kWe—enough to serve the needs of the building.

Some concern about fast reactor stability arose in 1955 after an experiment resulted in a partial meltdown of the MARK-II core in EBR-I.

Although the reactor was purposely being tested under extremely adverse conditions (wherein some fuel melting was expected), operator error allowed more fuel melting than had been anticipated. A subsequent core was built and tested at EBR-I, and these tests provided convincing evidence that stability problems of metal-fueled cores could be corrected by mechanical design.

In 1951, the federal government invited private industry to participate in reactor designs for producing civilian power. Of the four industrial groups responding, two were interested in pursuing the fast reactor concept. One group proposed a joint industrial/government enterprise that led eventually to the construction of the Enrico Fermi Fast Breeder Reactor (EFFBR). This 400 liter core, fueled with uranium metal and designed to produce up to 200 MWth and 65 MWe, attained criticality in August 1963. A local flow blockage in 1967 caused partial melting in several fuel assemblies. The reactor was repaired and returned to its authorized thermal power level of 200 MWth in 1972. It was decommissioned later that year.

The EBR-II facility was built at roughly the same time. This system was designed to include a complete fuel reprocessing plant, a fuel fabrication plant, and an electrical generating plant, in addition to the reactor. The decision to proceed was made in 1954, and dry criticality was attained in the fall of 1961. Wet criticality occurred in November 1963. EBR-II was the first fast reactor design to use the pool concept, in which all primary sodium components are located in a common sodium pool inside the reactor vessel. The reactor is currently used as a fuels and materials test facility while simultaneously generating electricity. Its sodium system has operated well; its double-tube-wall steam generator has operated since startup with no major repairs.

The Los Alamos Molten Plutonium Reactor Experiment (LAMPRE), conducted at LASL in the mid-1950s, was launched to further the study of plutonium as a breeder reactor fuel and, in particular, to investigate the feasibility of using plutonium in molten form. Dry criticality was achieved in 1961 and a second core was brought to full power in April 1962. A major problem with LAMPRE was the difficulty of containing the plutonium in liquid metal form.

Fundamental fast reactor physics work was augmented during the 1950s and 1960s by completion of the ZPR series of zero power critical facilities at ANL. These projects were followed by the large ZPPR facility.

Around 1960, emphasis in the design of fast reactors began to shift from the earlier emphasis on fuel conservation toward optimizing the economic potential of fast breeders. In both the United States and Western Europe, this shift led to an increased interest in uranium-plutonium oxide fuel for a fast breeder reactor. Experiments in the early 1960s demonstrated that oxide fuel could be irradiated to a specific energy value of 100 MWd/kg, a value far greater than that possible with metal fuel. The use of oxide fuel would lead to lower fuel costs, at the expense of decreasing the average neutron energy and, hence, the breeding ratio relative to metal-fueled breeders. Moreover, experience with uranium oxide fuel was being gained from light water reactor operation, and the Soviet Union had recently operated their fast reactor BR-5 with plutonium oxide fuel.

At the same time, a safety advantage to the use of oxide fuel was recognized. An inherent prompt-acting negative reactivity shutdown mechanism is needed in any reactor to terminate an unexpected transient. The principal prompt negative reactivity in a metal-fueled fast reactor is the axial expansion coefficient, and concern arose that this prompt effect might not be adequate for irradiated metal fuel. On the other hand, calculations indicated that a prompt, negative Doppler coefficient would be available in an oxide-fueled core, an effect which is much smaller in a metal core. In order to provide convincing experimental evidence of the effectiveness and reliability of this mechanism, an international partnership assembled under the leadership of K. P. Cohen, B. Wolfe, and W. Häfele and consisting of the USAEC, the General Electric Company, the Karlsruhe Nuclear Laboratory of the Federal Republic of Germany, Euratom, and a consortium of seventeen U.S. electrical utilities called the Southwest Atomic Energy Associates was formed to build the Southwest Experimental Fast Oxide Reactor (SEFOR). This endeavor was highly successful in demonstrating the efficacy of the Doppler effect in a mixed oxide (UO_2-PuO_2) fueled fast reactor. Excellent agreement was obtained between calculations and whole-core transient results, even when the reactor was intentionally made prompt critical.

The next major commitment in the United States was the Fast Flux Test Facility (FFTF), which was designed by Westinghouse Electric Corporation for the U.S. government. Although the long-range need for the breeder reactor had been clearly evident since the breeding process was discovered, the prevailing U.S. philosophy of the late 1960s was to establish a firm technological position, based on an extensive testing program, before proceeding with prototype plants. Therefore, it was decided to build a 400 MWth reactor, the FFTF, to provide both open and closed loop testing capability for fuels and materials in an irradiation environment typical of central station power reactors. Criticality was achieved in early 1980.

Although the FFTF should prove to be a highly valuable national and international facility for testing advanced fuels and materials, it does not provide an adequate test of large components, such as steam generators for a sodium/water system, or a complete test of the licensing process that must evolve for a maturing industry. To achieve this next step, a 350-MWe prototype plant called the Clinch River Breeder Reactor Plant (CRBRP), funded jointly by the U.S. government and U.S. utilities, was authorized and scheduled for operation in the early 1980s. Beginning in 1978, however, this project, together with the future course of fast breeder reactor development in the United States, became the subject of an extended national political debate. Start of construction and licensing activities of the CRBRP were deferred pending outcome of the U.S. energy policy debate, although the manufacturing of reactor components and design activities continued. The technology developed for the CRBRP represents a significant advance in U.S. LMFBR technology. Figure 1-1 and Table 1-2 reflect a schedule for CRBRP assuming a resolution of the energy debate which leads to a limited work authorization release during the last quarter of 1981.

One of the significant changes made in the CRBRP design during the deferral period was to incorporate a heterogeneous core configuration.* Since much of the early CRBRP technology was developed on the basis of a homogeneous core layout, several examples used in this book will be based on that earlier design. In order to avoid confusion, the terms "early CRBRP homogeneous core design" and "heterogeneous CRBRP core design" will be used in the text as appropriate. For those sections dealing with plant parameters unaffected by the core change, only the "CRBRP" designation will be used. Appendix A is based upon the heterogeneous core design. Appendix B is included to illustrate the principal differences in core parameters which resulted in converting from the homogeneous to the heterogeneous core design.

The U.S. supported a limited effort on the development of *gas cooled fast reactors* (GCFR). The General Atomic Company has designed helium-cooled fast breeder reactors of both prototype and demonstration size, but none have yet been built.

Development programs on several *thermal* breeder reactor concepts have been conducted in the United States, but only the light water seed-and-blanket reactor at Shippingport was supported through the 1970s. This reactor is a thermal breeder with a breeding ratio very close to unity. In the 1950s, a homogeneous thermal breeder reactor was built at the Oak Ridge National Laboratory (ORNL) in which uranyl sulfate was dissolved in heavy water. The core was surrounded by a thorium slurry. The difficulty with the design was the inability to contain the liquid solution. Later a molten salt thermal breeder reactor was also built at ORNL. The molten salt fluid consisted of fluorides of uranium, thorium, and other elements such as zirconium, sodium, and beryllium. The moderator was graphite.

---

*See Chapter 2-3 for a discussion of heterogeneous vs homogeneous core design options.

B. UNION OF SOVIET SOCIALIST REPUBLICS

Fast reactor work in the USSR dates from the early 1950s, when the BR-1 and BR-2 projects were initiated. BR-1 was a zero power critical assembly, fueled with plutonium metal, with a core volume of 1.7 liters. The reactor was upgraded in 1956, was renamed BR-2, and was operated at a thermal power level of 100 kWth. Mercury was used as the coolant.

BR-2 was soon replaced by BR-5, which attained criticality in 1958 and was operated at 5 MWth. This project was developed primarily to gain experience in operating a multi-loop, sodium-cooled reactor and to test fuel elements and equipment for subsequent fast reactors. BR-5 was the first reactor in the world to be operated with plutonium oxide fuel. Although several fuel elements developed leaks, the reactor was successfully operated until November 1964, when it was shut down and reloaded with a uranium carbide core. In 1972, the reactor was modified again, this time for 10 MWth operation with a new PuO_2 core. This upgraded facility has since been known as BR-10.

Construction for the 60 MWth BOR-60 reactor was started in May 1965, and wet criticality was attained in late 1969. Since it was built primarily to provide a materials test bed, the heat was originally rejected by air-dump heat exchangers. A steam generator was installed in 1970 and a second steam generator of a different design was put into service in 1973. The reactor could then generate electricity at the rate of 12 MWe.

In 1970, the Soviet Union completed the largest critical facility in the world, designated BFS-2. It can test full-scale assemblies up to the BN-1600 range.

The Soviet Union was the first nation to enter the prototype plant phase, with the design and construction of the BN-350, an equivalent 350 MWe (1000 MWth) UO_2 fueled reactor. BN-350 achieved criticality in November 1972 and was brought into power operation in mid-1973. Between 1973 and 1975, five of its six steam generators were repaired, and since 1976 it has operated at 650 MWth. This facility was designed for the dual purpose of producing electricity (150 MWe) and desalting water (120000 m^3 fresh water per day). BN-350 is a loop-type design, in which liquid sodium is piped to and from the core through six separate coolant loops.

Even before BN-350 achieved full power, construction began on a 600 MWe plant, BN-600. Unlike the BN-350, BN-600 is a pool-type reactor. Its size is intermediate between other prototype and demonstration plants. Construction was completed in 1979, and criticality was achieved in 1980.

A full size demonstration plant of 1600 MWe, called BN-1600, is being designed for construction in the 1980s. It is a pool design like the BN-600. The fuels in the BN-350 and BN-600 are not prototypical, since they start with UO_2 only. The BN-1600 will be the first Soviet fast reactor to use mixed oxide fuel initially.

In the gas cooled fast breeder reactor field, the USSR is in the advanced stages of designing an N_2O_4 gas-cooled plant, but no construction plans have been announced.

C. FRANCE

The French fast reactor program began in 1953 with development work on sodium systems. It was not until 1967 that their first fast reactor went into operation, but they have moved very rapidly since that time under the direction of G. Vendryes. Rapsodie began operation in 1967 at the Centre d'Études Nucléaires de Cadarache. This fuel test facility, originally rated at 20 MWth and soon raised to 24 MWth, has operated reliably with mixed oxide fuel and has provided fuel test data for the larger French demonstration reactors. In 1970, a "Fortissimo" core was loaded, raising the thermal power level of Rapsodie to 40 MWth.

Phénix, named for the mythological bird that was reborn from its own ashes, is a 250 MWe prototype plant located at Marcoule on the Rhône River. It achieved criticality in August 1973 and full power in March 1974. The reactor operated remarkably well until the summer of 1976 when problems in the intermediate heat exchanger required shutdown for their repair. The reactor was returned to full power in 1977 and operation has continued as before. The initial reactor core was half mixed oxide and half UO_2, but since 1977 it has been entirely mixed oxide.

Based on the success of the Rapsodie and Phénix programs, construction of the 1200 MWe demonstration plant, Super Phénix, was started in 1977 at Creys-Malville on the Rhône River. Criticality is scheduled for 1983. This reactor is supported financially by utilities from five countries—France (51%), Italy (33%), Germany (11%), and Belgium and the Netherlands (2.5% each). The lead designer is the French company Novatome, and it will be operated by the French national utility, Electricité de France. This utility has announced plans to begin construction on two commercial 1500 MWe LMFBR's in the mid-1980s. Moreover, in 1977 France signed major licensing agreements with a German-led consortium that involves that country as well as Belgium and the Netherlands. This is indicative of further joint European development of the LMFBR.

D. UNITED KINGDOM

Fast reactor work in the United Kingdom in the early 1950s was focused on reactor physics, as supported by the Zephyr and Zeus zero-power critical facilities at Harwell. The Zebra facility was built later for large critical assemblies.

Construction of the Dounreay Fast Reactor (DFR) began in 1955 and criticality was attained with uranium metal fuel in late 1959. In July 1963, Dounreay began operation at full power using redesigned fuel elements. From that time until 1977 (when it was decommissioned) this reactor served as a fuels and materials test reactor for the British fast reactor program. The test facility is well remembered by fast reactor designers as the place where the alarming rate of swelling of stainless steel by fast neutron irradiation was discovered in 1967, a phenomenon that has had an important influence on fast reactor core design. A unique feature of the Dounreay plant is that it is the only liquid metal (NaK) cooled reactor that incorporates downward flow through the core.

Construction of the Prototype Fast Reactor (PFR) in the UK was begun in 1967 on a site at the northern tip of Scotland adjacent to the Dounreay site. PFR employs a pool concept, and it was the first prototype plant to be initially fueled entirely with UO_2-PuO_2 fuel. Designed for 250 MWe operation, it attained criticality in March 1974. After early problems with leaks in the steam generator, it achieved full power in 1976.

Design is well advanced on the UK's Commercial Demonstration Fast Reactor (CDFR), which is rated at 1320 MWe. Construction of this plant is expected to begin during the 1980s. Like its prototype, CDFR will be a pool-type reactor with mixed oxide fuel.

E. FEDERAL REPUBLIC OF GERMANY

In the early 1960s German interest in the use of oxide fuel in fast breeder reactors and their recognition of the importance of the Doppler effect coincided with ideas that were gaining momentum in the United States. This led to the participation of the Karlsruhe Nuclear Laboratory in the joint U.S.-German SEFOR program described earlier.

Since that time, the West German fast reactor program has been conducted mostly in cooperation with INTERATOM and a consortium of Germany, Belgium, and the Netherlands on an approximately 70-15-15 percent basis. One of the early facilities built was the zero-power SNEAK critical facility. At Karlsruhe a sodium-cooled thermal reactor called KNK was in operation during the 1960s; later this facility was converted to a 58 MWth fast test facility entitled KNK-2. This conversion was essentially completed in 1975, and criticality was attained in 1977. The outer driver core is uranium oxide and the central test section is loaded with mixed oxide fuel of advanced reactor design. Considerable liquid metal experience has been gained from this facility.

The prototype plant in the German program is the SNR-300, designed for 300 MWe and fueled with UO_2-PuO_2. Construction of this loop-type reactor near Kalkar on the lower Rhine River is in the advanced stages, and operation is expected in the early 1980s.

Long-term planning is being focused on the design of the SNR-2 demonstration plant with a power level in the 1300 MWe range. Germany is an 11% partner in France's 1200 MWe Super Phénix demonstration plant, and in 1977 entered licensing agreements for joint development of commercial LMFBR's with France.

F. JAPAN

The fact that Japan imports almost 90% of the energy it uses has spurred the Japanese government into a very aggressive fast breeder reactor program. In 1967 the Power Reactor and Nuclear Fuel Development Corporation (PNC) was established to implement fast reactor development. The major companies engaged in LMFBR development include Fuji, Hitachi, Mitsubishi, and Toshiba.

The JOYO experimental fast reactor was designed for a power level of 100 MWth and was built primarily for testing advanced fuels and materials. Construction began in 1967 and criticality was attained in 1977. Initial operation was at 75 MWth; this will later be raised to 100 MWth with a second core.

MONJU is designed as a 300 MWe prototype plant. As in the case of JOYO, it employs a loop-type cooling system and is fueled with UO_2-PuO_2. Construction is scheduled to begin in the early 1980s, with operation expected before 1990.

Beyond MONJU, the Japanese have announced an ambitious LMFBR program. Design is underway for a demonstration plant in the 1000 to 1500 MWe range, with construction to start after MONJU is built. Serial construction of four "early-commercial" plants will be initiated in the 1990s.

G. ITALY, BELGIUM, THE NETHERLANDS, INDIA

In addition to the six countries discussed earlier with advanced LMFBR development programs, four other nations—Italy, Belgium, the Netherlands, and India—have significant fast reactor programs.

In Italy, the 118 MWth fuels and materials fast test reactor PEC has been under construction since 1973. Fast reactor component development and safety research programs are being carried out in Italy under a cooperative agreement signed in 1974 with France, and Italy is a 33% participant in France's 1200 MWe Super Phénix demonstration plant.

Belgium and the Netherlands have joined with the Federal Republic of Germany in a series of LMFBR development projects, each providing about 15% effort to Germany's 70%. They are thus participating in the SNR-300 and SNR-2 projects. They are both 2.5% participants in the Super Phénix plant, and they form part of the Germany-Belgium-Netherlands consortium involved in licensing agreements with France for joint development and commercialization of the LMFBR.

India declared its interest in LMFBR development by initiating the construction of a Fast Breeder Test Reactor in 1973. This reactor will have a steam generator and will generate electricity at the rate of 15 MWe. The reactor is unique in its use of thorium oxide in the blanket in order to demonstrate the use of thorium which India has in abundance.

REVIEW QUESTIONS

1-1. What is the principal reason for desiring a fast neutron spectrum for a breeder reactor?

1-2. Define the terms "breeding ratio" and "breeding gain."

1-3. What is meant by the term "doubling time"?

1-4. What are the principal reactor parameters which influence the doubling time?

1-5. Which countries have been actively pursuing the fast breeder reactor option?

1-6. How does the time span for fast breeder reactor development compare to that of the thermal reactor industry?

1-7. Why does the use of ^{239}Pu in a fast breeder reactor result in a higher breeding ratio than ^{233}U?

PROBLEMS

1-1. Suppose the initial fissile fuel inventory required for a 1000 MWe LMFBR is 3000 kg and the core power is 2200 MWth. If the breeding ratio is 1.4, can the reactor produce enough fuel to start up a new similar reactor in 8 years? (Assume a reasonable load factor. For α, make use of the average η in Table 1-3 together with a ν value for ^{239}Pu of 2.9.)

1-2. Suppose that in 1980 the installed U.S. nuclear capacity was 50 GWe and the total U.S. electrical capacity was 500 GWe. A number of hypothetical assumptions will be made regarding U.S. electricity generation for the 35-year period 1975-2010 for the purpose of utilizing various resource strategy calculations.

(a) For the assumptions listed below, plot

  (i) the annual U_3O_8 requirement between 1975 and 2010.
  (ii) the U_3O_8 committed between 1975 and 2010.
  (iii) the U_3O_8 consumed between 1975 and 2010.

Assumptions:

- Constant linear growth rate, K_N, of installed nuclear capacity of 10 GWe/y from 1975 to 2000, and 20 GWe/y from 2000 to 2010, and no capacity prior to 1975.

- No uranium or plutonium recycle.

- No introduction of FBR's before 2010.

- Decommissioned nuclear units are replaced with new nuclear units.

- The nuclear plant lifetime, τ, is 30 years.

(b) Referring to Table 1-5, and assuming that uranium resources greater than the $110/kg category in that table are prohibitively expensive for use in LWR's, discuss the relationship between uranium supply and new construction of LWR's beyond 2010 for the hypothetical conditions of part (a) of this problem.

(c) Assume the following exponential growth rates in total U.S. electrical capacity:

$$
\begin{array}{ll}
1980\text{--}1990: & 3.0\% \ (\text{i.e., } \lambda = 0.030 \ \text{y}^{-1}) \\
1990\text{--}2000: & 2.5\% \\
2000\text{--}2010: & 2.0\%
\end{array}
$$

Combining these with the hypothetical conditions of part (a), together with the fact that coal plants accounted for 50% of U.S. electrical capacity in 1980, fill in the table below.

	1980	2010
Non-Nuclear capacity (GWe)		
Coal capacity (GWe)		/////////
LWR capacity (GWe)		
Fraction nuclear		

    (d)   Assume that the FBR is commercialized in the year 2000 and that the nuclear capacity added between 2000 and 2010 is changed to the following:

2000–2005:	5 FBR's/y, 15 LWR's/y
2005–2010:	10 FBR's/y, 10 LWR's/y, and decommissioned plants are replaced with LWR's.
Beyond 2010:	Only FBR's and new LWR's operating on bred fissile material.

Assume now that uranium recycle *will* be used in all LWR's (retroactively to 1975) since introduction of breeder reactors requires the use of fuel reprocessing.

With these new assumptions, replot curve (ii) in part (a) (i.e., U_3O_8 committed) to the year 2020.

    (e)   Optional: Modify the input data to this problem to make them more consistent with current data, and repeat the calculations.

1-3.   (a)   Assume that Table 1-6 adequately reflects the world's economically available uranium resources; Table 1-4 correctly describes the uranium requirements for an LWR; and all reactors prior to FBR's are LWR's operating with uranium recycle only. Calculate the potential world LWR electrical capacity using economically available uranium.

    (b)   In 1980, the world nuclear electrical capacity was approximately 125 GWe. At that time about 200 GWe additional capacity was under construction, most of which was expected to be on-line by the mid-1980s, and another 75 GWe were on order (about 40% of these capacities were U.S., 60% non-U.S.). Comment on the influence of these values, together with those in part (a), on the timing for commercialization of FBR's.

REFERENCES

1. G. Robert Keepin, *Physics of Nuclear Kinetics*, Addison-Wesley Publishing Co., Inc., Reading, MA, 1965, 4.
2. DOE News Press Release, "DOE Reports Latest Estimate of U.S. Uranium Resources and 1979 U.S. Uranium Production and Exploration and Development Drilling," Grand Junction, Colorado Office, U.S. DOE (May 7, 1980).
3. *Energy: The Next Twenty Years*, Report by a study group sponsored by the Ford Foundation and Administered by Resources for the Future, H. H. Landsberg, Chairman, Ballinger Publishing Co., Cambridge, MA, 1979, 237.

CHAPTER 2

INTRODUCTORY DESIGN CONSIDERATIONS

2-1 INTRODUCTION

Before discussing the neutronics, systems, and safety considerations involved in designing fast breeder reactors, it is appropriate to follow the lead of Wirtz[1] in sketching the bases for fast breeder reactor designs. In this orientation, care will be taken to indicate the principal differences relative to thermal reactor systems with which the reader is more likely familiar.

This introduction to design begins with a brief discussion of major design objectives, followed by an overview of the mechanical and thermal systems designs of fast breeder reactors. Because of the position occupied by the sodium-cooled (liquid metal) fast breeder reactor (LMFBR) in the international breeder community, that system will be used for purposes of illustration. Chapter 17 focuses on another leading design, the Gas Cooled Fast Reactor (GCFR). This introductory description will be followed by considerations that have led to particular choices of fuels and major core parameters, coolant, and structural materials.

2-2 MAJOR DESIGN OBJECTIVES

Table 2-1 lists the principal goals of breeder reactor design and the specific reactor design objectives that accompany the goals.

Safe operation is mandatory for any nuclear reactor system. In terms of design objectives, safe operation translates into conscious awareness of safety at every stage of design, with the result of providing reliable components and adequate safety margins for potential accident situations. Part IV of this book (Safety) addresses these considerations in detail.

A high breeding ratio (BR) is needed to reduce fuel doubling time and thereby allow a sufficiently rapid rate of introduction of breeder reactors

TABLE 2-1   Goals and Design Objectives for the Breeder Reactor

Goals	Accompanying Design Objectives
Safe Operation	• Reliable Components • Adequate Containment Margin
High Breeding Ratio	• High Neutron Energy
Low Doubling Time	• High Breeding Gain • Low Fissile Specific Inventory
Low Cost	• High Burnup • Compatibility with LWR Cycle • Minimum Capital Costs

after uranium resources become scarce and U_3O_8 costs become excessive. As noted from Fig. 1-3 and Table 1-3, the best BR can be obtained with a fast neutron spectrum, although a reasonable value could be obtained in a thermal reactor if ^{233}U were used for fissile fuel.

A low doubling time is required in order for the breeder reactor to provide a significant fraction of the world's capacity for electricity generation during the first half of the next century. It is not sufficient simply to double the number of breeder reactors at the doubling rates of electrical demand. In order to continually increase the fraction of electrical capacity provided by breeder reactors, the doubling time must be shorter than the doubling time of electrical demand. The importance of high breeding gains and low fissile specific inventories in order to achieve low doubling time was illustrated by Eq. (1-11) in the previous chapter.

A very important objective of any energy generating system, of course, is low cost. From the standpoint of operational costs, overall fuel cycle considerations are of paramount importance. Hence, high burnup becomes an important objective, both for optimum utilization of high-fissile-content fuel and to minimize downtime for refueling. An additional cost objective is to mesh the breeder fuel cycle with that of the existing LWR cycle, which yields a considerable advantage for the ^{238}U-^{239}Pu system compared to the ^{232}Th-^{233}U system. Capital costs are directly related to hardware requirements and, therefore, put a premium on systems that require the least expensive capital equipment and still meet reliability criteria.

2-3   OVERVIEW OF MECHANICAL AND THERMAL SYSTEMS DESIGNS

This section provides an overview of some of the key features in a typical LMFBR system, as a prelude to the more detailed discussions in later chapters. Included are brief descriptions of core and blanket arrangements, the fuel configuration, vessel internals, and heat transfer systems. Discussions related to several of the design options under consideration are included as appropriate.

A. CORE AND BLANKET ARRANGEMENTS

Since the ability to breed fuel is the principal feature which distinguishes fast breeder reactors from thermal converter reactors, it is appropriate to ask how fertile and fissile fuels should be arranged to optimize breeding potential.

Two basic choices exist regarding where the breeding takes place. In the *external* breeding concept, all the fertile material is contained in the blanket surrounding the core; hence, all breeding takes place external to the core. In the *internal*, or *in-core* breeding concept, some fertile fuel is mixed with fissile fuel within the core fuel assemblies. Figure 2-1 illustrates the two basic breeding configuration concepts.

Some design consideration was given to the external breeding configuration for the small, early fast breeder reactor. Such an arrangement results in a very hard spectrum, a low fissile inventory, and a good breeding ratio. However, the lack of in-core breeding results in a rapid reactivity loss with burnup (requiring frequent fuel changes), low burnup, a small Doppler

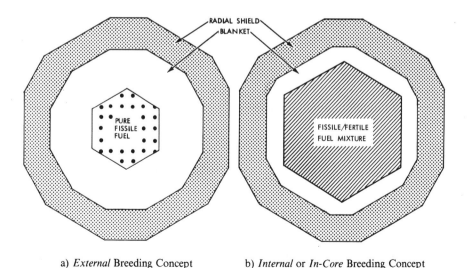

a) *External* Breeding Concept   b) *Internal* or *In-Core* Breeding Concept

FIGURE 2-1. External vs internal breeding core/blanket configurations.

effect, few fast fissions in fertile material, and requires a high fissile enrichment and a thick blanket. Consequently, all current designs allow for considerable internal or in-core breeding.

Whereas present reactors all allow internal breeding, they normally have an in-core breeding ratio less than unity. However, substantial breeding also takes place in the surrounding blankets, and this additional breeding raises the overall breeding ratio to well in excess of unity.

The configuration shown in Part (b) of Fig. 2-1 is known as a *homogeneous* core design because all assemblies containing pure fertile fuel are located in the radial and axial blanket regions. This leaves a relatively uniform or homogeneous mixture of fertile and fissile fuel spread throughout the core. An interesting variation of this internal breeding configuration is the *heterogeneous* core, in which blanket assemblies (containing pure fertile material) are distributed through the core region. These designs yield higher breeding ratios and reduced sodium void coefficients, but require higher fissile fuel inventories. While the early prototype reactors all utilized homogeneous-core designs, an advanced design of the Clinch River Breeder Reactor Plant incorporates a heterogeneous core.

A schematic top and side view of a typical homogeneous LMFBR core is given in Fig. 2-2a. The central region indicates the core itself (i.e., the region containing the initial load of fissile/fertile fuel), and the outer region represents the radial blanket and shielding arrangement. The core shown in this figure contains two enrichment zones. The outer zone incorporates a higher fissile fuel content than the inner zone in order to effect a flatter radial power distribution. Regions reserved for control are indicated on the figure. Boron-10, in the form of B_4C, is the absorber material most often used for this purpose. The reason for the triangular, or hexagon, shape of the core is given in Section 2-3B.

A heterogeneous core configuration is illustrated in Fig. 2-2b, with blanket assemblies at the center and in concentric rings. An early comparison[2] between 1000-MWe homogeneous and heterogeneous designs provided the results shown in Table 2-2.

The core of a fast reactor is smaller than that of a thermal reactor of comparable power. A thermal reactor is optimized at some particular fuel-to-moderator ratio, and any core smaller than optimum represents a less economic configuration. For a fast reactor, there are great incentives to minimize the size of the core, both for neutronics and cost considerations. No moderator is needed; in fact, a moderator is normally excluded by design in order to maintain a hard spectrum. By squeezing out as much coolant and structural material as possible to increase the fuel volume fraction, the neutron leakage is decreased, and the fissile fraction can be decreased. Viewed from a different perspective, reducing the core size of a fast reactor containing a given amount of fissile material increases reactivity.

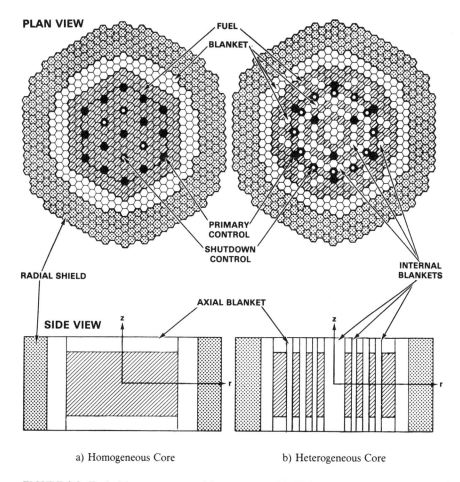

FIGURE 2-2. Typical homogeneous and heterogeneous LMFBR core/blanket arrangements wherein both schemes incorporate internal breeding.

B. FUEL LATTICE

Because of the incentive to minimize the specific fissile inventory in the breeder reactor, a "tight" fuel lattice which maximizes fuel volume fraction is desirable. A triangular lattice arrangement intrinsically allows a higher fuel volume fraction than a square lattice. The higher fuel volume fraction minimizes fissile loading mainly by reducing reactor leakage. Hence, the triangular lattice or hexagonal structure has been universally selected for breeder reactor design.

An LWR typically employs a square lattice because relatively larger spaces between fuel rods are needed to optimize the water-to-fuel ratio. The

TABLE 2-2 Comparison of Homogeneous and Heterogeneous Cores (1000-MWe Designs)[2]

	Homogeneous	Heterogeneous
Number of fuel assemblies*	276	252
Number of in-core blanket assemblies[†]	—	97
Number of ex-core blanket assemblies[†]	168	144
Number of control assemblies	19	18
Number of enrichment zones	3	1
Fissile mass (kg)	3682	4524
Pu fraction by zone (%)	13.5/14.9/20.3	20.2

*271 pins/fuel assembly, 1.22 m core height, 7.9 mm pin diameter, 0.36 m axial blankets

[†]127 pins/blanket assembly, 13 mm pin diameter

square lattice provides such space and also provides for easier mechanical assembly.

C. FUEL ASSEMBLY

Figure 2-3 is a simplified sketch of a typical LMFBR fuel assembly (also called a subassembly). For this standard triangular pitch lattice, the pins are separated by a spiral wire wrap and assembled as a cluster of 217 pins within the assembly duct. Alternatively, grid spacers can be used to separate the fuel pins. Fuel pellets make up the active core region, and blanket pellets provide the axial boundaries. For the design shown, the fission gas plenum is located above the upper blanket region, although there are substantial arguments in favor of a lower gas plenum region (below the lower blanket).

For neutronics calculations, it is necessary to define a unit cell, shown in Fig. 2-3, as a 1/12 segment of an assembly. The figure shows the geometric arrangement of the fuel, structure, and coolant. A table is also included to provide a ready reference for the relationship between concentric (hexagonal) rows of pins and the number of pins in an assembly. This same table can be used to determine the number of assembly positions in a core, given the number of assembly rows.

D. VESSEL INTERNALS

The core of an LMFBR is located on top of a core support structure, which is normally hung from the reactor vessel, or reactor tank, as shown in Fig. 2-4.

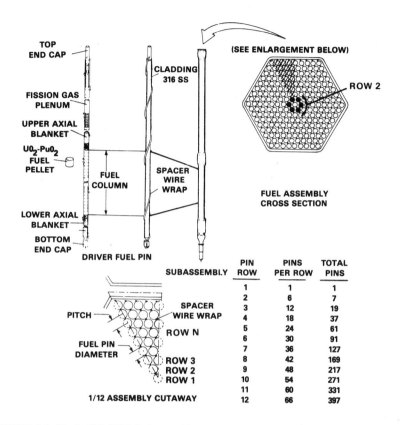

The following table appears within the figure:

SUBASSEMBLY	PIN ROW	PINS PER ROW	TOTAL PINS
	1	1	1
	2	6	7
	3	12	19
	4	18	37
	5	24	61
	6	30	91
	7	36	127
	8	42	169
	9	48	217
	10	54	271
	11	60	331
	12	66	397

FIGURE 2-3. Typical LMFBR fuel assembly system.

Because the LMFBR does not require pressurization to keep the coolant in a liquid state, outlet pressures are near atmospheric; consequently, the reactor vessel need only be thick enough to satisfy standard weight-bearing structural and safety requirements. A typical thickness may be the order of 30 mm, as compared to 300 mm for LWR systems.

Assemblies are slotted into positioning holes in the core support structure, and radial core restraint is normally provided at two axial locations (generally just above the active core and again near the top of the assemblies). Control rods enter from the top of the core, with the drive mechanisms located atop the vessel closure head. The vessel normally hangs from a support ledge, and the head is bolted to this structure. However, several variations in vessel support design have been used.

Sodium enters the lower inlet plenum, traverses upward through the core and blanket regions, and collects in a large reservoir above the core before exiting to the intermediate heat exchanger (IHX). An inert cover gas, usually argon, separates the sodium pool from the closure head.

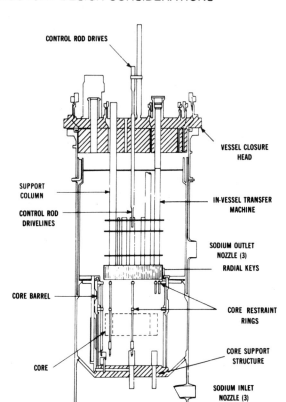

FIGURE 2-4. Typical loop-type LMFBR vessel internals.

E. SYSTEM HEAT TRANSFER

Neutron activation of the sodium coolant in the *primary* heat transport system of an LMFBR (discussed in Section 2-4B) requires that a *secondary*, or intermediate, heat transport system also be employed in an LMFBR. Hence, sodium coolant in a primary loop is pumped through the core and circulated through an intermediate heat exchanger and a secondary sodium loop transports heat from the IHX to the steam generator. The temperature of the sodium leaving the reactor (of the order of 550°C) is substantially higher than the coolant in an LWR (~300°C); hence, steam temperatures in the LMFBR are higher.

In contrast to the secondary sodium loop, the radioactive sodium in the primary loop must be shielded from plant personnel. Two different arrangements can be used to accomplish this; the IHX and the primary pump can be placed inside the reactor tank, or these two components can be located in adjacent hot cells with pipes connecting them to the reactor vessel. The first option is called the *pool system*, and the second is termed the *loop system*.

Simplified pool and loop design options are shown in Fig. 2-5. In the pool system, hot sodium flows directly into the IHX and then into a large pool that surrounds the vessel internals. A primary pump then recirculates this cool sodium into the core inlet plenum. In the loop system, two important design questions are the location of the primary pump (hot or cold leg) and the vessel penetration location for coolant inlet piping. The pros and cons of pool vs loop designs and the two pump locations will be covered in Chapter 12.

Figure 2-6 is a simplified schematic of the heat transport system for a pool design. With the exception of special considerations in steam generator

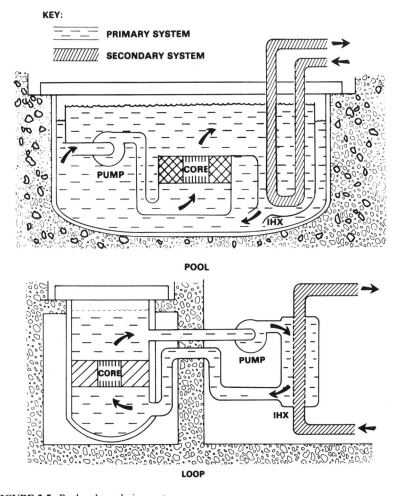

FIGURE 2-5. Pool vs loop design systems.

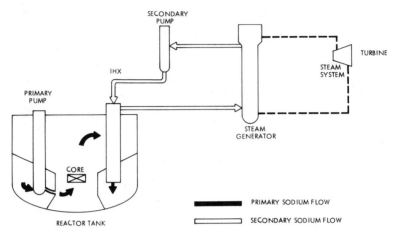

FIGURE 2-6. Pool-type reactor heat transport system.

design for a sodium/water system, the steam cycle side of the LMFBR system is quite similar to a thermal reactor or fossil-fuel power plant. It should be noted, however, that higher steam temperatures associated with the LMFBR lead to a higher overall plant thermal efficiency than for an LWR plant.

2-4 SELECTION OF CORE MATERIALS AND PARAMETERS

Having introduced the basic mechanical and thermal systems designs of an LMFBR, we can next address several obvious questions. What fuel is preferable for fast reactors? Why has sodium been so widely proposed as the coolant and are other alternatives attractive? What materials are used for cladding and other structures? Principal considerations that influence the answers to these questions are discussed briefly in this section. Chapter 11 deals with materials properties and selections in more detail.

In addition, we shall discuss basic core parameters related to the fuel—in particular the fuel fissile fraction, the fuel-pin size, and fuel burnup.

A. FUEL

Fuel Candidates and Properties

The prime candidates for FBR fuel are uranium-plutonium oxide, carbide, metal, and possibly nitride. Thorium-uranium-233 fuel can also be used in

FBR's, but higher breeding ratios and more favorable economics can be achieved with uranium-plutonium.

Generally desirable properties for fuels are the capability for irradiation to high exposures in terms of energy generated per unit mass (i.e., high burnup) and high values of linear power (i.e., kilowatts per unit length of fuel pin). As shown later in this section, high linear power implies either high thermal conductivity or high melting point.

Oxide fuel (i.e., a mixture of UO_2 and PuO_2) became the favored reference fuel in the early 1960s because of the demonstrated high-burnup capability (perhaps in excess of 100 MWd/kg). The large body of oxide fuel experience developed within the LWR program also provided considerable impetus to selecting this fuel for the base FBR program. Oxide fuel has a high melting temperature ($\sim 2750°C$), which largely compensates for its poor thermal conductivity. The neutron spectrum of an oxide core is softer than that of a metal-fueled breeder due to moderation by the oxygen; this softer spectrum allows a strong negative Doppler coefficient. Oxide fuel leads to lower power density compared to metal fuels, and this places less stringent requirements on the cooling system (indeed, allowing gas cooling to become a possibility). Apart from its low thermal conductivity, the principal disadvantage of oxide fuel is its somewhat lower breeding ratio (due to the softer neutron spectrum and lower density).

Carbide (UC-PuC) and nitride (UN-PuN) fuels are being given serious consideration as alternatives to oxide fuel, primarily because of their higher thermal conductivity. Their higher linear powers lead to a lower specific fissile inventory. The presence of only one moderator atom per heavy metal atom instead of two for oxide, plus a higher metal density due to the more compact lattice structure of UC, leads to a harder spectrum and, therefore, to higher breeding ratios. Higher breeding gain and lower specific fissile inventory both result in lower doubling times. However, the data base for these advanced fuels (particularly in the case of nitrides) is far from developed.

The early U.S. and U.K. experimental fast reactors utilized metal fuel. Metal fuel has a distinct advantage over oxide and slight advantage over carbide from the standpoint of breeding ratio. The neutron spectrum with metal fuel is hard since no moderation from oxygen, carbon, or nitrogen is present, and this leads to high breeding ratios. Also, the fissile concentration is high, which leads to a good doubling time. Furthermore, the thermal conductivity of metal fuel is high and this leads to a high linear power. Although its melting temperature is low, its high thermal conductivity offsets this to provide sufficiently high linear power. A serious disadvantage of metal fuel may be the difficulty in achieving high burnup. Metal fuel swells even at low burnups, and the only way to accommodate such swelling is to build in considerable porosity. Although this tends to negate some of its principal advantages, recent favorable burnup experience with metal fuel allows it to remain a contender for FBR fuel. Another disadvantage is that the neutron spectrum is so hard that the Doppler effect (important for safety reasons) is small. A strong negative Doppler coefficient is not crucial for metal fuels, however, since metal fuel is generally believed to exhibit a more reliable thermal axial expansion coefficient than ceramic fuels.

Fuel Fissile Fraction

The fissile fraction, or "enrichment," required to attain criticality in a fast reactor is considerably higher than that required for thermal reactors because of the low fission cross sections at high neutron energies. Plutonium fractions typical of commercial size LMFBR's are in the 12% to 30% range, depending on reactor size and configuration, with about 75% of the plutonium being fissile material (i.e., ^{239}Pu and ^{241}Pu). Hence, fissile fractions are in the range of 9% to 23%.

Figure 2-7 shows the fission cross sections of the principal fissile fuels as a function of energy. Note that σ_f for the fast reactor is approximately two barns, whereas the comparable cross section in a thermal reactor is of the order of 500 barns. The effect of this decrease in cross section at high neutron energies can be partially offset by fast fission in fertile fuel. (See Fig. 2-8.) It must also be recognized that all parasitic absorption cross sections are lower for a high energy spectrum. Nonetheless, the low fission

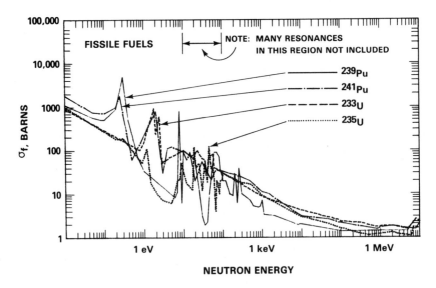

FIGURE 2-7. Fission cross sections vs energy for common fissile fuels (adapted from Ref. 3).

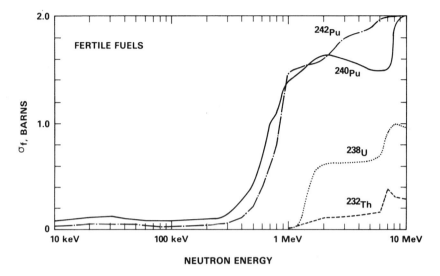

FIGURE 2-8. Fission cross sections vs energy for common fertile fuels (adapted from Ref. 3).

cross section results in a fast reactor requirement of four to five times the fissile fraction normally found in a thermal reactor.

Fuel-Pin Diameter

In LWR's the fuel-pin diameter is controlled by heat-transfer considerations. The same is not the case for FBR's, and it is useful to understand this important difference. One of two heat transfer considerations controls fuel-pin design—either maximum allowable heat flux or maximum center fuel temperature, and only the first of these affects pin diameter.

The surface heat flux can be expressed as

$$q = \frac{\chi}{\pi D},\qquad(2\text{-}1)$$

where

$$q = \text{heat flux } (W/m^2)$$

$$\chi = \text{linear power } (W/m)$$

$$D = \text{pin diameter (m)}.$$

For a water-cooled thermal reactor, this relationship defines a practical minimum to the pin diameter (to prevent cladding burnout). Because of the excellent cooling capabilities of sodium, however, this burnout limit does not hold for LMFBR's. Hence, the diameter of an LMFBR fuel pin can be appreciably less than that of a water-cooled reactor fuel pin.

The center fuel temperature is governed by the *linear power*, or linear heat rating, which can be determined from the expression:*

$$\chi = 4\pi \int_{T_s}^{T_0} k(T) \, dT \qquad (2\text{-}2)$$

where

$k(T)$ = thermal conductivity of fuel (watts/m·°C)

T_s = fuel surface temperature (°C)

T_0 = fuel center temperature (°C).

This expression limits the linear power in an FBR since T_0 must be below the fuel melting temperature. Note, however, that this expression is independent of pin diameter; heat extracted from the pin is a function only of fuel center and surface temperatures and thermal conductivity of the fuel.

It is for these reasons that heat transfer in FBR's does not control fuel-pin diameter. It is also of interest to observe from Eq. (2-2) that for a given power rating, the total length of pins in the core is to a great extent fixed by the fuel selected and the maximum T_0 allowed.

The pin diameter does, however, have a significant influence on economics; for example, it has an important effect on fissile specific inventory, and a primary design objective for a breeder reactor is to minimize this parameter. As the pin diameter is reduced, the fuel mass per pin is decreased and the required fissile fraction is increased. It turns out that the fissile mass per pin (i.e., the product of fuel mass and fissile fraction) is also reduced as the pin diameter is reduced. Hence, from the point of view of minimizing fissile specific inventory, a small pin diameter is desirable. Breeding ratio, however, decreases with decreasing pin diameter (e.g., see Table 7-3).

Other considerations are involved in placing a lower limit on fuel pin diameter, such as fabrication costs, pitch-to-diameter ratio, neutron leakage, and fuel volume fraction. These considerations typically lead to LMFBR mixed-oxide fuel-pin diameters from 6 to 9 mm, or approximately two-thirds those of water cooled reactor fuel pins. (See Chapter 8-2B for a more detailed discussion of fuel-pin diameter.)

---

*This expression will be derived in Chapter 9.

Fuel Burnup

Burnup requirements for an economical breeder reactor are considerably more demanding than for an LWR. This arises mainly from the difference in fissile enrichment involved. Whereas a burnup of 30 MWd/kg for a typical LWR is acceptable, and burnups as low as 7 MWd/kg can be accommodated for a natural uranium reactor, burnups of the order of 100 MWd/kg are required for economic operation of the LMFBR, resulting directly from the higher fissile fraction.

This high burnup requirement for an economical breeder is the principal reason for the world-wide shift in emphasis during the early 1960s from metal-fueled to oxide-fueled reactors. Relative to the LWR, this requirement translates into a need to get more energy out of each kilogram of heavy metal fabricated and loaded into an LMFBR before returning the fuel for reprocessing and refabrication.

B. COOLANT

The primary requirements for a breeder reactor coolant are as follows: (1) it must minimize neutron moderation, (2) it must remove heat adequately from a high-power-density system (approximately a factor of four higher than for an LWR), and (3) it must minimize parasitic neutron absorption. The first requirement automatically eliminates water and any organic coolant, but does allow gases or liquid metals of intermediate atomic number to be considered. Principal breeder coolant candidates, including a summary of their primary advantages and disadvantages, are discussed below. More detailed discussions of coolants appear in Chapters 9 and 11.

Principal Coolant Candidates

Liquid metals were the first candidates investigated for cooling a breeder reactor because they easily met the cooling requirements. However, helium and steam have also been given considerable thought, primarily in connection with the switch from metal to ceramic fuels, in which accompanying lower power densities allow the gas coolant to fulfill the basic heat transfer requirements.

Property Considerations for Coolants

With regard to liquid metals, sodium, NaK, mercury, bismuth and lead have all been considered. The latter three have generally been discarded because their high densities (and, consequently, high mass flow rates) require excessive pumping power. The sodium/potassium solution, NaK, was used in early designs because of its low melting point. Sodium melts at 98°C, requiring heaters somewhere in the system, whereas NaK is a liquid at room temperature. Potassium, however, is a fairly strong neutronic absorber. Hence, sodium is the only liquid metal presently given serious consideration for cooling central station breeder reactor plants.

Sodium has excellent thermal properties, has a high boiling point (880°C at one atmosphere pressure), and is quite compatible with the standard cladding material being employed. The high boiling point of sodium allows low pressure operation—in marked contrast to the pressurized systems required for LWR operation. The prime disadvantages of sodium are its neutronic activation properties (since ^{23}Na absorbs neutrons to form ^{24}Na, which decays with a 15-hour half life, and ^{22}Na, with a 2.6 d half life), its opaqueness, which somewhat complicates certain operating procedures such as refueling, and its chemical incompatibility with water and air. The 1.4- and 2.8-MeV gamma rays from the ^{24}Na decay process are sufficiently penetrating that an intermediate cooling loop is required to eliminate the possibility of radioactive contamination in the steam generator. This loop further protects the core from pressure surges or reactivity effects of hydrogen moderation should a steam generator leak result in a sodium-water reaction.

Helium gas can be used as a coolant for ceramic-fueled breeder reactors, with even less net degradation on the neutron spectrum than sodium. Although the atomic weight of He is lower than for Na, its atom density in gas form is much lower. Cladding compatibility with He is excellent since He is inert, and there is no neutron activation problem. Hence, the intermediate loop employed in a sodium cooled system would not be necessary. On the other hand, a He coolant system must be highly pressurized (~ 8 MPa, or 1200 psi) in order to meet cooling requirements, and considerably more pumping power is required than for a sodium system.

Steam received substantial attention in early designs, but was ruled out because of excess cladding corrosion and problems related to high pressure operation. (Fuel pins required internal pressurization.) Steam cooling also causes appreciable neutron spectrum degradation, which results in low breeding ratio and fuel cycle cost penalties.

At present, all countries actively pursuing the breeder reactor option have selected sodium as the reference coolant. Steam has essentially been discarded, but He is still actively being pursued as a backup, primarily in the

U.S., the U.K. (where considerable experience in gas cooling technology exists), and Germany. The Soviet Union has developed a fast breeder design utilizing N_2O_4 as a gas coolant.

C. STRUCTURE

The prime requirements for the structural material used in the core of a breeder reactor are (1) maintenance of integrity at high temperatures, (2) capability to withstand high fluence*, (3) compatibility with fuel and coolant, and (4) low parasitic neutron absorption. As in the case of coolant candidates, a summary of the major considerations is given below, with details left for coverage in Chapter 11.

Principal Structure Candidates

Stainless steel is the only material given serious consideration for breeder reactor cladding and assembly duct material, although considerable work has focused on specific variations in composition and heat treating to best match the particular structural function with material requirements. Zircaloy, the standard cladding material for LWR application, is not suitable because its strength is not compatible with the high-temperature fields in a fast breeder reactor.

Property Considerations for Structural Materials

Stainless steel has good strength and corrosion-resistant properties at high temperature and fluence and absorbs few neutrons. Most stainless steels swell significantly under the high fluences encountered in LMFBR operation, however, and a substantial developmental program has been directed toward solving this problem. Stainless steel 316, 20% cold worked, is the material most commonly used for cladding and assembly ducts because of its good high temperature properties and its resistance to swelling.

---

*Fluence is time integrated flux (ϕt or nvt), which represents a measure of the neutron bombardment that must be withstood by the structural material.

REVIEW QUESTIONS

2-1 What is meant by the term "fissile specific inventory" for the fissile fuel? Why is a low fissile specific inventory a major design objective for the breeder?

2-2 What is the difference between the "internal" and the "external" breeder? Which concept is favored and why?

2-3 Why are LMFBR pins smaller in radius than LWR pins?

2-4 How does the fissile fraction of an LMFBR core compare with that of an LWR?

2-5 Why is high burnup an economic requirement for the fast breeder reactor?

2-6 Why is a triangular (hexagonal) lattice used in fast breader reactors rather than the square lattice employed in LWR's?

2-7 What was the principal motivation for shifting from metal to oxide fuels in the fast breeder reactor program?

2-8 How does the shift from metal to ceramic fuels affect the choice of coolant?

2-9 Why is an LMFBR core smaller than a thermal reactor core of comparable power output?

2-10 What is the essential difference between the pool and loop type primary coolant system in an LMFBR?

2-11 What are the properties of sodium that make it so desirable as a coolant for a fast reactor?

REFERENCES

1. Karl Wirtz, *Lectures on Fast Reactors*, Kernforschungszentrum, Karlsruhe, 1973.
2. E. Paxon, Editor, *Radial Parfait Core Design Study*, WARD-353, Westinghouse Advanced Reactors Division, Waltz Mill, PA, June 1977.
3. D. I. Garber and R. R. Kinsey, *Neutron Cross Sections*, *Volume II*, *Curves*, BNL 325, Third Edition, Brookhaven National Laboratory, January 1976.

CHAPTER 3

ECONOMIC ANALYSIS OF NUCLEAR REACTORS

By P. S. Owen and R. P. Omberg*

...the ideas of economists and political philosophers, both when they are right and when they are wrong, are more powerful than is commonly understood. Indeed, the world is ruled by little else. Practical men, who believe themselves quite exempt from any intellectual influences, are usually the slaves of some defunct economist. I am sure the power of vested interests is vastly exaggerated compared to the gradual encroachment of ideas...soon or late, it is ideas, not vested interests, which are dangerous for good or evil.

John Maynard Keynes,
*The General Theory of
Employment, Interest,
and Money*

3-1 INTRODUCTION

Any enterprise, if it is to be viable for an extended period of time, must be able to sell a product for more than its cost. Or, insofar as the enterprise is concerned, revenues must be sufficient to cover expenditures. Since the very existence of the enterprise depends on this principle, an economic assessment of any project is an essential part of the decision to pursue it. This is particularly true for a project such as a power plant which entails a large investment.

The decision to construct a power plant involves both the disciplines of engineering analysis and economic analysis. Engineering analysis is used to determine the design of the plant, the efficiency of the plant, the life of components within the plant, and the time required to construct the plant. Economic analysis, on the other hand, is used to determine the cost of

---

*Westinghouse Hanford Company.

constructing the plant as well as the cost of operating the plant over an extended period of time.

Both engineering analysis and economic analysis should be employed in a balanced fashion in order to ensure the success of a large engineering project. For example, if engineering is emphasized at the expense of economics, the result will probably be a well-designed but economically uncompetitive plant. In contrast, if economics is emphasized at the expense of engineering, the result will probably be a plant which is initially inexpensive but which will subsequently incur large operating costs. Thus, a comprehensive understanding of both disciplines is desirable. Such an understanding of the engineering aspects is developed in other chapters; this chapter will develop a fundamental understanding of the economic aspects.

In the case of a nuclear power plant, the product which is ultimately sold in the marketplace is electricity. Consequently, it is important to know the average amount to charge for electricity since this determines revenues, and revenues must be sufficient to cover costs. Estimating the charge for electricity is particularly difficult in the case of a power plant because costs associated with the plant are incurred over several decades. The principal costs which are incurred are

(1) the *capital cost* which is incurred while constructing the plant,
(2) the *fuel cost* which is incurred while operating the plant,
(3) the *operation and maintenance cost* which is incurred over the life of the plant,
(4) *income taxes* which are incurred as a consequence of engaging in a private enterprise, and
(5) *other costs* which are incurred as a consequence of engaging in any commercial enterprise.

Capital and fuel costs are often the major costs and are also the costs which can be strongly influenced by engineering design. For example, there is a large difference in capital costs between a current reactor such as a light-water reactor (LWR) and an advanced reactor such as a fast breeder reactor (FBR). The standard LWR on the once-through cycle has a relatively low capital cost; however, the fuel cost will continually increase as the price of uranium increases. In contrast, the FBR has a relatively high capital cost but a fuel cost which is insensitive to uranium price. Thus, the LWR tends to be fuel intensive while the FBR tends to be capital intensive.

In order to calculate a single average charge for electricity for each of these plants, it is necessary to develop a method which handles capital and fuel expenditures in a consistent manner. Since capital expenditures occur during the construction period and fuel expenditures occur over the life of the plant, it is necessary to compare expenditures which occur at different times. A method for accomplishing this will be developed in the next section.

3-2 BASIC CONCEPT OF THE TIME VALUE OF MONEY

The stock which is lent at interest is always considered as a capital by the lender. He expects that in due time it is to be restored to him, and that in the meantime the borrower is to pay him a certain annual rent for the use of it. The borrower may use it either as a capital, or as a stock reserved for immediate consumption. If he uses it as a capital, he employs it in the maintenance of productive labourers, who reproduce the value with a profit. He can, in this case, both restore the capital and pay the interest without alienating or encroaching upon any other source of revenue. If he uses it as a stock reserved for immediate consumption, he acts the part of a prodigal, and dissipates in the maintenance of the idle, what was destined for the support of the industrious. He can, in this case, neither restore the capital nor pay the interest without either alienating or encroaching upon some other source of revenue, such as the property or the rent of land.

Adam Smith,
*An Inquiry Into
the Nature and Causes of
the Wealth of Nations*

Money is a valuable asset—so valuable, in fact, that individuals and organizations are willing to pay additional money in order to have it available for their use. This is indicated by the continual willingness of banks and savings institutions to pay for the use of money. These institutions provide a market which determines the interest rate. We will use this interest rate to develop the time value of money. For example, consider $100 borrowed from a bank for one year at an interest rate of 5 percent per year. Clearly, at the end of one year the bank will expect to receive $105. This sum includes both a *return of* the investment ($100) plus a *return on* the investment ($5).

More generally, consider an amount C which is borrowed for one year at an annual interest rate i, as shown in Fig. 3-1. At the end of one year, the investment C must be repaid. In addition, an amount iC representing a return on the investment must also be paid. Therefore, the total amount that must be paid back at the end of the year is

$$\text{Pay back} = C + iC = C(1+i). \tag{3-1}$$

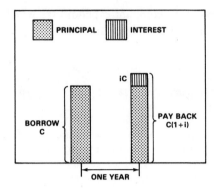

FIGURE 3-1. Borrowing money for one year.

Let us now consider a more complex example. Suppose it is known that an expense of $100 will be incurred one year in the future. In order to pay the expense in one year, one may wish to set aside the appropriate amount of money now. This amount could, of course, be invested such that the future expense of $100 is covered exactly. If the interest rate on the investment is 5 percent per year, then $95.24 must be invested now. That is, $95.24 (1+0.05)=$100.00.

This concept, of course, can be generalized for any amount of money, as shown in Fig. 3-2. If an expense C will be incurred one year in the future, an amount C' may be set aside such that the future expense C will be covered. If C' is invested at an interest rate i, the amount available to pay out at the end of one year is

$$\text{Pay out} = C' + iC' = C'(1+i). \tag{3-2}$$

Since we intend to use the amount $(1+i)C'$ to cover the expense C, we would like

$$C'(1+i) = C. \tag{3-3}$$

Therefore, the amount C' that must be invested at the beginning of the year in order to cover an expenditure C at the end of the year is

$$C' = \frac{C}{1+i}. \tag{3-4}$$

These two examples will now be extended to develop the concept of the *current value* of an expense or revenue occurring at any point in time. With

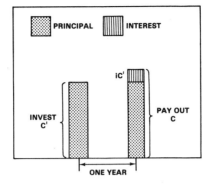

FIGURE 3-2. Meeting a known expense at the end of one year.

this concept, one can find an amount of money in any year which is equivalent to an amount in any other year. This concept, known as the *present value concept*, is very useful when analyzing the economics of a nuclear power plant.

It is useful to be able to determine the current value of either a cost which has occurred in the past or one which will occur in the future. Consider first the case of finding the present value of a cost which has occurred in the past. Assume for example, that a cost C was incurred two years ago, as shown in Fig. 3-3. If this cost had not been incurred, the amount C would

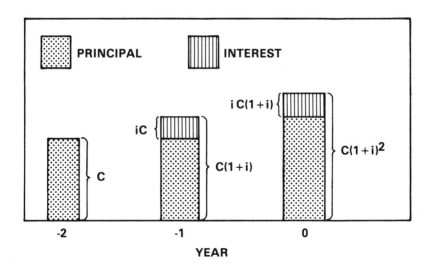

FIGURE 3-3. Present value of an expense in the past.

have been available for investment two years ago. If it had been invested, the amount C would increase in value as follows:

Available for investment two years ago: C
Value of investment one year ago: $C+iC=C(1+i)$
Value of investment today: $C(1+i)+i[C(1+i)]=C(1+i)^2$.

Thus, an *investment* of C two years ago has increased in value to the amount $C(1+i)^2$ today. Alternatively, we may say that the current value of an *expenditure* C which occurred two years ago is $C(1+i)^2$. More specifically, if we did not have that expenditure two years ago and invested it instead, we could have met an expenditure $C(1+i)^2$ today. This is known as the present value of a past expense C. This approach may be extended to an expense C which has occurred n years in the past. The reader should be able to verify that the present value of such a past expense is $C(1+i)^n$.

A similar technique can be used for finding the present value of an expense which will occur in the future. For example, assume that the amount C must be paid out in two years, as shown in Fig. 3-4. This expense could be met by investing an amount C' as follows:

Amount of investment today: C'
Value of investment in one year: $C'+iC'=C'(1+i)$
Value of investment in two years: $C'(1+i)+i[C'(1+i)]=C'(1+i)^2$.

Thus, an investment of C' today will increase in value to the amount

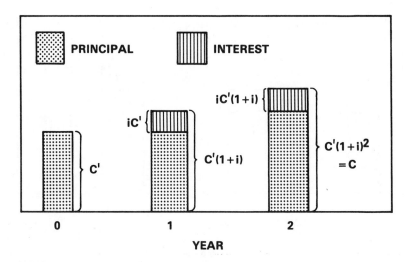

FIGURE 3-4. Present value of an expense in the future.

$C'(1+i)^2$ in two years. If we want an investment of C' today to cover an expenditure C in two years, then we would like $C'(1+i)^2$ to equal C. The amount C' is known as the present value of the future expense C, and

$$C' = \frac{C}{(1+i)^2}. \tag{3-5}$$

This approach may be extended to an expense C which will occur n years in the future. The reader should be able to verify that the present value of such a future expense is $C/(1+i)^n$.

To improve our understanding of the present value concept, let us consider a simple example involving nuclear fuel. The costs associated with this fuel involve (1) a purchase which occurs before the fuel is loaded into the reactor, and (2) an expenditure which occurs after the fuel is discharged from the reactor. The timing of a typical set of expenditures is shown in Fig. 3-5. Since the expenditures occur at different times, the complete cost of the fuel is determined by both the magnitude and the timing of the expenditures. In order to calculate the present value of the expenditures, it is necessary to choose a reference point in time. For convenience, we choose to present-value all expenditures to the time at which the fuel is loaded into the reactor. With this as a reference point, the front-end expenditure F has occurred one year in the past while the back-end expenditure B will occur four years in the future. The complete fuel cost is the sum of the present value of the expenditures, or $F(1+i)^1 + B/(1+i)^4$.

The present value concept can be used to find a single cost for fuel, as shown above. In addition, the present value concept may be applied to other components of the cost of a nuclear power plant. For example, with some additional analysis, it can be used to determine the amount which must be charged for electricity in order to cover the capital cost of the plant.

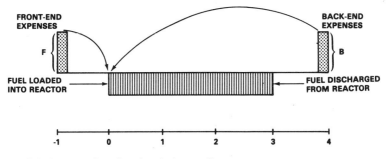

FIGURE 3-5. Present value of nuclear fuel expenditures.

However, before any further applications of the present value technique are developed, we will digress briefly to consider one of the key elements of any economic analysis.

3-3 COST OF MONEY

A key component of any economic analysis is the cost of money, or the interest rate i. This value is a function of the financial arrangements developed for the project, and this will vary from project to project. Therefore it is important to carefully select the most appropriate value for the project under consideration. Nuclear power plants are often financed with a combination of bonds (debt) and stocks (equity). When this is the case, a single effective interest rate or cost of money may be defined as the weighted mean rate of return on debt and equity. This may be expressed as

$$i=(b\times i_b)+(e\times i_e), \tag{3-6}$$

where

$i=$ effective interest rate
$b=$ fraction of funds obtained by debt (or bonds)
$i_b=$ debt (or bond) interest rate
$e=$ fraction of funds obtained by equity (or stocks)
$i_e=$ equity (or stock) interest rate.

Let us calculate an effective interest rate using typical values for bond and equity interest rates. For example, a *deflated* bond interest rate of 2.5 percent per year is often used in economic calculations. Deflated, in this case, implies a real rate of return of 2.5 percent per year and does not include the effect of inflation. Inflation, when included, simply raises the rate of return by the inflationary rate itself. A similarly deflated value for the equity interest rate would be 7 percent per year. It is common for utilities to finance nuclear power plants with a structure of 55 percent bonds and 45 percent stocks. In this case, the effective deflated interest rate would be

$$i=(0.55\times 0.025)+(0.45\times 0.07)=0.045, \text{ or } 4.5\%. \tag{3-7}$$

An effective interest rate calculated in this fashion will be used in the subsequent development of the major cost components, including capital cost and the taxes associated with capital expenditures, and fuel costs and the taxes associated with fuel expenditures.

3-4 CAPITAL COST

All production is for the purpose of ultimately satisfying a consumer. Time usually elapses, however—and sometimes much time—between the incurring of costs by the producer and the purchase of the output by the ultimate consumer. Meanwhile the entrepreneur has to form the best expectations he can as to what the consumers will be prepared to pay when he is ready to supply them after the elapse of what may be a lengthy period; and he has no choice but to be guided by these expectations, if he is to produce at all by processes which occupy time.

John Maynard Keynes,
*The General Theory of
Employment, Interest,
and Money*

A. CONSTRUCTION COST

The construction of a power plant occurs over an extended period of time and involves a sequence of nonuniform expenditures, as shown in Fig. 3-6. The complete capital cost of the plant is determined by both the magnitude and timing of these expenditures. In order to apply the present value concept, it is necessary to choose a point in time at which to value the

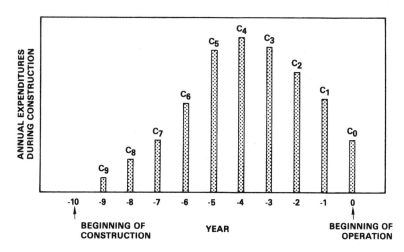

FIGURE 3-6. Annual construction costs.

expenditures. For convenience, we will choose the time at which the reactor begins to generate electricity as a reference point.

The total capital cost C is the sum of the present value of all expenditures. If the construction period extends for N years, the total capital cost including interest during construction may be expressed as

$$C = C_0 + C_1(1+i) + C_2(1+i)^2 + \cdots + C_{N-1}(1+i)^{N-1}, \qquad (3\text{-}8)$$

or

$$C = \sum_{k=0}^{N-1} C_k(1+i)^k, \qquad (3\text{-}9)$$

where we have assumed that the payment is made at the end of the year.

B. PAYING BACK A CAPITAL INVESTMENT

After the plant is constructed, it is necessary to recover revenues sufficient to repay the original investment. Let us develop the technique for calculating this revenue requirement with a simple example. For the sake of illustration, let us consider an initial investment of \$5000. Suppose we intend to pay back this investment over a five year period with an annual interest rate of 5 percent. There are several methods which could be employed to pay back the investment. One such method would involve five equal payments of \$1000 in order to return the principal. In addition to returning the principal, the interest on the outstanding balance must also be repaid each year. The total payment at the end of any year k is the sum of the two, and is given by

$$\text{Payment at end of year } k = \left(\begin{array}{c}\text{payment to}\\\text{principal}\end{array}\right) + \left(\begin{array}{c}\text{interest on out-}\\\text{standing balance}\end{array}\right)$$

$$= \frac{5000}{5} + 0.05\left[5000 - (k-1)\frac{5000}{5}\right]. \quad (3\text{-}10)$$

The annual payments for this method of paying back a capital investment are shown in Fig. 3-7. The reader will note that the total annual payment decreases as the outstanding balance decreases.

In some cases, a uniform—rather than a decreasing—set of payments is more convenient. If \$5000 is borrowed for five years at an interest rate of 5 percent, the original investment could be paid back with uniform payments,

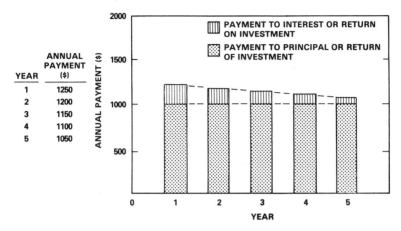

FIGURE 3-7. Repayment of a $5000 capital investment with equal payments to principal.

as shown in Fig. 3-8. The reader will note that, while the total annual payment is constant, the payment to principal increases in each succeeding year while the payment to interest simultaneously decreases.

Since uniform annual payments are often preferred, let us examine this method in more detail. Again, consider the previous example in which the total annual payment is $1155. This can be regarded as a uniform annual cost to the project, and let us call this quantity C_u. Since both principal and interest must be paid, each payment C_u is divided between return of the investment and return on the investment. The division between payments to principal and interest is shown in Table 3-1. This table was constructed by

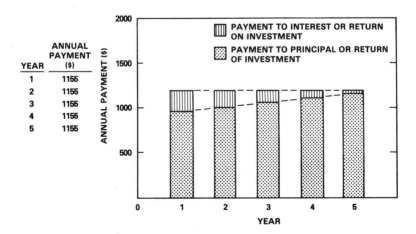

FIGURE 3-8. Repayment of a $5000 capital investment with equal annual payments.

first considering the outstanding debt in any year k, denoted here as D_k. In the same year k, the payment to interest I_k is $0.05 \times D_k$. The payment to principal P_k is then given by

$$P_k = C_u - I_k = C_u - 0.05 \times D_k. \tag{3-11}$$

Consequently, the outstanding debt at the beginning of year $k+1$ is

$$D_{k+1} = D_k - P_k = D_k - (C_u - 0.05 \times D_k), \tag{3-12}$$

and one must continue to make the annual payment C_u until the outstanding debt D_{k+1} is reduced to zero. In this case, the original capital investment will be completely paid back at the end of five years.

TABLE 3-1 Payments to Principal and Interest with the Uniform Annual Payment Method

Year (k)	Outstanding Debt at Beginning of Year (D_k)	Payment to Interest $(I_k = 0.05 \times D_k)$	Payment to Principal $(P_k = C_u - 0.05 \times D_k)$	Outstanding Debt at End of Year (D_{k+1})
1	5000	250	905	4095
2	4095	205	950	3145
3	3145	157	998	2147
4	2147	108	1047	1100
5	1100	55	1100	0

This example has verified that a uniform annual payment of $1155 will completely return the original capital investment over five years. However, it is not evident how one initially determines this payment. This can be accomplished by considering the problem from the viewpoint of the lender. Recall that the lender originally loaned $5000, which was to be returned in five equal annual payments. The lender, considering the time value of money, will require $5000 $(1.05)^5$ at the end of five years. This is because, by lending the money, he foregoes the opportunity to invest it. Thus, the sum of the present value of the payments C_u must be $5000 $(1.05)^5$ at the end of five years—or, as shown in Fig. 3-9,

$$5000 \, (1.05)^5 = C_u + C_u(1.05) + C_u(1.05)^2 + C_u(1.05)^3 + C_u(1.05)^4. \tag{3-13}$$

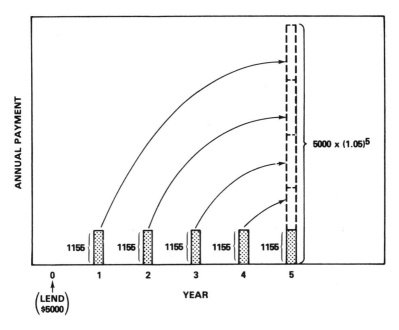

FIGURE 3-9. Uniform annual payment for a capital investment.

Solving for C_u,

$$5000 = \left(\frac{1}{(1+0.05)^5} + \frac{1}{(1+0.05)^4} + \frac{1}{(1+0.05)^3} \right.$$
$$\left. + \frac{1}{(1+0.05)^2} + \frac{1}{(1+0.05)} \right) C_u \qquad (3\text{-}14)$$

and the annual payment C_u is found to be $1155.

In more general terms, if an initial investment C is to be paid back uniformly over K years at an annual interest rate of i, then

$$C = \left[\frac{1}{(1+i)^K} + \frac{1}{(1+i)^{K-1}} + \cdots + \frac{1}{(1+i)^3} + \frac{1}{(1+i)^2} + \frac{1}{(1+i)} \right] \times C_u. \qquad (3\text{-}15)$$

Solving for the uniform annual payment C_u gives

$$C_u = \frac{C}{\displaystyle\sum_{k=1}^{K} \frac{1}{(1+i)^k}}. \qquad (3\text{-}16)$$

Regarding $1/(1+i)$ as a term in a geometric progression, one can show with elementary algebra that C_u can be expressed as

$$C_u = C \times \left[\frac{i \times (1+i)^K}{(1+i)^K - 1} \right].$$ (3-17)

This expression is commonly known as the *sinking-fund* repayment equation or the *amortization equation.*

The annual payment C_u, which represents a cost to the project, must be covered by revenues obtained from the sale of electricity. Revenues are determined by the amount charged for a unit of electricity multiplied by the number of units produced. If we define L_{cap} as the amount charged per unit of electrical energy in order to cover capital expenditures and E as the amount of electrical energy produced in any year, then

$$\text{Revenue} = L_{cap} \times E.$$ (3-18)

The revenue in each year will cover the expenditure in the year if $L_{cap} \times E = C_u$. Thus, the constant, or *levelized*, charge for electricity sufficient to cover capital expenditures is

$$L_{cap} = C_u / E,$$ (3-19)

and recalling the sinking fund repayment equation

$$L_{cap} = \frac{C}{E} \times \left[\frac{i \times (1+i)^K}{(1+i)^K - 1} \right].$$ (3-20)

C. FIXED CHARGES ASSOCIATED WITH A CAPITAL INVESTMENT

In the previous section, a method was developed for calculating the return of and the return on an investment. There are, however, other annual expenses associated with capital investments such as property insurance, property taxes, and replacement costs. These are collectively known as *fixed charges* and are usually expressed as a fixed percent of the original capital investment. In fact, since the value of the property decreases as the power plant ages, these annual costs may actually decrease with time. However, for simplicity we will consider that the fixed charges are a constant proportion, f, of the initial capital investment throughout the life of the plant.

For an original investment of C, the constant annual fixed charges may be expressed as

$$\text{Annual fixed charges} = f \times C.$$

While the value of f will vary with location, tax structure, and type of plant, a value of 5 percent is typically used for a nuclear power plant. Since annual fixed charges represent a cost to the project, they must be covered by revenues obtained from the sale of electricity. If we define L_{fc} as the levelized charge for a unit of electricity sufficient to cover fixed charges, then $L_{fc} \times E$ is the revenue collected in any single year. If this revenue is to cover expenditures, then

$$L_{fc} \times E = f \times C, \tag{3-21}$$

or

$$L_{fc} = f \times C / E. \tag{3-22}$$

D. TAXES ASSOCIATED WITH CAPITAL EXPENSES

If a power plant is owned by a private utility, then taxes must be paid on revenues collected to cover capital expenses. The calculation of taxes is complicated by provisions in the tax law which allow certain costs to be deducted from revenues before taxes are assessed. Although deductible costs are ultimately determined by the law, typical deductions for a nuclear power plant include items such as the interest paid on bonds and the amount of depreciation allowed on capital equipment. Since taxes are paid on revenues after deductions, the tax payment in any year may be represented as

$$\begin{pmatrix} \text{taxes on} \\ \text{capital} \end{pmatrix} = \begin{pmatrix} \text{tax} \\ \text{rate} \end{pmatrix} \times \left[(\text{revenues}) - \begin{pmatrix} \text{interest} \\ \text{paid on} \\ \text{bonds} \end{pmatrix} - (\text{depreciation}) \right].$$

$$\tag{3-23}$$

Revenues, in this case, include money collected to pay back the original capital investment. In addition, revenues must also be sufficient to pay the taxes associated with this investment.

As noted previously, the annual revenue required to cover a capital investment is given by $L_{cap} \times E$, where L_{cap} is the levelized charge for a unit of electricity necessary to cover the return of and the return on the investment. If we define L_{ctax} as the levelized charge for a unit of electricity sufficient to cover taxes, then the annual revenue which will be obtained to cover both capital investment and taxes is

$$\text{Revenue} = \left(L_{\text{cap}} + L_{\text{ctax}} \right) \times E. \qquad (3\text{-}24)$$

Note that, while we have defined L_{ctax}, we have yet to calculate it.

As indicated in Eq. (3-23), interest paid on bonds is a deductible cost. In general, it is somewhat difficult to calculate exactly the interest paid on bonds. It can, however, be calculated exactly with the sinking-fund repayment equation [Eq. (3-17)] which was developed in the previous section. This expression allowed us to calculate the annual cost to the project necessary to cover the return of and the return on a capital investment, where the return on a capital investment includes payment to both debt and equity.

In order to calculate the interest payment to debt alone, let us return to the example shown in Table 3-1. Since the total interest payment in that example is given by $0.05 D_k$, then the bond interest payment will be

$$\begin{array}{l} \text{Bond interest} \\ \text{payment} \end{array} = \frac{i_b b}{i_b b + i_e e} \times 0.05 D_k. \qquad (3\text{-}25)$$

Similarly, the equity interest payment would be

$$\begin{array}{l} \text{Equity interest} \\ \text{payment} \end{array} = \frac{i_e e}{i_b b + i_e e} \times 0.05 D_k. \qquad (3\text{-}26)$$

Therefore, if the debt and equity interest rates and fractions are known, a series of payments to bond interest can be constructed. This is shown in Table 3-2 for the example from the previous table. Note that the payment to bond interest varies year by year.

TABLE 3-2   Interest Payment to Debt and Equity

Year (k)	Payment to Interest $0.05 \times D_k$	Payment to Bond Interest $\dfrac{i_b b}{i_b b + i_e e} \times 0.05 \times D_k$	Payment to Equity Interest $\dfrac{i_e e}{i_b b + i_e e} \times 0.05 \times D_k$
1	250	76	.174
2	205	62	143
3	157	48	109
4	108	33	75
5	55	17	38
		$i_b b / (i_b b + i_e e) = 0.304$	$i_e e / (i_b b + i_e e) = 0.696$

In addition to bond interest, depreciation can also be deducted from revenues prior to the payment of taxes. Two commonly used methods, the straight-line method and the sum-of-years digits method, will be discussed here although any reasonable and consistent depreciation method can be used. The straight-line method assumes deductions due to depreciation will be taken in a uniform manner throughout the lifetime of the plant. If C represents the initial capital investment, K the plant lifetime, and $c\mathrm{dep}_k$ the depreciation allowance in year k, then

$$c\mathrm{dep}_k = \frac{C}{K}.\tag{3-27}$$

Alternatively, in the sum-of-years digits method, it is assumed that the largest deductions will be taken early in the plant life and that the depreciation deduction decreases with time. In this case, the depreciation allowance in any year k is given by

$$c\mathrm{dep}_k = \frac{C(K+1-k)}{\displaystyle\sum_{j=1}^{K} j}.\tag{3-28}$$

The depreciation allowance which results from these two methods of depreciating capital costs is shown in Fig. 3-10. The sum-of-years digits

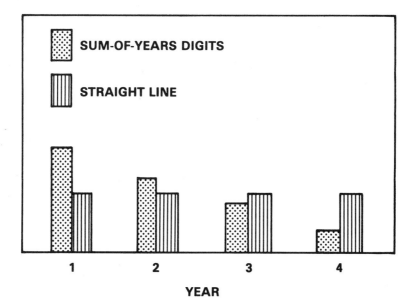

FIGURE 3-10. Comparison of the depreciation allowance using the straight line or sum-of-years digits method.

method is usually preferred since it reduces taxes in the early years of the project. Both methods, however, produce the same total depreciation over the life of the project. The advantage of the sum-of-years digits method is not that it foregoes tax payments, but rather that it defers them.

Since both the bond interest payment and the depreciation allowance vary throughout the lifetime of the plant, it should be clear that the tax payment will also vary. Thus, the revenue required to cover taxes on capital will vary from year to year. It is still desirable, however, to compute a single levelized value that can be charged for electricity over an extended period of time in order to cover taxes on capital. This can be done using a fundamental equation in economics which equates expenditures and revenues accounting for the time value of money in both cases. This may be expressed as

$$\sum_{k=1}^{K} \frac{\text{Exp}_k}{(1+i)^k} = \sum_{k=1}^{K} \frac{\text{Rev}_k}{(1+i)^k} \qquad (3\text{-}29)$$

where Exp_k is the expenditure in year k and Rev_k is the revenue allocated in year k to cover the expenditure. Note that it is not necessary for the revenue in year k to precisely equal the expenditure in that year. Rather, since money has a time value, it is only necessary that the sum of the present value of the revenues equal the sum of the present value of the expenditures.

Given this fundamental economic equality, suppose that the expenditures are taxes and that we wish to determine the revenues necessary to cover them. Recall from Eq. (3-23) that the expenditure for taxes in any year k is given by

$$\text{Exp}_k = t \times \left[\left(L_{\text{cap}} + L_{\text{ctax}} \right) \times E - c\text{bin}_k - c\text{dep}_k \right] \qquad (3\text{-}30)$$

where t is the tax rate and $(L_{\text{cap}} + L_{\text{ctax}}) \times E$ is the revenue required to cover both capital investment and the taxes. We have defined $c\text{bin}_k$ as the bond interest in any year k and $c\text{dep}_k$ as the depreciation allowance in any year k.

Although the expenditure for taxes will vary from year to year, the levelized charge for a unit of electricity L_{ctax} will produce the revenue $L_{\text{ctax}} \times E$ in any year. This revenue, when present-valued, should be sufficient to cover tax expenditures over the life of the plant. Or, in the form of an equation

$$\sum_{k=1}^{K} \frac{t \times \left[\left(L_{\text{cap}} + L_{\text{ctax}} \right) \times E - c\text{bin}_k - c\text{dep}_k \right]}{(1+i)^k} = \sum_{k=1}^{K} \frac{L_{\text{ctax}} \times E}{(1+i)^k}.$$

$$(3\text{-}31)$$

It is now possible to solve for the levelized charge for electricity sufficient to cover taxes $L_{c\text{tax}}$. Doing so gives

$$L_{c\text{tax}} = \frac{t}{1-t} \frac{\displaystyle\sum_{k=1}^{K} \frac{L_{\text{cap}} \times E - c\text{bin}_k - c\text{dep}_k}{(1+i)^k}}{E \times \displaystyle\sum_{k=1}^{K} \frac{1}{(1+i)^k}}. \qquad (3\text{-}32)$$

This complicated expression for taxes is often very difficult to evaluate. It can be simplified somewhat by rewriting the expression to separate the bond interest term as follows:

$$L_{c\text{tax}} = \frac{t}{1-t} \left[\frac{\displaystyle\sum_{k=1}^{K} \frac{L_{\text{cap}} \times E - c\text{dep}_k}{(1+i)^k}}{E \times \displaystyle\sum_{k=1}^{K} \frac{1}{(1+i)^k}} \right] - \frac{t}{1-t} \left[\frac{\displaystyle\sum_{k=1}^{K} \frac{c\text{bin}_k}{(1+i)^k}}{E \times \displaystyle\sum_{k=1}^{K} \frac{1}{(1+i)^k}} \right].$$

$$(3\text{-}33)$$

If we make the simplifying assumption that the depreciation allowance is constant each year, then the quantity $L_{\text{cap}} \times E - c\text{dep}_k$ is also constant. In this case, the numerator of the first term may be factored through the summation to give

$$L_{c\text{tax}} = \frac{t}{1-t} \frac{(L_{\text{cap}} \times E - c\text{dep})}{E} - \frac{t}{1-t} \left[\frac{\displaystyle\sum_{k=1}^{K} \frac{c\text{bin}_k}{(1+i)^k}}{E \times \displaystyle\sum_{k=1}^{K} \frac{1}{(1+i)^k}} \right]. \qquad (3\text{-}34)$$

We will now develop a method for evaluating the bond interest term. Recall that the payment to bond interest varies from year to year, and so it cannot be factored through the summation. It is possible, however, to develop a relatively simple expression for the sum of the present value of the payments to bond interest. The development is based on the fundamental principles involved in repaying a capital investment.

Recall that a capital investment may be repaid in a uniform fashion with an annual payment C_u, and this quantity is calculated using the sinking-fund repayment equation. The annual payment is divided between principal and interest, and in any year k

$$C_u = P_k + I_k, \quad \text{or}$$

$$P_k = C_u - I_k. \tag{3-35}$$

Recall also that the recursion relation for the outstanding debt, D_k, is

$$D_{k+1} = D_k - P_k. \tag{3-36}$$

We can now calculate the quantities D_k, I_k, and P_k for each year. Note that the outstanding debt at the beginning of the first year is the initial investment C, so

$$P_1 = C_u - I_1 = C_u - iC, \tag{3-37}$$

and

$$D_2 = D_1 - P_1 = C - (C_u - iC) = C(1+i) - C_u. \tag{3-38}$$

This sequence of calculations can be repeated for years two and three, as shown in Table 3-3, and for each additional year of the plant lifetime. It can be seen that payment to principal in any year k can be expressed in terms of the annual payment C_u and the initial investment C as follows:

$$P_k = C_u(1+i)^{k-1} - iC(1+i)^{k-1}.$$

We can now express the total payment to interest in a single year k as

$$I_k = C_u - P_k$$
$$= C_u - C_u(1+i)^{k-1} + iC(1+i)^{k-1}. \tag{3-39}$$

The fraction of this payment that is made to bond interest is given by

$$cbin_k = \frac{bi_b}{i} I_k$$
$$= \frac{bi_b}{i} \left[C_u - C_u(1+i)^{k-1} + iC(1+i)^{k-1} \right]. \tag{3-40}$$

We can now use this expression for $cbin_k$ to develop a simple formula for the sum of the present value of the payments to bond interest.

$$\sum_{k=1}^{K} \frac{cbin_k}{(1+i)^k} = \sum_{k=1}^{K} \frac{bi_b}{i} \frac{\left[C_u - C_u(1+i)^{k-1} + iC(1+i)^{k-1} \right]}{(1+i)^k}$$

$$= \sum_{k=1}^{K} \frac{bi_b}{i} \left[\frac{C_u}{(1+i)^k} - \frac{C_u}{(1+i)} + \frac{iC}{(1+i)} \right]$$

$$= \frac{bi_b}{i} \left[C - \frac{KC_u}{(1+i)} + \frac{KiC}{(1+i)} \right]. \tag{3-41}$$

TABLE 3-3 Calculation of Annual Payments to Principal and Interest

		Year			
		1	2	3	k
Outstanding debt at beginning of year	D_k	C	$C(1+i)$ $-C_u$	$C(1+i)^2$ $-C_u(2+i)$	—
Annual payment to interest	$I_k = iD_k$	iC	$iC(1+i)$ $-iC_u$	$iC(1+i)^2$ $-iC_u(2+i)$	—
Annual payment to principal	$P_k = C_u - I_k$	$C_u - iC$	$C_u(1+i)$ $-iC(1+i)$	$C_u(1+i)^2$ $-iC(1+i)^2$	$C_u(1+i)^{k-1}$ $-iC(1+i)^{k-1}$
Outstanding debt at end of year	$D_{k+1} = D_k - P_k$	$C(1+i)$ $-C_u$	$C(1+i)^2$ $-C_u(2+i)$	$C(1+i)^3$ $-C_u[(1+i)^2$ $+(2+i)]$	—

This expression may now be substituted into Eq. (3-34) to obtain the following equation for $L_{c\,\text{tax}}$

$$L_{c\,\text{tax}} = \frac{t}{1-t} \frac{(L_{\text{cap}} \times E - c\text{dep})}{E} - \frac{\left(\dfrac{t}{1-t}\right)\left(\dfrac{bi_b}{i}\right)\left[C - \dfrac{KC_u}{(1+i)} + \dfrac{KiC}{(1+i)}\right]}{E\left[\dfrac{(1+i)^K - 1}{i \times (1+i)^K}\right]}.$$

$$(3\text{-}42)$$

In summary, we have developed the three levelized charges necessary to cover all expenditures associated with capital. They are

L_{cap}: The levelized charge for electricity necessary to pay for the return on and return of a capital investment.

L_{fc}: The levelized charge for electricity necessary to pay for fixed charges, such as property taxes and insurance.

$L_{c\,\text{tax}}$: The levelized charge for electricity necessary to pay for taxes associated with a capital investment.

3-5 FUEL CYCLE EXPENSES

> ...the immense strength [of science] lies in its power of accurate prediction. There is a minimum of fumbling. Trial-and-error methods are reduced to a level which was quite unknown in former phases of human activity. An aero-engineer who, applying scientific principles, designs a new aircraft not only knows, even in the drawing board stage, that his aircraft will be capable of flying, but he also can forecast, with a fair degree of accuracy, its performance....The value of [scientific] prediction lies not so much in allowing us to see into the future, but in permitting us to exclude many avenues of progress which have no future. The scientific method directs human effort, not by helping us to see farther afield, but by limiting our progress along lines which have a reasonable chance of leading us where we hope to go....The ability to limit progress to profitable avenues carries in itself that economy of effort which has made science so successful.

> Kurt Mendelssohn,
> *The Secret of Western Domination*

In an era of increasing fuel costs, the ability to accurately assess the fuel cost of a power plant is often crucial to the success of the project. A plant which is destined to encounter ever-increasing fuel costs might be considered an avenue of progress with little future. In fact, the very health of the economy of the industrial nations is contingent upon identifying and avoiding such avenues. Nuclear power, with its relatively stable fuel costs, then becomes an attractive avenue of progress.

The evaluation of a nuclear fuel cost, however, is complex since the nuclear *fuel cycle* is composed of many processes that occur at different points in time. The processes which comprise the fuel cycle for both an LWR and an FBR are shown in Fig. 3-11. Consider, for a moment, a single fuel element or assembly which produces power within a nuclear reactor. This fuel element must first be prepared for use in the reactor—that is, fissile material must be purchased and must then be fabricated into a fuel pin. The time required for fuel preparation is known as the *lead time*, and is generally about one year. Many such fuel elements, which comprise a *batch* of fuel, are loaded into the reactor and generate power for a number of years. In light water reactors and fast breeder reactors, a fuel element typically resides in the reactor for a period of two to five years. Refueling generally occurs once a year when a new batch is loaded and an irradiated batch is discharged. The irradiated element is then stored in a cooling basin at the reactor site and remains there for approximately one year. The fuel element may then be shipped to a storage site or to a reprocessing plant. If it is reprocessed, the residual fissile material is extracted and recycled for

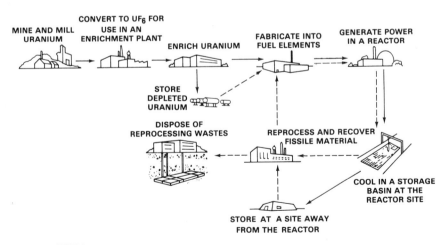

FIGURE 3-11. Nuclear fuel cycle.

use in another or in the same reactor. Waste products from the reprocessing stream are then permanently stored. The time between discharge from the reactor and either reprocessing or ultimate disposal is known as the *lag time*.

A. DIRECT EXPENSES FOR A SINGLE BATCH OF FUEL

In calculating the fuel expense incurred during the operation of a nuclear power plant, it is common to consider all the expenses associated with a single batch of fuel. Recall that the expenses associated with preparing and disposing of a single batch of fuel occur throughout a period of years. In a typical light water reactor with a three-year residence time and a one-year lead and lag time, the fuel expenses will occur over a period of five years. Thus, it is necessary to take into account the timing of each expense as well as the timing of the revenues generated by the fuel. The timing of expenses for a typical batch of fuel is shown in Fig. 3-12. In this figure, F represents all the front-end costs associated with preparing the fuel before it is loaded into the reactor. This includes the cost of purchasing uranium, enrichment, and fabrication. Similarly, B represents the back-end costs associated with disposing of the fuel after it is discharged. This includes the cost of shipping the spent fuel and the cost of permanent storage or reprocessing. The timing of these individual expenses is shown in Fig. 3-13.

A batch of fuel must produce revenues sufficient to cover fuel expenses. Each year, a single batch produces a fraction of the total energy produced by the reactor. For example, in a reactor with three batches, a single batch produces approximately one-third of the total energy production in each year. It is possible to calculate a levelized charge for electricity sufficient to cover fuel expenses from this fraction of the total energy production. The

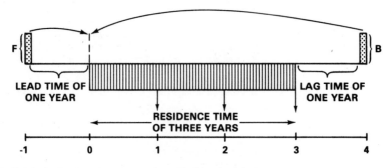

FIGURE 3-12. Timing of expenses of a typical batch of fuel.

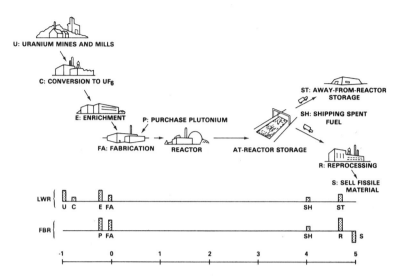

FIGURE 3-13. Timing of expenses for a batch of fuel.

levelized charge may be found by performing the following calculations:

(1) First, identify all revenues and expenses associated with the batch of fuel in each year. Revenues will, of course, depend upon the amount charged for electricity produced by the batch.

(2) Second, using the time at which the batch is loaded into the reactor as a reference point, calculate the present value of all expenses. Similarly, calculate the present value of all revenues. The choice of a reference point is arbitrary, but the start of operation is generally chosen for convenience.

(3) Next, equate the sum of the present value of the revenues to the sum of the present value of the expenses.

(4) Finally, solve this expression for the levelized charge for electricity sufficient to cover the expenses associated with the batch of fuel.

Let us consider a reactor which contains N batches of fuel. Assume that each batch resides in the reactor for a period of N years with a lead time ld and a lag time lg. Also assume that each batch generates the same fraction of the total energy produced by the reactor in each year. If we define L_b as the levelized charge for electricity necessary to cover the expenses associated with a batch of fuel, then the revenues obtained from the sale of electricity

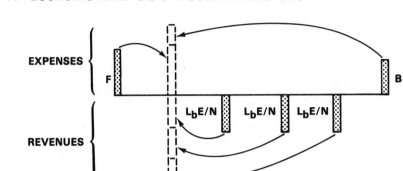

FIGURE 3-14. Present value of expenses and revenues for a batch of fuel.

each year will be $L_b E/N$. We can now equate the present value of the revenues to the present value of the expenses, as shown in Fig. 3-14.

$$\frac{L_b E/N}{(1+i)} + \frac{L_b E/N}{(1+i)^2} + \cdots + \frac{L_b E/N}{(1+i)^N} = F(1+i)^{ld} + \frac{B}{(1+i)^{lg+N}}. \quad (3\text{-}43)$$

Solving for the levelized charge for electricity necessary to cover the expenses associated with a batch of fuel L_b gives

$$L_b = \frac{F(1+i)^{ld} + \dfrac{B}{(1+i)^{lg+N}}}{\dfrac{E}{N}\left[\dfrac{1}{(1+i)} + \dfrac{1}{(1+i)^2} + \cdots + \dfrac{1}{(1+i)^N}\right]}. \quad (3\text{-}44)$$

B. TAXES ASSOCIATED WITH A SINGLE BATCH OF FUEL

In some nations, tax law requires fuel to be considered a capital investment rather than an operating expense. In this case, additional revenues must be collected in order to pay the taxes associated with fuel. Let us denote the levelized charge for electricity to cover taxes associated with a batch of fuel as $L_{b\text{tax}}$. The revenues obtained from the sale of electricity required to cover both direct expenses and taxes on a batch each year will then be $(L_b + L_{b\text{tax}})E/N$. In this case, we must again equate the present value of revenues and expenditures. If T represents the total tax payment present-valued to the beginning of energy production by the batch, then

$$\frac{(L_b + L_{b\text{tax}})E/N}{(1+i)} + \frac{(L_b + L_{b\text{tax}})E/N}{(1+i)^2} + \cdots + \frac{(L_b + L_{b\text{tax}})E/N}{(1+i)^N}$$

$$= F(1+i)^{ld} + \frac{B}{(1+i)^{lg+N}} + T. \quad (3\text{-}45)$$

Subtracting Eq. (3-43) from Eq. (3-45) gives

$$L_{b\text{tax}} E/N \left[\frac{1}{(1+i)} + \frac{1}{(1+i)^2} + \cdots + \frac{1}{(1+i)^N} \right] = T. \quad (3\text{-}46)$$

Our problem now becomes one of calculating the present value of the tax payment T. The present value of the tax payment T can be calculated by recalling that (1) taxes are paid on revenues, and (2) revenues can be reduced by depreciation prior to the payment of taxes. Thus, if t represents the tax rate and D the depreciation allowance in a single year, the present value of the tax payment is

$$T = t(\text{Revenues} - \text{Depreciation}) \quad (3\text{-}47)$$

$$= t \left[\frac{\{(L_b + L_{b\text{tax}})E/N\} - D}{(1+i)} + \frac{\{(L_b + L_{b\text{tax}})E/N\} - D}{(1+i)^2} \right.$$

$$\left. + \cdots + \frac{\{(L_b + L_{b\text{tax}})E/N\} - D}{(1+i)^N} \right]$$

where $(L_b + L_{b\text{tax}})E/N$ represents the revenue generated in any year. Collecting terms in this equation gives

$$T = t \left[\{(L_b + L_{b\text{tax}})E/N\} - D \right] \left[\frac{1}{(1+i)} + \frac{1}{(1+i)^2} + \cdots + \frac{1}{(1+i)^N} \right].$$

$$(3\text{-}48)$$

Recall that Eq. (3-46) equated the revenues required to pay taxes to the present value of the tax payment. Substituting the present value of the tax payment T from Eq. (3-48) into Eq. (3-46) gives

$$L_{b\,\text{tax}}E/N\left[\frac{1}{(1+i)}+\frac{1}{(1+i)^2}+\cdots+\frac{1}{(1+i)^N}\right]$$

$$=t\left[\{(L_b+L_{b\,\text{tax}})E/N\}-D\right]\left[\frac{1}{(1+i)}+\frac{1}{(1+i)^2}+\cdots+\frac{1}{(1+i)^N}\right].$$

$$(3\text{-}49)$$

Solving this equation for $L_{b\,\text{tax}}$ gives

$$L_{b\,\text{tax}}E/N=t\left[\{(L_b+L_{b\,\text{tax}})E/N\}-D\right], \qquad (3\text{-}50)$$

or

$$L_{b\,\text{tax}} = \frac{t}{1-t}(L_b-DN/E). \qquad (3\text{-}51)$$

We now have an explicit expression for the levelized charge for electricity necessary to cover the taxes associated with a single batch of fuel in terms of L_b and D. We have previously calculated the levelized charge L_b required to cover the direct expenses associated with a single batch, as shown in Eq. (3-44). We have yet to calculate the depreciation allowance D.

The depreciation allowance for a batch of fuel is often assumed to be directly proportional to the amount of energy produced by the batch in any year. A batch produces a cumulative amount of energy E throughout its residence time of N years. We will assume it will produce the same amount of energy E/N in any year, and the amount of depreciation which may be deducted from revenues in any year is

$$D=\text{Dep}\,(E/N)/E \qquad (3\text{-}52)$$

$$=\text{Dep}/N \qquad (3\text{-}53)$$

where Dep is the total depreciation allowance. This approach is often called the *unit of production method*.

The value of Dep depends upon current tax law, and hence may change with time. Currently all direct expenses associated with a batch can be depreciated. One must be very careful, however, when incorporating the time value of money into a calculation of depreciation. At present, the time value of money can only be considered when determining the value of expenses incurred either before or after operation. It cannot, as indicated in Fig. 3-15, be used to present-value all expenditures to a common reference

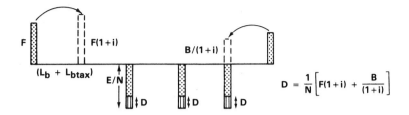

FIGURE 3-15. Calculation of depreciation for a batch of fuel.

time. With this approach,

$$\text{Dep} = F(1+i)^{ld} + \frac{B}{(1+i)^{lg}}. \tag{3-54}$$

We may now substitute Eq. (3-44) for L_b and Eqs. (3-53) and (3-54) for D into Eq. (3-51) to obtain $L_{b\,\text{tax}}$ which is

$$L_{b\,\text{tax}} = \frac{t}{1-t}\left[\frac{F(1+i)^{ld} + \dfrac{B}{(1+i)^{lg+N}}}{\dfrac{E}{N}\left(\dfrac{1}{(1+i)} + \dfrac{1}{(1+i)^2} + \cdots + \dfrac{1}{(1+i)^N}\right)} \right.$$

$$\left. - \frac{\left(F(1+i)^{ld} + \dfrac{B}{(1+i)^{lg}}\right)}{E} \right]. \tag{3-55}$$

Recall now we have developed two components associated with the charge for electricity necessary to cover fuel expenses. They are

L_b: The levelized charge for electricity necessary to cover the direct expenses associated with a batch of fuel.

$L_{b\text{tax}}$: The levelized charge for electricity necessary to cover the taxes associated with a batch of fuel.

The total levelized charge to cover all expenses associated with a batch of fuel is then $L_b + L_{b\text{tax}}$.

C. LEVELIZED CHARGE TO COVER FUEL EXPENSES

Often, the ability to calculate the levelized charge for electricity from a single but typical batch is adequate for design analysis. The levelized charge, however, can vary from batch to batch. In particular, the batches that comprise the initial core and the final core differ from equilibrium batches. If it is desirable to calculate an average charge for electricity over the life of the reactor, then all batches must be considered. Let us define L_{fuel} as the levelized charge for electricity necessary to cover all fuel expenses for all batches. Then equating the revenue for all batches to the sum of the revenues for each batch, as shown in Fig. 3-16, gives

$$
\sum_{m=1}^{M}\left[\left(\sum_{n=1}^{N} L_{\text{fuel}}\frac{E_{mn}}{(1+i)^n}\right)\frac{1}{(1+i)^{b_p}}\right]
$$

$$
= \sum_{m=1}^{M}\left[\left(\sum_{n=1}^{N}(L_b+L_{b\text{tax}})_m E_{mn}\frac{1}{(1+i)^n}\right)\frac{1}{(1+i)^{b_p}}\right], \quad (3\text{-}56)
$$

where M is the number of batches, E_{mn} is the amount of energy generated by batch m in year of residence n, and b_p is the time at which a batch begins

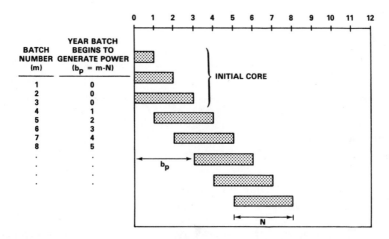

FIGURE 3-16. Calculation of a fuel cost considering all batches of fuel.

to generate power. Solving for L_{fuel} gives

$$L_{\text{fuel}} = \frac{\displaystyle\sum_{m=1}^{M}\left[\left(\sum_{n=1}^{N}(L_b+L_{b\,\text{tax}})_m E_{mn}\frac{1}{(1+i)^n}\right)\frac{1}{(1+i)^{b_p}}\right]}{\displaystyle\sum_{m=1}^{M}\left[\left(\sum_{n=1}^{N}\frac{E_{mn}}{(1+i)^n}\right)\frac{1}{(1+i)^{b_p}}\right]}. \qquad (3\text{-}57)$$

3-6 TOTAL POWER COST

The total levelized charge for electricity necessary to cover capital, fuel, and operating and maintenance expenditures may be expressed as

$$L_{\text{total}} = L_{c\,\text{tot}} + L_{b\,\text{tot}} + L_{om}$$

where

$$L_{c\,\text{tot}} = L_{\text{cap}} + L_{c\,\text{tax}} + L_{fc}$$

$$L_{b\,\text{tot}} = L_b + L_{b\,\text{tax}}$$

and L_{om} is the levelized charge necessary to cover operating and mainte-nance expenses.

3-7 AN ILLUSTRATIVE EXAMPLE

Figures have two important properties; they say a lot and they appear to be accurate. In both respects there is a real disadvantage connected with the apparent advantage. Figures say so much that without a detailed discussion it is impossible to understand them. When the discussion is finished, the substance is in the words rather than in the figures and the latter are only a device by which to remember the discussion. Figures are also much too accurate. The trouble is that with regard to the future this accuracy can never be attained.

Edward Teller,
Energy from Heaven and Earth

An example, unfortunately, has all the properties of a figure. An example has the ability to convey information in a very efficient manner while simultaneously appearing to be more accurate than the situation warrants. With this in mind, we will attempt to exploit the former while relying on the maturity of the reader to avoid the latter.

A. REACTOR CHARACTERISTICS

	Light-Water Reactor (Once-Through Cycle)	Homogeneous Oxide Fast Breeder Reactor
Power Level (MWe)	1300	1000
Capacity Factor (%)	70	70
Refueling Interval (years)	1	1
Plant Life Time (years)	30	30
Lead and Lag Time (years)	1	1

		Core	Axial Blanket	Radial Blanket
Number of Batches		3	2	2	5
Uranium Requirement* (short tons U_3O_8/yr)		240	0	0	0
Enrichment Requirement* (10^3 SWU/yr)		145	0	0	0
Plutonium (kg/yr)	charge	0	1480	0	0
	discharge	215	1410	160	135
Heavy Metal (kg/yr)	charge	33 300	12 600	9360	8480
	discharge	31 500	11 600	9150	8290

*Tails Composition of 0.2%

B. ECONOMIC PARAMETERS

Bond Fraction	0.55
Equity Fraction	0.45
Uninflated Bond Interest Rate	0.025
Uninflated Equity Interest Rate	0.07
Fixed Charge Fraction of Capital Cost	0.03

	Light-Water Reactor (Once-Through Cycle)	Homogeneous Oxide Fast Breeder Reactor
Tax Rate	0.50	0.50
U_3O_8 ($/lb)	40	
Enrichment ($/SWU)	100	
Plutonium ($/g)		30
Capital Cost of Plant* ($/kWe)	800	1000
Operation and Maintenance Costs		
($/kWe·yr) fixed	13	17
variable	1	1

		Core	Axial Blanket	Radial Blanket
Fabrication ($/kg)	150	1150	50	150
Spent Fuel Shipping and Disposal ($/kg)	150	—	—	—
Reprocessing, including Shipping and Waste Disposal ($/kg)	—	600	600	600

*Includes interest during construction

C. EFFECTIVE INTEREST RATE

$$i = (b \times i_b) + (e \times i_e)$$

$$= (0.55 \times 0.025) + (0.45 \times 0.07)$$

$$= 0.045 \ \$/\$\cdot yr.$$

D. RETURN OF AND RETURN ON A CAPITAL INVESTMENT

$$C_u = C \times \left[\frac{i \times (1+i)^K}{(1+i)^K - 1} \right],$$

$$L_{cap} = C_u / E.$$

LWR

$$C(\text{LWR}) = 800 \ \$/\text{kWe} \times 1300 \times 10^3 \ \text{kWe}$$

$$= 1.04 \times 10^9 \ \$$$

$$C_u(\text{LWR}) = 1.04 \times 10^9 \ \$ \times \left[\frac{0.045 \times (1.045)^{30}}{(1.045)^{30} - 1} \right] \$/\$ \cdot \text{yr}$$

$$= 1.04 \times 10^9 \ \$ \times 0.061 \ \$/\$ \cdot \text{yr} = 6.4 \times 10^7 \ \$/\text{yr}$$

$$E(\text{LWR}) = 1300 \times 10^3 \ \text{kWe} \times 0.70 \times 8760 \ \text{hr/yr}$$

$$= 8.0 \times 10^9 \ \text{kWh/yr}$$

$$L_{\text{cap}}(\text{LWR}) = \frac{6.4 \times 10^7 \ \$/\text{yr} \times 10^3 \ \text{mills}/\$}{8.0 \times 10^9 \ \text{kWh/yr}}$$

$$= 8.0 \ \text{mills/kWh.}$$

FBR

$$C(\text{FBR}) = 1000 \ \$/\text{kWe} \times 1000 \times 10^3 \ \text{kWe}$$

$$= 1.0 \times 10^9 \ \$$$

$$C_u(\text{FBR}) = 1.0 \times 10^9 \ \$ \times 0.061 \ \$/\$ \cdot \text{yr}$$

$$= 6.1 \times 10^7 \ \$/\text{yr}$$

$$E(\text{FBR}) = 1000 \times 10^3 \ \text{kWe} \times 0.70 \times 8760 \ \text{hr/yr}$$

$$= 6.1 \times 10^9 \ \text{kWh/yr}$$

$$L_{\text{cap}}(\text{FBR}) = \frac{6.1 \times 10^7 \ \$/\text{yr} \times 10^3 \ \text{mills}/\$}{6.1 \times 10^9 \ \text{kWh/yr}}$$

$$= 10.0 \ \text{mills/kWh.}$$

E. FIXED CHARGES ON A CAPITAL INVESTMENT

$$L_{fc} = \frac{f \times C}{E}.$$

LWR

$$L_{fc}(\text{LWR}) = \frac{0.03 \ \$/\$ \cdot \text{yr} \times 1.04 \times 10^9 \ \$ \times 10^3 \ \text{mills}/\$}{8.0 \times 10^9 \ \text{kWh}/\text{yr}}$$

$$= 3.9 \ \text{mills}/\text{kWh}.$$

FBR

$$L_{fc}(\text{FBR}) = \frac{0.03 \ \$/\$ \cdot \text{yr} \times 1.00 \times 10^9 \ \$ \times 10^3 \ \text{mills}/\$}{6.1 \times 10^9 \ \text{kWh}/\text{yr}}$$

$$= 4.9 \ \text{mills}/\text{kWh}.$$

F.   TAXES ON A CAPITAL INVESTMENT

$$L_{c\text{tax}} = \left(\frac{t}{1-t}\right) \frac{(L_{\text{cap}} \times E - c\text{dep})}{E} - \left(\frac{t}{1-t}\right)\left(\frac{bi_b}{i}\right)\left(\frac{1}{E}\right)$$

$$\times \left[\frac{i \times (1+i)^K}{(1+i)^K - 1}\right]\left[C - \frac{KC_u}{(1+i)} + \frac{KiC}{(1+i)}\right]$$

$$c\text{dep} = \frac{C}{K}.$$

LWR

$$c\text{dep}(\text{LWR}) = \frac{1.04 \times 10^9 \ \$}{30 \ \text{yr}}$$

$$= 3.5 \times 10^7 \ \$/\text{yr}$$

$$L_{c\text{tax}}(\text{LWR}) = \left(\frac{0.50}{1 - 0.50}\right)$$

$$\times \left(\frac{8.0 \ \text{mills}/\text{kWh} \times 8.0 \times 10^9 \ \text{kWh}/\text{yr} - 3.5 \times 10^7 \ \$/\text{yr} \times 10^3 \ \text{mills}/\$}{8.0 \times 10^9 \ \text{kWh}/\text{yr}}\right)$$

$$- \left(\frac{0.50}{1 - 0.50}\right)\left(\frac{0.55 \times 0.025}{0.045}\right)\left(\frac{0.061 \ \$/\$ \cdot \text{yr}}{8.0 \times 10^9 \ \text{kWh}/\text{yr}}\right)\left[1.04 \times 10^9 \ \$\right.$$

$$\left. - \frac{30 \ \text{yr} \times 6.4 \times 10^7 \ \$/\text{yr}}{1.045} + \frac{30 \ \text{yr} \times 0.045 \ \$/\$ \cdot \text{yr} \times 1.04 \times 10^9 \ \$}{1.045}\right](10^3 \ \text{mills}/\$)$$

$$L_{c\text{tax}}(\text{LWR}) = 3.6 \ \text{mills}/\text{kWh} - 1.3 \ \text{mills}/\text{kWh}$$

$$= 2.3 \ \text{mills}/\text{kWh}.$$

FBR

$$c\mathrm{dep}(\mathrm{FBR}) = \frac{1.0 \times 10^9 \ \$}{30 \ \mathrm{yr}} = 3.3 \times 10^7 \ \$/\mathrm{yr}$$

$$L_{c\,\mathrm{tax}}(\mathrm{FBR}) = \left(\frac{0.50}{1 - 0.50} \right)$$

$$\times \left(\frac{10.0 \ \mathrm{mills/kWh} \times 6.1 \times 10^9 \ \mathrm{kWh/yr} - 3.3 \times 10^7 \ \$/\mathrm{yr} \times 10^3 \ \mathrm{mills/\$}}{6.1 \times 10^9 \ \mathrm{kWh/yr}} \right)$$

$$- \left(\frac{0.50}{1 - 0.50} \right) \left(\frac{0.55 \times 0.025}{0.045} \right) \left(\frac{0.061 \ \$/\$\cdot\mathrm{yr}}{6.1 \times 10^9 \ \mathrm{kWh/yr}} \right) \left[1.0 \times 10^9 \ \$ \right.$$

$$\left. - \frac{30 \ \mathrm{yr} \times 6.1 \times 10^7 \ \$/\mathrm{yr}}{1.045} + \frac{30 \ \mathrm{yr} \times 0.045 \ \$/\$\cdot\mathrm{yr} \times 1.0 \times 10^9 \ \$}{1.045} \right] (10^3 \ \mathrm{mills/\$})$$

$$L_{c\,\mathrm{tax}}(\mathrm{FBR}) = 4.6 \ \mathrm{mills/kWh} - 1.7 \ \mathrm{mills/kWh}$$

$$= 2.9 \ \mathrm{mills/kWh}.$$

G. OPERATION AND MAINTENANCE

$$L_{om} = \frac{(\mathrm{fixed \ O\&M}) + [(\mathrm{variable \ O\&M}) \times (\mathrm{capacity \ factor})]}{E}.$$

LWR

$$L_{om}(\mathrm{LWR}) = \frac{\left[\begin{array}{l} (13\$/\mathrm{kWe}\cdot\mathrm{yr} \times 1300 \times 10^3 \ \mathrm{kWe}) \\ + (1 \ \$/\mathrm{kWe}\cdot\mathrm{yr} \times 0.70 \times 1300 \times 10^3 \ \mathrm{kWe}) \end{array} \right] \times 10^3 \ \mathrm{mills/\$}}{8.0 \times 10^9 \ \mathrm{kWh/yr}}$$

$$= 2.2 \ \mathrm{mills/kWh}.$$

FBR

$$L_{om}(\mathrm{FBR}) = \frac{\left[\begin{array}{l} (17\$/\mathrm{kWe}\cdot\mathrm{yr} \times 1000 \times 10^3 \ \mathrm{kWe}) \\ + (1\$/\mathrm{kWe}\cdot\mathrm{yr} \times 0.70 \times 1000 \times 10^3 \ \mathrm{kWe}) \end{array} \right] \times 10^3 \ \mathrm{mills/\$}}{6.1 \times 10^9 \ \mathrm{kWh/yr}}$$

$$= 2.9 \ \mathrm{mills/kWh}.$$

H.   FUEL CYCLE EXPENSES (for a single equilibrium batch)

Direct Expenses

$$L_b = \frac{F(1+i)^{ld} + \dfrac{B}{(1+i)^{lg+N}}}{\dfrac{E}{N}\left[\dfrac{1}{(1+i)} + \dfrac{1}{(1+i)^2} + \cdots + \dfrac{1}{(1+i)^N}\right]}.$$

LWR

$F(\text{LWR}) = (\text{uranium}) + (\text{enrichment}) + (\text{fabrication})$

$\qquad = (240 \text{ tons } U_3O_8/\text{yr} \times 2000 \text{ lb/ton} \times 40 \text{ \$/lb } U_3O_8)$

$\qquad\quad + (145 \times 10^3 \text{ SWU/yr} \times 100 \text{ \$/SWU}) + (33\,300 \text{ kg/yr} \times 150 \text{ \$/kg})$

$\qquad = 3.9 \times 10^7 \text{ \$/yr.}$

$B(\text{LWR}) = (\text{spent fuel shipping and disposal})$

$\qquad = 31\,500 \text{ kg/yr} \times 150 \text{ \$/kg}$

$\qquad = 4.7 \times 10^6 \text{ \$/yr}$

$$L_b(\text{LWR}) = \frac{\left[3.9 \times 10^7 \text{ \$/yr} \times 1.045 + \dfrac{4.7 \times 10^6 \text{ \$/yr}}{(1.045)^4}\right] \times 10^3 \text{ mills/\$}}{\dfrac{8.0 \times 10^9 \text{ kWh/yr}}{3}\left[\dfrac{1}{1.045} + \dfrac{1}{(1.045)^2} + \dfrac{1}{(1.045)^3}\right]}$$

$\qquad = 6.1 \text{ mills/kWh.}$

FBR

Fuel costs will be calculated individually for each of the three zones in the FBR. In addition, costs due to fabrication and reprocessing will be separated from those due to the purchase or sale of plutonium.

Fabrication and Reprocessing

— Core

$$L_b(C-F\&R)=$$

$$\left[12\,600\text{ kg/yr}\times1150\text{ \$/kg}\times1.045+11\,600\text{ kg/yr}\times600\text{ \$/kg}/(1.045)^3\right]\times10^3\text{ mills/\$}$$

$$\div\left[\frac{6.1\times10^9\text{ kWh/yr}}{2}\times\left(\frac{1}{1.045}+\frac{1}{(1.045)^2}\right)\right]$$

$$=3.7\text{ mills/kWh.}$$

— Axial Blanket

$$L_b(A-F\&R)=$$

$$\left[9360\text{ kg/yr}\times50\text{ \$/kg}\times1.045+9150\text{ kg/yr}\times600\text{ \$/kg}(1.045)^3\right]\times10^3\text{ mills/\$}$$

$$\div\left[\frac{6.1\times10^9\text{ kWh/yr}}{2}\left(\frac{1}{1.045}+\frac{1}{(1.045)^2}\right)\right]$$

$$=0.93\text{ mills/kWh.}$$

— Radial Blanket

$$L_b(R-F\&R)=$$

$$\left[8480\text{ kg/yr}\times150\text{ \$/kg}\times1.045+8290\text{ kg/yr}\times600\text{ \$/kg}/(1.045)^6\right]\times10^3\text{ mills/\$}$$

$$\div\left[\frac{6.1\times10^9\text{ kWh/yr}}{5}\left(\frac{1}{1.045}+\frac{1}{(1.045)^2}+\cdots+\frac{1}{(1.045)^5}\right)\right]$$

$$=0.96\text{ mills/kWh.}$$

— Subtotal

$$L_b(\text{FBR}-F\&R)=L_b(C-F\&R)+L_b(A-F\&R)+L_b(R-F\&R)$$

$$=3.7+0.93+0.96\text{ mills/kWh}$$

$$=5.6\text{ mills/kWh.}$$

Purchase and Sale of Plutonium

— Core

$$L_b(C-P)=$$

$$\frac{\left[1480 \text{ kg/yr}\times 1.045 - 1410 \text{ kg/yr}/(1.045)^3\right]\times 10^3 \text{ g/kg}\times 30 \text{ \$/g}\times 10^3 \text{ mills/\$}}{5.7\times 10^9 \text{ kWh/yr}}$$

$$= 1.6 \text{ mills/ kWh (cost)}.$$

— Axial Blanket

$$L_b(A-P)=\frac{\left[0-\dfrac{160 \text{ kg/yr}\times 10^3 \text{ g/kg}\times 30 \text{ \$/g}}{(1.045)^3}\right]\times 10^3 \text{ mills/\$}}{5.7\times 10^9 \text{ kWh/yr}}$$

$$= -0.74 \text{ mills/kWh (credit)}.$$

— Radial Blanket

$$L_b(R-P)=\frac{\left[0-\dfrac{135 \text{ kg/yr}\times 10^3 \text{ g/kg}\times 30 \text{ \$/g}}{(1.045)^6}\right]\times 10^3 \text{ mills/\$}}{5.4\times 10^9 \text{ kWh/yr}}$$

$$= -0.58 \text{ mills/kWh (credit)}.$$

— Subtotal

$$L_b(\text{FBR-}P)=L_b(C\text{-}P)+L_b(A\text{-}P)+L_b(R\text{-}P)$$

$$=1.6-0.74-0.58 \text{ mills/kWh}$$

$$=0.3 \text{ mills/kWh (cost)}.$$

Total cost for a Batch of FBR Fuel

$$L_b(\text{FBR})=L_b(\text{FBR-}F\&R)+L_b(\text{FBR-}P)$$

$$=5.6 \text{ mills/kWh}+0.3 \text{ mills/kWh}$$

$$=5.9 \text{ mills/kWh}.$$

Summary of FBR Fuel Costs (all values given in mills/kWh)

	Fabrication and Reprocessing	Plutonium Purchase or Sale	Subtotal
Core	3.7	1.6	5.3
Axial Blanket	0.9	−0.7	0.2
Radial Blanket	1.0	−0.6	0.4
Subtotal	5.6	0.3	5.9

Total Fuel Cost: 5.9 mills/kWh

Fuel Taxes

$$L_{b\,\text{tax}} = \left(\frac{t}{1-t}\right)\left[L_b - \frac{F(1+i)^{ld} + \dfrac{B}{(1+i)^{lg}}}{E}\right].$$

LWR

$$L_{b\,\text{tax}}(\text{LWR}) = \left(\frac{0.50}{1-0.50}\right)$$

$$\times\left[6.1 \text{ mills/kWh} - \frac{\left(3.9\times10^7 \text{ \$/yr}\times1.045 + \dfrac{4.7\times10^6 \text{ \$/yr}}{1.045}\right)\times10^3 \text{ mills/\$}}{8.0\times10^9 \text{ kWh/yr}}\right]$$

$$= 0.4 \text{ mills/kWh}.$$

FBR

$$L_{b\,\text{tax}}(\text{FBR}) = \left(\frac{0.50}{1-0.50}\right)$$

$$\times\left[5.9 \text{ mills/kWh} - \frac{\left(6.0\times10^7 \text{ \$/yr}\times1.045 - \dfrac{3.4\times10^7 \text{ \$/yr}}{1.045}\right)\times10^3 \text{ mills/\$}}{6.1\times10^9 \text{ kWh/yr}}\right]$$

$$= 1.0 \text{ mills/kWh}.$$

Total Levelized Power Cost

$$L_{\text{total}} = L_{\text{cap}} + L_{fc} + L_{c\text{tax}} + L_{om}L_b + L_{b\text{tax}}.$$

LWR

$$L_{\text{total}}(\text{LWR}) = (8.0 + 3.9 + 2.3 + 2.2 + 6.1 + 0.4)\text{mills/kWh}$$

$$= 22.9 \text{ mills/KWh}.$$

FBR

$$L_{\text{total}}(\text{FBR}) = (10.0 + 4.9 + 2.9 + 2.9 + 5.9 + 1.0)\text{mills/kWh}$$

$$= 27.6 \text{ mills/kWh}.$$

3-8 ECONOMIC ANALYSIS AND THE TRANSITION TO THE FBR

Commercialize: to cause something, having only a potential income-producing value, to be sold, manufactured, displayed, or utilized so as to produce income.

Webster's Dictionary

The principal goal of a civilian reactor development program is to develop a power plant which can supply electricity on a commercial basis. It is, therefore, important to understand thoroughly the factors which influence the time at which the FBR can be considered a commercial enterprise.

One factor which definitely affects the commercialization date is the size of the uranium resource base. In particular, the amount of electrical energy which can be produced by the LWR is ultimately limited by the amount of ^{235}U contained in that base. In contrast, the FBR is a technological advancement which will allow energy to be obtained from the ^{238}U in the uranium resource base. Since the amount of ^{238}U is considerably larger than the amount of ^{235}U, the amount of energy obtainable with the FBR is also considerably larger. Or, in other words, the energy recovered from the uranium resource base will be incomplete without the FBR.

It is important to note, however, that incomplete energy recovery is by no means unique to the LWR. In fact, it is not unusual to extract approximately one-third of the oil in any given oil reservoir, with the other two-thirds remaining in the ground because the cost of extraction is prohibitive. As oil prices increase, however, the employment of an enhanced recovery technique becomes increasingly attractive. In comparison, one can say that the FBR is an enhanced recovery technique which becomes increasingly attractive as U_3O_8 prices increase. Thus, the FBR is ultimately subject to the same requirements imposed upon any enhanced recovery technique—viz, the additional resource will be extracted when it is economically attractive.

The time at which the FBR, or any enhanced recovery technique for that matter, becomes economically attractive is determined by the price of the resource undergoing depletion as well as the additional capital investment required to obtain an additional amount of the resource. Thus, for the FBR, the time for commercialization is determined primarily by the price of the U_3O_8 and the capital cost differential between the FBR and the LWR. The price of U_3O_8 is in turn determined by the extent of the proven reserve as well as the cost of extracting U_3O_8 from this reserve. In addition, the price of U_3O_8 is affected by the costs associated with the additional exploration required to convert probable uranium resources into proven uranium reserves.

The price of U_3O_8 has a large influence on the cost of energy from the LWR, as illustrated in Fig. 3-17. In contrast, the price of U_3O_8 has little influence on the cost of energy from the FBR. While LWR energy costs will be lower than FBR energy costs when U_3O_8 prices are low, the reverse will be true when U_3O_8 prices are high. Thus, for some U_3O_8 price, a *point of*

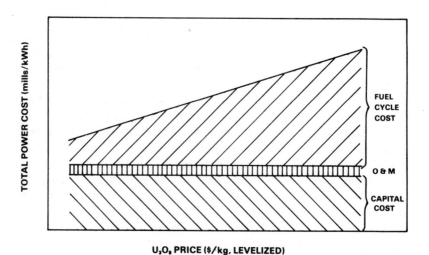

FIGURE 3-17. Total power cost components for an LWR as a function of U_3O_8 price.

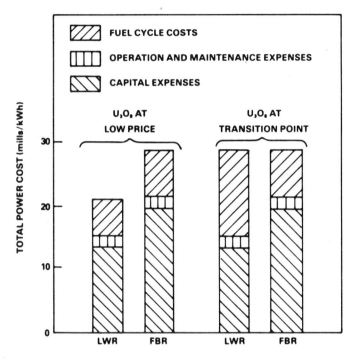

FIGURE 3-18. Comparison of power cost components for the LWR and the FBR at low U_3O_8 price and at the transition point.

economic indifference exists. This U_3O_8 price, at which the total cost of power from either an LWR or an FBR is equal, is often called the *transition point*. The components of the total cost of power both prior to, and at, the transition point are illustrated in Fig. 3-18. While the transition point is conceptually clear, it is often difficult to calculate uniquely. Thus, the timing of the breeder, and hence planning of an FBR program, is subject to uncertainty.

The reasons one has difficulty in calculating a unique economic transition point are several. First, the total cost of power from the FBR is a very strong function of its capital cost—and the capital cost is somewhat uncertain. It is possible that the capital cost of the FBR could be as low as 1.25 times that of an LWR, or it is possible that it could be as high as 1.75 times that of an LWR. This implies that the total cost of power from the FBR is commensurately uncertain. The effect of this uncertainty on the transition point is illustrated in Fig. 3-19.

The price of U_3O_8 in the future is a second uncertainty in an analysis of the timing of the breeder. Since uranium is a depletable resource, the price

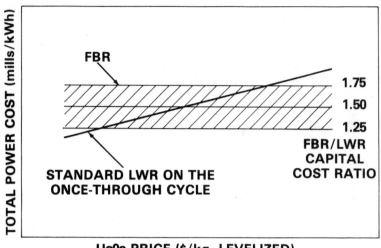

U₃O₈ PRICE ($/kg, LEVELIZED)

FIGURE 3-19. Transition from the standard LWR on the once-through cycle to the FBR. The horizontal lines represent FBR total power cost for the assumed ratio of FBR/LWR capital costs.

of U_3O_8 can be expected to increase as the supply is depleted. As with most minerals, the initial supply of U_3O_8 was probably obtained from rich deposits with a low marginal cost of production. As this supply is depleted, it becomes necessary to work poorer deposits with a higher marginal cost of production. Purchasers must then be willing to pay higher prices before a continuous supply of uranium will be available from these deposits. Thus, in order to accurately assess the rate at which U_3O_8 prices will increase, it is necessary to accurately assess the rate at which the uranium supply is consumed.

Since the price of U_3O_8 is an increasing function of the cumulative amount of uranium consumed, the ability to estimate future U_3O_8 prices is directly dependent upon the ability to estimate future uranium consumption. Future uranium consumption, in turn, depends directly upon the number of light water reactors. The number of light water reactors expected over the next two or three decades, often called the nuclear growth projection, is itself subject to uncertainty.

Although uncertainty exists, it is still possible to estimate a commercialization date for the FBR. Given a reasonable estimate for the nuclear growth projection, one can calculate the amount of U_3O_8 which would be committed without the FBR, as shown in Fig. 3-20. If the amount of both high and low grade uranium resources can be estimated, it is possible to

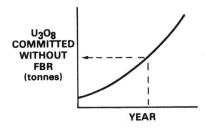

FIGURE 3-20. The amount of U_3O_8 committed as a function of time without the FBR.

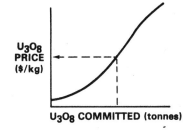

FIGURE 3-21. The price of U_3O_8 as a function of the amount committed.

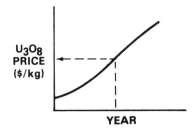

FIGURE 3-22. The price of U_3O_8 as a function of time obtained by combining Fig. 3-20 and 3-21.

estimate a U_3O_8 price as a function of the cumulative amount committed. One such estimate is shown in Fig. 3-21. When combined with Fig. 3-20 which indicates the cumulative amount committed as a function of time, one obtains an estimate of the price of U_3O_8 as a function of time, as shown in Fig. 3-22. As previously noted, the price of U_3O_8 strongly affects the total cost of power from the LWR, while the converse is true for the FBR. In contrast, the total cost of power from the FBR is strongly affected by the capital cost uncertainty. Both effects are illustrated in Fig. 3-23. If Fig. 3-23 is combined with Fig. 3-22, it is possible to estimate the total cost of power from both the LWR and the FBR as a function of time as shown in Fig. 3-24. The economic transition date is then the time at which the total cost of power from the FBR becomes less than that from the LWR.

Since the development of a commercial FBR requires decades—at least two, perhaps three—it is not surprising to find that different nations regard the FBR differently. Some nations, usually resource poor, are willing to run the risk of commercializing the FBR prior to the time of economic attractiveness. Other nations, usually resource rich, attempt to time the FBR program such that commercialization and economic attractiveness are coincident.

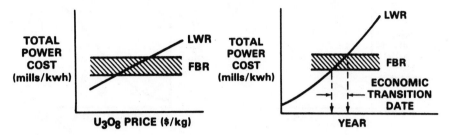

FIGURE 3-23. The total cost of power from the LWR and the FBR as a function of U₃O₈.

FIGURE 3-24. The time at which a transition would be economic.

3-9 SYMBOLS USED IN CHAPTER 3

Levelized Charges for Electricity

L_b = fuel costs for a single batch of fuel
L_{btax} = taxes associated with a single batch of fuel
L_{btot} = total fuel expenses
L_{cap} = return of and return on a capital investment
L_{ctax} = taxes associated with a capital investment
L_{ctot} = total capital expenses
L_{fc} = fixed charges associated with a capital investment
L_{fuel} = fuel costs based on all batches of fuel
L_{om} = operation and maintenance expenses
L_{total} = total power cost

Money Rates

b = fraction of funds obtained by bonds
e = fraction of funds obtained by equity
f = fixed charge fraction
i = effective interest rate or weighted mean rate of return
i_b = bond interest rate
i_e = equity interest rate
t = tax rate

Capital Investment

C = initial capital investment
$c$bin = payment to bond interest

$c$dep = depreciation allowance on a capital investment

C_u = uniform annual payment for the return of and return on a capital investment

D = outstanding debt on a capital investment

E = electrical output level

Exp = expenditures

I = payment to interest

K = number of years for payback of capital investment

P = payment to principal

Rev = revenues

Fuel

B = all back-end costs associated with a single batch of fuel

D = depreciation allowance in a single year for a batch of fuel

Dep = total depreciation allowance for a batch of fuel

F = all front-end costs associated with a single batch of fuel

ld = lead time for front-end costs

lg = lag time for back-end costs

N = residence time for a batch of fuel (also number of fuel batches)

T = total tax payment for a single batch of fuel

Subscript

k = year index

PROBLEMS

3-1. A buyer plans to pay for a $10 000 automobile in monthly installments over a three-year period. The annual interest rate is 15%. Calculate the monthly payments.

3-2. If the U_3O_8 price for a typical batch increases by a factor of three above the value assumed in Section 3-7, and no other changes occur, calculate the new total levelized power cost for the LWR and the FBR.

3-3. For the example given in Section 3-7, assume that the price for plutonium is doubled simultaneously with the U_3O_8 price increase assumed in Problem 3-2. Recalculate the total levelized power cost for the LWR and the FBR.

3-4. (a) Why might the price of plutonium be expected to rise as the price of U_3O_8 rises? (b) What major change would occur in the LWR fuel cycle assumed in Section 3-7 and Fig. 3-11 if FBR's were being built for commercial operation? (c) Despite the breeding of plutonium in the FBR in the example, the combined purchase and sale of plutonium was a net cost to the reactor. Why? (d) If the size of the FBR were increased above 1000 MWe, would you expect the net cost of the plutonium per kWh to increase or decrease, and why?

3-5. If the capital cost for the FBR in the example calculation increased to 1200 \$/kWe (and this were the only change), what would be the new total levelized power cost for the FBR?

PART II

NEUTRONICS

The neutron physics aspects of the fast breeder reactor deserve special attention because it is the hard neutron spectrum required for efficient breeding that provides a principal distinguishing feature from the more common thermal power reactors. Chapter 4 addresses nuclear design techniques, where multigroup diffusion theory is emphasized. The methodology for providing the multigroup cross sections is then included in Chapter 5. Kinetics features unique to the fast reactor are included in Chapter 6, and fuel management concepts—which are needed for the determination of breeding ratios and doubling times—are covered in Chapter 7.

CHAPTER 4

NUCLEAR DESIGN

4-1 INTRODUCTION

The nuclear design of a reactor plant involves defining the nuclear environment which will exist inside the reactor core. This phase of design work has a great bearing on thermal-hydraulic and mechanical analyses; hence, close coordination with these activities must be maintained. Safety and control requirements are closely tied to the nuclear design effort and must be considered for all phases of the fuel cycle. Power distributions are needed in order to determine peak-to-average power factors for thermal-hydraulics analysis. Other neutronics calculations are needed to obtain the following kinds of information: required fissile fraction and inventories, fuel cycle data, shielding data, and transient response.

The topics considered in this chapter include multigroup diffusion theory, geometrical considerations for obtaining input to the diffusion equation, and spatial power distributions and neutron flux spectra encountered in typical LMFBR designs. Problems at the end of the chapter emphasize neutronics design aspects of LMFBR's. Two sets of multigroup cross sections are supplied in Appendix C as reasonably good representations of effective cross sections for the typical LMFBR designs specified in the problems.

4-2 MULTIGROUP DIFFUSION THEORY

The principal nuclear design tools available to the fast reactor designer include diffusion theory, transport theory, and Monte Carlo techniques. Diffusion theory has been found to be adequate for most breeder design calculations. The primary reason for the validity of this relatively simple technique is that the mean free path of fast neutrons is generally long relative to the dimensions of the fuel pins and coolant channels. Hence, there is little localized spatial flux depression or flux peaking over most of the energy range. On the other hand, the core size is large relative to the mean free path.* Transport corrections (discussed in Chapter 5–6C) are often made for specialized problems, such as certain plate geometries in critical-facility analyses, but the general diffusion theory approach can still be successfully employed. Full transport theory or Monte Carlo methods are routinely used only for shielding calculations, where penetration analyses are more sensitive to directional aspects (discussed in Chapter 12-4). Therefore, the emphasis in this text and the problems at the end of this chapter will be concerned with diffusion theory.

A detailed treatment of the neutron energy spectrum is clearly very important in fast reactor neutronics analysis because of the wide range of neutron energies over which most capture and fission reactions occur in fast reactors as compared to thermal reactors. Multigroup analysis must begin, therefore, with many groups in the resonance and high energy regions. The starting point for the many groups is a complete data file of cross sections for fast reactor materials which is almost continuous in energy. One such file is ENDF-B[1], the Evaluated Nuclear Data File, which is commonly used in the U.S.

Multigroup cross sections are obtained from discrete group cross-section data by the use of conventional theories of resonance absorption and scattering. These theories are similar whether one is analyzing a fast reactor or a thermal reactor, although in some cases emphasis is placed on different methods due to differences in energy spectra of importance. Procedures peculiar to fast breeder design are now in common use, however, and these will be described in Chapter 5.

Spatial distributions of neutron flux and power density are as important in fast reactor analysis as in thermal reactor analysis. Fortunately, spatial analysis is easier for fast reactors as a result of the long mean free path. A further consequence of the long mean free path is that a fast power reactor is more tightly coupled neutronically than a large thermal reactor. Hence, the fundamental mode is more dominant, the neutron leakage from the core is higher, and spatial convergence of the diffusion equations is easier to achieve. Despite the general need for detailed energy analysis for fast reactors, many special space-dependent problems requiring two- and three-dimensional analyses can be solved adequately with the use of relatively few energy groups—even as few as four in some cases.

---

*Transport theory may be required for the analyses of small metal fast assemblies where the core dimensions are not large compared to the neutron mean free path.

The Multigroup Equation

For the reasons noted above, multigroup diffusion theory occupies a central role among the computational techniques employed for fast breeder reactor neutronics analyses. Because of this widespread applicability, it is important that the physical basis for each of the terms appearing in the multigroup equation be clearly understood.

The neutron diffusion equation is simply a neutron balance equation, with losses equal to sources. The equation is discussed here for steady state; a transient term will be added in the discussion of reactor kinetics in Chapter 6. Any imbalance between losses and sources is accounted for by the effective multiplication factor k_{eff}, an eigenvalue for this problem.

Since the multigroup equation is a neutron balance equation, the fluxes calculated from it are relative rather than absolute values; the fluxes are known only to within an arbitrary constant. The absolute value of the neutron flux can be calculated from the reactor power level. (See Section 4-6B.)

Suppose the energy spectrum is divided into G energy groups. Let us consider the gth group. The loss terms for group g, in terms of neutrons per unit volume per second, include leakage ($-D_g \nabla^2 \phi_g$), absorption ($\Sigma_{ag} \phi_g$ — which includes capture and fission since $\Sigma_{ag} = \Sigma_{cg} + \Sigma_{fg}$), removal by elastic scattering ($\Sigma_{erg} \phi_g$), and removal by inelastic scattering ($\Sigma_{irg} \phi_g$). Elastic or inelastic scattering *within* the group does not appear in the version of the multigroup equation presented in this book; hence, the total inelastic cross section is different from the value of Σ_{irg} used here. (Some versions of multigroup theory do include in-group scattering; the only requirement is that internal consistency be maintained between the scattering cross section and the scattering matrix.)

The *source terms* include neutrons produced by fission and neutrons scattered into group g by elastic and inelastic scattering at energies above group g. The rate of production of neutrons by fission (per cm^3 per second) is $\Sigma_{g'=1}^{G} (\nu\Sigma_f)_{g'} \phi_{g'}$. The summation is taken over all G energy groups since fission occurs at all energies; the subscript g' is used since groups other than group g are included in this summation. The fraction of those neutrons produced in fission that appear in group g (or are "born into" group g) is χ_g. Hence, the fission source for group g is $\chi_g \Sigma_{g'=1}^{G} (\nu\Sigma_f)_{g'} \phi_{g'}$. In the steady-state multigroup balance equation, the balance between production and loss is achieved by multiplying this source by $1/k_{\text{eff}}$ so that the effective production rate for group g from fission is $(1/k_{\text{eff}}) \chi_g \Sigma_{g'=1}^{G} (\nu\Sigma_f)_{g'} \phi_{g'}$. Hence, if the reactor is supercritical ($k_{\text{eff}} > 1$), so that the fission source is greater than the sum of the loss terms, multiplying the source by $1/k_{\text{eff}}$ reduces the production rate so that a neutron balance is achieved. The elastic and inelastic scattering source terms are $\sum_{g'=1}^{g-1} \Sigma_{eg' \to g} \phi_{g'}$ and $\sum_{g'=1}^{g-1} \Sigma_{ig' \to g} \phi_{g'}$, where $\Sigma_{g' \to g}$ represents scattering from g' into g. (In this book group 1 will denote the fastest energy group and group G the slowest; hence, the scattering source terms imply that only down-scattering from group g' into group g is allowed. Up-scattering is used in thermal reactor analysis in the thermal energy range, but in fast reactors no neutrons survive the slowing-down process to thermal energies.)

The multigroup equation can now be expressed as the following balance:

$$\underbrace{\left[-D_g \nabla^2 \phi_g\right]}_{\text{LEAKAGE}} + \underbrace{\left[\left(\Sigma_{ag} + \Sigma_{erg} + \Sigma_{irg}\right)\phi_g\right]}_{\text{REMOVAL}} =$$

$$\underbrace{\frac{1}{k_{\text{eff}}}\chi_g \sum_{g'=1}^{G} (\nu\Sigma_f)_{g'} \phi_{g'}}_{\text{FISSION SOURCE}} + \underbrace{\sum_{g'=1}^{g-1} \Sigma_{eg' \to g} \phi_g + \sum_{g'=1}^{g-1} \Sigma_{ig' \to g} \phi_{g'}}_{\text{SCATTERING IN}}. \quad (4\text{-}1)$$

4-3 SPATIAL SOLUTIONS TO THE MULTIGROUP EQUATION

One-dimensional, two-dimensional, and three-dimensional solutions to the multigroup diffusion equation are all regularly used in fast reactor design, always with appropriate compromises between the numbers of groups and dimensions and the economics of computer solutions for the problem of interest. Zero-dimensional (or fundamental mode) solutions are even used for special purposes, as for example in cross section and fuel cycle calculations.

In this book, our purpose is to develop an understanding of the most important features of the energy spectrum and spatial distributions in fast reactors and not to devote excessive attention to multi-spatial solution techniques. Hence, a one-dimensional computer formulation of the diffusion equation will be sufficient for our purposes. Even more useful as a tool for understanding the effects of energy spectrum in a fast reactor is a zero-dimensional solution, which the student can program on a computer in a reasonably short time. Hence, this technique will also be derived for use in problems at the end of the chapter. One two-dimensional topic will be briefly discussed, a two-dimensional triangular mesh, because this mesh is widely used for analysis of the hexagonal fast reactor geometry and is not used for light-water-reactor analysis.

First, we will discuss the approximation to be used for leakage in the directions perpendicular to the direction of the solution. This provides the basis for introducing the zero-dimensional solution. Next, a one-dimensional solution will be presented, and the section is then concluded with a discussion of the two-dimensional triangular mesh.

A. TRANSVERSE LEAKAGE APPROXIMATION

In one-dimensional diffusion theory, a technique for treating neutron leakage in the transverse directions is needed since spatial effects in these directions are not treated rigorously. In zero dimensions this technique must be applied for all directions. The method is not different from techniques used for thermal reactors.

To illustrate the technique, consider a single energy diffusion equation for a critical reactor:

$$-D\nabla^2\phi+\left(\Sigma_a-\nu\Sigma_f\right)\phi=0. \tag{4-2}$$

In terms of geometric buckling B^2, the equation is

$$\nabla^2\phi+B^2\phi=0. \tag{4-3}$$

We will consider cylindrical geometry since fast breeder reactors (like thermal power reactors) are normally calculated as right circular cylinders. Hence, Eq. (4-3) becomes

$$\frac{\partial^2\phi}{\partial r^2}+\frac{1}{r}\frac{\partial\phi}{\partial r}+\frac{\partial^2\phi}{\partial z^2}+B^2\phi=0. \tag{4-4}$$

By separation of variables,

$$\phi=R(r)Z(z), \tag{4-5}$$

where R is a function of r only and Z is a function of z only. Dividing Eq. (4-4) by Eq. (4-5) gives

$$\frac{1}{R}\left(\frac{d^2R}{dr^2}+\frac{1}{r}\frac{dR}{dr}\right)+\frac{1}{Z}\frac{d^2Z}{dz^2}=-B^2. \tag{4-6}$$

Since the sum of the two terms on the left-hand side is a constant, and the first term is a function of r only and the second term is a function of z only, and each term can be varied independently of the other, it follows that each term on the left-hand side is a constant. Hence,

$$\frac{1}{R}\left(\frac{d^2R}{dr^2} + \frac{1}{r}\frac{dR}{dr}\right) = -B_r^2 \tag{4-7}$$

$$\frac{1}{Z}\frac{d^2Z}{dz^2} = -B_z^2 \tag{4-8}$$

where

$$B_r^2 + B_z^2 = B^2; \tag{4-9}$$

and it follows that

$$\frac{\partial^2\phi}{\partial r^2} + \frac{1}{r}\frac{\partial\phi}{\partial r} = -B_r^2\phi, \tag{4-10}$$

and

$$\frac{\partial^2\phi}{\partial z^2} = -B_z^2\phi. \tag{4-11}$$

The simplest boundary conditions are (1) symmetry, and (2) disappearance of the flux at some extrapolated distance, δ, beyond the boundary; or

(1)
$$\frac{dR}{dr} = 0 \text{ at } r = 0, \text{ and } \frac{dZ}{dz} = 0 \text{ at } z = 0, \tag{4-12}$$

(2)
$$R = 0 \text{ at } r = R_0 + \delta_r, \text{ and } Z = 0 \text{ at } z = \frac{H}{2} + \delta_z. \tag{4-13}$$

The solutions to Eqs. (4-7) and (4-8) for these boundary conditions are a Bessel function and a cosine, respectively. Hence the solution to Eq. (4-2), to within an unspecified constant A, is

$$\phi = AJ_0(B_r r)\cos(B_z z), \tag{4-14}$$

where

$$B_r = \frac{2.405}{R_0 + \delta_r} \text{ and } B_z = \frac{\pi}{H + 2\delta_z}. \tag{4-15}$$

Now we can introduce the required approximations for transverse leakage. Equation (4-2) can be written

$$-D\left(\frac{\partial^2\phi}{\partial r^2} + \frac{1}{r}\frac{\partial\phi}{\partial r} + \frac{\partial^2\phi}{\partial z^2}\right) + \left(\Sigma_a - \nu\Sigma_f\right) = 0. \tag{4-16}$$

The terms with the r derivatives represent radial leakage; the $-D\partial^2\phi/\partial z^2$ term represents axial leakage.

For a one-dimensional solution in the r direction, the axial leakage is replaced by Eq. (4-11), or

$$-D\left(\frac{d^2\phi}{dr^2} + \frac{1}{r}\frac{d\phi}{dr}\right) + DB_z^2\phi + \left(\Sigma_a - \nu\Sigma_f\right)\phi = 0. \tag{4-17}$$

In the z direction the radial leakage is replaced by Eq. (4-10) so that

$$-D\frac{d^2\phi}{dz^2} + DB_r^2\phi + \left(\Sigma_a - \nu\Sigma_f\right)\phi = 0. \tag{4-18}$$

For a zero-dimensional solution, both r and z derivatives are replaced by Eqs. (4-10) and (4-11), so that

$$DB_r^2\phi + DB_z^2\phi + \left(\Sigma_a - \nu\Sigma_f\right)\phi = 0. \tag{4-19}$$

In order to use these approximations for leakage, it is necessary to evaluate δ_r and δ_z, which appear in the second boundary condition, Eq. (4-13). These parameters are similar to a reflector savings which, for a large reflector, is given approximately by $D_c M_r/D_r$, where D and M are the diffusion coefficient and migration length, respectively, and subscripts c and r refer to the core and reflector. In a breeder reactor the core is surrounded by a blanket for which the diffusion coefficient is about the same as that of the core. Hence the δ's would be about equal to M. The migration area, M^2, is approximately D/Σ_a. For a neutron energy of 100 keV, which is the order of magnitude for the median fission energy in an LMFBR, the diffusion coefficient in the blanket is about 1.3 cm and the absorption cross section is of the order of 0.003 cm^{-1}. This gives a value for M and for δ of the order of 20 cm.

The effective value of δ will vary with energy and with the buildup of fissile material in the blanket. Hence, appropriate values for δ for fast reactors must be determined by experience that can only be obtained by comparing actual multigroup spatial leakage calculations with the approximations represented by $DB^2\phi$ discussed here. This does, in fact, result in values for δ of about 15 to 20 cm over a relatively wide energy range, from several 10's of kilovolts to 1 MeV. Values of δ can be estimated from the flux shapes shown later in this chapter, in Section 4-6D.

B. ZERO-DIMENSIONAL SOLUTION

The zero-dimensional, or fundamental mode, solution is easier than the one-dimensional solution because of the simple treatment of leakage. This simplifying approximation results in a non-iterative solution for the effective criticality factor, k_{eff}, a situation not possible for the one-dimensional solution.

Replacing the leakage term, $-D_g \nabla^2 \phi_g$, in Eq. (4-1) with $D_g B_g^2 \phi_g$ as discussed in the previous part of this section, the multigroup equations can be written as

$$D_g B_g^2 \phi_g + \Sigma_{rg} \phi_g = \frac{1}{k_{eff}} \chi_g \sum_{g'=1}^{G} \left(\nu \Sigma_f \right)_{g'} \phi_{g'} + \sum_{g'=1}^{g-1} \Sigma_{eg' \to g} \phi_{g'} + \sum_{g'=1}^{g-1} \Sigma_{ig' \to g} \phi_{g'},$$

$$(4\text{-}20)$$

where
$$B_g^2 = B_{rg}^2 + B_{zg}^2, \qquad (4\text{-}21)$$

and
$$\Sigma_{rg} = \Sigma_{ag} + \Sigma_{erg} + \Sigma_{irg}. \qquad (4\text{-}22)$$

To solve for the fluxes, the first step is to specify that

$$\frac{1}{k_{eff}} \sum_{g'=1}^{G} \left(\nu \Sigma_f \right)_{g'} \phi_{g'} = 1. \qquad (4\text{-}23)$$

[Recall from an earlier discussion that the (ϕ_g)'s are relative values, a fact that is quite apparent from Eq. (4-23).]

As a result of Eq. (4-23), the equation for group 1 is

$$D_1 B_1^2 \phi_1 + \Sigma_{r1} \phi_1 = \frac{1}{k_{eff}} \chi_1 \sum_{g'=1}^{G} \left(\nu \Sigma_f \right)_{g'} \phi_{g'} = \chi_1. \qquad (4\text{-}24)$$

The only unknown in this equation is ϕ_1; hence it can be evaluated immediately.

The equation for group 2 is

$$D_2 B_2^2 \phi_2 + \Sigma_{r2} \phi_2 = \chi_2 + \Sigma_{e1 \to 2} \phi_1 + \Sigma_{i1 \to 2} \phi_1. \qquad (4\text{-}25)$$

Since ϕ_1 has by now been evaluated, we can solve for ϕ_2. We proceed in a similar manner until all (ϕ_g)'s are known.

Knowing all of the group fluxes, we can use Eq. (4-23) to solve for k_{eff}, or

$$k_{eff} = \sum_{g'=1}^{G} (\nu\Sigma_f)_{g'}\phi_{g'}. \tag{4-26}$$

Note that the solution is not iterative because this first value of k_{eff} is consistent with our original fission source defined in Eq. (4-23).

C. ONE-DIMENSIONAL SOLUTION

• Cylindrical Geometry (Radial Direction)

We present here the one-dimensional solution of the multigroup diffusion equation used in the two cross section codes, 1DX[2] and MINX,[3] as a typical example. Numerous alternate one-dimensional techniques are available.

The reactor is divided into N mesh intervals, as shown in Fig. 4-1. Note that for this formulation mesh points are not located on the boundaries. Equally valid methods are available with mesh points on the boundaries and at the center. The equation is solved numerically by approximating the axial leakage by the fundamental mode solution for $\nabla^2\phi(z)$ (i.e., $-B_z^2\phi$) and

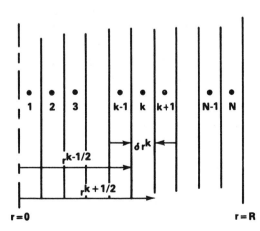

FIGURE 4-1. Mesh geometry for one-dimensional radial solution.

integrating over each mesh volume. Boundary conditions are needed at the center of the core and at the outside of the reactor.

In the following analysis, the superscript k is the mesh index, while the subscript g continues to be the energy group index.

Integration over a mesh volume around point k gives

$$-\int_k D_g^k \nabla^2 \phi_g^k \, dV + \int_k D_g^k B_{zg}^2 \phi_g^k \, dV + \int_k \Sigma_{rg}^k \phi_g^k \, dV$$

$$= \frac{1}{k_{\text{eff}}} \int_k \chi_g \sum_{g'=1}^{G} (\nu \Sigma_f)_{g'}^k \phi_{g'}^k \, dV + \int_k \sum_{g'=1}^{g-1} \Sigma_{g' \to g}^k \phi_{g'}^k \, dV,$$

$$(4\text{-}27)$$

where Σ_{rg} denotes all of the removal terms listed on the left-hand side of Eq. (4-1) and $\Sigma_{g' \to g}$ denotes the sum of both elastic and inelastic scattering from group g' to group g.

Using the divergence theorem,

$$\int D \nabla^2 \phi \, dV = \int D \nabla \phi \cdot \overrightarrow{dA}$$

and evaluating the derivatives halfway between the mesh points gives

$$-\int_k D_g^k \nabla \phi_g^k \cdot \overrightarrow{dA} = D_g^{k-1,k} A^{k-1,k} \frac{\phi_g^k - \phi_g^{k-1}}{r^k - r^{k-1}} - D_g^{k,k+1} A^{k,k+1} \frac{\phi_g^{k+1} - \phi_g^k}{r^{k+1} - r^k}.$$

$$(4\text{-}28)$$

The minus sign disappears in front of the first term on the right-hand side because $\nabla \phi \cdot \overrightarrow{dA}$ is negative (whereas $\nabla \phi \cdot \overrightarrow{dA}$ is positive in the second term). The difference $\phi_g^k - \phi_g^{k-1}$ in the first term is generally negative, in which case the first term would represent a source, or leakage, into the mesh interval at k. The second term would then be positive and would represent a loss, or leakage, out of the interval.

The coefficient $D_g^{k-1,k}$ is an effective diffusion coefficient between mesh points $k-1$ and k; it is based on the volume-averaged macroscopic transport cross section $\Sigma_{\text{tr}g}$ between mesh points. This volume-averaged value for $\Sigma_{\text{tr}g}$ is

$$\Sigma_{\text{tr}g}^{k-1,k} = \Sigma_{\text{tr}g}^{k-1} \frac{\delta r^{k-1}}{\delta r^{k-1} + \delta r^k} + \Sigma_{\text{tr}g}^k \frac{\delta r^k}{\delta r^{k-1} + \delta r^k}.$$

Since $D = 1/3\Sigma_{tr}$,

$$\frac{1}{D_g^{k-1,k}} = \frac{1}{D_g^{k-1}} \frac{\delta r^{k-1}}{\delta r^{k-1} + \delta r^k} + \frac{1}{D_g^k} \frac{\delta r^k}{\delta r^{k-1} + \delta r^k}$$

or

$$D_g^{k-1,k} = \frac{D_g^{k-1} D_g^k (\delta r^{k-1} + \delta r^k)}{D_g^{k-1} \delta r^k + D_g^k \delta r^{k-1}}. \qquad (4\text{-}29)$$

Values for the area and mesh volume are

$$A^{k-1,k} = 2\pi r^{k-1/2} H, \qquad (4\text{-}30)$$

$$V^k = \pi \left[(r^{k+1/2})^2 - (r^{k-1/2})^2 \right] H = \pi v^k H, \qquad (4\text{-}31)$$

where H is the axial dimension, and v^k is defined by Eq. (4-31) and is therefore a volume per unit height, divided by π.

Next we define a source S_g^k as

$$S_g^k = \frac{1}{k_{eff}} \chi_g \sum_{g'=1}^{G} (\nu\Sigma_f)_{g'}^k \phi_{g'}^k + \sum_{g'=1}^{g-1} \Sigma_{g' \to g}^k \phi_{g'}^k. \qquad (4\text{-}32)$$

Assuming that all cross sections and χ's are constant over the mesh volume, substituting Eqs. (4-28) through (4-32) into (4-27) gives

$$D_g^{k-1,k} 2 r^{k-1/2} \frac{\phi_g^k - \phi_g^{k-1}}{r^k - r^{k-1}} - D_g^{k,k+1} 2 r^{k+1/2} \frac{\phi_g^{k+1} - \phi_g^k}{r^{k+1} - r^k}$$

$$+ D_g^k B_{zg}^2 \phi_g^k v_k + \Sigma_{rg}^k \phi_g^k v^k = S_g^k v^k.$$

Rearranging gives

$$-D_g^{k-1,k} \frac{2 r^{k-1/2}}{r^k - r^{k-1}} \phi_g^{k-1} + \left(D_g^{k-1,k} \frac{2 r^{k-1/2}}{r^k - r^{k-1}} + D_g^{k,k+1} \frac{2 r^{k+1/2}}{r^{k+1} - r^k} \right.$$

$$\left. + D_g^k B_{zg}^2 v^k + \Sigma_{rg}^k v^k \right) \phi_g^k - D_g^{k,k+1} \frac{2 r^{k+1/2}}{r^{k+1} - r^k} \phi_g^{k+1} = S_g^k v^k. \qquad (4\text{-}33)$$

We next define α and β such that

$$\alpha_g^k = D_g^{k-1,k} \frac{2r^{k-1/2}}{r^k - r^{k-1}} \tag{4-34}$$

$$\beta_g^k = \alpha_g^k + \alpha_g^{k+1} + D_g^k B_{zg}^2 v^k + \Sigma_{rg}^k v^k, \tag{4-35}$$

so that Eq. (4-33) becomes

$$-\alpha_g^k \phi_g^{k-1} + \beta_g^k \phi_g^k - \alpha_g^{k+1} \phi_g^{k+1} = S_g^k v^k. \tag{4-36}$$

We next consider the boundary conditions for the two most usual cases:

(1) $$\frac{d\phi}{dr} = 0 \text{ at } r = 0$$

(2) $$\phi = 0 \text{ at } r = R + 0.71\lambda_{tr}$$

where R is the outermost boundary of the reactor and λ_{tr} is the transport mean free path.

For boundary condition (1), one can imagine a mesh interval ($k=0$) to the left of center and equal in width to mesh interval 1. The boundary condition is satisfied by setting $\phi^0 = \phi^1$. The first term and the first part of the second term of Eq. (4-33) cancel, so that Eq. (4-33) becomes

$$\left(D_g^{1,2} \frac{2r^{3/2}}{r^2 - r^1} + D_g^1 B_{zg}^2 v^1 + \Sigma_{rg}^1 v^1 \right) \phi_g^1 - D_g^{1,2} \frac{2r^{3/2}}{r^2 - r^1} \phi_g^2 = S_g^1 v^1,$$

or $\alpha_g^1 = 0$, and

$$\beta_g^1 \phi_g^1 - \alpha_g^2 \phi_g^2 = S_g^1 v^1. \tag{4-37}$$

For boundary condition (2), one can imagine a mesh point, $N+1$, a distance $0.71\lambda_{tr}$ to the right of the outermost boundary, R, where $\phi^{N+1} = 0.$* The distance between the mesh point N and $N+1$ is $\frac{1}{2}\delta r^N + 0.71\lambda_{tr}$.

The value of α_g^{N+1} (for use in β^N) is

$$\alpha_g^{N+1} = D_g^N \frac{2R}{\frac{1}{2}\delta r^N + 0.71\lambda_{trg}},$$

---

*Note that the extrapolation distance $0.71\lambda_{tr}$ is different from the fundamental mode distance δ at which the flux extrapolates to zero, as described in the first part of this section.

where $R = r^N + \frac{1}{2}\delta r^N$. Since $\phi_g^{N+1} = 0$, Eq. (4-33) becomes

$$-\alpha_g^N \phi_g^{N-1} + \beta_g^N \phi_g^N = S_g^N v^N. \qquad (4\text{-}38)$$

This system of multigroup equations can be written in matrix form as

$$A\phi = C, \qquad (4\text{-}39)$$

where A is a square matrix with $N \times G$ rows and columns:

$$
A = \begin{bmatrix}
\beta_1^1 & -\alpha_1^2 & & & & & & & \\
-\alpha_1^2 & \beta_1^2 - \alpha_1^3 & & & & & & & \\
 & & \ddots & & & & & & \\
 & & -\alpha_1^k & \beta_1^k - \alpha_1^{k+1} & & & 0 & & \\
 & & & & \ddots & & & & \\
 & & & & -\alpha_1^N & \beta_1^N & & & \\
 & & & & & \beta_2^1 - \alpha_2^2 & & & \\
 & & & & & -\alpha_2^2 & \beta_2^2 - \alpha_2^3 & & \\
 & & & & & & \ddots & & \\
 & & & & & & -\alpha_g^k & \beta_g^k - \alpha_g^{k+1} & \\
 & & & & & & & \ddots & \\
 & 0 & & & & & & -\alpha_G^{N-1} & \beta_G^{N-1} & -\alpha_G^N \\
 & & & & & & & & -\alpha_G^N & \beta_G^N
\end{bmatrix}
$$

and ϕ and C are the following vector matrices, each with $N \cdot G$ elements:

$$\phi = \left\{ \phi_1^1, \phi_1^2 \cdots \phi_1^N, \phi_2^1, \phi_2^2 \cdots \phi_g^k \cdots \phi_G^{N-1}, \phi_G^N \right\},$$

$$C = \left\{ S_1^1 v^1, S_1^2 v^2 \cdots S_1^N v^N, S_2^1 v^1, S_2^2 v^2 \cdots S_g^k v^k \cdots S_G^{N-1} v^{N-1}, S_G^N v^N \right\}.$$

This equation can be solved by inversion from the relation

$$\phi = A^{-1}C. \qquad (4\text{-}40)$$

The method of solution is iterative. To describe the method, we will first define another term, the fission source, denoted by F, and given by

$$F = \sum_{k=1}^{N} \sum_{g'=1}^{G} F_{g'}^k = \sum_{k=1}^{N} \sum_{g'=1}^{G} \left(\nu\Sigma_f\right)_{g'}^k \phi_{g'}^k v^k. \qquad (4\text{-}41)$$

In order to converge on the criticality factor, k_{eff}, it is necessary first to guess a flux distribution (ϕ_g^k) to be used in the expression for S_g^k and for F, and then to iterate until convergence is reached to within a specified criterion. A typical starting flux distribution is a fission spectrum, uniform in space. (The fission spectrum is discussed in Chapter 5-3.)

New fluxes are obtained from Eq. (4-40), and a new fission source, F, is calculated. Each new calculation of fluxes and the fission source is an iteration.

To see how convergence is obtained, let $F^{\nu+1}$ be the fission source obtained from the fluxes calculated in the νth iteration. A multiplication factor for the next iteration, $\lambda^{\nu+1}$, is defined as the ratio of the fission source of the $(\nu+1)$th iteration to that of the previous iteration, or

$$\lambda^{\nu+1} = \frac{F^{\nu+1}}{F^\nu}. \tag{4-42}$$

The effective multiplication factor $k_{eff}^{\nu+1}$ for the next iteration is then obtained by multiplying the previous iteration value by the new λ, or

$$k_{eff}^{\nu+1} = k_{eff}^\nu \lambda^{\nu+1}. \tag{4-43}$$

This value of k_{eff}, together with the new values of the neutron fluxes, are then used to obtain the new source terms, $(S_g^k)^{\nu+1}$, which are in turn used for the next iteration.

As the solution converges, the fluxes and, hence, the fission source F converge toward the same values on successive iterations. Thus, λ tends toward unity. A convergence criterion ε is specified such that, when

$$|1-\lambda| < \varepsilon, \tag{4-44}$$

the solution is said to have converged.

The actual convergence technique used in 1DX is more sophisticated than that presented above in that fission-source *over-relaxation* is employed to accelerate convergence. The procedure is as follows. After the new fission source is calculated (call it $F_1^{\nu+1}$), a second "new" value, $F_2^{\nu+1}$, is computed by magnifying the difference between the new and the old fission sources by a factor of β, the over-relaxation factor, as follows:

$$F_2^{\nu+1} = F^\nu + \beta(F_1^{\nu+1} - F^\nu). \tag{4-45}$$

$F_2^{\nu+1}$ is then normalized to give the same total source as $F_1^{\nu+1}$. The original value for β used in 1DX was 1.4.

• Slab Geometry

Slab geometry would be used for a one-dimensional calculation in the axial direction. The only differences between slab and cylindrical geometry arise in the equations for $A^{k-1,k}$ and V^k, Eqs. (4-30) and (4-31), and in replacing B_{zg}^2 with B_{rg}^2. Assuming the axial solution is desired for a cylindrical reactor, the equations for $A^{k-1,k}$ and V^k are

$$A^{k-1,k} = \pi R^2, \tag{4-30a}$$

$$V^k = \pi R^2 \delta z^k. \tag{4-31a}$$

Replacing δr^k by δz^k in the definition of $D^{k-1,k}$, Eq. (4-33) becomes

$$-\frac{D_g^{k-1,k}}{z^k - z^{k-1}}\phi_g^{k-1} + \left(\frac{D_g^{k-1,k}}{z^k - z^{k-1}} + \frac{D_g^{k,k+1}}{z^{k+1} - z^k} + D_g B_{rg}^2 \delta z^k + \Sigma_{rg}^k \delta z^k\right)\phi_g^k$$

$$-\frac{D_g^{k,k+1}}{z^{k+1} - z^k}\phi_g^{k+1} = S_g^k \delta z^k. \tag{4-33a}$$

In slab geometry,

$$\alpha_g^k = \frac{D_g^{k-1,k}}{z^k - z^{k-1}}, \tag{4-34a}$$

$$\beta_g^k = \alpha_g^k + \alpha_g^{k+1} + D_g^k B_{rg}^2 \delta z^k + \Sigma_{rg}^k \delta z_k, \tag{4-35a}$$

$$-\alpha_g^k \phi_g^{k-1} + \beta_g^k \phi_g^k - \alpha_g^{k+1} \phi_g^{k+1} = S_g^k \delta z^k, \tag{4-36a}$$

$$\beta_g^1 \phi_g^1 - \alpha_g^2 \phi_g^2 = S_g^1 \delta z^1, \tag{4-37a}$$

$$\alpha_g^{N+1} = \frac{D_g^N}{\frac{1}{2}\delta z^M + 0.71\lambda_{trg}}, $$

$$-\alpha_g^N \phi_g^{N-1} + \beta_g^N \phi_g^N = S_g^N \delta z^N. \tag{4-38a}$$

D. TWO-DIMENSIONAL TRIANGULAR MESH

A geometry often used for two-dimensional analysis in fast reactors but not encountered in light-water-reactor analysis is the triangular mesh. This mesh arrangement is useful for fast reactor hexagonal assemblies.* It is used in place of the conventional $x-y$ or $r-\theta$ mesh.

---

*This mesh is also used for analysis of the hexagonal assemblies in High Temperature Gas Reactors (HTGR).

In hexagonal geometry, a symmetric fast reactor can often be conveniently represented by one-sixth of the core and the blanket. Radial power distributions are frequently reported in this form, i.e., for each hexagonal assembly in a sextant of the core and blanket. Examples of this representation are given in Section 4-6.

An example of a 3×4 triangular mesh representing two hexagons is shown in Fig. 4-2. A mesh point of coordinates (i, j) is located at the center of each triangle. The mesh boundaries x_i and y_j are fixed by the flat-to-flat width, W, of the hexagons. The mesh spacings Δx and Δy are constant throughout the mesh.

The mesh height, Δy, is simply $W/2$, from Fig. 4-2. The mesh width, Δx, must be defined to conserve the area of the triangular mesh,

$$(\Delta x)(\Delta y) = \frac{1}{2} \cdot \frac{W}{2} \cdot \frac{W}{\sqrt{3}}, \tag{4-46}$$

so that

$$\Delta x = \frac{W}{2\sqrt{3}}. \tag{4-47}$$

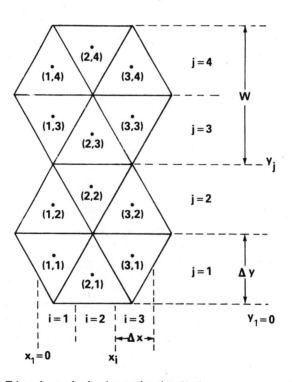

FIGURE 4-2. Triangular mesh, showing mesh points (i, j).

Setting x_1 and y_1 equal to zero, the mesh boundaries are

$$x_i = (i-1)\frac{W}{2\sqrt{3}} \qquad (4\text{-}48)$$

$$y_j = (j-1)\frac{W}{2}. \qquad (4\text{-}49)$$

4-4 VOLUME FRACTIONS AND ATOM DENSITIES

A starting point in nuclear design is the calculation of volume fractions, F, and atom densities, N. These must be calculated for each region in the core and blankets. Control assemblies will be distributed throughout the core, and for early scoping calculations these can be homogenized with the fuel assemblies.

The fuel pins are assembled into a triangular array in hexagonal fuel assemblies in a fast breeder reactor. The basic motivation for this arrangement is neutronics performance. The "tight" lattice offered by a triangular array, as opposed to a square lattice (Problem 4-1), allows a larger value for the fuel volume fraction and smaller core dimensions. As pointed out in Chapter 2 and illustrated later in Chapter 6-5, decreasing the core size with a given amount of fertile plus fissile fuel leads to a lower fissile fraction and lower fissile inventory in a fast reactor. Also, for a fixed size core, increasing the fuel volume fraction lowers the required fissile inventory and increases the breeding ratio; hence, increasing fuel volume fraction is generally desirable. Typical ranges for volume fractions for LMFBR cores are

Fuel	30–45%
Sodium	35–45%
Steel	15–20%
B_4C	1– 2%

The basic repeating cell for calculating volume fractions is the assembly. Volume fractions are calculated for fuel assemblies, radial blanket assemblies, and control assemblies. Fuel in the axial blankets is contained in the same pins as the fuel in the core; hence, volume fractions in axial blankets are the same as for the core. Core-wide average volume fractions can be approximated by weighting the volume fractions for fuel and control assemblies by the numbers of each type of assembly present. More detailed calculations can account for the spatial distribution of control assemblies by homogenizing an appropriate number of fuel assemblies with the control assemblies at a particular radial position to form an annular core region. More expensive but more precise calculations involving all three dimensions could be employed to allow individual modeling of each assembly in the reactor. Such a model would not require homogenization of adjacent assemblies.

Volume fractions are calculated from the geometrical cross-sectional dimensions, since the axial dimension of all materials is the same. A cross section of a fuel assembly was illustrated in Fig. 2-3. The repeating lattice is illustrated in the one-twelfth section shown in Fig. 4-3. As noted from this figure, the outer boundary of the repeating lattice is in the sodium halfway between adjacent assemblies. The fuel is assumed to occupy all the space inside the cladding. Structural material is present in the cladding and hexcan walls and in the spacers between fuel pins, either in the form of wire wrap or grid spacers. Often all the structural material is the same, generally some selected type of stainless steel. The sodium fills all of the remaining spaces inside the hexcan and is also present between the hexcans of adjacent assemblies. Useful geometric relationships for the flow areas in LMFBR fuel assemblies are given by Eq. (9-31).

When translating dimensions to volume fractions, a question arises regarding whether the dimensions are for the assemblies as-fabricated (hence, at room temperature) or at full power in the reactor. Here careful communication between the neutronics designer, the mechanical designer, and the

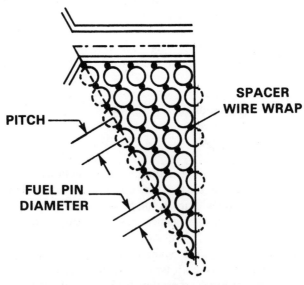

1/12 ASSEMBLY CUTAWAY

FIGURE 4-3. Repeating lattice.

thermal-hydraulics analyst is important. The neutronics designer wants to calculate neutronics parameters (hence, atom densities) for full-power conditions. Consider the fuel pin diameter, for example, for Type 316 stainless steel. The coefficient of thermal expansion $[\alpha = (1/L)(dL/dT)]$ between 20°C and 500°C (a typical cladding temperature at full power) is about $2 \times 10^{-5}/K$. The fractional change in the inside and outside diameter of the cladding between the pin as-fabricated and the pin at full power is about 1%. The cross sectional area of the structure (hence, the structure volume fraction) and the area inside the cladding (which controls the fuel volume fraction) would increase by $2\alpha \Delta T$, or 2% between 20°C and 500°C.

A parameter called *smear density* is used in calculating atom densities for the fuel. The smear density is the density of the fuel if it were uniformly spread or smeared throughout the inside of the cladding. During fabrication of fuel pins, pellets of a given density and at room temperature are loaded into the cladding, with a gap between the pellet and the cladding. For calculation of atom densities, the neutronics designer does not need to know the details of the pellet design—e.g., the exact diameter, whether the pellets are *dished* (i.e. have space between pellets), or how the pellets will crack and expand when the reactor is taken to power. He only needs to know the space available for the fuel and the mass of fuel in that space. He must, therefore, know the inside diameter of the cladding and the smear density of the fuel at full power.

The fuel fabricator is concerned with another type of density called the *percent of theoretical density*. Mixed oxide fuel with 20% to 25% PuO_2 in UO_2, as in LMFBR's, has a theoretical (crystalline) density of about 11 g/cm^3 with some variation depending on Pu/U ratio and stoichiometry (i.e., the average number of oxygen atoms in each molecule, which can vary 1% or 2% around two). Fuel pellets are designed to some percent of this theoretical density, usually around 95% or a little higher. The smear density generally lies in the range of 85% to 90% of theoretical density.

Atom densities for material m are calculated from the relation,

$$N_m = \frac{F_m \rho_m N_A}{M_m},$$ (4-50)

where

N_m = atom density (atoms/cm^3)*
F_m = volume fraction
ρ_m = density (g/cm^3)
M_m = molecular weight (g/g-mol)
N_A = Avogadro's number (6.023×10^{23} atoms/g-mol).

---

*Neutronics designers generally use cgs units for atom densities, cross sections, and neutron fluxes, and this procedure will be followed for these parameters in this book.

For each material in the metal alloys used for structures, such as iron, nickel and chromium in stainless steel, the ρ_m is replaced by the product of the alloy density and the weight percent of the material. The atomic weight of the particular material is used for M_m.

Care must be taken to be consistent in methods of calculating fuel atom densities. Different organizations develop their own methods. The method suggested here will be used throughout the fuel cycle analysis of Chapter 7. We use the smear density for ρ_f and a single molecular weight for the fuel, recognizing that the fuel is composed of isotopes from ^{235}U to ^{242}Pu plus even higher actinides for the U-Pu cycle. The single molecular weight suggested for the U-Pu cycle is 270, i.e., 238 for ^{238}U plus 32 for O_2. Though the average M may be slightly higher, additional accuracy is not generally warranted.

During irradiation of fuel in the reactor, some fuel atoms fission and others are transmuted to higher actinides by neutron capture. When fission occurs, a pair of fission products is formed. The number of fission-product pairs is exactly equal to the number of fissions. Hence, if a fission-product pair is considered to be one material (instead of two), the number density of heavy metal atoms ($\Sigma_i N_i$) plus fission-product pairs (N_{fp}) stays constant throughout the time the fuel remains in the reactor. The fact that the average molecular weight of this mixture is changing is irrelevant. Thus, at all times during fuel irradiation,

$$N_f = \frac{F_f \rho_f N_A}{270} = \sum_{\substack{\text{heavy} \\ \text{metals}, i}} N_i + N_{fp} = \text{constant}. \tag{4-51}$$

For a particular heavy metal isotope N_i the atom density is

$$N_i = I_i N_f, \tag{4-52}$$

where I_i is the fraction of fuel atoms which are isotope type i.

Let us next consider the units for the atom densities that are to be used in the multigroup equations. Macroscopic cross sections, Σ, have units of cm^{-1} and they are the product of an atom density and a microscopic cross section. Microscopic cross sections are given in barns, where 1 barn $= 10^{-24}$ cm^2. The barn is such a convenient unit that it is an unnecessary nuisance to convert it to cm^2. Moreover atom densities are in terms of atoms/cm^3, and there are generally 10^{23} or 10^{24} of them per cm^3. The logical solution is to calculate atom densities in "atoms per barn centimeter," so that the product $N\sigma$ is in cm^{-1} when σ is in barns. To do this, one uses for N_A the value 0.6023 instead of 0.6023×10^{24}. This way one never has to deal with 10^{24}'s and 10^{-24}'s again. "Well," as the Captain said in $H.M.S.$ *Pinafore*, "hardly ever." We will have to carefully reintroduce a 10^{-24} in a few places in the fuel cycle calculations. But that's a few chapters away.

4-5 NEUTRON BALANCE

Questions in nuclear analysis frequently arise that can be answered by examining the neutron balance for the core and blankets. A neutron balance for a large (1200 MWe) homogeneous LMFBR, with two core zones of equal volume, is provided in Table 4-1A.* The basis is 100 neutrons produced. Values are given for an equilibrium cycle, at mid-burnup. The captures in boron represent captures in the control rods. A neutron balance for a 1000 MWe heterogeneous design at the end of an equilibrium cycle is provided in Table 4-1B.*

Note that for these reactor designs the fraction of fissions in each isotope in the core fuel assemblies (not including the internal blankets) is apportioned as follows:

| | Fraction Fissions, % | |
Fuel Isotope	Homogeneous	Heterogeneous
^{235}U	1.5	0.6
^{238}U	14.7	10.6
^{238}Pu	—	0.6
^{239}Pu	64.3	70.2
^{240}Pu	5.6	5.4
^{241}Pu	13.3	12.1
^{242}Pu	0.6	0.5

These values are fairly typical for UO_2-PuO_2 LMFBR cores, except that the fraction of fissions in the ^{238}U in the homogeneous core may be slightly higher than typical. The core of this particular 1200 MWe design was larger than normal so that the fission fraction was somewhat lower than typical. The trend is correct, however; lower fissile fractions in homogeneous designs generally lead to higher ^{238}U fission fractions than in heterogeneous designs.

For this homogeneous core design, the fraction of fissions occurring in the core is 93.6%, with only 6.4% in the blankets. Neutron leakage from the outer boundaries of the blankets/shielding is very small (0.3%). However, the radial neutron leakage from the outer core enrichment zone to the radial blanket is approximately 10%, and the leakage to the axial blankets is 11%. Hence, the total neutron leakage from the core into the surrounding blankets is 21%.

---

*Compliments of C. L. Cowan, General Electric Company, Sunnyvale, California, 1979, 1981.

TABLE 4-1A   Neutron Balance for 1200 MWe Homogeneous LMFBR Design (Equilibrium Cycle, Mid-Burnup)

Event	Material	Zone 1	Zone 2	Axial Blanket	Radial Blanket	Shielding	Total
CAPTURES	U-238	19.15	11.79	7.65	7.13	0	45.72
	U-235	0.09	0.06	0.05	0.05	0	0.25
	Pu-239	3.46	2.64	0.25	0.23	0	6.58
	Pu-240	1.41	1.19	0.01	0.01	0	2.62
	Pu-241	0.43	0.04	0	0	0	0.83
	Pu-242	0.13	0.11	0	0	0	0.24
	F. P.	1.25	0.67	0.09	0.07	0	2.08
	O	0.10	0.07	0.02	0.02	0	0.21
	Na	0.14	0.10	0.07	0.06	0.07	0.44
	Fe	0.69	0.48	0.30	0.29	0.66	2.42
	Ni	0.29	0.20	0.09	0.09	0.24	0.91
	Cr	0.29	0.21	0.12	0.11	0.28	1.01
	B (control)	0.34	0.21	1.10	0	0	1.65
	TOTAL	27.77	18.13	9.75	8.06	1.25	64.95
FISSIONS	U-238	2.71	2.06	0.48	0.45	0	5.70
	U-235	0.29	0.20	0.14	0.14	0	0.77
	Pu-239	11.44	9.48	0.55	0.48	0	21.95
	Pu-240	0.91	0.90	0	0	0	1.81
	Pu-241	2.20	2.14	0	0	0	4.34
	Pu-242	0.09	0.09	0	0	0	0.18
	TOTAL	17.64	14.87	1.17	1.07	0.0	34.75
RADIAL LEAKAGE							0.03
AXIAL LEAKAGE							0.27
	TOTAL						100.00

4-6   POWER DENSITY

In the multigroup solutions discussed in Section 4-3, the neutron flux values were relative, not absolute. In order to obtain absolute fluxes, it is necessary to relate neutron flux to power density. Factors required to do this, together with representative values of neutron flux and power distribution, are presented in this section.

A.   ENERGY PER FISSION

The energy generated per fission that is ultimately transferred as heat to the coolant in a UO_2-PuO_2 fueled fast reactor is about 213 MeV. The source of this energy is given in Table 4-2. The neutrino energy ($\sim$9 MeV) is not included in the table, since this energy is not absorbed in the reactor.

TABLE 4-1B Neutron Balance for 1000 MWe Heterogeneous LMFBR Design (End of Equilibrium Cycle)

Event	Material	Core Fuel	Inner Blankets	Axial Blanket Extension Driver Fuel	Axial Blanket Extension Inner Blanket	Radial Blanket	Shielding	Total
CAPTURES	Oxygen	0.24	0	0.01	0.01	0.02	0	0.28
	Na	0.12	0.04	0.03	0.01	0.03	0.05	0.28
	Cr	0.42	0.14	0.09	0.03	0.09	0.17	0.94
	Mn	0.18	0.08	0.08	0.03	0.08	0.34	0.79
	Fe	1.07	0.35	0.23	0.07	0.24	0.50	2.46
	Ni	0.77	0.23	0.12	0.04	0.13	0.23	1.52
	Mo	0.32	0.12	0.09	0.03	0.10	0.27	0.93
	^{235}U	0.05	0.06	0.01	0.02	0.02	0	0.16
	^{238}U	16.57	12.32	5.31	2.61	8.42	0	45.23
	^{238}Pu	0.10	0	0	0	0	0	0.10
	^{239}Pu	5.34	0.81	0.15	0.07	0.34	0	6.71
	^{240}Pu	1.95	0.04	0	0	0.01	0	2.00
	^{241}Pu	0.61	0	0	0	0	0	0.61
	^{242}Pu	0.20	0	0	0	0	0	0.20
	F.P.	1.70	0.19	0.01	0.01	0.05	0	1.96
	B^{10}	0	0	0.65	0	0	0	0.65
	TOTAL	29.64	14.38	6.78	2.93	9.53	1.56	64.83
FISSIONS	^{235}U	0.17	0.12	0.07	0.03	0.10	0	0.49
	^{238}U	2.93	1.38	0.31	0.14	0.59	0	5.35
	^{238}Pu	0.16	0	0	0	0	0	0.16
	^{239}Pu	19.42	2.42	0.31	0.15	0.84	0	23.14
	^{240}Pu	1.50	0.02	0	0	0.00	0	1.52
	^{241}Pu	3.33	0.01	0	0	0	0	3.34
	^{242}Pu	0.14	0	0	0	0	0	0.14
	TOTAL	27.65	3.95	0.69	0.32	1.53	0	34.14
RADIAL LEAKAGE								0.16
AXIAL LEAKAGE								0.87
	TOTAL							100.00

The values in Table 4-2 vary slightly with the isotope fissioned. While ^{241}Pu gives about the same fission fragment energy as ^{239}Pu (~ 175 MeV), ^{238}U gives only about 169 MeV. On the other hand, the beta and gamma energies from fission products for ^{238}U are higher than for ^{239}Pu. Considering the distribution of fissions between ^{239}Pu, ^{241}Pu, and ^{238}U in a fast reactor, the results in Table 4-2 represent reasonable average values for the reactor. Useful reviews of fission energy for various fissionable isotopes are given in Refs. 4 through 7.

The gamma energy from (n, γ) reactions is equivalent to the binding energy of the target nucleus; note from Table 4-1 that the dominant target

TABLE 4-2 Approximate Energy Per Fission in a Fast Reactor

	Contribution	Energy (MeV)
PROMPT:	Fission fragment kinetic energy	174
	Neutron kinetic energy	6
	Fission gammas	7
	Gammas from (n, γ) reactions	13
DELAYED:	Betas from fission product decay	6
	Gammas from fission product decay	6
	Betas from ^{239}U and ^{239}Np decay*	1
	TOTAL	213

*Energy release due to this source is not always included in such tabulations.

nucleus for neutron capture is ^{238}U. There are about 1.9 (n, γ) reactions per fission in a fast reactor.

The higher value of 213 MeV for fast reactors relative to lower values generally used for thermal reactors results primarily from the difference in fission fragment kinetic energy for ^{239}Pu relative to ^{235}U. For ^{235}U, this energy is only ~ 169 MeV. Moreover, the large number of captures in hydrogen in a light water reactor leads to a lower value for the gamma energy from (n, γ) reactions.

Fission product kinetic energy and beta energy are absorbed in the fuel. Kinetic energy of the neutrons is transformed into gamma energy from inelastic scattering, and into kinetic energy of target nuclei from elastic scattering. Gamma rays are absorbed throughout the reactor, often distant from the source; relative absorption of gammas by each material in the core is approximately proportional to the mass of material present.

B. RELATION BETWEEN POWER DENSITY AND NEUTRON FLUX

There are 1.602×10^{-13} J/MeV. The relation between power density and absolute neutron flux is given by

$$p(\text{W/cm}^3) = \left(213 \frac{\text{MeV}}{\text{fission}}\right)\left(1.602 \times 10^{-13} \frac{\text{J}}{\text{MeV}}\right)\left(\sum_g \Sigma_{fg} \phi_g \frac{\text{fissions}}{\text{cm}^3 \cdot \text{s}}\right)$$

$$= \sum_g \Sigma_{fg} \phi_g \frac{\text{fissions}}{\text{cm}^3 \cdot \text{s}} \Big/ 2.93 \times 10^{10} \frac{\text{fissions}}{\text{W} \cdot \text{s}}, \qquad (4\text{-}53)$$

where ϕ is in n/cm$^2 \cdot$s, and Σ_f is in cm^{-1}.

In this relation the power density p is proportional to the fission distribution. From the previous discussion we know that the distribution of energy transferred to the coolant is slightly different from the fission distribution, due to diffusion and transport of neutrons and gamma rays.

C. POWER DISTRIBUTIONS

Power distributions can be shown in several ways. The fraction of total power transferred to the coolant in each region for the early CRBRP homogeneous core design is given in Table 4-3 for both the beginning and end of an equilibrium cycle.[8]

TABLE 4-3   Power Distribution for the Early CRBRP Homogeneous Core Design[8]

| | Fraction of Power in Region | |
Region	Beginning of Equilibrium Cycle	End of Equilibrium Cycle
Inner Core	0.50	0.46
Outer Core	0.38	0.38
Total Core	0.88	0.84
Radial Blanket	0.08	0.10
Axial Blanket	0.03	0.05
Beyond Blankets	0.01	0.01

The shift to the blankets during the cycle reflects the buildup of plutonium in the blankets. The small power beyond the blankets results from neutron leakage and gamma transport.

The radial power distribution calculated for a homogeneous 1200 MWe LMFBR design* at the core midplane at the middle of the equilibrium fuel cycle is plotted in Fig. 4-4. This sawtooth shape is typical of a two-zone homogeneous core design. The discontinuity between the inner and outer enrichment zones is a direct consequence of the higher fissile fuel content in the outer zone. The axial power distribution for this 1200 MWe LMFBR design is shown in Fig. 4-5.

The radial power distribution for a heterogeneous 1000 MWe LMFBR design[9] at the middle of the equilibrium cycle is plotted in Fig. 4-6.

---

*Data for the 1200 MWe homogeneous LMFBR design used for constructing Figs. 4-4, 4-5, 4-8, 4-9, and 4-11 were supplied by D. R. Haffner, R. W. Hardie, and R. P. Omberg of the Hanford Engineering Development Laboratory, 1978.

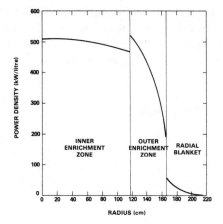

FIGURE 4-4. Radial power density at core midplane (1200 MWe homogeneous core design).

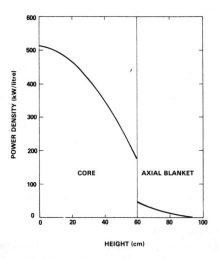

FIGURE 4-5. Axial power density at core midplane (1200 MWe homogeneous core design).

Discontinuities in the curves at assembly row boundaries result from different residence times during the fuel cycle. The fraction of power generated in each region of this core is

Central Blanket	0.016
Inner Driver	0.230
Internal Annular Blanket	0.062
Outer Driver	0.617
Radial Blanket	0.056
Axial Blanket	0.019

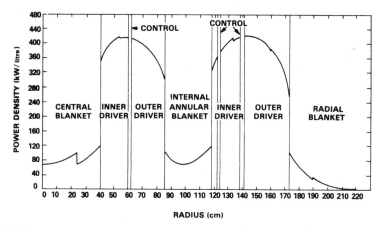

FIGURE 4-6. Radial power density for 1000 MWe heterogeneous core.[9]

A useful way to display the radial power distribution is illustrated in Fig. 4-7.[10] Here the CRBRP radial peaking factors for the heterogeneous core are listed in each hexagonal assembly, where the peaking factor is the ratio of the local power density to the average power density in the core or in the blanket. Values are listed for both the average pin and for the peak-power pin in each assembly. Values are shown in Fig. 4-7 for the beginning of Cycle 3 when peaking factors are quite high.

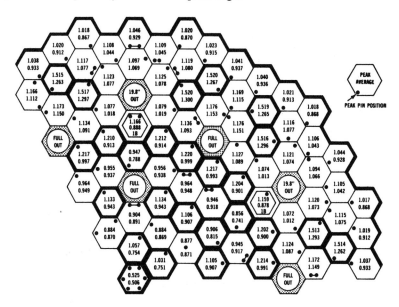

FIGURE 4-7. Assembly radial power factor and peak rod radial power factor at the beginning of Cycle 3 [CRBRP heterogeneous design[10]]. See Fig. B-2 Appendix B for a core map of the CRBRP. Heavy hexagons denote blanket assemblies.

D. NEUTRON FLUX DISTRIBUTIONS

Radial and axial flux distributions for a four-group calculation of an LMFBR are plotted in Figs. 4-8 and 4-9. The reactor is the same 1200 MWe homogeneous core for which power densities were plotted in Figs. 4-4 and 4-5. The radial fluxes are plotted at the core axial midplane, and the axial fluxes correspond to the core radial centerline. For this case, the total flux at the center of the core is 7.0×10^{15} n/cm$^2 \cdot$s. The energy group structures employed in these calculations are shown on the figures.

The axial fluxes follow a cosine shape relatively well, and group-dependent extrapolation distances for use in axial leakage approximations in radial diffusion theory calculations are relatively well defined. The use of two enrichment zones in the radial direction, however, causes considerable flattening of the radial flux in the center and, consequently, a large

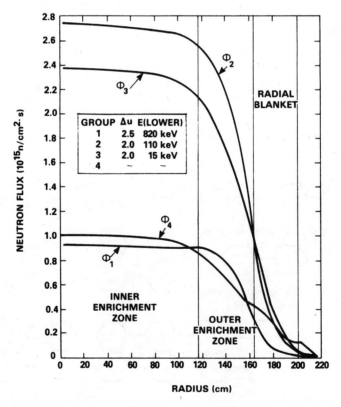

FIGURE 4-8. Radial four-group flux distribution, core midplane, 1200 MWe homogeneous core.

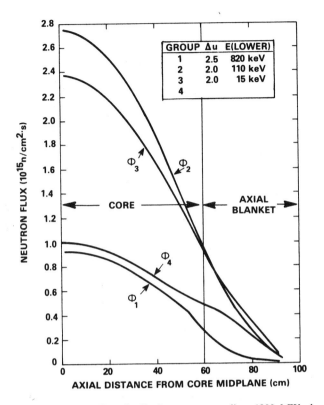

FIGURE 4-9. Axial four-group flux distribution, core centerline, 1200 MWe homogeneous core.

deviation from the uniform cylindrical Bessel function flux shape. This design measure is often employed to maximize total power production capabilities without exceeding the peak linear power constraint in the hottest fuel pins.

4-7 NEUTRON SPECTRUM

The neutron flux of most fast reactors tends to peak around 200 keV. However, substantial variations in the neutron spectrum do exist, especially for low-energy flux, depending upon coolant and fuel forms employed.

Neutron flux spectra for several fast reactor designs are compared in Fig. 4-10 as relative flux per unit lethargy.[11] The effect of oxygen and carbon in the fuel on softening the spectrum relative to metal-fueled LMFBR's is readily observed. The spectrum is softer for oxide than for carbide fuel

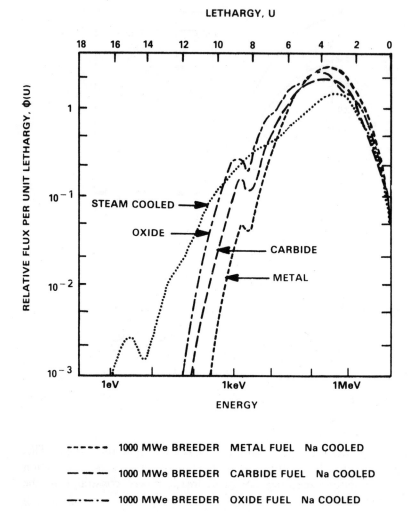

FIGURE 4-10. Neutron flux spectra of fast reactors.[11]

because of the higher density of oxygen relative to carbon; with M standing for metal, oxide fuel is MO_2, whereas carbide fuel is MC.

The depression in the flux at 3 keV is caused by the large sodium scattering resonance at this energy level.

The spectrum of a gas cooled fast reactor (not shown in Fig. 4-10, but included in Fig. 17-1) is slightly harder than a sodium cooled reactor due to the lack of slowing down by sodium.

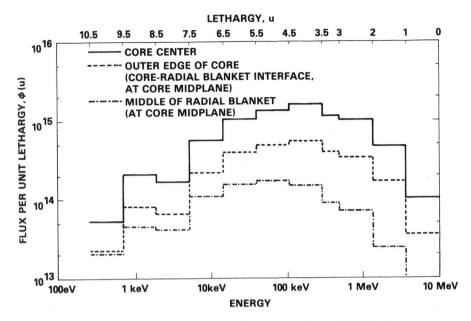

FIGURE 4-11. Twelve-group flux spectra at three radial locations, 1200 MWe homogeneous core (UO_2-PuO_2 fuel).

Twelve-group flux spectra are plotted in Fig. 4-11 at three locations of the 1200 MWe homogeneous core—at the core center, at the outer radial edge of the core at the core midplane, and at the middle of the radial blanket at the core midplane. We may note that there is little difference in the spectrum across this homogeneous core, but that the spectrum becomes significantly softer in the blanket. For a heterogeneous core, the internal blankets cause a more pronounced spatial variation in neutron spectra than in a homogeneous core.

A $1/E$ slowing down spectrum (discussed in Chapter 5-3B) would appear as a horizontal line in Figs. 4-10 and 4-11. In the oxide-fueled LMFBR of Fig. 4-11, the spectrum is fairly close to a $1/E$ spectrum between 10 keV and 1 MeV despite the large absorption rates in this energy range. Below 10 keV the flux drops rapidly. This represents a dramatic difference from a thermal reactor. Essentially no neutrons survive to thermal energies in the core of a fast reactor. Note also from Fig. 4-10 that the $1/E$ region is narrower for fast reactor designs with harder spectra, i.e., for carbide and metal-fueled reactors.

The fission spectrum, which dominates the spectrum above 1 MeV, is described in Chapter 5-3B.

Although radiation damage in structural materials such as cladding is caused by neutrons of all energies in a fast reactor, the neutron flux above

0.1 MeV is widely used in estimates of the lifetime of structural components. At the center of the 1200 MWe design for which the spectrum appears in Fig. 4-11, the neutron flux above 0.1 MeV is 3.7×10^{15} n/cm$^2 \cdot$s. This compares with the total flux at the center of 7.0×10^{15} n/cm$^2 \cdot$s.

An additional aspect of the neutron spectrum of interest is the energy at which fissions occur in a fast reactor. The fraction of fissions at each energy in a typical large oxide-fueled LMFBR is given in Table 4-4. The median fission energy is about 150 keV. The median fission energy in a metal-fueled LMFBR is significantly higher, e.g., several hundred keV. Moreover the relatively large fraction of fissions below 10 keV for the oxide reactor of Table 4-4 is not present in a metal-fueled fast reactor, and, as will be discussed in Chapter 6, this large fraction of low energy fissions is an important factor in contributing to the much larger Doppler coefficient in an oxide-fueled fast reactor relative to a metal-fueled reactor.

TABLE 4-4 Fraction of Fissions at Each Energy, 1200 MWe LMFBR

Group	Δu	Lower Energy of Group (E_l)	Fraction Fissions in Group	Fraction Fissions Above E_l
1	1.0	3.7 MeV	0.05	0.05
2	1.0	1.35	0.18	0.23
3	1.0	500 keV	0.12	0.35
4	0.5	300	0.05	0.40
5	1.0	110	0.14	0.54
6	1.0	41	0.13	0.67
7	1.0	15	0.11	0.78
8	1.0	5.5	0.08	0.86
9	1.0	2.0	0.03	0.89
10	1.0	740 eV	0.07	0.96
11	1.0	280	0.03	0.99
12	–	–	0.01	1.00

4-8 NUCLEAR PERFORMANCE PARAMETERS

It is interesting to summarize some of the characteristic nuclear parameters that vary with different LMFBR core designs. Values from Reference 12 are particularly illuminating in showing how nuclear design parameters vary as a function of fuel pin size for both homogeneous and heterogeneous cores.

The results reported in Table 4-5 are for 1200 MWe LMFBR designs, for which the thermal rating is about 3300 MWth. Common to all designs listed are peak linear power for fresh fuel (44.3 kW/m) and residence time of the fuel in the core (two years). Average burnup after two years is low for the

TABLE 4-5 Variation of Nuclear Design Parameters With Fuel Pin Diameter for Homogeneous and Heterogeneous Designs (As adapted from Reference 12)

Common Parameters:	Electrical rating	1200 MWe
	Thermal rating	~3300 MWth
	Linear power (fresh fuel)	44.3 kW/m
	Core fuel residence time	2 years
	Fuel smear density	88% TD

Pin Diameter		Homogeneous	Heterogeneous
6.35 mm	Pin pitch-to-diameter ratio	1.28	1.28
(0.25 in.)	Core fissile inventory* (kg)	3171	4041
	Fissile fraction, average (%)	15.9	22.8
	Fissile specific inventory (kg/MWth)	1.16	1.51
	Volume fractions		
	Fuel	0.3164	0.3863[†]
	Sodium	0.4593	
	Structure	0.2243	
7.62 mm	Pin pitch-to-diameter ratio	1.21	1.21
(0.30 in.)	Core fissile inventory* (kg)	3949	5992
	Fissile fraction, average (%)	12.8	18.8
	Fissile specific inventory (kg/MWth)	1.38	1.94
	Volume fractions		
	Fuel	0.3845	0.4436[†]
	Sodium	0.3848	
	Structure	0.2157	
8.38 mm	Pin pitch-to-diameter ratio	1.16	1.16
(0.33 in.)	Core fissile inventory* (kg)	4528	6704
	Fissile fraction	11.3	16.4
	Fissile specific inventory (kg/MWth)	1.51	2.09
	Volume fractions		
	Fuel	0.4375	0.4891[†]
	Sodium	0.3566	
	Structure	0.2059	

*Average during an equilibrium cycle.

[†] Fuel volume fraction after homogenizing core with internal blankets. Core assemblies in the heterogeneous cores are identical to homogeneous core assemblies, but the fuel volume fraction in the blanket assemblies is higher than in the core assemblies.

large pin designs; hence residence time might be increased for them. However, results in Reference 12 for these cases for 3-year residence times do not change much from the 2-year results, except for burnup.

The large variation of design parameters with fuel pin size is illustrated by these results. Also the large penalty in fissile mass for the heterogeneous design is apparent. Comparative results in Chapter 6 for sodium loss reactivity will show the compensating advantage of the heterogeneous designs that makes it attractive despite the high fissile mass. Results included in Chapter 7-8 will show little difference in doubling time since the higher breeding ratio in the heterogeneous design tends to offset the higher fissile mass requirement.

PROBLEMS

4-1. Fuel pins can be arranged in a triangular or a square lattice, as illustrated below.

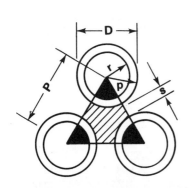 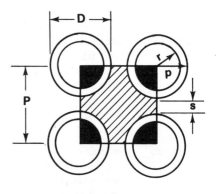

TRIANGULAR LATTICE   **SQUARE LATTICE**

It is advantageous to maximize the volume fraction of the fuel in an FBR. Compare the volume fractions of fuel, cladding, and sodium in an infinitely repeating (a) triangular lattice and (b) square lattice in which the pin diameter (i.e., outside diameter of the cladding) is 8 mm, the pitch-to-diameter ratio (P/D) is 1.20, and the cladding thickness is 0.3 mm. These values are reasonably typical for LMFBR's.

Why is the calculated fuel volume fraction for the triangular lattice so much higher than the range given in Section 4-4?

4-2. Assume plutonium in an LMFBR fuel is 25% ^{240}Pu. The plutonium is 22% of the fuel (the other 78% being ^{238}U). The fuel volume fraction is 0.40 and the fuel smear density is 9.5 g/cm³. There are 300 fuel assemblies and 20 control assemblies in the core.

Calculate the average ^{240}Pu atom density (averaged over the core, including fuel and control assemblies) in units of atoms/barn·cm.

4-3. The multigroup equation [Eq. (4-1)] did not include the ^{238}U $(n, 2n)$ reaction. Rewrite the equation to include this $n, 2n$ reaction, adding terms where appropriate.

4-4. A 1200 MWe LMFBR is characterized by the data listed below. Calculate the following quantities:

(1) Volume fractions for a fuel assembly (repeating cell).
(2) Core average volume fractions (including control rods).
(3) Pin pitch-to-diameter ratio.
(4) Effective core diameter and height-to-diameter ratio.
(5) Average core power density (kW/liter).
(6) Fissile specific inventory in the core (kg fissile/MWt), assuming a fissile atom fraction of 0.12.

DATA
- Materials
    Fuel: UO_2-PuO_2
            Smear density = 9.5 g/cm³
    Coolant: Sodium
    Cladding and Structure: Stainless Steel, Type 316
    Control: B_4C

- Geometry
    Core height = 1.0 m
    Fuel pin diameter = 7.5 mm
    Cladding thickness = 0.4 mm
    Wire wrap diameter = 1.5 mm
    Hexcan wall thickness = 3 mm
    Space between hexcans = 5 mm
    Average clearance between outer wire wrap and hexcan = 0.5 mm
    Rings of pins/assembly = 10 (with central pin as 1st ring)
    Assume that these are the dimensions during full-power operation.

- Power Information
    Thermal power = 3000 MW
    Fraction power generated in the blanket = 8%; remainder generated in the core

Assume all core power is generated in the fuel
Assume fraction core power generated by fertile fissions=20%
Peak linear power=50 kW/m
Peak-to-average power ratio=1.6

- Control Assemblies
  See Fig. 8-20 for control assembly design.
  Assume 20 control assemblies (with the same hexcan dimensions
  as the fuel assemblies) dispersed uniformly throughout the core.

  Assume the following volume fractions:
  B_4C=20%
  Sodium=50% (with rods inserted)
  Stainless Steel: Stationary hexcan=8%
                     Movable with rod=22%

  Assume that only one third of the control rods are in the core;
  the other two thirds are withdrawn in order to reach full power.

4-5. The 1200 MWe LMFBR in Problem 4-4 is loaded uniformly with
fresh (unirradiated) fuel. With the additional information provided
below and the 4 group cross sections provided in Appendix C and
using zero-dimensional diffusion theory,

(1) Calculate the fissile fraction that makes the reactor critical at full
     power to within 0.01% (i.e., $k=1.0000\pm0.0001$).

(2) (a) Calculate group dependent fluxes, ϕ_g.
     (b) Plot the neutron flux, $\phi(E)$, as a function of energy, where
          $\phi(E)$ can be approximated by a histogram with $\phi(E)=$
          $\phi_g/\Delta E_g$.
     (c) Plot $\phi(u)$, where $\phi(u)$ for group g is approximated by
          $\phi_g/\Delta u_g$, and compare the shape of your curve with Fig.
          4-11.

ADDITIONAL INFORMATION

- Use the following volume fractions and core dimensions based
  on values calculated in Problem 4-4 by assuming a symmetrical
  core of 340 assemblies (320 fuel and 20 control):

Fuel	0.392
Sodium	0.410
Steel	0.194
B_4C	0.004

Core Diameter	3.194 m
Core Height	1.0 m

- Use the following material densities:

Fuel:	9.5 g/cm^3 (smear density)
Sodium:	0.85 g/cm^3
Steel (assume iron):	8.0 g/cm^3
B_4C:	2.5 g/cm^3

- Use the following atomic or molecular weights:

Fuel:	270 kg/kg-mol
Sodium:	23.0 kg/kg-mol
Steel (assume iron):	55.8 kg/kg-mol
B_4C:	55.2 kg/kg-mol

- Assume that the fissile fuel is all ^{239}Pu and that the fertile fuel is all ^{238}U.

- Assume that the cross sections in the 4 group set in Appendix C are effective cross sections for this reactor composition at the temperatures associated with full power operation.

- The extrapolation distances are

$$\delta_z = 20 \text{ cm}$$
$$\delta_r = 20 \text{ cm}$$

Assume that B^2 is not group dependent.

- For stainless steel, assume for this calculation that it can be replaced with iron.

- The control rods are withdrawn two thirds of the way in order to achieve full power. (This is compatible with the 0.004 volume fraction B_4C). The boron in the B_4C control rods is natural boron.

4-6. For the reactor calculated in Problem 4-5:

(a) Calculate the fraction of fissions in ^{238}U in the core.
(b) Calculate the following contributions to a neutron balance for the core in terms of percent of neutrons produced in the core:
  (i) Leakage
  (ii) Fission
  (iii) Capture

(c) Estimate the median fission energy and calculate the fraction of fissions from neutrons with energies below 15 keV.

4-7. For the reactor calculated in Problem 4-5, obtain some first (crude) estimates of breeding ratios as follows:

(i) Internal breeding ratio:

$$IBR \sim \frac{\text{Capture rate in } ^{238}U \text{ in the core}}{\text{Absorption rate in } ^{239}Pu \text{ in the core}}$$

(ii) Potential reactor breeding ratio by assuming that 70% of the neutrons that leak from the core are captured in ^{238}U in the blankets:

$$BR \sim \frac{\text{Capture rate in } ^{238}U \text{ in the core and blankets}}{\text{Absorption rate in } ^{239}Pu \text{ in the core}}$$

REFERENCES

1. M. K. Drake, *Data Formats and Procedures for the ENDF Neutron Cross Section Library*, BNL-50274, Brookhaven National Laboratory, (April 1974 Revision).
2. R. W. Hardie and W. W. Little, Jr., *1DX, A One Dimensional Diffusion Code for Generating Effective Nuclear Cross Sections*, BNWL-954, Battelle Northwest Laboratory, March 1969.
3. C. R. Weisbin, P. D. Soran, R. E. MacFarlane, D. R. Harris, R. J. LaBauve, J. S. Hendricks, J. E. White, and R. B. Kidman, *MINX: A Multigroup Integration of Nuclear X-Sections from ENDF/B*, LA-6486-MS, Los Alamos Scientific Laboratory, August 1976.
4. M. F. James, "Energy Released in Fission," *J. Nuc. Energy*, *23*, (1969) 516-536.
5. J. P. Unik and J. E. Grindler, *A Critical Review of the Energy Release in Nuclear Fission*, ANL-7748, March 1971.
6. F. A. Schmittroth, *Decay Heat for the Fast Test Reactor (FTR)*, HEDL-TME 77-13, June 1977.
7. R. Sher and C. Beck, *Fission Energy Release for 16 Fissioning Nuclides*, EPRI NP-1771, March 1981.
8. *Preliminary Safety Analysis Report*, Clinch River Breeder Reactor Plant, Project Management Corporation, 1974.
9. *Liquid Metal Fast Breeder Reactor Conceptual Plant Design*, 1000 MWe, TID-27701-2, Vol. II, Rockwell International and Burns and Roe (May 1977).
10. *Preliminary Safety Analysis Report*, Clinch River Breeder Reactor Plant, Amendment #51, September 1979, 4.3–15.4.
11. W. Häfele, D. Faude, E. A. Fischer, and H. J. Laue, "Fast Breeder Reactors," *Annual Review of Nuc. Sci.*, Annual Reviews, Inc., Palo Alto, CA, 1970.
12. W. P. Barthold and J. C. Beitel, "Performance Characteristics of Homogeneous Versus Heterogeneous Liquid-Metal Fast Breeder Reactors," *Nuclear Technology*, *44*, (1979) 45, 50.

CHAPTER 5

MULTIGROUP CROSS SECTIONS

5-1 INTRODUCTION

Multigroup microscopic cross sections of the materials in a fast reactor depend on the particular composition of each region in the reactor; hence, a single set does not suffice for all designs. For detailed design, a multigroup cross section set of *effective cross sections* is developed for the particular reactor composition and design being analyzed. For scoping studies, however, less care is required and a single set can apply to a range of designs, provided the compositions are not too different from the composition for which the set was originally developed.

The main approach used in the United States to obtain fast reactor cross sections for design purposes is the *Bondarenko shielding factor* approach,[1,2] named for the lead scientist who developed the system first in the U.S.S.R. Shielding factors for a given material are factors that account for *energy self-shielding*, i.e., for the effects of other materials on the cross sections of the given material. The shielding factor approach as used for design among reactor vendors in the United States will be described in this chapter. An alternate approach, developed at Argonne National Laboratory, is represented by the MC^2 code;[3] and while this approach is expensive for ordinary design, it has been useful in providing a check on results obtained from the shielding factor method.

For both approaches, the starting point for calculating fast reactor cross sections is a reference set of cross sections and nuclear data, used and developed in the United States, called the Evaluated Nuclear Data Files (ENDF).[4] Most of the data used in the methods described in this chapter are contained in the part of this file called ENDF/B.

Two books related to neutron cross sections that cover some of the methods presented in this chapter are worth special reference. An early discussion of fast reactor cross sections appears in the first three chapters of

Reactivity Coefficients in Large Fast Power Reactors by Hummel and Okrent.[5] Second, Dresner's classic *Resonance Absorption in Nuclear Reactors*[6] provides an excellent background on resonance absorption.

5-2 SHIELDING FACTOR METHOD

The Bondarenko shielding factor approach consists of two parts. In the first, *generalized* multigroup cross sections are obtained for a given material, as a function of a *background cross section* σ_0 (to be defined quantitatively later), and for the abundant fuel isotopes, as a function of temperature T. These cross sections are generally obtained over a wide range of values of σ_0 for materials, and at three temperatures for fuel isotopes. The generalized multigroup cross sections need not relate to any particular reactor composition. Methods for obtaining these cross sections are presented in Sections 5-3 through 5-5.

In the second part of the shielding factor method, a cross section set is obtained from the generalized cross section set for a *particular* reactor composition at a given fuel temperature. The actual values of σ_0 are calculated for each material, and cross sections for the given composition are interpolated from the cross sections in the generalized set. For detailed design calculations, different cross sections are obtained for different regions of the core and the blanket. The second part of the shielding factor method is described in Section 5-6.

The most recent computer codes to perform this two-part calculation are MINX[7] and SPHINX;[8] the former was developed at Los Alamos Scientific Laboratory and the latter at Westinghouse (Advanced Reactors Division). The MINX-SPHINX technology is an extension of two earlier sets of codes developed for the shielding factor method during the late 1960s—the ETOX[9] and 1DX[10] codes developed at the Hanford Engineering Development Laboratory, and the ENDRUN[11] and TDOWN[12] codes developed at General Electric.

Shielding Factor, *f*

Self-shielding is an important effect in the resonance energy range where total cross sections become large enough to influence the local neutron flux and reaction rates. The explanation of the effect is given in Section 5-3 in the derivation of the slowing down flux, $\phi(E)$. The energy range where self-shielding begins to influence the results is between 100 keV and 1 MeV. Although effective cross sections must be obtained in both the high-energy range and the resonance region, the shielding factor method gets its name from the necessarily careful treatment of the resonance region.

The principal idea of the shielding factor method is the evaluation of effective cross sections in the resonance energy range by means of a *shielding factor*, f, for each material, m, and each cross section for which resonances make energy self-shielding important. The shielding factor is defined as the ratio of the *effective* cross section for reaction x, σ_{xmg}, to an *infinitely dilute* cross section, $\sigma_{xmg}(\infty)$. The effective cross section, and hence the shielding factor f, will be dependent on a background cross section, σ_0. Thus,

$$f_{xmg}(\sigma_0) = \frac{\sigma_{xmg}(\sigma_0)}{\sigma_{xmg}(\infty)}. \tag{5-1}$$

The background cross section and the methods for calculating $\sigma_{xmg}(\infty)$ and $\sigma_{xmg}(\sigma_0)$ are described in the next section.

5-3 GENERALIZED CROSS SECTIONS

We start with the generalized multigroup cross sections calculated in the first part of the Bondarenko shielding factor method (e.g., in the MINX, ETOX, and ENDRUN codes). In this part, the following infinitely dilute cross sections are calculated for each material and for each group: σ_t, σ_f, σ_c, σ_e and σ_i. Effective cross sections and shielding factors are calculated as functions of σ_0 and material temperature for σ_t, σ_f, σ_c and σ_e. Other quantities calculated by material and group include $\nu, \bar{\mu}, \xi, \chi$, and the inelastic scattering matrix.

A. GROUP CROSS SECTIONS AND GROUP FLUX

Group Flux

The group neutron flux, ϕ_g (n/cm$^2 \cdot$s), is defined as

$$\phi_g = \int_{E_g}^{E_{g-1}} \phi(E) \, dE = \int_g \phi(E) \, dE, \tag{5-2}$$

where $\phi(E)$ is the flux per unit energy, E_g is the lower energy of the group, and $\int_g$ refers to integration over the energy group.

Capture, Fission, and Scattering Cross Sections

For capture, fission and elastic and inelastic scattering reactions, the average cross section for a group is the cross section which, when multiplied by the group flux, will give the correct reaction rate. The reaction rate for material m and for group g [where x stands for capture (c), fission (f), elastic scattering (e), or inelastic scattering (i)] is

$$\int_g \sigma_{xm}(E)\phi(E)\,dE.$$

The reaction rate is said to be *conserved* when $\sigma_{xmg}\phi_g$ is equal to this rate. Hence, the cross section for material m for group g is

$$\sigma_{xmg} = \frac{\displaystyle\int_g \sigma_{xm}(E)\phi(E)\,dE}{\displaystyle\int_g \phi(E)\,dE}. \tag{5-3}$$

Transport and Total Cross Sections

The microscopic total cross section, σ_t, is used to calculate the transport cross section, σ_{tr} which is in turn used in evaluating the diffusion coefficient D. The different nature of leakage relative to other nuclear reactions requires that σ_t be averaged in a manner different from the other cross sections. The following relationships between D, σ_{tr} and σ_t are recalled:

$$D = \frac{\lambda_{tr}}{3} = \frac{1}{3\Sigma_{tr}} = \frac{1}{3\sum_m N_m \sigma_{trm}}, \tag{5-4}$$

$$\sigma_t = \sigma_a + \sigma_i + \sigma_e, \tag{5-5}$$

$$\sigma_{tr} = \sigma_a + \sigma_i + \sigma_e(1-\bar{\mu}),$$

$$= \sigma_t - \sigma_e\bar{\mu}, \tag{5-6}$$

where $\bar{\mu}$ is the average cosine of the scattering angle for elastic scattering in the laboratory system ($\bar{\mu}=2/3A$, where A is the atomic weight of the scattering isotope). The transport and total cross sections are nearly equal since $\bar{\mu}\ll1$ for materials in fast reactors. Hence, the total and transport cross sections should be averaged the same way.

For a proper treatment of leakage, the energy dependent transport mean free path, $\lambda_{\mathrm{tr}}(E)$, should be weighted with the flux in order to obtain an average transport mean free path for an energy group, or

$$\lambda_{trg} = \frac{\int_g \lambda_{\mathrm{tr}}(E)\phi(E)\,dE}{\int_g \phi(E)\,dE}. \tag{5-7}$$

In order to obtain this result, the microscopic transport cross section of each material should be weighted with the current density, $D\nabla\phi$, in obtaining average microscopic transport cross sections, or

$$\sigma_{trmg} = \frac{\int_g \sigma_{trm}(E)D(E)\nabla\phi(E)\,dE}{\int_g D(E)\nabla\sigma(E)\,dE}. \tag{5-8}$$

By assuming separability in space and energy, the *energy* dependence of $\nabla\phi(E)$ is the same as for $\phi(E)$. Hence, Eq. (5-8) can be written (replacing D with $\lambda_{\mathrm{tr}}/3$)

$$\sigma_{trmg} = \frac{\int_g \sigma_{trm}(E)\lambda_{\mathrm{tr}}(E)\phi(E)\,dE}{\int_g \lambda_{\mathrm{tr}}(E)\phi(E)\,dE}. \tag{5-9}$$

Since σ_t is close to σ_{tr}, we calculate σ_t also with Eq. (5-9). By replacing λ_{tr} in both the numerator and denominator of Eq. (5-9) by $1/\Sigma_{\mathrm{tr}}$ and then by $1/\Sigma_t$, since Σ_{tr} and Σ_t are approximately equal, we obtain

$$\sigma_{tmg} = \frac{\int_g \sigma_{tm}(E)\dfrac{\phi(E)}{\Sigma_t(E)}\,dE}{\int_g \dfrac{\phi(E)}{\Sigma_t(E)}\,dE}. \tag{5-10}$$

The crucial factor in calculating effective cross sections is the proper evaluation of the flux, $\phi(E)$. It is in the evaluation of the flux that the energy self-shielding appears.

B. ENERGY DEPENDENCY OF THE NEUTRON FLUX, $\phi(E)$

Fission Spectrum Flux

Two different flux energy spectra are used to evaluate cross sections. At high energy the flux follows a fission spectrum. At lower energy the flux spectrum is governed by slowing down, and it approaches the familiar $1/E$ spectrum. A cutoff energy is defined as the interface between the fission spectrum and the slowing down spectrum. A typical value of the cutoff energy is 2.5 MeV.

The fission spectrum for fissile isotope m has the form

$$\phi(E)=C\sqrt{\frac{E}{(kT_m)^3}}\ e^{-E/kT_m}, \qquad (5\text{-}11)$$

where T_m is a characteristic nuclear temperature for the fissile isotope m. For ^{239}Pu, kT is equal to ~ 1.4 MeV. The constant C provides the proper units and must have a numerical value such that the transition to the $1/E$ spectrum at the cutoff energy is continuous. Since $\phi(E)$ appears in both the numerator and denominator of the equations for the cross sections, however, this constant will eventually cancel out.

Slowing Down Flux

The relation for $\phi(E)$ below the fission spectrum is obtained from slowing down theory, and it is necessary to derive $\phi(E)$ in order to understand the shielding factor method. The derivation of $\phi(E)$ will be based on the *narrow resonance approximation* since the neutron energy in a fast reactor is confined to the energy range over which the narrow resonance approximation applies. *Wide resonances* exist below the energies of relevance to a fast reactor. In the narrow resonance approximation, the energy width of the resonance is considered to be so narrow that neutrons scattered into the resonance come from energies which are not influenced by the resonance.

The present derivation of the slowing down flux will include the influence only of elastic scattering since inelastic scattering becomes small at the low energies where the resonances are large enough to cause appreciable self shielding.

The cross sections involved for a given isotope with a resonance are illustrated schematically in Fig. 5-1.

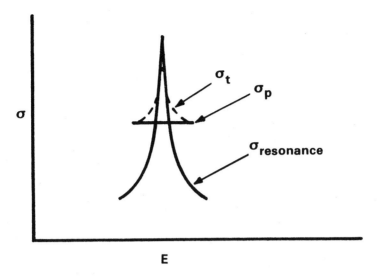

FIGURE 5-1. Schematic illustration of cross sections around a resonance.

The resonance can be a capture, fission, or scattering resonance. The *potential scattering cross section*, σ_p, varies slowly with energy and can be considered constant in the vicinity of the resonance. One would expect something dramatic to happen to the neutron flux around this resonance energy if the isotope with the resonance is present in the reactor to an appreciable extent. This, we shall see, is indeed the case.

Let us consider the energy picture of Fig. 5-2 where E describes the neutron energy. If a resonance occurs, we suppose that it occurs in the vicinity of energy E.

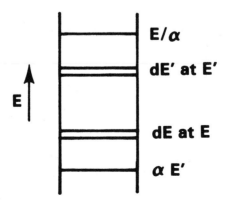

FIGURE 5-2. Energy diagram for scattering analysis.

The α in the figure is defined such that $(1-\alpha)$ is the maximum fraction of energy that can be lost in an elastic scattering collision. Thus, when a neutron of energy E suffers an elastic collision, the minimum energy of the neutron after the collision is αE. The quantity α is given by

$$\alpha = \left(\frac{A-1}{A+1}\right)^2, \tag{5-12}$$

where A is the atomic weight of the target isotope.

We define a *collision density*, $F(E)$ at E as the number of reactions per unit volume, or

$$F(E) = \Sigma_t(E)\phi(E). \tag{5-13}$$

At energies higher than those affected by the resonance, the macroscopic total cross section, Σ_t, and the macroscopic scattering cross section, Σ_s, are equal, so that outside the resonance,

$$F(E') = \Sigma_s(E')\phi(E').$$

We can write an expression for the number of scattering reactions occurring in dE' at E' on Fig. 5-2 as follows:

Neutrons scattered out of $dE' = \Sigma_s(E')\phi(E')\,dE' = F(E')\,dE'$.

For the purpose of deriving an expression for $\phi(E)$, we assume that the scattering is elastic and isotropic. Corrections can be made later for small deviations from this assumption. This assumption will be used to obtain the local neutron flux in the vicinity of the resonance in order to perform the integrations in Eqs. (5-3) and (5-10).

As a consequence of the assumption of isotropic elastic scattering, there is an equal probability that the energy of the neutron scattered at energy E' will lie in any given interval dE between E' and $\alpha E'$. Hence, the fraction of the neutrons scattered at E' that are scattered into dE at E is

$$\text{Fraction scattered into } dE = \frac{dE}{E' - \alpha E'}.$$

(The α is the value for the target scattering nucleus, and there may be several isotopes accounting for the scattering in a fast reactor. For the purpose of deriving the form of $\phi(E)$, it is not necessary to consider the details involved in obtaining an effective average value of α.)

Since neutrons elastically scattered into dE must have had energies ranging from E to E/α before the collision,

$$\text{Neutrons scattered into } dE = \int_{E}^{E/\alpha} \Sigma_s(E')\phi(E')\,dE'\,\frac{dE}{E'-\alpha E'}$$

$$= \int_{E}^{E/\alpha} \Sigma_s(E')\phi(E')\frac{dE'}{E'(1-\alpha)}\,dE.$$

A steady-state balance of neutrons in dE requires that the collision density in dE equal the rate that neutrons enter dE from scattering, or

$$\Sigma_t(E)\phi(E)\,dE = \int_{E}^{E/\alpha} \Sigma_s(E')\phi(E')\frac{dE'}{E'(1-\alpha)}\,dE. \qquad (5\text{-}14)$$

The dE on each side cancels, leaving

$$F(E) = \Sigma_t(E)\phi(E) = \int_{E}^{E/\alpha} \Sigma_s(E')\phi(E')\frac{dE'}{E'(1-\alpha)}. \qquad (5\text{-}15)$$

In the *absence* of resonances at E or anywhere in the energy range between E and E/α, $\Sigma_t(E) = \Sigma_s(E)$. We can also assume that $\Sigma_s(E)$ is constant in the energy range E to E/α since this is a small energy range for materials in fast reactors and Σ_s is slowly varying with energy. For these conditions the solution of the above integral equation is the familiar *asymptotic flux per unit energy*, $\phi_0(E)$:

$$\phi_0(E) = \frac{C_1}{E}. \qquad (5\text{-}16)$$

This solution can be verified by inserting C_1/E' for $\phi(E')$ into Eq. (5-15) and performing the integration.

With a resonance at E, the solution for $F(E)$ is

$$F(E) = \frac{C_2}{E},$$

which again can be verified by direct substitution, remembering that $\Sigma_s(E')\phi(E') = F(E')$ in the integral and that $F(E) = \Sigma_t(E)\phi(E)$ at the energy E of the resonance. The fact that $\Sigma_s(E')$ is used throughout the energy range of the integral in Eq. (5-15) despite the resonance at E, the tail of which actually will spread into the energy range between E and E/α, is a consequence of the narrow resonance approximation.

The collision density at E in the absence of a resonance at E is

$$F(E) = \Sigma_s(E)\phi_0(E). \tag{5-17}$$

If there *is* a resonance at E, $F(E)$ will be given by $\Sigma_t(E)\phi(E)$. The value of $F(E)$ for this case will be identical to that in Eq. (5-17), however, because $F(E)$ is equal to the integral on the right-hand side of Eq. (5-15) *regardless* of whether there is a resonance at E or not (as a consequence of the narrow resonance approximation). Thus, the actual flux and the asymptotic flux are related by

$$\Sigma_t(E)\phi(E) = \Sigma_s(E)\phi_0(E),$$

or

$$\phi(E) = \frac{\Sigma_s}{\Sigma_t}\phi_0(E). \tag{5-18}$$

Thus, there is a sharp decrease in the actual flux at the energy of the resonance relative to the flux without a resonance. Schematically the flux will appear as shown in Fig. 5-3.

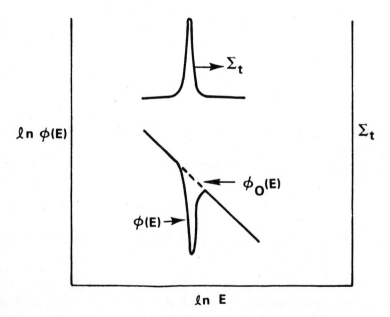

FIGURE 5-3. Schematic illustration of flux depression caused by a resonance.

The effect of the resonance is to depress the flux in the vicinity of the resonance by the ratio Σ_s/Σ_t. The effect of this flux depression on the microscopic cross sections is referred to as *energy self-shielding*.

C. SELF-SHIELDED AND BACKGROUND CROSS SECTIONS

We are now in a position to calculate this self-shielding effect. We can insert Eq. (5-18) for the flux into Eq. (5-3) to obtain the following equation for the effective microscopic cross section:

$$
\sigma_{xmg} = \frac{\displaystyle\int_g \sigma_{xm}(E)\frac{\Sigma_s}{\Sigma_t(E)}\phi_0(E)\,dE}{\displaystyle\int_g \frac{\Sigma_s}{\Sigma_t(E)}\phi_0(E)\,dE}
$$

$$
= \frac{\displaystyle\int_g \frac{\sigma_{xm}(E)}{\Sigma_t(E)}\frac{dE}{E}}{\displaystyle\int_g \frac{1}{\Sigma_t(E)}\frac{dE}{E}} , \tag{5-19}
$$

where we have assumed that Σ_s is independent of energy.

The total macroscopic cross section, Σ_t, is composed of the contribution from the resonance in material m plus the scattering cross section of all of the other materials that comprise the composition for which the effective cross section is being obtained, or

$$
\Sigma_t(E) = N_m\sigma_{tm}(E) + \sum_{\substack{\text{other} \\ \text{materials}}} N_{\text{other}}\sigma_{s,\,\text{other}} . \tag{5-20}
$$

Dividing by N_m gives

$$
\frac{\Sigma_t(E)}{N_m} = \sigma_{tm}(E) + \sigma_{om} , \tag{5-21}
$$

where

$$
\sigma_{om} = \frac{1}{N_m}\sum_{\substack{\text{other} \\ \text{materials}}} N_{\text{other}}\sigma_{s,\,\text{other}} . \tag{5-22}
$$

Hence, σ_o is the macroscopic scattering cross section of other materials per atom of material m per cm³, i.e., per N_m. This parameter, σ_o, is called the *background cross section*.

Equation (5-19) becomes, since $\Sigma_t(E)$ can be divided by N_m in both the numerator and denominator,

$$\sigma_{xmg}(\sigma_o) = \frac{\int_g \frac{\sigma_{xm}(E)}{\sigma_{tm}(E) + \sigma_{om}} \frac{dE}{E}}{\int_g \frac{1}{\sigma_{tm}(E) + \sigma_{om}} \frac{dE}{E}}, \tag{5-23}$$

where the $\sigma_{xmg}(\sigma_o)$ notation emphasizes the dependence of the effective cross section on the background cross section.

Similarly, Eq. (5-10) can be written as follows:

$$\sigma_{tmg}(\sigma_o) = \frac{\int_g \frac{\sigma_{tm}(E)}{[\sigma_{tm}(E) + \sigma_{om}]^2} \frac{dE}{E}}{\int_g \frac{1}{[\sigma_{tm}(E) + \sigma_{om}]^2} \frac{dE}{E}}. \tag{5-24}$$

Equations (5-23) and (5-24) are the expressions used in the three codes— MINX, ETOX, and ENDRUN—to calculate effective microscopic cross sections.

D.   INFINITELY DILUTE CROSS SECTIONS

The infinitely dilute cross section is the cross section of a material that is present in such small quantities that its resonances do not affect either its own effective cross section or that of other materials present. This cross section is denoted by $\sigma_{xm}(\infty)$. As a consequence of the narrow resonance approximation and the infinitely dilute amount of material present, the flux at the resonance of material m is equal to the flux without resonances, or $\phi_0(E)$, which was defined by Eq. (5-16). Using this flux in Eq. (5-3) gives for $\sigma_{xmg}(\infty)$

$$\sigma_{xmg}(\infty) = \frac{\int_g \sigma_{xm}(E)\, dE/E}{\int_g dE/E}. \tag{5-25}$$

This cross section applies for the total cross section as well as capture, fission, elastic scattering and inelastic scattering.

We now have all of the information needed to obtain the shielding factor f, as defined in Eq. (5-1).

E. INELASTIC SCATTERING

In an inelastic scattering collision, the nucleus is excited to a higher energy state by an incident neutron of energy E' and then returns to a ground state by emitting a neutron of energy E and a gamma ray. Inelastic scattering generally occurs at high energies where resonances are small and self-shielding has little effect. The inelastic scattering cross section can therefore be calculated by Eq. (5-3) with the fission spectrum or the asymptotic flux. The inelastic scattering cross section $\sigma_i(E)$ is a part of the total cross section $\sigma_t(E)$, as given by Eq. (5-5).

The matrix for inelastic scattering from group g' to group g is obtained from the relation:

$$
\sigma_{img' \to g} = \frac{\int_g \int_{g'} \sigma_{im}(E')W(E' \to E)\phi(E')\,dE'\,dE}{\int_g \int_{g'} W(E' \to E)\phi(E')\,dE'\,dE}, \tag{5-26}
$$

where the dE' integration is over group g' and the dE integration is over group g. The evaluation of $W(E' \to E)$ from the ENDF/B files is complex and will not be reproduced here; details are given in References 7, 9, and 11.

Another reaction that is generally included in the inelastic scattering matrix is the $(n,2n)$ reaction. The σ_{n2n} cross section is added to the σ_{irg} cross section and $2\sigma_{n2ng' \to g}$ is added to the appropriate scattering matrix.

There is no need to calculate an elastic scattering matrix in the generalized cross section part of the multigroup cross section calculation (even though it is done in some of the codes). The reason for this is that the elastic removal cross section, σ_{ermg}, is dependent on the neutron flux spectrum, which is dependent on the reactor composition. The procedure for evaluating σ_{er} is described in Section 5-6B.

F. CALCULATION OF $\chi, \nu, \bar{\mu}, \xi$

Values for the fission spectrum fraction, χ, the number of neutrons emitted per fission, ν, the average cosine for elastic scattering, $\bar{\mu}$, and the average logarithmic energy decrement, ξ, are calculated for the generalized cross section set from the ENDF/B files.

An accurate expression for χ_{mg} is

$$\chi_{mg} = \frac{\int_g \int \nu_m(E')\sigma_{fm}(E')\chi_m(E' \to E)\phi(E')\,dE'\,dE}{\int_{\nu_m}(E')\sigma_{f_m}(E')\phi(E')\,dE'}, \qquad (5\text{-}27)$$

where $\chi(E' \to E)$ is obtained from the ENDF/B files and $\phi(E')$ is the fission spectrum or the asymptotic flux. The spectrum $\chi(E' \to E)$ is a slowly varying function of E' over the important energy range. Therefore, the MINX code allows the use of a simple approximation for χ,

$$\chi_{mg} = \int_g \chi_m(E^* \to E)\,dE, \qquad (5\text{-}28)$$

where $E^* = 1$ MeV.

The integrals for ν, $\bar{\mu}$, and ξ are

$$\nu_{mg} = \frac{\int_g \nu_m(E)\sigma_{fm}(E)\phi(E)\,dE}{\int_g \sigma_{fm}(E)\phi(E)\,dE}, \qquad (5\text{-}29)$$

$$\bar{\mu}_{mg} = \frac{\int_g \bar{\mu}_m(E)\sigma_{em}(E)\phi(E)\,dE}{\int_g \sigma_{em}(E)\phi(E)\,dE}, \qquad (5\text{-}30)$$

$$\xi_{mg} = \frac{\int_g \xi_m(E)\sigma_{em}(E)\phi(E)\,dE}{\int_g \sigma_{em}(E)\phi(E)\,dE}, \qquad (5\text{-}31)$$

where $\phi(E)$ is the fission spectrum or the asymptotic flux.

5-4 RESONANCE CROSS SECTIONS

In the resonances, cross sections vary so rapidly with energy that it is unreasonable to include values at enough energy points in a data file such as ENDF/B to evaluate the integrals in Eqs. (5-23) and (5-24). Instead analytical expressions for resonance cross sections are used in these integrals.

Resonance cross sections of a material are dependent on the material temperature. The source of this dependence is the relative motion between target nuclei and neutrons, the phenomenon referred to as the *Doppler effect*. As the temperature of a material is raised, the thermal motion of the material nuclei increases, thus modifying the relative motion and changing the effective cross section.

The Doppler effect is very important in a fast reactor with ceramic fuel, as will be discussed in Chapter 6. The most important cross section with regard to the Doppler effect is the capture cross section of the fertile isotope, such as ^{238}U in the U-Pu fuel cycle. Of lesser importance are the capture and fission cross sections of ^{239}Pu and the capture cross section of ^{240}Pu. The fuel isotopes are important because of the magnitude of the resonances and also because the fuel temperature changes over a wide range both between zero power and full power and in the event of an accidental power transient.

Temperature has only a small effect on the capture cross section of materials other than fuel and on resonance scattering cross sections. However, the temperature dependence of these cross sections can also be evaluated in fast reactor cross section codes such as MINX.

The starting point for calculating absorption and scattering resonance cross sections is the *Breit-Wigner single-level formula*.* The *resonance parameters* in this relation are provided in the ENDF/B file. This relation is summarized in this section. (For an understanding of such parameters as wave numbers, statistical factors, angular momentum wave numbers, etc., the reader must go to more basic texts in reactor theory or nuclear physics.) The method used to modify the Breit-Wigner relation to account for the Doppler effect for the fuel isotopes is presented in Section 5-5.

Resonances can be subdivided into two groups—*resolved* and *unresolved*. For the resolved resonances the resonance parameters have been determined (or "resolved") experimentally. The unresolved resonances occur at higher energies, where the peak resonance cross sections are lower and the resonance parameters are not well established. For uranium-238, capture resonances are resolved up to ~ 10 keV; for plutonium-239, fission and capture resonances are resolved only to a few tenths of a keV. Methods for obtaining effective cross sections in both the resolved and unresolved regions are described in this section.

---

*Multilevel theory can also be used, and this method is available in addition to the single-level formulation in the MINX code. Single-level theory is usually adequate, however.

A. RESOLVED RESONANCES

The expressions for resonance cross sections are written here first for s-wave interaction only, for which the angular momentum quantum number, l, is zero, and then for the general case that includes higher angular momentum values. Below 100 keV, interactions with $l>0$ contribute little. The U.S. codes MINX, ETOX, and ENDRUN have adopted the Breit-Wigner formulas presented in Reference 13, which include higher-order angular-momentum terms.

For absorption (capture plus fission) the Breit-Wigner single-level formula for the resonance cross section for $l=0$ is

$$\sigma_a(E) = \frac{\pi}{k^2} g_J \frac{\Gamma_n(\Gamma_\gamma + \Gamma_f)}{(E-E_0)^2 + \left(\frac{\Gamma}{2}\right)^2} \tag{5-32}$$

$$= \sigma_0 \frac{\Gamma_\gamma + \Gamma_f}{\Gamma} \frac{1}{1+x^2}, \tag{5-33}$$

where

$k=$ neutron wave number (cm^{-1})

$\quad = 2\pi/\lambda = 0.002197 \ \mu\sqrt{E}$

$\lambda =$ DeBroglie wavelength (cm)

$\mu =$ reduced mass $= mA/(A+1)$, where m is the neutron mass ($m=1$ in the expression above for k) and A is the mass number of the target nucleus

$g_J = \dfrac{2J+1}{2(2I+1)}$ (g_J is a statistical spin factor)

$J=$ spin quantum number of the compound nucleus (target + neutron)

$I=$ spin quantum number of the target nucleus

$E=$ neutron energy (eV)

$E_0 =$ resonance energy (eV)

$\Gamma_n =$ neutron width of the resonance level (eV)

$\Gamma_\gamma =$ radiative capture width (eV)

$\Gamma_f =$ fission width (eV)

$\Gamma =$ total width (eV) ($\Gamma = \Gamma_n + \Gamma_\gamma + \Gamma_f$)

$\sigma_0 = \dfrac{4\pi}{k^2} g_J \dfrac{\Gamma_n}{\Gamma}$

$x = \dfrac{E-E_0}{\Gamma/2}.$

Note that the cross section is symmetrical around the resonance energy E_0. The symmetrical function $1/(1+x^2)$ is called the *natural line shape*. The peak absorption cross section at the center of the resonance (where $x=0$) is $\sigma_0(\Gamma_\gamma+\Gamma_f)/\Gamma$.

For fissile isotopes, the fission resonance cross section is $\sigma_a(E)\Gamma_f/(\Gamma_\gamma+\Gamma_f)$ and the capture cross section is $\sigma_a(E)\Gamma_\gamma/(\Gamma_\gamma+\Gamma_f)$.

The capture width Γ_γ is fairly constant for all resonances, the fission width Γ_f varies somewhat, and the neutron width Γ_n is approximately proportional to $\sqrt{E}$. The neutron width, therefore, varies significantly between resonances, although it can be considered constant over the small energy range of a single resonance.

In order to include contributions for higher-order angular-momentum numbers (i.e., $l>0$), the cross section is

$$\sigma_a(E) = \sum_l \sigma_a^l(E), \tag{5-34}$$

where

$$\sigma_a^l(E) = \frac{\pi}{k^2} \sum_J g_J \sum_{r=1}^{NR_J} \frac{\Gamma_{nr}(\Gamma_{\gamma r}+\Gamma_{fr})}{(E-E_r')^2 + \frac{(\Gamma_r)^2}{2}}. \tag{5-35}$$

The summation on J extends over all possible J-states for a particular l-state. The r-summation limit, NR_J, is the number of resonances for a given pair of l and J values. Details of the evaluation of the Γ_{nr}'s and Γ_r's and the difference between E_0 and E_r are given in References 7, 9, 11, and 13.

For $l=0$, the elastic scattering cross section is given by

$$\sigma_e(E) = \frac{\pi}{k^2} g_J \frac{\Gamma_n^2}{(E-E_0)^2 + \left(\frac{\Gamma}{2}\right)^2} + \frac{4\pi R}{k} g_J \frac{\Gamma_n(E-E_0)}{(E-E_0)^2 + \left(\frac{\Gamma}{2}\right)^2} + 4\pi R^2 \tag{5-36}$$

$$= \sigma_0 \frac{\Gamma_n}{\Gamma} \frac{1}{1+x^2} + \left(\sigma_0 \sigma_p g_J \frac{\Gamma_n}{\Gamma}\right)^{1/2} \frac{2x}{1+x^2} + \sigma_p, \tag{5-37}$$

where σ_p is the potential scattering cross section (which is energy independent) and R is the *potential scattering radius*. For fissile and fertile isotopes, the value of R is about 9×10^{-13} cm, so σ_p is about 10 barns. The expression $2x/(1+x^2)$ is called the *interference line shape*. Note that this line shape is *asymmetrical*; in fact, for most scattering resonances it results in a large *negative* contribution to the cross section at energies below E_0.

The elastic scattering cross section evaluated in MINX, ETOX and ENDRUN for all l-states is

$$\sigma_e(E) = \sum_l \sigma_e^l(E), \qquad (5\text{-}38)$$

where

$$\sigma_e^l(E) = \frac{\pi}{k^2} \sum_J g_J \sum_{r=1}^{NR_J} \frac{\Gamma_{nr}^2 \cos 2\theta_l - 2\Gamma_{nr}(\Gamma_{\gamma r} + \Gamma_{fr}) \sin^2 \theta_l + 2(E - E_r')\Gamma_{nr}\sin 2\theta_l}{(E - E_r')^2 + \frac{(\Gamma_r)^2}{2}}$$

$$+ (2l+1)\frac{4\pi}{k^2}\sin^2\phi_l, \qquad (5\text{-}39)$$

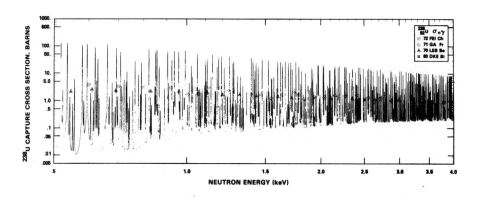

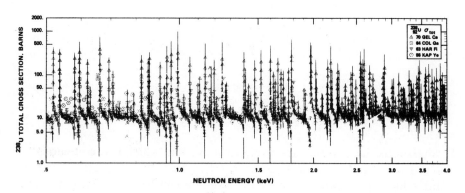

FIGURE 5-4. Capture and total resonance cross sections for ^{238}U in the low keV energy range.[14]

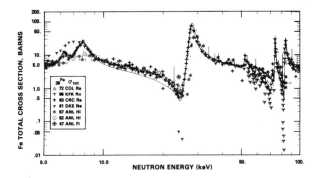

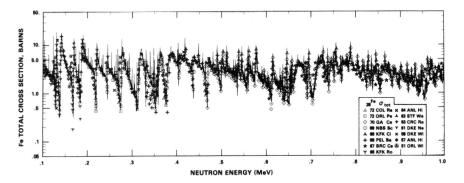

FIGURE 5-5. Resonance cross sections for Fe.[14]

and where, again, detailed expressions for θ, ϕ, Γ_{nr}, Γ_r and E'_r are given in References 7, 9, 11, and 13. The second and third terms in the summation account for *interference* between resonance and potential scattering. A further effect called interference scattering is sufficiently small that it is not included in Eqs. (5-36) and (5-39).

Examples of resonance cross sections are given in Figs. 5-4 through 5-6 over resonance ranges of particular interest for the LMFBR. Capture and total resonance cross sections are shown in Fig. 5-4 for ^{238}U. Note that the capture resonances are symmetrical, according to Eq. (5-33). The scattering cross sections often contribute more to the total cross sections than the capture; hence, the large negative contribution from the interference line shape can be observed. Figure 5-5 shows results for iron. Figure 5-6 gives resonances for sodium; note that the low energy scattering resonances at 3 keV and 52 keV do not exhibit the large negative contribution from the second term while resonances at higher energies do.

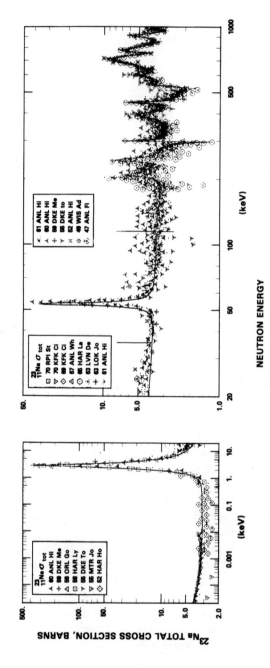

FIGURE 5-6. Total cross section for sodium.[14]

B. UNRESOLVED RESONANCES

For the energy range for which resonance parameters (Γ_n, Γ_γ, Γ_f) have not been measured, i.e., the unresolved resonance region, it is necessary to estimate these parameters in order to calculate cross sections. The estimates are based on properties observed for the resolved resonances.

The resolved resonances provide information on the number of resonances in a given energy range, or the *average spacing*, D (in units of energy), between resonances. They also provide estimates of the resonance parameters. Studying the statistical distribution of resolved resonance parameters, Porter and Thomas[15] found that the parameters followed a chi-squared distribution in energy with varying degrees of freedom. The radiative capture widths are almost constant with energy, which is equivalent to a chi-squared distribution with infinite degrees of freedom. The

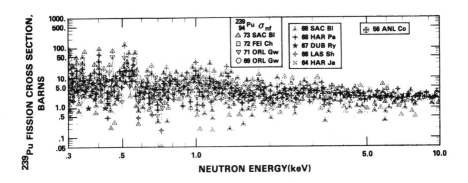

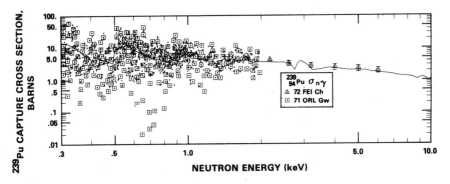

FIGURE 5-7. Fission and capture cross sections in the unresolved resonance region for ^{239}Pu.[14]

fission widths vary from resonance to resonance according to a chi-squared distribution with three or four degrees of freedom. The neutron widths are about proportional to $\sqrt{E}$, so that a new parameter called the *reduced neutron width*, Γ_n^0 (equal to $\Gamma_n / \sqrt{E}$), is about constant in energy. There is some statistical variation in Γ_n^0, nevertheless, and Porter and Thomas showed that Γ_n^0 follows a chi-squared distribution with one degree of freedom.

A method for approximating these distributions, introduced by Greebler and Hutchins,[16] is used in U.S. codes for the unresolved region. A constant value for Γ_γ is assumed. The fission width distribution is represented by five values for Γ_f of equal probability. The reduced neutron width is represented by ten values of Γ_n^0 of equal probability. Cross sections are calculated for all combinations of these resonance parameters and are then averaged to obtain the cross sections for use in Eqs. (5-23) and (5-24).

Fission and capture cross sections for ^{239}Pu in the unresolved resonance range of principal interest are shown in Fig. 5-7. Below ~ 0.3 keV, the resonances are resolved. The difficulty of measuring α, the ratio of σ_c to σ_f, in this energy range led for a long time to a rather large uncertainty in the breeding ratio for a fast oxide-fueled reactor; in recent years, however, this uncertainty has been substantially reduced.

5-5 DOPPLER BROADENED CROSS SECTIONS

A. GENERAL THEORY

As pointed out in the previous section, the effective cross section of a material will be influenced by the thermal motion of the resonance nuclei. Previously we defined a neutron reaction rate for a stationary target material (per atom per cm^3 of the material) as $\sigma_x(E)\phi$, where x refers to capture, fission, absorption, or elastic scattering and where $\phi = nv$. Here n is the neutron density and v is the neutron velocity in the laboratory system. We must now redefine a cross section, called the *Doppler broadened* cross section, such that, when multiplied by the neutron flux ϕ, it gives the correct reaction rate for nuclei in thermal motion. We will label this cross section $\sigma_{x,\text{Dop}}$.

The target nuclei are in thermal motion with a distribution of velocities. The relative velocity between the target nucleus with velocity vector $\vec{V}$ in the laboratory system and the neutron velocity vector $\vec{v}$ is equal to $\vec{v} - \vec{V}$, as illustrated in Fig. 5-8.

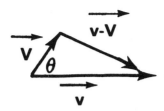

FIGURE 5-8. Relative motion between neutron and target.

Only the component of $\vec{V}$ parallel to the neutron velocity $\vec{v}$ will have an influence on the Doppler broadened cross section. Since we can assume any direction for $\vec{v}$, let us consider that $\vec{v}$ is in the z direction. Hence, only the distribution of the z-component of the target velocities, V_z, will influence the result.

The Doppler broadened cross section is

$$\sigma_{x,\text{Dop}}\,\phi = \sigma_{x,\text{Dop}}\,nv = n\int_0^\infty |v-V|\sigma_x(v-V_z)P(V_z)\,dV_z \qquad (5\text{-}40)$$

because σ_x will vary with the z-component of the target velocity. Here $|v-V|$ is the magnitude of the relative velocity. $P(V_z)$ represents the Maxwellian distribution* of the z-component of the target velocities so that the fraction of target nuclei with velocities in dV_z is

$$P(V_z)\,dV_z = \sqrt{\frac{M}{2\pi kT}}\,\exp\!\left(-\frac{MV_z^2}{2kT}\right)dV_z, \qquad (5\text{-}41)$$

where M is the mass of the target nucleus, T is the temperature, and k is Boltzmann's constant.

Since $V\ll v$, $|v-V|$ is close enough to v that the v and $|v-V|$ on the two sides of Eq. (5-40) can be cancelled. Hence,

$$\sigma_{x,\text{Dop}} = \int_0^\infty \sigma_x(v-V_z)P(V_z)\,dV_z. \qquad (5\text{-}42)$$

We will next switch to an energy variable because we want $\sigma_{x,\text{Dop}}(E)$ for use in Eqs. (5-23) and (5-24), where E is the neutron energy in the

---

*The Maxwellian distribution holds strictly only for a gas, whereas the resonance absorber is a solid. Above a temperature referred to as the Debye temperature (based on Debye's model for solids) this distribution is satisfactory for solids. The Debye temperature for most solids is of the order of 300 K; hence, the distribution holds for the entire range of fast reactor temperatures.

laboratory system. Hence, we must consider the relationship between E, E_c (the neutron energy in the center-of-mass system) and the velocity component V_z. The energy E_c is

$$E_c = \tfrac{1}{2}\mu|v - V|^2, \tag{5-43}$$

where μ is the reduced mass [as defined after Eq. (5-33)], and

$$|v - V|^2 = v^2 - 2\vec{v}\cdot\vec{V} + V^2,$$

$$\cong v^2 - 2vV\cos\theta \text{ (where } \theta \text{ is given in Fig. 5-8)} \tag{5-44}$$

$$\cong v^2 - 2vV_z \text{ (since } \vec{v} \text{ is postulated to be in the } z \text{ direction).}$$

The energy E is

$$E = \tfrac{1}{2}\mu v^2. \tag{5-45}$$

Substituting these relations into Eq. (5-43) gives

$$E_c = E - \sqrt{2\mu E}\, V_z, \tag{5-46}$$

and

$$dV_z = \frac{dE_c}{\sqrt{2\mu E}}. \tag{5-47}$$

Substituting Eqs. (5-41), (5-46), and (5-47) into (5-42) gives

$$\sigma_{x,\text{Dop}}(E) = \int_0^\infty \sigma_x(E_c)\sqrt{\frac{M}{2\pi kT}}\,\exp\left[-\frac{M}{4\mu kTE}(E - E_c)^2\right]\frac{dE_c}{\sqrt{2\mu E}}. \tag{5-48}$$

When E stands alone, it can be approximated by a constant in the integral and can be replaced by E_0. Only energy differences [such as $(E - E_0)$ and $(E_c - E_0)$] vary significantly with E_c inside the integral. Hence,

$$\sigma_{x,\text{Dop}}(E) = \frac{1}{\sqrt{\pi}}\sqrt{\frac{M}{4\mu kTE_0}}\int_0^\infty \sigma_x(E_c)\exp\left[-\frac{M}{4\mu kTE_0}(E - E_c)^2\right]dE_c. \tag{5-49}$$

B. ABSORPTION CROSS SECTIONS

Next, let us consider the Doppler broadened absorption cross section, $\sigma_a(E)$. (For compactness, we drop the subscript Dop in the remainder of this section.) Here we will consider only s-wave interactions ($l = 0$).

Substituting Eq. (5-33) for $\sigma_a(E_c)$ into Eq. (5-49) gives

$$\sigma_a(E) = \sigma_0 \frac{\Gamma_\gamma + \Gamma_f}{\Gamma} \frac{1}{\sqrt{\pi}} \sqrt{\frac{M}{4\mu k T E_0}}$$

$$\times \int_0^\infty \frac{1}{1 + \left(\dfrac{E_c - E_0}{\Gamma/2}\right)^2} \exp\left[-\frac{M}{4\mu k T E_0}(E - E_c)^2\right] dE_c. \quad (5\text{-}50)$$

We next introduce the following definitions:

$$\Delta \cong \sqrt{\frac{4\mu k T E_0}{M}} = \text{Doppler width (eV)}, \quad (5\text{-}51)$$

$$\theta = \Gamma/\Delta, \quad (5\text{-}52)$$

$$y = \frac{E_c - E_0}{\Gamma/2}. \quad (5\text{-}53)$$

Recognizing that $(\Gamma/2)^2 [(E - E_0 + E_0 - E_c)/(\Gamma/2)^2]$ is equivalent to $(E - E_c)^2$, we can rewrite $(E - E_c)^2$ in Eq. (5-50) as

$$(E - E_c)^2 = \left(\frac{\Gamma}{2}\right)^2 \left(\frac{E - E_0}{\Gamma/2} - \frac{E_c - E_0}{\Gamma/2}\right)^2 = \left(\frac{\Gamma}{2}\right)^2 (x - y)^2,$$

where x was defined by Eq. (5-33) and y by Eq. (5-53).

Since from Eq. (5-53) $dE_c = \frac{\Gamma}{2} dy$, Eq. (5-50) can be written as

$$\sigma_a(E) = \sigma_0 \frac{\Gamma_\gamma + \Gamma_f}{\Gamma} \frac{\theta}{2\sqrt{\pi}} \int_{-\infty}^\infty \frac{\exp\left[-\frac{1}{4}\theta^2(x - y)^2\right]}{1 + y^2} dy, \quad (5\text{-}54)$$

where the limits of integration of y (corresponding to the limits for E_c) have been extended to $\pm\infty$.

We define the function $\psi(\theta, x)$, called the *Doppler broadened line shape*, to be

$$\psi(\theta, x) = \frac{\theta}{2\sqrt{\pi}} \int_{-\infty}^\infty \frac{\exp\left[-\frac{1}{4}\theta^2(x - y)^2\right]}{1 + y^2} dy. \quad (5\text{-}55)$$

Hence Eq. (5-54) can be written as

$$\sigma_a(E) = \sigma_0 \frac{\Gamma_\gamma + \Gamma_f}{\Gamma} \psi(\theta, x). \quad (5\text{-}56)$$

By comparing this result with Eq. (5-33), we note that the Doppler broadened line shape has replaced the natural line shape. Like the natural line shape, the Doppler broadened line shape is symmetrical around the resonance energy E_0. Its value at E_0, however, is less than the value of the natural line shape at E_0. On the other hand its value far from E_0 is greater than the natural line shape. Thus, the effect of Doppler broadening on $\sigma_a(E)$ is to decrease the peak cross section at the resonance energy while increasing the cross section away from the resonance, i.e., "in the wings" of the resonance. This is illustrated schematically in Fig. 5-9.

A number of properties of Doppler broadened cross sections are described by Dresner[6] and others. For example, as the temperature, T, approaches zero, the Doppler broadened line shape, ψ, does indeed approach the natural line shape, as it should.

Another interesting property is that the integral over a resonance of a Doppler broadened absorption cross section weighted with an asymptotic energy spectrum, i.e., $\int \sigma_a(E)dE/E$—called the *resonance integral*—is the same as the resonance integral for the non-Doppler broadened cross section. The result is curious because, since dE/E is the weighting factor for the infinitely dilute cross section, this result says that the infinitely dilute group cross section is unaffected by Doppler broadening and, hence, by the temperature of the absorbing material. On the other hand, the important integral for effective cross sections, $\int \sigma_a(E)\phi(E)\,dE$, where $\phi(E)$ is the self-shielded flux given by Eq. (5-18), *does* depend on the temperature. The actual absorption rate is higher for the Doppler broadened cross section because the Doppler cross section is higher in the wings (where the flux, $\phi(E)$, is not greatly depressed), whereas *both* the Doppler and the non-Doppler cross sections are so high at the center of the resonance that the absorber is essentially black to neutrons regardless of Doppler broadening.

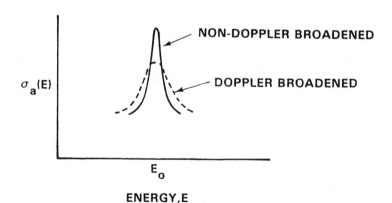

FIGURE 5-9. Effect of Doppler broadening on $\sigma_a(E)$.

In other words, the lower peak Doppler cross section is just about compensated by a higher effective neutron flux at the center of the resonance, whereas the reverse is not the case out in the wings. The important result of this is that self-shielding of the neutron flux is responsible for the variation of the effective group cross section with temperature and for Doppler reactivity effects which will be discussed in Chapter 6. Without self-shielding, there would be no Doppler reactivity effect from temperature changes.

As a result of the above discussion, the influence of a temperature change on the effective cross section of a resonance absorber depends not only on the magnitude of the temperature change but also on the concentration of the absorber. This is because the self-shielding depends on the *macroscopic* cross section, Σ_t, and hence the atom density of the absorber. If there is very little of the resonance absorber present, the effective group cross section will hardly change with temperature. For this reason the change in the effective capture cross sections with temperature for the fertile material ^{240}Pu will be much lower than for the fertile material ^{238}U. Further subtleties in the behavior and effects of Doppler broadening appear when the effective cross sections are combined with atom densities and adjoint fluxes for the calculation of Doppler reactivities. That will be the subject of Chapter 6.

C. RESONANCE OVERLAP

A complication in the calculation of effective cross sections from Eqs. (5-23) and (5-24) not discussed in Section 5-3 is that there may be more than one resonance (i.e., from different materials) at about the same energy. Thus, parts of the two resonances may *overlap* and the assumption that σ_{om} is constant across a resonance in material m is, therefore, inaccurate.

The influence of resonance overlap on the absolute magnitude of an effective cross section is not great. However, the influence of overlap on the temperature dependence of cross sections, and hence on the Doppler effect, is quite important because here one is concerned with small changes in effective cross sections with temperature. This problem has been extensively analyzed by contributors to the theory of the Doppler effect in fast reactors, including Greebler and Hutchins, Nicholson, and Hwang and Hummel in the U.S. and Fischer, Rowlands, and Codd and Collins in Europe. These investigators showed, for example, that it was necessary to account for resonance overlapping in order to calculate the contribution of the fissile isotopes to the Doppler effect. A review of resonance overlapping is given by Hummel and Okrent.[5] The analysis is too complex to describe in this text, but it is accounted for in the generalized cross section codes—MINX, ETOX, and ENDRUN.

D. ELASTIC SCATTERING CROSS SECTION

The elastic scattering cross section is also Doppler broadened. The result (for $l=0$) is

$$\sigma_e = \sigma_0 \frac{\Gamma_n}{\Gamma} \psi(\theta, x) + \left(\sigma_0 \sigma_p g_J \frac{\Gamma_n}{\Gamma} \right)^{1/2} \chi(\theta, x) + \sigma_p, \qquad (5\text{-}57)$$

where
$$\chi(\theta, x) = \frac{\theta}{2\sqrt{\pi}} \int_{-\infty}^{\infty} \frac{\exp\left[-\frac{1}{4}\theta^2(x-y)^2\right]}{1+y^2} 2y \, dy. \qquad (5\text{-}58)$$

Here $\chi(\theta, x)$ is the Doppler broadened counterpart of the interference line shape, $2x/(1+x^2)$, which appeared in Eq. (5-37). Like the interference line shape, $\chi(\theta, x)$ is asymmetrical around E_0.

5-6 CROSS SECTIONS FOR A SPECIFIC REACTOR COMPOSITION

A. GENERAL METHOD

The second part of the shielding factor method is the calculation of effective cross sections for each material from the generalized cross sections obtained in the previous part. This is done in the SPHINX,[8] 1DX,[10] and TDOWN[12] codes. To do this, it is necessary first to calculate the background cross section, σ_0, for each material and then to interpolate for the appropriate shielding factor from the values of $f(\sigma_0)$ found in the generalized cross section calculation.

Evaluation of the effective cross section for each material is an iterative process because the correct σ_0 for a given material depends on the effective total cross sections of all the other materials present. The iterative scheme proceeds as follows.

First iteration:

- Make a first guess for σ_{tmg} for each material in each energy group. The infinitely dilute value represents a reasonable first choice.
- Calculate σ_{omg} for all materials, using the first guesses for σ_{tmg}.
- Obtain the shielding factor for the total cross section for each material, f_{tmg}, from the generalized cross section correlations of f_{tmg} vs σ_0.
- Calculate σ_{tmg} for each material and each group from Eq. (5-1).

Second and subsequent iterations:

- With the new values of σ_{tmg}, recalculate all σ_{omg}'s.
- Obtain new values of f_{tmg}.
- Obtain new values of σ_{tmg}.
- Compare each σ_{tmg} with the σ_{tmg} from the previous iteration. Continue the iteration procedure until the fractional change in every value of σ_{tmg} between iterations is less than a selected convergence criterion.

After convergence is achieved, we have the final values of both σ_{tmg} and σ_{omg} for each m. At this time all other f_{xmg} factors (i.e., those for capture, fission, and elastic scattering) and σ_{xmg} values are calculated. The process usually converges in a few iterations.

The inelastic scattering removal cross section, together with the inelastic matrix, is carried over directly from the generalized cross section set; it is not composition dependent. (Recall that inelastic scattering occurs at high enough energies that self-shielding is not important.)

In the SPHINX and 1DX codes the interpolation between σ_o's for which cross sections are tabulated is accomplished by recognizing that graphs of f factors versus $\ln \sigma_o$ have a hyperbolic tangent shape. An example is shown in

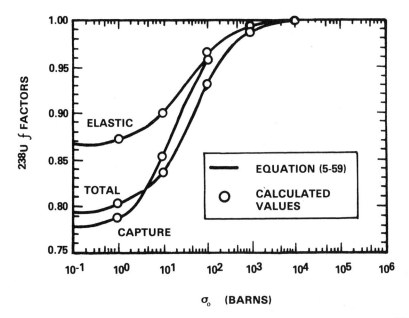

FIGURE 5-10. Example of f factors vs background scattering cross section for ^{238}U (from Ref. 2).

Fig. 5-10. The following four-parameter expression is used to calculate f:

$$f(\sigma_o) = A \tanh B(\ln \sigma_o + C) + D. \qquad (5\text{-}59)$$

The parameters A and D are obtained by inspection of the range and magnitude of the shielding factors. Parameters B and C are solved by utilizing the tabulated σ_o points nearest the σ_o value for which f is being determined. In the TDOWN code, the interpolation is done by a Taylor series expansion of f around neighboring values of $\ln \sigma_o$ for which f has been tabulated.

B.   ELASTIC REMOVAL CROSS SECTION

For most cross sections, the flux distribution $\phi(E)$ used in the generalized calculations is adequate for obtaining the specific composition cross sections. The elastic removal cross section, however, may be sensitive to the energy variation in flux across an energy group, particularly if the energy width of the group is wide relative to the maximum energy loss in an elastic collision—a situation frequently encountered in fast reactor analysis.

The expression for σ_{ermg} can be obtained with the aid of Fig. 5-11. From arguments presented earlier for energy lost by isotropic elastic scattering, σ_{ermg} can be expressed as

$$\sigma_{ermg}\phi_g = \int_{E_g}^{E_g/\alpha} \sigma_{em}(E)\phi(E)\frac{E_g - \alpha E}{E - \alpha E} dE \qquad (5\text{-}60)$$

where $(E_g - \alpha E)/(E - \alpha E)$ is the fraction of neutrons scattered at E that are scattered out of the group. It is more convenient to use lethargy, u,

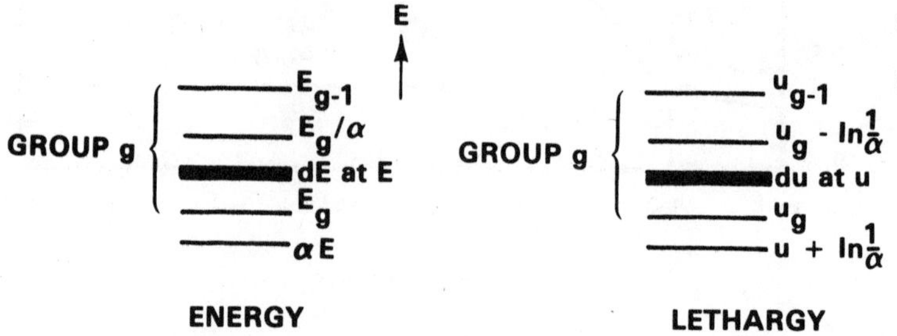

FIGURE 5-11. Lethargy and energy variables for group g.

where

$$E = E_0 e^{-u},$$

$$\phi(E)\, dE = -\phi(u)\, du, \tag{5-61}$$

$$u = u_g - \ln\frac{1}{\alpha} \text{ at } E = E_g/\alpha.$$

Substituting these values into Eq. (5-60) gives

$$\sigma_{ermg}\phi_g = \int_{u_g - \ln\frac{1}{\alpha_m}}^{u_g} \sigma_{em}(u)\phi(u)\frac{e^{-(u_g-u)} - \alpha_m}{1 - \alpha_m}\, du. \tag{5-62}$$

For the asymptotic flux [see Eq. (5-16)] $\sigma_{em}(u)$ and $\phi(u)$ are constant. For this case, we label σ_{ermg} as $(\sigma_{ermg})_0$, and

$$(\sigma_{ermg})_0 = \frac{\xi_m \sigma_{emg}}{\Delta u_g}, \tag{5-63}$$

where ξ_m is the logarithmic energy decrement, or

$$\xi_m = 1 + \frac{\alpha_m \ln \alpha_m}{1 - \alpha_m}. \tag{5-64}$$

To account for the deviation from the asymptotic flux, the elastic removal cross section can be modified in the following way. We define a quantity $\psi_m(u)$ proportional to a slowing down density as

$$\psi_m(u) = \xi_m \sigma_{em}(u)\phi(u). \tag{5-65}$$

It is assumed that $\psi_m(u)$ varies linearly in lethargy between groups, and that this product has its average value for the group at the center (in lethargy) of this group. Hence, ψ_m has the group value ψ_{mg} at the mid-lethargy of group g.

From this point on, the exact treatment of the correction to σ_{er} becomes complex algebraically. The exact treatment is used in the TDOWN code and is derived in Reference 12.

A simpler approximate correction for σ_{er} is used in the 1DX and SPHINX codes, and this result is given here. The corrected value of σ_{ermg} is calculated using the flux at lethargy point u'_{mg}, where

$$u'_{mg} = u_g - 0.66\xi_m. \tag{5-66}$$

The quantity $\psi_m(u)$ is assumed to vary linearly between mid-lethargies of each group, and the expression for σ_{ermg} becomes

$$\sigma_{ermg} = (\sigma_{ermg})_0 \left[1 + \frac{u'_{mg} - u_g}{u_{g+1} - u_g} \left(\frac{\psi_{g+1}}{\psi_g} - 1 \right) \right]. \qquad (5\text{-}67)$$

The evaluation of σ_{ermg} is an iterative process because it depends on the group fluxes (through ψ), which depend on the σ_{er} values. The iteration is started with the asymptotic flux values, $(\sigma_{ermg})_0$. Convergence is obtained when the following fractional change between iterations is less than a convergence criterion, ε, that is

$$\frac{\left(\dfrac{\phi_g}{\phi_{g+1}} \right)^{\nu+1} - \left(\dfrac{\phi_g}{\phi_{g+1}} \right)^{\nu}}{\left(\dfrac{\phi_g}{\phi_{g+1}} \right)^{\nu}} < \varepsilon, \qquad (5\text{-}68)$$

where ν refers to the iteration number.

A final note on σ_{ermg} concerns elastic removal at energies above E_c, the interface between the fission spectrum for $\phi(E)$ and the asymptotic flux. Since values of ξ_{mg} are available for these high energies, it would be useful to use the simple expression $\xi\sigma_{eg}/\Delta u_g$ for σ_{erg}. However since ϵ was based on a $1/E$ spectrum, a correction factor must be introduced into the simple expression above to account for the fission spectrum. Such a correction is defined in Reference 11.

C. CORRECTION FOR HETEROGENEOUS GEOMETRY

The expressions derived thus far have assumed a homogeneous reactor composition. Little error results from this assumption for most fast reactor calculations because the mean free paths are large relative to the fuel-rod and coolant-channel dimensions. In the resonance region, however, mean free paths are shorter where the resonance material is located—for example, in the fuel pin when a ^{238}U or Pu resonance is being considered. Hence, heterogeneity will have some influence in the resonance region.

A simple correction to account for the effect of heterogeneity on self-shielding for the fuel isotopes is based on the *rational approximation* (sometimes called the Wigner or canonical approximation). This method is based on an approximation of the neutron leakage probability from a fuel rod. For repeating lattices of rods separated by moderator or, in the case of fast reactors, by cladding and sodium, a modification of the rational

FIGURE 5-12. Geometry for heterogeneity correction.

approximation provided by Bell[17] provides an appropriate correction. This correction is available in all three computer codes discussed here—1DX, TDOWN, and SPHINX.

The Bell formulation of the rational approximation can be applied to the fuel isotopes in a fast reactor lattice by modifying the background cross section, σ_o.

Consider the following two-region cylindrical geometry, which approximates fuel in region 1 surrounded by cladding and coolant in region 2 (Fig. 5-12).

Let the value of the background scattering cross section for fuel isotope m modified for heterogeneity be called σ_{ohetm}, so that

$$\frac{\Sigma_t}{N_m} = \sigma_{tm} + \sigma_{ohetm}, \qquad (5-69)$$

which is an equation of the same form as Eq. (5-21). In calculating effective cross sections in the second part of the shielding factor method, σ_{ohetm} is used in exactly the same way that the homogeneous equivalent σ_{om} is used.

Our task, then, is to evaluate σ_{ohetm}. We begin with some careful definitions, where all energy dependent quantities are at energy E for which Σ_t is being evaluated:

P = probability that an arriving neutron at energy E suffers its next collision in region 1 (in which a uniform spatial source of moderated neutrons arriving at E is assumed)

S = surface between regions 1 and 2 = $\pi D_1 L$ (cm^2)

V_1 = fuel volume in the cell = $(\pi/4) D_1^2 L$ (cm^3)

V_2 = cladding-coolant volume in the cell = $(\pi/4)(D_2^2 - D_1^2)L$ (cm^3)

V_{cell} = $V_1 + V_2$ (cm^3)

Σ_1 = macroscopic total cross section in region 1

Σ_2 = macroscopic total cross section in region 2

Σ_t = effective macroscopic total cross section for the homogenized cell

$N_{m,1}$ = atom density of fuel isotope m in region 1

N_m = atom density of fuel isotope m in the homogenized cell

$N_{other,1}$ = atom density of other materials in region 1

N_{other} = atom density of other materials present in region 1 only, averaged over the homogenized cell

ϕ_1 = neutron flux in region (n/cm$^2 \cdot$s).

The neutron reaction rate in region 1 per homogenized cell volume is $\Sigma_1\phi_1 V_1/V_{\text{cell}}$. The neutron reaction rate in both regions 1 and 2 per homogenized cell volume is $(\Sigma_1\phi_1 V_1/V_{\text{cell}})(1/P_1)$. This reaction rate is approximately equal to $\Sigma_t\phi_1$; although the average for the cell differs slightly from ϕ_1, most of the reactions are occurring in region 1 at the resonance energy. Therefore,

$$\Sigma_t\phi_1 = \left(\Sigma_1\phi_1 \frac{V_1}{V_{\text{cell}}}\right)\left(\frac{1}{P_1}\right). \tag{5-70}$$

The Bell-modified rational approximation provides the expression for P_1:

$$P_1 = \frac{\Sigma_1}{\Sigma_1 + \tau_1}, \tag{5-71}$$

where

$$\tau_1 = \frac{\Sigma_2 \dfrac{S}{4V_1}}{\Sigma_2 + \dfrac{S}{4V_1}\dfrac{V_1}{V_2}}. \tag{5-72}$$

Substituting into Eq. (5-70) gives

$$\Sigma_t = (\Sigma_1 + \tau_1)\frac{V_1}{V_{\text{cell}}} \tag{5-73}$$

$$= \left(N_{m,1}\sigma_{tm} + \sum_{\substack{\text{other} \\ \text{in 1}}} N_{\text{other},1}\sigma_{s,\text{other}}\right)\frac{V_1}{V_{\text{cell}}} + \tau_1\frac{V_1}{V_{\text{cell}}}. \tag{5-74}$$

Since $N_{m,1} = N_m V_{\text{cell}}/V_1$, and $N_{\text{other},1} = N_{\text{other}} V_{\text{cell}}/V_1$,

$$\Sigma_t = N_m\sigma_{tm} + \sum_{\substack{\text{other} \\ \text{in 1}}} N_{\text{other}}\sigma_{s,\text{other}} + \tau_1\frac{V_1}{V_{\text{cell}}},$$

and

$$\frac{\Sigma_t}{N_m} = \sigma_{tm} + \frac{1}{N_m}\sum_{\substack{\text{other} \\ \text{in 1}}} N_{\text{other}}\sigma_{s,\text{other}} + \frac{\tau_1}{N_m}\frac{V_1}{V_{\text{cell}}}. \tag{5-75}$$

Defining σ'_{om} by the second term on the right-hand side of this equation, Σ_t/N_m can be written as

$$\frac{\Sigma_t}{N_m} = \sigma_{tm} + \sigma'_{om} + \frac{\tau_1}{N_m}\frac{V_1}{V_{\text{cell}}}. \tag{5-76}$$

By comparing Eq. (5-76) with Eq. (5-69),

$$\sigma_{\text{ohet}m} = \sigma'_{om} + \frac{\tau_1}{N_m} \frac{V_1}{V_{\text{cell}}}. \tag{5-77}$$

The difference between σ_{om} of Eq. (5-22) and σ'_{om} of Eq. (5-77) should be emphasized; σ'_{om} has the same form as σ_{om} (including N's based on the homogenized cell), but σ'_{om} includes *other materials* in the *fuel only* (e.g., oxygen and other fuel isotopes in UO_2-PuO_2 fuel). The cross sections of the other materials in the cladding and coolant do enter into the $\sigma_{\text{ohet}m}$ calculation, but they enter through the Σ_2 appearing in the τ_1.

5-7 COLLAPSING OF MULTIGROUP CROSS SECTIONS

Frequently it is useful to use few-group cross sections, or even, for fuel cycle analysis, one-group cross sections. The ENDF/B file contains almost continuous data on cross sections as a function of energy. The generalized cross sections may involve from 20 to 100 energy groups, and in certain regions of rapidly varying cross sections, even more groups. For one-dimensional design calculations for specific reactor compositions, it may be sufficient to use 10 to 30 groups, and for two- or three-dimensional analysis even fewer groups. For fuel cycle analysis, relative reaction rates are important and sufficient accuracy can be achieved with one-group capture and absorption cross sections—although a single effective one-group value may have to be changed for different stages in the burnup cycle. For these reasons it is necessary to have techniques for collapsing cross sections from many groups to fewer groups.

Two approaches to collapsing cross sections can be followed. The first conserves reaction rates. The second conserves reactivity worths. The first is more often used, and it will be described here.

Consider the group structure in Fig. 5-13.

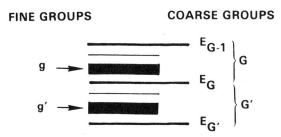

FIGURE 5-13. Group structure for group collapsing.

Suppose several *fine groups*, with index g, are to be collapsed into one *coarse group* with index G, as shown in Fig. 5-13. The lower energy of the new coarse group is E_G. For capture, fission and elastic scattering for each material, the collapsed cross section is (omitting the material index, m)

$$\sigma_{xG}\phi_G = \int_{E_G}^{E_{G-1}} \sigma_x(E)\phi(E)\,dE \qquad (5\text{-}78)$$

$$= \sum_{g\epsilon G} \sigma_{xg}\phi_g, \qquad (5\text{-}79)$$

where $g\epsilon G$ means that the fine group g is contained in the larger coarse group G. Since $\phi_G = \sum_{g\epsilon G}\phi_g$,

$$\sigma_{xG} = \frac{\displaystyle\sum_{g\epsilon G} \sigma_{xg}\phi_g}{\displaystyle\sum_{g\epsilon G} \phi_g}. \qquad (5\text{-}80)$$

For σ_{tr}, leakage is to be conserved. Hence,

$$\sigma_{trG} = \frac{\displaystyle\int_{E_G}^{E_{G-1}} \sigma_{tr}(E)\lambda_{tr}(E)\phi(E)\,dE}{\displaystyle\int_{E_G}^{E_{G-1}} \lambda_{tr}(E)\phi(E)\,dE}, \qquad (5\text{-}81)$$

$$= \frac{\displaystyle\sum_{g\epsilon G} \sigma_{trg}D_g\phi_g}{\displaystyle\sum_{g\epsilon G} D_g\phi_g}. \qquad (5\text{-}82)$$

Note that the collapsed value of σ_{tr} depends on composition dependent quantities, not only through ϕ but also through the diffusion coefficients D_g for the composition.

Generally the collapsed cross sections are sufficiently wide in energy that elastic removal involves transfer to only one group below G, i.e., to group G' in Fig. 5-13. For this case,

$$\sigma_{erG} = \sigma_{eG\to G'} = \sum_{g\epsilon G} \phi_g \sum_{g'\epsilon G'} \sigma_{eg\to g'} \Big/ \sum_{g\epsilon G} \phi_g. \qquad (5\text{-}83)$$

Here $\sigma_{eg\to g'}$ is a scattering matrix for which there will be more than one entry if elastic scattering across the narrow groups g is possible. If elastic scattering across collapsed group G' is possible, the collapsed elastic scattering cross section and collapsed matrix are calculated in the same manner as the inelastic cross section.

The collapsed inelastic scattering cross section matrix is

$$\sigma_{iG \to G'} = \sum_{g \epsilon G} \phi_g \sum_{g' \epsilon G'} \sigma_{ig \to g'} \bigg/ \sum_{g \epsilon G} \phi_g, \qquad (5\text{-}84)$$

where G' is any group below G. The inelastic removal cross section for group G is

$$\sigma_{irG} = \sum_{G' > G} \sigma_{iG \to G'}. \qquad (5\text{-}85)$$

The total cross section for the collapsed group is simply

$$\sigma_{tG} = \sigma_{trG} + \sigma_{eG} \bar{\mu}_G. \qquad (5\text{-}86)$$

The collapsed values for χ, ν, $\bar{\mu}$, and ξ are

$$\chi_G = \sum_{g \epsilon G} \chi_g, \qquad (5\text{-}87)$$

$$\nu_G = \sum_{g \epsilon G} \nu_g \sigma_{fg} \phi_g \bigg/ \sum_{g \epsilon G} \sigma_{fg} \phi_g, \qquad (5\text{-}88)$$

$$\bar{\mu}_G = \sum_{g \epsilon G} \bar{\mu}_g \sigma_{eg} \phi_g \bigg/ \sum_{g \epsilon G} \sigma_{eg} \phi_g, \qquad (5\text{-}89)$$

$$\xi_G = \sum_{g \epsilon G} \xi_g \sigma_{eg} \phi_g \bigg/ \sum_{g \epsilon G} \sigma_{eg} \phi_g. \qquad (5\text{-}90)$$

PROBLEMS

5-1. For the reactor calculated in Problem 4-5, obtain one-group capture and fission cross sections for materials to be used in fuel cycle problems in Chapter 7, i.e., ^{238}U, ^{239}Pu, ^{240}Pu, ^{241}Pu, ^{242}Pu, and fission product pairs.

5-2. Suppose there is a large ^{238}U capture resonance at 500 eV but no ^{232}Th or other resonances there. Schematically draw curves for $\phi(E)$ versus energy near 500 eV for the following conditions:

(a) A lattice that contains 35% steel, 25% sodium, 40% $^{232}ThO_2$, by volume.

(b) A lattice that contains 35% steel, 25% sodium, and 40% $^{238}UO_2$-PuO_2 fuel.

5-3. Suppose a file of generalized multigroup cross sections as a function of background scattering cross section has been developed and is available to you. For a specific LMFBR core composition, it is desired to obtain the cross sections for each energy group for that composition. List in detail the steps required to obtain the ^{238}U capture cross section for group g for that composition.

5-4. Calculate the total cross sections for U, Pu, Fe, Na, and O, the fission cross section for ^{239}Pu, and the capture cross section for ^{238}U, for a group g covering the energy range 1–5 keV and for a reactor composition (atom densities) given below.

The following data for this energy range are needed, where σ's are in barns:

Material	m	$\sigma_{tmg}(\infty)$	σ_o	f_{tmg}	f_{cmg}	f_{fmg}
^{238}U	1	14	10	0.5	0.3	
			10^3	0.9	0.9	
^{239}Pu	2	14	10	0.5		0.7
			10^3	0.9		1.0
Fe	3	12	10	0.7		
			10^3	1.0		
Na	4	6	10	0.3		
			10^3	0.9		
O	5	8	10	0.8		
			10^3	1.0		

^{238}U: $\sigma_{clg}(\infty)=2$
^{239}Pu: $\sigma_{f2g}(\infty)=1$

Assume that f varies linearly with $\log_{10}\sigma_o$ above $\sigma_o=10$, as shown for a typical case below

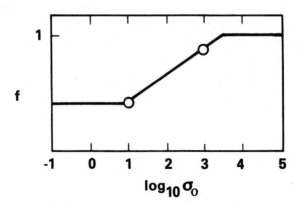

Atom Densities: (atoms / barn · cm)

Fuel	0.02
Fe	0.02
Na	0.01

$$N(^{238}\text{U})/N(\text{fuel}) = 0.8$$

An iterative solution is needed such that for each material,

$$\frac{\sigma_{tmg}^{(r+1)} - \sigma_{tmg}^{(r)}}{\sigma_{tmg}^{(r)}} < \varepsilon$$

for convergence, where superscript r represents the iteration number.
You will need to select your own ε.

REFERENCES

1. I. I. Bondarenko, et al., *Group Constants for Nuclear Reactor Calculations*, Translation—Consultants Bureau Enterprises, Inc., New York, 1964.

2. R. B. Kidman, R. E. Schenter, R. W. Hardie, and W. W. Little, "The Shielding Factor Method of Generating Multigroup Cross Sections for Fast Reactor Analysis," *Nucl. Sci. and Eng.* 48 (1972) 193–201.

3. H. Henryson, B. J. Toppel, and C. G. Stenberg, $MC^2 - 2$: *A Code to Calculate Fast Neutron Spectra and Multigroup Cross Sections*, ANL-8144, Argonne National Laboratory 1976.

4. M. K. Drake, *Data Formats and Procedures for the ENDF Neutron Cross Section Library*, BNL-50274, Brookhaven National Laboratory, (April 1974 Revision).

5. H. H. Hummel and D. Okrent, *Reactivity Coefficients in Large Fast Power Reactors*, American Nuclear Society, 1970.

6. L. Dresner, *Resonance Absorption in Nuclear Reactors*, Pergamon Press, New York, 1960.

7. C. R. Weisbin, P. D. Soran, R. B. MacFarlane, D. R. Morris, R. J. Lebauve, J. S. Hendricks, J. E. White, and R. B. Kidman, *MINX: A Multigroup Integration of Nuclear X-Sections from ENDF/B*, LA-6486-MS, Los Alamos Scientific Laboratory, August 1976.

8. W. J. Davis, M. B. Yarborough, and A. D. Bortz, *SPHINX: A One-Dimensional Diffusion and Transport Nuclear Cross Section Processing Code*, WARD-XS-3045-17, Westinghouse Advanced Reactors Division, August 1977.

9   R. E. Schenter, J. L. Baker, and R. B. Kidman, *ETOX, A Code to Calculate Group Constants for Nuclear Reactor Calculations*, BNWL-1002, Battelle Northwest Laboratory, 1969.

10. R. W. Hardie and W. W. Little, Jr., *1DX, A One-Dimensional Diffusion Code for Generating Effective Nuclear Cross Sections*, BNWL-954, Battelle Northwest Laboratory, 1969.

11. B. A. Hutchins, C. L. Cowan, M. D. Kelly, and J. B. Turner, *ENDRUN-II, A Computer Code to Generate a Generalized Multigroup Data File from ENDF/B*, GEAP-13704, General Electric Company, March 1971.

12. C. L. Cowan, B. A. Hutchins, and J. E. Turner, *TDOWN—A Code to Generate Composition and Spatially Dependent Cross Sections*, GEAP-13740, General Electric Co., August

1971. (See also R. Protsik, E. Kujawski, and C. L. Cowan, *TDOWN-IV*, GEFR-00485, September 1979.)

13. K. Gregson, M. F. James, and D. S. Norton, *MLBW—A Multilevel Breit-Wigner Computer Programme*, UKAEA Report AEEW-M-517, March 1965.

14. D. I. Garber and R. R. Kinsey, *Neutron Cross Sections, Vol. II, Curves*, 3d ed., BNL 325, Brookhaven National Laboratory, January 1976.

15. C. E. Porter and R. G. Thomas, "Fluctuation of Nuclear Reaction Widths," *Phys. Rev.*, *104* (1956) 483.

16. P. Greebler, B. A. Hutchins, and J. R. Sueoka, *Calculation of Doppler Coefficient and Other Safety Parameters for a Large Fast Oxide Reactor*, GEAP-3646, General Electric Company, March 1961.

17. G. I. Bell, "A Simple Treatment of the Effective Resonance Absorption Cross Sections in Dense Lattices," *Nucl. Sci. and Eng.*, *5* (1958) 138.

CHAPTER 6

KINETICS, REACTIVITY EFFECTS, AND CONTROL REQUIREMENTS

6-1 INTRODUCTION

Values for reactivity effects are required both for transient safety analysis and for control requirements during normal operation. Reactivity effects of importance in fast reactor design and safety include (1) effects of dimensional changes in core geometry, (2) the Doppler effect, (3) effects of sodium density changes or loss of sodium, and (4) long-term reactivity loss from fuel burnup.

The reactor control system must compensate for these reactivities during normal operation and provide sufficient margin to handle off-normal situations.

We begin this chapter with a review of the reactor kinetics equations (Section 6-2). We then proceed to discuss adjoint flux and perturbation theory (Section 6-3) since these are needed for an understanding of reactivity effects. Kinetics parameters $\bar{\beta}$ and l, the effective delayed neutron fraction and the neutron lifetime respectively, are then discussed and the differences in these values between fast and thermal reactors are presented (Section 6-4). Sections 6-5 through 6-7 cover the first three categories of reactivity effects. Section 6-8 is addressed to reactivity worth distribution, and the final section discusses the control requirements for a fast reactor. A detailed discussion of the fourth category of reactivity, that associated with fuel burnup, will be delayed until Chapter 7, although sufficient information will be summarized in this chapter to define the control requirements.

6-2 REACTOR KINETICS

Reactor kinetics equations for both fast and thermal reactors are identical. However, point kinetics approximations can be used more effectively for fast reactors than for thermal reactors because fast reactors are more tightly coupled neutronically. Tighter coupling implies that the neutron flux is more nearly separable in space and time, which is a necessary condition for point kinetics approximations to be valid. Fast reactor safety codes to date have therefore generally employed point kinetics.* Inevitably, as commercial fast reactors reach the 1000 to 2000 MWe range, and particularly if they employ heterogeneous cores, problems involving space-time kinetics will arise. Such considerations, however, are beyond the scope of this text.

A. POINT KINETICS EQUATIONS

The point kinetics equations can be written as

$$\frac{dn}{dt} = \frac{\rho - \bar{\beta}}{\Lambda} n + \sum_{i=1}^{6} \lambda_i C_i \qquad (6\text{-}1)$$

$$\frac{dC_i}{dt} = \frac{\bar{\beta}_i}{\Lambda} n - \lambda_i C_i \qquad (i = 1 \text{ to } 6) \qquad (6\text{-}2)$$

where n = neutron density (neutrons/cm^3),
$\quad C_i$ = delayed neutron precursor concentration for the ith group (precursors/cm^3),
$\quad \rho = (k-1)/k = \delta k/k$ = reactivity,
$\quad \bar{\beta}$ = effective delayed neutron fraction,
$\quad \bar{\beta}_i$ = effective delayed neutron fraction for the ith group,
$\quad \Lambda$ = neutron generation time (s),
$\quad \lambda_i$ = decay constant for the ith delayed neutron group (s^{-1}).

The neutron generation time, Λ (which measures the neutron birth-to-birth time) is related to the neutron lifetime, l (which measures neutron birth-to-death) as follows:

$$l = k\Lambda. \qquad (6\text{-}3)$$

This relationship implies that l exceeds Λ when the power is rising ($k > 1$) and that Λ exceeds l when the power is dropping ($k < 1$). Because k is always very close to unity, even for large postulated accident conditions, both terms are loosely referred to as neutron lifetime.

---

*Though the point kinetics formulation is normally employed in fast reactor safety codes, the reactivity feedback terms due to reactor temperature variations and material motion generally contain spatial effects.

As in the case of thermal reactors, reactivity for fast reactors is often reported in units of the *dollar** which is obtained by dividing the absolute value of ρ by $\bar{\beta}$. For reactivities less than 1 dollar (i.e., $\rho < \bar{\beta}$), the first term on the right hand side of Eq. (6-1) is small relative to the second term and the reactor kinetics equations are controlled by delayed neutrons. This is as valid for fast reactors as it is for thermal reactors. Hence, normal startup and shutdown and power level changes proceed in a fast reactor exactly as in a thermal reactor.

It is only when the net reactivity approaches or exceeds $\bar{\beta}$ that the behavior of a fast reactor differs significantly from that of a thermal reactor. This difference arises because of the very much smaller prompt neutron lifetime for the FBR ($\sim 4 \times 10^{-7}$ s) relative to that of the LWR ($\sim 2 \times 10^{-5}$ s). Even for prompt critical conditions, however, the transient behavior of an FBR is smooth and predictable. This can be shown by observing the solution to Eqs. (6-1) and (6-2) for neutron density (using a one-delayed group precursor approximation):

$$\frac{n}{n_0} = \frac{\bar{\beta}}{\bar{\beta} - \rho} \exp\left(\frac{\lambda \rho}{\bar{\beta} - \rho} t\right) - \frac{\rho}{\bar{\beta} - \rho} \exp\left(-\frac{\bar{\beta} - \rho}{\Lambda} t\right) \qquad (6-4)$$

where λ represents a weighted decay constant for all six delayed neutron groups. For $\rho < \bar{\beta}$, the second term drops out rapidly, leaving the first term to express the long term transient response. Within this region, the reactor period, τ, is approximated by

$$\tau \cong \frac{\bar{\beta} - \rho}{\lambda \rho} \qquad \left(\text{for } \rho < \bar{\beta}\right).$$

For $\rho > \bar{\beta}$ the second term in Eq. (6-4) becomes positive in sign and large in magnitude (due to the very small value for Λ). The reactor period is then approximated by

$$\tau \cong \frac{\Lambda}{\rho - \bar{\beta}} \qquad \left(\text{for } \rho > \bar{\beta}\right).$$

A plot of reactor period vs reactivity is shown in Fig. 6-1. Note that for a reactivity of less than about $.90 the period is unaffected by neutron lifetime (i.e., fast and thermal reactors behave identically). Even for prompt critical conditions, behavior is smooth and reactor response can be accurately predicted—as graphically demonstrated in the SEFOR program,[1,2]

---

*Reactivity is sometimes reported in units of cents. One cent is 0.01 dollars.

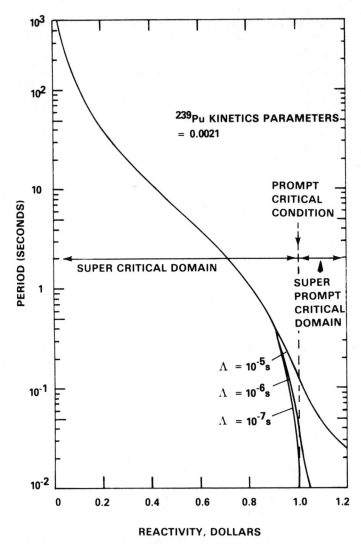

FIGURE 6-1. Plot of reactor period vs reactivity for ^{239}Pu fueled FBR.

in which the reactor was purposely made superprompt critical to demonstrate the capability of the Doppler effect to arrest a transient.

The absolute value for the reactivity corresponding to one dollar is lower for a fast reactor (using plutonium as the principal fissile material) relative to a thermal reactor (using principally enriched uranium) since β for ^{239}Pu is only 0.00215 compared to 0.0068 for ^{235}U. Delayed neutron data for fast fission in these and other relevant isotopes are listed in Table 6-1. Methods for calculating β and $\bar{\beta}$ are presented later in Section 6-4.

TABLE 6-1A   Delayed Neutron Parameters

Isotope	Delayed[a] Neutrons Per Fission	Total[b] Neutrons Per Fission, ν	β	Group	$\lambda_i^{[c]}$ (s^{-1})	$\beta_i/\beta^{[c]}$
^{232}Th	0.0531 ±0.0023	2.34	0.0227	1	0.0124	0.034
				2	0.0334	0.150
				3	0.121	0.155
				4	0.321	0.446
				5	1.12	0.172
				6	3.29	0.043
^{233}U	0.00731 ±0.00036	2.52	0.0029	1	0.0126	0.086
				2	0.0334	0.274
				3	0.131	0.227
				4	0.302	0.317
				5	1.27	0.073
				6	3.13	0.023
^{235}U	0.01673 ±0.00036	2.45	0.0068	1	0.0127	0.038
				2	0.0317	0.213
				3	0.115	0.188
				4	0.311	0.407
				5	1.40	0.128
				6	3.87	0.026
^{238}U	0.0439 ±0.0010	2.77	0.0158	1	0.0132	0.013
				2	0.0321	0.137
				3	0.139	0.162
				4	0.358	0.388
				5	1.41	0.225
				6	4.02	0.075

[a]   Ref. 3. Delayed neutrons per fission rather than β is the parameter measured. These values are for fast fission, meaning fissions in the few keV to 4 MeV range. This parameter is not very sensitive to the neutron energy causing the fission.

[b]   The average value of ν (to be used with delayed neutrons per fission to obtain β) varies slightly with the FBR neutron spectrum. The values here are consistent with the values in Appendix C and the neutron spectrum for which the Appendix C cross sections were obtained. Since β is the ratio of the first two columns, β will differ slightly for a neutron spectrum different from that assumed here.

[c]   Ref. 4. Uncertainties in λ_i and β_i/β are reported in that reference.

TABLE 6-1B Delayed Neutron Parameters

Isotope	Delayed[a] Neutrons Per Fission	Total[b] Neutrons Per Fission, ν	β	Group	$\lambda_i^{(c)}$ (s^{-1})	$\beta_i/\beta^{(c)}$
^{239}Pu	0.00630 $\pm$0.00016	2.93	0.00215	1	0.0129	0.038
				2	0.0311	0.280
				3	0.134	0.216
				4	0.331	0.328
				5	1.26	0.103
				6	3.21	0.035
^{240}Pu	0.0095 $\pm$0.0008	3.07	0.0031	1	0.0129	0.028
				2	0.0313	0.273
				3	0.135	0.192
				4	0.333	0.350
				5	1.36	0.128
				6	4.04	0.029
^{241}Pu	0.0152 $\pm$0.0011	2.95	0.00515	1	0.0128	0.010
				2	0.0299	0.229
				3	0.124	0.173
				4	0.352	0.390
				5	1.61	0.182
				6	3.47	0.016
^{242}Pu	0.0221 $\pm$0.0026	3.05	0.0072	1	0.0128	0.004
				2	0.0314	0.195
				3	0.128	0.161
				4	0.325	0.412
				5	1.35	0.218
				6	3.70	0.010

[a]   Ref. 3. Delayed neutrons per fission rather than β is the parameter measured. These values are for fast fission, meaning fissions in the few keV to 4 MeV range. This parameter is not very sensitive to the neutron energy causing the fission.

[b]   The average value of ν (to be used with delayed neutrons per fission to obtain β) varies slightly with the FBR neutron spectrum. The values here are consistent with the values in Appendix C and the neutron spectrum for which the Appendix C cross sections were obtained. Since β is the ratio of the first two columns, β will differ slightly for a neutron spectrum different from that assumed here.

[c]   Ref. 4. Uncertainties in λ_i and β_i/β are reported in that reference.

For many safety calculations, it is convenient to change the principal variable in the kinetics equations from neutron density, n, to reactor power density, $p(\text{W}/\text{cm}^3)$. This can be done by introducing the following definitions and approximations:

$$p = \frac{\Sigma_f n v}{3.0 \times 10^{10}},$$

$$y_i = \frac{\Sigma_f C_i v}{3.0 \times 10^{10}}.$$

With these variables, the kinetic equations become*

$$\frac{dp}{dt} = \frac{\rho - \bar{\beta}}{\Lambda} p + \sum_{i=1}^{6} \lambda_i y_i \tag{6-5}$$

$$\frac{dy_i}{dt} = \frac{\bar{\beta}_i}{\Lambda} p - \lambda_i y_i \quad (i = 1 \text{ to } 6). \tag{6-6}$$

B. PROMPT JUMP APPROXIMATION

A very convenient approximation that can be used to estimate the change in power density resulting from a change in reactivity (for reactivities less than about $0.9) is called the prompt jump approximation. For small values of reactivity ($\rho < \$0.9$), the rate of change of power is sufficiently slow that the derivative dp/dt is negligible compared to the two terms on the right hand side of Eq. (6-5). Hence, the power density can be approximated from Eq. (6-5) as

$$p = \frac{\Lambda \sum_{i=1}^{6} \lambda_i y_i}{\bar{\beta} - \rho}. \tag{6-7}$$

Prior to the reactivity change, the derivative dy_i/dt in Eq. (6-6) is zero and the power density is the initial or steady-state value, p_0. Hence, $\lambda_i y_i$ at the time of the reactivity change is $(\bar{\beta}_i/\Lambda)p_0$, from Eq. (6-6). Inserting this expression into Eq. (6-7) gives the value of p immediately after the reactivity

---

*This formulation ignores the difference in timing between prompt and delayed energy generation.

change, or

$$p \simeq \frac{p_0 \sum\limits_{i=1}^{6} \bar{\beta}_i}{\bar{\beta} - \rho} = \frac{p_0 \bar{\beta}}{\bar{\beta} - \rho} = \frac{p_0}{1 - \left(\dfrac{\rho}{\bar{\beta}} \right)} \, .$$

Hence,

$$p \simeq \frac{p_0}{1 - \rho(\$)}, \qquad \text{(for } \rho < \$0.9) \tag{6-8}$$

where the reactivity is in dollars. Equation (6-8) represents the almost instantaneous change (or "jump") in power density due to prompt neutrons when a reactivity change less than \$0.9 is made, yet the result does not contain the neutron generation time Λ. It is further noted that this approximate value for p/p_0 is simply the coefficient of the first term in Eq. (6-4), i.e., the term that remains after the rapidly decaying second term of Eq. (6-4) dies away. This is yet another way of recognizing that the transient behavior of an FBR is essentially independent of the prompt neutron lifetime for reactivities less than the prompt critical condition.

6-3 ADJOINT FLUX AND PERTURBATION THEORY

The adjoint flux, ϕ^*, is introduced in this chapter because of its importance to the calculation of reactivity effects. We shall not take the time here to derive the adjoint flux or the perturbation expression for reactivity; the theory is identical for fast and thermal reactors and appears in standard reactor theory textbooks. We will, however, discuss the physical meaning of the adjoint flux and will use the adjoint flux in multigroup perturbation expressions for reactivity effects.

A. ADJOINT FLUX

The adjoint flux is often called the importance function, a name that describes its physical meaning. The adjoint flux is proportional to the importance of the neutron in continuing the neutron chain reaction. A neutron of high importance has a high chance of causing a fission, and the neutrons resulting from the fission have a high chance of causing other fissions, and so on. Spatially, a neutron near the center of the core has a higher importance than one near the edge because the center neutron has less chance to leak from the core—hence, more chance to cause fission. In fact, we may recall that for a bare core the spatial shapes of the flux and the adjoint flux are identical, being highest at the center.

In a fast reactor, the *spatial* shapes of the flux and the adjoint flux in the most important energy ranges are generally quite similar, even though the core is surrounded by blankets.

Of particular interest in a fast reactor is the *energy* dependence of the adjoint flux, and here we find a large difference in behavior between flux and adjoint flux. We also find significant differences in behavior between fast and thermal reactors in this regard. In a fast reactor, the neutron importance tends to increase with energy above some energy in the low keV range. This is primarily because of the increase in η of the fissile isotopes with increasing energy—the same property that led to high breeding ratios for fast reactors fueled with plutonium. Above the fission threshold energy of the fertile materials, the importance also rises because of fission by these materials.

At low energies, below the few-keV range, the importance function begins to rise again. This rise is caused by the increasing fission cross section of the

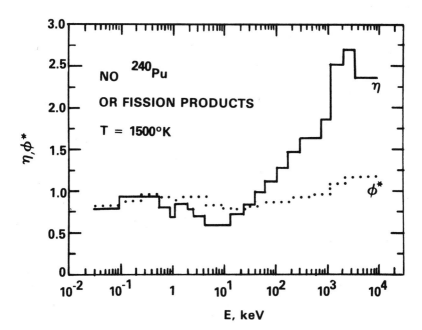

FIGURE 6-2. Energy dependence of adjoint flux and effective η in a large fast oxide reactor. (Adapted from Ref. 5.)

fissile material relative to other competing cross sections; we are referring here to resonance rather than thermal cross sections because neutrons do not survive to thermal energies in a fast reactor. For a large fast reactor with oxide fuel, the neutron energy of minimum importance is just above 10 keV.

The energy dependence of the adjoint flux is illustrated in Fig. 6-2 for a large oxide-fueled fast reactor. Also plotted in the figure is η versus neutron energy, which shows the rise above 10 keV. The η plotted is an effective η for the fissile and fertile materials. The sharp rise above 1 MeV reflects the effect of fertile material fission, while the rise at low energy reflects the rise in fissile material fission cross section relative to fertile capture cross section.

We will soon see that the energy dependence of the adjoint flux plays an important role in all fast reactor safety parameters, including Doppler, sodium, and fuel compaction reactivities and the effective delayed neutron fraction.

B. REACTIVITY FROM PERTURBATION THEORY

A reactivity effect based on multigroup first order perturbation theory is calculated as follows:

$$
\rho = \frac{\left[\int \sum_g \phi_g^* \chi_g \sum_{g'} \delta(\nu\Sigma_f)_{g'} \phi_{g'}\, dV + \int \sum_g \nabla\phi_g^* \,\delta D_g\, \nabla\phi_g\, dV \right.}{\int \sum_g \phi_g^* \chi_g \sum_{g'} (\nu\Sigma_f)_{g'} \phi_{g'}\, dV}
$$

$$
\left. - \int \sum_g \phi_g^* \,\delta\Sigma_{rg}\phi_g\, dV + \int \sum_g \phi_g^* \sum_{g'<g} \delta\Sigma_{g'\to g}\phi_{g'}\, dV \right]
$$

(6-9)

where the integrals are taken over the entire reactor. Here reaction rates are being weighted with adjoint fluxes to determine the effect on reactivity. The second and third terms in the numerator represent loss terms from leakage and from absorption, elastic scattering, and inelastic scattering; increases in these loss terms lead to negative reactivity. The first and fourth terms represent source terms from fission and scattering; increases in these terms provide positive reactivity. The denominator is a weighted fission source term for the unperturbed reactor.

6-4 EFFECTIVE DELAYED NEUTRON FRACTION AND NEUTRON LIFETIME

A. EFFECTIVE DELAYED NEUTRON FRACTION

The delayed neutron fraction, β, can be estimated from a zero-dimensional neutronics calculation by weighting the delayed neutron fraction of each fissionable isotope with the number of neutrons produced from fissions by that isotope, as follows:

$$\beta = \frac{\sum\limits_{m}\left(\beta_m \sum\limits_{g}\left(\nu\Sigma_f\right)_{mg}\phi_g\right)}{\sum\limits_{g}\left(\nu\Sigma_f\right)_g\phi_g}. \tag{6-10}$$

A more accurate space-dependent calculation must include the adjoint function in the weighting procedure and the energy spectrum of the delayed neutrons, as described below [Eq. (6-11)]. Values of β for fast fission for the principal fissionable isotopes that may be in fast reactors were given in Table 6-1.

A typical value for β for a fast oxide reactor can be estimated from values of β_m and ν_m in Table 6-1 by assuming a typical distribution of fissions between the fissionable isotopes. For the fission fractions in the core of the 1200 MWe homogeneous design with the neutron balance in Table 4-1A, an approximate value of β is 0.0046. For the 1000 MWe heterogeneous design of Table 4-1B, the approximate β is 0.0040.

An accurate calculation of β for use in the kinetic equations must account for both the spatial distribution of fissions and, more important, for the difference in energy spectra of delayed neutrons and prompt neutrons. The resulting expression for β is called the *effective beta*, $\bar{\beta}$:

$$\bar{\beta} = \frac{\sum\limits_{m}\sum\limits_{i=1}^{6}\beta_{im}\int\sum\limits_{g}\phi_g^*\chi_{dg}\sum\limits_{g'}\left(\nu\Sigma_f\right)_{mg'}\phi_{g'}\,dV}{\int\sum\limits_{g}\phi_g^*\chi_g\sum\limits_{g'}\left(\nu\Sigma_f\right)_g\phi_{g'}\,dV}. \tag{6-11}$$

The summation over m includes all fissionable materials m, for which β and $\nu\Sigma_f$ will differ. We assume that the spectrum of delayed neutrons, χ_d, from each material is identical. The summation over i refers to the six delayed-neutron groups. Note too that the weighted source term in the denominator is identical to the denominator of Eq. (6-9).

In a fast reactor, $\bar{\beta}$ is less than β by about 10%, a condition that differs from a thermal reactor. Delayed neutrons are born at energies lower than the energy of prompt neutrons. Since the adjoint flux in a fast reactor rises with energy (as in Fig. 6-2), a delayed neutron is worth less in a fast reactor than a prompt neutron, so that $\bar{\beta}/\beta < 1$. In a thermal reactor a delayed neutron is worth more than a prompt neutron because the delayed neutron has less chance to leak from the reactor, while there is little chance (relative to a fast reactor) that either a delayed or a prompt neutron will cause a fission at high energy; hence, $\bar{\beta}/\beta > 1$ for a thermal reactor.

For the estimates of β given above for the homogeneous and heterogeneous designs of Table 4-1A and Table 4-1B, the values of $\bar{\beta}$ are: $\bar{\beta} \simeq 0.0042$ for the homogeneous design and $\bar{\beta} \simeq 0.0036$ for the heterogeneous design. These values would represent the reactivity dollar for these designs. The values are fairly typical for FBR's although the homogeneous design value is slightly higher than those usually reported. The reason for the difference between the two designs is the larger fraction of ^{238}U fissions in the homogeneous design, which results from the considerably lower fissile fraction in that design.

B. NEUTRON LIFETIME

The neutron lifetime is given by the following relation for multigroup theory:

$$l = \frac{\int \sum_g \phi_g^* \frac{1}{v_g} \phi_g \, dV}{\int \sum_g \phi_g^* \chi_g \sum_{g'} (v\Sigma_f)_{g'} \phi_{g'} \, dV}. \tag{6-12}$$

A reasonable value for v_g for group g can be obtained by setting

$$\frac{1}{v_g} \int_{E_{lg}}^{E_{ug}} \phi(E) \, dE = \int_{E_{lg}}^{E_{ug}} \frac{1}{v(E)} \phi(E) \, dE, \tag{6-13}$$

where subscripts u and l refer to the upper and lower energy boundaries of the group. Letting $\phi(E) = C/E$, $v(E) = \sqrt{2E/m}$, and solving for v_g gives

$$v_g = \frac{v_{ug} v_{lg} \ln(E_{ug}/E_{lg})}{2(v_{ug} - v_{lg})}, \tag{6-14}$$

although at low energy, $\phi(E)$ decreases with energy faster than $1/E$.

A typical value for the neutron lifetime of an LMFBR is 4×10^{-7} seconds.

6-5 REACTIVITY FROM DIMENSIONAL CHANGES

Changes in reactor core dimensions occur during normal operation of an LMFBR and might occur during off-normal transients. Fuel will expand axially when its temperature is increased, as in normal startup from zero to full power. The structure that supports the core will expand radially if the inlet sodium temperature increases, thus causing a radial expansion of the core. If fuel melts in a severe accident, fuel might slump downward tending to compact the core, or it might be swept upward or forced outward causing an expansion of the core. For these reasons, it is important to understand the reactivity effects caused by changes in core dimensions.

One of the most important differences between a fast and a thermal power reactor is that fuel compaction in a fast reactor leads to a large positive reactivity effect, while in a thermal reactor it does not. A fast reactor is not designed to operate in an optimum configuration from a reactivity point of view; the fissile mass could always be reduced, for example, if a way were found to reduce the coolant volume fraction. A light water reactor, on the other hand, is designed to operate at an optimum moderator-to-fuel ratio.

The reason that compaction of a fast reactor leads to a positive reactivity can be illustrated by a simple argument, based on simplified but fundamental reactor theory concepts.* We represent the criticality factor by

$$k = k_\infty P_{NL} = k_\infty e^{-B^2 M^2}, \qquad (6\text{-}15)$$

where k is the product of the criticality factor for an infinite reactor, k_∞, and the fast neutron nonleakage probability, $e^{-B^2 \tau}$ (a form often used in the early days of thermal reactor analysis, except that τ is replaced by a migration area M^2).

Now, let us examine the principal factors that change significantly on compaction and identify those that do not. In a fast reactor, compaction has little effect on k_∞ because the coolant plays a minor role in both moderation and absorption. This situation is fundamentally different from a thermal, water-moderated reactor for which moderator-to-fuel ratio has a great influence on k_∞.

---

*This argument was introduced to one of the present authors in the early days of fast oxide reactor development by B. Wolfe of the General Electric Company.

Next, let us consider B^2. Suppose we have a spherical reactor of radius R. Then,

$$B^2 = \left(\frac{\pi}{R}\right)^2 \tag{6-16}$$

B^2 will increase with compaction through the reduction in R.

According to conventional reactor theory, the migration area varies with transport and total cross sections as

$$M^2 = \frac{1}{\Sigma_{\text{tr}}\Sigma_t} = \frac{1}{\displaystyle\sum_m N_m\sigma_{\text{tr}m}\sum_m N_m\sigma_{tm}}, \tag{6-17}$$

where Σ_{tr} and Σ_t are effective transport and total cross sections that provide the correct nonleakage probability, and the summations are over the materials present. Microscopic cross sections σ_{tr} and σ_t do not change much with compaction (since the change in energy spectrum with reduction in coolant volume is of secondary importance). Neglecting the effect of loss of coolant atoms (since their effect is small compared to the remaining fuel and structure atoms), we can consider that M^2 depends only on the atom densities that remain in the core after compaction, or

$$M^2 \propto \frac{1}{N^2}. \tag{6-18}$$

For a fixed number of fuel and structure atoms,

$$N \propto \frac{1}{\text{volume}} \propto \frac{1}{R^3}. \tag{6-19}$$

Combining Eqs. (6-18) and (6-19) gives

$$M^2 \propto R^6. \tag{6-20}$$

Substituting Eqs. (6-16) and (6-20) into the expression for k gives

$$k = k_\infty e^{-CR^6/R^2} = k_\infty e^{-CR^4}, \tag{6-21}$$

where k_∞ and C are both constants (essentially independent of radius).

When the core is compacted, R decreases, e^{-CR^4} (and, hence, P_{NL}) increases, and the criticality factor increases. The neutron leakage has been reduced by decreasing the radius (in what may at first appear to be a paradoxical situation), thereby causing an increase in reactivity.

This simple development can be carried one step further to provide insight into another important phenomenon in fast reactor safety—namely, axial fuel expansion. For a cylindrical core, Eq. (6-15) can be written as

$$k = k_\infty e^{-(B_z^2 + B_r^2)M^2} = k_\infty e^{-[(\pi/Z)^2 + (2.405/R)^2]M^2}. \qquad (6-22)$$

Now let us suppose that fuel expands axially as a result of increasing fuel temperature during the rise to full power or in a power transient. Radial fuel expansion may increase fuel pin diameters slightly but will have relatively little effect on radial expansion of the core. Bulk radial core expansion is governed primarily by the structure and, hence, the coolant temperatures, together with the influence of the radial restraint system. Hence, the primary core expansion from increased fuel temperature is in the axial direction. Again M^2 will be proportional to $1/N^2 \propto V^2$, and $V^2 \propto R^4 Z^2$, but now Z is the only dimension in V that changes during axial fuel expansion. Hence, the criticality factor can be presented in terms of the only parameter that is varying, Z, as

$$k = k_\infty P_{\mathrm{NL,axial}} P_{\mathrm{NL,radial}} = k_\infty e^{-C_1' R^4 Z^2 / Z^2} e^{-C_2' R^4 Z^2 / R^2}$$

$$= k_\infty e^{-C_1} e^{-C_2 Z^2}. \qquad (6-23)$$

This result indicates that the axial leakage does not change with axial fuel expansion, but that the radial leakage does increase (due to the increase in M^2 even while R and B_r^2 remain constant), thereby producing a negative reactivity effect.*

Obviously, the nuclear designer would use multigroup calculations to obtain design values for these compaction and expansion effects. Also, the change in axial leakage with fuel expansion is not absolutely zero. Nevertheless, this approximate analysis clearly emphasizes the most important phenomena that occur.

Fuel motion plays an important role in fast reactor safety. The most serious adverse consequence would be compaction of molten fuel in a postulated unprotected transient, either by gravity or external pressure forces. Fortunately, as discussed in the chapters on safety, experimental

---

*An interesting historical application of the effect of axial fuel expansion in fast reactors was provided by the SEFOR reactor. The reactor was constructed for the specific purpose of measuring the Doppler effect. Consequently, it was desirable to eliminate the axial fuel expansion effect entirely. This was successfully done by inserting a gap in each fuel pin at just the right axial position (about 2/3 of the way up the core) so that the natural increase in radial leakage with fuel expansion was exactly offset by closing the gap.

evidence in recent years appears to demonstrate that, should an unprotected transient occur (discussed in Chapter 15), fuel motion would be predominantly away from the middle of the core instead of the reverse.

A role of axial expansion of fuel in the normal solid fuel pin geometry is to provide a prompt negative reactivity feedback at the start of a power transient. This mechanism is the principal prompt negative feedback available in a metal-fueled fast reactor. (Doppler feedback is also available for the metal-fueled system, but it is relatively small due to the hard neutron spectrum.) For a ceramic-fueled reactor, lack of structural integrity (due to cracking) in high-burnup fuel renders the axial expansion mechanism somewhat unreliable so that more dependence is placed on the Doppler effect for a prompt negative feedback.

6-6 DOPPLER EFFECT

A. COMPUTATIONAL METHODS

As indicated at the end of the previous section, it is important to have a prompt negative reactivity feedback that reverses a power transient if the reactor becomes prompt critical. Mechanical action by control rods is too slow after prompt criticality is reached, whether one is talking about thermal or fast reactors. A prompt negative feedback is particularly important for fast reactors because two mechanisms, fuel compaction and sodium loss, have the potential to make the reactor superprompt critical.

The Doppler effect provides this prompt negative reactivity feedback for a fast reactor fueled with ceramic fuel. In a power excursion, excess fission energy quickly raises the fuel to a high temperature. As we observed in Chapter 5, the temperature rise in the principal fertile isotope (^{238}U or ^{232}Th) results in a relatively large increase in the effective parasitic capture cross section of this isotope. This provides a large negative reactivity effect. The existence of the Doppler effect in fast oxide-fueled reactors and our ability to predict its magnitude and its feedback effect on transients with reactivities above prompt critical was fully demonstrated by the SEFOR reactor experiment in the U.S. in the late 1960s.[1,2]

The feedback response is prompt only if the fissile and fertile materials are mixed intimately together. Most of the fission energy comes from slowing down of the fission products; hence, the fertile isotopes must be close enough to the fissile isotopes that the loss in kinetic energy by the fission products will promptly cause an increase in temperature and thermal motion of the fertile nuclei. This requires that the distance between the fertile and fissile nuclei be no greater than the order of the slowing-down distance, or range, of the fission products. This range is of the order of 10 μm. This intimate mixture of fertile and fissile materials is achieved by the fabrication processes generally used for making mixed oxide and carbide fuels.

As discussed in Chapter 5, the Doppler reactivity comes predominantly from the capture of low-energy neutrons. Energy self-shielding is required for an increase in the temperature of the fertile material to increase the effective cross section of that material. At high energy there is little self-shielding. Ceramic-fueled reactors, due to the presence of oxygen or carbon in the fuel, have a soft enough neutron spectrum to have a large Doppler effect. Reactors with hard spectra, particularly the early small metal test reactors, have small Doppler effects. Even large metal-fueled fast reactor designs have significantly smaller Doppler effects than ceramic-fueled reactors. This represents one of the important advantages of ceramic over metal fuel.

The Doppler reactivity effect can be calculated from perturbation theory, or it can be obtained from two successive criticality calculations with effective cross sections at different temperatures. A multigroup perturbation expression for the Doppler reactivity is

$$\rho = \frac{-\int \sum_g \phi_g^* \delta\Sigma_{ag}\phi_g \, dV + \int \sum_g \phi_g^* \chi_g \left[\sum_{g'} \delta(\nu\Sigma_f)_{g'}\phi_{g'} \right] dV}{\sum_g \phi_g^* \chi_g \left[\sum_{g'} (\nu\Sigma_f)_{g'}\phi_{g'} \right] dV}, \qquad (6\text{-}24)$$

where the $\delta\Sigma_a$ and $\delta(\nu\Sigma_f)$ terms reflect changes in the absorption and fission cross sections due to heating. The integrals are over the entire reactor volume. The first integral in the numerator accounts for the increase in the effective absorption cross sections of the fertile and fissile isotopes (including both capture and fission); this is a negative reactivity effect for an increase in fuel temperature (i.e., $\delta\Sigma_a$ is positive for this case). The second term accounts for the positive contribution of the increase in the effective fission cross section of the fissile isotopes with an increase in fuel temperature. Note that in the first term the increased absorption rate, $\delta\Sigma_{ag}\phi_g$, is weighted with the importance, ϕ_g^*, of the same energy group in which the absorption occurs because the neutron is lost from the reactor at that energy. On the other hand, the neutrons produced in the second integral are weighted with the importance of the energy group into which they are born.

The importance of the shape of the adjoint flux curve in Fig. 6-2 can be seen from Eq. (6-24). In a ceramic-fueled fast reactor, the main contribution to $\delta\Sigma_{cg}\phi_g$ for the principal fertile material occurs at low energy, i.e., in the

0.1–10 keV range. From Fig. 6-2, we can see that the adjoint flux levels out at about 10 keV and rises again below that energy.

Most of the Doppler effect comes from the fertile isotope. For the fissile isotopes, the effects of $\delta\Sigma_c$ and $\delta\Sigma_f$ almost cancel, an effect first observed experimentally and later explained by the application of resonance-overlap theory (Chapter 5-5C). Also, the minor fertile isotopes, such as ^{240}Pu, contribute little to the Doppler effect because of their small quantities (hence, small atom densities, N). Furthermore, as a result of their small quantities, their energy self-shielding is small and the effective $\delta\sigma_{cg}$ values are consequently small.

The criticality factor for a ceramic-fueled reactor as a function of absolute fuel temperature has the shape illustrated in Fig. 6-3. Whereas the absolute value of the variation in k about 1.0 is small, this difference can be large in terms of reactivity increments. The slope of the curve at any temperature is the *Doppler coefficient, dk/dT*, at that temperature. The coefficient is always negative, but dk/dT is not constant; it is a function of temperature, decreasing in magnitude with increasing T.

It has been found that, for oxide-fueled fast reactors, the temperature dependence of dk/dT is almost exactly

$$\frac{dk}{dT} = \frac{K_D}{T} \qquad (6\text{-}25)$$

where K_D is a constant. For this reason, the constant K_D has taken on the title *Doppler constant*, where

$$K_D = T\frac{dk}{dT}. \qquad (6\text{-}26)$$

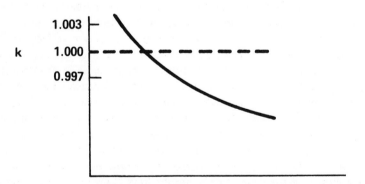

AVERAGE ABSOLUTE FUEL TEMPERATURE

FIGURE 6-3. Decrease of k with increasing fuel temperature.

This constant is the value that characterizes the Doppler effect for fast oxide reactors. (Frequently this constant is referred to in the literature as the Doppler coefficient, but we prefer to reserve the term "coefficient" for the derivative dk/dT.)

For a uniform change in fuel temperature from T_1 to T_2, the Doppler reactivity effect can be calculated from Eq. (6-25) as follows:

$$\rho = \int_{T_1}^{T_2} \frac{dk}{dT} dT = \int_{T_1}^{T_2} \left(\frac{K_D}{T} \right) dT = K_D \ln(T_2/T_1). \qquad (6\text{-}27)$$

In the reactor there is a fuel temperature distribution that must be taken into account in calculating the reactivity effect. This can enter Eq. (6-24) through the changes in cross sections $\delta\Sigma_{ag}$ and $\delta\Sigma_{fg}$, which are space-dependent since the temperature changes are space-dependent. Moreover, one must be careful to consider the type of power change one is calculating. One type of change is from one steady-state power level to another. This type of change must be analyzed to determine steady-state power coefficients or the required reactivity compensation by the control rods for a given change in power level. Another type of change would be a postulated accidental power excursion. Here the spatial change in fuel temperature for a given change in power level would differ from the spatial temperature variation between the same power levels at steady-state.

These temperature variations are often accounted for by the expression

$$\rho = \sum_i \frac{K_{Di}}{V_i} \int_{V_i} W(r,z) \ln \frac{T(r,z)}{T_0(r,z)} dV, \qquad (6\text{-}28)$$

where the summation is over regions i of the reactor (including the blankets), K_{Di} is the contribution to the Doppler constant for that region, and $T_0(r,z)$ is the initial temperature distribution. The weighting factor $W(r,z)$ is a normalized spatial variation of the Doppler effect, such that

$$\frac{1}{V_i} \int_{V_i} W(r,z) dV = 1. \qquad (6\text{-}29)$$

Studies of the effect of temperature variation within an actual fuel rod indicate that the use of an effective temperature equal to or slightly less than the mean gives results of adequate accuracy.[5]

While accounting for subtleties is a requirement for final design and final safety analysis, these differences are generally second-order effects. For steady-state power changes, calculational experience indicates that the magnitude of the actual reactivity change from a power level with a fuel temperature distribution $T_1(r)$ for which the average fuel temperature is $\overline{T}_1$

to a power with fuel at $T_2(r)$ with average fuel temperature $\overline{T}_2$ is about 10% higher than the calculated value for a uniform temperature change from $\overline{T}_1$ to $\overline{T}_2$, i.e.,

$$\rho\left[T_1(r) \rightarrow T_2(r)\right] \approx 1.1 K_D \ln\left(\overline{T}_2/\overline{T}_1\right). \tag{6-30}$$

The reason that dk/dT varies so closely with $1/T$ for fast oxide reactors is due more to convenient good luck than to some guiding law of nature. For a reactor with a much harder neutron spectrum, such as a metal-fueled fast reactor, the theory for the effective Doppler cross section indicates that dk/dT should vary as $1/T^{3/2}$. For a reactor with a much softer spectrum, such as a light water reactor, resonance theory indicates that dk/dT should vary as $1/T^{1/2}$. The spectrum of the fast oxide reactor lies in between, and by chance, $1/T$ appears to describe the variation remarkably well.

B. EFFECT OF COOLANT LOSS ON DOPPLER FEEDBACK

We have learned that the magnitude of the Doppler coefficient, or Doppler constant, is relatively large for a ceramic-fueled fast reactor because of the large neutron flux below 10 keV. This soft spectrum results from the combination of the moderating effect of sodium and the low-mass material (oxygen or carbon) in the fuel. In an unprotected accident of the type described in Chapter 15, however, the liquid sodium may have boiled and been expelled from all or part of the core, leaving behind only low-density sodium vapor. In this event the neutron spectrum will be hardened and the magnitude of the Doppler coefficient will be reduced. Figure 6-4 illustrates the cause of the reduced Doppler coefficient due to the loss of neutrons in the Doppler resonance region. This reduced Doppler coefficient must be used in safety analyses when sodium is lost from the core. Values of the Doppler constant for the CRBRP design will be given later in this section in order to indicate the influence of the loss of sodium on K_D.

One might also argue that the change in spectrum should alter the form of the temperature variation of dk/dT, since the exponent m in $1/T^m$ is spectrum-dependent. The spectrum change due to sodium loss is insufficient to change the variation significantly, however, and values of K_D (with the implied $1/T$ variation in dk/dT) are quoted for oxide fueled designs both with and without sodium.

In calculating the Doppler reactivity with sodium removed, it is important to account for the change in all effective cross sections due to the change in self-shielding caused by the absence of sodium scattering. The

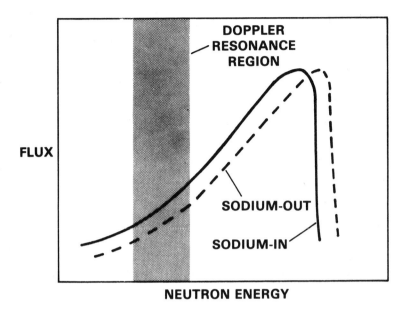

FIGURE 6-4. Flux reduction in the Doppler resonance region when sodium is lost from the core.

most important change, of course, is the modification of the Doppler-broadened cross sections for the fuel as a function of temperature.

C. CHANGES IN THE DOPPLER EFFECT DURING THE FUEL CYCLE

The Doppler coefficient changes during the fuel cycle. The magnitude and even the direction of the change, however, are difficult to estimate a priori because a number of subtle changes are occurring simultaneously. First, boron absorption is being removed with the withdrawal of the shim control rods. Second, fission products are being produced. Each of these effects modifies both ϕ and ϕ^* at low energies, where the Doppler is important, relative to ϕ and ϕ^* at high energies. Both types of absorber reduce the low-energy ϕ's and ϕ^*'s more than the high-energy ones, but there is competition between the two as to which has the dominant influence. The fuel plays two roles. The small decrease in ^{238}U concentration decreases the Doppler slightly through a decrease in $\delta\Sigma_c$, but more importantly the internal conversion ratio influences the extent to which boron absorber must be removed during the fuel cycle, and hence, changes the relative influence between boron and fission products on ϕ and ϕ^*. The internal conversion ratio is strongly influenced by reactor size (or rated power); because of this, the influence of burnup on the Doppler coefficient may vary between small and large fast reactors.

D. SAMPLE RESULTS

Typical values of the Doppler constant for the early CRBRP homogeneous design[6] are listed below, together with the influences of reactor size (i.e., power level), sodium loss, fuel cycle, and heterogeneous core design. All results in this section are for UO_2-PuO_2 fueled LMFBR's. Values of K_D for various reactors appear in Appendix A, and additional values appear in Table 6-3 in the section on sodium loss reactivity.

Effect of Reactor Size (or Power Level). Values can be compared for the early CRBRP homogeneous design[6] and a 1200 MWe homogeneous design[7]

$$K_D(\text{CRBRP, 350 MWe}) = -0.0062$$

$$K_D(\text{1200 MWe}) = -0.0086.$$

An increase in K_D with size is expected predominantly because the neutron spectrum for a large reactor is softer (due to decreased leakage).

Effect of Sodium Loss. At the beginning of the equilibrium fuel cycle, the Doppler constants with and without sodium present are

$$K_D \text{ (with Na)} = -0.0062$$

$$K_D \text{ (without Na)} = -0.0037.$$

Note that the reduction in Doppler due to sodium loss is quite large.

Effect of Fuel Cycle. The Doppler constants with sodium present at the beginning and the end of a cycle are

$$K_D \text{ (beginning of cycle)} = -0.0062$$

$$K_D \text{ (end of cycle)} = -0.0070.$$

We observe that, for the CRBRP-size design, the Doppler effect increases slightly during the cycle. This was also the case for the Doppler effect without sodium present.

Effect of Heterogeneous Core Design. Doppler constants are reported in Reference 7 for a 1200 MWe homogeneous core and for two heterogeneous core designs. The results are

$$K_D \text{ (homogeneous, 7.26 mm pin)} = -0.0086$$

$$K_D \text{ (heterogeneous, 7.26 mm pin)} = -0.0088$$

$$K_D \text{ (heterogeneous, 5.84 mm pin)} = \sim -0.008.$$

Thus, the heterogeneous design appears to have little influence on the Doppler constant. However, it should be noted that blankets respond more slowly than driver fuel; thus, the effective Doppler feedback is less for the heterogeneous design—especially for rapid transients.

Doppler Contribution by Region. The integrations in Eq. (6-24) are over the entire reactor. By separating the integrations in the numerator into regions, one can find the contributions to the Doppler constant from individual regions. A complete summary of these contributions for the CRBRP is given in Table 6-2.

TABLE 6-2 Doppler Constants by Region for the Early CRBRP Homogeneous Design[6]

| | Contribution to $-K_D$ | | | |
| | Beginning of Cycle | | End of Cycle | |
Region	With Na	Without Na	With Na	Without Na
Inner core	0.0034	0.0016	0.0037	0.0019
Outer core	0.0011	0.0006	0.0013	0.0008
Radial blanket	0.0011	0.0010	0.0012	0.0012
Axial blanket	0.0006	0.0005	0.0008	0.0005
Total	0.0062	0.0037	0.0070	0.0044

6-7 SODIUM LOSS REACTIVITY

The loss of sodium from a large fast reactor can result in a large positive reactivity effect. Sodium might be expelled from the core in the unlikely event of an unprotected transient, as described in Chapter 15, in which sodium boiling results from undercooling the reactor. This condition presents an important safety problem for an LMBFR, a problem not present in thermal reactors. This problem is also not present in gas-cooled fast reactors, which is one of the major advantages of gas cooling.

The sodium loss reactivity effect is exceedingly space-dependent. Sodium loss from the center of the core yields a highly positive reactivity effect; sodium loss from near the edge gives a negative effect. An understanding of such behavior is revealed by observing individually the four phenomena

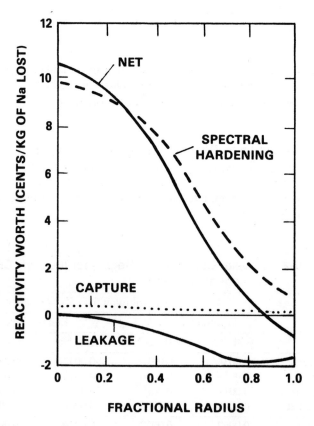

FIGURE 6-5. Components of sodium void coefficient for a small cermet-fueled LMFBR. (Adapted from Ref. 8.)

which contribute to the overall sodium void effect:

(1) spectral hardening
(2) increased leakage
(3) elimination of sodium capture
(4) change in self-shielding.

The first two effects are large and of opposite sign. The last two are small. Hence, most of the sodium loss reactivity results from the difference of two large numbers, as illustrated in Fig. 6-5, and this situation makes an accurate calculation of the net effect difficult. Each of these contributions to the sodium loss reactivity will be discussed later. Before doing so, however, we will discuss briefly the methods used to evaluate the sodium loss effect.

A. METHOD FOR CALCULATING SODIUM LOSS REACTIVITY

The method generally used to calculate sodium loss reactivity is to perform successive two-dimensional multigroup calculations (one with sodium present and the second with sodium removed from the zone of interest) and to compare the criticality factors. Care must be taken to use the proper self-shielded cross sections in the zone from which sodium is removed; all effective cross sections of the fuel and structure will change when the scattering effect of the sodium is removed. This technique does not yield the four individual contributions listed above; consequently, it is easy to lose sight of the fact that the result is still, in effect, the difference between two large reactivity effects. Large errors could accrue unless extraordinary care is taken.*

An alternate method for calculating the sodium loss effect is to use perturbation theory and calculate the four contributions separately. The calculation is more difficult, but it provides useful insight into the physical processes. It also provides a convenient data base for use in conducting transient calculations where only small localized perturbations occur during the time scale of interest. The appropriate perturbation expression for each contribution is given in the following discussion.

Spectral Hardening. The spectral hardening effect is positive. The loss of sodium from the core results in decreased moderation of the neutrons so the

---

*While the zero-dimensional calculation introduced in Chapter 4 can provide reasonable results for many fast reactor problems and is useful as an educational tool, it gives misleading results for a process as spatially dependent as sodium loss reactivity.

average neutron energy increases. This produces a positive reactivity effect because of the increase in neutron importance with increasing energy, as illustrated in Fig. 6-2.

The denominator of the perturbation theory expression for each of the four contributions is the same weighted source term that has appeared previously. Hence, it will be replaced in each expression by the integral S, where

$$S = \int \sum_g \phi_g^* \chi_g \sum_{g'} \left(\nu \Sigma_f \right)_{g'} \phi_{g'} \, dV. \tag{6-31}$$

The perturbation theory expression for the spectral hardening contribution can be written as

$$\rho = \frac{1}{S} \int \sum_g \phi_g^* \left[\left(\sum_{g'<g} \delta\Sigma_{g'\to g} \phi_{g'} \right) - \left(\delta\Sigma_{erg} + \delta\Sigma_{irg} \right) \phi_g \right] dV. \tag{6-32}$$

An alternate, but equivalent, expression also appears in the literature:

$$\rho = \frac{1}{S} \int \sum_g \sum_{g'>g} \left(\phi_{g'}^* - \phi_g^* \right) \delta\Sigma_{g\to g'} \phi_g \, dV. \tag{6-33}$$

Note that this contribution to the sodium loss effect is proportional to the product of ϕ and ϕ^*, both of which peak at the center of the core. Hence, this contribution would be highest per unit volume of sodium lost near the center of the core and lowest near the edge of the core.

Increased Leakage. Sodium loss results in increased leakage, which in turn yields a negative contribution to sodium loss reactivity. The strong spatial dependence of the sodium loss effect is due primarily to the different spatial behavior of the leakage component relative to that of the spectral hardening component. Sodium loss from a unit volume of core near the center of the core adds little to the leakage contribution because the flux gradient is low, and hence, leakage per cubic centimeter is low. The effect of sodium loss near the center of the core is highly positive, since the large spectral hardening effect is not counterbalanced by a large leakage contribution. The same sodium loss near the edge of the core, however, increases the leakage considerably and adds much to the negative leakage contribution to the sodium loss effect and little to the positive spectral hardening contribution.

The source of the spatial dependence of the leakage contribution can be deduced from the perturbation expression for this contribution:

$$\rho = \frac{1}{S} \int \sum_g \nabla \phi_g^* \, \delta D_g \, \nabla \phi_g \, dV. \tag{6-34}$$

Elimination of Sodium Capture. The elimination of sodium capture results in a small positive reactivity effect, given by

$$\rho = -\frac{1}{S} \int \sum_g \phi_g^* \, \delta \Sigma_{cg,\mathrm{Na}} \, \phi_g \, dV. \tag{6-35}$$

Change in Self-Shielding. The elimination of sodium scattering results in a small reactivity effect, given by the following expression, where the changes in the cross sections are those due only to the change in self-shielding:

$$\rho = \frac{1}{S} \int \sum_g \phi_g^* \left[\chi_g \sum_{g'} \delta(\nu \Sigma_f)_{g'} \phi_{g'} - \delta \Sigma_{ag} \phi_g \right] dV. \tag{6-36}$$

B. SODIUM LOSS REACTIVITY VALUES AND METHODS TO REDUCE THIS REACTIVITY

For a commercial-size LMFBR with UO_2-PuO_2 fuel and a homogeneous core, the positive reactivity from loss of sodium from the core is large. Values for a number of designs are presented in Table 6-3, which shows reactivities in the \$5 to \$7 range for commercial-size homogeneous cores. Comparison of the early CRBRP homogeneous design with the commercial-size plants shows the increase in sodium loss reactivity with size. The reason for this trend is that, as the core becomes larger, the (negative) leakage effect becomes less pronounced, while the (positive) spectral hardening effect is less affected.

TABLE 6-3 Selected Reported Values of Sodium Loss Reactivity

Reactor	Geometry	Core Fuel	Core Sodium Loss Reactivity $	Doppler Constant $\left(T\dfrac{dk}{dT}\right)$
CRBRP	homogeneous[6]	UO_2-PuO_2	3.3	−0.0062
(350 MWe)	heterogeneous[*]	UO_2-PuO_2	2.3	−0.008
1000 MWe[9]	heterogeneous	UO_2-PuO_2	2.9 (core only) 4.1 (core plus internal blankets)	
1000 MWe[10]	homogeneous	UO_2-PuO_2	6.9	
		ThO_2-PuO_2	2.3	
		$^{238}UO_2$-$^{233}UO_2$	1.5	
1200 MWe[7]	homogeneous	UO_2-PuO_2	5.0	−0.0086
	heterogeneous, tightly coupled	UO_2-PuO_2	1.5	−0.0088
	heterogeneous, tightly coupled, small pin[†]	UO_2-PuO_2	1.7	∼ −0.008
	heterogeneous, less tightly coupled	UO_2-PuO_2	1.5	—
	heterogeneous, modular island	UO_2-PuO_2	0.25	−0.0060
	homogeneous, pancake height/ diameter=0.1	UO_2-PuO_2	1.75	−0.004
	homogeneous, BeO moderated	UO_2-PuO_2	2.2	−0.0116

*See Appendix B, Table B-3.
†Small pin diameter is 5.84 mm. Pin diameter for other cases in Ref. 7 is 7.26 mm.

Numerous attempts have been made to reduce the sodium loss reactivity by design. Most design modifications have focused on increasing the leakage component of the sodium loss reactivity. The most promising appears to be a concept first proposed by the French,[11] generally referred to as the heterogeneous core, and described earlier in Chapters 2 and 4. In this method, annular rings of blanket assemblies are placed in the core (as in Fig. 2-2) to achieve a high neutron leakage rate from the core to the blankets.

The effect of the heterogeneous core design on sodium loss reactivity can be seen in Table 6-3. For sodium loss from the core assemblies only, the reactivity drops to the $2 range. In an accident involving loss of sodium, sodium expulsion from the blanket assemblies would likely lag the core assemblies if it occurred at all, so that the core reactivity would control the accident sequence. This is a significant safety advantage for the heterogeneous design. On the other hand, we saw earlier the disadvantage of a higher fissile inventory for this design (Table 4-5).

Two other methods to decrease the sodium loss reactivity by increasing leakage were explored early in LMFBR development; these were the "pancake core" and the "modular core." As seen by the example results for these concepts in Table 6-3, the sodium loss reactivity is indeed reduced, but the economic penalties for both concepts are greater than for the heterogeneous core.

Given a particular fuel type, the spectral component of the sodium loss reactivity is difficult to modify. One early suggestion to accomplish this was to deliberately soften the neutron spectrum by adding diluents such as BeO. The introduction of such a light element moderator reduces the dependency of the spectrum on the presence of sodium; when sodium is lost, the BeO keeps the spectrum soft. This softened neutron spectrum also enhances the magnitude of the Doppler coefficient, but economic penalties and a large drop in breeding ratio made this design unacceptable.

The sodium loss reactivities for a core utilizing thorium for the fertile material (even with plutonium as the fissile material) and for a core utilizing uranium-233 as the fissile material (even with uranium-238 as the fertile material) are both lower than for a $^{238}UO_2$-PuO_2 reactor. The lower value for the thorium fuel results from the lower fast fission fraction in thorium relative to uranium-238; hence, the spectral hardening component in the thorium reactor is smaller. The lower value for the uranium-233 fuel results from the lower rate of increase of η with energy for ^{233}U than for ^{239}Pu below 1 MeV, as can be observed from Fig. 1-3.

Finally, it is of interest to observe how the different designs used to modify the sodium loss reactivity affect the Doppler coefficient. Note from Table 6-3 that little difference exists in the Doppler constant, $T dk/dT$, between the homogeneous and the heterogeneous cores.

6-8 REACTIVITY WORTH DISTRIBUTIONS

In performing safety calculations, it is often of interest to know the reactivity implications of moving a particular reactor material such as fuel, steel, or sodium from its original location to another. A direct way to compute such an effect is to perform two eigenvalue calculations, one for the original core configuration and one for the new configuration. However, this computation is expensive and impractical to perform for many transient domains of interest.

As long as the amount of material which moves is relatively small, perturbation calculations can be performed to provide a good approximation of the expected reactivity change. Given the flux and adjoint flux distributions, the perturbation worth curves for any material of interest can be determined using Eq. (6-9). The unit of the perturbation worth calculations is usually given as ρ per kg of material (although ρ per unit volume of material is sometimes used). Figures 6-6 through 6-11 are included to illustrate typical results [shown in this case for the FFTF[(12)]].

Figures 6-6 and 6-7 show the flux and power distributions, respectively, for the central core assembly. They are included to provide a frame of reference. Figures 6-8 through 6-11 show respectively the fuel, sodium void, steel and Doppler weighting perturbation curves for the same central assembly. It should be noted that the FFTF is relatively small core (1000 litres) and has axial reflectors, rather than blankets. The slight axial skew in the flux and power is due to the beginning-of-life core configuration in which the control rods are banked about halfway into the core. Of particular interest is the sodium void worth, which shows the typical positive region in the central regions of the core, and the negative region near the core periphery where the leakage component dominates. The steel worth curve

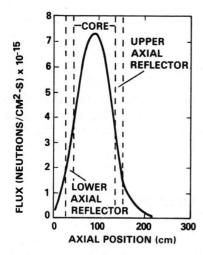

FIGURE 6-6. FTR total flux, central assembly.

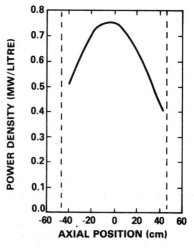

FIGURE 6-7. FTR power profiles, central assembly.

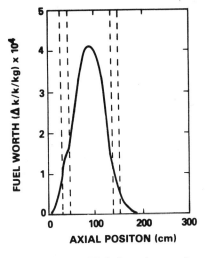

FIGURE 6-8. FTR fuel worth, central assembly.

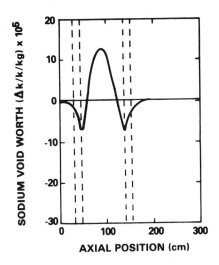

FIGURE 6-9. FTR sodium void worth, central assembly.

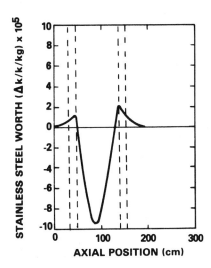

FIGURE 6-10. FTR stainless steel worth, central assembly.

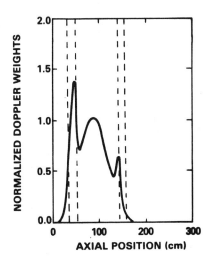

FIGURE 6-11. FTR Doppler weight, central assembly.

has a shape similar to that of sodium (note that Fig. 6-9 is for the sodium *void* worth), due to the same processes. The Doppler weighting distribution (Fig. 6-11) is determined by altering the cross section of ^{238}U in Eq. (6-24) to reflect a temperature increase. This curve indicates the increased effectiveness of Doppler feedback due to heating of the fuel near the axial reflectors, where the neutron spectrum is softer and more neutrons are in the Doppler resonance region. A curve of this type can be used in transient computer codes to assess the Doppler feedback due to non-uniform heating of the core.

6-9 REACTIVITY CONTROL REQUIREMENTS

Reactivity is controlled by control assemblies, usually containing rods of enriched B_4C, as described in Chapters 2 and 8. The reactivity control systems are part of the Plant Protective System, which is described in Chapter 14. For our purposes here, it is sufficient to know that two independent reactivity control systems—a primary and a secondary system —are used in fast reactors in order to assure that fuel design limits are not exceeded. Independent operation means that each system must operate in a manner such that the capability of either system to fulfill its control function is not dependent on the operation of the other system. The design of the two systems must be diverse as well as independent in order to reduce the potential for common mode failure. These concepts are discussed further in Chapter 14.

Although detailed design criteria for the two control systems continue to evolve, the criteria are well enough established to indicate the function of each system. The primary system must be able to shut down the reactor from any operating condition to the refueling temperature, with the most reactive control assembly inoperative (i.e., failed, or stuck out of the core); this requirement is commonly called the "stuck rod criterion." Any operating condition, in this context, means an *overpower condition** together with a *reactivity fault*. Such a fault is a positive reactivity introduced by some accident situation. The basis often used to quantify this reactivity fault is the maximum worth of a control assembly because the uncontrollable withdrawal of that assembly could be the fault that must be overridden. The primary system also serves to compensate for the reactivity effect of the fuel cycle, or fuel burnup. Moreover, reactivities associated with uncertainties in criticality and fissile inventory are accommodated by the primary control system.

---

*LMFBR's are generally designed for $\sim 115\%$ overpower operation (i.e., 15% above nominal full power)—to be discussed in Chapter 10-4.

The secondary system must be able to shut down the reactor from any operating condition to the *hot standby* condition, also with the most reactive assembly inoperative. In the hot standby condition, the reactor power is zero but the coolant is at the inlet temperature for full power operation. The fuel, consequently, is at the coolant temperature. The secondary system does not have to duplicate the primary system's ability to hold down the excess reactivity for the fuel cycle because this excess reactivity does not represent additional reactivity to be overridden in an accident. The reasoning here is that, although the secondary system must shut down the reactor without insertion of the primary control assemblies, it is not necessary to assume that the primary assemblies are removed from the core during an accident situation. The secondary system should, however, be able to override the uncontrollable withdrawal of one primary control assembly which is being used for burnup control. Hence, the reactivity fault is included in the secondary requirements. Since uncertainties in fuel loading are accommodated by the primary system, these reactivity uncertainties are not a part of the secondary system requirements.

Reactivity worth requirements for both the primary and secondary control systems are listed in Table 6-4 for a representative large (1000 MWe) LMFBR plant[10] and for the Clinch River Breeder Reactor Plant design.[6] Note especially the large difference between the fuel cycle reactivity requirements. This difference results from the difference in size and, hence, core leakage between the two plants—with the resulting larger internal breeding ratio for the larger plant.

TABLE 6-4 Reactivity Worth Requirements ($) (Equilibrium Cycle)

	Representative 1000 MWe LMFBR[10]		Early CRBRP Homogeneous Design[6]	
	Primary	Secondary	Primary	Secondary
1. Temperature defect:				
a. Full power to hot standby	1.6	1.6	2.4±0.8	2.4±0.8
b. Hot standby to refueling	0.8	—	0.8±0.3	—
2. Overpower	0.3	0.3	0.2±0.1	0.2±0.1
3. Reactivity fault	3	3	2.8±0.4	0.2±0.4
4. Fuel cycle excess reactivity	5	—	18.2	—
5. Allowance for refueling uncertainties	—	—	0.8±0.4	—
6. Uncertainties				
a. Criticality	+1	—	—	—
b. Fuel tolerances	+0.8	—	0.4	—
Subtotal	11	5	25.2	5.4
Propagated uncertainties	1.2	—	1.3	0.9
Maximum requirement	12	5	26.5	6.3

In Table 6-5, the control requirements are compared for these two reactors with the reactivity available in the control systems. The shutdown margins shown for the primary systems are considered adequate. Little margin is needed for the secondary system because uncertainties are already accounted for in the maximum requirements.

TABLE 6-5 Comparison of Control Requirements and Available Reactivity Worths (Representative 1000 MWe LMFBR and Early CRBRP Homogeneous Design)

	Representative 1000 MWe LMFBR[10]		CRBRP[6]	
	Primary	Secondary	Primary	Secondary
Number of control assemblies	13	6	15	4
Reactivity worth of control system ($)	20	8	31.8	8.4
Worth of 1 stuck rod ($)	3	2.5	2.8	2.0
Reactivity worth available ($)	17	5.5	29.0	6.4
Maximum requirements ($) (From Table 6-4; includes uncertainties)	12	5	26.5	6.3
Shutdown margin ($)	5	0.5	2.5	0.1

The *temperature defect* is the reactivity effect due to changes in coolant and fuel temperatures. This temperature defect appears in Table 6-4 as the reactivity gained in cooling the reactor from full power to the refueling temperature. The contributions to the temperature defect for the CRBRP design are listed in Table 6-6.

TABLE 6-6 Temperature Defect, Early CRBRP Homogeneous Design[6]

Contribution	Full Power to Hot Standby	Hot Standby (316°C) to Refueling (205°C)
Doppler Effect ($)	1.9	0.3
Axial Fuel Expansion ($)	0.15	0.05
Radial Core Expansion ($)	0.3	0.4
Coolant Density ($)	0.01	0.01
TOTAL ($)	2.4	0.8

PROBLEMS

6-1. For the reactor calculated in Problem 4-5, calculate $\bar{\beta}$, assuming $\bar{\beta}/\beta = 0.9$ and $\beta(^{239}\text{Pu}) = 0.00215$ and $\beta(^{238}\text{U}) = 0.0158$.

6-2. Conduct the following calculations based on the results of Problem 4-5.

   (a) Using the Doppler cross sections in the 4-group cross section set, calculate the Doppler constant, K_D, assuming that all of the Doppler effect is due to ^{238}U.

   (b) Calculate the reactivity effect of the withdrawal of a single average B_4C control assembly.

6-3. For a particular LMFBR design, the reactivity due to the Doppler effect for a change in average fuel temperature from 1400 K to 700 K is $2.5. The effective delayed neutron fraction, $\bar{\beta}$, is 0.0034. Neglecting any effect of spatial temperature variation, calculate the Doppler constant, K_D.

6-4. For a 1600 MWe LMFBR, suppose that the number of primary system control assemblies is 16. The reactivity worth of the primary system is $16; the worth of the most reactive assembly is $1.5. The temperature defect from full power to refueling temperature is $2, of which $1.2 covers full power to hot standby and $0.8 is from hot standby to refueling. Overpower and reactivity fault requirements add up to another $2. The control requirement for the fuel cycle is $3. Propagated uncertainties total $2.

   (a) Show whether the primary system has enough control assemblies to meet the requirements, as defined in Section 6-9.

   (b) The secondary system has 8 assemblies, worth a total of $6.7. The most reactive assembly is worth $1.3. Show whether another secondary assembly should be added.

REFERENCES

1. W. Häfele, K. Ott, L. Caldarola, W. Schikarski, K. P. Cohen, B. Wolfe, P. Greebler, and A. B. Reynolds, "Static and Dynamic Measurements on the Doppler Effect in an Experimental Fast Reactor," *Proc. of the Third Int. Conf. on the Peaceful Uses of Atomic Energy*, Geneva, 1964.

2. L. D. Noble, G. Kassmaul, and S. L. Derby, "SEFOR Core I Transients," GEAP-13837, General Electric Co., August 1972, and "Experimental Program Results in SEFOR Core II," GEAP-13833, June 1972.

3. R. J. Tuttle, "Delayed Neutron Yields in Nuclear Fission," *Proc. of the Consultants' Meeting on Delayed Neutron Properties*, IAEA, Vienna, March 1979.

4. R. J. Tuttle, "Delayed-Neutron Data for Reactor-Physics Analysis," *Nucl. Sci. and Eng., 56* (1975) 70.

5. H. H. Hummel and D. Okrent, *Reactivity Coefficients in Large Fast Power Reactors*, American Nuclear Society, LaGrange Park, IL (1970) 87–88, 148.

6. *Preliminary Safety Analysis Report*, Clinch River Breeder Reactor Plant, Project Management Corp. (1974) 43–61.

7. H. S. Bailey and Y. S. Lu, "Nuclear Performance of Liquid-Metal Fast Breeder Reactors Designed to Preclude Energetic Hypothetical Core Disruptive Accidents," *Nucl. Tech., 44* (1979) 81.

8. D. Okrent, "Neutron Physics Considerations in Large Fast Breeder Reactors," *Power Reactor Tech., 7* (1964) 107.

9. E. Paxson, ed., *Radial Parfait Core Design Study*, WARD-353, Westinghouse Electric Corp., Madison, PA, June 1977, 79.

10. B. Talwar, *Preconceptual Design Study of Proliferation Resistant Homogeneous Oxide LMFBR Cores*, GEFR-00392 (DR)-Rev. 1, General Electric Co., Sunnyvale, CA, September 1978, 5.14.

11. J. C. Mougniot, J. Y. Barre, P. Clauzon, C. Giacomette, G. Neviere, J. Ravier, and B. Sicard, "Gains de Regeneration des Reacteurs Rapides á Combustible Oxyde et á Réfrigerant Sodium," *Proc. European Nuclear Conf., 4* (April 1975) 133.

12. J. V. Nelson, R. W. Hardie, and L. D. O'Dell, *Three Dimensional Neutronics Calculations of FTR Safety Parameters*, HEDL-TME 74-52, Hanford Engineering Development Laboratory, Richland, WA, August 1974.

CHAPTER 7

FUEL MANAGEMENT

7-1   INTRODUCTION

Fuel management deals with the irradiation and processing of fuel. An analysis of the fuel cycle is necessary to estimate fuel costs and to define operational requirements such as initial fuel compositions, how often to refuel, changes in power densities during operation, and reactivity control.

Fuel costs represent one contribution to the total power costs, as discussed in Chapter 3. Unlike the LWR, fuel costs for an LMFBR are insensitive to U_3O_8 price. Hence, this contribution to total power cost is predicted to be lower for a fast breeder reactor than for a thermal reactor as the price of U_3O_8 rises. Since more fissile material is produced in a breeder than is consumed, the basic (feed) fuel for the LMFBR is depleted uranium which is available for centuries without further mining of uranium ore.

In this chapter we describe methods to calculate isotopic compositions in fuel being loaded into the reactor and fuel discharged from the reactor. These figures are needed in order to calculate fuel costs. Methods will also be presented for calculating breeding ratios and doubling times. These parameters are not required for fuel cost analysis, but they provide convenient figures of merit for comparison of various fast breeder concepts and designs.

7-2   FUEL MANAGEMENT CONCEPTS

There are a number of choices regarding the loading and removal of fuel from a reactor. Consider first the concept of the *equilibrium cycle* versus the *first-core cycle*. Initially, the entire reactor is loaded with unirradiated fuel, also called *fresh*, or *new fuel*, in preparation for the first-core cycle. After an initial period of operation, some irradiated fuel is removed and replaced with more fresh fuel.

After several years, a transition is made to an equilibrium fuel cycle. The equilibrium cycle is said to be reached when, after each refueling, the composition of materials remaining in the core after spent fuel has been discharged and fresh fuel reloaded is the same as the composition of materials at the start of the previous irradiation cycle. The details of transition to the equilibrium cycle are complex and will not be considered here. Instead we will only be concerned with analysis of the equilibrium cycle itself.

When fuel is irradiated, a net decrease in heavy atoms occurs due to fission and the fuel is said to be *burned*. The term *burnup* is used as the measure of either the energy obtained from the burned fuel or the fraction of fuel that has fissioned (i.e., burned). Units of burnup are described in Section 7-3.

We will use the term *fuel cycle* in a specific way in the remainder of this chapter; the term will refer to the irradiation process that begins at the time of reactor startup after refueling and ends when the reactor is shut down for subsequent refueling. This use of fuel cycle is in contrast to the broader use of the term to describe all the processes involving fuel, from mining to fabrication, irradiation, reprocessing, and storage as described in Chapter 3. The duration of the fuel cycle, i.e., the time between refuelings, is called the *refueling interval*.

Fuel can be loaded into nuclear reactors either *continuously* or *batchwise*. In batch loading, some fraction of the irradiated fuel is replaced by a fresh batch of fuel at periodic shutdowns for refueling. Fast breeder reactors, like light water reactors, always utilize batch loading. If one-third of the fuel is replaced, there would be three batches of fuel in the reactor, each with its separate irradiation history. Generally the refueling interval for proposed commercial fast breeder reactors is one year, primarily because that is convenient for utilities; however, shorter refueling intervals have generally been used in the prototype plants. A typical residence time for fuel in the regions of highest power in an LMFBR is two years. The combination of a one-year refueling interval and a two-year residence time corresponds to a two-batch loading.

Another fuel management choice involves *scatter loading* versus *zone loading*. In scatter loading, fuel assemblies in a batch are loaded at various positions throughout the core. In zone loading, irradiated fuel is removed from one zone of the core only, while all fresh fuel is loaded in a different zone. Before loading the fresh fuel, it is necessary to *shuffle* partially irradiated fuel from the space into which the fresh fuel is to be placed to the positions from which the fully irradiated fuel was removed. Light water reactors use a combination of these two loading schemes. For LMFBR's it is generally planned to use scatter loading. This has the disadvantage that highly irradiated fuel with lower fissile inventories and power densities will be located adjacent to fresh fuel with higher power densities, but the time-consuming step of shuffling fuel will not be required.

The fresh fuel loaded into the core can come either from thermal reactors or from other fast breeder reactors. In the early stages of a mixed fast/thermal reactor economy the fissile plutonium (or ^{233}U) produced in a thermal reactor can be more economically utilized as fuel for a fast reactor than as recycled fuel for a thermal reactor. As we will determine later from Section 7-8, it would take about 12 to 15 light water reactors operating for one year to generate enough plutonium to start up a fast breeder reactor of comparable electrical rating. Once a breeder reactor economy is established, the breeders develop enough fissile materials both to feed themselves and to start up new breeder reactors, at a rate depending on doubling time (or, of course, demand for new electrical generating capacity). Eventually these breeders may produce more fuel than is needed for new fast reactor power plants, in which case the excess fissile material bred can be used in thermal reactors. Thermal reactors will presumably always be less expensive to build; hence, it may be economical to continue to build and operate thermal converter reactors to run on fuel from breeders long after the mining of uranium ore becomes too expensive.

One can imagine an equilibrium fuel cycle in which the feed from a breeder reactor comes from another breeder just like it, and each cycle is exactly like the previous one. Computer codes are available to calculate such a cycle; iteration is required to find a feed isotope composition that is consistent with the discharge composition. The problem is complicated by having the choice of either recycling fuel from the core only and using fuel from the blanket to start up new reactors, or mixing all the fuel discharged from both core and blankets before using it for recycle and new startups. As one considers these and other options, the possibilities multiply and the computer codes and logic correspondingly become very complex.

Examples of fast reactor fuel cycle codes used widely in the United States are FUMBLE,[1] REBUS,[2] 2DB[3] and 3DB.[4]

In this book we will restrict ourselves to a single type of relatively simple fuel cycle calculation. We will suppose that we know the isotopic composition of the plutonium feed. We will then develop a technique for calculating the plutonium fraction required to allow the reactor to operate for a refueling interval and will calculate the composition of fuel discharged from the reactor. We will *not* calculate the buildup of plutonium in the blanket, although the technique to do this if we were using a space-dependent calculation should be apparent. The proposed calculation will demonstrate most of the basic logic in fuel cycle analysis and will illustrate the approximate discharge compositions from the core. Yet it will be simple enough for the student to program in a reasonable length of time. (See Problem 7-5 for a way to approximate the buildup of plutonium in the blankets.) Our objective will then have been accomplished—to illustrate basic fuel management logic and concepts. To consider further details of any of the numerous actual codes in use in the industry would be too tedious for our purposes here.

Other aspects of the fuel cycle that must be considered in more detailed analyses involve disposition of fuel outside the reactor. After fuel is removed from the core and blankets, it is generally allowed to remain in the reactor vessel to allow the short-lived fission products to decay. It may then be placed in an ex-vessel storage pool for further decay before being sent to a reprocessing plant. The time required for storage, shipping, reprocessing, fabrication, and return to the reactor all affect fuel cost analysis and doubling time. Moreover, there will inevitably be a small wastage rate of fissile material in reprocessing and fabrication, which also influences doubling time and makes the accountability of fissile inventory more difficult.

7-3 FUEL BURNUP

The two units most widely used for reporting burnup of fuel, MWd/kg (or MWd/tonne) and atom percent, are measures of fundamentally different quantities. MWd/kg measures the energy obtained from irradiating fuel; it is most frequently used, at least in the United States, by design engineers performing fuel cycle and fuel cost analyses. Atom percent calculations measure the fraction of heavy atoms in the fuel that undergo fission. This unit is more useful to a research engineer investigating fuel damage or behavior as a function of irradiation although it is widely used in Europe for fuel cycle design calculations. We will define both units quantitatively and will show how the two units can be related to one another. We will, however, use the specific energy unit (MWd/kg) in the rest of the chapter.

When burnup is reported in MWd/kg, the fuel mass in the denominator includes the heavy atoms only. It does not include the oxygen in oxide fuel; hence, the oxide fuel mass must be multiplied by the ratio of the fuel atomic weight to the oxide molecular weight, i.e., 238/270 for UO_2-PuO_2, in order to obtain the heavy atom mass for use in calculating burnup.

In calculating burnup, one must be careful to distinguish between *chronological time* and *time at rated power*. Time at rated power is the product of a load factor (or capacity factor), f, and chronological time. The refueling interval is given in chronological time. Suppose the thermal power of an LMFBR core containing 30000 kg of oxide fuel is 3000 MWth, the load factor is 0.7, and the refueling interval t_c is 1 year. The average burnup, B, during a fuel cycle is

$$B = \frac{P(\text{MW}) ft_c(\text{d})}{m(\text{kg oxide}) \dfrac{\text{mass metal}}{\text{mass oxide}}}$$

$$= \frac{(3000 \text{ MW})(0.7)(1 \text{ yr})(365 \text{ d/y})}{(30\,000 \text{ kg oxide})(238 \text{ kg metal}/270 \text{ kg oxide})}$$

$$= 29 \text{ MWd/kg}. \tag{7-1}$$

Burnup in MWd/kg can be related to neutron flux, fission rate, and chronological time, t, as follows:

$$B(t) =$$

$$\frac{\Sigma_f \phi \left(\dfrac{\text{fissions}}{\text{cm}^3 \text{ core} \cdot \text{s}} \right) ft(\text{d}) \cdot 10^3 (\text{g/kg})}{2.9 \times 10^{16} \left(\dfrac{\text{fissions}}{\text{MW} \cdot \text{s}} \right) F_f \left(\dfrac{\text{cm}^3 \text{ oxide}}{\text{cm}^3 \text{ core}} \right) \rho_{\text{oxide}} \left(\dfrac{\text{g oxide}}{\text{cm}^3 \text{ oxide}} \right) \dfrac{238}{270} \left(\dfrac{\text{g metal}}{\text{g oxide}} \right)},$$

$$\tag{7-2}$$

where F_f is the fuel volume fraction and ρ_{oxide} is the fuel smear density. This relation assumes a constant $\Sigma_f \phi$ with time. In reality, $\Sigma_f \phi$ is both strongly space-dependent and slightly time-dependent. Even the average core value of $\Sigma_f \phi$ [and P in Eq. (7-1)] is slightly time-dependent due to the increase in power in the blankets during a fuel cycle.

Burnup in atom percent as a function of chronological time is

$$B(\text{atom \%}) = 100 \frac{\sum_m N_m \sigma_{fm} \phi ft}{\sum_m N_{m,0}}, \tag{7-3}$$

where sums are over all heavy atoms and $N_{m,0}$ are the atom densities at $t = 0$.

The two burnup units are related to one another; the ratio of the specific energy value to the percent fission value is

$$\frac{B\left(\dfrac{MWd}{kg}\right)}{B(\text{atom }\%)} = \frac{B\left(\dfrac{MWd}{kg\ \text{heavy metal}}\right)}{B\left(\dfrac{\%\ \text{fissions}}{\text{atoms heavy metal}}\right)}$$

$$= \frac{6.023\times10^{26}\left(\dfrac{\text{atoms heavy metal}}{\text{kg-mol heavy metal}}\right)}{\left[100\left(\dfrac{\%}{\text{absolute}}\right)\cdot2.9\times10^{16}\left(\dfrac{\text{fissions}}{\text{MW}\cdot\text{s}}\right)\cdot0.864\times10^{5}\left(\dfrac{s}{d}\right)\cdot238\left(\dfrac{\text{kg heavy metal}}{\text{kg-mol heavy metal}}\right)\right]}$$

$$= 10. \tag{7-4}$$

Fuel exposure in the reactor is usually limited by fuel damage or by cladding or structural damage or swelling, not by reactivity control. Design exposure goals for LMFBR's with oxide fuel are of the order of 100 MWd/kg, although these may be reduced in early LMFBR's due to limitations on the cladding or hexagonal ducts.

The time that fuel must remain in the core in order to reach the design burnup depends on the specific power (MWth/kg metal), which in turn

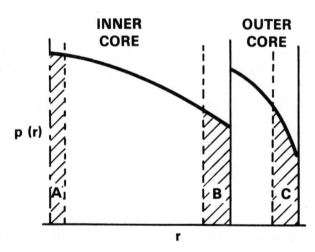

FIGURE 7-1. Radial power distribution and location of assemblies A, B and C.

depends on the spatial distribution of the power density. In order to consider the chronological time that fuel can remain in the core, consider the radial power density distribution illustrated in Fig. 7-1 and the following numerical example.

Suppose that the radial peak-to-average power ratios for assemblies A, B, and C are 1.3, 0.8, and 0.65. Suppose that the axial peak-to-average ratio is 1.25. Assume typical values for specific inventories in the inner and outer cores of 1.2 and 1.6 kg fissile/MWth, respectively, and for fissile atom fractions of 0.11 and 0.15 for the inner and outer cores. Finally, assume a load factor of 0.7. The peak burnup for assembly A after a chronological time in the core of 2 years is

$$\left(\frac{1}{1.2}\frac{\text{MWth}}{\text{kg fissile}}\right)\left(0.11\frac{\text{kg fissile}}{\text{kg metal}}\right)(2 \text{ yr})(365 \text{ d/yr})(0.7)(1.3)(1.25)$$

$$= 76 \text{ MWd/kg}.$$

Average and peak exposures for assemblies A, B, and C as a function of residence time are shown in Table 7-1.

TABLE 7-1 Typical Exposures for Various Residence Times at Core Positions A, B, and C in Figure 7-1

	Burnup (MWd/kg)					
	Average			Peak		
Residence Time	A	B	C	A	B	C
2 years	61			76		
3 years	91	56	47	114	70	58
4 years		75	62		94	78

For an allowable peak burnup of 80 MWd/kg, most of the fuel would stay in the core only 2 or 3 years, with the outermost assemblies remaining for 4 years. Residence time for radial blanket assemblies would be longer, varying perhaps from 4 to 6 years.

7-4 BURNUP EQUATIONS

The central calculation in a fuel management code is the depletion and production of isotopes as a function of time. The differential equations governing the burnup of fuel are called the *burnup equations*. These equations and their solutions are described in this section.

Either an analytical solution or a numerical integration procedure can be used to calculate atom densities from the burnup equations. An analytical solution assumes constant neutron fluxes and cross sections during the period of the calculation. The numerical integration technique assumes the same thing during a time step. The constant-flux cross section approximation applies for longer periods of time in a fast reactor than in a thermal reactor because the cross sections are not as sensitive to energy changes and fission product buildup. In the iterative calculation illustrated later in Fig. 7-4, the flux and cross sections are constant throughout the fuel cycle.

In this section the burnup equations will first be presented. The two methods of solution (numerical and analytical) will then be described.

The two potential fertile-to-fissile conversion chains for the U-Pu cycle and the Th-U cycle were given in Chapter 1. We will assign material numbers to each of the isotopes in the chain according to Table 7-2.

TABLE 7-2 Material Numbers for Conversion Chains

Material Number	U-Pu Cycle	Th-U Cycle
1	^{238}U	^{232}Th
2	^{239}Pu	^{233}U
3	^{240}Pu	^{234}U
4	^{241}Pu	^{235}U
5	^{242}Pu	^{236}U
6	^{243}Am	^{237}Np
7	^{241}Am	—
8	f.p. pairs	f.p. pairs

In commercial codes all isotopes are accounted for, such as ^{239}U, ^{239}Np, ^{237}Np, ^{238}Pu, ^{235}U, ^{236}U, and the isotopes of Am and Cm. The more complete conversion chains are shown in Figs. 7-2 and 7-3. Inclusion of ^{239}U and ^{239}Np makes little difference since their half-lives are so short (23.5 min and 2.35 days, respectively). Inclusion of ^{233}Pa in the Th-U cycle is more important because of its relatively long half-life (27.4 days). In the derivations that follow, we will use the U-Pu cycle as typical because of its more widespread interest.

Plutonium-238 is frequently included in fuel cycle results (e.g., see Table 7-5 near the end of the chapter). Two sources of ^{238}Pu are (1) neutron capture by ^{237}Np followed by beta decay of ^{238}Np, and (2) alpha decay of ^{242}Cm which comes from beta decay of ^{242}Am, which in turn results from neutron capture by ^{241}Am. There are also two principal sources of ^{237}Np: (1) $(n,2n)$ reaction by ^{238}U with subsequent beta decay of ^{237}U, and (2) (n,γ) reactions in ^{235}U and ^{236}U, followed by beta decay of ^{237}U.

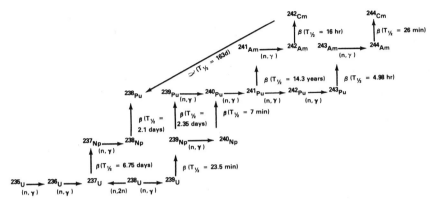

FIGURE 7-2. ^{238}U-^{239}Pu conversion chain.

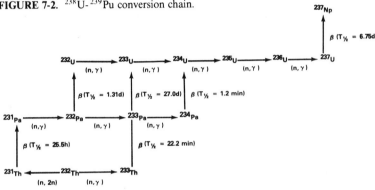

FIGURE 7-3. ^{232}Th-^{233}U conversion chain.

A. THE DIFFERENTIAL EQUATIONS

The ^{238}U (and ^{232}Th) concentration (or mass) is obtained from the equation:

$$\frac{dN_1}{dt} = -N_1 \sigma_{a1} \phi f \tag{7-5}$$

where

N_1 = concentration or mass of ^{238}U in any convenient unit such as atoms per barn·cm or per cm^3, or kg

t = chronological time (s)

σ_{a1} = ^{238}U absorption cross section (cm^2)

f = load factor (fraction of time at full power)

ϕ = one-group absolute value of neutron flux (neutrons/cm^2·s).

It is noted that σ must be in cm^2 here because it will always be multiplied by ϕ; hence, cross sections in barns must be multiplied by 10^{-24} cm^2/barn for use in the burnup equations.

The variable $f\phi t$ (or $fnvt$) appears repeatedly in the solution of the burnup equations; this variable is called the *fluence* and will be denoted by z:

$$z = f\phi t. \tag{7-6}$$

Equation (7-5) can be written in terms of fluence as

$$\frac{dN_1}{dz} = -N_1\sigma_{a1}. \tag{7-7}$$

The ^{239}Pu (and ^{233}U) concentration is obtained from the equation:

$$\frac{dN_2}{dz} = -N_2\sigma_{a2} + N_1\sigma_{c1}. \tag{7-8}$$

Note that only the capture rate of fertile material 1 is included, whereas the absorption rate, $N_1\sigma_{a1}\phi$, was used in Eq. (7-5).

The equations for ^{240}Pu, ^{242}Pu, and ^{243}Am are similar to the equation for ^{239}Pu, so that

$$\frac{dN_m}{dz} = -N_m\sigma_{am} + N_{m-1}\sigma_{c,m-1}\ (m=2,3,5,6). \tag{7-9}$$

Depletion of ^{243}Am will be ignored in the model described here, so that $\sigma_{a6} = 0$.

The equation for ^{241}Pu must include the β decay of ^{241}Pu, for which the half-life is 14.7 years. The equation is written first in terms of chronological time, t, and then in terms of fluence, z.

$$\frac{dN_4}{dt} = -N_4\sigma_{a4}\phi f - \lambda_4 N_4 + N_3\sigma_{c3}\phi f. \tag{7-10}$$

$$\frac{dN_4}{dz} = -N_4\left(\sigma_{a4} + \frac{\lambda_4}{\phi f}\right) + N_3\sigma_{c3}. \tag{7-10a}$$

For the model described here, we will ignore all depletion of ^{241}Am. Hence, the equation for ^{241}Am is

$$\frac{dN_7}{dz} = \frac{\lambda_4}{\phi f}N_4, \tag{7-11}$$

A fission-product pair is produced every time a fission occurs. Therefore, the fission product concentration can be written easily if destruction of fission products by absorption is ignored;

$$\frac{dN_8}{dz} = \sum_{m=1}^{5} N_m \sigma_{fm} \tag{7-12}$$

where only materials 1 through 5 are considered fissionable in the present model.

B. NUMERICAL SOLUTION OF THE BURNUP EQUATIONS

Several numerical integration procedures are available for solving the burnup equations. The simplest is the Euler method. This is an explicit method, in which the value of each atom density, N, at time step $j+1$, N_{j+1}, is expressed explicitly in terms of its value, N_j, at time j and its time derivative, $\dot{N}_j$, at j and the time step δt:

$$N_{j+1} = N_j + \dot{N}_j \delta t. \tag{7-13}$$

A more accurate integration method is an implicit one, such as the following used in the 2DB and 3DB codes:

$$N_{j+1} = N_j + \frac{\delta t}{2} \left(\dot{N}_j + \dot{N}_{j+1} \right). \tag{7-14}$$

Since $\dot{N}_{j+1}$ is a function of the unknown N_{j+1}, $\dot{N}_{j+1}$ is also an unknown. The solution for N_{j+1} must therefore be iterative.

C. ANALYTICAL SOLUTION OF THE BURNUP EQUATIONS

We will present analytical solutions of the burnup equations used in some fuel cycle codes such as FUMBLE. One-group collapsed cross sections for capture and fission are used to obtain reaction rates. For fast reactor analysis, the effective one-group cross section does not change very rapidly with irradiation during a fuel cycle, so reasonably accurate results can be obtained by using a single set of one-group cross sections for the entire cycle. For more detailed analysis (as in FUMBLE), new one-group cross sections, together with new fluxes, would be obtained at various times during the cycle.

The solutions are given in terms of the initial values of material m, $N_{m,0}$, loaded into the fresh fuel, i.e., at $z=0$. The solution of Eq. (7-5) for ^{238}U is

$$N_1 = N_{1,0} e^{-\sigma_{a1} z}. \tag{7-15}$$

Equation (7-8) can be solved by using an integrating factor, $p = \exp(\int \sigma_{a2}\, dz)$ $= \exp(\sigma_{a2} z)$, to give

$$N_2 = \frac{1}{p} \int p N_1 \sigma_{c1}\, dz,$$

$$= e^{-\sigma_{a2} z} N_{1,0} \sigma_{c1} \left[\frac{e^{(\sigma_{a2} - \sigma_{a1}) z}}{\sigma_{a2} - \sigma_{a1}} + C \right]. \tag{7-16}$$

To solve for the constant C, we use the initial condition that at $z = 0$, $N_2 = N_{2,0}$. Thus,

$$C = \frac{N_{2,0}}{N_{1,0} \sigma_{c1}} - \frac{1}{\sigma_{a2} - \sigma_{a1}},$$

and

$$N_2 = N_{2,0} e^{-\sigma_{a2} z} + N_{1,0} \frac{\sigma_{c1}}{\sigma_{a2} - \sigma_{a1}} \left(e^{-\sigma_{a1} z} - e^{-\sigma_{a2} z} \right). \tag{7-17}$$

In general, the solutions to the equations for materials 1 through 6 as defined in Table 7-2, have the following form which can be written as *recursion relationships*:

$$N_1 = A_{11} e^{-\sigma_{a1} z}$$

$$N_2 = A_{21} e^{-\sigma_{a1} z} + A_{22} e^{-\sigma_{a2} z}$$

$$N_3 = A_{31} e^{-\sigma_{a1} z} + A_{32} e^{-\sigma_{a2} z} + A_{33} e^{-\sigma_{a3} z}$$

$$\vdots$$

$$N_m = \sum_{n=1}^{m} A_{mn} e^{-\sigma_{an} z} \qquad (m \leqslant 6) \tag{7-18}$$

where

$$A_{mn} = \frac{\sigma_{c, m-1}}{\sigma_{am} - \sigma_{an}} A_{m-1, n} \qquad (1 \leqslant n \leqslant m-1) \tag{7-19}$$

and A_{mm} is determined from the relation

$$N_{m,0} = \sum_{n=1}^{m} A_{mn}. \tag{7-20}$$

In these equations [and in all subsequent equations in this chapter except Eqs. (7-31) and (7-32)], the "σ_{a4}" for ^{241}Pu must be the sum of the actual σ_{a4}

and $\lambda_4/\phi f$. The solution of Eq. (7-11) for ^{241}Am does not follow the recursion formula of Eq. (7-18). Using Eq. (7-18) for N_4, the solution for N_7 becomes

$$N_7 = \frac{\lambda_4}{\phi f} \sum_{n=1}^{4} A_{4n} \int_0^z e^{-\sigma_{an} z'} dz'$$

$$= \frac{\lambda_4}{\phi f} \sum_{n=1}^{4} \frac{A_{4n}}{\sigma_{an}} (1 - e^{-\sigma_{an} z}). \tag{7-21}$$

The solution of Eq. (7-12) for fission-product pairs, N_8, is

$$N_8 = \int_0^z \sum_{m=1}^{5} N_m(z') \sigma_{fm} dz'. \tag{7-22}$$

Substituting Eq. (7-18) for N_m gives

$$N_8 = \sum_{m=1}^{5} \left[\sigma_{fm} \sum_{n=1}^{m} A_{mn} \int_0^z e^{-\sigma_{an} z'} dz' \right]$$

$$= \sum_{m=1}^{5} \left[\sigma_{fm} \sum_{n=1}^{m} \frac{A_{mn}}{\sigma_{an}} (1 - e^{-\sigma_{an} z}) \right]. \tag{7-23}$$

Equation (7-23) has the same form as Eq. (7-21), except that (7-23) is summed over all five fissionable isotopes whereas the source in (7-21) is only one isotope, $N=4$.

Accounting for all heavy isotopes provides a useful internal check on the correct operation of a fuel cycle computer program. Since it is assumed that there is no destruction of ^{241}Am, ^{243}Am, or fission-product pairs, the total number of heavy isotopes is $\sum_{m=1}^{8} N_m$, where we have included fission-product pairs in this total. This total must remain constant throughout the irradiation; and, in fact, the total is the initial fuel atom density, N_f, as calculated by the methods of Chapter 4. For commercial fuel cycle codes, the same type of check is available; the difference is that higher-order actinides [and fission-product (n, γ) progenies] become the last isotopes in the chain.

D. AVERAGE ATOM DENSITIES DURING THE FUEL CYCLE

For various nuclear design calculations it is useful to know average atom densities at various stages during a fuel cycle—e.g., at the beginning or end of an equilibrium cycle or at the middle or some average point in the cycle.

Furthermore, for more accurate fuel cycle calculations, it may be important to recalculate multigroup neutron fluxes and one-group collapsed cross sections at various stages in the cycle.

For these reasons there is a need to calculate average atom densities during a fuel cycle. The average atom density for material m during an irradiation to fluence z is

$$\bar{N}_m(z) = \frac{1}{z} \int_0^z N_m(z')\,dz'. \tag{7-24}$$

For materials 1 through 6, the result is

$$\bar{N}_m(z) = \frac{1}{z} \sum_{n=1}^m \frac{A_{mn}}{\sigma_{an}} (1 - e^{-\sigma_{an}z}) \qquad (1 \leqslant m \leqslant 6). \tag{7-25}$$

To be certain that the reader is correctly understanding the relationship between all of these N_m's, the following exercise in logic may be useful at this point. One can reason that the atom density for the fission-product pairs, N_8, at fluence z should be the sum of the products of the $\bar{N}_m$'s, σ_{fm}'s and z, or

$$N_8(z) = \sum_{m=1}^5 \bar{N}_m \sigma_{fm} z. \tag{7-26}$$

Substituting Eq. (7-25) for N_m into Eq. (7-26) shows that Eqs. (7-26) and (7-23) are indeed equivalent.

For materials 7 and 8:

$$\bar{N}_7(z) = \frac{1}{z} \frac{\lambda_4}{\phi f} \sum_{n=1}^4 \frac{A_{4n}}{\sigma_{an}} \left[z - \frac{1}{\sigma_{an}} (1 - e^{-\sigma_{an}z}) \right] \tag{7-27}$$

$$\bar{N}_8(z) = \frac{1}{z} \sum_{m=1}^5 \sigma_{fm} \sum_{n=1}^m \frac{A_{mn}}{\sigma_{an}} \left[z - \frac{1}{\sigma_{an}} (1 - e^{-\sigma_{an}z}) \right]. \tag{7-28}$$

7-5 INITIAL AND DISCHARGE COMPOSITIONS AND FUEL CYCLE REACTIVITY

In this section an approximate method will be presented to calculate the required plutonium fraction in the fresh fuel, the composition of the discharge fuel, the reactivity loss during a fuel cycle, and a basis for estimating the breeding ratio. This simplified model is based on a zero-dimensional core calculation; hence, only the core will be treated in detail. Space-dependent calculations would obviously be required for analysis of the blankets. The logic for fuel cycle calculations will be effectively presented, however, by this simplified model.

Our primary task is to calculate an initial plutonium fraction that will lead to a criticality factor of unity at the end of the fuel cycle with all control rods withdrawn. Let us begin with some definitions:

$E =$ atom fraction of Pu in the fuel

$Q =$ number of batches; also equal to the number of cycles for the discharge fuel

$q =$ cycle index, with $q=0$ for fresh fuel, $q=1$ for fuel after 1 cycle, etc., $q=Q$ for discharge fuel

$N_m(q) =$ atom density of material in a batch after q cycles if all of the core were composed of this batch

$N_{m,b} =$ average atom density (i.e., averaged over all batches) at the beginning of the cycle

$N_{m,e} =$ average atom density (i.e., averaged over all batches) at the end of the cycle

$\overline{N}_m =$ average atom density during the cycle (i.e., averaged over all batches *and* over the duration of the cycle)

$k_b =$ criticality factor at the beginning of the cycle, with control rods withdrawn

$k_e =$ criticality factor at the end of the cycle, with control rods withdrawn

$E^{(\nu)}, N^{(\nu)}, k^{(\nu)} =$ variables for the νth iteration

$\Sigma_{c,nf} =$ macroscopic capture cross section for the non-fuel material

$DB^2 =$ product of one-group diffusion coefficient and buckling

$t_c =$ refueling interval, or duration of one cycle, in chronological time in seconds

$B =$ burnup (MWd/kg metal).

The average atom densities at the beginning and end of the cycle are

$$N_{m,b} = \frac{1}{Q} \sum_{q=0}^{Q-1} N_m^{(q)} \tag{7-29}$$

$$N_{m,e} = \frac{1}{Q} \sum_{q=1}^{Q} N_m^{(q)}. \tag{7-30}$$

The criticality calculation is approximated by a one-group neutron balance, whereas in an actual design fuel cycle code, the calculation would be returned to a spatial multigroup code for each criticality calculation;

$$k_b = \frac{\sum\limits_{m=1}^{5} \nu_m N_{m,b}\sigma_{fm}}{\sum\limits_{m=1}^{8} N_{m,b}\sigma_{am} + \Sigma'_{c,nf} + DB^2}, \tag{7-31}$$

$$k_e = \frac{\sum\limits_{m=1}^{5} \nu_m N_{m,e}\sigma_{fm}}{\sum\limits_{m=1}^{8} N_{m,e}\sigma_{am} + \Sigma_{c,nf} + DB^2}, \tag{7-32}$$

where the cross sections and D are all one-group collapsed values. Also, it should be noted that σ_{a4} (for ^{241}Pu) in these two equations are the actual σ_{a4} values, not $(\sigma_{a4} + \lambda_4/\phi f)$.

We assume that the isotopic composition of the feed plutonium fuel is known, so that all initial fuel atom densities are known once the plutonium atom fraction is specified. We assume that the refueling interval, t_c, and the load factor, f, are given; t_c will generally be one year.

The flow sheet for the iterative calculation of the required plutonium fraction in the fresh fuel is given in Fig. 7-4, where the equations in the BURNER module are those given in Section 7-4. We begin with an initial and a second guess for the plutonium fraction, $E^{(1)}$ and $E^{(2)}$, for use in the first two iterations. For the third and subsequent iterations, $E^{(\nu+1)}$ will be interpolated (or extrapolated) from comparsion of $E^{(\nu)}$ and $E^{(\nu-1)}$ with $k_e^{(\nu)}$ and $k_e^{(\nu-1)}$ according to a user-selected convergence scheme. When $k_e^{(\nu)}$ is sufficiently close to unity (i.e., to within a convergence criterion, ε), the iterative procedure is terminated.

The neutron flux should be recalculated for each iteration, even though its change between iterations will be small:

$$\phi^{(\nu+1)} = \frac{2.9 \times 10^{16} P_{\text{core}}(\text{MW})}{V_{\text{core}} \sum\limits_{m=1}^{5} \bar{N}_m^{(\nu)}\sigma_{fm}}. \tag{7-33}$$

Note that the average atom density, $\bar{N}_m$, is used in the denominator; hence, the flux is an average flux during the cycle. (Little error would be introduced by using the arithmetic average of $N_{m,b}$ and $N_{m,e}$ instead of $\bar{N}_m$.) For the first iteration, $\phi^{(1)}$ is calculated with the guessed values of $N_{m,0}$ instead of $\bar{N}_m$.

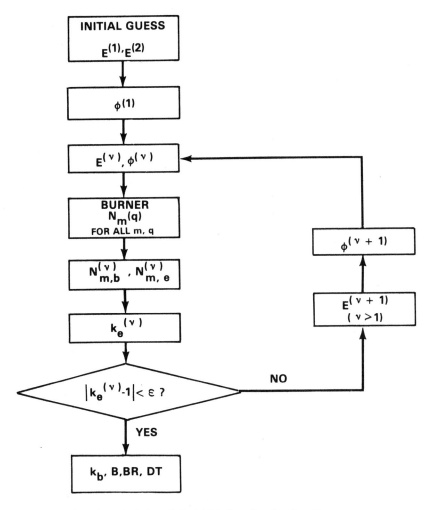

FIGURE 7-4. Iterative calculation of the initial plutonium fraction, E.

The fluence for the cycle, needed for the BURNER calculations, is

$$z^{(v)} = \phi^{(v)} f t_c. \qquad (7\text{-}34)$$

At the conclusion of the iterative calculation, the initial criticality factor, burnup, and breeding ratio (discussed in Section 7-6) can be calculated. The discharge compositions, $N_{m,d}$, are already available from the BURNER calculation as $N_m^{(Q)}$.

The reactivity for the cycle is

$$\Delta k_{\text{cycle}} = k_b - 1. \qquad (7\text{-}35)$$

We are interested in the burnup of the discharge fuel, in MWd/kg metal. The number of days at full power for the discharge fuel is

$$Qft_c \times \frac{1}{86\ 400\ \text{s/d}}.$$

The mass of heavy atoms (U+Pu) in the core is

$$\rho_f F_f V_{\text{core}} \times 238/270$$

with ρ_f in kg/m³ and V_{core} in m³. Thus, the burnup of the discharge fuel is given by

$$B\left(\frac{\text{MWd}}{\text{kg}}\right) = \frac{P_{\text{core}}(\text{MW})Qft_c/86\ 400}{\rho_f F_f V_{\text{core}} 238/270}. \qquad (7\text{-}36)$$

7-6 BREEDING RATIO

A number of special quantities will be needed to calculate both breeding ratio and doubling time. These quantities are defined at the beginning of this section:

FP = fissile material produced per cycle
FD = fissile material destroyed per cycle
FG = fissile material gained per cycle
FBOC = fissile material in the core and blankets at the beginning of the fuel cycle
FEOC = fissile material in the core and blankets at the end of the fuel cycle
FLOAD = fissile material loaded into the core as fresh fuel at the beginning of the fuel cycle
FDIS = fissile material in the fuel discharged from the core and blankets at the end of the cycle
FE = fissile material external to the reactor per cycle
FL = fissile material lost external to the reactor per cycle
FPL = fissile material lost in processing per cycle
EF = out-of-reactor (or ex-reactor) factor
PLF = processing loss fraction
RF = refueling fraction
^{241}PuED = out-of-reactor decay of ^{241}Pu per cycle
^{241}PuBOC = ^{241}Pu in the core and blankets at the beginning of the cycle
^{241}PuEOC = ^{241}Pu in the core and blankets at the end of the cycle
C = cycles per year ($=1/t_c$).

The units for the fissile material are the same as for N_m in the previous sections; the units can be, for example, kg, or atoms/barn·cm. Expressions for breeding ratio and doubling time always contain ratios of fissile material so that these units eventually cancel themselves.

The breeding ratio was defined in Chapter 1 as

$$BR = \frac{FP}{FD}.$$ (7-37)

Despite the apparent simplicity of the definition, the breeding ratio has been calculated by several different methods, as reviewed for example, by Wyckoff and Greebler.[5] We will adopt their proposed definition and method for calculating breeding ratio. The definitions here apply to an equilibrium fuel cycle.

The breeding ratio is to be averaged over a fuel cycle. The numerator includes all fissile material produced during the cycle, not the net material produced. Therefore, the fissile material that is produced but later destroyed during the cycle is included in the numerator. Hence, the numerator can be written as

$$FP = FD + FEOC - FBOC.$$ (7-38)

We digress for a moment at this point to discuss another commonly used basis for the breeding ratio. A different expression for fissile material produced, FP′, based on the fuel entering and leaving the reactor is

$$FP' = FD + FDIS - FLOAD.$$

Some would argue that this is a more logical basis for defining breeding ratio; Wyckoff and Greebler, however, prefer the definition based on all of the fuel in the reactor at the beginning and end of each cycle. The reason for the slight difference in breeding ratio calculated by the two methods is that ^{239}Pu and ^{241}Pu production rates are not linear with fluence.

Before returning to Eq. (7-38), we can add one further complicating factor by noting yet another attempt to improve the consistency of the definition of breeding ratio. This concept introduces a provision for the difference in the value of the various fissile isotopes. For example, ^{241}Pu is a better fuel than ^{239}Pu; it has a higher reactivity value and, hence, less is required for criticality. Moreover, the ^{240}Pu and the ^{242}Pu that are bred have some value, since these are fissionable. Hence, Ott suggested the concept of a breeding ratio based on *equivalent* ^{239}Pu.[6] In this way greater credit could be given for ^{241}Pu, instead of simply lumping all the fissile isotopes together, as in Eq. (7-38). In Reference 5, ^{241}Pu is worth 1.5 times as much as ^{239}Pu with regard to reactivity value primarily because of its higher fission cross section.

With Eq. (7-38), the breeding ratio is

$$BR = 1 + \frac{FEOC - FBOC}{FD}. \tag{7-39}$$

We further observe that, since breeding gain (G) is defined as (BR-1), the breeding gain is given by the second term on the right-hand side of Eq. (7-39). The difference between FEOC and FBOC is the net increase in fissile material (i.e., the fissile material *gained*) during a fuel cycle, so that

$$G = \frac{FEOC - FBOC}{FD} = \frac{FG}{FD}. \tag{7-40}$$

To calculate the terms in Eq. (7-39), we introduce the subscript k as a region index because the quantities in Eq. (7-39) must be summed over all parts of the core and the radial and axial blankets. With the region volume expressed as V_k and with the quantities in Eqs. (7-29) and (7-30) in atoms or mass per unit volume, the beginning and end-of-cycle fissile inventories are

$$FBOC = \sum_k (N_{2,b} + N_{4,b})_k V_k, \tag{7-41}$$

$$FEOC = \sum_k (N_{2,e} + N_{4,e})_k V_k, \tag{7-42}$$

where subscripts 2 and 4 refer to ^{239}Pu and ^{241}Pu (or ^{233}U and ^{235}U).

With $z_k^{(Q)}$ signifying the fluence for the discharge fuel in region k, the total fissile destroyed in one cycle is

$$FD = \sum_k \sum_{m=2,4} \left(\int_0^{z_k^{(Q)}} N_m \sigma_{am} \, dz \right)_k V_k / Q_k$$

$$= \sum_k \left[\sigma_{a2} \sum_{n=1}^{2} \frac{A_{2nk}}{\sigma_{an}} \left(1 - e^{-\sigma_{an} z_k^{(Q)}} \right) \right.$$

$$\left. + \sigma_{a4} \sum_{n=1}^{4} \frac{A_{4nk}}{\sigma_{an}} \left(1 - e^{-\sigma_{an} z_k^{(Q)}} \right) \right] V_k / Q_k, \tag{7-43}$$

where N_{mk} and A_{mnk} are in atoms or mass per unit volume in region k as if all of the region were composed of this composition.* The breeding radio is then computed directly by substituting Eqs. (7-41), (7-42), and (7-43) into Eq. (7-39).

---

*One can integrate over the fluence from zero to discharge here (i.e., for Q cycles) for a single batch as long as N_m is calculated as if all the region were composed of this batch. This is equivalent to the sum of the fissile material destroyed in each of the Q batches in the region for one cycle.

7-7 DOUBLING TIME

The concept of doubling time was introduced in Chapter 1. Doubling time is one of the figures of merit used to compare breeder reactor designs, different fast reactor fuels, and fuel cycle systems involving many breeder reactors. Considerable confusion has existed regarding this parameter, however, because it can be defined in many different ways. Following a long history of proliferation of increasingly complex definitions, some of which were quite ingenious, Wyckoff and Greebler attempted to standardize the doubling-time concept by providing three useful definitions.[5] The first (reactor doubling time, or RDT) is the one discussed in Chapter 1. In the remainder of this section, all three of these definitions will be applied to an equilibrium fuel cycle. Despite the definitions here and in Reference 5, other expressions are still widely used, e.g., Reference 7. An expression for a compound inventory doubling time that includes the approach to equilibrium has also been developed,[8] but treatment of non-equilibrium cycles is beyond the scope of this text.

Wyckoff and Greebler's three proposed definitions of doubling time are given below, followed by a mathematical description of each:

RDT = reactor doubling time
= time required for a specific reactor to produce enough fissile material in excess of its own fissile inventory to fuel a new, identical reactor;

SDT = system doubling time*
= time required for a specific reactor to produce enough fissile material in excess of its own fissile inventory to supply its requirements of fissile material *external to the reactor* and to fuel a new, identical reactor;

CSDT = compound system doubling time
= time required for a system of identical breeder reactors to double the fissile material in the system, assuming that the number of reactors is increasing at a rate such that all of the fissile material is being utilized.

A. REACTOR DOUBLING TIME

The reactor doubling time is given by

$$RDT = \frac{FBOC}{(FG)(C)},$$ (7-44)

---

*This doubling time is called fuel-cycle-inventory doubling time (IDT) in Ref. 5.

where FBOC was defined in Section 7-6 and FG is the net increase in fissile material during one fuel cycle as defined in Eq. (7-40). The fuel cycles per year, C, is equivalent to $1/t_c$, where t_c was defined in Section 7-5 as the refueling interval. This result for RDT is equivalent to $M_0/\dot{M}_g$ in Eq. (1-9). This definition for doubling time is sometimes referred to as a *simple* or *linear* reactor doubling time, to distinguish it from definitions that consider the fuel cycle external to the reactor and compounding growth rates of fissile material.

As shown in Chapter 1 [by Eq. (1-11)], the RDT can be approximated by

$$RDT \approx \frac{2.7 \, FBOC}{GPf(1+\alpha)}, \qquad (7\text{-}45)$$

where FBOC here is mass, in kg, and P is power, in MW. This relation shows the strong dependence of RDT on breeding gain, G, and on fissile specific inventory, FBOC/P.

B. SYSTEM DOUBLING TIME

The Reactor Doubling Time does not account for the processes in the fuel cycle external to the reactor. These processes are part of the breeder reactor *system*. Fissile inventory is tied up in these processes and this inventory must be produced in the reactor. The time required for external processing also affects doubling time. In addition, there are two types of losses of fissile material external to the reactor that affect doubling time. There is loss of material in fabrication and reprocessing, and (for the U-Pu cycle) there is loss from the β decay of ^{241}Pu.

It is estimated in Reference 5 that the out-of-reactor fissile inventory is in the range of 60% of the in-reactor inventory. This includes fuel that is (a) being fabricated, (b) being shipped to the plant, (c) in storage at the plant awaiting loading or in a cool-down period, (d) in shipment for reprocessing, and (e) being reprocessed. By denoting the fissile inventory in these processes as *fissile external*, FE, the effect of this inventory is accounted for through a factor (the *ex-reactor factor*) defined as

$$EF = \frac{FBOC + FE}{FBOC} \qquad (7\text{-}46)$$

Its value is approximately 1.6.

The *processing losses*, FPL, are usually based on a *processing loss fraction*, PLF, which is the fraction of material lost in fabrication and reprocessing operations; this fraction may range from 1% to a few percent. With *refueling fraction*, RF, being defined as the fraction of core fuel assemblies replaced at each refueling (and, therefore, equal to $1/Q$), the processing losses per cycle are

$$FPL = \tfrac{1}{2}(FBOC + FEOC)(RF)(PLF). \qquad (7\text{-}47)$$

The ^{241}Pu decay external to the reactor during one fuel cycle is

$$^{241}PuED = \tfrac{1}{2}\left(^{241}PuBOC + {}^{241}PuEOC\right)(EF - 1)\left(1 - e^{-\lambda_4 t_c}\right) \qquad (7\text{-}48)$$

where $(1 - e^{-\lambda_4 t_c})$ is the fraction of the ^{241}Pu in the fissile material external to the reactor that decays during one fuel cycle. The expression for ^{241}PuED assumes that the average level of ^{241}Pu in an out-of-reactor fuel assembly can be represented by the average level of ^{241}Pu in an in-reactor fuel assembly.

With the results of Eqs. (7-47) and (7-48), we can define all of the losses external to the reactor per cycle as

$$FL = FPL + {}^{241}PuED. \qquad (7\text{-}49)$$

Finally, with Eqs. (7-46) and (7-49), we are in a position to define the System Doubling Time, SDT:

$$SDT = \frac{(FBOC)(EF)}{(FG - FL)(C)}. \qquad (7\text{-}50)$$

C. COMPOUND SYSTEM DOUBLING TIME

If the excess fissile material produced in a system of breeder reactors were used to start up a new breeder as soon as the fuel from each cycle was reprocessed and fabricated into new fuel assemblies, the fissile inventory in the system of reactors would grow exponentially, as is the case with compound interest. The governing equation for this growth rate is

$$\frac{dM}{dt} = \lambda M \qquad (7\text{-}51)$$

where M = fissile mass in the system as a function of time (kg), and λ = fractional increase in fissile mass in the system per unit time (y^{-1}). The constant λ can be expressed in terms previously defined as

$$\lambda = \frac{(FG - FL)(C)}{(FBOC)(EF)}. \qquad (7\text{-}52)$$

It is recognized from Eq. (7-50) that λ is the inverse of the System Doubling Time, or

$$\lambda = 1/\text{SDT}. \qquad (7\text{-}53)$$

Eq. (7-51) can be integrated to give

$$\frac{M}{M_0} = e^{\lambda t}, \qquad (7\text{-}54)$$

or, from Eq. (7-53),

$$\frac{M}{M_0} = e^{t/\text{SDT}}, \qquad (7\text{-}55)$$

where M_0 = initial fissile mass in the system of reactors. [M_0 could also be thought of as the number of reactors in the system at time zero $\times$ (FBOC) $\times$ (EF).]

The compound system doubling time is the time required for M/M_0 to equal 2. Substituting 2 for M/M_0 and CSDT for t in Eq. (7-55) gives

$$\text{CSDT} = \text{SDT} \ln 2 = 0.693 \, \text{SDT}. \qquad (7\text{-}56)$$

Even though the concept of CSDT represents an idealization that will never be realized in practice, it is a useful figure of merit for a potential breeder system. Occasionally the 0.693 factor has been applied to the simple RDT, but a compound doubling time for a single reactor is misleading since it ignores the time and other requirements of the fuel cycle external to the reactor. The three doubling times defined here — RDT, SDT, and CSDT — all have merit in comparisons of breeder reactors, but only when it is clearly understood which doubling time is being considered.

7-8 FUEL CYCLE RESULTS

A. U-PU CYCLE AND COMPARISON WITH LIGHT WATER REACTORS

Selected results from LMFBR and LWR fuel cycles reported in the literature are provided in this section, both to gain an idea of the numbers for fast breeder reactors and to allow some comparison with water reactors.

Results for the designs described in Chapter 4-8 are given in Table 7-3. These results show the extent to which breeding ratio and doubling time can

TABLE 7-3 Fuel Cycle Results as a Function of Fuel Pin Diameter for Homogeneous and Heterogeneous Designs (Adapted from Reference 9)

Common Parameters:

Fuel	UO_2–PuO_2
Electrical rating	1200 MWe
Thermal rating	3300 MWth
Residence times	
core and internal blankets	2 y
radial blanket	5 y
axial blanket	1 y
Fuel cycle interval	1 y
Capacity factor	0.7

Configuration	Homogeneous						Heterogeneous					
	6.35		7.62		8.38		6.35		7.62		8.38	
Fuel pin diameter, mm	BOEC	EOEC	BOEC	EOEC	BOEC	EOEC	BOEC	EOEC	BOEC	EOEC	BOEC	EOEC
Fissile inventory (kg)												
core	3319	3023	4019	3878	4551	4504	4292	3791	6233	5751	6908	6500
internal blanket							255	710	278	797	258	743
radial blanket	436	637	391	574	350	516	410	596	456	670	453	667
axial blanket	114	325	120	349	121	353	65	190	77	226	81	238
Total	3869	3985	4530	4801	5022	5373	5022	5287	7044	7444	7700	8148
Discharge burnup (MWd/kg)												
core average	82		56		46		77		52		43	
core peak	116		80		66		109		78		70	
Fraction power (%)												
core	93	89	94	91	94	93	83	74	87	81	88	84
internal blanket							10	18	7	13	6	10
radial blanket	4	6	3	4	3	3	5	6	4	4	4	4
axial blanket	3	5	3	5	3	4	2	2	2	2	2	2
Fuel cycle reactivity (absolute)	−0.058		−0.023		−0.009		−0.019		−0.023		−0.020	
Breeding ratio	1.11		1.26		1.33		1.24		1.35		1.39	
Doubling time, CSDT (y)	43.5		19.3		16.5		22.3		20.7		20.2	

TABLE 7-4 1000-MWe Reactor Charges, Discharges, and Inventories of Plutonium (Adapted from Reference 10)

Reactor	Fuel fraction replaced per charge	Average residence time, days	Plutonium discharged,† kg	Plutonium charged,† kg	Maximum plutonium inventory, kg	Average amount of plutonium, kg		Average burnup of core fuel, MWd/kg
						Discharged per year	Charged per year	
PWR (U-fueled)	1/3	1100	256		512	256		33
PWR (fueled with Pu from U-fueled PWR)	1/3	1200	442	800	2042	403	730	33
Atomics International* (AI) reference oxide LFMBR (fueled with Pu from U-fueled PWR)								
Core and axial blanket	1/2	540	1270	1380	2740	1716	1865	80
Radial blanket	0.28	970	223		560	302		
Total			1493	1380	3300	2018	1865	
General Electric (GE)* LMFBR (fueled with Pu from U-fueled BWR)								
Core and axial blanket	0.46	796	1304	1094	2713	1304	1094	100
Radial blanket	0.29	1260	157		356	157		
Total			1461	1094	3069	1461	1094	

*The AI and GE designs are the 1000-MWe *follow-on* designs developed for the USAEC in 1968.

†Refers to plutonium charged or discharged at an actual refueling. Refueling occurred annually for the PWR (U-fueled) and GE LMFBR reactors, so that these numbers agree with the average annual amounts for these reactors. Refueling was not annual for the PWR (Pu-fueled) and AI LMFBR reactors.

be varied by core design and fuel pin diameter. The breeding ratio is larger for the heterogeneous design; despite the larger fissile inventories required for the heterogeneous design, the doubling times (for the larger fuel pins) are only slightly higher than for the homogeneous design. Due to the breeding ratio advantage of the heterogeneous design, fuel management optimization studies have been carried out (by Dickson and Doncals of Westinghouse) that go beyond the general comparative studies reported in Table 7-3. Such studies show that doubling times for oxide fuel heterogeneous designs can be reduced to the 15 to 20 year range.

Plutonium inventories and fuel management data for a uranium-fueled Pressurized Water Reactor (PWR) and a plutonium-fueled PWR and for two LMFBR designs are given in Table 7-4.[10] The LMFBR results are based on early 1000-MWe designs reported by Atomics International and General Electric. The isotopic composition of plutonium in fuel discharged from LMFBR's and LWR's (Pressurized and Boiling Water Reactors) is given in Table 7-5.[10]

TABLE 7-5 Isotopic Composition of Plutonium in Discharge Fuels (wt. %): Plutonium-Fueled Reactors[†] (Adapted from Reference 10)

	Burnup MWd/kg	Weight Percent				
		^{238}Pu	^{239}Pu	^{240}Pu	^{241}Pu	^{242}Pu
PWR (U-fueled)	33.0	1.8	58.7	24.2	11.4	3.9
BWR (U-fueled)	27.5	1.0	57.2	25.7	11.6	4.5
PWR (Pu-fueled)	33.0	2.7	39.3	25.6	17.3	15.1
LMFBR	80.0[‡]					
(AI reference oxide)*						
Core and						
axial blanket		0.9	61.5	26.0	7.2	4.5
Radial blanket		0.02	97.6	2.33	0.04	
Core plus blankets		0.8	66.8	22.5	6.2	3.8
(average)						
LMFBR	100.0[‡]					
(GE design)*						
Core and						
axial blanket			67.5	24.5	5.2	2.8
Radial blanket			94.9	4.9	0.2	
Core plus blankets			70.5	22.4	4.6	2.5
(average)						

[†]Plutonium from U-fueled PWR.
[‡]Core.
*The AI and GE designs are the 1000-MWe *follow-on* designs developed for the USAEC in 1968.

Several observations from Tables 7-4 and 7-5 are of interest. The plutonium discharged annually from an LMFBR is five to ten times the amount discharged from a uranium-fueled LWR, yet the net plutonium produced (i.e., Pu discharged minus Pu charged) is not very different for the two. The plutonium discharged from a uranium-fueled LWR each year will supply about 10% of the initial plutonium needed to start up an LMFBR and from 15 to 25% of the annual needs thereafter. The excess plutonium produced in two LMFBR's of the GE design would provide approximately enough fuel for a plutonium-fueled PWR.

The fissile fraction of the plutonium discharged from the core and axial blanket of an LMFBR is 0.7, which is similar to the value for fuel discharged from an LWR. The split between ^{239}Pu and ^{241}Pu is different, however, between an LMFBR and an LWR, with more ^{241}Pu in the LWR discharge. The ^{240}Pu content is similar in both LMFBR's and LWR's.

The isotopic composition of plutonium in the radial blanket of the LMFBR is considerably different from that of LWR discharge fuel; namely, the ^{239}Pu content is high and the ^{240}Pu content is low. The principal reasons for this result are the low capture-to-fission ratio for ^{239}Pu in a fast neutron spectrum and the relatively low reaction rate in the radial blanket of an FBR. For the GE and AI designs shown in Table 7-5, the radial blanket was to be removed before a large amount of ^{240}Pu was produced. The axial blanket plutonium in the outer part of the axial blanket is also low in ^{240}Pu, but this fuel is present in the same fuel pins as the core fuel. Hence, it would likely be processed and mixed directly with the core fuel during reprocessing.

Breeding ratios and doubling times for some of the operating and proposed LMFBR's are listed in Appendix A, where available, even though it is difficult to be sure of the basis for the results. The breeding ratio value of 1.16 for Phénix is probably the most reliable because it is the only one reported as being measured (from irradiated fuel). All of the demonstration and prototype reactors listed in Appendix A use or plan to use UO_2-PuO_2 fuel.

Predicted breeding ratios of carbide, nitride, and metal-fueled LMFBR's are significantly higher than for oxide-fueled LMFBR's, and doubling times are correspondingly shorter. Results of a study of fuel cycles for an oxide, a carbide, and a metal-fueled 1000-MWe LMFBR are given in Table 7-6. While calculated breeding ratios for the oxide reactor were in the 1.2 to 1.3 range, breeding ratios for the carbide reactor were ~ 1.4 and for the metal reactor ~ 1.6. Doubling times (simple reactor doubling times—RDT) are reported as being reduced from 15 years for oxide fuel to 9 years for carbide fuel and 6 years for metal fuel.

TABLE 7-6 Summary of Breeding Performance for Oxide, Carbide and Metal Fueled LMFBR's Under Various Fuel Cycle Options[11]

Cycle		Pu-U	Pu fueled-Th	$^{233}U/^{238}U$-Th	^{233}U-Th
Core Fissile/Fertile		LWR Pu/Dep U	LWR Pu/Th	$^{233}U/^{238}U$	^{233}U-Th
Blanket Fertile		Dep U	Th	Th	Th
Oxide	Doubling Time*	16	29	23	112
	Breeding Ratio	1.28	1.20	1.16	1.041
Carbide	Doubling Time*	9	20	15	91
	Breeding Ratio	1.42	1.23	1.23	1.044
Metal	Doubling Time*	6	12	12	43
	Breeding Ratio	1.63	1.38	1.30	1.11

*Reactor doubling time, years (75% capacity factor).

B. THORIUM-URANIUM CYCLE

The results in Table 7-6 also provide predicted breeding ratios and doubling times for LMFBR's that utilize the thorium-uranium cycle and mixed Th-U-Pu cycles. The breeding ratio is considerably reduced, down to 1.04, for the ^{233}U-Th oxide cycle, and the doubling time is therefore increased.

PROBLEMS

7-1. Prove that the following recursion relations are valid for ^{240}Pu, for which $m=3$:

$$N_m = \sum_{n=1}^{m} A_{mn} e^{-\sigma_{an}z}$$

where

$$A_{mn} = \frac{\sigma_{c,m-1}}{\sigma_{am}-\sigma_{an}} A_{m-1,n} \qquad 1 \leqslant n \leqslant m-1$$

and

$$N_{m,0} = \sum_{n=1}^{m} A_{mn}.$$

7-2. In an LMFBR utilizing the ^{233}U-Th fuel cycle, Protactinium-233 is formed by β-decay of ^{233}Th. Protactinium-233 decays to ^{233}U with a 27.4d half-life. It also has a relatively high absorption cross section that cannot be ignored in fuel burnup calculations.

Write an expression for dN_2/dz (i.e., a burnup equation) for ^{233}Pa, where material 2 is ^{233}Pa and material 1 is ^{232}Th. The decay of ^{233}Th can be assumed to occur instantaneously since its half life is only 22 minutes.

7-3. Suppose the fissile specific inventory of an LMFBR is 1.25 kg/MWth fissile at the beginning of the cycle. The load factor is 0.7. The ratio of σ_c to σ_f (i.e., α) is 0.2. The System Doubling Time (SDT) for the system of which this reactor is a part, is 1.5 times the Reactor Doubling Time (RDT).

If all LMFBR's in a fast breeder reactor system are identical to this reactor and all of the fissile material is being utilized to the fullest, estimate the minimum time required to double the fissile material in the reactor system.

7-4. An LMFBR is operating on an equilibrium fuel cycle, with the following conditions:
- 3-batch loading for fuel at every position in the core
- total core inventory of oxide fuel = 25 000 kg oxide
- average core power = 3000 MWth
- product of radial and axial peak-to-average power factors = 1.6
- refueling interval = 1 year
- load factor = 0.7.

Calculate the peak burnup for the fuel discharged from the core, in MWd/kg.

Approximately what is the peak burnup in atom percent?

7-5. The LMFBR described in Problem 4-5 is operating on an equilibrium fuel cycle. A 3-batch loading scheme is used, and fuel is removed and reloaded uniformly throughout the reactor. Refueling is performed annually and the load factor is 0.7. Assume uniform burnup throughout the core, i.e., neglect power peaking factors. The fuel volume fraction and smear density are identical to those in Problem 4-5 (i.e., $F_f = 0.392$ and $\rho_f = 9.5$ g/cm^3).

With the additional data provided below, calculate the following by means of an iterative computer code as illustrated in Fig. 7-4.

(a) Plutonium atom fraction in the fresh fuel required to give $k = 1$ with all control rods withdrawn at the end of the cycle. (Assume that $\overline{N}_m = (N_{mb} + N_{me})/2$ for the calculation of ϕ.)
(b) Isotopic composition of the plutonium discharged from the core *and* atom densities of the discharge fuel.
(c) Burnup in MWd/kg for fuel discharged from the core.
(d) Reactivity requirement per cycle.
(e) Breeding ratio.
(f) Reactor Doubling Time (RDT).

Additional Data for Problem 7-5

- Pu composition in the fresh fuel (typical of Pu discharged from an LMFBR, including the blankets):

^{239}Pu	0.70
^{240}Pu	0.22
^{241}Pu	0.05
^{242}Pu	0.03

- One-group microscopic heavy-metal cross sections:

	σ_c	σ_f	ν
^{238}U	0.29	0.044	2.77
^{239}Pu	0.50	1.82	2.93
^{240}Pu	0.50	0.36	3.07
^{241}Pu	0.46	2.52	2.96
^{242}Pu	0.35	0.28	3.01
fission-product pairs	0.47	0	—

- One-group diffusion coefficient: $D = 1.6$ cm

- One-group non-fuel capture cross section (with control rods withdrawn):

$$\Sigma_{c,nf} = 6 \times 10^{-4} \text{ cm}^{-1}$$

- Buckling: $B^2 = 7 \times 10^{-4} \text{ cm}^{-2}$

- Blanket inventories and fissile destruction values needed for the BR and RDT calculations:

$$\text{BOC}: {}^{239}\text{Pu} = 240 \text{ kg}$$

$$\text{EOC}: {}^{239}\text{Pu} = 660 \text{ kg}$$

(Neglect ^{235}U in both the core and the blankets)

(Neglect plutonium isotopes other than ^{239}Pu in the blanket)

- $\dfrac{\text{fissile destroyed in blankets}}{\text{fissile destroyed in core}} = 0.08.$

REFERENCES

1. P. Greebler and C. L. Cowan, *FUMBLE: An Approach to Fast Power Reactor Fuel Management and Burnup Calculations*, GEAP-13599, General Electric Co., November 1970.
2. J. Hoover, G. K. Leaf, D. A. Meneley, and P. M. Walker, "The Fuel Cycle Analysis System, REBUS," *Nucl. Sci. and Eng.*, 45 (1971) 52–65.
3. W. W. Little, Jr., and R. W. Hardie, *2DB User's Manual*, BNWL-831, Pacific Northwest Laboratory, Richland, Washington, 1968.
4. R. W. Hardie and W. W. Little, Jr., *3DB, Three-Dimensional Diffusion Theory Burnup Code*, BNWL-1264, Pacific Northwest Laboratory, Richland, Washington, March 1970.
5. H. L. Wyckoff and P. Greebler, "Definitions of Breeding Ratio and Doubling Time," *Nuclear Technology*, 21 (1974) 158–164.
6. K. Ott, "An Improved Definition of the Breeding Ratio for Fast Reactors," *Trans. Am. Nucl. Soc.*, 12 (1969) 719.
7. W. P. Barthold and Y. I. Chang, "Breeding Ratio and Doubling Time Definitions Used for Advanced Fuels Performance Characterization," *Trans. Am. Nucl. Soc.*, 26 (1977) 588.
8. D. R. Marr, R. W. Hardie, and R. P. Omberg, "An Expression for the Compound System Doubling Time Which Explicitly Includes the Approach to Equilibrium," *Trans. Am. Nucl. Soc.*, 26 (1977) 587.
9. W. P. Barthold and J. C. Beitel, "Performance Characteristics of Homogeneous Versus Heterogeneous Liquid-Metal Fast Breeder Reactors," *Nucl. Tech.*, 44 (1979)45, 50–52.
10. C. A. Erdman and A. B. Reynolds, "Radionuclide Behavior During Normal Operation of Liquid-Metal-Cooled Fast Breeder Reactors, Part I: Production," *Nuclear Safety, 16* (1975) 44–46.
11. Y. I. Chang, C. E. Till, R. R. Rudolph, J. R. Deen, and M. J. King, *Alternative Fuel Cycle Options: Performance Characteristics and Impact on Nuclear Power Growth Potential*, ANL-77-70, Argonne National Laboratory, September 1977.

PART III

SYSTEMS

A fast breeder reactor differs substantially from a thermal reactor with regard to core design and the heat transport system. This difference arises mainly from the much higher flux and power densities encountered in the fast breeder reactor, the high energy neutron spectrum, and the desire to take advantage of higher possible coolant temperatures to improve the overall thermal efficiency.

Five chapters are included to provide coverage for the principal systems design questions of interest. Chapters 8 through 10 address the mechanical and thermal aspects of LMFBR core design. Although actual core design procedures involve a close coupling of these topics, an artificial separation is helpful for instructional purposes. Hence, Chapter 8 is focused on the mechanical aspects of fuel pin and assembly design, Chapter 9 is addressed to the thermal aspects of a single fuel pin, and Chapter 10 is then broadened to discuss full assembly and overall thermal hydraulics aspects. Whereas certain core materials properties must be introduced in the earlier chapters, the more detailed materials questions are concentrated in Chapter 11. Overall LMFBR plant systems, including the heat transport system, are included in Chapter 12.

CHAPTER 8

FUEL PIN AND ASSEMBLY DESIGN

8-1 INTRODUCTION

This chapter deals with the mechanical designs of fuel pins and assemblies. These core components must be designed to withstand the high temperature, high flux environment of a fast breeder reactor for a long irradiation exposure time. In this chapter we will describe many of the factors that influence this design, and we will examine in some detail the stress analysis of the fuel pin.

We begin in Section 8-2 with the basic geometric and heat transfer relationships for the fuel pin, and then discuss some topics related to fuel and fission gas that must be considered in analysis of steady-state fuel-pin performance. The discussion of fuel-pin design is continued in Section 8-3, in which failure criteria and stress analysis are presented. Discussion is then shifted in Section 8-4 to grouping the pins into a fuel assembly. This will include discussion of mechanical design problems such as fuel-pin spacing and duct swelling. Limited attention is then directed to the design of other assemblies, including blanket, control and shielding assemblies. The final section of the chapter will provide a description of methods to restrain the assemblies and to control duct bowing.

8-2 FUEL PIN DESIGN CONSIDERATIONS

Fuel pin design is a complex process that involves an integration of a wide range of phenomena. The design procedure must integrate the thermal analysis of the pin with an assessment of the characteristics of the fuel and cladding as a function of temperature and irradiation history and with the stress analysis of the fuel-cladding system. Many of the phenomena that affect fuel pin performance are illustrated schematically in Fig. 8-1. Each of these will be treated in Chapters 8 and 9, with additional discussions in

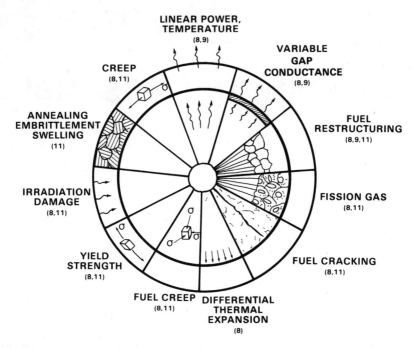

FIGURE 8-1. Phenomena affecting fuel pin performance. (Numbers in parentheses denote chapters where the indicated phenomena are discussed.)

Chapter 11. It should be readily apparent that there is no single way to order the examination of so many interacting phenomena. Each sequence selected to cover these tightly coupled processes will inherently suffer from the need for input from processes not yet discussed. In actual design practice, all of the governing processes are integrated in large time-dependent pin analysis codes such as LIFE.[1]

Our presentation will begin with the geometrical arrangement of the various sections of a fuel pin, including the function of each section. We will then extend the discussion of pin geometry to the selection of pin diameter. This will involve the concept of linear power. Fuel restructuring is then introduced, followed by fission gas release and the associated length of the gas plenum to establish the basis for the cladding loading function. Such discussions lead directly to the stress analysis of the fuel-cladding system, the subject of Section 8-3. These structural considerations are required to establish the requirements for cladding thickness, fuel porosity, and initial gap size (wherein the latter two items define the fuel smear density, given the theoretical density of the fuel material, at steady-state operating conditions). In addition, it is necessary to discuss fuel pin lifetime or design criteria in order to understand how the results of the stress analysis are used in the pin design.

The cladding is stressed by fission gas pressure when the fuel-cladding gap is open, and it may be stressed directly by the fuel if the gap is closed. A quantitative determination of this fuel-loading effect requires a thermal analysis of the fuel and cladding in order to calculate thermal expansions. (This thermal analysis is the subject of Chapter 9.) It also requires knowledge of the creep strength of the fuel, since for very slow transients the fuel can deform sufficiently to prevent significant loading on the cladding. Once the fuel pin temperature field is known, the temperature-dependent stress-strain relations can be incorporated.* We will introduce the basic factors required for the fuel-pin stress analysis, including some aspects of material behavior, in this chapter, but a more detailed discussion of materials properties will be delayed for Chapter 11.

Emphasis in this chapter will be placed on the behavior of mixed-oxide fuel since this is the fuel used in most FBR systems. Certain aspects of other fuels are discussed in Chapter 11.

A. MATERIALS AND GEOMETRICAL ARRANGEMENT

The fuel pin, or element, for a ceramic fueled FBR consists of an axial stack of cylindrical fuel pellets encased in a metal tube called the cladding. The cladding provides structural integrity for the fuel pin and serves to separate the fuel from direct contact with the coolant, thereby preventing the fission products from entering the primary coolant. Typical values for the outside diameter of the cladding are 6 to 8 mm. Each fuel pellet is approximately 7 mm in length and 6 mm in diameter. The pellet diameter is fabricated to be slightly less than the inside cladding diameter, so that there is an initial gap between the fuel and the cladding.

The fuel for the current generation of fast reactors is mixed uranium-plutonium oxide (UO_2-PuO_2). A second ceramic fuel, mixed carbide (UC-PuC), is also being investigated for possible use in future designs. Designs are also being considered that use metallic fuel and mixed thorium-^{233}U ceramic fuel. The mechanical designs described in this chapter apply principally to the oxide fuel designs, although the same general configuration is used for carbide fuels. All the fuels, including the metallic form, are described in more detail in Chapter 11.

---

*It is of interest to note that the temperature distribution in the fuel may be dependent upon the stress-strain relationships between fuel and cladding. The reason for this is that gap conductance is a function of the fuel-cladding interface pressure. As we shall see later, this gap conductance directly affects the fuel surface temperature, which in turn influences the temperature profile within the fuel pin.

The axial configuration of an FBR fuel pin is quite different from an LWR pin. In the FBR, the zone containing the fissile fuel (which will form the reactor core) is only about a third of the pin axial length, whereas the fuel region of an LWR typically represents 80% of the overall pin length. The FBR core is generally about one metre long, while the entire fuel pin is about three metres long. Because of the much higher core height for a similarly powered LWR, overall pin lengths for the LWR and FBR systems are comparable.

Cutaway views of two FBR pins are shown in Fig. 8-2. Axial blanket pellets, which are fabricated from depleted uranium oxide, are located above and below the core pellets. Typical heights for the axial blankets are 0.3 to 0.4 m. The spring above the pellet stack is used to hold the pellets in place during shipment and prevent pellet separation during insertion of the fuel assembly into the core; this spring provides no structural function once the pin has been brought to power.

A fission gas plenum is located in the pin as a reservoir for gaseous fission products produced during irradiation. The fission gas plenum is normally long, approximately the height of the core. As noted in Fig. 8-2, the plenum can be located either above or below the core. For example, in the early U.S. designs (FFTF and CRBRP), the plenum is above the core; the French Phénix and Super Phénix designs contain a small upper and a large lower plenum. The advantage of the above-core location is that a cladding rupture in the plenum region would not allow fission gas to pass through the core (since the sodium flow is upward). The disadvantage is that the coolant is at its highest temperature above the core and the plenum length (or volume) required to accommodate the fission gas pressure (since the fission gas is at the coolant temperature) is larger than would be the case for a plenum in the cooler region below the core. The longer plenum length adds to the total length of the fuel pin, which in turn leads to a higher pressure drop across the fuel assembly, higher required pumping power, and potentially a larger reactor vessel and related vessel internal components.*

Figure 8-3 contains a schematic diagram of the lengths of fuel pins for several fast reactors. The pins are drawn to scale, with the core centerline

---

*It is necessary to conduct a total systems evaluation before deciding the best location for the fission gas plenum. It is possible, for example, that a below-core plenum location could actually lead to an increase in the overall length of the assembly. This possibility arises from the general desire to keep the outlets of all core assemblies at about the same elevation. Since control rods normally enter from the top of the core, the control ducts extend well above the core to accommodate withdrawn absorber positions. That vertical distance, plus the added fuel assembly distance below the core to accommodate a below-core fission gas plenum location, may result in a longer fuel assembly length than if the above-core ducting were utilized to house a somewhat larger fission gas plenum.

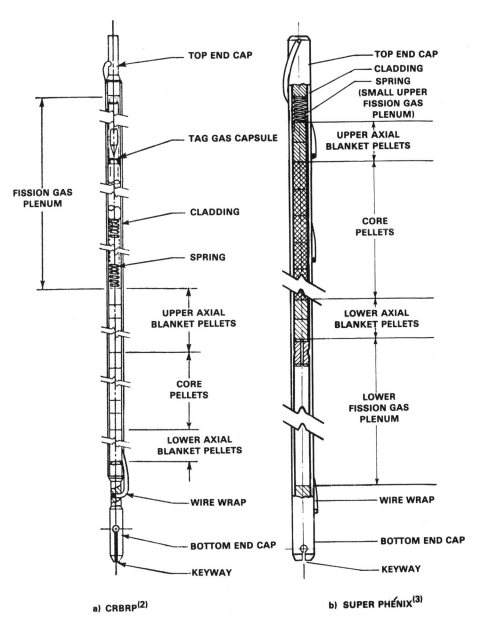

FIGURE 8-2. Typical fast reactor fuel pins.

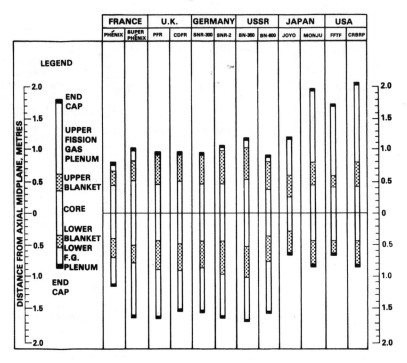

FIGURE 8-3. Intercomparison of fuel pin lengths for selected fast reactor systems. (Updated from Ref. 4)

used for vertical normalization. The various design options are readily apparent, particularly with regard to the placement of the fission gas plenum. It can be observed from this figure that the length of the plenum is approximately the same as that of the core.

Tag gas capsules are sometimes included in the plenum region to assist in the location of a failed fuel pin. These capsules in the fuel pins contain an isotopic blend of inert gases that is unique to that assembly. At the time of final fabrication, the capsule is punctured so that the gas mixture enters the fission gas plenum. Should a break occur in the cladding during irradiation, the unique blend of gases would escape with the fission product gases into the primary loop where they could be collected and analyzed to identify the assembly containing the failed pin. The effectiveness of this technique has been demonstrated in EBR-II. (See Chapter 12-6B and Fig. 12-25 for a more complete presentation of this technique.)

Above and below the plenum and blankets are solid end caps. In designs employing a wire wrap pin spacing technique, the wire is normally pulled through a hole in each end cap and then welded at that point. For some designs the wire is merely laid along the side of the end cap and then welded.

Cladding in the current generation of U.S. LMFBR's is 20% cold worked austenitic type 316 stainless steel. This steel alloy contains 16–18 weight percent Cr, 10-14% Ni, 2-3% Mo, and the balance Fe (65%) except for small amounts of C, Nb, Si, P, and S. Characteristics of this and other candidate cladding and structural materials, definition of cold working, and reasons for the selection of 316 stainless steel are discussed in Chapter 11-3.

B.   LINEAR POWER AND FUEL-PIN DIAMETER

Linear power (also called linear power density and linear heat rate) governs the total length of all the fuel pins in the core—or, given a specified core height and a total core thermal power level, the total number of pins in the core. The linear power was introduced briefly in Chapter 2, through Eq. (2-2), as

$$\chi = 4\pi \int_{T_s}^{T_0} k(T)\, dT,$$

where T_s and T_0 are the fuel surface and center temperatures and k is the fuel thermal conductivity. This equation will be derived in Chapter 9 when the detailed thermal analysis is presented. Note that the fuel pin diameter does not enter this equation. The fuel surface temperature does enter, however. This temperature is influenced strongly by the thickness of the gap between the fuel and the cladding and, hence, by the pin design, at least during early operation of the reactor before the gap closes. An objective of any fast reactor design is to make the average linear power as high as possible without allowing the center temperature to exceed the melting point of the fuel. Selection of average and peak values of linear power is a major step in pin design, and considerable attention to the thermal analysis required for this selection is given in Chapter 9.

For now, let us suppose that an average linear power has been selected. Our next step is to see how this selection influences the choice of a pin diameter. The term *pin diameter* generally refers to the outside diameter of the cladding. In the next few paragraphs we will look at the fuel pellet radius, R_f. This radius, together with the fuel-cladding gap and the cladding thickness, defines the pin diameter. Values of pin diameters for several fast breeder reactors, together with peak linear powers, are listed in Table 8-1.

In selecting a pin diameter, we seek to minimize *fissile specific inventory*, M_0/P, where M_0 is the fissile mass in the core and P is the core power.

TABLE 8-1 Pin Diameter and Linear Power in FBR's

Reactor	Country	Pin Diameter (mm) Core	Pin Diameter (mm) Radial Blanket	Peak Linear Power* (kW/m)
BN-350	USSR	6.1	14.2	44
Phénix	France	6.6	13.4	45
PFR	UK	5.8	13.5	48
SNR-300	Germany	6.0	11.6	36
FFTF	US	5.8	—	42
BN-600	USSR	6.9	14.2	53
Super Phénix	France	8.5	15.8	47
MONJU	Japan	6.5	11.6	36
CDFR	UK	6.7	13.5	42
SNR-2	Germany	7.6	11.6	42
CRBRP	US	5.8	12.8†	42

*Values for linear power change slightly during the burnup cycles. The CRBRP (heterogeneous core) and FFTF values quoted are for the beginning of the first cycle.
†Radial and Internal Blankets

Minimizing the fissile specific inventory minimizes the fissile inventory and, as pointed out in Chapter 1, Eq. (1-11), minimizes doubling time.

The fissile specific inventory is related to the linear power through the relation

$$\chi = e\rho_f \pi R_f^2 \frac{P}{M_0}, \qquad (8\text{-}1a)$$

where χ is the linear power, e is the fissile mass fraction in the fuel, ρ_f is the fuel density in the pellet, and R_f is the fuel pellet radius. Rearranging gives for the fissile specific inventory:

$$\frac{M_0}{P} = \frac{\pi \rho_f e R_f^2}{\chi}. \qquad (8\text{-}1b)$$

Having selected χ, we can minimize M_0/P by modifying ρ_f, e, and R_f. Little flexibility is available for varying ρ_f since several considerations limit ρ_f to a narrow range. It is desirable to make ρ_f as near the theoretical density (TD) as possible to enhance the thermal conductivity (and hence, maximize χ) and to decrease the required fissile inventory [note in Eq. (8-1b) that increasing ρ_f results in a sufficient reduction in neutron leakage to cause a larger decrease in e than the relative increase in ρ_f]. On the other hand, ρ_f must be low enough to accommodate fuel swelling during steady-state

operation and volumetric expansion during thermal upset conditions.*
These considerations lead to a typical pellet density for mixed oxide
LMFBR's in the range of 85% to 95% TD and a smear density in the range
of 75% to 90% TD.

The parameters e and R_f are tied together. Reducing R_f will increase the
required fissile fraction, e, but this will not offset the reduction in R_f^2 as long
as the fuel volume fraction does not have to be decreased significantly.
Therefore, reducing R_f reduces eR_f^2 and reduces M_0/P. These trends are
illustrated in Table 4-5. Consequently, reducing R_f is a desirable objective
and this is the reason that fast reactor fuel pins have such a small diameter.

One lower limit to an acceptable fuel pin diameter might be the ability of
the coolant to handle the ever-increasing heat flux. The heat flux, $q(W/m^2)$,
diameter, D, and heat transfer coefficient, h, are related by

$$q = \frac{\chi}{\pi D} = h(T_{co} - T_b),\tag{8-2}$$

where T_{co} and T_b are the outside cladding and bulk coolant temperatures,
respectively. With sodium as the coolant, h remains high even for low flow
velocity without the problems of burnout which are of concern in an LWR.
Hence, sodium heat transfer is not the limiting factor in reducing the pin
diameter.†

There must eventually be some limiting factor on how small the pin
diameter can be. One important limitation is that beyond some point small
pins become increasingly expensive to fabricate. A second limitation involves pitch-to-diameter ratio. As pin diameter is reduced, the point is
reached eventually when further reduction requires an increase in P/D ratio
which leads to a significant reduction in fuel volume fraction and increased
fissile inventory. The P/D ratio must eventually increase as D gets small for
two reasons. First, there is a lower limit on spacing between pins due to hot
spots with wire wrap spacers or fabrication with grid spacers; when this
spacing is reached, further reduction in D necessarily increases P/D.
Second, as D is lowered, P/D must eventually be increased in order to hold
axial pressure drop (and hence, pumping power) down.

An important incentive for a large pin diameter is breeding ratio. The
breeding ratio is reduced with reduced pin diameter (see Table 7-3). Other

---

*UO_2 expands approximately 10% upon melting.
† The arguments relating to Eq. (8-1b) apply also to a gas-cooled fast reactor, so small pins are
  proposed for GCFR's; however, the pumping power and system pressure for the gas coolant
  must both be far greater than for an LMFBR to achieve the required heat transfer coefficient.
  (See Table 11-7 and Chapter 17 for further discussion of these points.)

more subtle factors also come into play. For example, the increased fissile fraction, e, that accompanies reduced R_f would reduce the Doppler coefficient and require greater atom percent burnup in the fuel for a given burnup in MWd/kg of total fuel. Reducing the pin size increases the neutron flux, leading to higher fluences for a specific irradiation time.*

In summary, Eq. (8-1) identifies the principal factors which must be balanced in determining the fuel pin diameter, but a detailed optimization procedure involves several secondary effects which can become quite important.

C. FUEL RESTRUCTURING

Analysis of a fuel pin at high temperature and during its irradiation history is strongly influenced by the structure of the fuel. Like all ceramic and metallic materials, oxide fuel is composed of *grains*. A grain is a single crystal within which all the atoms are arranged in some particular pattern characterized by a *unit cell*. When the fuel is fabricated, the shapes of the grains are irregular, with the shape of each grain being controlled by the surrounding grains. The *grain boundary* is a zone of transition between crystalline alignments in adjacent grains. For curved boundaries, the atoms are more stable on concave than on convex surfaces; hence, thermal agitation at high temperatures will cause atoms to transfer from grains with convex surfaces to those with concave surfaces. As illustrated in Fig. 8-4, small grains have more convex surfaces, so atoms from small grains tend to move to larger grains. This process is called *grain growth*. Since an increase in temperature increases the thermal agitation of the atoms, which in turn facilitates grain growth, a large amount of grain growth might be expected at the high temperatures encountered by the fuel in a reactor.

Immediately upon reactor startup and attainment of full-power fuel temperature distributions, substantial alterations in the structure of the fuel do indeed take place. Above a certain temperature band, called the *equiaxed* temperature, the grains grow very rapidly and eventually reach many times their original size. These grains are called *equiaxed grains*. No radiation field is required to induce this change; it is an effect of temperature only. At an even higher temperature closer to the radial centerline of the fuel pin, the

---

*The flux increase results from the relation $\Sigma_f \phi V_{core} = 2.9 \times 10^{16} P_{core}(MW) = $ constant. As the pin diameter is reduced, V_{core} decreases faster than Σ_f increases; hence, ϕ increases. This is of benefit to the FFTF, which is a test reactor with the objective of maximizing ϕ in order to maximize fluence. It has smaller diameter pins (5.8 mm) than a typical power LMFBR.

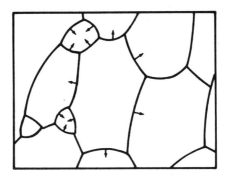

FIGURE 8-4. Grain growth. The boundaries move toward the center of curvature (arrows). As a result the small grains eventually disappear.[5]

temperature gradient causes the internally fabricated *pores** to collect and migrate up the temperature gradient (i.e., toward higher temperatures). Very long grains, called *columnar grains*, begin to emanate from the center like spokes of a wheel. The flow of pores to the radial center of the pellet leaves a more fully dense fuel material behind and this can result in a large hole in the center of the pellet. Most of this *restructuring* occurs in a matter of hours at full power.

Figure 8-5 is a cross section of a mixed-oxide fuel pellet irradiated at a linear power of 56 kW/m to a burnup of ~27 MWd/kg. The void in the center of the element is clearly visible, as well as the very long grains associated with the columnar region and the large grains of the equiaxed region. The microstructure near the cooler edges of the pin remains unaltered from its fabricated condition; this fuel is called *unrestructured*. It should be noted that the large black traces extending from the central void all the way to the inner cladding wall are cracks that most likely occurred during the cooldown from operating temperature. These cracks were probably not present during most of the fuel lifetime. Small cracks in the unrestructured and equiaxed fuel do appear during irradiation, however, due to stress in the fuel.

Another feature illustrated by Fig. 8-5 is the absence of the gap at the fuel-cladding interface in an irradiated pin. Fuel swelling and cracking cause the gap to disappear after a short irradiation time.

Since restructuring patterns are directly dependent upon temperature and temperature gradients, it is expected that the differing axial temperature distributions in an FBR pin would result in a corresponding axial variation

---

*Pores are tiny void spaces purposely fabricated into the fuel to allow space for fuel swelling and expansion during steady-state and thermal upset conditions. *Porosity* is the fraction of the fuel volume that is void.

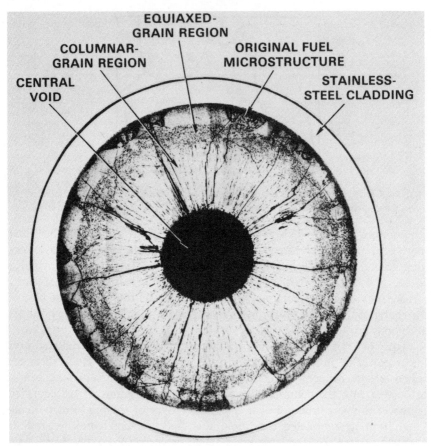

CENTRAL VOID — COLUMNAR-GRAIN REGION — EQUIAXED-GRAIN REGION — ORIGINAL FUEL MICROSTRUCTURE — STAINLESS-STEEL CLADDING

FIGURE 8-5. Cross section of mixed-oxide fuel pin irradiated to 27 MWd/kg. No melting.[6]

in fuel microstructure. This indeed is the case, as graphically illustrated by Fig. 8-6. The fuel pin shown was irradiated in EBR-II to a burnup of 50 MWd/kg at a peak linear power of 36 kW/m. Even though EBR-II has an active core length of only 34.3 cm, there is sufficient axial temperature variation to produce substantial differences in local fuel microstructure. We shall see in the later safety chapters that such differences play an important role in determining the transient response of the fuel pin to thermal upset conditions.

These microstructural differences affect fission gas retention patterns. Tracks left by pore migration within the columnar region allow fission gas to be easily vented to the central void region. Hence, it is of considerable interest to predict where such restructuring takes place, and how to assess the position of the boundaries physically separating the various

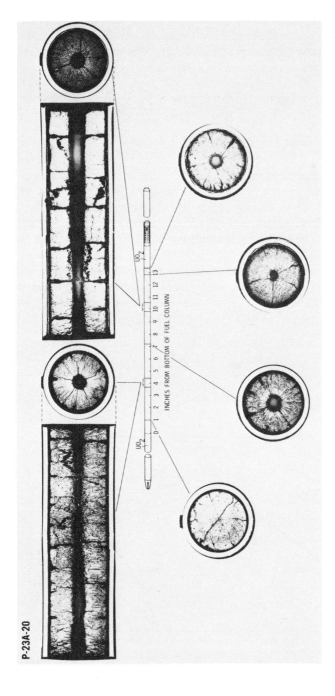

Figure 8-6. Cutaway view of microstructure patterns observed in mixed-oxide fuel irradiated in ERB-II. Compliments of L. A. Lawrence, J. W. Weber, and J. L. Devary, Hanford Engineering Development Laboratory, 1979.

263

microstructures. This is done by estimating the temperature boundaries which separate (1) the equiaxed from the unrestructured fuel (T_{eq}), and (2) the columnar from the equiaxed fuel (T_{col}). Approximate expressions for these temperatures can be written as (Ref. 7)

$$T_{eq} = \text{Equiaxed Temperature} = \frac{62\,000}{2.3 \log_{10} t + 26}, \qquad (8\text{-}3)$$

$$T_{col} = \text{Columnar Temperature} = \frac{68\,400}{2.3 \log_{10} t + 28}, \qquad (8\text{-}4)$$

where t is time measured in hours, and T is the absolute temperature (K). Figure 8-7 shows the drop in columnar temperature T_{col} as a function of time, as well as the fractional radius of columnar grain growth for three different centerline temperatures T_0. The equiaxed temperature, from Eq. (8-3), lies below the columnar temperature by 58 K at 1 hour, and extends to 78 K less at 10 000 hours.

Whereas the above formulation contains an explicit time dependence for the restructuring temperatures, more recent experience developed in the formulation of Reference 8 suggests that in-pile mixed oxide fuel data can be fit quite adequately by using the following simplified forms, which are time-independent:

$$T_{eq} = 1500°C, \qquad (8\text{-}3a)$$

$$T_{col} = 1900°C - 5\chi, \qquad (8\text{-}4a)$$

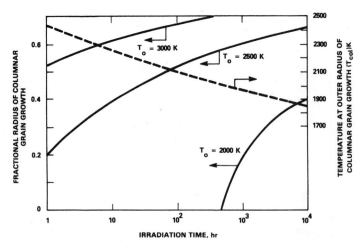

FIGURE 8-7. Location and temperature of the outer boundary of the columnar-grain region in pure UO_2 as a function of irradiation time.[7]

where χ is the linear power (kW/m). The reason these simplified forms apply is that the fuel-cladding gap closes very early in irradiation, causing a sufficient drop in the fuel pin temperature gradient to offset continued pore migration.

D. FISSION GAS RELEASE AND GAS PLENUM LENGTH

Fission product gas is produced in the fission process, both directly from fission and from the decay of radionuclides to stable gases, as in the decay of iodine to xenon. Most of the fission gas is xenon; the second most abundant gas is krypton. The yield of stable fission product gas, i.e., the fraction of fission products that become stable gas atoms, for a mixed UO_2-PuO_2 fast reactor is about 0.27.

Much of the fission gas produced in the unrestructured fuel, where temperatures are lowest, is retained in the fuel grains during normal operation. In the equiaxed and columnar grain regions, most of the fission gas escapes to the central void region or through cracks to the fuel-cladding interface. (The gas locked in the unrestructured fuel could be released in the event of fuel melting, a phenomenon that will influence the safety analysis of unprotected transients, as discussed in Chapter 15.)

The processes by which fission gas bubbles nucleate, grow, diffuse, and eventually collect and condense at grain boundaries are extremely complex. A description of these processes is beyond the scope of this text. Suffice it to say that below a temperature of about 1300 K, fission gas mobility is very low and there is essentially no gas escape. Between 1300 and 1900 K, atomic motion allows some diffusion to take place such that an appreciable amount of gas can reach escape surfaces over a long period of time. Above 1900 K, thermal gradients can drive gas bubbles and pores over distances comparable to grain sizes in days or months. Gas release then occurs when cavities reach a crack or other surface connected directly with free volume. Attempts to simulate the physics of these processes have been made with the FRAS2[9] and POROUS[10] computer codes.

A relatively simple mathematical correlation that effectively describes fission gas release from a large number of solid oxide pins irradiated in EBR-II follows.[11] The fraction of fission gas released, F, is

$$F = F_r A_r + F_u A_u, \tag{8-5}$$

where F_r and F_u are the release fractions for the restructured and unrestructured regions and A_r and A_u are the fractional fuel areas associated with these regions. The fit for F_r is a function only of burnup, B. The fit for F_u

includes the linear power, χ, in addition to burnup. Useful updated expressions from Reference 8 are

$$F_r = 1 - \frac{4.7}{B}\left(1 - e^{-B/5.9}\right),\qquad (8\text{-}6a)$$

$$F_u = 1 - \frac{25.6}{B - 3.5}\left[1 - e^{-(B/3.5-1)}\right]e^{-0.0125\chi}F'(B),\qquad (8\text{-}6b)$$

where
$$F'(B) = 1\ (B < 49.2)$$
$$= e^{-0.3(B-49.2)}\ (B \geqslant 49.2)$$
$$B = \text{local burnup (MWd/kg)}$$
$$\chi = \text{local linear power (kW/m)}$$

and F_u is set equal to zero whenever Eq. (8-6b) yields a negative number. A plot of fission gas release vs burnup and linear power using the model is shown in Fig. 8-8. For the illustration, the 23 kW/m case was assumed to

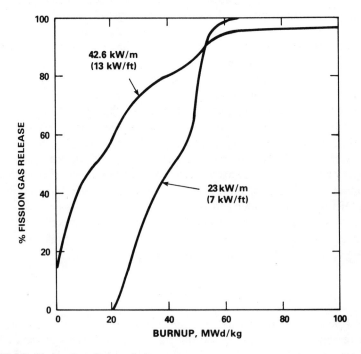

FIGURE 8-8. Fission gas release vs burnup.

have zero restructuring (i.e., $A_u = 1.0$, $A_r = 0.0$), whereas for the 42.6 kW/m case the values $A_u = 0.3$ and $A_r = 0.7$ were used. It should be noted that, whereas this figure illustrates the character of the simplified fission gas release model just presented, a determination of the fission gas released from a full fuel pin during irradiation would show a smoother set of curves (wherein the actual restructuring areas associated with an axially varying power distribution are taken into account).

Fission gas, once released from the fuel matrix and vented to open collecting zones (such as the central void or the fuel-to-cladding gap), will not reenter the solid form. The gas is collected in the fission gas plenum above or below the fuel stack. The pressure in the fission gas plenum is normally in approximate equilibrium with pressures throughout the fuel pin. This pressure is manifest in the fuel-cladding gap or in the cracks in the fuel adjacent to the cladding if the fuel there is well cracked and has little structural integrity left. Under these conditions the fission gas plenum pressure provides the loading pressure on the inside of the cladding.

For pin design, a plenum pressure at the end of pin life, P_p, is specified, consistent with the pin failure criteria (as described in Section 8-3). Typical fission gas plenum pressures at discharge are in the range of 6 to 10 MPa. The next problem to be treated here is to develop a relationship between plenum length, L_p, and P_p. A common technique for calculating the plenum pressure is to calculate the volume, V_0, of fission gas at a standard temperature $T_0 = 273$ K and pressure $P_0 = 1$ atm $= 1.013 \times 10^5$ Pa and to use the perfect gas relation

$$\frac{P_p V_p}{T_p} = \frac{P_0 V_0}{T_0}, \tag{8-7}$$

where subscript p refers to plenum conditions at rated power at the end of life.

A new parameter α_0 is defined as

$\alpha_0 =$ volume of fission gas released per cubic meter of fuel at standard temperature and pressure (273 K, 1 atmosphere).

The use of this parameter will, conveniently, introduce the active fuel length, L_f, (i.e., the core height) into the analysis so that the plenum length will ultimately be expressed in terms of the core height. The volume V_0 is related to the active fuel volume, V_f, as

$$V_0 = \alpha_0 V_f. \tag{8-8}$$

Hence,

$$\frac{P_p V_p}{T_p} = \frac{\alpha_0 V_f P_0}{273} . \tag{8-9}$$

Since the gas plenum is contained within the same cladding as the fuel,

$$\frac{V_p}{V_f} = \frac{L_p}{L_f}$$

so that

$$P_p L_p = \alpha_0 L_f \frac{T_p P_0}{273} . \tag{8-10}$$

The value of α_0 is related to the fission gas release fraction, F, and burnup, B, as follows:

$$\alpha_0 = \frac{F n \mathbf{R} T_0}{P_0} , \tag{8-11}$$

where

$n =$ kg-mol fission gas produced/m^3 fuel

$\mathbf{R} =$ universal gas constant $(8317 \text{ J/kg-mol} \cdot \text{K})$.

For a fuel theoretical density of 11 g/cm^3 ($=11\,000$ kg/m^3) and a smear density of 85% TD, and for a fission gas yield of 0.27,

$$n = B \frac{\text{MWd}}{\text{kg heavy metal}} \times \frac{2.93 \times 10^{16} \dfrac{\text{fissions}}{\text{MWs}} 86\,400 \dfrac{\text{s}}{\text{d}}}{6.023 \times 10^{26} \dfrac{\text{molecules}}{\text{kg-mol}}} \times (11\,000)(0.85) \frac{\text{kg oxide}}{\text{m}^3 \text{ fuel}}$$

$$\times \frac{238}{270} \frac{\text{kg heavy metal}}{\text{kg oxide}} \times 0.27 \frac{\text{kg-mol fission gas}}{\text{kg-mol fissioned}} = 0.94 \times 10^{-2} B.$$

This gives for α_0,

$$\alpha_0 = \frac{(0.94 \times 10^{-2})(8317)(273)}{1.013 \times 10^5} FB$$

$$= 0.21 \; FB \; (\text{m}^3 \text{ fission gas at STP/m}^3 \text{ fuel}), \tag{8-12}$$

where B is in MWd/kg, and F can be obtained as a function of the linear power at which the irradiation occurred from Eqs. (8-5) and (8-6).

8-3 FAILURE CRITERIA AND STRESS ANALYSIS FOR PIN DESIGN

The primary objectives of fuel pin design are to design a pin that will maintain its structural integrity during its lifetime in the reactor and will not expand or swell beyond the geometrical limits required for adequate cooling. A pin is considered to have failed if a breach occurs in the cladding that allows fission products to escape into the primary coolant.

Two approaches which have been used to predict cladding failure are described in this section. These predictions require the calculation of cladding strain and hoop stress. Hence, cladding stress analysis methods are presented after the introduction of the failure criteria.

A. STRAIN LIMIT APPROACH

A method commonly used to determine cladding failure for early fast reactor designs was the *strain limit* approach. This method was based on the observation that a combination of internal fuel pin gas pressure and/or mechanical interaction between the fuel and the cladding would at some point cause sufficient permanent cladding strain to cause rupture.

Though straightforward in concept, several difficulties exist in attempting to provide pin failure criteria in such simple terms. The principal difficulty is that the actual allowable cladding strain which can be accommodated prior to cladding rupture is a strong function of temperature, fluence and strain rate. Hence, failure conditions are strongly dependent upon the location of the cladding in the core, as well as steady-state vs accident conditions. Nonetheless, conservative bounds can be established to envelop worst-case conditions. The cladding strain limits established for the FFTF design provide such an example: 0.2% inelastic strain for steady state at goal exposure and 0.3% inelastic strain for thermal upset conditions. An inelastic strain limit of 0.7% was established as the integrity limit for cladding during emergency conditions (cf. Table 14-3).

B. CUMULATIVE DAMAGE FUNCTION

A somewhat more sophisticated guideline for pin lifetime more recently developed is the *cumulative damage function* (CDF), which utilizes the linear *life fraction rule* and includes damage for both steady-state and transient operation. The steady-state portion of the cumulative damage function is based on *time-to-rupture*, t_r, (also called time-to-failure) at the operating stress and temperature. If the cladding is subjected to a given stress at a constant temperature, the cladding will rupture at t_r. The life fraction rule assumes that if the cladding is subjected to that stress at that temperature for a short time, δt, the ratio of this duration to the time-to-rupture will produce a damage fraction, D, or the life fraction at that stress and temperature, i.e.,

$$D = \delta t / t_r. \tag{8-13}$$

The fractional damage is assumed to accumulate linearly. During the lifetime of a pin, the cladding may be subjected to a series of different values of stress and temperature for times, δt_i, for which the time-to-failure is t_{ri}. The cumulative damage function for irradiation at varying conditions is

$$\mathrm{CDF} = \sum_i \left(\frac{\delta t}{t_r} \right)_i. \tag{8-14}$$

The life fraction rule states that failure will occur when the cumulative damage fraction reaches some value, generally unity, or

$$\sum_i \left(\frac{\delta t}{t_r} \right)_i \simeq 1. \tag{8-15}$$

This rule is entirely empirical, but high-temperature design experience indicates that satisfactory correlations for t_r are often obtained. Additional experience will be necessary before it can be determined whether the effect of irradiation hardening influences the linear addition process generally employed.

The designer must in addition consider two transient conditions: thermal cycling during normal anticipated startups and shutdowns, and off-normal, or accidental, transients. It is possible to treat off-normal transients by a life fraction rule based on fraction of time-to-failure at a given stress level and temperature during the transient. Hence, the $\sum_i (\delta t / t_r)_i$ term in Eq. (8-15) would include the off-normal transients. The appropriate time-to-failure relations for transients are determined from cladding temperature tests which duplicate the transient thermal conditions.

A proposed method for treating fatigue from thermal cycling is to determine the number of cycles to failure, N_r, for a particular type of cycle, say type j, and to define a cycle-fraction equal to the number of this type of cycle expected during the cladding lifetime, N, to the allowable cycles, or $(N/N_r)_j$. The cyclic damage fractions could then be accumulated linearly along with the steady-state damage fractions, thus defining a new value for the cumulative damage function and life fraction rule as follows:

$$\text{CDF} = \sum_i \left(\frac{\delta t}{t_r}\right)_i + \sum_j \left(\frac{N}{N_r}\right)_j = 1. \qquad (8\text{-}16)$$

It might be noted that a strain-fraction rule, such as that mentioned in Section 8-3A above, could be added to the life fraction rule by adding a third item $\Sigma(\delta\varepsilon/\varepsilon_r)$, where $\delta\varepsilon$ is the cladding strain increment during a transient and ε_r is the allowable strain to rupture.

C. CORRELATIONS FOR TIME-TO-FAILURE

The time-to-failure must account for stress level, temperature, fluence, and any other mechanism degrading to cladding strength. Whereas it would be desirable to account for such mechanisms in a deterministic fashion, the present state-of-the-art requires t_r to be determined from correlations of stress rupture and transient tests performed on fuel cladding. These correlations may be based on one of several stress-rupture parameters or they may be simply a fit of the time-to-rupture data as a function of temperature, fluence and stress. One of the commonly used correlation parameters is the *Larson-Miller parameter* (LMP),

$$\text{LMP} = T(\log_{10} t_r + A), \qquad (8\text{-}17)$$

where

$t_r = $ failure time (hours)
$T = $ temperature of the stressed cladding (K)
$A = $ constant.

Another correlation commonly employed is the *Dorn parameter*, θ, given by

$$\theta = t_r e^{-Q/\mathbf{R}T}, \qquad (8\text{-}18)$$

where

$t_r = $ failure time (hours)
$Q = $ activation energy (J/kg-mol)
$\mathbf{R} = $ universal gas constant, 8317 J/kg-mol·K.

For steady-state unirradiated performance of 20% cold worked 316 stainless steel cladding without fuel, the time-to-rupture correlation utilizing the Dorn stress-rupture parameter is given by[12]

$$\ln \theta = A + \frac{1}{\lambda} \ln \ln \frac{\sigma^*}{\sigma},\tag{8-19}$$

where λ^{-1} is the slope of an experimental correlation of $\ln\theta$ vs $\ln\ln\sigma^*/\sigma$. Combining Eqs. (8-18) and (8-19) gives

$$\ln t_r = A + \frac{B}{T} + C\ln\ln \frac{\sigma^*}{\sigma},\tag{8-20}$$

where[12]

$$A = -42.980$$
$$B = \frac{Q}{\mathbf{R}} = 42\,020 \text{ (K)}$$
$$C = \frac{1}{\lambda} = 9.5325$$
$$\sigma^* = 930 \text{ MPa}$$
$$\sigma = \text{cladding hoop stress (MPa) [See Eq. (8-29)]}.$$

Extensive radiation experience with mixed oxide fuel clad in 20% cold worked 316 stainless steel has shown that the fuel pin will fail sooner than predicted with the above correlation, especially at low irradiation temperatures.[13] This fuel pin failure at substantially lower cumulative damage fractions is illustrated in Fig. 8-9. In this figure each data point represents the calculated value of the CDF for an irradiated fuel pin, i.e., calculated from Eqs. (8-20) and (8-15) using hoop stress values from time-dependent pin analysis codes. Some of these pins failed; some did not. At 800°C the fuel pin failures correspond to a calculated damage fraction of nearly one, as should be the case. However, at a lower irradiation temperature, for example, at 600°C, the calculated damage fractions of the failed pins are only about 0.005 instead of 1. This means that there is a further reduction in time-to-failure at these low irradiation temperatures. The effect may be attributed to cladding damage caused by chemical reactions between fission products and the cladding, but the precise mechanisms for this effect are still being determined. This phenomena has been termed the *Fuel Adjacency Effect* (FAE). In order to provide a quantitative basis for relating this observed in-pile fuel failure experience to the CDF methodology, the following expression has been superimposed on the data:

$$\text{FAE} = M + \frac{1}{1 + 10^{(5 - \alpha N)}},\tag{8-21}$$

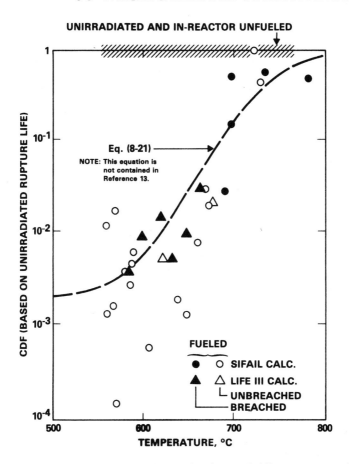

FIGURE 8-9. Calculated stress rupture damage fraction vs cladding temperature for cold worked 316 breached and selected unbreached EBR-II test fuel pins. (Adapted from Ref. 13).

where

$$M = \text{minimum CDF at failure} \simeq 2 \times 10^{-3}$$
$$\alpha = 0.017$$
$$N = T - 450$$
$$T = \text{temperature } (°C).$$

Therefore, the actual steady-state lifetime may be predicted by multiplying the time-to-failure from Eq. (8-20) and the FAE term from equation (8-21) for utilization in the CDF summation equation (8-16) as follows:

$$\text{CDF} = \sum_i \left(\frac{\delta t}{\text{FAE} \cdot t_r} \right)_i + \sum_j \left(\frac{N}{N_r} \right)_j = 1. \qquad (8\text{-}22)$$

The reader should be cautioned that the state of the art regarding steady-state time-to-failure is still evolving; hence, the details of this analysis will certainly change and perhaps even some of the bases of the approach will be modified. Nevertheless, the methods presented were the most current at the time of writing and they introduce many of the important concepts in an extremely complex and important field.

For transient analysis the failure correlation should be based on test data acquired under temperature transient conditions which are similar to those of the reactor transients involved. As with the steady-state fuel pin performance, there is also a Fuel Adjacency Effect exhibited under transient conditions.[14, 15] Both the Larson-Miller parameter and the Dorn parameter, as well as other correlation approaches, have been used to describe the dependence of fast reactor irradiated 20% cold worked 316 stainless steel cladding on the irradiation and transient conditions. One such correlation utilizing the Dorn parameter is given by[16]

$$\ln t_r = \frac{42\,800}{T} - 37.4201 (\dot{T})^{0.00743} + 8.8754 \ln \ln \frac{\sigma^*}{\sigma} + F \cdot A, \quad (8\text{-}23)$$

where

$T = $ temperature (K)
$\dot{T} = $ heating rate (K/s)
$\sigma = $ cladding hoop stress (MPa)
$\sigma^* = $ reference cladding hoop stress (930 MPa)

$$F = \left\{ \exp\left[-0.44 \left(\ln \ln \frac{\sigma^*}{\sigma} - 0.2 \right)^2 \right] \right\}^{0.8}$$

$$A = -(0.03677 + 1.53366\,X + 1.79437X^2$$

$$-0.47756\,X^3 + 0.039902\,X^4$$

$$-0.001099\,X^5)/(\ln \dot{T})^{0.74}$$

$X = $ neutron fluence, in units of 10^{22} n/cm^2 > 0.1 MeV.

This equation incorporates the influence of the Fuel Adjacency Effect as well as the temperature and stress dependence of cladding failure under transient heating conditions.

D. STRESS ANALYSIS

In order to calculate the cumulative damage fraction, it is necessary to analyze the stress-strain history of the cladding. The analysis involves the time-dependent elastic-plastic response of both fuel and cladding to loading pressures, and must include creep deformations and swelling. The analysis is performed with computer codes in time steps during the irradiation and at various axial positions of the fuel pin. The damage fraction $(\delta t/t_r)_i$ is calculated at each time step and position, and is added to the previous value of the cumulative damage fraction.

Accurate stress-strain analyses are extremely complex. The most sophisticated U.S. codes for this analysis are the LIFE[1] and PECT[17] codes. Simplified codes are also available, such as PECS[18] and SIEX-SIFAIL.[8] These codes perform thermal as well as stress analyses since the two always have to be coupled.

We will present here a simplified model for stress analysis that can be readily understood and used by the student to demonstrate the major principles of design analysis. While this model is somewhat similar to the simplified design codes, especially PECS, the reader must recognize that the model is only a beginning, and that fuel-pin design methods are continually being revised as data are obtained and evaluated.*

A flow chart for the simplified stress-strain analysis is shown in Fig. 8-10. It is assumed that the fission gas pressure, P_p, is known at the beginning of the time step. At each time step the stresses in the cladding are calculated. These are used later to calculate elastic strains. In addition, inelastic strains from creep and swelling are calculated.

An evaluation of the stresses depends on the nature of the loading on the cladding. As discussed in Section 8-2, the cladding can be loaded either by fission gas or by the fuel cladding mechanical interaction (FCMI) pressure, P_{fc}, of the fuel pushing against the cladding, depending on whether the fuel-cladding gap is closed. Hence, the first step in the calculation is to determine whether the gap is closed.

It is possible for the fuel-cladding gap to be closed but for the fuel to be so cracked that it is incapable of exerting a large force on the cladding. In this case the fission gas will continue to provide the loading on the cladding. For such conditions, only during a rapid startup early in the history of the

---

*In the United States, most of the early experimental data came from EBR-II, which is a relatively low-flux fast reactor. The FFTF will provide a step increase in the U.S. ability to understand cladding behavior. The Europeans—with Phénix, BN-350, and PFR—have been obtaining these types of data since the early 1970s.

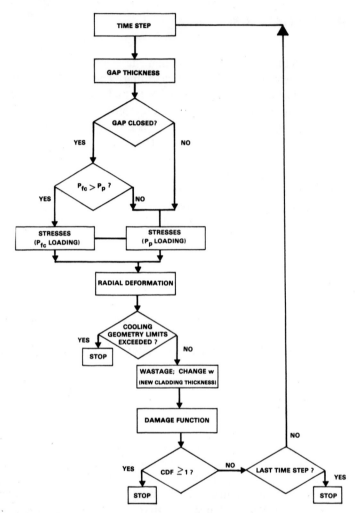

FIGURE 8-10. Flow chart of simplified fuel pin stress analysis.

irradiation will the fuel maintain enough structural integrity for the fuel-cladding interface pressure to be of dominant importance.

Calculation of Gap Thickness

It is necessary to calculate the relative positions of the outer fuel surface and the inner cladding surface at rated power in order to determine the hot gap thickness. There are two steps to the calculation.

The first step is to determine the *residual* gap thickness, G_c. This is the cold gap that would exist if the irradiated fuel pin were removed from the reactor and cooled to room temperature. Initially, G_c is simply G_0, the gap in the original fuel as fabricated. In fast reactor design the following correlation is widely used[19]

$$G_c = G_0 \left\{ 1 - 4.851 \times 10^{-4} \chi (\chi - 19.1)(1 - e^{-C}) \right.$$

$$-0.365 \left[1 - \exp(-2.786 \times 10^{-2} \chi B) \right]$$

$$\left. -8.81 \times 10^{-5} \frac{B}{G_0} \right\}, \tag{8-24}$$

where

$$G_c = \text{residual gap thickness (mm)}$$
$$G_0 = \text{fabricated gap thickness (mm)}$$
$$\chi = \text{linear power (kW/m)}$$
$$B = \text{burnup (MWd/kg)}$$
$$C = \text{power cycles.*}$$

The second step is to calculate the thermal expansion of the fuel and cladding from room temperature up to operating temperatures. The thermal expansion of uncracked UO_2-PuO_2 is used here. If the fuel is cracked, it may close the gap. Letting G refer to the hot gap,

$$G = G_c - \left(\frac{\delta R_f}{R_f} R_f - \frac{\delta R_{ci}}{R_{ci}} R_{ci} \right), \tag{8-25}$$

where R_f and R_{ci} are the cold pellet and inner cladding radii.

The fractional change in the inner cladding radius is obtained using the mean linear thermal expansion coefficient α_c as follows:

$$\frac{\delta R_{ci}}{R_{ci}} = \alpha_c (T_c - 298), \tag{8-26}$$

---

*One power cycle represents zero to full power and back to zero power. NOTE: If $\chi < 19.1$, the term in Eq. (8-24) containing the power cycle correction should be set to zero.

where for type 316 stainless steel,[20]

$$\alpha_c(T) = 1.789 \times 10^{-5} + 2.398 \times 10^{-9} T_c + 3.269 \times 10^{-13} T_c^2,$$

and

$$T_c = \text{average cladding temperature (K)}.$$

The thermal expansion of the fuel surface can be approximated by using a simplified model which neglects the internal forces due to differential thermal growth. Such a model, determined by summing up the thermal expansion of the radial fuel increments, can be mathematically expressed as

$$\frac{\delta R_f}{R_f} = \frac{\int_{R_0}^{R_f} [\alpha_f(T_f - 273)] r \, dr}{\int_{R_0}^{R_f} r \, dr}, \qquad (8\text{-}27)$$

where for 75% UO_2-25% PuO_2 mixed oxide fuel,[20] the mean linear thermal expansion coefficient is

$$\alpha_f = 6.8 \times 10^{-6} + 2.9 \times 10^{-9}(T_f - 273),$$

and T_f = fuel temperature (K), where T_f is a function of r as calculated by methods explained in Chapter 9.

If the calculated value of G is negative, the gap is closed. Since physically the gap cannot be negative, the fuel and cladding must deform to accommodate this situation, and this gives rise to fuel cladding mechanical interaction.

Plenum Gas Loading

When the plenum gas provides the loading on the cladding, the cladding may be conveniently approximated by a *thin-walled* tube *closed* at both ends. The geometry for the stress-strain analysis is shown in Fig. 8-11.

The stresses are calculated from thin-walled vessel theory, for which $w \ll R$ and the difference between the inside cladding radius and the mean cladding radius is ignored. The thin-walled vessel stresses can be obtained from Fig. 8-12.

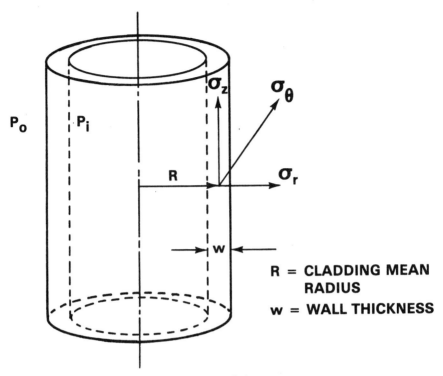

FIGURE 8-11. Geometry for cladding stress analysis.

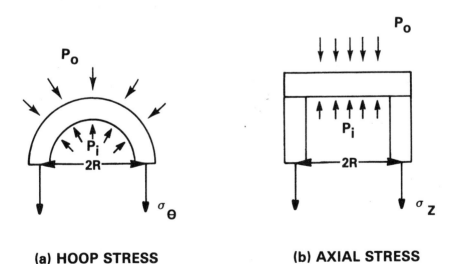

(a) HOOP STRESS   **(b) AXIAL STRESS**

FIGURE 8-12. Geometry for hoop stress and axial stress.

The hoop stress, σ_θ, is obtained from the following force balance in the tangential (θ) direction

$$(P_i - P_o)2RL = \sigma_\theta \cdot 2wL, \tag{8-28}$$

where L is the length of the thin-walled vessel of thickness w. Solving for the hoop stress gives

$$\sigma_\theta = (P_i - P_o)\frac{R}{w}. \tag{8-29}$$

For a closed-end cylindrical vessel there is an axial stress σ_z. The geometry is shown in Fig. 8-12(b). An axial force balance gives

$$(P_i - P_o) \cdot \pi R^2 = \sigma_z \cdot 2\pi Rw,$$

or

$$\sigma_z = (P_i - P_o)\frac{R}{2w}. \tag{8-30}$$

This expression for σ_z is equal to $\sigma_\theta/2$ if P_i and P_o each are the same in Eqs. (8-29) and (8-30).

The radial stress, σ_r, is compressive (hence, negative). The maximum value of σ_r equals P_i at the inside of the cladding or P_o at the outside of the cladding. Some analysis codes assume that $\sigma_r = -P_i$; others ignore σ_r altogether, whereas others use the approximation

$$\sigma_r = -\tfrac{1}{2}(P_i + P_o). \tag{8-31}$$

The radial stress is always used with the hoop or axial stress and is much smaller than either; hence, the various approximations for σ_r have little effect on the final results.

For the calculation of deformation with multiaxial stresses, the various stresses combine to give a single *effective stress*, $\bar\sigma$, equal to

$$\bar\sigma = \frac{1}{\sqrt{2}}\left[(\sigma_\theta - \sigma_r)^2 + (\sigma_r - \sigma_z)^2 + (\sigma_z - \sigma_\theta)^2\right]^{1/2}. \tag{8-32}$$

However, for the determination of time-to-failure in Eqs. (8-20) and (8-23), the maximum principal stress is used (i.e., hoop stress). The effective stress from Eq. (8-32) will be used later in the Prandtl-Reuss flow rule Eq. (8-66).

For plenum gas loading of the cladding,

$$P_i = P_p, \text{ and } P_o = P_b,$$
(8-33)

where $P_p =$ fission gas plenum pressure and $P_b =$ bulk coolant pressure at the axial position of interest.

Fuel-Cladding Mechanical Interaction Loading

When the gap is closed and the fuel-cladding mechanical interaction pressure is greater than the fission gas pressure, the interface pressure P_{fc} is the pressure that loads the cladding.

Neglecting the axial stress due to fission gas and fuel growth, the hoop and radial stresses in the cladding are again given by Eqs. (8-29) and (8-31), with

$$P_i = P_{fc}, \text{ and } P_o = P_b.$$
(8-34)

Given the assumption of zero axial stress, i.e.

$$\sigma_z = 0,$$
(8-35)

Eq. (8-32) becomes

$$\bar{\sigma} = \frac{1}{\sqrt{2}} \left[(\sigma_\theta - \sigma_r)^2 + \sigma_r^2 + \sigma_\theta^2 \right]^{1/2}.$$
(8-36)

When the gap computed from Eq. (8-25) is a negative quantity, the fuel applies a pressure $P_i = P_{fc}$ to the cladding. Defining u_c and u_f as the elastic deformations of the cladding and the fuel, as shown in Fig. 8-13, the magnitude $|G|$ is given as

$$|G| = u_c - u_f.$$
(8-37)

The fuel displacement u_f is negative (i.e., in the opposite direction from u_c) so that the magnitudes of the two deformations add together. In order to determine P_{fc}, we must express u_c and u_f in terms of P_{fc} and substitute the resulting expressions into Eq. (8-37).

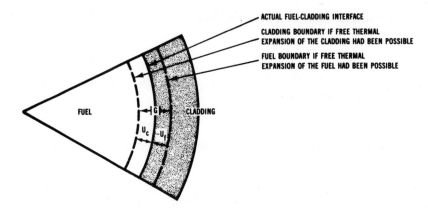

FIGURE 8-13. Schematic of elastic fuel and cladding deformations which occur when gap closure leads to interface pressure loading.

Cladding Deformation (Thin-Walled Cylinder). In order to obtain u_c in terms of P_{fc}, we write the hoop strain, ε_θ, in terms of the hoop stress and radial stress as

$$\varepsilon_\theta = \frac{1}{E_c}(\sigma_\theta - \nu_c\sigma_r), \qquad (8\text{-}38)$$

where E_c = Young's modulus for the cladding

ν_c = Poisson's ratio for the cladding.

The hoop strain is defined as:

$$\varepsilon_\theta = \frac{\delta(\text{circumference})}{\text{circumference}} = \frac{\delta R}{R}, \qquad (8\text{-}39)$$

where R is the radius at the fuel-cladding interface (essentially equal to R_{ci}). Combining Eqs. (8-38) and (8-39), and replacing δR by the radial cladding displacement u_c gives

$$u_c = \varepsilon_\theta R = \frac{R}{E_c}(\sigma_\theta - \nu_c\sigma_r). \qquad (8\text{-}40)$$

Substituting Eqs. (8-29) and (8-31) for σ_θ and σ_r, together with Eq. (8-34), yields

$$u_c = \frac{R}{E_c}\left[(P_{fc} - P_b)\frac{R}{w} + \frac{\nu_c}{2}(P_{fc} + P_b)\right]. \qquad (8\text{-}41)$$

Fuel Deformation (Thick-Walled Cylinder). The next step is to express the fuel displacement, u_f, in terms of P_{fc}. The relation between u_f and P_{fc} involves the use of thick-walled vessel theory because the fuel must be considered as a thick cylinder. Since thick-walled theory may not be familiar to the reader, we shall divert our attention briefly from the flow chart of Fig. 8-10 to a derivation of stresses in a thick-walled cylinder needed to relate u_f to P_{fc}. The derivation appears in various textbooks, e.g., Reference 21, which we follow closely here.

The geometry is shown in Fig. 8-14. We assume that $\sigma_z = 0$, and the axial strain in the fuel is constant. Using the nomenclature in Fig. 8-14, we can write the following force balance in the tangential direction on a ring of thickness dr (recalling that σ_r is the negative of pressure):

$$2\sigma_\theta \, dr + 2\sigma_r r = 2(\sigma_r + d\sigma_r)(r + dr). \tag{8-42}$$

Neglecting higher order terms gives the equilibrium equation for a thick-walled cylinder:

$$r\frac{d\sigma_r}{dr} + \sigma_r - \sigma_\theta = 0. \tag{8-43}$$

Our objective is to solve this equation for the two unknowns, σ_r and σ_θ, in terms of the boundary condition, $P_o = P_{fc}$, and to relate these stresses at the fuel cladding interface to the fuel displacement u_f.

A second equation is needed to evaluate the two unknowns. This equation is supplied by the axial strain, ε_z, which is independent of radius and is related to σ_θ and σ_r through Poisson's ratio, ν, and Young's modulus, E, for

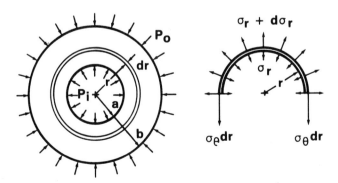

FIGURE 8-14. Geometry for a thick-walled cylinder.

the thick-walled material:

$$\varepsilon_z = -\frac{\nu}{E}(\sigma_\theta + \sigma_r). \tag{8-44}$$

The negative sign is needed because σ_θ is a tensile stress and, hence, will be positive and will tend to decrease the axial length, while σ_r is compressive (negative) and will tend to increase the length. The axial strain, ε_z, is also unknown, but it will be evaluated in terms of the pressure boundary condition P_i at $r=a$ and P_o at $r=b$.

The hoop stress can be eliminated from Eq. (8-43) by means of Eq. (8-44) to give

$$r\frac{d\sigma_r}{dr} + 2\sigma_r = -\frac{E}{\nu_f}\varepsilon_z. \tag{8-45}$$

Multiplying this equation by r gives

$$\frac{d(r^2\sigma_r)}{dr} = -\frac{E}{\nu_f}\varepsilon_z r. \tag{8-46}$$

Integrating gives

$$\sigma_r = -\frac{E}{\nu_f}\frac{\varepsilon_z}{2} + \frac{C}{r^2}. \tag{8-47}$$

The first term on the right-hand side is a constant. This constant and C are evaluated simultaneously from the two boundary conditions:

$$\sigma_r = -P_i \text{ at } r=a, \tag{8-48}$$

$$\sigma_r = -P_o \text{ at } r=b.$$

Hence,

$$C = \frac{a^2 b^2(P_o - P_i)}{b^2 - a^2}, \tag{8-49}$$

$$-\frac{E}{\nu_f}\frac{\varepsilon_z}{2} = \frac{P_i a^2 - P_o b^2}{b^2 - a^2}. \tag{8-50}$$

From Eqs. (8-44) and (8-47),

$$\sigma_\theta = \frac{P_i a^2 - P_o b^2 - a^2 b^2 (P_o - P_i)/r^2}{b^2 - a^2}, \qquad (8\text{-}51)$$

$$\sigma_r = \frac{P_i a^2 - P_o b^2 + a^2 b^2 (P_o - P_i)/r^2}{b^2 - a^2}. \qquad (8\text{-}52)$$

The radial displacement of the outer surface of the thick-walled cylinder is

$$u(b) = \varepsilon_\theta b = \frac{b}{E} \left[\sigma_\theta(b) - \nu \sigma_r(b) \right]$$

$$= \frac{b}{E} \left[P_i \frac{2a^2}{b^2 - a^2} - P_o \left(\frac{b^2 + a^2}{b^2 - a^2} - \nu \right) \right]. \qquad (8\text{-}53)$$

Letting $P_i = P_p$, $P_o = P_{fc}$, $a = R_o$ (where R_o is the radius of the central void as shown in Figs. 8-5 and 9-4), and $b = R$ (R being satisfactorily approximated by R_{ci}), the fuel displacement can be obtained from Eq. (8-53) as

$$u_f = \frac{R}{E_f} \left[P_p \frac{2R_0^2}{R^2 - R_0^2} - P_{fc} \left(\frac{R^2 + R_0^2}{R^2 - R_0^2} - \nu_f \right) \right]. \qquad (8\text{-}54)$$

Determination of P_{fc}. Substituting the expressions for u_f and u_c into Eq. (8-37) gives

$$|G| = \frac{R}{E_c} \left[(P_{fc} - P_b) \frac{R}{w} + \frac{\nu_c}{2} (P_{fc} + P_b) \right]$$

$$- \frac{R}{E_f} \left[P_p \frac{2R_0^2}{R^2 - R_0^2} - P_{fc} \left(\frac{R^2 + R_0^2}{R^2 - R_0^2} - \nu_f \right) \right]. \qquad (8\text{-}55)$$

Solving for P_{fc} gives the desired result

$$P_{fc} = \frac{\dfrac{|G|}{R} + \dfrac{P_p}{E_f} \dfrac{2R_0^2}{R^2 - R_0^2} + \dfrac{P_b}{E_c} \left(\dfrac{R}{w} - \dfrac{\nu_c}{2} \right)}{\dfrac{1}{E_f} \left(\dfrac{R^2 + R_0^2}{R^2 - R_0^2} - \nu_f \right) + \dfrac{1}{E_c} \left(\dfrac{R}{w} + \dfrac{\nu_c}{2} \right)}, \qquad (8\text{-}56)$$

where $|G|$ is the absolute value of the negative gap calculated by Eq. (8-25). We now have a value for the interface pressure that can be used to calculate cladding stresses when the gap is closed and $P_{fc} > P_p$.

Radial Deformation

Cladding. The *radial cladding deformation*, δR_c, is needed in order to determine if the cladding has expanded beyond the geometrical limits required for adequate cooling. [This deformation is different from the radial displacement, u_c, of Eq. (8-41).]

The radial cladding deformation is related to the creep strain, ε_θ^c, and the swelling strain, ε_θ^s, by

$$\delta R_c = R_c \left(\varepsilon_\theta^c + \varepsilon_\theta^s \right), \qquad (8-57)$$

where R_c is the mid-wall cladding radius. The creep and swelling strains are calculated from a combination of the Prandtl-Reuss flow rule and correlations for swelling and thermal and irradiation creep. (In some structural codes, e.g., PECS, a radial velocity, or *deformation rate*, $\dot{R}_c$, is calculated from *strain rates*, and deformation is obtained by integration of $\dot{R}_c$.)

As will be discussed in Chapter 11, the swelling of 20% cold worked 316 stainless steel is a function of the fluence and irradiation temperature. Using the relationship for volumetric swelling $\Delta V / V$, given in Chapter 11, the swelling strain is

$$\varepsilon_\theta^s = \frac{1}{3} \frac{\Delta V}{V}. \qquad (8-58)$$

The Prandtl-Reuss flow rule evaluates creep strain by means of *equivalent stresses and strains* (as described, for example, in Reference 22). For the case of interest here, the flow rule relates creep strain in the cladding to calculated stresses [as in Eqs. (8-29)–(8-31)], an effective stress [Eq. (8-32) or (8-36)], and an empirical effective inelastic strain, $\bar{\varepsilon}$, as follows:*

$$\varepsilon_\theta^c = \frac{\bar{\varepsilon}}{\bar{\sigma}} \left(\sigma_\theta - \frac{\sigma_r + \sigma_z}{2} \right), \qquad (8-59)$$

where

$$\bar{\varepsilon} = \tfrac{2}{3} \left(\varepsilon_{\text{th creep}} + \varepsilon_{\text{ir creep}} \right), \qquad (8-60)$$

and $\varepsilon_{\text{th creep}}$ and $\varepsilon_{\text{ir creep}}$ are thermal and irradiation creep strains.

---

*By comparison with the more complete Prandtl-Reuss flow rules of Eqs. (8-66)–(8-68), it is observed that Eqs. (8-59) and (8-60) represent a special case for which only inelastic hoop strains are non-zero.

Correlations for thermal and irradiation creep strain for 20% cold worked 316 stainless steel cladding are given below. These correlations tend to change as more data become available. In fact, there are some recent indications that it may be appropriate to combine the thermal and irradiation creep relationships, rather than to maintain a separate-effects approach.

Thermal Creep[23]

$$\varepsilon_{th\ creep} = C_1 \sigma_\theta \cosh^{-1}(1 + C_2 t) + C_3 \sigma_\theta^n t^m + C_4 \sigma_\theta^n t^{2.5}, \qquad (8\text{-}61)$$

where $\qquad \sigma_\theta$ = hoop stress (MPa)

$\qquad\qquad\quad t$ = time (hr)

and the remaining constants (C_1 through C_4) and coefficients (n and m) are supplied by Reference 23 for the particular temperature ranges of interest.

Irradiation Creep[24]

$$\varepsilon_{ir\ creep} = 10^{-6} \sigma_\theta \left[0.67F + 5.8 \times 10^4 e^{-8000/T} (F - 8.5 \tanh F/8.5) \right], \quad (8\text{-}62)$$

where $\quad \sigma_\theta$ = hoop stress (MPa)

$\qquad\quad F$ = displacements per atom (dpa)

$\qquad\quad T$ = temperature (K).

Displacements per atom can be related to neutron fluence above 0.1 MeV for a particular reactor design. In EBR-II, where testing was done to obtain this equation, one dpa corresponds to $\sim 2 \times 10^{21}$ n/cm^2 (above 0.1 MeV).

Fuel. Displacement of the fuel surface is also generally calculated in pin-design codes. One use of this calculation is to determine the amount of porosity required during fuel fabrication. A phenomenological treatment of fuel swelling and fuel thermal and irradiation creep (as, for example, in the LIFE code) is complex and will not be introduced in the present simplified model. It should be noted, however, that fuel deformation is just as important as cladding deformation as a cladding stress relief mechanism. Discussions of fuel swelling and fuel creep are given in Chapter 11.

Cladding Wastage

Corrosion of the cladding by sodium and chemical reactions of fuel and fission products with the cladding slowly decrease the thickness of the cladding. This phenomenon is called *wastage*. Wastage is accounted for in the stress-strain analysis through the calculation of the hoop and axial stresses. The reduction in cladding thickness due to wastage is calculated at each time step and the cladding thickness w in the equations for the stresses σ_θ and σ_z is modified accordingly.

E. LIFE CODE APPROACH

Field Equations

The field equations for the stress analysis in the LIFE code[1] are listed below. They apply to both fuel and cladding.

(a) *Equilibrium**

$$r\frac{\partial \sigma_r}{\partial r}+\sigma_r-\sigma_\theta=0. \qquad (8\text{-}63)$$

(b) *Kinematics*

$$\varepsilon_r^T=\frac{\partial u}{\partial r}, \ \varepsilon_\theta^T=u/r, \ \varepsilon_z^T=\text{constant}, \qquad (8\text{-}64)$$

where

$$\varepsilon^T=\text{total strain}$$
$$u=\text{radial displacement (m)}.$$

(c) *Constitutive relations*

$$\varepsilon_r^T=\frac{1}{E}\left[\sigma_r-\nu(\sigma_\theta+\sigma_z)\right]+\alpha T+\varepsilon_r^c+\varepsilon_r^s,$$

$$\varepsilon_\theta^T=\frac{1}{E}\left[\sigma_\theta-\nu(\sigma_r+\sigma_z)\right]+\alpha T+\varepsilon_\theta^c+\varepsilon_\theta^s, \qquad (8\text{-}65)$$

$$\varepsilon_z^T=\frac{1}{E}\left[\sigma_z-\nu(\sigma_r+\sigma_\theta)\right]+\alpha T+\varepsilon_z^c+\varepsilon_z^s,$$

where

$$\varepsilon^c=\text{creep strain}$$
$$\varepsilon^s=\text{swelling strain}$$
$$\alpha T=\int \alpha\, dT, \text{ linear thermal expansion}.$$

---

*Same as Eq. (8-43).

(d) *Prandtl-Reuss flow rule*

$$\Delta\varepsilon_r^c = \frac{\Delta\varepsilon^c}{\bar{\sigma}}\left(\sigma_r - \frac{\sigma_\theta + \sigma_z}{2}\right),$$

$$\Delta\varepsilon_\theta^c = \frac{\Delta\varepsilon^c}{\bar{\sigma}}\left(\sigma_\theta - \frac{\sigma_z + \sigma_r}{2}\right), \qquad (8\text{-}66)$$

$$\Delta\varepsilon_z^c = \frac{\Delta\varepsilon^c}{\bar{\sigma}}\left(\sigma_z - \frac{\sigma_r + \sigma_\theta}{2}\right),$$

where $\Delta\varepsilon^c = $ change in ε^c in a time step, and

$$\bar{\sigma} = \frac{1}{\sqrt{2}}\left[(\sigma_r - \sigma_\theta)^2 + (\sigma_r - \sigma_z)^2 + (\sigma_\theta - \sigma_z)^2\right]^{1/2}, \qquad (8\text{-}67)$$

$$\Delta\varepsilon^c = \frac{\sqrt{2}}{3}\left[(\Delta\varepsilon_r^c - \Delta\varepsilon_\theta^c)^2 + (\Delta\varepsilon_r^c - \Delta\varepsilon_z^c)^2 + (\Delta\varepsilon_\theta^c - \Delta\varepsilon_z^c)^2\right]^{1/2}. \qquad (8\text{-}68)$$

These equations are solved for several radial nodes in both the fuel and the cladding at various axial nodes. Cracking of fuel is accounted for in the fuel stress analysis. Attaining a solution for the pressure loading when the fuel-cladding gap is closed is considerably more complex than for the simplified model of Section 8-3D.

Example of Calculated Pressure Loadings and Cladding Strains

A sample calculation of long-term fuel-pin deformation, as performed by the LIFE code, is shown in Fig. 8-15. Shown on the graph are the cladding swelling and the inelastic strain (creep) at the core midplane for a peak power LMFBR fuel pin irradiated at full power to design burnup, together with the cladding loading pressures.

Although the fuel and cladding are not initially in contact, the gap closes soon after operation begins and the fuel cladding mechanical interaction pressure is calculated by the LIFE code to load the cladding to 6 MPa. After 5000 hours, the gap is predicted to open again if fuel maintained its structural integrity, i.e., did not crack. Beyond about 7000 hours, the plenum gas pressure provides the primary cladding loading. Also beyond this time, the two inelastic strain contributions control the radial deformation of the cladding.

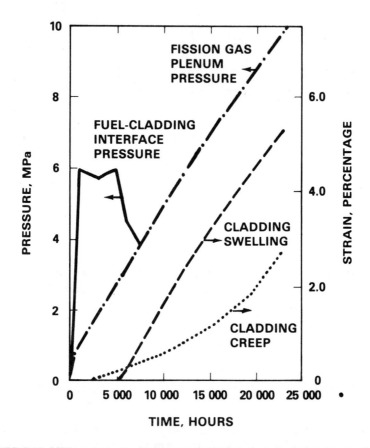

FIGURE 8-15. LIFE analysis of a high power LMFBR fuel pin to design burnup. Compliments of D. S. Dutt, Hanford Engineering Development Laboratory, 1978.

In an actual fuel irradiation, the gap does not reopen as predicted in this calculation. Instead, fuel cracking due to repeated thermal stresses from shutdown and startup, together with fuel swelling, keeps the fuel pressed against the cladding. The fuel has little strength at this stage, however. Hence, the fuel-cladding interface pressure is essentially equal to the fission gas plenum pressure.

The complexity of fuel pin design, first illustrated in Fig. 8-1, is now summarized further in the diagram of Fig. 8-16. By now we have referred to

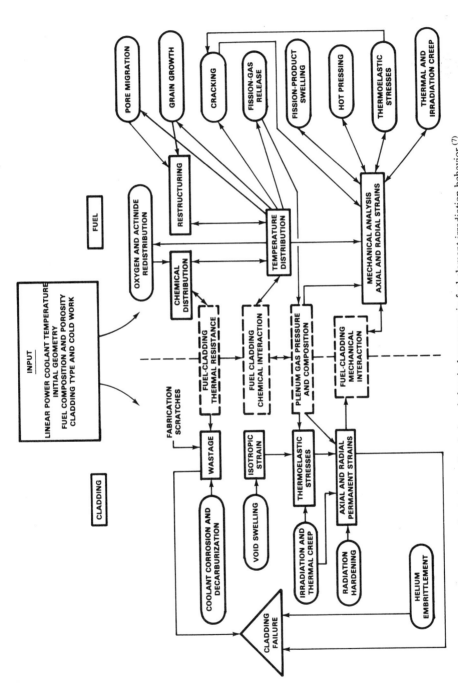

Figure 8-16. Interrelation of mechanical, metallurgical, and chemical processes in fuel-element irradiation behavior.[7]

291

most of the phenomena on the chart, although many will be discussed further in Chapters 9, 10 and 11. The relevant physical processes are shown in the ovals and the observable consequences on the fuel pin are shown in rectangles.

8-4 FUEL ASSEMBLY DESIGN

The method universally adopted for assembling FBR fuel pins into manageable clusters is to collect them into a hexagonal duct, as depicted in Fig. 2-3. These ducts, together with the fuel pins and the associated end hardware, are referred to in the literature as both assemblies and subassemblies. They are assembled into an overall core according to the possible arrangements denoted by Fig. 2-2. Control assemblies are normally dispersed throughout the fueled core region whereas blanket assemblies surround the core.* Shielding assemblies are normally included outside the radial blanket. The purpose of the present section is to provide some background on the geometric and material considerations involved in designing such assemblies.

The bulk of the reactor power is generated in the fuel assemblies. A typical homogeneous LMFBR generates from 85 to 95% of the power in the active core, 3 to 8% in the radial blanket, and about 3 to 6% in the axial blanket. Hence, fuel assemblies constitute an area of very high power density, and the geometric arrangement of the pins within an assembly is of considerable importance.

A. DUCT CONSIDERATIONS

The ducts serve a number of functions in an LMFBR, including the following:

(1) The ducts force the sodium to flow past the fuel pins and not bypass the high-flow-resistance path within the pin bundle.
(2) By enclosing the flow through an individual pin bundle, the ducts allow individual assembly orificing, thus giving the reactor designer positive control of the power-to-flow ratio throughout the core and radial blanket.
(3) The ducts provide structural support for the pin bundle.

---

*Blanket assemblies are also included within the core for the heterogeneous design.

(4) The ducts provide a mechanical means to load the pins into the core as a unit, with the assembly of ducts constrained by the core restraint system.

(5) The ducts provide a barrier to the potential propagation to the rest of the core of a possible accident initiated by the rupture of a few pins in an assembly.

Given the decision to employ a duct in which to house a pin bundle, the next question is the geometrical arrangement of the pins. Whereas LWR's normally employ a square lattice, the desire to maximize the fuel volume fraction in an FBR results in a considerable advantage to the triangular array, as discussed in Chapter 4. Also, the tighter triangular pitch improves the overall heat transfer characteristics. Hence, the hexagonal geometry has been universally chosen for LMFBR duct designs.

Although rigid ducts have become generally accepted for the early LMFBR designs, the amount of space occupied by the steel duct is well recognized to reduce the fuel volume fraction and, as such, directly reduce the breeding ratio. Three approaches to minimize the steel content in the duct walls are (1) to use a non-uniform duct wall thickness, (2) to slot the ducts to reduce the pressure differential across the duct wall, and (3) to use thinner duct walls. The first approach employs a mechanical deformation analysis to assess where duct thinning can be accomplished without an undue loss of structural strength. The second idea is to cut slots, as shown in

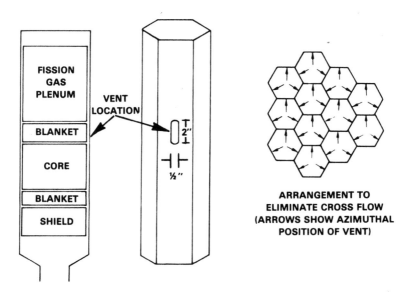

FIGURE 8-17. Slotted duct design.

Fig. 8-17, to reduce the internal duct pressure (and, hence, the stress). An axial vent location above the active core region results in a substantial drop for the in-core pressure without allowing coolant crossflow in the high heat generation regions. The crossflow that would occur could be minimized by a duct arrangement shown in the inset to Fig. 8-17. The third approach, to use thinner duct walls, would allow the ducts to touch in the core region during operation. This would necessitate refueling at as low a temperature as practicable, in order to allow thermal contraction to reduce the loads required to withdraw the fuel assemblies.

Efforts to minimize the swelling characteristics of duct material are equally important to that of reducing wall thickness. In fact, it is quite possible that duct swelling may be more of a limitation to core lifetime than material degradation or stress effects in individual pins.

B. FUEL ASSEMBLY SIZE

An important design decision is to determine the number of pins per assembly. Included in the factors for consideration are the following: [7]

(1) Reactivity Worth per Assembly—There is a desire to keep this to a minimum for the peak worth assemblies to minimize the reactivity swing during refueling; an argument exists to keep the maximum worth less than one dollar to assure a mild transient response even in the extremely unlikely event of dropping a fuel assembly into a just critical core.

(2) Decay Heat Removal—Spent fuel discharged from the reactor must be cooled, and the total heat removal capability required in the refueling equipment is directly proportional to the number of pins involved for a single movement.

(3) Assembly Weight—The heavier the assembly, the more difficult the handling procedures during fabrication and refueling.

(4) Mechanical Performance—Bowing and dilation problems become magnified with total size.

(5) Cost—Larger assemblies generally result in lower overall core cost.

(6) Refueling Time—Larger assemblies require fewer steps, hence less time, to refuel.

(7) Criticality During Shipment—Safety precautions normally require that the possibility of flooding with water be considered during ex-vessel shipments. Reactivity clearly increases with the size of the assembly.

Current practice is to cluster about 9 or 10 rows of fuel pins, which results in a total assembly size of 217 and 271 pins, respectively. The total number of pins as a function of pin rows is shown in tabular form in Fig. 2-3.

C. FUEL PIN SPACER DESIGN

The two basic design choices commonly considered for radially separating the pins within an assembly are the use of wire wraps or grid spacers. The prototype and demonstration plants in the United Kingdom and Germany and the U.S. Fermi reactor use grid spacers; other plants use wire wrap. In the wire wrap spacer design, a round steel wire is welded at one axial pin extremity and then spirally wound around the pin at a specified pitch or lead (the axial distance separating the repeat pattern) up to a weld at the other pin extremity. Grid spacers, on the other hand, consist of a cross-lacing of steel webbing which is normally anchored to the assembly duct wall* at specified axial levels. The two principal variations of the grid concept are the staggered grid, wherein the grid mesh is offset at alternate axial levels, and the honeycomb concept. Figure 8-18 contains a sketch of the wire wrap vs the two grid pin spacer concepts.

Wire wraps have enjoyed the most popularity to date, mainly because fabrication is relatively easy and inexpensive. Mechanical vibration problems are minimized because even for a pitch of one foot, wire wrap contact with cladding occurs every two inches (because of the six nearest neighbors—each with their own wire wrap). It is possible, however, for the wire wrap to move slightly during handling. Thus, the precise location of support points for the pins is not known. This adds some uncertainty to the thermal hydraulic and fuel pin bundle structural calculations.

Perhaps the greatest stimulus for considering grid spacers is that such devices can be installed with appreciably less steel volume allocated to the pin spacer component. This potentially allows the fuel volume fraction of the core to be increased, which directly improves the breeding ratio and lowers the overall doubling time. The overall pressure drop tends to be lower† since the coolant has a straight-through pass (as opposed to swirling) and the required pump size is correspondingly lower. An additional advantage of the grid spacers is that the hot channel factor (the statistical factor which must be added to nominal temperature calculations to be assured that design limits are not exceeded) may be reduced. For a wire wrap design, the wire necessarily contacts the hottest cladding in the core (which occurs just below the top of the core); and local flow reduction, possibly induced by eddy currents behind the point of contact, could aggravate the local hot spot. When employing grid spacers, some design flexibility is available to locate the grids axially away from the position of

---

*Some grid designs employ either three or six rods in duct corners to support and space the grids.

†Actual pressure drop is a function of (1) the number of grids required to maintain acceptable bundle deformation grid and (2) the grid design itself.

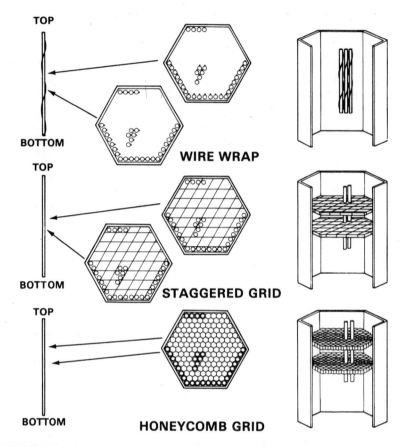

FIGURE 8-18. Pictorial intercomparison of three primary pin separation concepts.

the peak cladding temperature. However, it should be noted that whereas lower peak cladding temperatures result from using grid spacers, this directly increases the duct temperature and, as such, reduces duct lifetime.

The principal concern with both grid concepts is the fabrication cost, since carefully aligned physical indentures must be made in the assembly duct wall to axially affix the grids. Perhaps the biggest uncertainty is simply how well such designs will perform for high burnup conditions. Very little long-term experience has been gained to fully assess overall design implications. Also, the safety considerations have yet to be fully assessed. One such concern is that associated with a low probability unprotected transient overpower accident. For a wire wrap design, molten fuel entering the coolant channel could be hydraulically removed from the active core region without mechanical obstruction (other than any blockage that might be formed by molten fuel or cladding refreezing), whereas the cross-lacing of a grid design would form a possible barrier to the flow of particulate material.

D. FUEL ASSEMBLY LENGTH

The overall length of the fuel assembly is determined by the active fuel length, the upper and lower blankets, the fission gas plenum, the orifice-shielding block, the orifice assembly, and the handling socket. The considerations involved in specifying the fission gas plenum length were discussed earlier, and the blanket lengths are fixed from neutronics considerations.

There is not a unique criterion for establishing the length of the active fuel column. Several factors, including pressure drop, coolant temperature rise, and sodium void worth, are considered for determining an approximate height. Fine tuning of the axial core length is then generally accomplished by molding the total core power requirements with average linear heat ratings to arrive at a symmetric core layout.

Both the pressure drop and the rise in coolant temperature increase with fuel bundle length. Hence, there is a general incentive to keep the core height low, resulting in height-to-diameter (H/D) ratios below unity. As shown in Chapter 6, any design parameters that can be adjusted to increase the neutron leakage will reduce the magnitude of the positive reactivity effect resulting from sodium voiding. Hence, there have been attempts to flatten the core (greatly reduced H/D) or modularize it to enhance leakage. A heterogeneous core layout, as discussed briefly in Chapter 2, also increases neutron leakage and reduces the sodium void effect (cf. Chapter 6). Current designs tend to have an active core height of about one metre.

Figure 8-19 shows a cutaway of a typical fuel pin assembly for an LMFBR. Coolant enters the assembly from the coolant inlet plenum via slotted ports that completely surround the assembly nozzle. Such a design is intended to prohibit accidential assembly blockage by allowing for a multitude of multi-directional coolant entry ports. From there, the coolant winds through an orifice-shielding block, sized to effect the degree of pressure drop (hence, flow reduction) desired. It then enters the pin bundle region.

As shown by the enlargement on the right side of the figure, the fuel pins are typically affixed to a lower rail by locking pins. Since the fuel pins are floating free at the top end, it is primarily this locking device which keeps the fuel bundle in position during its useful life.

The coolant then flows up past the lower fission gas plenum (if present below the core), through the lower blanket region, through the active core, past the upper blanket region and on past the upper fission gas plenum (if

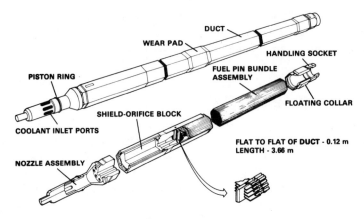

FIGURE 8-19. Typical full length LMFBR assembly.

present above the core). Coolant then collects above the fuel bundle and is ejected into the upper coolant plenum via slots just below the assembly cap, which constitutes a mechanical grapple device for fuel handling purposes. The assembly shown in Fig. 8-19 is, of course, only typical of an LMFBR assembly. Several variations exist for particular designs, such as for the FFTF in which the axial blankets are replaced by reflectors and the coolant leaves the assembly out the top where it directly traverses an instrumentation tree prior to ultimate ejection into the upper coolant plenum region.

8-5 OTHER ASSEMBLIES

The other principal components that make up a fast reactor core are the blanket, control, and shielding assemblies. The principal features of these key FBR components will be outlined below; a more complete discussion can be found in Reference 25.

A. BLANKET ASSEMBLIES

The primary function of the blanket assemblies is to allow the efficient conversion of fertile fuel (e.g., depleted UO_2) to fissile fuel via capture of stray neutrons. Blanket assemblies located at the radial periphery of the core also provide some shielding for structures beyond the blanket.

From their outward appearance, blanket assemblies look very similar to fuel assemblies. Both assemblies are hexagonally shaped, are the same approximate length, and are normally made of the same material. Since some fissioning occurs within the blanket pins, a fission gas plenum is normally present.

However, there are appreciable differences—in addition to the obvious change in fuel (a pure fertile loading in the blanket rather than a mixture of fertile and fissile fuel). Due both to low fissile fraction and low neutron flux,* the volumetric heat generation rate in the blanket fuel is much lower than in the core fuel. Hence, blanket pins can be larger in diameter than core pins while maintaining linear powers, χ, no greater than linear powers in the core. For this reason, and since it is more economical to incorporate the fertile fuel into large fuel pins, the blanket pins are substantially larger than core pins. Blanket pins are typically twice the diameter of fuel pins. As a result of the increased pin size, there are far fewer pins per assembly, and the coolant volume fraction is smaller. Blanket pin diameters are compared with core pin diameters in Table 8-1.

Because of the large pin diameter for blanket assemblies, blanket pins are considerably stiffer than core fuel pins. For blanket assemblies located in the radial periphery of the core, or for in-core positions (heterogeneous design), a pronounced power gradient within an assembly can occur. This combination of large temperature gradients and stiffer structure can lead to the development of higher stresses between pins—resulting in a potentially greater pin/duct interaction and stress problem than found in fuel assemblies.

Blanket assemblies located within the core proper, i.e., characteristic of the heterogenous core design (Fig. 2-2), do not have the extreme radial power and temperature gradients typical of the radial blanket assemblies, but they tend to suffer more rapid power and temperature changes during normal operations. As a consequence, the lifetime of in-core blanket pins may be more limited by transient conditions than by cladding damage resulting from long-term steady-state irradiation.

B. CONTROL ASSEMBLIES

Control assemblies are required to perform three primary functions: (1) reactivity compensation during the fuel burn cycle, (2) neutronic startup and shutdown capability for normal operations, and (3) rapid shutdown during off-normal conditions.

Whereas several control materials can be employed in a fast reactor (discussed in Chapter 11), boron carbide, B_4C, is the most commonly used substance. During reactor operation, the neutrons absorbed in ^{10}B results in

---

*The neutron flux can be quite high for in-core blankets in heterogeneous designs, but the volumetric heat generation rate is still low in comparison to fuel assemblies.

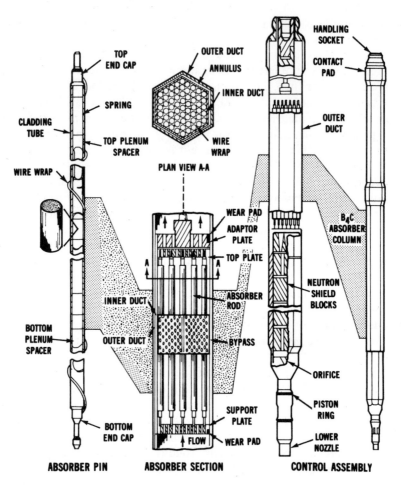

FIGURE 8-20. Typical FBR control assembly composition.

an (n, α) reaction. This process both liberates heat and produces helium gas. Consequently, cooling must be provided for control rods, and a gas plenum is normally incorporated into the design to accommodate the pressure buildup inside the control pins.

The control material is normally fabricated in pellet form and assembled in steel tubing, as illustrated in Fig. 8-20. Pin diameters are typically about twice that of the fuel (i.e., comparable to the blanket pin diameter). The pins are clustered into the usual hexagonal array within a double-walled hexagonal duct assembly. When control rods are inserted or withdrawn, the inner ducted structure moves within the stationary outer duct. In some designs (e.g., Fig. 8-20), the inner hexagonal duct is perforated by thousands of holes, both to allow coolant crossflow within the control pins and to relieve excessive pressure buildup during a scram.

C. SHIELDING ASSEMBLIES

Radial shielding assemblies surround the core and blanket assemblies for all FBR designs. The principal function of these assemblies is to provide neutron and gamma shielding for the reactor vessel and the major components within the vessel. An elevation view of the shielding assemblies for the CRBRP design is shown in Fig. 8-21. The assemblies fit into the regular hexagonal lattice with the core and blanket assemblies. They also serve to transmit the load between the core and the core restraint system. Similar removable radial shielding assemblies are used for pool designs, but there are several important differences. The height of the radial shielding is greater for the pool design in order to prevent activation of the secondary sodium in the IHX's which are located in the reactor tank (e.g., see Fig. 12-6).

Shielding assemblies are normally designed in two types: removable and fixed. The removable assemblies, as the word implies, can be withdrawn and replaced during the core lifetime, whereas the fixed assemblies cannot. The

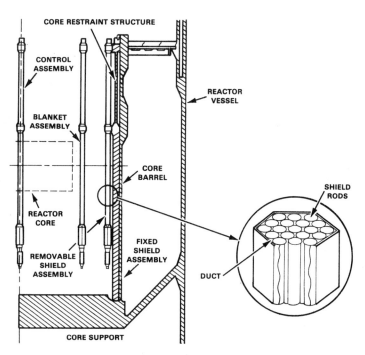

FIGURE 8-21. Elevation view of CRBRP reactor system showing relation of removable shielding assemblies to the fixed shield assembly and the core barrel and core restraint structure.

removable positions are situated closest to the core, where the fluence levels are highest. Further discussion of both removable and fixed shielding appears in Chapter 12-4.

8-6 CLUSTERED ASSEMBLY BEHAVIOR

Earlier sections of this chapter have been addressed to the principal design features and behavior associated with individual pins and pin assemblies. In this concluding section we turn our attention to duct assembly interaction effects that came into play as fuel, blanket, control and shielding assemblies are brought together to form a reactor. In this context we will discuss duct swelling, duct bowing, and core restraint systems. Duct swelling could have been treated in Section 8-4, in conjunction with fuel assembly design considerations, but it has been placed in the present context because (1) duct swelling is common to all types of ducts in the reactor, and (2) accommodation of duct swelling is an important function of the core restraint system.

A. DUCT SWELLING

The same mechanism that causes void swelling in the fuel pin cladding leads to appreciable swelling in the ducts. It is possible that assembly duct swelling may limit core lifetime rather than the onset of rupture in individual fuel pins.

Substantial swelling in both the axial and radial directions can take place, as illustrated in Figs. 8-22 and 8-23. Both fluence and temperature are responsible for the spatial problems depicted on the two illustrations. For instance, the relatively smooth drop in the axial elongation pattern noted from Fig. 8-22 clearly follows the fluence level; the peaks in the radial blanket and reflector result simply from the larger exposure time (hence, larger fluence even with lower flux). The magnitudes of the radial dilations can be better appreciated when one recognizes that the 1.3 mm separating the vertical lines in Fig. 8-23 would be quite adequate to satisfy mechanical requirements if there were no swelling. Because of swelling, initial gaps up to 7 mm may be required in this case. If these gaps were not supplied, the core would dilate radially during irradiation and would soon cause unacceptable misalignment of control rods and assembly handling heads. Refueling, which must be done in opaque sodium for an LMFBR, would then become very difficult.

Such gaps designed into the core to accommodate swelling result in an increase in fissile inventory and an appreciable reduction in the attainable breeding ratio. Hence, there is considerable incentive to reduce the swelling problem. Table 8-2 contains a qualitative assessment of potential design solutions to the duct swelling problem. The development of improved alloys to mitigate void swelling would appear the most desirable.

TABLE 8-2   Possible Design Changes to Accommodate Assembly Duct Swelling

Design Change	Implications
Mechanically open up grid	Lower breeding ratio; increased fissile inventory
Lower burnup	Increased fuel cycle cost
Reduce outlet temperature	Thermal efficiency reduced
Periodically rotate ducts by 180°	More down time
Remove ducts for high temperature annealing	Need more on-site fuel assemblies
Permit significant duct-to-duct contact	Requires refueling at lowest practical temperature
Develop alloys with improved swelling resistance	R&D effort

B. DUCT BOWING

Whereas pure void swelling in ducts leads to a uniform dilation pattern of the type illustrated in Figs. 8-22 and 8-23, there are additional forces which lead to deformation in the duct walls. In particular, thermal expansion,

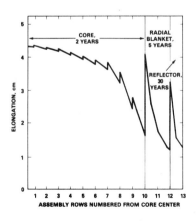

FIGURE 8-22. Axial elongation of assembly duct walls due to void swelling at the nominal lifetime of each component.[26]

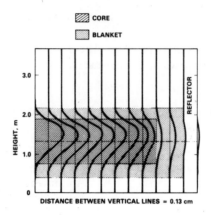

FIGURE 8-23. Radial dilation of assembly duct due to void swelling at end-of-life.[27]

irradiation creep, and mechanical interaction with neighboring assemblies occur simultaneously. The interaction of the above forces in combination with the mechanical tie-down points imposed by the core restraint system (to be discussed below) leads to a phenomenon called bowing.

Because of the complicated interties that can accrue due to the forces outlined, bowing is difficult to predict in a precise fashion. Each core configuration has to be individually assessed. Complex code systems such as CRASIB,[28] AXICRP,[29] and BOW-V[30] have been developed to analyze such systems.

Figure 8-24 contains a calculational sample for CRASIB which qualitatively illustrates the bowing pattern emerging from various clamped and unclamped configurations (where clamping occurs at three axial levels: core support structure, top of the core, and top of duct). Figure 8-24(a) depicts the effect of thermal bowing alone, and (b) represents pure swelling effects. Note that swelling results in a permanent deformation, even for the unclamped, power off condition. Case (c) illustrates the combined effects of thermal and swelling bowing. For the clamped core at power condition, this case results in a sharp duct curvature which produces large bending stresses in the duct wall. However, these high stress points are effectively relaxed by irradiation creep, as illustrated in case (d). The importance of irradiation creep on eventual bowing patterns is dramatically illustrated by observing cases (c) and (d) for the unclamped, power off condition. Creep is seen to actually change the *direction*, as well as the magnitude, of duct bowing for unrestrained conditions. This underscores both the importance and complexity of determining duct bowing patterns for a particular reactor of interest.

In addition to the long-term structural problems arising from the bowing phenomena, it is also possible that bowing could cause a net positive power

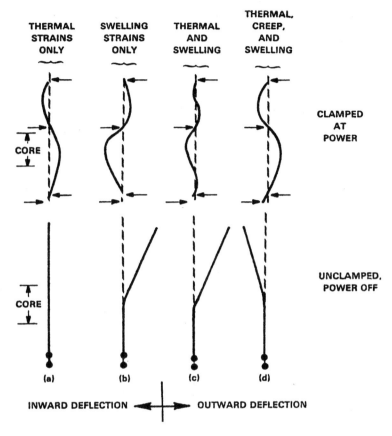

FIGURE 8-24. Bowing of fuel assemblies with core clamps fixing the wrappers at the spacer-pad elevations.[26]

coefficient* over certain ranges of reactor operation. Since this generally represents an undesirable performance characteristic, considerable design analysis is normally given to assessing the bowing potential for reactor systems of interest.

C. CORE RESTRAINT

Given the requirements (a) to allow clearance between assembly ducts to accommodate swelling, and (b) to constrain the core to resist bowing (due to

---

*The power coefficient is defined as the change in reactivity resulting from an increase in core power, while holding all other variables fixed.

swelling and thermal gradients), it is necessary to provide a core restraint system. Such a system should fulfill three functions:

(1) provide a calculable and reproducible structural response of the core within the limits imposed by reactivity insertion considerations (during both long-term irradiation or transient conditions),
(2) maintain the tops of the reactor assemblies in a position such that handling heads can be remotely located and grappled by the refueling machine, and
(3) provide clearance for duct insertion and removal, with minimal vertical friction, during shutdown refueling conditions.

Uncertainties in areas such as (a) fast-flux irradiation effects on core structural materials, and (b) fluence and temperature conditions that affect duct bowing have led to several design concepts for core restraint devices. Early prototype systems employed an *active* device, wherein radial yokes could be repositioned to provide varying degrees of lateral force as conditions warranted. For commercial use, however, designers prefer to use *passive* devices, to minimize any maintenance problems which may be associated with hydraulic (or other) systems within the core barrel.

The three passive systems being used or under development are briefly described below.[25]

Leaning Post.   Fig. 8-25 illustrates the leaning post concept, which is used in PFR. Transverse support pads are located in the lower axial blanket region of the fuel assembly. These support pads are forced into contact with a leaning post via a cantilever beam spring in the fuel assembly nozzle, which is forced into a deflected configuration during fuel assembly insertion. The fuel assemblies are supported in groups of six, each centered on a leaning post used as a control assembly. Thus, the large uncertainties associated with core restraint system design are more localized in a cluster of six assemblies, rather than being additive throughout an entire core. Principal limitations of the leaning post concept appear to be in the areas of seismic load capability and impact on core breeding capability.

Free Standing.   Figure 8-26 illustrates the free standing core restraint concept, which allows the cantilever-supported fuel and blanket assemblies to bow freely outward until contact is made with shield assemblies at the core periphery. The large number of peripheral shield assemblies act as a soft spring, restraining further outward motion of the core assemblies. Pressed dimple-type spacer pads are located in or just above the active core region. Rapsodie, Phénix and EBR-II use this type of core restraint system which features simplicity, but there are uncertainties associated with its extrapolation potential to larger cores and higher fluence levels.

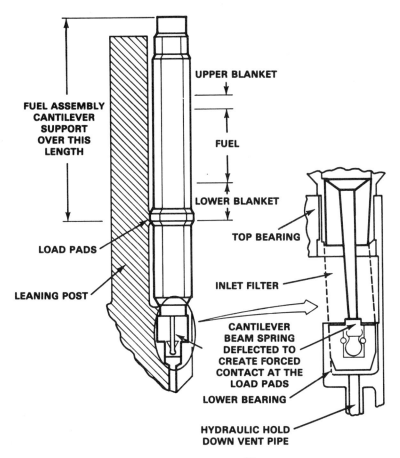

FIGURE 8-25. The leaning post core restraint concept.[25]

Limited Free Bow. Figure 8-27 illustrates the limited free bow core restraint system concept, wherein lateral restraint of the fuel assembly ducts is provided at three locations: one at either end of the assembly and a third in the low swelling region just above the active core. This corresponds to the clamping system depicted in Fig. 8-24. Such a support configuration results in the active core region of the fuel assemblies bowing radially outward from the core centerline in response to the temperature gradients generated in the ducts at power. Whereas this concept appears to have high burnup potential, it has only limited testing for reactor operating conditions. It was designed to accommodate the anticipated effects of irradiation swelling and creep as defined in Chapter 11. The limited free bow core restraint concept has been adopted in the FFTF and CRBRP design.

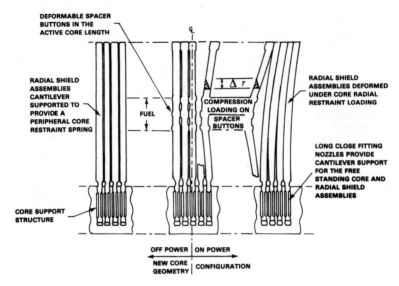

FIGURE 8-26. Free-standing core restraint concept.[25]

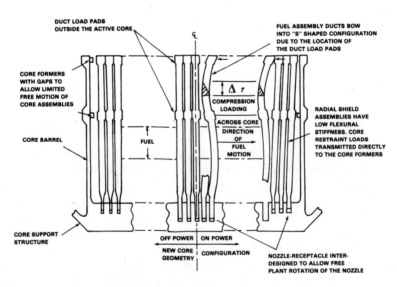

FIGURE 8-27. Limited free bow core-restraint concept.[25]

PROBLEMS

8-1 Calculate the plenum length, L_p, required to limit the plenum pressure, P_p, to 8 MPa at a burnup of 80 MWd/kg and a temperature of 800 K. The linear power throughout the irradiation is 33 kW/m. The length of the active fuel, L_f, is 1.0 m. The fuel is UO_2-PuO_2 with a theoretical density, TD, of 11 g/cm³ and a smear density of 85% TD. The fission gas yield is 0.27. Assume Ar=0.4 and Au=0.6 at this linear power.

8-2 (a) Calculate the cumulative damage function for steady-state operation for 800 full-power days (i.e. $800 = f \times$ chronological time) at the following conditions:

- The cladding is loaded only by plenum gas.
- The calculation can be made in four time steps of 200 days each, using an average plenum pressure during each step.
- The average plenum pressures are

0–200 days	$P_p = 1$ MPa
200–400 days	$P_p = 3$ MPa
400–600 days	$P_p = 5$ MPa
600–800 days	$P_p = 7$ MPa

- Coolant pressure = 0.4 MPa
- Cladding temperature = 800 K
- $\phi(>0.1 \text{ MeV}) = 2 \times 10^{15}$ n/cm²·s
- Cladding radius, $R = 3.5$ mm
- Cladding thickness, $w = 0.4$ mm.

Use the Dorn-parameter correlation as given by Eq. (8-20) to calculate time-to-rupture.

(b) Calculate the product FAE·t_r for $P_p = 7$ MPa at a cladding temperature of 900 K instead of 800 K. Compare your result with the corresponding value at 800 K obtained in part (a), and comment.

8-3 (a) The theory for a thick-walled cylinder is needed for the calculation of some of the stresses when the cladding is loaded directly by fuel, whereas it is not needed when the cladding is loaded by fission gas. Explain the reasons for this difference.

(b) What is the purpose of the Prandtl-Reuss flow rule?

(c) Explain in words what the "equilibrium" equation [Eq. (8-63)] in the LIFE code represents.

REFERENCES

1. V. Z. Jankus and R. W. Weeks, "LIFE-II—A Computer Analysis of Fast Reactor Fuel-Element Behavior as a Function of Reactor Operating History," *Nuc. Eng. and Des.*, *18* (1972) 83–96.

2. *Clinch River Breeder Reactor Plant, Preliminary Safety Analysis Report*, Amendment 51, September 1979, 4.2–4.78.

3. P. Delpeyroux, "Super-Phénix 1 Fuel Element Optimization–First Check of Overall Behavior," *Proc. from Conference on Optimisation of Sodium-Cooled Fast Reactors*, London, November 28–December 1, 1978.

4. G. Karsten, *The Fuel Element of the Sodium Breeder*, EURFNR-1418, October 1976, Karlsruhe, Germany.

5. L. H. Van Vlack, *Elements of Materials Science*, Addison-Wesley Publishing Co., Reading, MA (1959) 99.

6. D. R. O'Boyle, F. L. Brown, and J. E. Sanecki, "Solid Fission Product Behavior in Uranium-Plutonium Oxide Fuel Irradiated in a Fast Neutron Flux," *J. Nucl. Mat.*, *29*, No. 1 (1969) 27.

7. D. R. Olander, *Fundamental Aspects of Nuclear Reactor Fuel Elements*, TID-26711-P1, U.S. ERDA, 1976.

8. D. S. Dutt and R. B. Baker, *SIEX—A Correlated Code for the Prediction of Liquid Metal Fast Breeder Reactor (LMFBR) Fuel Thermal Performance*, HEDL-TME 74–75, Hanford Engineering Development Laboratory, 1975.

9. E. E. Gruber and L. W. Deitrich, "Dispersive Potential of Irradiated Breeder Reactor Fuel During a Thermal Transient," *Trans. ANS*, *27* (1977), 577.

10. J. R. Hofmann and C. C. Meek, "Internal Pressurization in Solid Mixed Oxide Fuel Due to Transient Fission Gas Release," *Nuc. Sci. Eng.*, *64* (1977) 713.

11. D. S. Dutt, D. C. Bullington, R. B. Baker and L. A. Pember, "A Correlated Fission Gas Release Model For Fast Reactor Fuels," *Trans. ANS*, *15* (1972) 198–199.

12. G. D. Johnson, J. L. Straalsund and G. L. Wire, "A New Approach to Stress-Rupture Data Correlation," *Mat. Sci. and Eng.*, *28* (1977) 69–75.

13. A. J. Lovell, B. Y. Christensen and B. A. Chin, "Observations of In-Reactor Endurance and Rupture Life for Fueled and Unfueled FTR Cladding," *Trans. ANS*, *32* (1979) 217–218.

14. C. W. Hunter and G. D. Johnson, "Fuel Adjacency Effects on Fast Reactor Cladding Mechanical Properties," *Int. Conf. Fast Breeder Reactor Fuel Performance*, ANS/AIME, Monterey, CA, March 1979, 478–488.

15. C. W. Hunter and G. D. Johnson, "Mechanical Properties of Fast Reactor Fuel Cladding for Transient Analysis," *ASTM Symposium on the Effects of Radiation on Structural Materials*, ASTM-STP-611, 1976.

16. G. D. Johnson and C. W. Hunter, "Mechanical Properties of Transient-Tested Irradiated Fast-Reactor Cladding," *Trans. ANS*, *30* (November 1978) 195–196.

17. F. E. Bard, G. L. Fox, D. F. Washburn and J. E. Hanson, "Analytical Models for Fuel Pin Transient Performance," *Proc. of Int. Meeting on Fast Reactor Safety and Related Physics*, CONF-761001, *III*, Chicago, October 1976, US ERDA (1977) 1007.

18. W. S. Lovejoy, M. R. Patel, D. G. Hoover and F. J. Krommenhock, *PECS-III: Probabilistic Evaluation of Cladding Lifetime in LMFBR Fuel Pins*, GEFR-00256, General Electric Co., Sunnyvale, CA (1977).

19. D. S. Dutt, R. B. Baker and S. A. Chastain, "Modeling of the Fuel Cladding Postirradiation Gap in Mixed-Oxide Fuel Pins," *Trans. ANS*, *17* (1973) 175.

20. L. Leibowitz, E. C. Chang, M. G. Chasanov, R. L. Gibby, C. Kim, A. C. Millunzi, and D. Stahl, *Properties for LMFBR Safety Analysis*, ANL-CEN-RS-76-1, Argonne National Laboratory, March 1976.

21. J. E. Shigley, *Mechanical Engineering Design*, McGraw-Hill Co., 2d ed., New York, 1972, 73–76.

22. A. Mendelson, *Plasticity: Theory and Applications*, MacMillan Co., New York, 1968.

23. E. R. Gilbert and L. D. Blackburn, "Creep Deformation of 20% Cold Worked Type 316 Stainless Steel," *J. of Eng. Mat. and Tech.*, *99* (April 1977) 168–180.

24. E. R. Gilbert and J. F. Bates, "Dependence of Irradiation Creep on Temperature and Atom Displacements in 20% Cold Worked Type 316 Stainless Steel," *J. of Nuc. Mat.*, *65* (1977) 204–209.

25. Y. S. Tang, R. D. Coffield, Jr., and R. A. Markley, *Thermal Analysis of Liquid Metal Fast Breeder Reactors*, American Nuclear Society, La Grange Park, IL, 1978, 19–21, 47.

26. P. R. Huebotter, "Effects of Metal Swelling and Creep on Fast Reactor Design and Performance," *Reactor Technology*, *15* (1972) 164.

27. P. R. Huebotter and T. R. Bump, "Implications of Metal Swelling in Fast Reactor Design," *Radiation-Induced Voids in Metals*, CONF-710601, June 9, 1971 AEC Symposium Series, No. 26, J. W. Corbett and L. C. Ianniello, eds., New York, 84–124.

28. W. H. Sutherland and V. B. Watwood, Jr., *Creep Analysis of Statistically Indeterminate Beams*, BNWL-1362, Battelle Northwest Laboratory, 1970.

29. W. H. Sutherland, "AXICRP-Finite Element Computer Code for Creep Analysis of Plane Stress, Plane Strain and Axisymmetric Bodies," *Nuc. Eng. Des.*, *11* (1970) 269.

30. D. A. Kucera and D. Mohr, *BOW-V: A CDC-3600 Program to Calculate the Equilibrium Configurations of a Thermally-Bowed Reactor Core*, ANL/EBR-014, Argonne National Laboratory, 1970.

CHAPTER 9

FUEL PIN THERMAL PERFORMANCE

9-1 INTRODUCTION

Fast breeder reactor design requires the simultaneous application of mechanical and thermal-hydraulics analysis methods. We reviewed some aspects of mechanical analysis in Chapter 8; we will discuss mechanical design further in Chapter 12. Chapters 9 and 10 will deal with thermal-hydraulics analysis. In the present chapter we will investigate methods of determining temperature distributions within fuel pins. We will then extend these methods to assembly and core-wide temperature distributions in Chapter 10.

There are several ways to approach thermal-hydraulics analysis. We could begin with coolant inlet flow and temperature conditions, work up through the core axially, and then go back to determine the pin temperatures for each axial level, working radially inward. This is the sequence followed by most thermal-hydraulic computational schemes and is the one we will follow in discussing hot channel factor applications in Chapter 10. For purposes of instruction, however, we have chosen to begin with the temperature distribution in the fuel itself, and then work radially out to the coolant. This allows us to begin with the heat source and physically follow the flow of heat in a decreasing temperature gradient through the system's several heat transfer barriers outward to the heat sink.

This chapter deals with steady-state fuel pin thermal performance, but a reactor designer is also interested in transient fuel pin behavior—both for anticipated transients (e.g., normal adjustments in power/flow operations and scrams) and off-normal or accident conditions. The same basic techniques apply to analysis of transient conditions as to steady state, but the need to understand the interaction of a high-temperature environment with materials properties becomes more acute. These concerns are discussed in the chapters on core materials (Chapter 11) and on safety (Chapters 13–16).

Only liquid metal coolants are considered in the discussion of hydraulics and coolant heat transfer in Chapters 9 and 10. Discussion of the thermal hydraulics of gas-cooled fast reactors will be delayed until Chapter 17. A more detailed presentation on LMFBR thermal hydraulics, including the methods developed during the design of the FFTF and CRBRP, appears in the book by Tang, Coffield, and Markley, *Thermal Analysis of Liquid-Metal Fast Breeder Reactors.*[1]

9-2 FUEL AND CLADDING THERMAL ANALYSIS

A. RADIAL FUEL TEMPERATURES

The steady-state temperature distribution in cylindrical rods with an internal heat source is obtained from the heat conduction equation, which can be derived with the aid of Fig. 9-1. We assume that the heat source in the fuel is uniform, i.e., there is no neutronic spatial self-shielding within the pin.*

Neglecting heat transfer in the axial direction,[†] a one-dimensional heat balance on the cylindrical element dr gives, per unit length of the cylinder,

$$-2\pi rk\frac{dT}{dr}+Q2\pi r\,dr=-2\pi\left[rk\frac{dT}{dr}+\frac{d}{dr}\left(rk\frac{dT}{dr}\right)dr\right],$$

where $T=$ temperature (°C)
$k=$ thermal conductivity (W/m·°C)
$Q=$ uniform volumetric heat source (W/m³).

Simplifying gives the steady-state heat conduction equation:

$$\frac{1}{r}\frac{d}{dr}\left(rk\frac{dT}{dr}\right)+Q=0. \tag{9-1}$$

Note that the thermal conductivity remains in the argument for the derivative because thermal conductivity is a strong function of temperature and, hence, of radius.

---

*Neglect of neutronics spatial self-shielding is an excellent assumption for fast reactors. The spatial variation of the heat source is complicated by restructuring of the fuel, however, as in Eq. (9-9).

[†]This is an acceptable approximation in most cases, since the radial temperature gradient is much greater than the gradient in the axial direction.

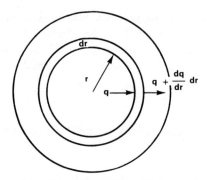

q = HEAT TRANSFER RATE AT RADIUS r (W)

FIGURE 9-1. Geometry for the cylindrical heat conduction equation.

The two boundary conditions needed to solve Eq. (9-1) are

(1) $$\frac{dT}{dr}=0 \text{ at } r=0,$$

(2) $$T=T_s \text{ at } r=R_f,$$

where $R_f=$ radius at the outer surface of the fuel
$T_s=$ temperature at the outer surface of the fuel.

Integrating Eq. (9-1) yields

$$rk\frac{dT}{dr}+Q\frac{r^2}{2}=C_1.$$

From boundary condition 1, $C_1=0$. We incorporate the second boundary condition directly into the second integration to give

$$\int_{T(r)}^{T_s} k\, dT+\frac{Q}{2}\int_r^{R_f} r\, dr=0.$$

Reversing the limits on the first integral, the result is

$$\int_{T_s}^{T(r)} k\, dT=\frac{Q}{4}\left(R_f^2-r^2\right). \tag{9-2}$$

Given the surface temperature T_s and thermal conductivity as a function of temperature, the fuel temperature at any radius r can be determined from the above expression.*

If we are interested only in the fuel centerline temperature T_{CL}, then $r=0$ and we have

$$\int_{T_s}^{T_{CL}} k\, dT = \frac{QR_f^2}{4}.$$ (9-3)

We can now evaluate the linear power, or linear heat rate, χ, originally introduced in Chapter 2, Eq. (2-2). Linear power and volumetric heat source are related by

$$\chi = Q\pi R_f^2.$$ (9-4)

From Eqs. (9-3) and (9-4), one obtains

$$\chi = 4\pi \int_{T_s}^{T_{CL}} k\, dT.$$ (9-5)

We shall use the most logical SI unit for χ, kW/m. Units also used frequently for χ are W/cm and, in the earlier U.S. literature, the hybrid unit kW/ft. Values for linear power in fast breeder reactors were given in Table 8-1.

One method to evaluate $\int k\, dT$ is to use an analytical expression for k. A plot of k vs T is given in Fig. 9-2 for mixed oxide fuel (80% U, 20% Pu) at 95% theoretical density and for an O/M ratio of 2.00, using the expression[2]

$$k = (0.042 + 2.71 \times 10^{-4} T)^{-1} + 6.9 \times 10^{-11} T^3 \quad (95\% TD),$$ (9-6)

where k is in W/m·K and T is in degrees Kelvin.

The 95% TD case was chosen since that is the porosity of mixed oxide fuel typically employed in breeder reactor fuel systems. Porosity is the fraction of fuel pellet volume that is void. For fuel with porosity different from 5%, the thermal conductivity can be obtained by modifying the 95% TD value according to

$$k_p = k \frac{1 - 2.5p}{0.875} \quad \text{if} \quad p \leqslant 0.1,$$ (9-7a)

$$k_p = k \frac{1 - p}{0.875(1 + 2p)} \quad \text{if} \quad p \geqslant 0.1,$$ (9-7b)

---

*Alternatively, Eq. (9-2) can be obtained by defining $\theta = \int_{T_s}^{T(r)} k\, dT$ so that $d\theta/dr = k\, dT/dr$ (cf. Problem 9-2). Eq. (9-1) can then be written as $d^2\theta/dr^2 + (1/r)(d\theta/dr) + Q = 0$. The solution of this equation is $\theta(r) = -Qr^2/4 + C\ln r$.

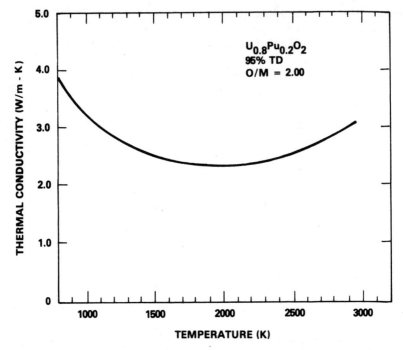

FIGURE 9-2. Thermal conductivity for 80% U, 20% Pu—mixed oxide fuel.[2]

where k_p = thermal conductivity of fuel with porosity p and k is determined from Eq. (9-6) for 95% TD (or 5% porosity).*

Another way to evaluate center temperature, given linear power and surface temperature, is to plot $\int_{T_{ref}}^{T} k_{ref} \, dT$ against fuel temperature T, where k_{ref} is the conductivity for a specified (or reference) fuel density or porosity, and T_{ref} is an arbitrary reference temperature. Such a plot is shown in Fig. 9-3 for mixed oxide fuel at 95% theoretical density. The first step is to find $\int_{T_s}^{T_{CL}} k \, dT$ from the linear power [by means of Eq. (9-5)]. If the actual density of the fuel is the same as that for which $\int k_{ref} \, dT$ has been provided, we start with the pin surface temperature and obtain the center temperature by using the calculated $\int_{T_s}^{T_{CL}} k \, dT$ as the increment in $\int k \, dt$ between T_s and T_{CL} on the graph, i.e.,

$$\int_{T_s}^{T_{CL}} k_{ref} \, dT = \int_{T_{ref}}^{T_{CL}} k_{ref} \, dT - \int_{T_{ref}}^{T_s} k_{ref} \, dT. \qquad (9-8)$$

---

*Laboratory measurements indicate that the thermal conductivity of mixed oxide fuel varies with the oxygen-to-metal (O/M) ratio. Equation (9-6) and Fig. 9-2 correspond to a stoichiometric O/M ratio of 2.00. The precise effect of O/M ratio on fuel conductivity and temperature during actual reactor operation, however, is still in question.

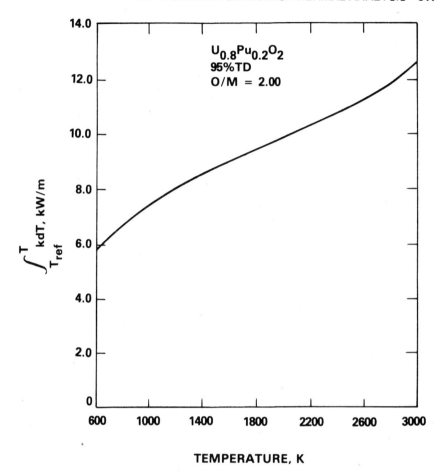

FIGURE 9-3. Integral thermal conductivity data for (80% U, 20%Pu) mixed oxide fuel [constructed to be consistent with Eq. (9-6) and Fig. 9-2].

(Conversely, if we know the surface temperature and the fuel melting temperature, we can use the graph to determine the linear power that would lead to centerline melting.)

If the density of the fuel is different from the value for which $\int k_{ref} \, dT$ has been provided, it is necessary to modify the increment in $\int k \, dT$ from the graph. Let k_p represent the fuel conductivity at a given porosity, so that the integral for the fuel is $\int k_p \, dT$. One then multiplies $\int k_p \, dT$ by the ratio k_{ref}/k_p obtained from Eq. (9-7) and uses this result as the increment on the graph to find T_{CL}.

The problem for oxide fuel is actually more complicated because (as described in Chapter 8-2C), the oxide fuel restructures, forming a void in the center of the pin. (See Problem 9-1 for the simplified case of a pin of only one structure but with a void in the center.)

B. EFFECTS OF RESTRUCTURING

Metal, carbide, and nitride fuels do not restructure, so no voids form at their centers.* Hence, radial temperature distributions for these fuels can be determined by using the simple expressions derived above. Mixed oxide fuels, however, *do* restructure, as described in Chapter 8-2C. With regard to fuel temperature distribution, the most important results of restructuring are the development of a central void and changes in thermal conductivity, density, and volumetric heat generation rates in the columnar and equiaxed regions.

Figure 9-4 is a sketch of the salient features of Fig. 8-5 and identifies the region and radius numbers used in the model for the restructured fuel pin. Given the density and the thermal conductivity of the three microstructure regions of Fig. 9-4, we can determine the temperature profile.

Due to porosity loss in the equiaxed and columnar grains, the local fuel density will increase. Hence, given the linear power χ (which does not change with restructuring), the actual volumetric heat generation rates in each of the three zones will be

$$Q_3 = \frac{\chi}{\pi R_f^2}, \tag{9-9a}$$

$$Q_2 = \frac{\chi}{\pi R_f^2} \frac{\rho_2}{\rho_3}, \tag{9-9b}$$

$$Q_1 = \frac{\chi}{\pi R_f^2} \frac{\rho_1}{\rho_3}, \tag{9-9c}$$

where the densities, ρ, are straightforward reflections of the porosity loss.

With these new values of Q, the heat conduction equation is solved in each region, with the limits of $\int k \, dT$ being the surface temperatures bounding the different fuel structures. Between regions, e.g., between regions 1 and 2 with a boundary at R_1, continuity of heat flux is a boundary condition in addition to continuity of temperature, i.e.,

$$k_1 \left(\frac{dT}{dr} \right)_1 = k_2 \left(\frac{dT}{dr} \right)_2, \quad \text{and} \quad T_1 = T_2, \quad \text{at} \quad r = R_1. \tag{9-10}$$

---

*See Chapter 11 for a discussion of the thermal properties of these fuels.

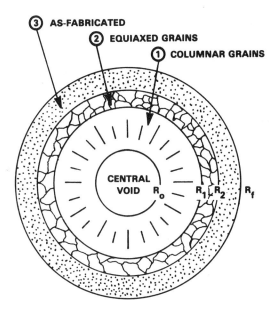

FIGURE 9-4. Geometry for restructured fuel pin analysis.

Furthermore, the radii and densities are interrelated. For example, regions 1 and 2 contain material that was at density ρ_3 before restructuring. A mass balance inside radius R_2 gives

$$\rho_1\left(R_1^2 - R_0^2\right) + \rho_2\left(R_2^2 - R_1^2\right) = \rho_3 R_2^2. \tag{9-11}$$

Applying the boundary conditions [Eq. (9-10)] and the mass continuity relation Eq. (9-11), a solution to the heat conduction equation Eq. (9-1) yields the following results for the three regions, where T_0, T_1, T_2 and T_s refer to temperatures at R_0, R_1, R_2 and R_f:*

Region 3 (unrestructured fuel):

$$\int_{T_s}^{T_2} k_3 \, dT = \frac{\chi}{4\pi}\left[1 - \left(\frac{R_2}{R_f}\right)^2\right]. \tag{9-12}$$

Region 2 (equiaxed fuel):

$$\int_{T_2}^{T_1} k_2 \, dT = \left(\frac{\chi}{4\pi}\right)\left(\frac{\rho_2}{\rho_3}\right)\left(\frac{R_2}{R_f}\right)^2\left[1 - \left(\frac{R_1}{R_2}\right)^2 - 2\left(1 - \frac{\rho_3}{\rho_2}\right)\ln\frac{R_2}{R_1}\right]. \tag{9-13}$$

---

*Additional details in the solution are given in Ref. 3. Also see Problem 9-1.

Region 1 (columnar fuel):

$$\int_{T_1}^{T_0} k_1 \, dT = \left(\frac{\chi}{4\pi}\right)\left(\frac{\rho_1}{\rho_3}\right)\left(\frac{R_1}{R_f}\right)^2\left[1 - \left(\frac{R_0^.}{R_1}\right)^2 - 2\left(\frac{R_0}{R_1}\right)^2 \ln\frac{R_1}{R_0}\right]. \quad (9\text{-}14)$$

The center temperature, T_0, can be obtained from a knowledge of χ and using Fig. 9-3 and the above equations as follows. Start with Eq. (9-12) and Fig. 9-3 to find the temperature T_2. If the density of the unrestructured fuel is different from the density for which $\int k_{\text{ref}} \, dT$ is provided, the $\int k \, dT$ increment must be modified in the manner described in Section 9-2A. Then, calculate T_1 using Eq. (9-13) and Fig. 9-3; and, finally, calculate T_0 using Eq. (9-14) and Fig. 9-3.

In Fig. 9-5, the radial fuel temperature distribution is plotted for a typical mixed-oxide fuel pin before and after restructuring. Note that the development of the central void and the enhanced thermal conductivity in the

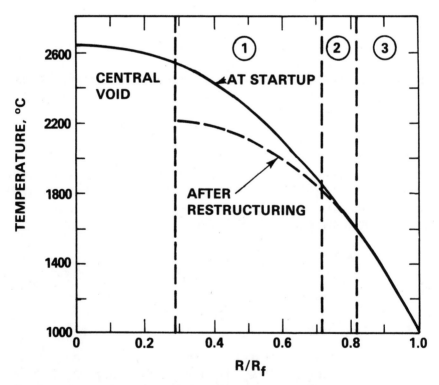

FIGURE 9-5. Temperature distribution in a mixed-oxide fuel pin before and after restructuring. $\chi = 50$ kW/m; $T_s = 1000°C$; Initial Density = 85% Theoretical Density; $T_1 = 1800°C$, $\rho_1 = 98\%$ *TD*; $T_2 = 1600°C$, $\rho_2 = 95\%$ *TD* (Ref. 3).

restructured regions decreases the hottest temperature by about 400°C, even though the linear power remains the same. Hence, restructuring in mixed oxide fuels partially offsets the disadvantages of low inherent thermal conductivity. A further reduction in center temperature occurs after the initial fuel-cladding gap closes, as explained below.

C. FUEL-CLADDING GAP

Fuel temperatures determined in Section 9-2A and 9-2B were based on a knowledge of the fuel surface temperature, T_s. Additional heat transfer barriers that must be traversed before reaching the bulk coolant include the fuel-cladding gap, the cladding, and the coolant film. Of these three barriers, the fuel-cladding gap provides the greatest resistance to heat flow. Initially, when the gap is open, the temperature drop across the gap is an order of magnitude greater than the drop across either the cladding or the coolant film. Even after the hot gap closes, the temperature drop across the contacting surfaces remains larger than the other two.

Most experimental evidence with irradiated fuel pins indicates that the hot gap closes after a short time due to fuel swelling and thermal cracking in addition to the differential thermal expansion. At this time the gap resistance decreases significantly, which leads to a large reduction in fuel center temperature. Methods to calculate the thickness of the hot gap, G, as a function of irradiation were presented in Chapter 8-3D.

In order to calculate the temperature drop across the gap, $(T_s - T_{ci})$, where T_{ci} is the inside cladding temperature, a gap conductance, h_g, is defined such that

$$\chi = h_g 2\pi R_f (T_s - T_{ci}).$$ (9-15)

Methods are available to estimate gap conductances (equivalent to a heat transfer coefficient) for open and closed gaps.[4] The method generally used for fast reactor analysis follows the method developed for thermal reactors by Ross and Stoute.[5] Uncertainties in gap conductance are large because experimental measurements of this parameter must be unraveled from integral fuel pin experiments that involve linear power and $\int k \, dT$ values for the fuel in addition to gap conductance.

For an open gap, heat transfer can occur across the fuel-cladding gap by conduction, radiation, and convection. However, surface temperatures are seldom high enough to allow the radiation component to be very appreciable and convection currents in the gap result in negligible heat transfer. Hence, the heat flow is mainly governed by conduction through the gas that fills the gap. The gap conductance for an open gap is given approximately by

$$h_g(\text{open}) \sim \frac{k_m}{G}, \qquad (9\text{-}16)$$

where k_m = thermal conductivity of the gas mixture in the gap (W/m·K)
G = gap(m)*.

More accurate expressions for the open gap conductance take into account the roughness of the fuel and cladding surfaces and a quantity called the jump distance. Roughness is illustrated for an open gap in Fig. 9-6(a). Reported values of roughness vary over a wide range of the order of 10^{-4} to 10^{-2} mm for both fuel and cladding.[4] These values are to be compared with initial hot gaps of the order of 0.1 mm.

The jump distance is an extrapolated distance used to account for incomplete energy exchange by gas molecules at each wall. The concept is illustrated schematically in Fig. 9-7.

By adding the jump distance to the gap, an effective gap thickness is obtained that provides the correct temperature drop across the gap when combined with the gas conductivity, k_m. Calculated values for jump distance are of the order of 10^{-4} mm or less, although higher values have been reported;[5] hence, they have little influence on the open gap.

Equation (9-16) can now be modified to include roughness and jump factors, as follows:

$$h_g(\text{open}) = \frac{k_m}{G + (\delta_f + \delta_c) + (g_f + g_c)}. \qquad (9\text{-}17)$$

The gas thermal conductivity diminishes rapidly with burnup because the thermal conductivities of the fission product gases Xe and Kr are considerably less than that of the fill gas (normally helium).† On the other hand, the increasing gas pressure with burnup partially offsets this effect by increasing the molecular density. Moreover, gap closure after extended operation further increases the conductance.

For a closed gap, the gap coefficient is not reduced to zero because surface roughness remains, as illustrated in Fig. 9-6(b). The gap conductance now becomes the sum of two parts, one due to the material contact

---

*Throughout the literature the word "gap" is used for the width or thickness of the gap in addition to referring to the region between the fuel and the cladding.

† The deleterious effect of Xe can be observed by comparing Figs. 9-10 and 9-11, which are included in the last part of this section in conjunction with linear-power-to-melting discussions.

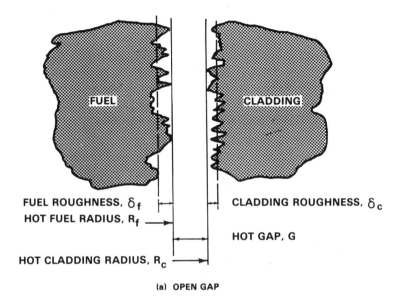

FUEL ROUGHNESS, δ_f

HOT FUEL RADIUS, R_f

HOT CLADDING RADIUS, R_c

CLADDING ROUGHNESS, δ_c

HOT GAP, G

(a) OPEN GAP

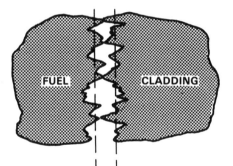

(b) CLOSED GAP

FIGURE 9-6. Geometry for gap conductance for open and closed gaps.

surfaces and the other due to remaining gas caused by the roughness. The conductance due to the contact surfaces is proportional to the contact pressure, P_{fc}, (defined earlier in Chapter 8-3D) and an effective thermal conductivity of the fuel cladding surface materials. It is inversely proportional to the Meyer hardness of the softer material and the square root of an effective roughness. This effective roughness is defined as

$$\delta_{\text{eff}} = \sqrt{\left(\delta_f^2 + \delta_c^2\right)/2}\,.$$

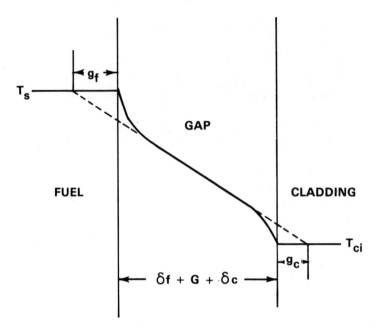

FIGURE 9-7. Schematic temperature profile across an open gap to illustrate jump distances, g_f and g_c.

The conductance due to gas conduction remains the same as for the open gap with $G=0$ and with roughnesses assumed to be unchanged. Hence, the closed-gap conductance is

$$h_g(\text{closed}) = \frac{Ck_s P_{fc}}{H\sqrt{\delta_{\text{eff}}}} + \frac{k_m}{(\delta_f + \delta_c) + (g_f + g_c)}, \qquad (9\text{-}18)$$

where $C =$ empirical constant $(m^{-1/2})$

$k_s =$ effective conductivity of surface materials $\left(= \dfrac{2k_f k_c}{k_f + k_c} \right)$

$H =$ Meyer hardness of the softer material (Pa).

The net results of Eqs. (9-17) and (9-18) are qualitatively illustrated in Fig. 9-8, which shows the gradual increase in the effective heat transfer coefficient as the gap is reduced, and then the sharp increase upon physical contact. Further increases occur as the contact pressure becomes more intense up to the yield stress of one of the materials.

The influence of the fuel-cladding gap on the radial temperature distribution and the effect of uncertainties in the hot gap are illustrated in Fig. 9-9.

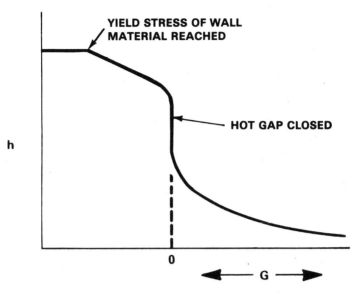

FIGURE 9-8. Qualitative representation of the fuel-cladding coefficient.

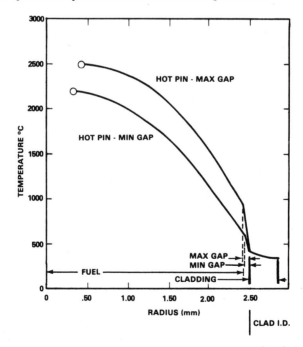

FIGURE 9-9. Influence of a beginning-of-life gap on the hot pin temperature distribution. Curves are for the FFTF hot pin at the axial mid-plane at 115% overpower with hot channel factors included, one with the maximum gap and one with the minimum.[6]

Both curves represent peak temperatures (at 115% overpower and after the effect of hot channel factors have been included*) at beginning-of-life conditions at the axial midplane of the hottest FFTF fuel pin. The upper curve shows the effect of the maximum hot gap, G (max); the lower curve is for the minimum hot gap, G (min), but still at beginning-of-life. The 280°C difference in temperature drop across the gap results in virtually the same difference in center fuel temperature.

Further illustrations of the influence of gap on fuel temperature appear in Part E of this section and in Section 9-4 (Figs. 9-20 and 9-21).

D.   TEMPERATURE DROP ACROSS THE CLADDING

The temperature drop across the cladding is relatively small. The temperature difference between the inside and outside cladding surfaces, $(T_{ci} - T_{co})$, is obtained from Fourier's law for the heat flux, $q(W/m^2)$:

$$q = -k_c \frac{dT}{dr}.$$

In terms of the linear power, χ, this becomes

$$\chi = -k_c 2\pi r \frac{dT}{dr}.$$

Since the thermal conductivity of the cladding can normally be assumed constant, integrating across the cladding gives

$$T_{ci} - T_{co} = \frac{\chi}{2\pi k_c} \ln \frac{R_{co}}{R_{ci}}. \tag{9-19}$$

A note of caution should be added at this time. Throughout the thermal analysis so far we have tacitly assumed that all heat generation contributing to the linear power was in the fuel. This is not exactly correct, and detailed design codes must account for the deviation from this assumption. Gamma radiation, accounting for about 14% of the heat generated, is absorbed throughout the core—particularly in the more massive materials such as fuel, cladding, and structure. Also, a small amount of energy is transferred directly to the steel and the sodium through elastic scattering of neutrons and beta deposition. In all, about 2 or 3% of the fission energy is absorbed in the cladding.

---

*See Section 10-4 for a discussion of overpower and hot channel factors.

E. INTEGRAL MEASUREMENTS OF LINEAR-POWER-TO-MELTING

A major concern of the thermal analyst is to demonstrate that fuel does not melt. This is discussed at greater length in Chapter 10-4 on hot channel factors. At this point, however, it is useful to introduce the types of experimental results that are important for this demonstration.

From the preceding discussion of fuel thermal performance, it is evident that an elaborate calculation is needed to obtain the center temperature of the fuel as a function of linear power. This calculation has numerous potential sources of uncertainty. It would be desirable to take a fuel pin for a particular fast reactor and actually irradiate it in a fast test reactor to determine experimentally what linear power is required to cause centerline melting. This value is called *linear-power-to-melting* and is denoted by χ_m. The measurement of χ_m is called an integral measurement because it incorporates a series of phenomena into a single overall result. Such an experiment would provide a direct and independent demonstration of whether center melting would occur for the design peak linear power without having to depend on numerous intermediate calculations.

This type of experiment was performed for the FFTF and CRBRP design fuel pins by the Hanford Engineering Development Laboratory[7, 8] and in another experiment by the General Electric Co.[9] In the HEDL-P-19 experiment, nineteen fresh mixed oxide ($75\% \ UO_2 - 25\% \ PuO_2$) fuel pins with cold gap sizes varying from 0.09 to 0.25 mm (3.4 to 10 mils) and pin diameters of 5.84 mm (0.23 in.), corresponding to the FFTF driver fuel and the CRBRP fuel pin designs, and 6.35 mm (0.25 in.) diameter pins were irradiated in EBR-II. Peak linear powers varied from 55.4 to 57.7 kW/m (16.9 to 17.6 kW/ft) for the 5.84-mm pins and 65.6 to 70.2 kW/m (20.0 to 21.4 kW/ft) for the 6.35-mm pins. These linear powers were sufficient to produce centerline fuel melting in some, but not all, of the pins. Helium fill gas was used in all of the P-19 pins.

Later, a second similar experiment, called HEDL-P-20, was performed with preirradiated fuel pins. Preirradiation varied from 3.7 to 10.9 MWd/kg. In addition a few fresh pins were run in the P-20 experiment. Some of these pins were filled with 18% xenon and 82% helium to assess the effect of xenon tag gas.

Figure 9-10 shows the results of the P-19 experiment, together with results from the P-20 fresh fuel pins. Lower-bound points (Figs. 9-10 and 9-12) were obtained from values of pins that did not reach incipient melting (i.e., χ_m lies above these values). Upper-bound points were obtained from pins that experienced some melting throughout the axial length of the pins; two such points are plotted on Fig. 9-11. For pins that experienced melting over

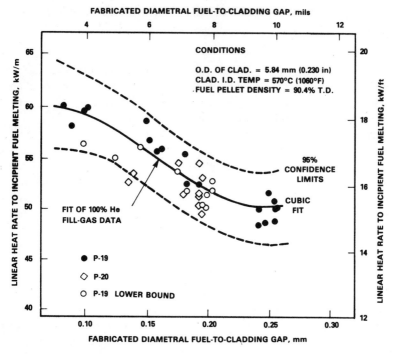

FIGURE 9-10. Cubic fit of χ_m data from fresh pins normalized to FFTF conditions with helium fill-gas.[8] Lower bound data refer to pins in which no fuel melting occurred.

part of the axial length, the χ for the axial position of incipient melting could be accurately determined.

The following correlation was found for linear-power-to-melting, χ_m, for fresh FFTF and CRBRP pins with 100% He fill gas:[8]

$$\chi_m = 53.51 + 199.3G_0 - 1.795 \times 10^3 G_0^2 + 3.764 \times 10^3 G_0^3, \quad (9\text{-}20)$$

$$(0.076 \leqslant G_0 \leqslant 0.254)(100\% \text{ He fill gas})$$

where G_0 is the cold gap in millimeters and χ_m is in kW/m. This fit and the corresponding 3σ level of confidence (discussed in Chapter 10-4) are plotted in Fig. 9-10.

A correlation was also developed for 10% Xe and 90% He fill gas.[8] The correlation is plotted on Fig. 9-11 together with the results of the two P-20 tests with 18% Xe–82% He fill gas in which melting occurred throughout the axial length of the pins. The correlation is

$$\chi_m = 56.7 + 100.7G_0 - 1.449 \times 10^3 G_0^2 + 3.424 \times 10^3 G_0^2 \quad (9\text{-}21)$$

$$(0.076 \leqslant G_0 \leqslant 0.254)(10\% \text{ Xe}, 90\% \text{ He fill gas}).$$

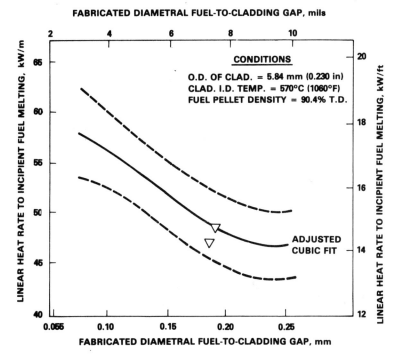

FIGURE 9-11. Cubic fit of χ_m data from fresh pins adjusted to FFTF conditions with 10% Xe tag gas (8). The two data points refer to χ_m for pins which contained 18% Xe and 82% He fill gas. Values of χ_m for 10% Xe, as in the FFTF pins, should be higher than for 18% Xe.

Lower-bound results of the P-20 experiment for preirradiated pins are plotted in Fig. 9-12. These lower-bound results can be represented by the linear relation,[8]

$$\chi_m = 66.3 - 0.41 G_0 \quad (0.076 \leqslant G_0 \leqslant 0.254) \tag{9-22}$$

(for burnup from 3.7 to 10.9 MWd/kg).

The P-20 test demonstrated that higher values of linear power were required to cause irradiated fuel to melt than fresh fuel because the gaps had closed during the preirradiation and fuel restructuring had matured. These results suggest that it may be possible to limit operation to lower values of linear power during the early stages of FBR operation and to operate at higher linear powers later.

A similar concept linear-power-to-melt test was performed by General Electric, called the F20 experiment. It included a range of variables complementary to those run in the HEDL tests. The results agreed with the HEDL tests and extended the variable range for these type of data.

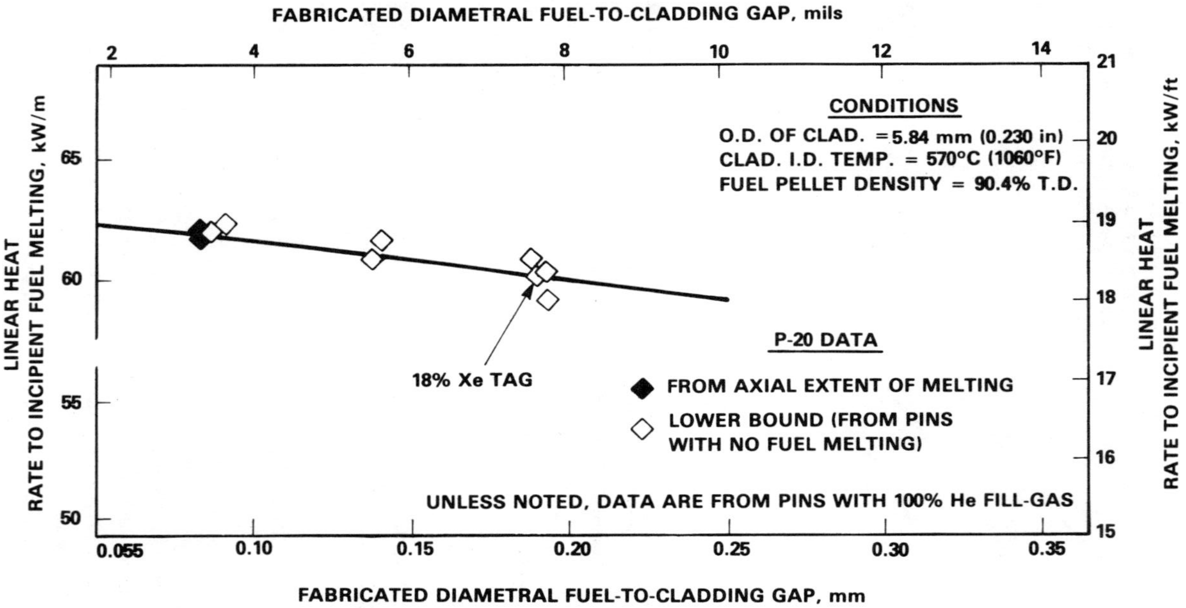

FIGURE 9-12. Fit of lower bound χ_m data for fuel with burnups to 10.9 MWd/kg (1.12 atom %).[8]

9-3 COOLANT HEAT TRANSFER

A. GENERAL ENERGY AND CONVECTION RELATIONS

Energy is transferred from the fuel pins by convection to the coolant. The coolant, either liquid metal (sodium) or gas (usually He), flows between the pins in the assembly geometry. Flow channels (sometimes called subchannels or cells) are designated as the spaces between pins and between pins and duct walls, as shown in Fig. 9-13. There are three types of channels—interior, edge, and corner. In the repeating interior lattice of the assembly, there are two interior channels per fuel rod. Designs with wire-wrap spacers have a wire wrap present in each interior channel half the time.* As noted in Fig. 9-13, the axial distance between each complete spiral of the wire is called the *lead* (sometimes referred to as the axial pitch).

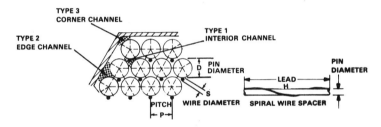

FIGURE 9-13. Definition of flow channels and of wire spacer lead.

The heat flux from the cladding to the coolant is given by

$$q = h(T_{co} - T_b) = \frac{\chi}{2\pi R_{co}},$$ (9-23)

where h = heat transfer coefficient (W/m² · °C),
T_{co} = outer cladding temperature (°C),
T_b = coolant bulk temperature (°C), and
R_{co} = outer radius of the cladding (m).

With the geometry defined by Fig. 9-14, this energy is convected away according to the following expression:

$$h(T_{co} - T_b)2\pi R_{co}\,dz = \chi\,dz = \dot{m}c_p\,dT_b,$$ (9-24)

---

*See Section 9-3C and Eqs. (9-31) and (9-32) for further details of geometry for each type of channel.

where the coolant flow rate, $\dot{m}$, is the flow associated with one fuel pin. The model defined by Eq. (9-24) and Fig. 9-14 is highly simplified and will be used to describe the main features of the axial temperature distribution in Section 9-4B. In Chapter 10, however, a more detailed analysis will be presented which is required for actual LMFBR design. The mass flow rate in Eq. (9-24) is associated with one fuel pin, rather than an individual channel as in Fig. 9-13. The flow is shown upward for this calculation, as is the case for an LMFBR. For a gas-cooled fast reactor, the flow may be downward or upward (discussed in Chapter 17).

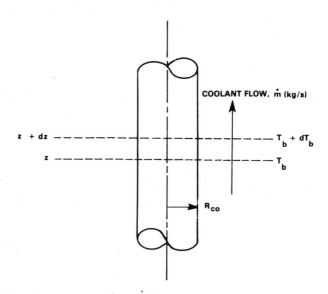

FIGURE 9-14. Geometry for axial coolant flow (simplified).

Before using Eqs. (9-23) and (9-24), we need to evaluate the heat transfer coefficient h. For this reason we shall digress from pin analysis in the remainder of this section in order to consider liquid metal thermal characteristics and heat transfer coefficient correlations. We will return to the use of Eqs. (9-23) and (9-24) in Section 9-4.

B. DIFFERENCE BETWEEN LIQUID METALS AND OTHER FLUIDS

As discussed in Chapter 8-2, a fast breeder reactor needs a high heat transfer coefficient in order to exploit the advantage of a small fuel-pin diameter. Heat transfer coefficients are higher for liquid metals than for other fluids at relatively low flow velocities and pressures; hence, liquid metals are prime candidates for use as FBR coolants. The high heat transfer coefficients for liquid metals result from their high thermal conductivities compared to other fluids.

Before presenting the correlations available for obtaining heat transfer coefficients for LMFBR core design, it is useful to consider some of the general differences between liquid metals and ordinary fluids that follow from the large difference in thermal conductivity.

The Prandtl number, Pr, is an important dimensionless grouping of properties that influence convection heat transfer:

$$\mathrm{Pr} = \frac{c_p \mu}{k} = \frac{\nu}{\alpha}, \tag{9-25}$$

where ν is the kinematic viscosity (μ/ρ) and α is the thermal diffusivity $(k/\rho c_p)$. Kinematic viscosity is related to the rate of momentum transfer in a fluid; thermal diffusivity is related to the rate of heat transfer by conduction. Hence, for a Prandtl number of unity, ν and α are equal, and the mechanism and rate of heat transfer are similar to those for momentum transfer. For many fluids, including water, Pr lies in the range from 1 to 10. For gases, Pr is generally about 0.7.

For liquid metals the Prandtl number is very small, generally in the range from 0.01 to 0.001. This means that the mechanisms of conductive heat transfer in liquid metals dominate over those of momentum transfer. The low Prandtl number results from the exceptionally high thermal conductivity of metals; the values for viscosity and specific heat for sodium are not appreciably different from those of water, whereas the thermal conductivity of sodium is about a factor of 100 greater than that of water. The high thermal conductivity for liquid metals results from the high mobility of free electrons in metal lattices. The Prandtl number for sodium at a typical mid-core temperature of 500°C is 0.0042.

Two classical calculations related to liquid metal heat transfer provide some insight into the physical consequences of a low Prandtl number. The first involves the difference in boundary layer behavior in laminar flow over a flat plate; the second is Martinelli's analysis of temperature distribution in turbulent flow in tubes.

Flow Over a Flat Plate

Laminar flow of liquid metals over a flat plate is analyzed in textbooks on heat transfer (e.g., Reference 10) and the analysis will not be reported here. Physically, though, the problem can be described as follows:

When a fluid flows over a flat plate and heat is simultaneously transferred between the fluid and the plate, two boundary layers develop along the plate surface: a hydrodynamic boundary layer and a thermal boundary layer. The velocity u and temperature T vary in the boundary layers and attain the free stream values u_∞ and T_∞ at the edges of the hydrodynamic and thermal boundary layers, respectively. The temperature at the wall is T_w. For fluids in which the Prandtl number does not differ greatly from unity, the thicknesses of the thermal and hydrodynamic boundary layers (δ_t and δ_h) are about equal, as illustrated in Fig. 9-15(a). For liquid metals, however, the thickness of the thermal boundary layer is substantially larger than the thickness of the hydrodynamic boundary layer as a consequence of the large value of α relative to ν; the high thermal conductivity allows heat to be transported far out into the fluid with relatively little resistance, whereas the normal magnitude for viscosity allows the fluid to attain the free stream velocity relatively quickly. This situation is depicted in Fig. 9-15(b).

The consequences of these differences is a difference in the heat transfer correlation for the Nusselt number ($Nu = hx/k$, where x is the distance down the flat plate). This difference also appears in correlations for turbulent flow in tubes. For liquid metals, the Reynolds (Re) and Prandtl (Pr) numbers in the correlations for both flat plates and tubes appear to the same power, so that the Nusselt number is correlated directly with their product, which is called the Peclet number, Pe. For a flat plate,

$$Pe = Re\,Pr = \left(\frac{\rho V x}{\mu} \right) \left(\frac{\mu c_p}{k} \right) = \frac{\rho V x c_p}{k} . \tag{9-26}$$

Note that viscosity does not appear in the correlation at all. For ordinary fluids, Re and Pr appear to different powers, so that viscosity remains in the correlation and even becomes an important part of the correlation since viscosity is such a strong function of temperature.

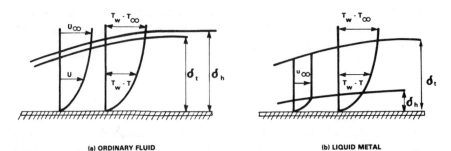

(a) ORDINARY FLUID **(b) LIQUID METAL**

FIGURE 9-15. Comparison of thermal (δ_t) and hydrodynamic (δ_h) boundary layers for ordinary fluids vs liquid metals.

Heated Turbulent Flow in Tubes

Turbulent flow is characterized by high Reynolds number (Re>2300 for tube flow), where

$$\mathrm{Re} = \frac{\rho V D}{\mu}. \tag{9-27}$$

In fully developed tube flow, the hydrodynamic boundary layer occupies the entire tube; one can imagine the boundary layers from the walls all meeting at the centerline of the tube. For turbulent flow, the hydrodynamic boundary layer consists of three sublayers: a laminar sublayer next to the wall, then a buffer layer, and finally a turbulent core that fills the center of the tube. Special equations for the velocity distributions in each sublayer are available for turbulent flow in tubes, known in fluid mechanics as the *law of the wall*.

Martinelli[11] analyzed turbulent flow through a tube with heat transfer by using velocity profiles through the three boundary layers but accounting for the differences in thermal conductivity and Prandtl number between liquid metals and ordinary fluids, and accounting for eddy diffusivity of heat and momentum. His results for a Reynolds number of 10 000 are reproduced in Fig. 9-16, in which the normalized temperature difference in the tube, $[T_w - T(y)]/(T_w - T_{CL})$, is plotted against the normalized distance from the tube wall y/r_o, where r_o is the tube radius, $y/r_o = 1$ at the centerline, and T_w and T_{CL} are the wall and centerline temperatures. (Typical Reynolds numbers in LMFBR cores are 50 000; Martinelli's results for Re = 100 000 are also similar to Fig. 9-16.) It is observed that for ordinary fluids (Pr~1) there is a high resistance to heat transfer in the laminar sublayer and the buffer layer, whereas for liquid metals (Pr<0.01) the resistance to heat transfer is distributed throughout the tube, i.e., well into the turbulent core.

From this result one would expect that the temperature profile in a coolant channel in an LMFBR would be similar to the curves in Fig. 9-16 for Pr=0.01 and 0.001; hence, there would not be a thin film adjacent to the fuel pin across which most of the temperature drop would occur. This is indeed the case and represents one difference between LMFBR's and LWR's. In the water flow channels in an LWR, most of the temperature drop occurs across a thin film adjacent to the fuel rod. The LMFBR literature continues to use the term "film heat transfer coefficient;" we, too, will use the term in Chapter 10, even referring to a region as a "film" region as if it physically existed between the coolant and the cladding. A hydrodynamic film of low-velocity flow adjacent to the tube wall does exist even in liquid metal tube flow, however.

Figure 9-16 also illustrates why viscosity plays a large role in correlations for heat transfer coefficient in turbulent flow in tubes for fluids with normal

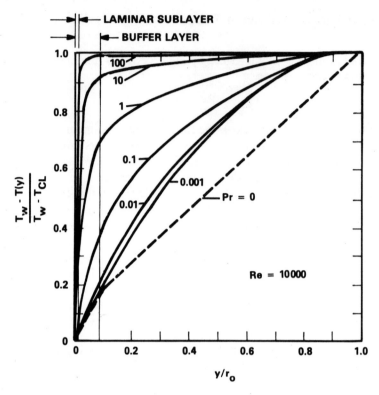

FIGURE 9-16. Influence of Prandtl number on turbulent flow in tubes with heat transfer.[11]

or high Prandtl number, whereas it does not enter into the liquid metal correlations. Although viscosity has a strong influence on velocity distributions in laminar and buffer layers in liquid metal tube flow, it does not affect temperature distributions; hence, heat transfer correlations are based on the viscosity-independent Peclet number.

C. LIQUID-METAL HEAT TRANSFER CORRELATIONS FOR ROD BUNDLES

Heat transfer coefficients for sodium flow through fuel-pin bundles with pitch-to-diameter ratios encountered in the LMFBR have been measured extensively, and numerous correlations have been proposed. Uncertainty limits on the heat transfer correlations are still rather large, but the importance of this uncertainty must be assessed with regard to the influence of the uncertainty on core and fuel pin design. This subject will be explored in Chapter 10 on hot channel factors. It will be observed that even a large uncertainty in heat transfer coefficient will have a relatively minor effect on maximum cladding or fuel temperatures. Hence, from the point of view of LMFBR design, knowledge of the sodium heat transfer coefficient is reasonably adequate.

Experimenters try to find a correlation that best fits existing data. The reactor designer must use a method of analysis that provides reasonable assurance that allowable cladding and fuel temperatures are not exceeded. One way to accomplish this is to use a correlation that gives heat transfer coefficients at some specified lower limit of uncertainty when compared to existing reliable experimental data. Another method is to use the best-fit correlation and provide for the uncertainties through appropriate hot-channel factors. For the FFTF and CRBRP analyses, a correlation was used that gives heat transfer coefficients below the best-fit values, and further allowance for uncertainties was added through hot channel factors. Both the FFTF/CRBRP correlations and several best-fit correlations are discussed in this section.

Liquid metal convection-heat-transfer correlations are generally presented in terms of Nusselt number versus Peclet number, where for tubes and rod bundles,

$$\text{Nu} = \frac{hD_e}{k}, \tag{9-28}$$

$$\text{Pe} = \text{Re} \, \text{Pr} = \frac{\rho V D_e c_p}{k}, \tag{9-29}$$

and D_e is an effective hydraulic diameter [cf. Eq. (9-30)]. Once the Nusselt number is determined, the heat transfer coefficient, h, can be obtained and used in Eqs. (9-23) and (9-24).

When the experimental data for log(Nu) are plotted against log(Pe), the curve generally has a dog-leg* shape; the Nusselt number is constant up to some value of Peclet number called the critical Peclet number, Pe_c, and gradually rises for values greater than Pe_c. The value of Pe_c is about 200. Experimental data of this type are summarized in Fig. 9-17 for sodium flow through rod bundles of various pitch-to-diameter (P/D) ratios.[12]

The difference in behavior above and below Pe_c results from eddy conduction. Below Pe_c, heat transfer is by molecular conduction (which is proportional to the thermal conductivity k), while above Pe_c eddy conduc-

---

*A boomerang has a dog-leg shape. If the reader is still uncertain about this shape, he can ask anyone who plays golf, for there are always several dog-leg holes on a golf course.

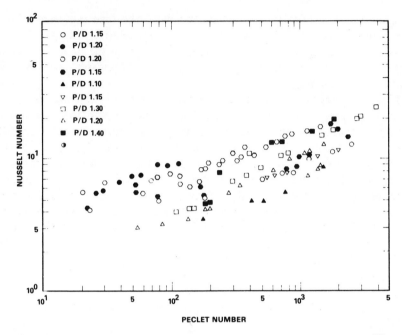

FIGURE 9-17. Heat transfer coefficient correlation data for sodium in rod bundles.[12]

tion (which is proportional to effective eddy conductivity k_e) becomes important. Eddy conduction begins to make a contribution well above the laminar/turbulent boundary. For example, for sodium at 500°C (Pr= 0.0042) and Pe=200, the Reynolds number is 50 000, which is well into the turbulent regime.

For flow through rod bundles the diameter in the Re, Pe, and Nu numbers is an effective hydraulic diameter, D_e. For the channel types shown in Fig. 9-13, the value of D_e for channel i is

$$D_{ei} = \frac{4A_i}{P_{wi}}, \tag{9-30}$$

where A_i is the flow area of channel i and P_{wi} is the wetted perimeter. For fuel assemblies with wire spacers, as shown in Fig. 9-13, the flow area in the basic channel ($i=1$) is the area between the fuel pins less half the cross-sectional area of the wire wrap, because wire wrap is in the channel for half the time on the average as the wires spiral around the three fuel pins adjacent to the channel*. In the edge channel ($i=2$) the flow area is again

---

*This can be observed from Fig. 10-4 in Chapter 10.

the area between the pins and the wall less half the wire-wrap area. In the corner channel ($i=3$), the flow area is the area between the pin and the corner less one sixth of the wire-wrap area. The wetted perimeter P_{wi} is the sum of the perimeters of the fuel-pin and wall surfaces in each channel plus one half the circumference of the wire wrap in channels 1 and 2 and one sixth of the wire-wrap circumference in channel 3. Therefore,

$$A_1 = \frac{\sqrt{3}}{4} P^2 - \frac{\pi D^2}{8} - \frac{\pi s^2}{8}, \qquad (9\text{-}31a)$$

$$A_2 = P\left(\frac{D}{2} + s\right) - \frac{\pi D^2}{8} - \frac{\pi s^2}{8}, \qquad (9\text{-}31b)$$

$$A_3 = \frac{1}{\sqrt{3}}\left(\frac{D}{2} + s\right)^2 - \frac{\pi D^2}{24} - \frac{\pi s^2}{24}, \qquad (9\text{-}31c)$$

$$P_{w1} = \frac{\pi D}{2} + \frac{\pi s}{2}, \qquad (9\text{-}32a)$$

$$P_{w2} = P + \frac{\pi D}{2} + \frac{\pi s}{2}, \qquad (9\text{-}32b)$$

$$P_{w3} = D + 2s + \frac{\pi D}{6} + \frac{\pi s}{6}, \qquad (9\text{-}32c)$$

where D, P, and s are defined in Fig. 9-13.

The vertical flow velocity, V_i, in the Re and Pe numbers is governed by the average assembly velocity $\bar{V}$ and the channel flow area and hydraulic radius through the flow distribution factor, X_1, given in Chapter 10-2A, Eq. (10-8).

An early review of liquid metal heat transfer correlations in rod bundles was given by Dwyer.[13] A useful recent comparison of correlations and more recent data has been provided by Kazimi and Carelli.[14] This reference shows experimental data (Nu versus Pe) for pitch-to-diameter ratios from 1.04 to 1.30. These data were compared with four proposed correlations, which we will now discuss. The pitch-to-diameter ratios of principal interest for LMFBR design are indicated by the values listed in the table of Appendix A. These values are around 1.2 (varying from 1.15 to 1.32) for fuel, and lower (1.05 to 1.1) for control and radial blanket assemblies. Typical Peclet numbers for normal operation are from 150 to 300 in the fuel assemblies.

Borishanskii, Gotovskii, and Firsova proposed the following correlation: [15]

$$Nu = 24.15 \log_{10}\left[-8.12 + 12.76(P/D) - 3.65(P/D)^2\right] \quad (9\text{-}33a)$$

for $1.1 \leqslant P/D \leqslant 1.5$ and $Pe \leqslant 200$, and

$$Nu = 24.15 \log_{10}\left[-8.12 + 12.76(P/D) - 3.65(P/D)^2\right] \quad (9\text{-}33b)$$
$$+ 0.0174\left[1 - e^{-6(\frac{P}{D} - 1)}\right](Pe - 200)^{0.9}$$

for $1.1 \leqslant P/D \leqslant 1.5 \leqslant$ and $200 \leqslant Pe \leqslant 2000$.

The critical Peclet number is 200. Reference 14 shows that this correlation agrees well with most of the experimental data over the ranges of P/D and Pe indicated.

Graber and Rieger proposed the following correlation: [16]

$$Nu = 0.25 + 6.2(P/D) + \left[0.32(P/D) - 0.007\right](Pe)^{[0.8 - 0.024(P/D)]}$$

for $\quad 1.25 \leqslant P/D \leqslant 1.95$ and $150 \leqslant Pe \leqslant 3000$. $\qquad (9\text{-}34)$

This correlation fits the data for Graber and Rieger's experiments ($P/D = 1.25$) and other data at $P/D = 1.2$, but overestimates values of Nu for P/D ratios below 1.2.

FFTF Analysis. The FFTF correlation is [6]

$$Nu = 4.0 + 0.16(P/D)^{5.0} + 0.33(P/D)^{3.8}(Pe/100)^{0.86} \quad (9\text{-}35)$$

for $20 \leqslant Pe \leqslant 1000$.

CRBRP Analysis. The correlation used for CRBRP[14] utilized the FFTF correlation in the P/D range of 1.2 and 1.3 and the Schad correlation as modified by Carelli[14] for the P/D range 1.05 to 1.15. The modified Schad correlation has two parts, one below Pe_c and one above Pe_c, where $Pe_c = 150$. The correlations used were

(a) $1.2 \leqslant P/D \leqslant 1.3$; the FFTF correlation, Eq. (9-35).
(b) $1.05 \leqslant P/D \leqslant 1.15$; modified Schad:

$$Nu = 4.496\left[-16.15 + 24.96(P/D) - 8.55(P/D)^2\right] \quad (9\text{-}36a)$$

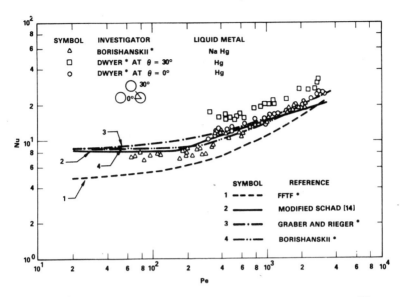

FIGURE 9-18. Comparison of predicted and experimental results for $P/D=1.3$.[14] (*References to investigators are found in Ref. 14.)

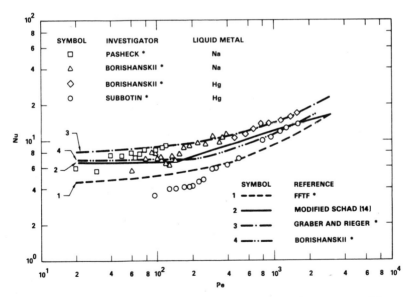

FIGURE 9-19. Comparison of predicted and experimental results for $P/D=1.2$.[14] (*References to investigators are found in Ref. 14.)

for $Pe \leqslant 150$, and

$$Nu = 4.496 \left[16.15 + 24.96(P/D) - 8.55(P/D)^2 \right] \left(\frac{Pe}{150} \right)^{0.3} \quad (9\text{-}36b)$$

for $150 \leqslant Pe \leqslant 1000$.

Values of Nu determined from experimental data are compared against values predicted by the various correlations in Figs. 9-18 and 9-19. Figure 9-19 also shows how the value of Nu can be affected by circumferential position within the core; data are from Dwyer and co-workers at Brookhaven National Laboratory.[17] The conservatism of the FFTF-CRBRP correlation can be readily discerned.

9-4 FUEL PIN TEMPERATURE DISTRIBUTION

Having determined the means to compute the temperature distribution from the fuel center to the outer surface of cladding (Section 9-2) and to evaluate the coolant heat transfer coefficient (Section 9-3), we are now ready to complete the thermal analysis of the fuel pin.

A. RADIAL TEMPERATURE DISTRIBUTION

Given the bulk coolant temperature at a specified core axial position, we can use Eq. (9-23) to determine the temperature rise to the outer surface of the cladding. This allows us to compute the absolute value of the radial temperature profile across the fuel pin.

Illustrative results of radial temperature calculations at the core midplane of the FFTF at beginning-of-life are shown in Fig. 9-20. The *peak* pin distribution is the nominal distribution for the peak power pin. The *hot* pin distribution is the distribution at 115% overpower and with the effects of all hot channel factors included. The *average* distribution is the nominal average-power fuel pin. Note that the center temperature for the average pin is below the value required for fuel restructuring so that no central void appears.

B. AXIAL TEMPERATURE DISTRIBUTION

In determining the axial temperature distribution in the coolant, it is tempting to simply integrate Eq. (9-24) across the assembly to obtain the coolant temperature rise in a single flow channel as the coolant flows

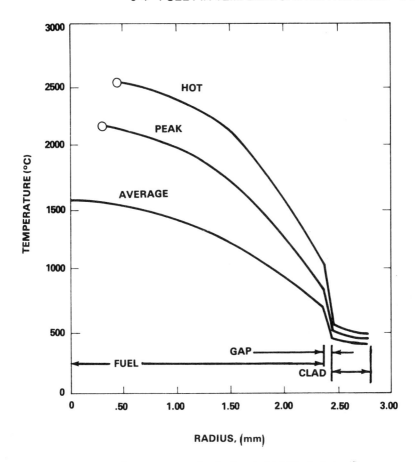

FIGURE 9-20. Typical radial temperature distributions in LMFBR fuel pins.[6]

through the assembly. Using average values for an assembly, one can obtain an average temperature rise, $\overline{\Delta T_b}$, for the assembly. As will be discussed in Chapter 10, however, there is actually a wide variation in ΔT_b between flow channels within an assembly. For the time being, defining the parameters as average values for an assembly, the average $\overline{\Delta T_b}$ for the assembly is

$$\overline{\Delta T_b} = \frac{1}{\dot{m}c_p} 2\pi R_{co} h \int \left[\overline{T}_{co}(z) - \overline{T}_b(z) \right] dz = \frac{1}{\dot{m}c_p} \int \chi(z)\, dz, \quad (9\text{-}37)$$

where the integration is over the assembly, including the axial blankets.

Note that the average coolant temperature rise across an assembly is directly proportional to linear power and inversely proportional to mass flow rate through the assembly. It is desirable to have a uniform $\overline{\Delta T_b}$ for all assemblies in the core (and to some extent in the radial blanket). Since χ

varies radially in the core according to the radial power distribution, the achievement of a uniform $\overline{\Delta T_b}$ requires that $\dot{m}$ be varied between assemblies so that the ratio $\chi/\dot{m}$ is the same for each. This is done by orificing the flow rate through each assembly, meaning that an orifice is physically placed in each assembly near the coolant inlet. Exact $\overline{\Delta T_b}$ matching cannot be achieved since the radial power distribution varies with the fuel cycle, so the designer must optimize on a reasonable orificing scheme. (Gas cooled fast reactor designs have been proposed with variable orificing during the fuel cycle.)

Axial temperature distributions for FFTF fuel, cladding, and coolant are plotted in Fig. 9-21 for fresh fuel and for fuel irradiated to 60 MWd/kg. Since $\chi(z)$ normally has a chopped cosine shape, the coolant and cladding temperatures are S-shaped curves and their shapes hardly change during the irradiation. The FFTF does not have axial blankets; otherwise there would be a coolant temperature rise in the blankets. Because of the large temperature drop across the fuel-cladding gap for gas-bonded pins and the closing of the gap with burnup, the axial temperature distribution in the fuel pin can change appreciably during irradiation. This effect is evident from Fig. 9-21. For the initial core conditions the fuel-cladding gap is open everywhere, and the smooth coolant temperature profile is reflected throughout

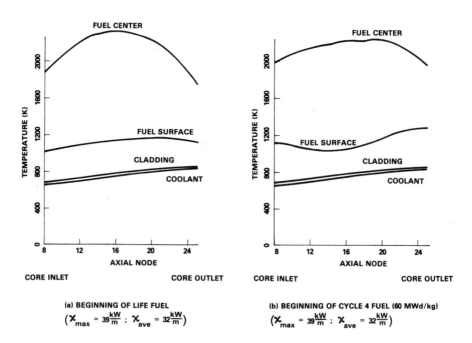

(a) BEGINNING OF LIFE FUEL
$$\left(\chi_{max} = 39\frac{kW}{m} \; ; \; \chi_{ave} = 32\frac{kW}{m}\right)$$

(b) BEGINNING OF CYCLE 4 FUEL (60 MWd/kg)
$$\left(\chi_{max} = 39\frac{kW}{m} \; ; \; \chi_{ave} = 32\frac{kW}{m}\right)$$

FIGURE 9-21. Axial temperature distribution if FFTF fuel pins.[18]

the fuel pin. For later burnup conditions, however, fuel/cladding contact is made over the central two-thirds of the pin and the substantially enhanced gap conductance is responsible for the large dip observed in fuel surface temperature. The lower surface temperature leads to a drop in center fuel temperature as well.

REVIEW QUESTIONS

9-1. Why is burnout of so little concern in an LMFBR relative to an LWR?

9-2. If burnout does not limit the minimum size of the pin diameter in an LMFBR as it does in an LWR, then why not make the pins of smaller diameter, say 4 or 5 mm diameter?

9-3. With regard to Question 9-2, what is the incentive that makes one want to consider reducing the diameter to a small value?

PROBLEMS

9-1. Some FBR's (e.g., BN-350, BN-600, and PFR) use fuel pellets which are fabricated with a central hole (i.e., annular pellets). Like ordinary pins, heat is transferred to the coolant only at the outer surface of the pin.

(a) Derive an expression for the fuel temperature in a pin with annular pellets as a function of the radius, assuming constant thermal conductivity. The radius of the central hole and the outer radius of the fuel are R_o and R_f.

(b) Next, consider the same fuel pin with variable thermal conductivity. We wish to compare the linear power, χ, for pins with annular and solid pellets. Assume for the fuel in the two pins equal values of outer fuel radius, peak and surface temperatures, and fuel density; and assume no restructuring of the fuel. Show that the ratio of χ for the annular fuel to that of the solid fuel is

$$\left[1 - \frac{2\ln\dfrac{R_f}{R_o}}{\left(\dfrac{R_f}{R_o}\right)^2 - 1} \right]^{-1} .$$

(c) Discuss (qualitatively) some of the implications of the result of Part (b) on fuel pellet design. For example, Consider as a function of R_o/R_f such factors as the above ratio and the fuel mass. Consider also the effect on fuel volume fraction, the fissile/fertile ratio, the breeding ratio, and the influence on fabrication costs.

9-2. By letting $\theta = \int_{T_{ref}}^{T(r)} k\, dT$, [so that $d\theta/dr = k\, dT/dr$, and $d^2\theta/dr^2 = d/dr(k\, dT/dr)$], Eq. (9-1) can be written

$$\frac{d^2\theta}{dr^2} + \frac{1}{r}\frac{d\theta}{dr} + Q = 0,$$

for which the solution is

$$\theta = -\frac{Qr^2}{4} + C\ln r.$$

Using this definition of θ, derive Eqs. (9-12), (9-13), and (9-14).

9-3. Calculate the center temperature for a $(U_{0.8}Pu_{0.2})O_2$ fuel pin for the following operating conditions:

Linear power, $\chi = 50$ kW/m.
Sodium temperature $= 400°C$ at the axial position of the calculation.
$h = 100$ kW/m² · °C.
$k_c = 20$ W/m · °C (cladding).
Fuel pin outer diameter $= 8.8$ mm (i.e., outer cladding diameter).
Cladding thickness $= 0.35$ mm.
$G = 0.05$ mm [neglect δ's and g's in Eq. (9-17)].
k_m(gas in the gap) $= 0.3$ W/m · °C.
$\rho_3 = 90\%$ T.D. ($p = 0.10$).
$\rho_2 = 95\%$ T.D. ($p = 0.05$).
$\rho_1 = 100\%$ T.D. ($p = 0$).
$k(p) = k(95\%$ T.D.)$[(1 - 2.5p)/0.875]$.
$R_3 = 4.0$ mm (surface).
$R_2 = 3.5$ mm.
$R_1 = 3.0$ mm.
$R_0 = 1.3$ mm.

9-4. Based on the approximations that the equiaxed and columnar temperatures are

$$T_{eq} = 1500°C$$
$$T_{col} = 1900°C - 5\chi \text{ (where } \chi \text{ is in kW/m)}$$

together with the $\int k\,dT$ data of Fig. 9-3, sketch roughly (i.e., without estimating R_i's, but indicating the different fuel structures) the cross section of a fuel pellet operated at linear powers of

(a) 50 kW/m

(b) 20 kW/m.

9-5. For the basic (triangular) channel for the LMFBR specified in Problem 4-4, Chapter 4, calculate the heat transfer coefficient at the fuel pin surface with sodium flowing with a velocity of 6 m/s, using the following two correlations:

(i) Borishanskii, Gotovskii, and Firsova.

(ii) CRBRP correlation.

Assume that sodium properties are independent of temperature and that

$$\rho = 850 \text{ kg/m}^3$$
$$c_p = 1.2 \text{ kJ/kg} \cdot °\text{C}$$
$$\mu = 3 \times 10^{-4} \text{ kg/s} \cdot \text{m}$$
$$k = 70 \text{ W/m} \cdot °\text{C}.$$

9-6. In a radial blanket assembly of an LMFBR, the fuel pins are still in a triangular lattice, but both the heat generation rate and sodium flow rate are much reduced below the values in the core. Also the blanket pins are larger in diameter.

For the following conditions, calculate the cladding surface temperature at the top of the radial blanket, using the Borishanskii, Gotovskii, and Firsova correlation:

Blanket height $= 1$ m
D (pin) $= 12$ mm
$P/D = 1.3$
Flow area/channel $= 4.9 \times 10^{-5} \text{ m}^2$
$V = 0.4$ m/s (sodium flow velocity)

Na properties: $\rho = 800 \text{ kg/m}^3$
$c_p = 1.2 \text{ kJ/kg} \cdot K$
$k = 70 \text{ W/m} \cdot K$
$\mu = 3 \times 10^{-4} \text{ kg/s} \cdot \text{m}$

T (blanket inlet) $= 400°\text{C}$
$\bar{\chi}$ (average over blanket) $= 3$ kW/m
χ (at top of blanket) $= 2$ kW/m
Flow is fully developed, steady state, and incompressible.

9-7. Using the Borishanskii correlation and the parameters given below, calculate the temperature difference between the outer cladding surface and the bulk sodium.

$$\chi = 30 \text{ kW/m}$$
$$Pe = 150$$
$$k(\text{sodium}) = 70 \text{ W/m} \cdot K$$
$$D_e = 3 \text{ mm (effective diameter of coolant channel)}$$
$$D = 7 \text{ mm (pin diameter)}$$
$$P/D = 1.2 \text{ (pitch-to-diameter).}$$

REFERENCES

1. Y. S. Tang, R. D. Coffield, Jr., and R. A. Markley, *Thermal Analysis of Liquid Metal Fast Breeder Reactors*, The American Nuclear Society, La Grange Park, Illinois, 1978.
2. A. B. G. Washington, *Preferred Values for the Thermal Conductivity of Sintered Ceramic Fuel for Fast Reactor Use*, TRG-Report-2236, September 1973.
3. D. R. Olander, *Fundamental Aspects of Nuclear Reactor Fuel Elements*, Chapter 10, TID-26711-P1, Office of Public Affairs, U.S. ERDA, 1976.
4. R. B. Baker, *Calibration of a Fuel-to-Cladding Gap Conductance Model for Fast Reactor Fuel Pins*, HEDL-TME 77-86, 1978, Hanford Engineering Development Laboratory, Richland, Washington.
5. A. M. Ross and R. L. Stoute, *Heat Transfer Coefficient between UO₂ and Zircalloy 2*, AECL-1552, June 1962.
6. G. J. Calamai, R. D. Coffield, L. Jossens, J. L. Kerian, J. V. Miller, E. H. Novendstern, G. H. Ursim, H. West, and P. J. Wood, *Steady State Thermal and Hydraulic Characteristics of the FFTF Fuel Assemblies*, FRT-1582, June 1974.
7. R. D. Leggett, E. O. Ballard, R. B. Baker, G. R. Horn, and D. S. Dutt, "Linear Heat Rating for Incipient Fuel Melting in UO₂-PuO₂ Fuel," *Trans. ANS, 15,* 1972, 752.
8. R. B. Baker, *Integral Heat Rate-to-Incipient Melting in UO₂ − PuO₂ Fast Reactor Fuel*, HEDL-TME 77-23, 1978, Hanford Engineering Development Laboratory, Richland, Washington.
9. W. H. McCarthy, *Power to Melt Mixed-Oxide Fuel—A Progress Report on the GE F20 Experiment*, GEAP-14134, September 1976.
10. J. P. Holman, *Heat Transfer*, McGraw Hill Co., New York, 4th Edition, 1976.
11. R. C. Martinelli, "Heat Transfer to Molten Metals," *Trans. ASME, 69* (1947) 949–959.
12. J. Muraoka, R. E. Peterson, R. G. Brown, W. D. Yule, D. S. Dutt, and J. E. Hanson, *Assessment of FFTF Hot Channel Factors*, HEDL-TI-75226 November 1976, Hanford Engineering Development Laboratory, Richland, Washington.
13. O. E. Dwyer, "Heat Transfer to Liquid Metals Flowing In-Line Through Unbaffled Rod Bundles: A Review," *Nucl. Eng. Des.*, (1969)) 3–20.
14. M. S. Kazimi and M. D. Carelli, *Heat Transfer Correlation for Analysis of CRBRP Assemblies*, CRBRP-ARD-0034, November 1976.
15. V. M. Borishanskii, M. A. Gotovskii and E. V. Firsova, "Heat Transfer to Liquid Metal Flowing Longitudinally in Wetted Bundles of Rods," *Sov. At. Energy 27* (1969) 1347–1350.
16. H. Graber and M. Reiger, "Experimental Study of Heat Transfer to Liquid Metals Flowing In-Line Through Tube Bundles," *Progress in Heat and Mass Transfer, 7,* Pergamon Press, New York, (1973) 151–166.

17. O. E. Dwyer, H. Berry and P. Hlavac, "Heat Transfer to Liquid Metals Flowing Turbulently and Longitudinally Through Closely Spaced Rod Bundles," *Nucl. Eng. Des.*, *23* (1972) 295–308.

18. A. E. Waltar, N. P. Wilburn, D. C. Kolesar, L. D. O'Dell, A. Padilla, Jr., L. N. Stewart, and W. L. Partain, *An Analysis of the Unprotected Transient Overpower Accident in the FTR*, HEDL-TME 75-50, June 1975, Hanford Engineering Development Laboratory, Richland, Washington.

CHAPTER 10

CORE THERMAL HYDRAULICS DESIGN

10-1 INTRODUCTION

In the previous chapter we explored the methodology for determining the temperature field for a single fuel pin. Since a typical fast breeder reactor core comprises several thousand fuel pins clustered in groups of several hundred pins per assembly, a complete thermal-hydraulic analysis requires knowledge of coolant distributions and pressure losses throughout the core. This chapter will address these determinations.

We focus first on the coolant velocity and temperature distributions within an assembly. Then the discussion is broadened to include coolant velocity and pressure distributions throughout the reactor vessel. Having established these temperature and flow fields, the important design question of hot channel factor determination is then addressed.

10-2 ASSEMBLY VELOCITY AND TEMPERATURE DISTRIBUTION

In Chapter 9 we presented a simple relation for the average axial coolant temperature rise, based on an average bundle mass flow rate [Eq. (9-37)]. The problem is actually more complicated, for several reasons. First, the flow areas for the three types of channels in Fig. 9-13 are different, so the velocity and mass flow rates in each are different. Second, there is a radial power distribution, or *power skew*, across individual fuel assemblies. Third, there is *crossflow*, i.e., flow transverse to the axial direction, between assembly channels. In the first part of this section a simplified approximate velocity distribution will be derived that accounts for the variation in flow area between channels. In the second part the more complex set of equations accounting for crossflow and turbulent mixing between channels will be presented.

A. APPROXIMATE VELOCITY DISTRIBUTION

An approximate velocity distribution, or flow split, between flow channels in Fig. 9-13 can be obtained from the physical fact that the axial pressure drop across each channel of a fuel assembly must be the same since the flow begins and ends in common inlet and outlet plenum regions. This technique was used by Novendstern[1] in his method for evaluating pressure drop in a fast reactor fuel assembly and is similar to the technique developed earlier by Sangster.[2] The approximate velocity distribution is useful for calculating assembly pressure drops and for use as a term in the Peclet number to calculate heat transfer coefficients. It is not sufficiently accurate for detailed design calculations of outlet temperature distribution or maximum cladding temperatures.

Novendstern's flow-split model was later improved by Chiu, Rohsenow, and Todreas.[3] The method of Novendstern is presented here, after which the final result for the Chiu-Rohsenow-Todreas model is given.

Novendstern Model

The pressure drop across channel i is

$$\Delta p_i = f_i \frac{L}{D_{ei}} \frac{\rho V_i^2}{2}, \tag{10-1}$$

where f_i is the effective friction factor in channel i (to be defined further in Section 10-3C), V_i is the velocity, and D_{ei} is the effective hydraulic diameter [Eq. (9-30)]. Setting the three pressure drops equal gives

$$f_1 \frac{L}{D_{e1}} \frac{\rho V_1^2}{2} = f_2 \frac{L}{D_{e2}} \frac{\rho V_2^2}{2} = f_3 \frac{L}{D_{e3}} \frac{\rho V_3^2}{2}. \tag{10-2}$$

The velocities V_i must account for the way in which the mass flow is divided between the three types of channels. To relate V_i to the geometry, we first define an average axial velocity, $\bar{V}$, for the fuel assembly, which is related to the total mass flow rate $\dot{m}_T$ through the assembly by

$$\dot{m}_T = \rho A_T \bar{V}, \tag{10-3}$$

where A_T is the sum of the flow areas of all the channels.

We let N represent the total number of channels in the assembly and N_i the number of channels of type i. The number of type-3 channels per assembly is always 6 (Fig. 9-13). For a 217-pin assembly there are 384 basic channels and 48 edge channels. For a 271-pin assembly there are 486 basic channels and 54 edge channels. Letting A_i represent the flow area in channel i averaged over one wire lead (where lead is defined in Fig. 9-13),

$$A_T = N_1 A_1 + N_2 A_2 + N_3 A_3. \tag{10-4}$$

The flow areas A_1, A_2, and A_3 are given by Eq. (9-31).

Assuming incompressible flow (i.e., constant coolant velocity), continuity gives

$$N_1 V_1 A_1 + N_2 V_2 A_2 + N_3 V_3 A_3 = \bar{V} A_T. \tag{10-5}$$

Flow distribution is governed by Eq. (10-2). The friction factor for each channel can be approximated by

$$f_i = \frac{C}{Re_i^m} = \frac{C}{\left(\rho V_i D_{ei} / \mu \right)^m}, \tag{10-6}$$

where D_{ei} is given by Eq. (9-30), together with the associated Eqs. (9-31) and (9-32). For turbulent flow the recommended value for m is the exponent that appears in the Blasius relation for the friction factor [Eq. 10-36)], where $m = 0.25$.

Substituting Eq. (10-6) into Eq. (10-2) and assuming that C and m are the same for each channel gives

$$\frac{V_1^{(2-m)}}{D_{e1}^{(1+m)}} = \frac{V_2^{(2-m)}}{D_{e2}^{(1+m)}} = \frac{V_3^{(2-m)}}{D_{e3}^{(1+m)}}. \tag{10-7}$$

Using Eq. (10-7) to eliminate V_2 and V_3 from Eq. (10-5) gives

$$V_1 \left[N_1 A_1 + N_2 A_2 \left(\frac{D_{e2}}{D_{e1}} \right)^{\left(\frac{1+m}{2-m} \right)} + N_3 A_3 \left(\frac{D_{e3}}{D_{e1}} \right)^{\left(\frac{1+m}{2-m} \right)} \right] = \bar{V} A_T,$$

or

$$V_1 = \frac{\bar{V} A_T}{\displaystyle\sum_{j=1}^{3} N_j A_j \left(\frac{D_{ej}}{D_{e1}} \right)^{\left(\frac{1+m}{2-m} \right)}} = X_1 \bar{V}, \tag{10-8}$$

where X_1 is a *flow distribution factor*, or *flow-split parameter*, for channel 1.

For turbulent flow, for which $m=0.25$,

$$V_1 = \frac{\bar{V}A_T}{N_1 A_1 + N_2 A_2 \left(\dfrac{D_{e2}}{D_{e1}}\right)^{0.714} + N_3 A_3 \left(\dfrac{D_{e3}}{D_{e1}}\right)^{0.714}}. \qquad (10\text{-}9)$$

In general, for the ith channel,

$$V_i = \frac{\bar{V}A_T}{\displaystyle\sum_{j=1}^{3} N_j A_j \left(\dfrac{D_{ej}}{D_{ei}}\right)^{\left(\frac{1+m}{2-m}\right)}} = X_i \bar{V}. \qquad (10\text{-}10)$$

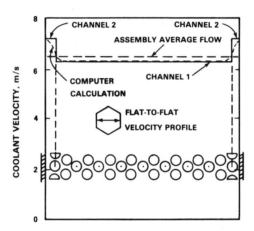

a) Assembly transverse (flat-to-flat)

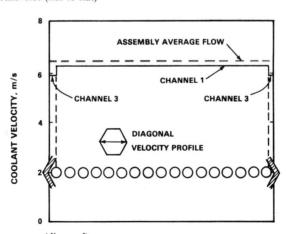

b) Assembly transverse (diagonal)

FIGURE 10-1. Coolant velocity distribution in the maximum power FFTF assembly based on Eq. (10-10).

The mass flow rates and flow velocities through the FFTF driver fuel assemblies are typical of LMFBR power reactors. The mass flow rate through the highest-power FFTF fuel assembly is 23.4 kg/s, and the average velocity, $\bar{V}$, is 6.4 m/s. The distribution according to Eq. (10-10) is plotted in Fig. 10-1 and compared to a more detailed computer prediction (which includes radial momentum exchange).

Chiu-Rohsenow-Todreas (CRT) Model

The CRT flow-split model represents an improvement over the simpler Novendstern model. The principal difference is that the CRT model divides the pressure drop across the channel into two components, one due to skin friction losses, and one due to form losses from flow perpendicular to the wire wrap. The bases for the CRT flow-split model, including the two pressure components, are given in Section 10-3C. Flow split parameters are derived for channels 1 and 2 only; values for the corner channels (channel 3) are assumed to be the same as for the edge channels (channel 2) since the corner channels have so little influence on the flow. Results are given here for turbulent flow only. Results for laminar flow appear in Reference 4. The resulting velocities for channels 1 and 2 for turbulent flow are

$$V_1 = \cfrac{\bar{V} A_T}{N_1 A_1 + (N_2 A_2 + N_3 A_3)\left(\cfrac{D_{e2}}{D_{e1}}\right)^{0.714}\left[\cfrac{C_1 \cfrac{D_{e1}}{H} \cfrac{A_{r1}}{A_1'} \cfrac{P^2}{(\pi P)^2 + H^2} + 1}{C_3\left\{1 + \left[C_2 n\left(\cfrac{V_T}{V_2}\right)_{gap}\right]^2\right\}^{1.375}}\right]^{0.571}},$$

$$\tag{10-11}$$

$$V_2 = \cfrac{\bar{V} A_T}{N_1 A_1 \left(\cfrac{D_{e1}}{D_{e2}}\right)^{0.714}\left[\cfrac{C_3\left\{1 + \left[C_2 n\left(\cfrac{V_T}{V_2}\right)_{gap}\right]^2\right\}^{1.375}}{C_1 \cfrac{D_{e1}}{H} \cfrac{A_{r1}}{A_1'} \cfrac{P^2}{(\pi P)^2 + H^2} + 1}\right]^{0.571} + N_2 A_2 + N_3 A_3},$$

$$\tag{10-12}$$

where C_1, C_2, and C_3 are dimensionless constants that must be determined experimentally; N_i, A_i, D_{ei}, A_T, and $\overline{V}$ are defined the same way as for the Novendstern method; V_T is a transverse velocity (defined in Fig. 10-14); and

P = pitch, as defined in Fig. 9-13
H = lead, as defined in Fig. 9-13
A_{r1} = projected area for one wire wrap in channel 1 over one lead,

$$= \frac{1}{6}\left[\frac{\pi}{4}(D+2s)^2 - \frac{\pi D^2}{4}\right],$$

where D and s are defined in Fig. 9-13,

A_1' = flow area in channel 1 without the wire,

$$= \frac{\sqrt{3}}{4}P^2 - \frac{\pi D^2}{8},$$

$$n = \frac{s}{\left[\left(\frac{D}{2}+s\right)\frac{P}{2} - \frac{\pi D^2}{16}\right]/\frac{P}{2}},$$

$$\left(\frac{V_T}{V_2}\right)_{\text{gap}} = 10.5\left(\frac{s}{P}\right)^{0.35}\frac{P}{\sqrt{(\pi P)^2 + H^2}}\left(\frac{A_{r2}}{A_2'}\right)^{0.5},$$

where A_{r2} = projected area for one wire wrap in channel 2 over one lead, i.e.,

$$A_{r2} = \frac{1}{4}\left[\frac{\pi}{4}(D+2s)^2 - \frac{\pi D^2}{4}\right],$$

$$A_2' = P\left(\frac{D}{2}+s\right) - \frac{\pi D^2}{8}.$$

Based on experimental data from seven references, values for the constants C_1, C_2 and C_3 were determined to be[3]

$$C_1 = 2200$$
$$C_2 = 1.9$$
$$C_3 = 1.2.$$

Comparison of the CRT flow split with the Novendstern result, as given in Eq. (10-9), shows that the only significant difference is the large bracketed term in the denominators of the CRT values.

Circumferential Velocity and Temperature Distributions

Average flow velocities in the channels have been assumed throughout the above discussion. In reality, there is a circumferential variation of velocity around a fuel pin, with the axial component of velocity a minimum at the minimum gap between pins. This circumferential velocity variation leads to a circumferential cladding temperature variation, which in turn causes a hot spot in the cladding temperature at the minimum gap. For the CRBRP and FFTF designs, the ratio of the peak-to-average ΔT between the cladding temperature and the bulk coolant temperature due to the circumferential velocity distribution is ~ 1.5. This appears in the hot channel factors for the maximum cladding temperature calculation in the CRBRP design (discussed in Section 10-4C). Obviously, circumferential velocity and temperature distributions for each LMFBR fuel assembly design must be carefully considered.

B. EFFECTS OF CROSSFLOW ON VELOCITY AND TEMPERATURE DISTRIBUTIONS

Although it is possible to obtain approximate values for the temperature and velocity distributions across a fuel assembly from the use of simple analytical models, the geometry of the reactor lattice is so complicated that it is virtually impossible to obtain exact solutions. This difficulty arises because the flow channels in a reactor lattice (defined in Fig. 9-13) are coupled to one another so tightly by lateral mass, momentum, and energy transport that it is not possible to analyze one channel independently of the others. The reactor designer, however, needs a better solution than is provided by the simple models because the fuel must be designed for a maximum cladding temperature, and this temperature is tied closely to the coolant temperature which varies significantly between flow channels within a particular fuel assembly. A severe economic penalty would be imposed if the reactor designer were forced to set the maximum cladding temperature on the basis of the simple models. Accurate temperature distributions are also needed for the analysis of bowing behavior of assembly hexcans and fuel pins and, in safety-related transients, for the determination of the onset of coolant boiling.

For these reasons most LMFBR thermal hydraulics problems must be solved by complex numerical methods rather than either analytical means or simple models. Many computer programs have been developed to analyze the thermal hydraulic performance of LMFBR fuel assemblies. Some of these codes, such as COBRA, were orginally developed for water reactor applications but have been modified for application to LMFBR's. Recent U.S. LMFBR codes include COBRA,[5] THI3D,[6] TRITON,[7] and ORRI-BLE.[8] A simple and faster code developed specifically for the LMFBR is ENERGY,[9,10] and later SUPERENERGY.[11] The SABRE code,[12] developed in the U.K., is also widely used. The *control volume approach* described below is used in each of the codes listed. These codes are continually being developed further for application to new transient conditions as well as for improvement in describing the steady state.

The Control Volume Approach

The fuel lattice geometry imposes constraints on the flow field that cannot be handled conveniently with a conventional coordinate system. While some reactor analysts have attempted to obtain a detailed resolution of the flow field within the boundaries of each individual flow channel by using the differential equations of fluid mechanics, the more useful approaches have involved systems of approximate equations which rely on simple physical arguments to describe the conservation of mass, energy, and momentum within each channel.

One such approach that has been particularly successful is the control volume approach developed by Meyer[13] and extended to three-dimensional reactor geometry by Rowe[5] (the author of the original COBRA series of codes) and later by others. In this approach the flow channels form control volumes which are allowed to communicate laterally with one another through the gaps between the fuel pins. A basic assumption of the approach is that any flow entering or leaving a control volume loses its sense of direction after crossing the control-volume boundary. Another important feature of the method is that the equations must ultimately be normalized to experiments for the particular lattice geometry and flow conditions.

We present below the conservation equations used in the control volume approach. We shall limit this discussion to the steady-state case. Although transient solutions are needed for safety analysis, a satisfactory understanding of the physical problems can be obtained from the steady-state equations. We further limit the presentation to single-phase flow even though advanced codes like COBRA can, in some instances, be applied to two-phase flow to include boiling that may occur during transients. Finally, we will not discuss the numerical methods used to solve these equations since they are quite sophisticated and beyond the scope of this book. Moreover, new schemes are still being devised to reduce computer running time and simultaneously to provide more detailed information about coolant behavior, e.g., Reference 14.

The objective of the calculation is to determine the velocities at both the inlet and outlet of each flow channel and the temperature (or enthalpy) and density at each channel outlet. The problem is usually defined in terms of four conservation equations, together with boundary conditions and an appropriate equation-of-state for the coolant. The four conservation equations generally used are conservation of mass, energy, axial momentum, and transverse momentum. It is usually assumed that the temperature (or enthalpy), density, and pressure at the inlet of the fuel assembly, and the pressure at the outlet of the assembly, are known and are the same for every flow channel in the assembly.

Conservation of Mass

Consider a control volume to be the space occupied by channel i in an axial section of length dz. Channel i is adjacent to J channels (with $J=3$ for a repeating triangular lattice). The continuity equation is written for channel i adjacent to channel j, as defined by Fig. 10-2; a plan view of the geometry appears in Fig. 10-3.

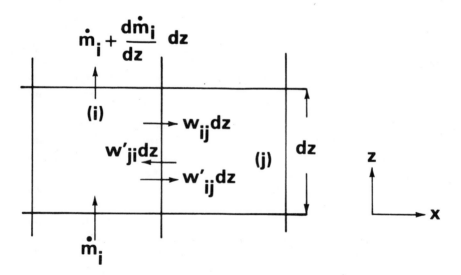

FIGURE 10-2. Geometry for continuity equation.

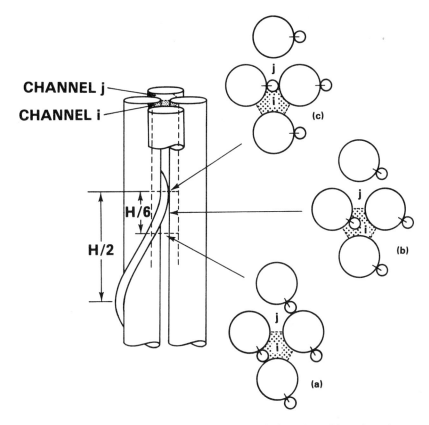

FIGURE 10-3. Wire wrap orientation as it passes through channel i and into channel j.

The upward flow rate $\dot{m}_i(\text{kg/s})$ changes with axial length dz by $(d\dot{m}_i/dz)dz$.

The quantity $w_{ij}dz$ is the net *diversion crossflow*, i.e. the net transverse mass flow rate, from channel i to channel j. Consequently w_{ij} is the crossflow rate per unit length of channel (kg/m·s). Several sources of crossflow interact to produce the net diversion crossflow, including: (1) directional flow induced across the gap between fuel pins by the presence of the wire wrap, (2) non-directional flow induced by objects such as grid spacers, and (3) flow between channels caused by radial (transverse) pressure gradients.

The quantities w'_{ij} and w'_{ji} are *turbulent crossflow rates per unit length* (kg/m·s), and are associated with transport by eddy diffusion. These quantities are sometimes called the *turbulent mixing rates*. No net transport of mass occurs by this mechanism; hence, $w'_{ij}=w'_{ji}$. The phenomenon is important for energy and momentum transfer, however. For example, energy transport from turbulent mixing tends to reduce temperature gradients in the coolant.

Considering for the moment only one adjacent channel as shown in Fig. 10-2, a mass balance on the control volume gives, at steady state,

$$\dot{m}_i + w'_{ji}\,dz = \dot{m}_i + \frac{d\dot{m}_i}{dz}\,dz + w_{ij}\,dz + w'_{ij}\,dz. \qquad (10\text{-}13)$$

Since $w'_{ij} = w'_{ji}$, this reduces to

$$\frac{d\dot{m}_i}{dz} = -w_{ij}. \qquad (10\text{-}14)$$

Expanding the analysis to J channels around channel i,

$$\frac{d\dot{m}_i}{dz} = -\sum_{j=i}^{J} w_{ij}. \qquad (10\text{-}14a)$$

Interchannel mixing has been described in several publications, e.g., References 5, 15 and 16. Diversion crossflow induced by wire wrap can be described with the aid of Figs. 10-3 and 10-4. Flow channels i and j are illustrated in Fig. 10-3 and a wire is shown sweeping across channel i as it advances an axial distance $H/6$ and turns through an angle $\pi/3$ radians (60°). In the plan views the locations of the wire in channel i are illustrated at three axial positions. As the wire approaches channel j, then passes through the gap and on into channel j, it carries a fraction of the flow in channel i into channel j. The maximum crossflow occurs just below and above the wire as it passes through the gap (position c on Fig. 10-3). It is assumed that there is no crossflow between i and j when the wire enters channel i (position a), although crossflow enters channel i at that point from a channel other than j. At an axial height $H/2$ above and below position c, the wire from the pin on the right side of channel i will turn π radians around the pin and pass through the gap between channels i and j, and the crossflow between i and j will be a maximum in the reverse direction.

The crossflow between channels i and j consequently tends to pulse (in opposite directions) at axial intervals of $H/2$ (corresponding to angular wire intervals of π radians) when wire wrap spacers are used. This pulsing-type flow for w_{ij} is illustrated qualitatively in Fig. 10-4. Channel i is also adjacent to two other flow channels in the typical fast-reactor triangular lattice, shown as channels k and l in Fig. 10-4. The total crossflow from channel i is the sum of the crossflows to all adjacent channels and is oscillatory in nature, as also illustrated qualitatively in Fig. 10-4. It can be further noted from Fig. 10-4 that a wire is present in channel i half the time; e.g., no wire passes through channel i between positions a and b while a wire does pass through the channel between b and c.

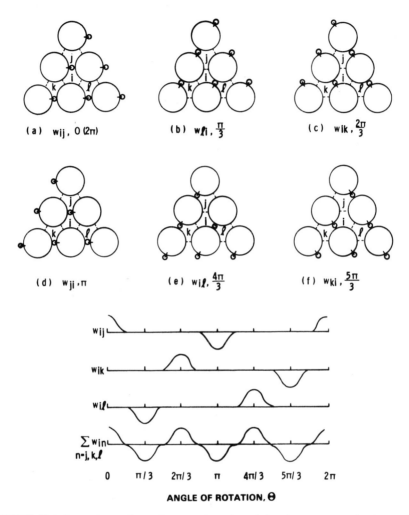

FIGURE 10-4. Dependence of crossflow on orientation of the wire wrap. Listed below each configuration *a* through *f* is the positive crossflow which is maximum for the specific wire orientation followed by the angle of rotation of the wire, θ, relative to *a*. Plots of crossflows versus θ are qualitative; exact curves must be normalized to experiments. The short lines between the wires and the pins indicate the pin to which the wire is attached.

The functional form of the crossflow rate per unit length, w_{ij}, can be related to the channel mass flow rate, $\dot{m}_i$, and the geometry as follows. Consider the portion of the wire in Fig. 10-5 that carries crossflow from channel *i* to channel *j*.

Flow is carried from channel *i* to channel *j* as the wire advances a distance of roughly $H/6$, i.e. $H/12$ below and above the position where the wire passes through the gap which marks the boundary between the

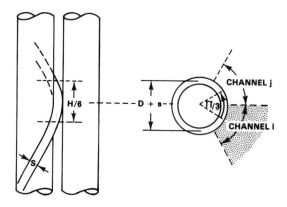

FIGURE 10-5. Geometry for crossflow w_{ij}.

channels. Near the gap, the flow is swept by the wire in a direction approximately parallel to the wire. Hence, the flow through the gap has two components, an axial component u_i and a transverse component v. As the wire advances an axial distance $H/6$, the projection of the crossflow follows an arc of length $(\pi/6)(D+s)$ as illustrated in Fig. 10-5. Therefore, for this approximate model, the ratio of the velocity components where the wire passes through the gap is

$$\frac{v}{u_i} = \frac{\pi(D+s)}{H}. \tag{10-15}$$

The diversion crossflow per unit length can be expressed by the following proportionality:

$$w_{ij} \propto \rho_i s v. \tag{10-16}$$

Replacing v by Eq. (10-15) and u_i by $\dot{m}_i/\rho_i A_i$, where A_i is the average channel flow area [defined by Eq. (9-31)], gives

$$w_{ij} = F\pi \frac{D+s}{H} \frac{s}{A_i} \dot{m}_i, \tag{10-17}$$

where F is a proportionality constant which must be obtained from experiments and which may be a function of Re, P/D, and H/D.

Conservation of Energy

The energy flows are shown in Fig. 10-6. The quantity h^* refers to the enthalpy in the channel from which the diversion crossflow is coming. If the actual direction of the crossflow is from i to channel j (hence, w_{ij} positive),

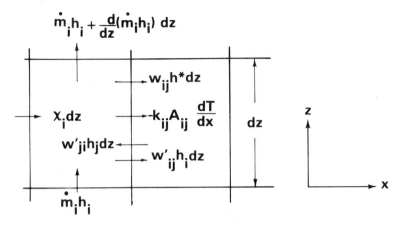

FIGURE 10-6. Geometry for the energy equation.

h^* is h_i; if the flow is from j to i, h^* is h_j. The $\chi_i dz$ term refers to energy from the fuel pins bounding the flow channel; χ_i is the linear power per channel. The $-k_{ij}A_{ij}dT/dx$ term refers to heat conduction to adjacent channel j, where the x-direction is the direction perpendicular to the gap between flow channels i and j. This conduction term is more important in fast reactors than in water reactors because of the high conductivity of sodium and the tight spacing of the lattice.

The energy balance is

$$\dot{m}_i h_i + \chi_i\,dz + w'_{ji}h_j\,dz = \dot{m}_i h_i + \frac{d(\dot{m}_i h_i)}{dz}\,dz$$

$$-k_{ij}A_{ij}\frac{dT}{dx} + w_{ij}h^*\,dz + w'_{ij}h_i\,dz. \quad (10\text{-}18)$$

We expand the derivative in the second term on the right as

$$\frac{d(\dot{m}_i h_i)}{dz} = \dot{m}_i\frac{dh_i}{dz} + h_i\frac{d\dot{m}_i}{dz} \quad (10\text{-}19)$$

and replace $d\dot{m}_i/dz$ by the continuity equation, so that

$$\frac{d(\dot{m}_i h_i)}{dz} = \dot{m}_i\frac{dh_i}{dz} - w_{ij}h_i. \quad (10\text{-}20)$$

The conduction term is written as

$$-k_{ij}A_{ij}\frac{dT}{dx} = -k_{ij}\left(\frac{s}{\Delta x}\right)_{ij}dz(T_j - T_i), \quad (10\text{-}21)$$

where s is the crossflow width ($A_{ij} = s_{ij} dz$) and Δx_{ij} is a representative distance between channels.

Recalling that $w'_{ji} = w'_{ij}$, substitution of the above relations gives for the energy equation

$$\frac{dh_i}{dz} = \frac{1}{\dot{m}_i}\left[\chi_i - k_{ij}\left(\frac{s}{\Delta x}\right)_{ij}(T_i - T_j) + w_{ij}(h_i - h^*) - w'_{ij}(h_i - h_j)\right].$$

(10-22)

For J channels adjacent to channel i,

$$\frac{dh_i}{dz} = \frac{1}{\dot{m}_i}\left\{\chi_i + \sum_{j=1}^{J}\left[-k_{ij}\left(\frac{s}{\Delta x}\right)_{ij}(T_i - T_j) + w_{ij}(h_i - h^*) - w'_{ij}(h_i - h_j)\right]\right\}.$$

(10-22a)

Conservation of Axial Momentum

Axial forces acting on the control volume are shown in Fig. 10-7. The $F_i dz$ represents shear forces, $\rho_i g A_i dz$ is the gravitational force, $\dot{m} u_i$ is the axial momentum flux, $p_i A_i$ is the pressure force in the axial direction, and

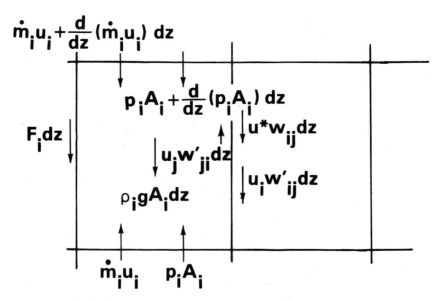

FIGURE 10-7. Geometry for the axial momentum equation. The arrows indicate the direction of the force on the control volume.

the $uw\,dz$ terms represent axial momentum exchanges between adjacent channels. The velocity u^* is the axial velocity in the channel from which the diversion crossflow is coming; thus, u^* is u_i for positive w_{ij} and u_j for negative w_{ij}. The flow area A_i is a function of z in the present context (varying with the changing position of the wire wrap) rather than being an average area as in Eqs. (9-31) and (10-17). Hence, A_i is included in the axial derivative of the pressure term.

The axial momentum equation* for one channel i adjacent to channel j is

$$\dot{m}_i u_i + p_i A_i + u_j w'_{ji}\,dz = \dot{m}_i u_i + \frac{d}{dz}(\dot{m}_i u_i)\,dz$$

$$+ p_i A_i + \frac{d}{dz}(p_i A_i)\,dz + F_i\,dz + \rho_i g A_i\,dz$$

$$+ u^* w_{ij}\,dz + u_i w'_{ij}\,dz. \tag{10-23}$$

The shear term is divided into frictional and form loss terms as follows:

$$F_i\,dz = \left(f_i \frac{dz}{D_i} \frac{\rho_i u_i^2}{2} + K'_i\,dz \frac{\rho_i u_i^2}{2} \right) A_i, \tag{10-24}$$

where f_i is a friction factor and K'_i is a form loss coefficient per unit length. Since $\dot{m}_i = \rho_i u_i A_i$, this term can be written as

$$F_i\,dz = \left(\frac{f_i}{2 D_i \rho_i} + \frac{K'_i}{2 \rho_i} \right) \left(\frac{\dot{m}_i}{A_i} \right)^2 A_i\,dz. \tag{10-25}$$

---

*A form of the momentum equation perhaps more familiar to some is

$$\vec{F_S} + \vec{F_B} = \int_{CS} \vec{V} \rho \vec{V} \cdot d\vec{A},$$

where $\vec{F_S}$ and $\vec{F_B}$ are the surface and body forces acting on the control volume, and the momentum flux is integrated over the control surface (CS). The axial component equation [which is equivalent to Eq. (10-23)] is

$$F_{Sz} + F_{Bz} = -\dot{m}_i u_i + \dot{m}_i u_i + \frac{d}{dz}(\dot{m}_i u_i)\,dz + u^* w_{ij}\,dz + u_i w'_{ij}\,dz - u_j w'_{ji}\,dz.$$

The change in momentum flux can be expressed, using $\dot{m}=\rho u A$ and replacing $d\dot{m}/dz$ by the continuity equation, as

$$\frac{d}{dz}(\dot{m}_i u_i)\,dz = \left(\dot{m}_i \frac{du_i}{dz} + u_i \frac{d\dot{m}_i}{dz}\right) dz$$

$$= \left\{\dot{m}_i\left[\frac{1}{\rho_i A_i}\frac{d\dot{m}_i}{dz} - \frac{\dot{m}_i}{(\rho_i A_i)^2}\frac{d(\rho_i A_i)}{dz}\right] + u_i\frac{d\dot{m}_i}{dz}\right\} dz$$

$$= \left[-2u_i w_{ij} - \left(\frac{\dot{m}_i}{\rho_i A_i}\right)^2 \frac{d(\rho_i A_i)}{dz}\right] dz. \tag{10-26}$$

With these substitutions, the axial momentum equation can be written as

$$\left[-2u_i w_{ij} - \left(\frac{\dot{m}_i}{\rho_i A_i}\right)^2 \frac{d}{dz}(\rho_i A_i)\right] dz + \frac{d}{dz}(\rho_i A_i)\,dz$$

$$+ \left(\frac{f_i}{2D_i\rho_i} + \frac{K_i'}{2\rho_i}\right)\left(\frac{\dot{m}_i}{A_i}\right)^2 A_i\,dz + \rho_i g A_i\,dz + u^* w_{ij}\,dz + (u_i - u_j)w_{ij}'\,dz = 0$$

or

$$\frac{d}{dz}(p_i A_i) = -\left(\frac{\dot{m}_i}{A_i}\right)^2 \left[\frac{f_i A_i}{2D_i\rho_i} + \frac{K_i' A_i}{2\rho_i} - \frac{1}{\rho_i^2}\frac{d}{dz}(\rho_i A_i)\right] - \rho_i g A_i$$

$$+ (2u_i - u^*)w_{ij} - (u_i - u_j)w_{ij}'. \tag{10-27}$$

For J channels adjacent to channel i,

$$\frac{d}{dz}(p_i A_i) = -\left(\frac{\dot{m}_i}{A_i}\right)^2 \left[\frac{f_i A_i}{2D_i\rho_i} + \frac{K_i' A_i}{2\rho_i^2} - \frac{1}{\rho_i^2}\frac{d}{dz}(\rho_i A_i)\right] - \rho_i g A_i$$

$$+ \sum_{j=i}^{J} \left[(2u_i - u^*)w_{ij} - (u_i - u_j)w_{ij}'\right]. \tag{10-27a}$$

Conservation of Transverse Momentum

The control volume for the equation to calculate momentum transfer between channels i and j is illustrated in Fig. 10-8. The control volume is the gap region between channel i and j in which s is the gap thickness (the

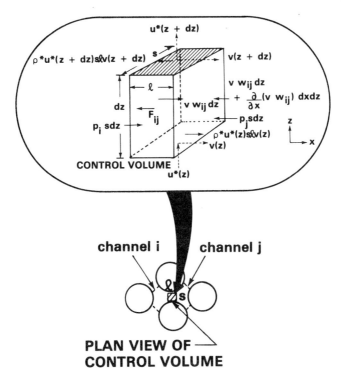

FIGURE 10-8. Geometry for the transverse momentum equation. The arrows associated with the forces indicate the direction of the force on the control volume. The dashed arrows indicate the direction of the flow velocities, v and u^*.

minimum distance between fuel pins) and l is a characteristic distance between channels (a parameter which must be normalized to experiments).

The transverse momentum entering the control volume with the flow upward through the bottom is $\rho^*u^*(z)slv(z)$, where ρ^* and u^* represent values in the channels from which the flow is coming. The momentum $\rho^*u^*(z+dz)slv(z+dz)$ is lost out the top. A second transverse momentum flux entering the control volume across the face of area $s\,dz$ is $vw_{ij}\,dz$, and the momentum flux leaving is $[vw_{ij} + \partial/\partial x(vw_{ij})\,dx]\,dz$. The force balance is

$$\rho^*u^*(z)slv(z)+vw_{ij}\,dz+p_is\,dz=\rho^*u^*(z+dz)slv(z+dz)$$

$$+\left[vw_{ij}+\frac{\partial}{\partial x}(vw_{ij})\,dx\right]dz+p_js\,dz+K_{ij}\frac{\rho^*v^2}{2}s\,dz, \quad (10\text{-}28)$$

where K is a coefficient that accounts for both frictional and form losses for the crossflow. Since $\rho^* s v = w_{ij}$ in the gap, the equation can be written as

$$l\frac{\partial}{\partial z}\left(u^* w_{ij}\right) dz + \frac{\partial}{\partial x}\left(v w_{ij}\right) l\, dz = \left(p_i - p_j\right) s\, dz - K_{ij}\frac{w_{ij}^2}{2\rho s}dz,$$

or

$$\frac{\partial}{\partial z}\left(u^* w_{ij}\right) + \frac{\partial}{\partial x}\left(v w_{ij}\right) = \left(p_i - p_j\right)\frac{s}{l} - K_{ij}\frac{w_{ij}^2}{2\rho s l}. \qquad (10\text{-}29)$$

The two momentum flux terms contribute little during normal steady-state flow. Both terms were omitted in early thermal-hydraulic codes, and the second one is sometimes still omitted; they are left in computer codes more for reasons of numerical stability for transient solutions than for the magnitude contributed to the momentum equation. Setting the left hand side of Eq. (10-29) equal to zero provides a simple relationship between $(p_i - p_j)$ and w_{ij} which is accurate for the steady-state case.

Calculated Results

Codes that solve these equations numerically have achieved considerable success in comparisons with experimental data. It is necessary, however, to obtain some of the coefficients in the equations from a combination of physical arguments and normalizations to experiments—quantities such as F, w_{ij}', K_{ij}, l, $k(s/\Delta x)_{ij}$, and K_i'. The friction factor f_i in Eq. (10-24) can be determined by the methods described in Section 10-3C.

Examples of detailed temperature distributions calculated with transverse mixing models are shown in Figs. 10-9 and 10-10. The curves in Fig. 10-9 were calculated for FFTF using the COBRA code. Outlet temperatures for a fuel assembly in Phénix are shown for calculations with and without accounting for transverse mixing. The asymmetry in Fig. 10-10 results from the radial gradient in the power distribution across the assembly.

The ENERGY Model

The solution of the flow equations for every channel in an assembly requires a large amount of computer time. A useful and simplified model called ENERGY, which calculates a reasonably good exit temperature distribution for forced convection problems in a short computer time, was developed specifically for the LMFBR.[9, 10]

The simplicity of the model results from the replacement of the exact momentum coupling between channels with approximations appropriate for

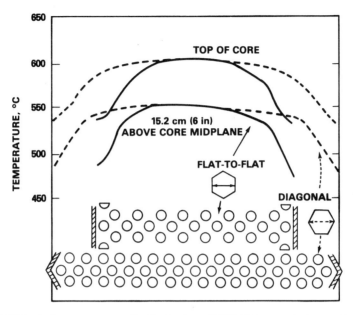

FIGURE 10-9. Sodium temperature distribution in an FFTF high power fuel assembly at two axial locations (217-pin bundle; flat radial power profile).[17]

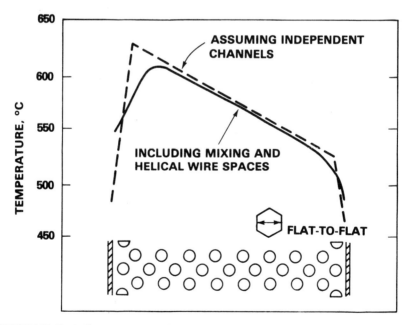

FIGURE 10-10. Sodium temperature distribution in a Phénix fuel assembly at the top of the core (217-pin bundle; radial power skew).[18]

LMFBR's. To accomplish this task, the fuel assembly is radially divided into only two regions instead of the entire number of subchannels. The central region includes most of the central flow channels. The outer region includes the flow channels near and adjacent to the hexcan.

The only differential equations required for the model are the energy equations in the two regions. Three velocities are defined, as shown in Fig. 10-11; these are the axial velocity in the central region $[(U_I)_z]$, the axial velocity in the outer region $[(U_{II})_z]$, and the circumferential velocity in the outer region $[(U)_s]$. The two axial velocities are obtained from the analysis described in Section 10-2A. The effect of crossflow is represented by turbulent diffusion through an eddy diffusivity ε. The circumferential velocity $(U)_s$ and the eddy diffusivity ε are obtained by fitting the solution to experimental data.

The energy equations solved in ENERGY are

$$\rho c_p (U_I)_z \frac{\partial T}{\partial z} = \left(\rho c_p \varepsilon_I + \zeta k \right) \left(\frac{\partial^2 T}{\partial x^2} + \frac{\partial^2 T}{\partial y^2} \right) + Q(x, y, z) \quad (10\text{-}30)$$

$$\rho c_p (U)_s \frac{\partial T}{\partial s} + \rho c_p (U_{II})_z \frac{\partial T}{\partial z} = \left(\rho c_p \varepsilon_n + \zeta k \right) \frac{\partial^2 T}{\partial n^2}$$

$$+ \left(\rho c_p \varepsilon_s + \zeta k \right) \frac{\partial^2 T}{\partial s^2} + Q(s, n, z). \quad (10\text{-}31)$$

The terms on the left represent energy transport by convection. The first terms on the right side of the equations represent energy transport by eddy

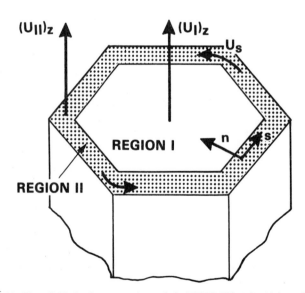

FIGURE 10-11. Flow fields in the two regions of the ENERGY code. (Adapted from Ref. 9).

diffusion and conduction. The heat source from the fuel is given by Q. The factor $\zeta(<1)$ that multiplies the conductivity k accounts for the winding (and hence lengthened) path between fuel pins followed by the sodium as energy is conducted in the direction transverse to the bulk flow; this factor is estimated from physical arguments. Good results are obtained by assuming that all eddy diffusivities are equal.

The boundary conditions for this system of equations include continuity of heat flux and temperature between zones, inlet temperature distribution, and a thermally insulated boundary at the hexcan wall. A finite difference method is used to solve the equations.

The model is normalized to experimental data through the two parameters ε and $(U)_s$. Nondimensional forms of these parameters are expressed as functions of Re, P/D, and H/D.[19] The authors of the ENERGY code report that temperature distributions obtained with that code for forced convection problems for rod bundles of normal geometry are as accurate as those obtained with larger codes, with a considerable saving in computation time. However, the model is not as effective as the more detailed codes for analyzing natural convection, transient effects, flow blockages, and other conditions out of the range of data to which the values of ε and $(U)_s$ have been normalized.

10-3 COOLANT FLOW DISTRIBUTIONS AND PRESSURE LOSSES

Having discussed coolant velocity and temperature fields within assembly bundles, we can now broaden our knowledge of core thermal-hydraulics by addressing the flow patterns and pressure drops within the reactor vessel as a whole. We will then conclude this section by focusing on the pressure drop across the fuel pins in a fuel bundle.

A. IN-VESSEL COOLANT FLOW DISTRIBUTION

Before discussing the flow patterns within the reactor vessel, it is important that we define the portion of the primary loop under discussion. We recall from Fig. 2-5 that for a loop-type system, the primary sodium is circulated through pumps and intermediate heat exchangers (IHX) outside the reactor vessel. For a pool-type system all the heavy primary components are contained within the main reactor vessel. As a consequence, the pump and IHX characteristics and the accompanying pressure distribution exterior to the inlet and outlet plenum are substantially different between these two basic systems. These considerations are discussed in Chapter 12. Between the inlet and outlet plenum, i.e., in the immediate environs of the core, the general flow characteristics are similar for both loop- and pool-type designs.

This discussion will focus on flow and pressure distributions as coolant is driven from the inlet to the outlet plenum. The example given is based on a loop-type design, but the methods would be similar for a pool-type design. This particular example is taken from the CRBRP design.[20] Numbers would differ for other designs, of course, but the flow division shown is instructive because it is typical of most breeder designs.

In the CRBRP design, flow enters an inlet plenum from three loops. The inlet flows must be directed so that mixing during a transient does not result in excessive thermal stresses in the core support structure and the reactor vessel when combined with other pressure and mechanical load stresses.

From the inlet plenum, the flow for the homogeneous CRBRP core was calculated to be distributed according to the following pattern:

Fuel assemblies	80%
Control assemblies	1.6%
Radial blanket assemblies	12%
Radial shielding (blanket modules)	0.3%
Periphery of core support structure	4.7%
Leakage through seals	1.4%

The flow through the fuel, control and radial blanket assemblies passes from the inlet plenum through core support modules. Each module contains seven assemblies. The center inlet module has the highest flow rate; this flow dictates the core pressure drop. Flow distribution through the fuel and control assemblies is controlled by orificing in the individual assemblies rather than in the inlet module. For the radial blanket assemblies, the orificing is in the inlet module instead of the individual assemblies to allow radial blanket assemblies to be shuffled. There are some radial shielding assemblies adjacent to the radial blanket that form seven-assembly clusters with blanket assemblies, and hence are fed by inlet modules. These account for the 0.3% flow given in the list above.

From the inlet module, the sodium flows through the inlet nozzle at the bottom of each assembly. The sodium then flows through an orifice in the fuel and control assemblies. The orifice is designed to equalize maximum cladding midwall temperatures at the equilibrium cycle end-of-life conditions. In the fuel assembly, the sodium then flows past the neutron shielding below the core and upward into and through the rod bundle itself, i.e., the lower axial blanket, the core, the upper axial blanket, and the fission gas plenum. As described later in this section, a large part of the pressure drop occurs across the rod bundle.

The sodium exits from the assemblies into channels that guide the flow up to the vessel outlet plenum. From there it passes to the outlet pipes of the three primary system loops.

The maximum hydraulic load is on the core support plate because it is subjected to the core pressure drop. For the CRBRP the design pressure above the sodium pool, in the cover gas, is only 104 kPa absolute, or 2.5 kPa above atmospheric (0.36 psig). The maximum pressure in the inlet plenum below the core support plate is 950 kPa absolute (123 psig).

A hydraulic balance system is used to assure holddown of the fuel assemblies. A significant portion of the bottom of each assembly is exposed to the low pressure of the outlet plenum so that the net force downward on the fuel assembly is gravity and is equal to the buoyant weight of the assembly.

B. PRESSURE LOSSES

The pressure drop in the flow paths described above are considered to be either *form* or *friction* pressure losses. Form pressure losses are given by the equation

$$\Delta p = K \frac{\rho V^2}{2} \qquad (10\text{-}32)$$

and friction pressure losses are given by

$$\Delta p = f \frac{L}{D} \frac{\rho V^2}{2}, \qquad (10\text{-}33)$$

where the form factors K are either determined experimentally for the particular design or, like the friction factors, obtained from analysis.

Complete results are available for the CRBRP design.[20] It is instructive to see the detail with which the contributions to the pressure drop are calculated by studying the CRBRP results in Table 10-1 for the maximum-power fuel assembly. The *rod bundle* includes the active core region, the axial blankets, and the fission gas plenum.

All pressure drops except the rod bundle loss were reported to be known to $\pm 20\%$ uncertainty with 95% confidence. The rod-bundle pressure drop was known to $\pm 14\%$ uncertainty with 95% confidence. These uncertainties, when applied to the results in Table 10-1, lead to a maximum pressure drop of 850 kPa between the inlet and outlet plenums.

TABLE 10-1   Pressure Losses in the CRBRP Reactor Vessel for
Maximum-Power Fuel Assembly[20]

Component Sub-Component	Pressure Drop		Loss Coefficients
	Component	Sub-Component	
	(kPa)	(kPa)	
Inlet Plenum	32		$K=1.29$
Module	62		
Slots		39	$K=2.01$
Strainer		3	$K=0.363$
Expansion		8	$K=0.41$
Manifold		6	$K+f\dfrac{L}{D}=1.0$
Stalk		6	$K+f\dfrac{L}{D}=1.0$
Inlet Nozzle	105		
Transition		19	$f\dfrac{L}{D}=0.022$
Inlet loss (form)		59	$K=0.098$
Inlet loss (frictional)		4	$f\dfrac{L}{D}=0.0024$
Nozzle friction (1)		4	$f\dfrac{L}{D}=0.000082$
Nozzle friction (2)		19	$f\dfrac{L}{D}=0.00035$
Shield, Orifice	155		Not given
Rod Bundle Inlet	7		$K=0.37$
Rod Bundle	305		$f\dfrac{L}{D}$ from Novendstern correlation
Assembly Exit	21		$K=0.79$
Outlet Plenum	33		$K=1.32$
Total	720		

C.   PRESSURE DROP IN THE FUEL ASSEMBLY

Early United States, French, USSR, and Japanese prototype LMFBR designs utilized wire wrap to space the fuel pins in the hexagonal assembly channel, as discussed in Chapter 8-4C. The pressure drop for this method of spacing is presented here. An alternate method that employs grid spacers at various heights along the fuel pins was used in early British and German designs and the U.S. Fermi Reactor, but the pressure drop caused by spacers is sensitive to the particular grid spacer design. Therefore, no general correlations for grid spacer pressure drop are given in this text.

Two pressure drop correlations for an LMFBR fuel assembly with wire wrap spacers are presented here, the first due to Novendstern[1] and the second due to Chiu, Rohsenow, and Todreas.[3]

The earliest correlation used extensively in the U.S. for determining pressure losses in LMFBR pin bundles was developed by de Stordeur[21] for the Fermi Reactor in the early 1960s (a reactor which used grid spacers instead of wire wrap). In the late 1960s Sangster[2] presented a correlation for wire wrap spacers. Later Novendstern,[1] adopting several of Sangster's principal features, developed a correlation which was used for the FFTF and CRBRP designs. Rehme[22] also contributed an improved method at about the same time as Sangster. Novendstern's analysis takes advantage of all the older methods, together with experimental data of Reihman,[23] Rehme,[22] and Baumann, et al.,[24] and his method agrees closely with pressure loss measurements for an FFTF fuel assembly.[25] Finally, Chiu, Rohsenow, and Todreas extended the analysis further to include drag across the wire wrap more explicitly.[3]

Prior to the development of the Chiu-Rohsenow-Todreas model, the pressure drop was obtained from a modified form of the skin friction pressure drop equation, Eq. (10-33). Chiu, et al., introduced separate expressions to account for the following two components of the pressure drop:

(1) form drag pressure drop induced by flow over the wires,
(2) skin friction pressure drop characterized by the resultant velocity of both the axial and transverse velocity components of the flow.

Hence, the Chiu-Rohsenow-Todreas model uses modified forms of both Eqs. (10-32) and (10-33).

Novendstern Model

In this model the influence of the wire wrap is accounted for by means of an effective friction factor in Eq. (10-33). A friction factor multiplier, M, is introduced such that the effective friction factor, f_1, for channel 1 (Fig. 9-13) is given by

$$f_1 = M f_{smooth}. \tag{10-34}$$

The multiplier M is obtained from a correlation based on the dimensionless parameters pitch-to-diameter (P/D), lead-to-diameter (H/D), and Reynolds number (Re); and f_{smooth} is a function of Re. Hence, M accounts for the increase in f above the smooth tube value caused by the wire spacer.

The correlation for M proposed by Novendstern is

$$M=\left[\frac{1.034}{(P/D)^{0.124}}+\frac{29.7(P/D)^{6.94}\mathrm{Re}_1^{0.086}}{(H/D)^{2.239}}\right]^{0.885}, \qquad (10\text{-}35)$$

where $\mathrm{Re}_1=\rho V_1 D_{e1}/\mu.$

In the expression for Re_1 and in the skin friction pressure drop equation, V_1 is given by Eq. (10-9) and D_{e1} by Eq. (9-30).

The Blasius relation for the friction factor, f_{smooth}, is applicable for the flow conditions in LMFBRs; hence,

$$f_{\mathrm{smooth}}=\frac{0.316}{\mathrm{Re}_1^{0.25}}. \qquad (10\text{-}36)$$

Note that only the value of f_1 is given. It is sufficient that the final calculated pressure drop is based on channel type 1 because the definition of the flow distribution factor, X_i, assures that Δp is the same for all channels (see Section 10-2A).

The final value for Δp across the fuel pins comes from combining Eq. (10-33) and Eq. (10-34) to obtain

$$\Delta p=Mf_{\mathrm{smooth}}\frac{L}{D_{e1}}\frac{\rho V_1^2}{2}. \qquad (10\text{-}37)$$

Calculated results are compared[1] with experimental pressure drop data for a 217-pin FFTF fuel test assembly[25] in Fig. 10-12. These data were not used for the development of the correlation; hence, they provide an independent check of the correlation. The $\pm 14\%$ error curves (also from Reference 1) include most of the data points.

The friction factor multiplier, M, is plotted as a function of P/D and of H/D in Fig. 10-13 for a Reynolds number of 50000, a fairly typical value for an LMFBR. For the FFTF the parameters are

$$D=5.84 \text{ mm } (0.230 \text{ in.})$$
$$P=7.31 \text{ mm } (0.288 \text{ in.})(P/D=1.25)$$
$$H=305 \text{ mm } (12 \text{ in.}) \quad (H/D=52)$$
$$\mathrm{Re}=58000$$

for which $M=1.05$. Hence, the multiplier that accounts for the wire wrap deviates only slightly from unity for designs of most practical interest.

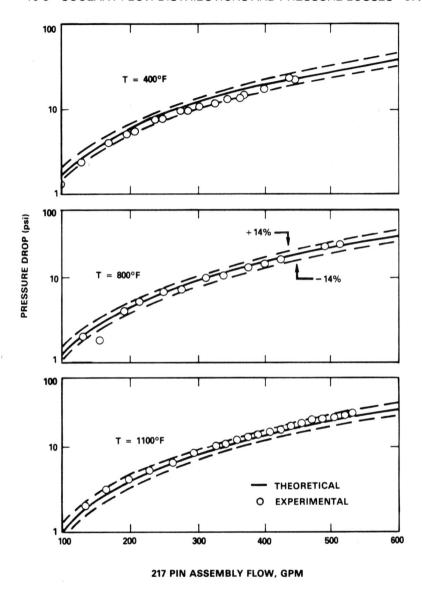

FIGURE 10-12. Comparison of Novendstern model with experimental sodium pressure drop data.[1]

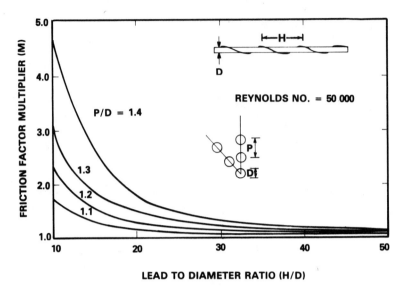

FIGURE 10-13. Friction factor multiplier, M.[1]

Chiu-Rohsenow-Todreas (CRT) Model

While the Novendstern model provides reasonably accurate results, as illustrated in Fig. 10-12, the CRT model treats the mechanisms responsible for the pressure drop in greater detail. Thus the CRT model has greater potential for application over a wider range of flow conditions.

In the CRT model the pressure drop is modeled as the sum of two terms:

$$\Delta p = \Delta p_s + \Delta p_r, \qquad (10\text{-}38)$$

where

$\Delta p_s =$ skin friction losses
$\Delta p_r =$ form drag losses from the component of velocity in the direction perpendicular to the wire.

For the form drag term, the form coefficient K in Eq. (10-32) is expressed as $C_D A_r / A'$, where C_D is a drag coefficient and A_r and A' are defined in Eq. (10-11). The drag coefficient and the velocity component perpendicular to the wire, V_p, are related through a dimensionless constant, C_1', and the skin friction factor based on the pressure drop in the flow channel without the wire, f_s, such that the drag pressure loss component over one lead of fuel pin is

$$\Delta p_r \,(\text{one lead}) = C_1' f_s \frac{A_r}{A'} \frac{\rho V_p^2}{2}. \qquad (10\text{-}39)$$

For a fuel pin length L, there are L/H leads, so that Δp_r for the entire pin is

$$\Delta p_r = C_1' f_s \frac{A_r}{A'} \frac{L}{H} \frac{\rho V_p^2}{2}. \tag{10-40}$$

The velocity component V_p is expressed in terms of the more useful axial, transverse, and resultant velocities V_A, V_T, and V_R according to the geometrical considerations in Fig. 10-14. Thus,

$$V_p^2 = V_R^2 \sin^2 \theta = V_A^2 \left[1 + \left(\frac{V_T}{V_A} \right)^2 \right] \sin^2 \theta. \tag{10-41}$$

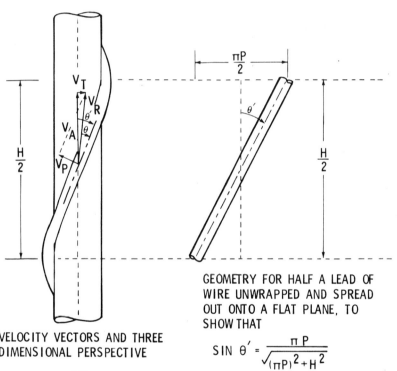

GEOMETRY FOR HALF A LEAD OF WIRE UNWRAPPED AND SPREAD OUT ONTO A FLAT PLANE, TO SHOW THAT

VELOCITY VECTORS AND THREE DIMENSIONAL PERSPECTIVE

$$\text{SIN } \theta' = \frac{\pi P}{\sqrt{(\pi P)^2 + H^2}}$$

PLAN VIEW
(PROJECTION OF WIRE WRAP)
LENGTH OF DASHED LINE = $\pi P/2$

FIGURE 10-14. Geometry for the Chiu-Rohsenow-Todreas model. The axial and transverse velocity components V_A and V_T are equivalent to u and v in Section 10-2B.

The form drag pressure loss is, therefore,

$$\Delta p_r = C_1' f_s \frac{A_r}{A'} \frac{L}{H} \frac{\rho V_A^2}{2} \left[1 + \left(\frac{V_T}{V_A}\right)^2\right] \sin^2 \theta. \tag{10-42}$$

The skin friction component of the pressure drop is

$$\Delta p_s = f_R \frac{L_R}{D_{eR}} \frac{\rho V_R^2}{2}, \tag{10-43}$$

where the subscript R denotes values related to the resultant velocity V_R. Geometrical arguments led Chiu, et al., to replace the values related to V_R as follows:

$$\Delta p_s = f_s \frac{L}{D_e} \frac{\rho V_A^2}{2} \left[1 + \left(C_2 \frac{V_T}{V_A}\right)^2\right]^{1.375}, \tag{10-44}$$

where f_s is the same as in Eq. (10-42) and C_2 is a constant to be determined empirically.

Chiu, et al., next provide reasons why the skin friction component is controlling for the edge channel pressure drop to the extent that the ΔP_r component can be dropped for channel 2. They argue further that $V_T/V_A \ll 1$ for the interior channels so that V_T/V_A can be made zero for channel 1. Moreover, when $V_T/V_A = 0$, θ and θ' on Fig. 10-14 become congruent so that $\sin \theta$ in Eq. (10-42) can be replaced by $\sin \theta'$ as given in Fig. 10-14. Finally the π^2 in the numerator of the expression for $\sin^2 \theta'$ is absorbed into the constant C_1' to give a new constant C_1. Thus, the form drag pressure drop in channel 1 is

$$\Delta p_{r1} = C_1 f_{s1} \frac{A_{r1}}{A_1'} \frac{L}{D_{el}} \frac{D_{el}}{H} \frac{P^2}{(\pi P)^2 + H^2} \frac{\rho V_1^2}{2}, \tag{10-45}$$

where V_A in channel 1 has been replaced by V_1.

The final expressions for the total pressure drop in channels 1 and 2 are

$$\Delta p_1 = f_{s1} \frac{L}{D_{el}} \frac{\rho V_1^2}{2} \left[1 + C_1 \frac{A_{r1}}{A_1'} \frac{D_{el}}{H} \frac{P^2}{(\pi P)^2 + H^2}\right], \tag{10-46}$$

$$\Delta p_2 = f_{s2} \frac{L}{D_{e2}} \frac{\rho V_2^2}{2} \left\{1 + \left[C_2 n \left(\frac{V_T}{V_2}\right)_{gap}\right]^2\right\}^{1.375}, \tag{10-47}$$

where V_A in channel 2 has been replaced by V_2 and V_T/V_A has been replaced by $n(V_T/V_2)_{gap}$, where n and $(V_T/V_2)_{gap}$ are defined in Eq. (10-11).

The velocities V_1 and V_2 to be used in Eqs. (10-46) and (10-47) are those given by Eqs. (10-11) and (10-12). These velocities (and flow split parameters X_1 and X_2, equal to $V_1/\overline{V}$ and $V_2/\overline{V}$) were obtained by forcing the above expressions for Δp_1 and Δp_2 to be equal, together with the continuity equation, Eq. (10-5).

10-4 HOT CHANNEL FACTORS

The first three sections of this chapter have addressed the methods for calculating nominal temperatures, where "nominal" refers to values calculated for rated full-power operation without accounting for uncertainties. We noted in Section 10-2 that substantial variations in temperature can exist within an assembly at a given axial position. These are all predictable values, however, and these determinations constitute only the first part of the designer's job. The next part is the treatment of uncertainties in the analysis. The impacts of uncertainties in theoretical and experimental analyses, instrumentation accuracy, manufacturing tolerances, physical properties, and correlation uncertainties must be considered in order to assure safe and reliable reactor operation. Moreover, the influence of uncertainties on design will indicate the importance of increasingly sophisticated thermal-hydraulics experiments in the development of fast reactor technology. This section deals with treatment of these uncertainties.

The thermal-hydraulics design of a fast reactor must conform to a set of *design bases*, or *design criteria*. Many of these relate to fuel, cladding, and coolant exit temperatures under various conditions. One fuel design criterion for FFTF and CRBRP, for example, was that no fuel melting should occur at some specified *overpower*; for both reactor designs this overpower was set at 115% of rated power. The maximum allowable cladding temperature must assure fuel pin integrity. The maximum allowable exit coolant temperature must assure structural integrity above the core. Other thermal-hydraulic design bases refer to other parameters such as allowable pressure drop and coolant flow velocities.

In order to satisfy the design bases, *design limits* must be specified. This is often done by establishing a statistical *level of confidence* with which selected parameters must be known. An uncertainty in a parameter, or a limiting value, is then associated with the specified level of confidence.

Uncertainties and limiting values are treated through the use of *hot channel factors*, or *hot-spot factors*. The hot channel factor, F, for a particular parameter is the ratio of the maximum value of that parameter to its nominal value. It is therefore a number greater than unity, and the decimal part (i.e., $F - 1$) represents the fractional uncertainty in the parameter. Hot channel factors must be based on a combination of experimental data and experimentally verified analytical methods; justification of these factors is one of the most important and challenging of the reactor designer's tasks.

The exact definitions and values of hot channel factors and the way they are applied to the analysis of uncertainties will change and evolve with each new fast reactor design. Methods used are strongly influenced by precedents used in the licensing process; indeed, LWR precedents have already influenced fast reactor methods. At this time we have available the extensive uncertainty analyses for FFTF[26] and CRBRP;[27, 28] these are used here as guides to the methodology for calculating uncertainties. Analyses for later designs will inevitably become more elaborate, particularly as more experimental data become available and as incentives grow to decrease design margins. The important objective of this section, therefore, is to introduce the methodology involved in the use of hot channel factors, not to spell out the details of the particular analyses currently available.

The parameter multiplied by a hot channel factor is generally a temperature difference. Consequently, the result is independent of the temperature units being used.

A. STATISTICAL METHODS

Two types of uncertainties might influence a particular parameter—*direct uncertainties*, or *biases*, and *random uncertainties*. A direct uncertainty in a parameter represents influence on the parameter by a variable that is not subject to random variation, but for which the exact value cannot be predicted in advance. A random uncertainty represents influence on the parameter by a variable that can be represented by a frequency distribution of occurrence. Several sources of each type of uncertainty might be present. Individual uncertainties propagate to yield an overall uncertainty for the parameter; the method of propagation differs for the two types of uncertainties, as will be described.

The general measure of uncertainty, or dispersion in data, is the *variance*, σ^2. If the statistical distribution of experimental data around a mean value (or of calculated values around the true value) is normal, the square root of the variance, σ, will represent a *standard deviation*. In this case, the probability is 67% that the measured or calculated value will lie within $\pm\sigma$ of the mean or true value, 95% that it will be within $\pm2\sigma$, and 99.73% that it will lie within $\pm3\sigma$. For the 3σ limit, this probability means that there is a 0.13% probability that the measured or calculated value will lie more than

3σ above the mean or true value and an equal probability that it will lie more than 3σ below the mean or true value. Hence, there is a 99.87% probability that the measured or calculated value will lie below the mean or true value plus 3σ.

Random uncertainties in thermal-hydraulics design are assumed to follow a normal distribution. If the design basis requires that the true value of a parameter for which the uncertainty is random be less than a specified design limit for that parameter with 99.9% confidence, then the nominal value of the parameter must be 3σ less than the design limits. This level of confidence is frequently required for LMFBR thermal-hydraulics designs, although for some parameters a 2σ level (97.5% confidence) is considered satisfactory.

If a 3σ level of confidence is required for a particular variable, the hot channel factor for that variable would be

$$F = 1 + 3\sigma. \tag{10-48}$$

It is not always possible to characterize an uncertainty by a standard deviation, especially in the case of direct uncertainties. In such a case it is necessary to find another justification to provide the appropriate level of confidence so that uncertainty factors will not allow design limits to be exceeded.

Combination of Hot Channel Factors

Numerous uncertainties influence each design parameter, and a hot channel factor is associated with each uncertainty. Individual hot channel factors combine to yield an overall hot channel factor for the design parameter.

Hot channel factors based on direct uncertainties multiply nominal values directly; hence, their combination is straightforward:

$$F_d = \prod_{k=1}^{D} F_{d,k}, \tag{10-49}$$

where

$F_{d,k}$ = value of a direct hot channel factor for hot channel effect k,
D = number of direct hot channel factors.

Hot channel factors based on random uncertainties are combined statistically before multiplying a nominal value. The propagated uncertainty is the square root of the sum of the squares of the individual random uncertainties. Each statistical hot channel factor, $F_{s,k}$, is related to the uncertainty, $n\sigma$, in a manner similar to Eq. (10-48), with the number of standard deviations n depending on the level of confidence required. Hence, the part of the statistical hot channel factor that is propagated as the sum of the squares is $(F_{s,k} - 1)$. Therefore,

$$F_s = 1 + \left[\sum_{k=1}^{S} (F_{s,k} - 1)^2 \right]^{1/2}, \qquad (10\text{-}50)$$

where S = number of statistical hot channel factors.

The overall hot channel factor, F, is the product of the combined direct and statistical uncertainty factors:

$$F = F_d F_s. \qquad (10\text{-}51)$$

B. CRBRP AND FFTF HOT CHANNEL FACTORS

The hot channel factors used for the early CRBRP homogeneous core design are listed in Table 10-2.[27] Factors are listed for five regions (coolant, film, cladding, gap, and fuel) and for the heat flux. Hot channel factors are combined by columns in the table so that a total hot channel factor for each region and for the heat flux is obtained; the reason for this will become clear in Section 10-4C where the method for applying hot channel factors is illustrated.

The heat flux factor accounts for the uncertainties in power generation.* It is used in evaluating uncertainties in the film, cladding, gap, and fuel temperature differences. It does not apply to the coolant enthalpy rise (and, hence, to the temperature rise). Other factors, averaged along the flow channel, account for the coolant enthalpy uncertainties.

The FFTF hot channel factor analysis[26] was similar to the early CRBRP analysis; we focus here on the early CRBRP approach since it is the more recent and applies to a power reactor design.

The individual direct and statistical factors are discussed briefly below.

---

*In FFTF, the heat flux factor is included in the total hot channel factor for each region.

TABLE 10-2 Hot Channel Factors (Early CRBRP Homogeneous
Core Design, 3σ Level of Confidence) [27]

	Coolant	Film	Cladding	Gap	Fuel	Heat Flux
A. DIRECT						
Power measurement and control system dead band	1.03					1.03
Inlet flow maldistribution	1.05	1.035				
Assembly flow maldistribution	1.08					
Cladding circumferential temperature variation		1.0 (1.7)*				
Direct combination (F_d)	1.17	1.035 (1.76)*				1.03
B. STATISTICAL						
Inlet temperature variation	1.02					
Reactor ΔT variation	1.04					
Nuclear Data (power distribution)	1.06					1.07
Fissile fuel maldistribution	1.01					1.04
Wire wrap orientation	1.01					
Coolant properties	1.01					
Subchannel flow area	1.03	1.0				
Film heat transfer coefficient		1.12				
Pellet-cladding eccentricity		1.15	1.15			
Cladding thickness and conductivity			1.12			
Gap conductance				1.48†		
Fuel thermal conductivity					1.10	
Statistical Combination (F_s)	1.08	1.19	1.19	1.48†	1.10	1.08
TOTAL (F)	1.26	1.23 (2.10)*	1.19	1.48	1.10	1.11

*This factor affects the maximum cladding temperature only; it does not affect the maximum fuel temperature.
†Used for fresh fuel only.

Direct Factors

Power Level Measurement and Control System Dead Band. The calibration error in power measurement instrumentation is $\pm 2\%$, based mainly on water flow rate and feedwater temperature measurement uncertainties in the steam cycle.* An additional allowance of 1%, called a *dead band allowance*, is incorporated in the design of the control system to prevent excessive exercising of the control rod drives for small changes in reactor power. The combined hot channel factor for power measurement and dead band is 1.03; this was considered a direct uncertainty factor for the early CRBRP design.

Inlet Flow Maldistribution. Due to flow and pressure distribution within the inlet plenum, and due to the potential accumulation of fabricated dimensional tolerances in the orifice, fuel pin, wire wrap, and duct, the total flow to any one fuel assembly might be lower than nominal. Flow reduction affects the coolant enthalpy rise (and, hence, coolant temperature rise) directly. Allowance for a 5% flow reduction yields the hot channel factor of 1.05. The flow reduction influences the ΔT across the film, changing the Peclet number in the heat-transfer-coefficient correlation. The lower flow is also associated, however, with a lower hydraulic diameter for the flow, an effect that raises the heat transfer coefficient. The hot channel factor of 1.035 for inlet flow maldistribution for the film accounts for both reduced flow and reduced hydraulic diameter, as well as inlet and assembly flow maldistributions.

Assembly Flow Maldistribution. Flow and coolant temperature distributions within an assembly due to differences between interior and edge flow channels were discussed in Section 10-2. The nominal value of the temperature rise in the hot channel of an assembly takes into account the ratio of the rise in the hot channel to the average value for the assembly as calculated by the methods of Section 10-2. The assembly flow maldistribution hot channel factor accounts for the uncertainty in this ratio.

Cladding Circumferential Temperature Variation. The coolant velocity and temperature distribution in the channel formed by three adjacent fuel pins is not uniform, but varies circumferentially around the rod, as discussed at the end of Section 10-2A. The maximum cladding temperature

---

*One might suppose that this uncertainty is statistical and should be propagated as such rather than as a direct factor. This uncertainty is considered as a bias, however, in LWR licensing; hence, this is an example where LWR precedent dictates FBR methodology.

occurs at the minimum gap between the rods. The presence of wire wrap in the gap further increases the cladding surface temperature. The calculated increase in ΔT across the film at the minimum gap was 1.5 times the average ΔT. Another 20% was added for calculation method uncertainties, to give a direct hot channel factor of 1.7. An alternate procedure would be to use the factor of 1.5 as a known design factor and include only the 20% uncertainty in the hot channel factor analysis.

This circumferential temperature factor is applied only to the cladding temperature limit. It does not affect the fuel-cladding gap or fuel temperature differences; hence, this hot channel factor is unity when calculating the film temperature rise for evaluation of maximum fuel temperature.

Statistical Factors

Inlet Temperature Variation. A statistical analysis of the integrated reactor and plant system, including primary and secondary sodium and steam systems, indicated an uncertainty in inlet coolant temperature of $\pm 5°C$. This uncertainty is included in the uncertainty in the coolant temperature rise across the core. For a nominal temperature rise of $\sim 280°C$, the $\pm 5°C$ uncertainty corresponds to a $\pm 2\%$ variation in the temperature rise.

Reactor ΔT Variation. The same system analysis indicated a 3σ uncertainty in coolant flow of $\pm 4\%$, which would lead to the same ($\pm 4\%$) uncertainty in ΔT across the core.

Nuclear Data (Power Distribution). This factor represents uncertainties in the radial, axial, and local power peaking factors. It reflects uncertainties in nuclear data and nuclear design methods, after adjusting for comparisons with critical experiments. For the early CRBRP homogeneous core design, it was estimated that this uncertainty has a 6.5% influence on the heat flux factor and a 6% uncertainty on the coolant enthalpy rise.

Fissile Fuel Maldistribution. This factor is based on statistical measurements of fuel made for the FFTF. It is a 3.5% uncertainty in the fuel and cladding temperature limit analyses for an individual fuel pin, but only a 1% uncertainty for the coolant enthalpy rise since this rise is influenced by several pins.

Wire Wrap Orientation. This small contribution to the hot channel factor is associated with uncertainties due to swirl flow induced by the wire wrap in channels near the assembly ducts.

Coolant Properties. This reflects the effect of uncertainties in the specific heat and density of sodium.

Subchannel Flow Area. This factor represents the effect of propagated tolerances in fuel pin diameter, pitch-to-diameter ratio, and pin bowing. It does not affect the film hot channel factor because the decrease in hydraulic diameter associated with the 2.8% flow reduction overrides the influence of the Peclet number in calculating the film heat transfer coefficient.

Film Heat Transfer Coefficient. When calculating cladding and fuel temperature limits, one must be sure not to use too high a value for the heat transfer coefficient since the higher the coefficient, the lower the calculated cladding and fuel temperature. Comparison of the CRBRP (or FFTF) correlation with experimental data (e.g., Fig. 9-19) indicated that the minimum value of the actual heat transfer coefficient would not be more than 12% below the value calculated with the correlation.

Pellet-Cladding Eccentricity. An eccentric position of the fuel pellet within the cladding will result in an increased heat flux in the area of minimum fuel-cladding gap, with consequent increases in temperature rise through the coolant film and cladding. Calculations indicated that a hot channel factor of 1.15 is appropriate for the film and peak cladding temperatures. This factor is not applied to the peak fuel temperature because pellet eccentricity actually reduces this temperature.

Cladding Thickness and Conductivity. A 10% uncertainty in cladding conductivity (mostly to account for irradiation effects) was propagated statistically with a 6.7% uncertainty due to a 3σ cladding thickness tolerance of ± 1 mil (0.03 mm) to give a hot channel factor of 1.12.

Gap Conductance. Uncertainties in beginning-of-life (BOL) gap conductance are due to scatter in integral measurements (such as in the P-19 data, as illustrated earlier in Fig. 9-10), and to the difficulty of separating gap uncertainties from uncertainties in integral experiments resulting from fuel thermal conductivity and from tolerances on cladding inner diameter and fuel pellet outer diameter. For the fuel pin with the highest center fuel temperature, the gap hot channel factor was 1.48. (The cold gap nominal value for the early CRBRP design was 6.5 mils, with a tolerance of 2.5 mils on the pellet diameter and 0.5 mils on the cladding inner diameter.)

This hot channel factor is valid only for BOL fuel. After some period of irradiation and thermal cycling, fuel swelling and cracking cause the hot fuel-cladding gap to close; thus increasing the gap conductance, increasing the linear power-to-melting, and decreasing the peak fuel temperature. This was observed in the HEDL-P-20 experiments and can be seen by comparing Figs. 9-10 and 9-11 with Fig. 9-12.

Fuel Thermal Conductivity.   Several sources of uncertainty in the fuel thermal conductivity propagate to an estimated uncertainty of $\pm 10\%$ in the fuel temperature rise from the surface to the center of the fuel pellet, for any given value of linear power. The uncertainty in fuel thermal conductivity may actually be greater than this value, especially at high temperatures; but this uncertainty, when combined with the 1.48 factor for gap conductance, gives an appropriate overall uncertainty for the temperature change across gap plus fuel for BOL fuel.

Special attention must be given to applying the gap conductance and fuel thermal conductivity hot-spot factors, because the parameters to which they are applied are always interrelated in experimental measurements.

C.   APPLICATION OF HOT CHANNEL FACTORS

The use of hot channel factors will be illustrated first by applying them to calculations of peak exit coolant temperature and maximum cladding temperature. Next, the method used to assure that fuel melting does not occur will be described.

The nomenclature used for the combined hot channel factors is listed below. Numerical values appearing in Table 10-2 for these factors are shown in parentheses.

Hot Channel Factor	Parameter Affected
F_q	Heat flux (1.11)
F_b	Coolant enthalpy and temperature rise (1.26)
F_{film}(cladding)	Temperature rise across film at hot spot beneath the wire wrap, to be used in obtaining peak cladding temperatures (2.10)
F_{film}(fuel)	Temperature rise across film to be used in obtaining peak fuel temperature (1.23)
F_c	Temperature rise across cladding (1.19)
F_{gap}	Gap conductance (1.48)
F_{fuel}	Fuel conductivity (1.10)

Further nomenclature used includes:*

q_{OP} = Product of the nominal hot-channel heat flux at the outer surface of the pin cladding and the overpower factor (hence, heat flux at overpower conditions)

χ_{OP} = Product of nominal hot-channel linear power and overpower factor

h = Heat transfer coefficient

k = Thermal conductivity

D_i, D_o = Inside and outside diameter of the cladding

$T_{b,0}$(inlet) = Nominal value of the inlet coolant temperature

$T_{ci,m}$ = Maximum value of the inner cladding temperature

$T_{co,m}$ = Maximum value of the outer cladding temperature

$\Delta T_{b,OP}$ = Product of nominal hot channel coolant temperature rise and overpower factor.

As the starting point for the temperature calculations we will use the nominal value of the inlet coolant temperature, $T_{b,0}$(inlet), since the uncertainty in the inlet temperature is included in the temperature rise.

The maximum coolant temperature rise across a fuel assembly is

$$\Delta T_{b,m}(\text{exit}) = \Delta T_{b,OP}(\text{exit})F_b. \tag{10-52}$$

Hence, the maximum exit temperature is

$$T_{b,m}(\text{exit}) = T_{b,0}(\text{inlet}) + \Delta T_{b,m}(\text{exit}). \tag{10-53}$$

In order to calculate the peak temperature at the outer surface of the cladding, the first step is to locate the axial position of the peak nominal cladding temperature, z_{max}. The peak outer surface temperature is the sum of the inlet coolant temperature, the maximum rise in coolant temperature to that axial position $[\Delta T_{b,m}(z_{max})]$, and the maximum temperature rise across the film at that axial position, or

$$T_{co,m} = T_{b,0}(\text{inlet}) + \Delta T_{b,m}(z_{max}) + \Delta T_{film,m}(\text{cladding}), \tag{10-54}$$

where

$$\Delta T_{film,m}(\text{cladding}) = \frac{q_{OP}F_q}{h_{film}/F_{film}(\text{cladding})}, \tag{10-55}$$

---

*Subscript m refers to maximum values; subscript 0 refers to nominal values.

and q_{OP} and h_{film} are calculated at the axial position of peak cladding temperature.

The maximum temperature rise across the cladding is

$$\Delta T_{c,m} = \frac{F_q \chi_{OP} \ln(D_o/D_i)}{2\pi k_c/F_c}. \qquad (10\text{-}56)$$

Hence, the maximum inner surface cladding temperature would be the sum of Eq. (10-54) and Eq. (10-56), with all terms evaluated at the axial level in question.

The third problem discussed here is to demonstrate that centerline fuel melting does not occur. Two approaches can be used to show this:

(1) the fuel centerline temperature can be calculated and compared with the fuel melting temperatures; or
(2) the operating linear power can be calculated and compared with experimentally observed values of linear-power-to-melting.

In the first approach, we find the axial location of nominal peak fuel temperature and then calculate the maximum coolant temperature at that position. Next, we calculate the maximum temperature rise across the film, but we use the film hot channel factor for the average cladding $F_{film}(fuel)$, instead of the one that applies to the cladding hot spot beneath the wire wrap, or

$$\Delta T_{film,m}(fuel) = \frac{q_{OP} F_q}{h_{film}/F_{film}(fuel)}. \qquad (10\text{-}57)$$

From this point, the calculation of maximum fuel temperature becomes more complicated because of the experimental interrelation between gap conductance, fuel conductivity, and linear power. The procedure used for the early CRBRP homogeneous core is presented here.

The maximum temperature differences across the fuel and gap are calculated together. It is reasonable to propagate the uncertainties due to fuel conductivity and gap conductance randomly (i.e., as the sum of squares), since they cannot be isolated experimentally. This can be done as follows, recognizing that the uncertainty from the fuel conductivity is $\Delta T_{fuel,0}(F_{fuel}-1)$ and the uncertainty due to gap conductance is $\Delta T_{gap,0}(F_{gap}-1)$.

The nominal temperature differences across the fuel and gap are both multiplied by the heat flux hot channel factor. To this is added the statistically propagated uncertainties due to gap conductance and fuel conductivity. Hence,

$$\Delta T_{(\text{fuel}+\text{gap}),\,m} = \left(\Delta T_{\text{fuel},0} + \Delta T_{\text{gap},0} \right) F_q$$

$$+ \sqrt{\left[\Delta T_{\text{fuel},0} \left(F_{\text{fuel}} - 1 \right) \right]^2 + \left[\Delta T_{\text{gap},0} \left(F_{\text{gap}} - 1 \right) \right]^2},$$

$$(10\text{-}58)$$

where $\Delta T_{(\text{fuel}+\text{gap}),\,m} = T_{\text{fuel},\,m} - T_{ci,\,m}$. Hence, the maximum fuel temperature is

$$T_{\text{fuel},\,m}(z) = T_{b,0}(\text{inlet}) + \Delta T_{b,\,m}(z) + \Delta T_{\text{film},\,m}(\text{fuel}) + \Delta T_{c,\,m}(z)$$

$$+ \Delta T_{(\text{fuel}+\text{gap}),\,m}. \tag{10-59}$$

The second approach for showing that fuel melting will not occur is based on integral experiments of linear power required for incipient fuel melting, as described in Section 9-2E. By comparing the operating linear power with the experimentally observed linear-power-to-melting, a *margin-to-melting* is obtained. The desired result is the margin after uncertainties have been taken into account. The procedure to do this is illustrated in Fig. 10-15.[28] The *minimum linear-power-to-melting* is calculated by starting with the experimentally measured value for the nominal gap, decreasing this value to the result for the maximum (3σ level) cold gap, and decreasing it further by the experimental uncertainties at the -3σ confidence level. Such a procedure leads to line A on Fig. 10-15. This value must be greater than the

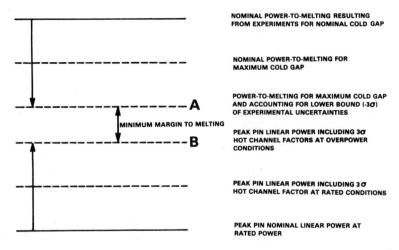

FIGURE 10-15. Scheme of accounting for uncertainties in assessing margin-to-melting.[28]

maximum operating linear power, which is the nominal linear power, increased by the 3σ level of confidence (i.e., multiplied by F_q) and increased again by the overpower factor (i.e., 115% for FFTF and early CRBRP design). This result is line B on the figure. The amount by which line B is lower than line A is the design margin-to-melting.

PROBLEMS

10-1. The following parameters apply for the 217 pin FFTF fuel assembly:

Using the Novendstern method, calculate the pressure drop across a rod bundle three metres in length for which $Re = 58\,000$.

$D = 5.84$ mm	$A_T = 4344$ mm^2
$P = 7.31$ mm	$A_1 = 8.97$ mm^2
$H = 305$ mm	$A_2 = 17.94$ mm^2
$s = 1.47$ mm	$A_3 = 6.39$ mm^2
$N_1 = 384$	$D_{e1} = 3.15$ mm
$N_2 = 48$	$D_{e2} = 3.84$ mm
$N_3 = 6$	$D_{e3} = 2.90$ mm

a) Calculate the flow distribution factors X_1 and X_2 by the Novendstern model.

b) Show that the bracketed terms in Eqs. (10-11) and (10-12) are 0.83 and 1.21, respectively, assuming $C_1 = 2200$, $C_2 = 1.9$, and $C_3 = 1.2$.

c) Calculate X_1 and X_2 by the Chiu-Rohsenow-Todreas model.

10-2. Suppose that the average velocity, $\overline{V}$, for an LMFBR fuel assembly is 6 m/s. The flow distribution factor for channel 1 of Fig. 9-13 is 0.97. Assume the following parameters typical of an LMFBR:

$D_{e1} = 3.5$ mm (effective hydraulic diameter)
$D = 7$ mm (fuel pin diameter)
$P/D = 1.25$ (pitch-to-diameter)
$H/D = 50$ (lead-to-diameter)

Using the Novendstern method, calculate the pressure drop across a rod bundle three metres in length for which $Re = 58\,000$.

10-3. Show how the individual terms in the energy equations, Eqs. (10-30) and (10-31), correspond to the terms in Eq. (10-22).

10-4. The fuel assemblies for a particular LMFBR are designed for coolant temperatures at 115% overpower with a 99.9% level of confidence. The nominal temperature rise of the coolant across the fuel assembly is 170°C. Based on hot channel factors listed in Table 10-2, what is the maximum ΔT for which the assembly is designed?

10-5. Suppose a linear power is to be specified for an LMFBR based on the data of Fig. 9-10. The fabricated fuel-to-cladding gap is 0.15 mm, and we shall assume for this problem that there is *no* uncertainty in the gap. The design must allow for 115% overpower and for a 2σ level of confidence.

Find the maximum value for the linear power for normal operation (i.e., at 100%, or rated, power) that would assure that incipient melting at 115% overpower would not occur, with 2σ confidence.

REFERENCES

1. E. H. Novendstern, "Turbulent Flow Pressure Drop Model for Fuel Rod Assemblies Utilizing a Helical Wire-Wrap Spacer System," *Nucl. Eng. and Des.*, 22 (1972) 19–27.
2. W. A. Sangster, "Calculation of Rod Bundle Pressure Loss," Paper 68-WA/HT-35, ASME, 1968.
3. C. Chiu, W. M. Rohsenow, N. E. Todreas, *Flow Split Model for LMFBR Wire Wrapped Assemblies*, COO-2245-56TR, Massachusetts Institute of Technology, Cambridge, April 1978.
4. J. T. Hawley, C. Chiu, W. M. Rohsenow, and N. E. Todreas, "Parameters for Laminar, Transition, and Turbulent Longitudinal Flows in Wire Wrap Spaced Hexagonal Arrays," *Topical Meeting on Nuclear Reactor Thermal Hydraulics*, Saratoga, NY, 1980.
5. D. S. Rowe, *COBRA-IIIC: A Digital Computer Program for Steady State and Transient Thermal Hydraulic Analysis of Rod Bundle Nuclear Fuel Elements*, BNWL-1695, Battelle Pacific Northwest Laboratories, March 1973. See also T. L. George, K. L. Basehore, C. L. Wheeler, W. A. Prather, and R. E. Masterson, *COBRA-WC: A Version of COBRA for Single-Phase Multiassembly Thermal Hydraulic Transient Analysis*, Pacific Northwest Laboratory, PNL-3259, July 1980.
6. W. T. Sha, R. C. Schmitt, and P. R. Huebotter, "Boundary-Value Thermal Hydraulic Analysis of a Reactor Fuel Rod Bundle," *Nuc. Sci. and Eng.*, 59 (1976) 140–160.
7. M. D. Carelli and C. W. Bach, "LMFBR Core Thermal Hydraulic Analysis Accounting for Interassembly Heat Transfer," *Trans. ANS*, 28 (June 1978) 560–562.
8. J. L. Wantland, "ORRIBLE—A Computer Program for Flow and Temperature Distribution in 19-Rod LMFBR Fuel Subassemblies," *Nuc. Tech.*, 24 (1974) 168–175.
9. E. U. Khan, W. M. Rohsenow, A. A. Sonein, and N. E. Todreas, "A Porous Body Model for Predicting Temperature Distribution in Wire-Wrapped Fuel Rod Assemblies," *Nuc. Eng. and Des.*, 35, (1975) 1–12.
10. E. U. Khan, W. M. Rohsenow, A. A. Sonein, and N. E. Todreas, "A Porous Body Model for Predicting Temperature Distribution in Wire Wrapped Rod Assemblies in Combined Forced and Free Convection," *Nuc. Eng. and Des.*, 35 (1975) 199–211.

11. B. Chen and N. E. Todreas, *Prediction of Coolant Temperature Field in a Breeder Reactor Including Interassembly Heat Transfer*, COO-2245-20TR, Massachusetts Institute of Technology, Cambridge, 1975.

12. J. N. Lillington, *SABRE-3—A Computer Program for the Calculation of Steady State Boiling in Rod-Clusters*, AEEW-M-1647, United Kingdom Atomic Energy Authority, 1979.

13. J. E. Meyer, *Conservation Laws in One-Dimensional Hydrodynamics*, WAPD-BT-20, Westinghouse Electric Corp., Bettis Atomic Power Laboratory, Pittsburgh, PA, September 1960.

14. R. E. Masterson and L. Wolf, "An Efficient Multidimensional Numerical Method for the Thermal-Hydraulic Analysis of Nuclear Reactor Cores," *Nuc. Sci. Eng.*, *64* (1977) 222–236.

15. J. T. Rogers and N. E. Todreas, "Coolant Interchannel Mixing in Reactor Fuel Rod Bundles Single-Phase Coolants," *Symp. Heat Transfer in Rod Bundles*, ASME (1965), 1–56.

16. T. Ginsberg, "Forced-Flow Interchannel Mixing Model for Fuel Rod Assemblies Utilizing a Helical Wire-Wrap Spacer System," *Nuc. Eng. and Des.*, *22* (1972) 28–42.

17. M. W. Cappiello and T. F. Cillan, *Core Engineering Technical Program Progress Report, Jan-March 1977*, HEDL-TME 77-46 (July 1977), Hanford Engineering Development Laboratory, Richland, WA.

18. Chaumont, Clauzon, Delpeyroux, Estavoyer, Ginier, Marmonier, Mougniot, "Conception du Coeur et des Assemblages d'une Grande Centrale a Neutrons Rapides," *Conf. Nucleaire Europeene*, Paris (April 1975).

19. S. F. Wang and N. E. Todreas, *Input Parameters to Codes Which Analyze LMFBR Wire-Wrapped Bundles*, Rev. 1, COO-2245-17TR, Massachusetts Institute of Technology, Cambridge, (May 1979).

20. *Preliminary Safety Analysis Report, Clinch River Breeder Reactor Plant*, Project Management Corporation (1974).

21. A. N. deStordeur, "Drag Coefficients for Fuel-Element Spacers," *Nucleonics*, *19* (1961) 74–79.

22. K. Rehme, *The Measurement of Friction Factors for Axial Flow Through Rod Bundles with Different Spacers, Performed on the INR Test Rig*, EURFNR-142P, November 1965.

23. T. C. Reihman, *An Experimental Study of Pressure Drop in Wire Wrapped FFTF Fuel Assemblies*, BNWL-1207, September 1969.

24. W. Baumann, V. Casal, H. Hoffman, R. Moeller, and K. Rust, *Fuel Elements with Spiral Spacers for Fast Breeder Reactors*, EURFNR-571, April 1968.

25. R. A. Jaross and F. A. Smith, *Reactor Development Program Progress Report*, Argonne National Laboratory, ANL-7742 (1970) 30.

26. G. J. Calamai, R. D. Coffield, L. J. Ossens, J. L. Kerian, J. V. Miller, E. H. Novendstern, G. H. Ursin, H. West, and P. J. Wood, *Steady State Thermal and Hydraulic Characteristics of the FFTF Fuel Assemblies*, ARD-FRT-1582, Westinghouse Electric Corp. June 1974.

27. M. D. Carelli and R. A. Markley, "Preliminary Thermal-Hydraulic Design and Predicted Performance of the Clinch River Breeder Reactor Core," *Nat. Heat Transfer Conf.*, ASME Paper 75-HT-71, ASME (1975).

28. Y. S. Tang, R. K. Coffield, Jr., and R. A. Markley, *Thermal Analysis of Liquid Metal Fast Breeder Reactors*, The American Nuclear Society, La Grange Park, IL, 1978.

CHAPTER 11

CORE MATERIALS

11-1 INTRODUCTION

The most hostile environment to be found in any nuclear reactor system is inside the core. Relative to a thermal reactor, the high flux, high burnup, and high temperature conditions encountered in a fast breeder reactor (FBR) place severe requirements on the materials selected for core design. Hence, considerable effort has been devoted to understanding and improving the performance of fuels and structural component candidates for FBR use.

This chapter is included to provide a more complete materials treatment of several of the general observations offered in Chapter 2 and of the design discussions included in Chapters 8, 9, and 10. However, because so much study has been given to the materials that comprise the primary building blocks of the FBR, it is not possible in an introductory text of this type to treat this subject with the degree of detail that a materials-oriented student would wish. Much of the information presented in this chapter, particularly that relating to stainless steel clad mixed-oxide fuel, is taken from Olander,[1] which represents an exceptionally clear and comprehensive treatment for the student interested in the materials phase of FBR technology.

This chapter is organized in four sections to deal with the four primary ingredients in the core, namely, the fuel, cladding and duct, coolant, and control material. Each section opens with general requirements and continues with general property considerations and the specific characteristics of the various material candidates. A general intercomparison is then included as appropriate. Materials of direct interest to the LMFBR are emphasized, in recognition of the central FBR role occupied by the LMFBR.

11-2 FUEL

A. REQUIREMENTS

Perhaps the most demanding requirement placed on the fuel for an FBR is that of high burnup ($\simeq 100$ MWd/kg goal exposure). Such an exposure level, about a factor of three higher than for an LWR, leads directly to a high fission product generation. This requires particular design emphasis on accommodating larger fuel swelling and fission gas release consequences than those associated with an LWR.

The fuel must also be capable of sustaining specific powers about a factor of four higher than for an LWR, as well as a higher temperature gradient, due to the smaller fuel pin diameter.

Finally, the inherent dynamic response of the core during accident conditions plays an important role in fuel selection. A desired feature of overall plant safety is that a temperature rise in the fuel should lead promptly and directly to a negative reactivity feedback effect, such that the accident condition causing the original temperature rise would be arrested rather than augmented. A negative Doppler coefficient and axial fuel expansion upon heating constitute two safety coefficients that could offer the desired response function.

B. DEFINITIONS

In searching for fuel materials capable of rendering high performance within the environment outlined above, the materials scientist is continually measuring and intercomparing certain characteristics. Hence, before exploring the features of each of the major fuel candidates under study for FBR application, it is constructive to define some of the terminology commonly used to discuss nuclear fuel performance.

Burnup

As noted in Chapter 7, fuel *burnup* can be defined either in terms of energy yield, i.e., MWd/kg, or as the fraction of heavy atoms fissioned, i.e., atom percent burned. The latter definition is usually the more meaningful to the fuel pin designer since it conveys most directly the potential damage done to the fuel matrix during irradiation. These two definitions can be quantitatively related to each other, as discussed in Chapter 7-3.

Dimensional Stability

Dimensional stability (or instability) is the term generally used to denote volumetric changes in the fuel matrix shape as a result of irradiation. The material distortion caused by radiation damage to fuel is extremely important because of the direct implications to pin and lattice design. There are two fundamental processes which contribute to dimensional instability. First, for every fission event that takes place, two fission fragments are formed. These new particles are each normally larger than the original atom, which results in a fractional increase in matrix volume about three times the burnup percent taking place.

Second, since the atom undergoing fission no longer exists, a vacancy (the place originally occupied by that atom) is left in the fuel matrix. Additional vacancies are caused by fission fragments knocking other atoms away from their original locations. Both the fission products and the displaced knock-on atoms eventually come to rest at interstitial locations (i.e., at positions between those atom sites which are arrayed in order).* The net effect of such vacancy/interstitial patterns on the dimensional stability of a particular fuel material depends heavily on the properties of the original, unirradiated lattice structure.

Microstructural Changes

Actual changes in grain size and orientation can be effected by even the slightest alloying of certain materials. Grain shapes and characteristics can also be grossly modified by the presence of strong temperature gradients, as noted in Chapter 8-2. Such structural changes can have a profound effect on subsequent material behavior. *Sintering* is a term used to describe the process of redistributing *porosity*† at high temperatures, where porosity is the fraction of void volume in the fuel. Thermal conductivity is an example of a property that could be altered significantly by changes in porosity.

Strength

Material mechanical properties such as hardness, yield and ultimate strength do not play nearly as significant a role in fuel selection as in the

---

*It is possible, of course, that such slowing down atoms could relocate in a vacancy position. This is called vacancy annihilation.

† This normally causes densification.

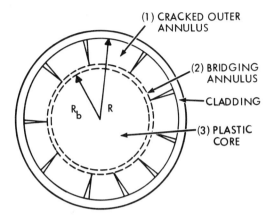

FIGURE 11-1. Model of the mechanical state of a fuel pin under irradiation.

case of cladding materials. This is because fuel breakup can be accommodated with the cladding jacket. An understanding of fuel cracking processes, however, is important in order to predict how fuel pins will perform under both steady-state and accident conditions. Figure 11-1 illustrates the possible mechanical state of a fuel material under irradiation. The hot interior exhibits plastic behavior whereas the cooler outer region, beyond R_b, is relatively brittle and is characterized by radial cracks.

Creep

Creep is time-dependent strain occurring under constant stress over long periods of time. Whereas elastic or instantaneous plastic deformation can occur the moment that stress is applied, creep strain is a longer time-frame phenomenon. One source of constant stress that could cause creep in a fuel matrix is thermal stress resulting from a steep temperature gradient.

When a material is initially exposed to a stress field, the elastic and instantaneous plastic deformations which occur exhaust all the mechanisms readily available for allowing dislocation motion. Hence, any additional dislocations which allow creep to occur must result from thermal or irradiation activation. Since the probability per unit time of supplying the energy, E, needed to start a dislocation moving is proportional to the Boltzman factor, $\exp(-E/kT)$, the *thermal creep rate $\dot{\varepsilon}$*, can be generally expressed in the form:

$$\dot{\varepsilon} \propto \sigma^m \exp\left\{\frac{-E}{kT}\right\}. \tag{11-1}$$

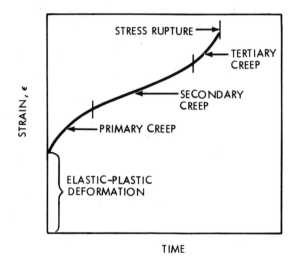

FIGURE 11-2. Typical creep curve.

Thus, the creep rate rapidly increases with stress level (since the exponent m is well in excess of unity) and, for a given stress level, increases with temperature.

Creep that occurs at elevated temperatures, but in the absence of an irradiation field, is called *thermal creep*. Creep occurring due to the presence of a radiation field where dislocations are the direct result of irradiation damage is called *irradiation creep*.

Figure 11-2 is included to illustrate the general deformation characteristics involved during a typical constant load creep test. In particular, the three general stages that occur prior to stress rupture are shown. Immediately after the elastic and instantaneous plastic deformations occur, the material undergoes primary creep (assuming the temperature is at least one-third to one-half the melting temperature). This monotonic process becomes approximately constant (secondary creep) during the intermediate time frame and then accelerates (tertiary creep) just before failure.

Swelling

A goal burnup of 10 atom percent ($\simeq 100$ MWd/kg) in a fast breeder reactor implies a total inventory of 20% fission products somewhere in the fuel system. Most of these fission products lodge within the fuel matrix as solid particles (as discussed in the above section on Dimensional Stability) and contribute to an overall volumetric increase, known as *fission product fuel swelling*. Swelling due to this effect can be defined as:

$$\left(\frac{\Delta V}{V}\right)_{\text{solid fp}} = \frac{V - V_0}{V_0}, \qquad (11\text{-}2)$$

where

V_0 = volume of a region of fresh fuel,
V = volume of the same region after irradiation.

The unit cell region being measured is assumed to envelop all solid fission products generated during the irradiation.

During the fission process, about 15% of the fission products occur as the rare gases xenon and krypton. Since these gases are insoluble in fuel, they tend to be rejected from the fuel matrix and contribute only indirectly to Eq. (11-2). These gases initially collect as small bubbles within the grains (intragranular) or as large bubbles along the grain boundaries (intergranular) as illustrated in Fig. 11-3. In either case, the net material density is appreciably less than if such space were filled by the original fuel. Hence, an additional volumetric increase, swelling due to fission gases, takes place.

Fission Gas Release

Not all of the fission product gases remain confined within the fuel. Depending largely on the nature of the fuel matrix structure and tempera-

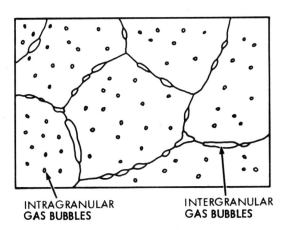

INTRAGRANULAR
GAS BUBBLES

INTERGRANULAR
GAS BUBBLES

FIGURE 11-3. A schematic distribution of fission gas bubbles in the fuel matrix.

ture, some of the gas can diffuse to the grain boundaries and from there much of it can flow via interconnected porosity to major cracks or open regions of the fuel. This released fission gas, unless intentionally bled off, then pressurizes the overall fuel system and applies stress to the cladding. A fission gas plenum is normally included in the fuel pin to keep the pressure buildup due to long-term irradiation within acceptable limits (discussed in Chapter 8-2). An alternate design option which has been considered is the use of vented* pins, which release filtered gas directly to the coolant system. The rate of fission gas release as a function of burnup for oxide fuel was shown in Fig. 8-8.

C. URANIUM BASED FUELS

The uranium-plutonium fuel system was shown in Chapter 1 to represent the best potential for high breeding performance in a fast breeder reactor. Of the several U-Pu fuel matrix forms available, the most widely used and tested fuel to date is mixed oxide. Hence, this candidate will be given the most attention in the present section. Other fuels, however, have enjoyed certain success and have the potential to be employed in advanced breeder reactor designs. These include uranium based metals, carbides and nitrides, and thorium based fuels. Cermets, a form in which fissile and fertile material is dispersed in a matrix such as steel, will not be treated because the minimum amount of non-heavy metal required is normally high enough to render this fuel system unattractive for large-scale breeder application.

In discussing materials properties of the uranium based fuels, the reader is cautioned that much of the early data was obtained for pure uranium as opposed to the uranium/plutonium system. Some property changes occur with Pu present, and detailed fuel design procedures must take these changes into account.

Oxide Fuel

Mixed-oxide fuels (UO_2-PuO_2) have provided the reference program fuel material for all nations pursuing the breeder option for at least the last decade. Interest in this fuel system logically followed from the wealth of experience gained from oxide fuels in the water-cooled power reactors. Principal factors that motivated interest in mixed oxides for the FBR were

---

*See Chapter 17-5 for an example of a vented fuel pin design.

the recognition of the high burnup potential and the existence of an established industry for manufacturing oxide fuels.

For LWR applications, oxide fuels have demonstrated very satisfactory dimensional and radiation stability, as well as chemical compatibility with cladding and coolant materials. However, the environment in an FBR is considerably more hostile (higher temperatures and much larger exposure) and some alterations, beyond enrichment, must be made to the fuel to maintain satisfactory performance. For instance, whereas the oxygen-to-metal ratio (O/M) is maintained at almost precisely 2.00 for LWR oxide fuel, FBR fuel is purposely fabricated with a deficiency of oxygen (i.e., hypostoichiometric).* This is done primarily to reduce the propensity for irradiated fuel to oxidize the cladding (which leads to "thinning" or "wastage").

The principal disadvantages of mixed-oxide fuel for FBR use are its low thermal conductivity and low density. The former property leads to high temperature gradients in the fuel and low values for linear power, and the latter property is undesirable from a breeding-ratio point of view.

Fabrication. Because of the high melting temperature of mixed-oxide fuel ($\simeq 2800°C$ for UO_2), fabrication is normally accomplished by powder metallurgy techniques. Uranium and plutonium oxides are mixed in the proportions required to achieve the desired fissile content and then the mixture is cold-compacted into a pellet. Sintering at temperatures around $1600°C$ is then performed to achieve the desired degree of densification. Although very high theoretical densities (TD) can be attained by this method (and arc casting techniques can be employed to reach full TD), it is normally desired to fabricate pellets in the range of 85 to 95% TD. This allows sufficient internally distributed porosity to accommodate burnup and also provides for substantial melting accommodation under accident conditions.[†]

Microstructure. The principal alterations in the morphology of oxide fuel as it heats up are described in Chapter 8-2C. Steep radial temperature profiles cause columnar and equiaxed grains to develop on the time scale of a few hours. Such restructuring patterns provide a strong interaction with fission gas retention patterns (discussed in chapter 8-2D) and overall thermal performance (discussed in Chapter 9-2B).

---

*A *stoichiometric* material contains an atomic ratio exactly equal to its chemical formula. *Hypostoichiometric* refers to a deficiency in the non-heavy metal constituent, and *hyperstoichiometric* refers to an excess in the non-heavy metal constituent.

[†] Mixed-oxide fuels undergo a volumetric increase of approximately 10% upon melting.

Physical Properties. The thermal conductivity for UO_2 is presented in Chapter 9-2A. Although the relatively low value for this physical property represents one of the principal disadvantages of UO_2 as a fuel, the densification of the columnar grains up to nearly 100% TD improves the thermal conductivity in that region and allows a significant increase in linear power.

The melting temperature of oxide fuel is high. Figure 11-4 contains the melting point of mixed-oxide fuel as a function of PuO_2 content. Note the substantial ΔT between the solidus and liquidus ($\simeq 50$ K for 20 mol percent PuO_2). This high melting point compensates somewhat for the low thermal conductivity in allowing adequate values of linear power [Eq. (2-2)].

Uranium dioxide has a fluorite crystalline structure, wherein the oxygen ions are arrayed in a simple cubic structure and the heavy metal ions form a face centered cubic sublattice. This arrangement results in an unoccupied interstitial position in the body center of the oxygen cubic structure, a situation which directly contributes to the relatively low density of mixed oxide fuel. Uranium carbide, on the other hand, exhibits an interlaced, face centered cubic structure which leads to a higher concentration of heavy metal atoms and, thus, higher material density.

Mixed-oxide, like most ceramics, is a relatively brittle material at temperatures less than half the melting point. Plastic deformation only occurs at the higher temperature levels. Consequently, the steep thermal gradients lead to rapid crack formation upon startup and cooldown. During steady-state power, cracks are believed to exist only near the radial periphery, as

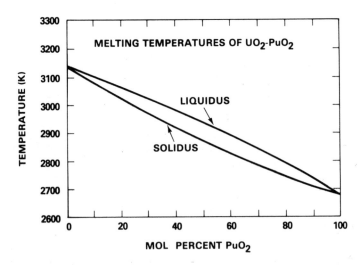

FIGURE 11-4. Melting point of mixed uranium-plutonium oxide fuel.[2] (*Note*: The present data uses a UO_2 melting point of 3138 K rather than 3113 K used in Ref. 2.)

schematically illustrated in Fig. 11-1, the larger cracks having largely been annealed out. However, the thermal cycling that necessarily occurs in a power reactor leaves a cracked structure that is not well defined. As a result, it is difficult to know how much of a fuel continuum exists for thermal expansion during a thermal upset condition.

Swelling. As noted earlier, some porosity is intentionally incorporated into mixed-oxide fuel to accommodate fuel swelling. Some of this swelling is due to solid fission products. Olander[1] estimates that swelling due to this source may range from 0.15 to 0.45% per atom percent burnup. A number close to 0.2% per a/o burnup is estimated for the oxygen deficient mixed-oxide fuel generally used. This implies a solid fission product swelling contribution of approximately 2% for a typical goal exposure of 100 MWd/kg.

Substantially greater swelling could result from fission gases. The situation is complicated, however, because the net swelling is derived from the balance between fission gas retention versus release, grain structure, porosity distribution, temperature, and temperature gradient. For example, most of the fission gas generated from the relatively cool periphery of the pin is retained in the fuel mixture. Since the temperature is low, swelling in this region occurs in a manner similar to solid fission product swelling.

Further up the temperature gradient, in the equiaxed region,* even though a substantial fraction of the gas is still retained, the temperature is high enough to allow gas migration and collection along the grain boundaries; hence, substantial swelling results. In the still higher temperature columnar region, most of the fission gas is vented to the central void region. Consequently, very little swelling due to fission gas occurs in that locale.

In addition to built-in internal porosity, a physical gap between the fuel pellets at the cladding wall can be adjusted to yield the smear density desired for accommodating fuel swelling. Too large a gap is undesirable, however, since it greatly increases the temperature drop from the fuel surface to the cladding and also reduces the fission density (which directly lowers the breeding ratio). Pellet densities of approximately 90% *TD* and a smear density of 85% *TD* are sufficient to allow a typical mixed-oxide fueled LMFBR to reach an exposure of 100 MWd/kg. As of 1977,[3] over 50 000 mixed-oxide fuel pins have been successfully irradiated (failure rates less than 1%) and, in some cases, burnup levels in excess of 150 MWd/kg have been achieved.

---

*See Chapter 8-2 for a discussion of equiaxed and columnar grain development.

Fission Gas Release. Fission gas, once released from the fuel matrix and vented to open collecting zones (such as the central void or the fuel-cladding gap), does not re-enter the fuel matrix during continued irradiation. The normal procedure for collecting such gas is to provide a fission gas plenum immediately above or below the fuel stack. It should be noted that since Xe and Kr are chemically insoluble in fuel, the buildup of pressure in the fission gas plenum (which is normally in approximate equilibrium with pressure through the fuel pin) does not affect the rate of gas released from the fuel matrix.

The processes by which fission gas bubbles nucleate, grow, diffuse, and eventually collect and condense at grain boundaries are extremely complex, though fuel temperature appears to provide a useful index for establishing the relative rates of fission gas mobility.* Below a temperature of about 1300 K, fission gas mobility is very low and there is essentially no gas escape. Between 1300 K and 1900 K, atomic motion allows some diffusion to take place such that an appreciable amount of gas can reach escape surfaces over a long period of time. Above 1900 K, thermal gradients can drive gas bubbles and closed pores over distances comparable to grain sizes in days or months. Gas release occurs when the cavities reach a crack or other surface that communicates directly with free volume.

Approximate numerical values for fission gas release in the restructured and unrestructured regions as a function of burnup can be obtained using Eq. (8-6).

It should be noted that solid fission products tend to migrate toward the cooler region of the pin (radially toward the cladding and axially toward the ends of the pin). A particular concern is cesium which, if sufficiently concentrated near the pin periphery, can cause appreciable degradation of the cladding.

Carbide Fuel

Several compounds of uranium and carbon exist, but UC has been given the most attention because of its high uranium density. As indicated earlier, UC is a densely packed, face centered cubic structure and contains 4.8 weight percent carbon for a stoichiometric composition. Like UO_2, UC can be fabricated by compacting and sintering powders to achieve the degree of porosity desired. The primary factor of interest in UC as a fast reactor fuel, beyond its relatively high density, is its good thermal conductivity.

---

*While most of the factors governing fission gas retention and redistribution are reasonably well understood, disagreement among ceramicists still remains over some of them.[4] Fuel temperature is only one of several governing factors.

Microstructure. Because of its high thermal conductivity, peak temperature and temperature gradients are smaller in a UC pin relative to UO_2. Although some cracking does occur upon thermal cycling, such effects are inherently smaller than for UO_2. There is a tendency for pores to migrate up the temperature gradient, but the overall magnitude is far less than in UO_2 because of the much lower temperature gradient. Hence, no appreciable fuel restructuring occurs and no central void region develops.

Properties. Table 11-1 contains a qualitative comparison of mixed-oxide and carbide fuel materials properties. Table 11-2 indicates the principal differences in irradiation performances which follow from the properties of these two fuel systems. By comparison with the highly successful irradiation of over 50 000 mixed-oxide fuel pins, the irradiation results of some 500

TABLE 11-1   Qualitative Comparison of Mixed Oxide and Carbide Fuel Materials Properties[3]

Materials Property	(U, Pu)O$_2$	(U, Pu)C
Thermal Conductivity		Higher
Thermal Expansion	Higher	
Thermal Creep	Similar*	
Irradiation Creep	Higher	
Fuel-Sodium Compatibility		Better
Fuel-Cladding Chemical Compatibility		Better

*At equivalent temperatures the thermal creep rates are similar although there are differences in the stress dependencies.

TABLE 11-2   Comparison of Irradiation Performance of Mixed-Oxide and Carbide Fuels (Adapted from Ref. 3)

Mixed-Oxide Fuel	Mixed-Carbide Fuel
High centerline fuel temperature and low beginning-of-life (BOL) power capability (~ 50 kW/m)	Low centerline fuel temperature and high beginning-of-life (BOL) power capability (> 100 kW/m)
Fuel restructuring	Little or no fuel restructuring
High fission gas release and significant fission gas pressure effects	Low fission gas release and significant fuel swelling
High plasticity fuel and low levels of fuel cladding mechanical interaction (FCMI)	Low plasticity fuel and high levels of fuel cladding mechanical interaction (FCMI)
Release of volatile fission products and fuel-cladding chemical interaction	Little or no release of volatile fission products and cladding carburization

mixed-carbide pins, up to 1977, indicated high failure rates.[3] However, recent experience indicates that goal exposure (i.e.$\simeq$100 MWd/kg) can be realized with carbide fuel.

Swelling. Control of stoichiometry during the fabrication of carbide fuel appears to be *very* important with regard to swelling characteristics. Hypo-stoichiometric mixtures, i.e., carbon deficient $UC_{(1-x)}$, leaves free uranium in the matrix. This material tends to migrate and collect as a metallic phase at the grain boundaries. Hence, the swelling characteristics of metal fuel become manifest and lead to intolerable swelling rates.[5] In addition to this fabrication problem,* it is essential to ensure that no mechanism (such as fission product generation) could preferentially lead to a carbon deficiency.

On the other hand, a carbon excess leads to a precipitation of higher carbides, and these can be transferred to the cladding to cause cladding carburization. Whereas the last entry of Table 11-2 indicated manageable conditions for helium-bonded fuel, this problem can be particularly acute with sodium bonding (sodium residing in the fuel-cladding gap). Sodium bonding is preferable to gas bonding from a thermal efficiency viewpoint, since the high temperature drop across a gas bond negates much of the advantage of the high-conductivity carbide fuel. One possible design solution to this problem is the inclusion of a sacrificial steel sheath, immersed within the sodium bond, to protect the cladding from chemical attack.

Fission Gas Release. For operating temperatures of interest, the fission gas released from stoichiometric UC is far less than that obtained from UO_2. Release rates in excess of 50% have been observed for specimens around 1350°C, but such release rates are normally associated with hypo-stoichiometric mixtures, where the uranium phase has led to appreciable swelling. Figure 11-5 indicates that fission gas release, particularly for the carbon-deficient composition, can become significant above $\simeq$ 1300°C.

Mixed Uranium-Plutonium Carbides. Whereas appreciably less in-pile experience has been accumulated for UC-PuC mixtures, data to date indicate that such mixtures have properties very similar to UC. Excess carbon tends to be precipitated as Pu_2C_3, and a carbon deficiency results in a U-Pu alloy deposition. The metallic alloy results in swelling characteristics very similar to hypostoichiometric UC.

---

*Another difficulty of the carbide fuel fabrication process is the necessity to keep oxygen concentrations at very low levels. Uranium will preferably oxidize, and the presence of oxygen could produce unwanted UO_2 as well as pose a fire hazard.

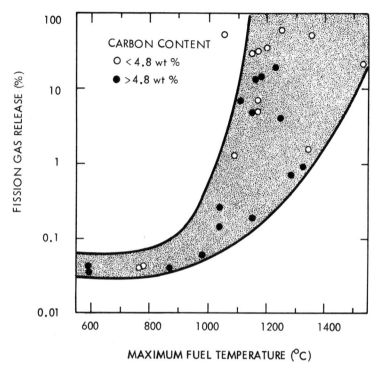

FIGURE 11-5. The release of fission-product gases from UC as a function of the irradiation temperature.[6]

Metal Fuel

All the early, low-power fast reactors employed metallic uranium fuel. It is relatively easy to fabricate, has excellent thermal conductivity (as shown in Fig. 11-6 in the section on general intercomparison), and has a high density (19.0 g/cm³ at room temperature). The principal problem with metal fuel is its highly anisotropic growth patterns upon irradiation and the large associated dimensional change. If fully dense metallic fuel is used, along with a small initial fuel-cladding gap in order to achieve a high smear density, burnup is limited to the order of 10 MWd/kg, far too low to be economical. However, at least two techniques have been proposed to improve metal fuel burnup capability. One scheme[7] is to fabricate an axial hole in the fuel to accommodate fuel swelling internally. The second technique, developed in association with the EBR-II project,[8] involves increasing the initial fuel-cladding gap to accommodate such swelling.

Success in this approach with metal alloy fuels is described below in the paragraph on Recent EBR-II Experience.

Crystalline Structure. Metallic uranium exhibits three crystalline structural phases (alpha, beta and gamma) over low, intermediate, and high temperature ranges. The alpha phase (orthorhombic structure) is exceptionally anisotropic. Some of this anisotropy is suppressed in bulk material, wherein the grains are randomly oriented, but even the slightest amount of material cold working will preferentially orient the grains and allow highly nonuniform growth patterns. The gamma phase (body centered cubic) demonstrates the most isotropic behavior of the three phases.

Physical Properties. As in all fuel materials, thermal conductivity of metal fuel tends to decrease with burnup due to the material density loss. This is particularly aggravated when uranium swelling occurs. Both hardness and yield strength increase appreciably, even at low levels of irradiation, but there are conflicting data regarding the effects of radiation on ultimate tensile strength and ductility. The melting point of uranium metal is 1132°C.

Swelling. Both uranium and uranium-plutonium metallic fuels have extremely low resistance to swelling* over the temperature ranges of interest, i.e., the 400 to 600°C range. Below 400°C, no appreciable swelling occurs. In the range from 400 to 500°C, tears occur along the grain boundaries and rapid swelling results. Deformation is believed to occur primarily due to grain-boundary sliding and cavitation, in preference to plastic flow within the grains. The fission gas diffuses to the grain boundaries and causes rapid grain boundary swelling, called "break-away" swelling. Beyond 500°C a self-annealing process appears to take place that minimizes the effects of swelling.

Alloying. Because of the pronounced fuel swelling observed, much of the research effort on metallic fuels has focused on a search for alloying materials to mitigate the problem. In particular, efforts to stabilize the gamma phase, which has the most isotropic properties, have received primary attention.

---

*The term fuel "swelling" is somewhat ambiguous when applied to uranium metal because much of the observed growth over certain temperature ranges is due to the highly anisotropic behavior of uranium instead of fission product generation. However, the net effects of temperature and burnup are often referred to as swelling in the literature.

Molybdenum has perhaps been the most successful material used in alloying uranium to retard the swelling problem. Both the Dounreay and Fermi reactors successfully employed an uranium molybdenum cermet fuel containing approximately 10 wt% Mo. However, the burnup for these reactors was substantially less than desired for commercial plants. Achievement of high burnup with fissium and zirconium alloys is described in the next paragraph.

Recent EBR-II Experience. The EBR-II reactor has been fueled with a uranium-5 wt% Fs* metal driver fuel, U-Fs alloy, which has operated with burnups as high as 10.5 a/o for the past decade. (With the increased role of EBR-II as an irradiation test facility, a large fraction of the core has been loaded with mixed-oxide test fuel.)

The primary reason such high burnups have been achieved with metallic fuel is that substantial porosity has been designed into the fuel element. Since swelling cannot be suppressed, the fuel was fabricated to assure a fuel-cladding gap large enough to result in a cold smear density of approximately 75%. Initial swelling quickly expands the fuel to close the gap, but by the time contact is made with the cladding, sufficient swelling has occurred to open numerous paths for fission gas to escape to the fission gas plenum. Subsequent swelling is then mitigated.

Research with a U-Pu-Zr alloy[8] suggests that performance even better than that achieved with the U-Fs alloy is possible. The principal improvement is the potential for higher temperature operation. A high eutectic temperature ($\simeq 810°C$) exists between this fuel and the stainless steel cladding, as well as a solidus temperature appreciably higher ($\simeq 1150°C$) than previous metal alloys. (The comparable U-Fs alloy temperatures are 705°C and 1000°C, respectively.) It is argued[8] that even with the low smear densities necessary to achieve economic burnup levels for such fuels, the fissile density is still better than other fuel candidates and the breeding potential is thereby high.

One of the uncertainties associated with the low density metal alloy fuels involves safety parameters. The lack of a strong Doppler coefficient for metal fuels (caused by the characteristic hard neutron spectrum) has historically been compensated for by the presence of a reliable axial thermal expansion coefficient. There is some question, however, about the presence or efficacy of such an axial expansion effect for a high porosity fuel matrix. Efforts to resolve this uncertainty are underway.

---

*Fissium (Fs) is an equilibrium concentration of fission product elements left by the pyrometallurgical reprocessing cycle designed specifically for EBR-II. It consists of 2.4 wt% Mo, 1.9 wt% Ru, 0.3 wt% Rh, 0.2 wt% Pd, 0.1 wt% Zr, and 0.01 wt% Nb.

Other Uranium Compounds

In principle, many uranium compounds could serve as a fast breeder reactor fuel. However, of the numerous other compounds, only uranium nitride (UN) has received appreciable attention. It has physical properties[9] quite similar to UC. Uranium nitride is more compatible with cladding than UC (no carburization), but much less in-pile experience exists upon which to judge its overall merits. Fabrication problems are somewhat complicated, particularly if an arc casting process is used, because of the need for a nitrogen atmosphere to prevent nitrogen loss. Thermal decomposition of UN occurs above 2000°C, which may raise certain safety questions but should not affect steady-state performance.

Uranium sulphide, US, likewise has many properties similar to UC, but its density is no better than that of UO_2. Hence, there is little incentive for active development. U_3Si has been studied as promising among the possible uranium-silicon systems because it yields a very high uranium density. Again, however, very little in-pile data are available upon which to judge its overall merits. A final uranium compound that has been studied to a limited extent as a potential fuel is the phosphide, UP.

D. THORIUM BASED FUELS

As noted from the discussion of Chapter 1, the ^{232}Th-^{233}U fuel system could be employed in a breeder reactor (thermal or fast), although with an inherently lower performance with regard to breeding ratio and doubling time. Because the potential to breed does exist for such a system, it is appropriate to provide some background[10] on the potential of thorium based fuel systems for fast reactor application.

Relative to uranium, thorium metal has both a higher thermal conductivity and a lower coefficient of thermal expansion. Both these properties tend to reduce the thermal stress in the fuel elements, but the latter property would tend to reduce the negative reactivity feedback during a power excursion. Thorium has a considerably lower density than uranium.

Perhaps the most significant difference with regard to in-pile performance is that Th has an isotopic cubic crystalline structure (face centered cubic) and, therefore, undergoes appreciably less dimensional change upon thermal cycling and irradiation than does the anisotropic uranium. Thorium also has higher irradiation creep resistance, higher ductility, and a higher melting point (1700°C). The latter property somewhat complicates the fabrication process since melting and casting are more difficult.

Cast thorium has a large grain size and easily cracks upon cold working. However, hot Th can actually be rolled to a thickness of 0.025 mm at room temperature without intermediate annealing. Overall, thorium metal is substantially easier to work with than uranium metal.

Th-U Oxides

For thorium to be used as a fuel, it must be mixed with uranium. In the case of a mixed thorium-uranium oxide system, it is possible to maintain a single phase structure over the entire mixture range of interest. The thorium-uranium oxide system can be fabricated by pressing and sintering in an air atmosphere, but sintering in a hydrogen atmosphere significantly reduces the oxygen excess.

Columnar grain growth, with associated lenticular pores, has been observed for thorium dioxide. The growth is very similar to that observed in UO_2. However, the vapor pressure of ThO_2 is less than for UO_2 and, as a result, such grain formation does not occur until higher temperatures (approximately 350°C higher) are reached. Equiaxed grain growth occurs in a manner very similar to UO_2. Thermal stress cracking patterns are also similar to UO_2. Pure ThO_2 has a thermal conductivity approximately 10% higher than pure UO_2.

Irradiation experience that is available for ThO_2-UO_2 fuel yields little indication of fuel swelling. This fuel system should perform in a manner comparable to the UO_2-PuO_2 system. However, considerably more in-pile data needs to be collected for the ThO_2-UO_2 fuel system before a high degree of confidence can be attained regarding its long-term performance.

Th-U Carbides

ThC is a face centered cubic (FCC) structure with a melting point of 2625°C and a density of 10.65 g/cm^3. One of the positive aspects of this carbide system is that the carbon content can vary from 3.8 to 4.9 wt% without changing the crystalline structure (in sharp contrast to the UC system). Very little information has yet been gathered on the ThC-UC system.

Th-U Metal

Metallic fuels of thorium and uranium can be readily obtained, but the uranium fraction must be kept below a certain level if reasonable dimensional stability is to be maintained. Only 1% of U can be dissolved into solid solution with thorium for temperatures up to 1000°C. At higher content, U precipitates predominantly at the grain boundaries. For uranium additions up to 20 wt% the U can be regarded as a dispersion in the FCC thorium matrix, but beyond that point the U may form a continuous metallic network.

This solubility characteristic is extremely important from the standpoint of dimensional stability. Good dimensional stability has been maintained in Th-U fuels for mixtures with low U content. Based on early in-pile work, it was reasoned[11] that the principal reasons for the encouraging results were (1) most fission fragments were deposited outside the U precipitates and directly in the Th metal, (2) the fine dispersion of U particles helped to anchor gas-filled pores, and (3) the higher Th melting point might have resulted in lower gas diffusion rates at a given temperature (hence, less propensity for fission gas induced swelling). Unless significant cracking occurs, fission gas retention is believed to be very high for Th-U fuel.

E. GENERAL INTERCOMPARISON

From the discussion above, it is apparent that several fuel systems exist that could be employed in fast breeder reactors. Because of the hostile environment in which such fuels must perform, however, considerable high-burnup experience is critical to the confident prediction of long-term performance. Such data are generally quite expensive and time-consuming to accumulate; hence, the backlog of experience with mixed uranium-plutonium oxide fuels currently gives that fuel system a considerable advantage.

In addition to in-pile performance, other considerations such as neutronics and safety are important in the fuel selection process. The reactor spectrum is one such factor since breeding ratio increases as the neutron spectrum is hardened. The high density metal reactor, having no moderator in the fuel matrix, yields the hardest neutron spectrum, but the flux in the principal Doppler resonance region (0.1–10 keV) is appreciably less than for the ceramic fueled reactor. Consequently, both oxide and carbide fueled fast reactors possess much higher Doppler coefficients than metal fueled reactors.

Table 11-3 lists selected physical properties for uranium and plutonium fuels. Metals are clearly superior with regard to density but have substantially lower melting points than the ceramic fuels. Figure 11-6, which shows the thermal conductivities of stoichiometric fuels, indicates that full density metal fuel has substantially higher values than either the carbide or the oxide fuel form. On the other hand, as we saw from earlier discussions, fully dense metal U-Pu fuel cannot be used for even moderately high burnup applications in an FBR. A reduction in smear density to approximately 75%, to allow reasonable burnup levels, directly reduces the effective thermal performance of the pin.

TABLE 11-3 Properties of Uranium and Plutonium Fuels

	Melting Point (K)	Heat of Fusion (kJ/kg)	Density 100% TD (g/cm³)	Thermal Conductivity (W/m·K)		Specific Heat (c_p) (kJ/kg·K)	at melting point	
				500°C	1500°C	298K	solid	liquid
Metal				30*				
U†	1408	38.2	19.0			0.12	0.16	0.20
Pu	913		19.9			0.18[b]		
Oxide**								
UO$_2$	3138	277	10.97			0.24	0.77	0.50
(U$_{0.8}$Pu$_{0.2}$)O$_2^†$	3023(sol) 3063(liq)	277	11.08	4.0[a]	2.3[a]	0.26	0.64	0.50
PuO$_2$	2670		11.46			0.35[c]		
Carbide				16*	17*			
UC	2780	184	13.6			0.20	0.35	0.28
(U$_{0.8}$Pu$_{0.2}$)C†	2548(sol) 2780(liq)	186	13.5			0.19	0.43	0.28
PuC*	1920		13.6			0.24[c]		
Nitride*				12	16			
UN	2870		14.3			0.27[c]		
PuN	2770		14.2			0.26[c]		
Phosphide*				17				
UP	2880		10.2					
PuP	2870		9.9					
Sulfide*				14				
US	2750		10.9					
PuS	2620		10.6					

*Adapted from page 169 of Ref. 12
**The vapor pressures for UO$_2$ and mixed oxide fuel are discussed in Chapter 15-7B.
†Ref. 13
[a]95% TD, O/M=2.00 (Cf. Fig. 9-2).
[b]at 500°C
[c]at 1500°C

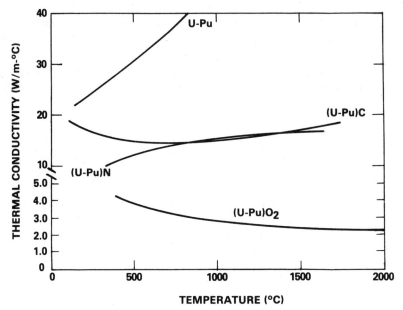

FIGURE 11-6. Thermal conductivity of mixed-oxide, nitride, carbide, and metal fuels (from page 33 of Ref. 12).

11-3 CLADDING AND DUCT

The fuel cladding provides the basic structural integrity of the fuel element. It prevents fission gas from entering the primary coolant system* and provides physical separation of the fuel from the coolant. The duct is likewise a fundamental part of the fuel assembly design, as delineated in Chapter 8-4A. Since both the cladding and duct provide structural support for the fuel and experience a similar temperature and radiation environment, the general materials background for both components will be provided in the same section (though specific references will generally be made to the cladding, since it tends to experience the more hostile environment).

A. REQUIREMENTS

High temperature strength and long exposure irradiation resistance provide the principal requirements for fast breeder reactor cladding. For an economical fuel cycle, FBR cladding must be capable of residing in the core

---

*Controlled venting could be purposely designed into the system.

TABLE 11-4 Absorption Cross Sections of the More Important Metals*

Group	Metal	$\sigma_{n,\gamma}$ (100 keV) [millibarns]	$\sigma_{n,\gamma}$ (thermal) [barns]
	Al	4	0.230
	Ti	6	5.8
1	Fe	6.1	2.53
	Cr	6.8	3.1
	V	9.5	15.1
	Si	10.0	0.16
	Co	11.5	37.0
	Ni	12.6	4.8
2	Zr	15.1	0.18
	Cu	24.9	3.77
	Mn	25.6	13.2
	Mo	71.0	2.7
	Nb	100.0	1.15
3	W	178.0	19.2
	Ta[†]	325.0	21.0

*From page 184 of Ref. 12.
[†]Ta is an absorber in control rods.

for up to three years at peak temperatures approaching 700°C. Whereas the peak exposure in an LWR may be around 10^{22} n/cm^2, the fluence requirements for FBR cladding are about 2×10^{23} n/cm^{2}[†] (a factor of 20 higher!). An important limitation to meeting the latter requirement is the phenomenon of void swelling since accommodation of diametral cladding strains beyond about 3% begins to impose severe economic penalties. Also, cladding rupture or breach is an important life limiting phenomenon.

Since the cladding is a parasite from the standpoint of neutron economy, it is desirable to select a material with a low neutron capture cross section, $\sigma_{n,\gamma}$ (or σ_c). Table 11-4 contains values of $\sigma_{n,\gamma}$ for several metals in thermal and fast spectrum environments. When comparing these values, however, it should be kept in mind that neutron absorption in the cladding is considerably less important for an FBR than for an LWR. This situation arises because (1) the metal-to-fuel neutron cross section ratio is considerably lower in a fast system, relative to a thermal spectrum, and (2) there is

---

[†]Material damage is normally correlated with neutron fluence above 0.1 MeV; this is sometimes referred to as a *damage fluence*. Therefore, all fluences quoted in this section are for neutron energies above 0.1 MeV. (See Chaper 4-7 for the fraction of flux above 0.1 MeV relative to total flux.)

relatively less cladding material in an FBR. The latter fact is a consequence of a strong desire to maximize the fuel/cladding ratio to enhance the FBR breeding ratio.

It is noted from Table 11-4 that zirconium based alloys required in light water reactors for their low neutron capture cross section would not be attractive for FBR use due to their relatively high cross section at higher energies. Another reason for bypassing the standard LWR zircaloy cladding material in the FBR is that zircaloy does not possess sufficient strength at the high temperature operations required. Though only of secondary importance, neutronics considerations deter the use of high nickel bearing alloys in the FBR, even though some high nickel alloys may demonstrate generally superior irradiation properties.

Austenitic 316 stainless steel, normally cold worked to approximately 20%, has been selected as the reference cladding material for the LMFBR. Austenitic steel is a face centered cubic form of iron (as opposed to ferritic steel, which is body centered cubic). This austenitic form is the natural structure of iron between 910 and 1400°C, and this structure is stabilized at room temperature by the addition of nickel. The austenitic steels have advantages over most ferritic steels in LMFBR's because of their better high-temperature creep strength and corrosion resistance. A few ferritic steels are still being considered for cladding, however, and even more serious consideration is being given to ferritics for use in ducts.

B. RADIATION EFFECTS

The four major categories of mechanical behavior that must be investigated for cladding selection are summarized in this section. These include radiation hardening, embrittlement and fracture, void swelling, and irradiation creep.

Radiation Hardening

A material *hardens* when the matrix becomes entangled, i.e., when interstitial atoms become lodged in glide planes. Such a process, called *work hardening*, occurs during plastic deformation, as shown in Fig. 11-7. Once the elastic strain limit is passed, an increased stress causes permanent deformation. These increasing stress levels add to the deformation and entanglement process, leading to greater levels of hardening and material strength.

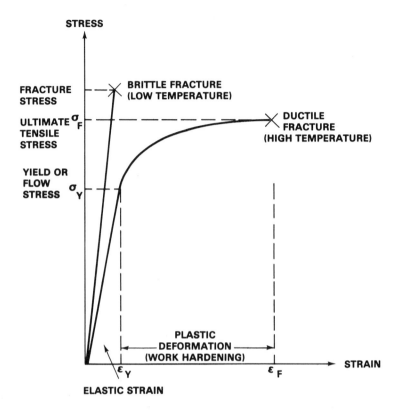

FIGURE 11-7. Typical stress-strain curves at temperatures above and below the brittle-ductile transition temperature.

One of the standard techniques employed to modify the strength characteristics of a structural metal is to apply *cold working*. This is a process of reducing the cross-sectional area of the cladding by drawing the tubing under room temperature conditions. The 20% cold working used for LMFBR cladding and ducts results in a 20% reduction in area (which is nearly equivalent to a 20% reduction in cladding or duct thickness). A mechanical deformation process of this type introduces dislocation tangles into the matrix and leads to a greater yield strength and ultimate strength, while decreasing the ductility.

A similar hardening effect occurs in metals undergoing irradiation. Interstitial atoms caused by neutron displacement damage form dislocation loops which considerably strengthen the metal. However, the overall effect is strongly dependent upon the irradiation temperature because, if the lattice is hot enough, a substantial amount of thermal annealing can occur simultaneously to mitigate the hardening effect of neutron irradiation. Figures 11-8

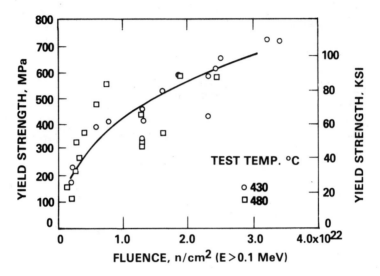

FIGURE 11-8. Effect of fluence on the low temperature (430–480°C) yield strength (0.2% offset) of Type 316 stainless steel. (Ref. 14.)

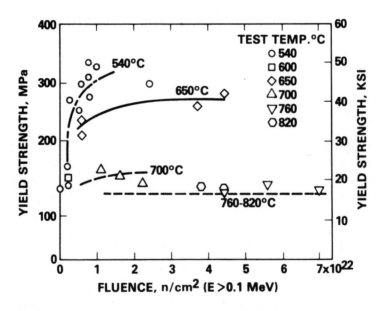

FIGURE 11-9. Effect of fluence on the high temperature (>540°C) yield strength (0.2% offset) of Type 316 stainless steel. (Ref. 14.)

and 11-9 compare typical yield strengths characteristic of irradiated and unirradiated metal at various temperatures.* At cold conditions, the hardening effect is very pronounced, whereas thermal annealing removes most of the difference at temperatures around 800°C.

Source hardening is a term used to describe the increased stress required to start a dislocation moving on its glide plane. *Friction hardening* measures the increased resistance to plastic deformation due to radiation-produced obstacles once movement along the slip plane occurs.

The relative dominance of the various mechanisms responsible for radiation hardening vary with the fluence level. For low fluence ($\phi t < 10^{21}$ n/cm^2), most of the resistance to plastic flow results from depleted gases. For the higher fluence levels of principal interest for FBR's, dislocation loops constitute the dominant hardening mechanism.

It was mentioned earlier that radiation hardening is similar in net effect to cold working. Because of this similarity, work hardening of irradiated materials is not nearly as effective as work hardening of unirradiated material. In irradiated material, there are already so many radiation-produced obstacles that any additional effect of work hardening causes only a small increase in frictional stress.

Embrittlement and Fracture

When introducing the concept of embrittlement for cladding materials, it is useful to review the general stress-strain relationships involved. Fig. 11-10 represents the relationship for ferritic steel. As stress is originally applied to a ferritic specimen, elastic strain occurs according to Hooke's law up to the proportional limit (PL) which is generally below the yield point. The yield point is attained at point U and the specimen will strain to point L, corresponding to the flow stress level. For such a material, an additional strain (called Lüders strain) will occur at this lower yield stress point. If the stress level is subsequently increased, additional strain will occur, and the work hardening that occurs in this area strengthens the material (by the mechanism described above).

This process continues until the ultimate tensile strength (UTS) level is reached, beyond which necking predominates and fracture follows (unless the load is reduced). The dashed line on Fig. 11-10 represents the true stress-strain curve, wherein actual area reductions and incremental strains in the specimen under test are properly taken into account.

---

*Refer ahead to Fig. 11-11 for an explanation of the 0.2% offset concept.

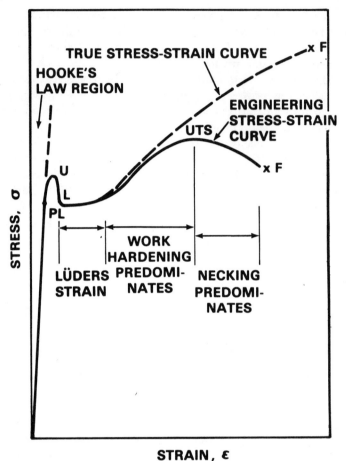

STRAIN, ε

FIGURE 11-10. Stress-strain curve for ferritic steel.

Whereas Fig. 11-10 is representative of the behavior of ferritic steel, the austenitic steels behave somewhat differently, as depicted in Fig. 11-11. The principal difference in the stress-strain relationship is that there is no well defined yield point, i.e., there is no unique stress level at which plastic flow begins. As a consequence, it is common practice in the case of such steels to arbitrarily label the yield stress level as that corresponding to a 0.2% permanent strain, i.e., where the 0.2% offset line on Fig. 11-11 intersects the stress-strain curve. This is called the *0.2% offset yield strength* of the metal.

The term *ductility* is defined as the strain between the ultimate tensile strength (UTS) and the yield stress ($\varepsilon_{UTS} - \varepsilon_Y$), or more commonly as the total uniform strain up to the UTS. *Embrittlement* refers to the reduction in

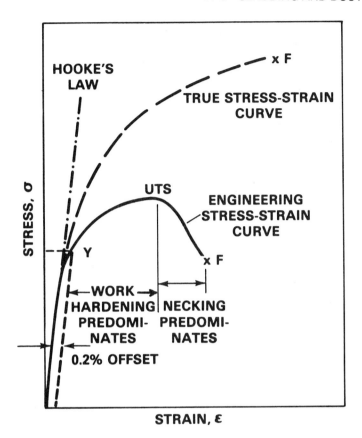

FIGURE 11-11. Stress-strain curve for austenitic steel.

either measure of ductility. Fast neutron irradiation normally renders a metal less ductile than unirradiated metal. A *brittle* material fails prior to the yield point or before the 0.2% offset strain is reached.

Whereas Figs. 11-10 and 11-11 represent the general stress-strain relationship for the common cladding material candidates, the actual magnitudes of the curves are strain rate dependent. The yield strength is strain rate dependent at higher temperatures, and at such elevated temperatures, time-dependent plastic deformation (creep) also occurs at low strain rates $(<10^{-4}/\text{min})$. Such strain rates are characteristic of long term pressure on the cladding due to fuel swelling. When such tests are conducted under high temperature conditions typical of an in-pile environment, they are called creep rupture tests. Data for higher strain rates $(\simeq 10^{-2}/\text{min})$ are also collected to ascertain the cladding performance expected for typical startup, shutdown, and power cycling operations. Data are also collected under high strain rate and transient heating conditions for application to reactor transient conditions.[15]

Due to the obvious necessity to ensure long-term structural integrity of cladding at high burnup and temperature levels, the effect of neutron irradiation on creep rupture strength is of considerable interest. Generally speaking, neutron irradiation leads to considerable matrix displacement damage and, at temperature conditions low enough to prevent thermal annealing, such damage impedes creep. Consequently, the strain rates at a given stress level are reduced for irradiated material as shown in Fig. 11-12. For the irradiated annealed type 304 stainless steel in the example, the irradiation temperature must be raised to nearly 800°C before the strain rate equals that of unirradiated material.

Another interesting factor revealed by Fig. 11-12 is that radiation reduces the elongation to fracture, i.e., radiation significantly reduces the ductility of the metal. This embrittlement is graphically illustrated in Fig. 11-13 for the case of 304 stainless steel, where the elongation decreases exponentially to a saturation level at a fluence of 3 to 5×10^{22} n/cm^2, and then does not further decrease.

It was mentioned earlier that cold working is often done to improve the strength characteristics of a metal. Figure 11-14 illustrates the effect of various degrees of cold working on the creep rupture strength for type 316 stainless steel. Cold working up to about 30% is seen to allow substantially higher stress levels for short-term rupture, but just the opposite is true for

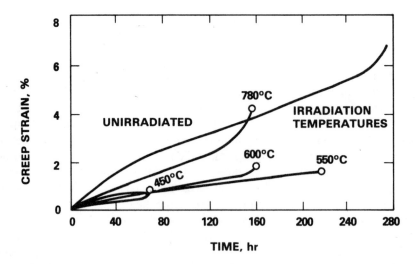

FIGURE 11-12. Effect of irradiation temperature on the creep rupture of annealed type 304 stainless steel irradiated to 1.9×10^{22} neutron/cm^2 (>0.1 MeV) and tested at 550°C under a stress of 3×10^5 kN/m^2. [16]

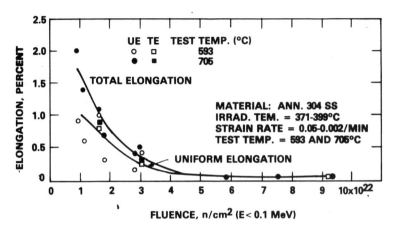

FIGURE 11-13. Effect of fluence on high-temperature ductility of EBR-II type 304 stainless steel.[17]

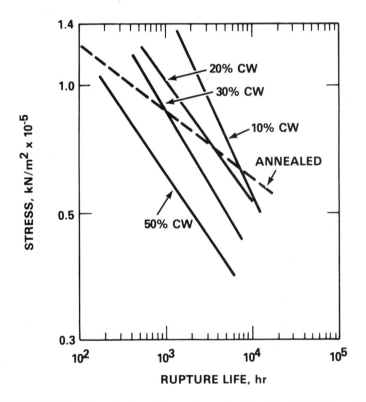

FIGURE 11-14. The effect of cold working on the rupture life of type 304 stainless steel, tested at 700°C.[18]

long-term strength. These data indicate that a fully annealed material would be preferred if the only criterion for cladding selection was long-term rupture life at low stress. However, as will be seen later, the principal reason for introducing cold work is to suppress void formation and swelling.

Other than cold working or irradiation induced matrix distortions, the primary phenomenon causing metal embrittlement is helium production during irradiation. Helium is produced by the (n, α) reaction occurring with the boron impurity that always exists to some degree in stainless steel. While this reaction tends to burn out most of the boron impurity relatively early in life, helium is continuously produced by (n, α) reactions with nickel. Boron is normally found interchanged with carbon at the grain boundaries, and it is precisely in this location that helium causes the most damage in terms of embrittlement potential.

Helium implantation into the cladding results from ternary fission in the fuel. As shown in Fig. 11-15, both calculations[19] and helium measurements show that helium from ternary fission can penetrate up to 0.13 mm into fuel cladding, which causes additional degradation of cladding adjacent to fuel.

Embrittlement due to helium is more damaging than that from cold work or displacement damage. With the latter, an increase in strength always occurs to compensate for the loss in ductility. Helium embrittlement, however, can decrease the strength and ductility simultaneously by producing premature fracture at the grain boundaries. Below 500°C, austenitic stainless steel cannot fracture at the grain boundaries and is therefore insensitive to helium embrittlement.[17] Above this temperature, however, the material can be severely embrittled, with strengths reduced to 50% of the unirradiated value and ductilities less than 0.1%.[19] A higher strain rate loading can prevent this grain boundary embrittlement until temperatures above 650°C are attained.

Although helium embrittlement (and attendant ductility loss) is a classic problem for austenitic stainless steels, this phenomenon is completely absent in the more open body centered cubic ferritic steels.

Void Swelling

During the early to mid-1960s, the key cladding problem for the LMFBR envisioned by designers was embrittlement. In 1967, however, designers were surprised by the discovery of another key problem. It was observed from irradiations at the Dounreay Fast Reactor that considerable *swelling* of the cladding took place under certain high fluence conditions. The data indicated that such swelling only occurred for fast fluence levels above a

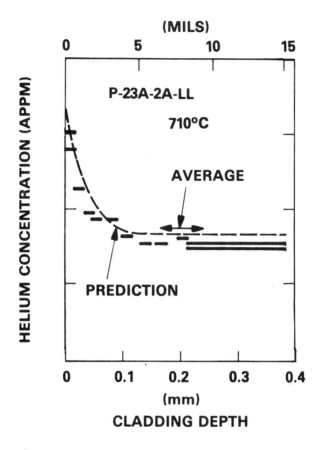

FIGURE 11-15. Helium profile through the cladding wall in specimen from the fuel column region of fuel pin. The inner surface of the cladding is at wall position "zero." [20]

threshold of about 10^{22} n/cm², but it increased rapidly beyond that point. Since it was known that fluence levels of about 2×10^{23} n/cm² would be required for successful cladding performance, considerable attention was drawn to these early Dounreay results.

Close examination of the irradiated cladding specimens revealed that very small voids or cavities had been formed within the grains. These voids, ranging from the smallest observable with an electron microscope up to 0.1 μm, did not contain sufficient gas to be called bubbles. They formed only within the temperature range of 350°C to 700°C, but this is precisely the operating temperature domain for LMFBR application.

The magnitude of volumetric increases observed for the high fluence levels was much higher than expected. Subsequent work demonstrated that most metals are susceptible to swelling by this mechanism for temperatures between 0.3 and 0.55 of their absolute melting temperature. As it turns out, stainless steel is one of the most resistant alloys to this process. Nonetheless, the implications of having to provide for cladding growths of the magnitude implied by direct extrapolation of the existing data was sufficient to launch an extensive program toward understanding the basic phenomenon and searching for ways to mitigate the problem.

Investigations revealed that four conditions must be satisfied for void swelling to occur[1]

1) Both interstitials and vacancies must be mobile in the solid. Whereas interstitials are always mobile in metals even at low temperatures, vacancies become mobile only at relatively high temperatures. Since immobile vacancies will be readily annihilated by a cloud of moving interstitials, void swelling cannot occur at low temperatures.

2) Point defects must be capable of being removed at sinks provided by structural defects in the solid, in addition to being destroyed by recombination. Furthermore, there must be a preference for interstitials at one sink in order for the vacancy population excess necessary for void formation to exist.

3) The supersaturation of vacancies must be large enough to permit voids and dislocation loops to be nucleated and grow. At high enough temperatures, however, the thermal equilibrium concentration of vacancies at the void surfaces becomes comparable to that sustained within the matrix by irradiation. Hence, void nucleation and growth ceases at high temperatures.

4) Trace quantities of insoluble gas must be present to stabilize the embryo voids and prevent their collapse.* Helium provides this requirement, although O_2, N_2, or H_2 impurities can fill the same function. Since He is introduced into the cladding materials of interest only via neutron-induced transmutation, the reason for the fluence threshold for void swelling could be explained by the incubation period required to produce sufficient quantities of the He catalyst. Although such a He concentration could stabilize the growing voids, this concentration is far too low to classify these cavities as bubbles.

A macroscopic effect of void swelling can be measured as an overall incremental increase in volume. After the fluence threshold of 10^{22} n/cm^2 is attained, early experience characterized the increase in terms of an

---

*This last requirement is a controversial point among materials specialists.

exponential rise, i.e.,

$$\left(\frac{\Delta V}{V}\right)_{\text{swelling}} \propto [\phi t]^n, \tag{11-3}$$

where n is greater than unity. (In early correlations, the exponent n tended to increase from unity at 400°C to nearly 2 at higher temperatures.)

More exact expressions for void swelling include three characteristic parameters: the incubation parameter, τ, the swelling rate parameter, R, and the curvature parameter, α. As shown in Fig. 11-16, swelling is very low up to a certain fluence and then tends to rise fairly rapidly at a roughly linear rate. Hence, the incubation parameter is used as a measure of the threshold effect, the swelling rate parameter is a measure of the linear increase beyond the threshold, and the curvature parameter characterizes the transition from the low to a high swelling rate.

A form of the stress-free void swelling relationship that has received widespread usage is as follows:

$$\frac{\Delta V}{V} = \frac{V_f - V_0}{V_0} \cong (0.01)R\left[\phi t + \frac{1}{\alpha}\ln\left(\frac{1 + \exp[\alpha(\tau - \phi t)]}{1 + \exp(\alpha\tau)}\right)\right], \tag{11-4}$$

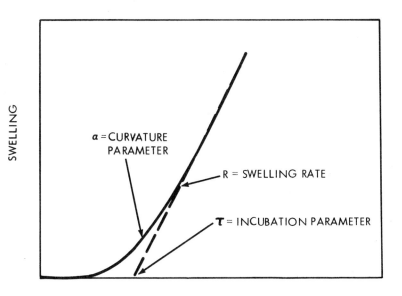

FLUENCE

FIGURE 11-16. Schematic representation of void swelling as a function of fluence.

where

V_f = final specimen volume
V_0 = initial specimen volume
R = swelling rate parameter in units of % per 10^{22} n/cm² ($E > 0.1$ MeV)
ϕt = neutron fluence in units of 10^{22} n/cm² ($E > 0.1$ MeV)
α = curvature parameter in units of $(10^{22}$ n/cm²$)^{-1}$
τ = incubation parameter in units of 10^{22} n/cm² ($E > 0.1$ MeV).

Whereas the actual values for the relevant parameters change as the data base increases, one useful set of consistent data for 20% cold-worked 316 stainless steel is as follows: [21]

$$\alpha = 0.75$$
$$R = \exp(0.497 + 0.795\beta - 0.948\beta^2 + 0.908\beta^3 - 1.49\beta^4)$$
$$+ 1.3 \exp[-8(\beta - 1.35)^2]$$
$$\tau = 6.58 - 0.566\beta; \quad T < 575°C$$
$$\tau = 4.3105 + 2.46\beta; \quad T \geqslant 575°C$$
$$\beta = (T - 500)/100,$$

where T is temperature, in units of °C.

Figure 11-17 shows the swelling relationship when these data are employed in Eq. (11-4). A statistical analysis was employed to determine the 3σ confidence level* resulting from scatter in the data.

The results shown in Fig. 11-17 were obtained for a stress-free environment on the structural material. For most applications of cladding and duct material, however, a stress field will exist. Data for such stress fields indicate a reduction in the incubation period, τ, preceding the onset of steady-state swelling. [22]

It is possible that saturation in the swelling effect may occur at high goal exposures. Very little in-pile data are available to provide statistically meaningful results at this time, but limited data from high energy bombardments [23] indicate that saturation may not occur until the fluence level is around 10^{24} n/cm². The present expectation is that commercial LMFBR cladding will swell from 0 to 15% at goal fluence conditions.

It was mentioned earlier that swelling is quite sensitive to irradiation temperature. Figure 11-18 illustrates this behavior for type 316 stainless steel. For this case, maximum swelling occurs for a temperature of about 550°C.

The effect of metal cold working was mentioned previously in conjunction with improving early rupture strength characteristics. Figure 11-19

---

*See chapter 10-4A for a discussion of $\pm 3\sigma$ confidence limits.

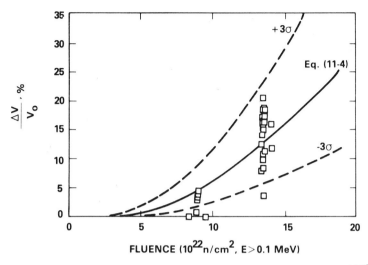

FIGURE 11-17. Swelling vs fluence for 20% cold worked 316 stainless steel at 500°C.[21]

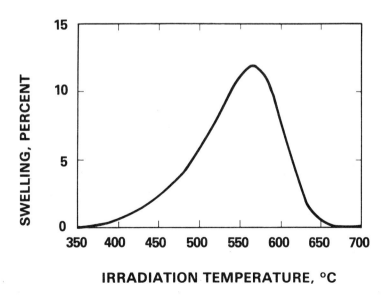

FIGURE 11-18. Influence of temperature on void swelling for 20% cold worked 316 stainless steel irradiated to a fluence of 1×10^{23} n/cm². (Adapted from Ref. 24)

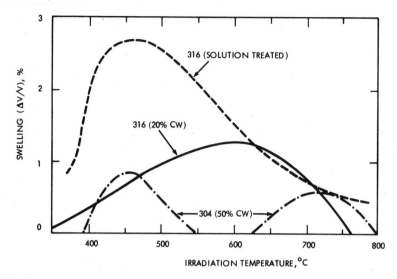

FIGURE 11-19. Effect of cold work (CW) on the swelling behavior of austenitic stainless steels. The curves are all for a fluence of approximately 5×10^{22} neutrons/cm² (from Refs. 1 and 25).

indicates the improvement in swelling resistance offered by 20% cold-worked 316 stainless steel, as compared to the fully annealed material. Whereas some cold working is clearly advantageous with regard to inhibiting swelling, there is an apparent limit on the value of such a procedure. This is suggested by the 50% CW type 304 stainless steel swelling curves, also shown in Fig. 11-19. This material, differing from type 316 SS only by its reduced molybdenum content, exhibits two swelling humps. The second hump is believed to be the result of an instability of the dislocation network introduced by the large degree of cold work. At high temperatures, large segments of such a network become free of dislocations and provide a ready environment for the formation of voids and subsequent void growth.

The materials used for alloying steel can appreciably influence the swelling characteristics. Figure 11-20 illustrates the reduction in swelling effected by using a low purity nickel relative to high purity nickel. Apparently, the impurities present in low grade nickel precipitate and act as recombination sites for vacancies and interstitials, thus removing the swelling potential. As noted in Fig. 11-20, Inconel®* actually densifies slightly during irradiation. Unfortunately, it becomes severely embrittled because of its high nickel content.

---

*Inconel® is a registered trademark of International Nickel Company.

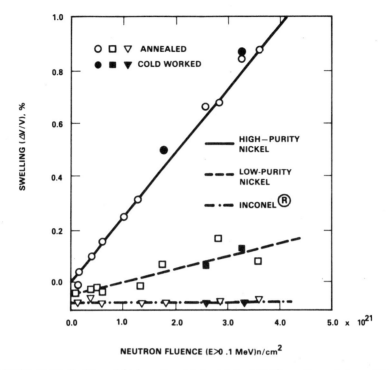

FIGURE 11-20. Swelling of high-purity nickel, nickel of 99.6% purity, and Inconel® (73% Ni-17% Cr-8% Fe) at 425°C. [26] Values for swelling below zero indicate densification.

Irradiation Creep

Irradiation creep, as defined earlier, refers to a long term plastic flow induced by an irradiation field under stress. In order to be properly classified as irradiation creep, however, the applied stress must cause deformation, and the deformation rate must change when the fast neutron flux changes.

For cladding material considerations, this classification is subdivided according to the role that thermal creep plays. Irradiation *enhanced* creep is an augmentation of thermal creep due to irradiation, whereas irradiation *induced* creep refers to the development of creep under conditions in which thermal creep is absent.

Measurements of irradiation creep in pressurized tubes of 20% cold worked type 316 stainless steel have shown that above a fluence of 10^{22} n/cm^2 the creep rate increases with temperature. Up to about 500°C, irradiation creep has been shown[27] to have essentially the same temperature dependence as void swelling. Also, above 500°C the creep rate increases

TABLE 11-5 Nominal Compositions of Selected Structural Materials*

Type		C	Fe	Cr	Ni	Mn	Mo	Nb	Al	Ti	Si	W	P	S	Cu
Stainless Steels†	304	0.08 max	base	18.0 to 20.0	8.0 to 12.0	2.0 max	—	—	—	—	1.0 max	—	0.045 max	0.030 max	—
	316	0.08 max	base	16.0 to 18.0	10.0 to 14.0	2.0 max	2.0 to 3.0	—	—	—	1.0 max	—	0.045 max	0.030 max	—
	321	0.08 max	base	17.0 to 19.0	9.0 to 12.0	2.0 max	—	—	—	5X carbon min	1.0 max	—	0.045 max	0.030 max	—
	347	0.08 max	base	17.0 to 19.0	9.0 to 13.0	2.0 max	—	10X carbon (min)	—	—	1.0 max	—	0.045 max	0.030 max	—
	Incoloy® 800	0.04	46.0	20.5	32.0	0.75	—	—	—	—	0.35	—	—	—	0.30
Nickel Base Alloys	Inconel® 600	0.15	10.0	17.0	base	1.0	—	—	—	—	0.5	—	—	—	—
	Inconel® X750	0.08	9.0	17.0	base	1.0	—	1.2	1.0	2.75	0.5	—	—	—	—
	Hastelloy X	0.1	18.5	22.0	base	1.0	9.0	—	—	—	0.75	0.6	—	—	—
	Inconel® 718	0.1	base	21.0	55.0	0.35	3.3	5.5	0.8	1.15	0.35	—	—	—	—
	Inconel® 625	0.1	5.0	23.0	base	0.5	10.0 •	4.15	0.4	0.4	0.5	—	—	—	—

*Taken from pages 194 and 201 of Ref. 12.

†ASTM A271-64T specifies somewhat different compositions for heavy-walled stainless steel tubing and pipe. For example: 0.040 percent maximum phosphorus, 0.75 percent maximum silicon, and 8.0 to 11.0 percent and 9.0 to 13.0 percent nickel for Type 304 and Types 321 and 347, respectively. The suffix H applied to tubing material indicates a carbon content in the range from 0.04 to 0.10 percent and defines certain heat-treating temperatures and times for the various alloys.

with fluence, while below 500°C the rate is not affected by fluence. Irradiation creep, $\varepsilon_{ir\,creep}$, to fluences of 6×10^{22} n/cm² may be described by Eq. (8-62).

C. CLADDING MATERIALS

Given the intense high temperature and high fluence requirements imposed on FBR cladding candidates, much effort is being concentrated on the selection of special alloys that will provide the best overall balance of design. As in any complex engineering application, optimal properties cannot be consistently obtained from any one material over the full range of parameters desired. Table 11-5 provides a composition listing for the stainless steels of most interest for FBR applications, as well as for the nickel base alloys.

Type 316, 20% cold worked stainless steel is the reference material used for most LMFBR in-core cladding and structural applications. Table 11-6

TABLE 11-6   Thermophysical Properties of 316 Stainless Steel[28] (all temperatures in kelvins)

Melting Point $= 1700$ K
Heat of Fusion $= 2.70 \times 10^5$ J/kg
Boiling Point $= 3090$ K
Heat of Vaporization $= 7.45 \times 10^6$ J/kg
Specific Heat:

$c_p = 462 + 0.134\,T$	(J/kg·K)	solid region
$c_p = 775$	(J/kg·K)	liquid region

Thermal Conductivity:

$k = 9.248 + 0.01571\,T$	(W/m·K)	solid region
$k = 12.41 + 0.003279\,T$	(W/m·K)	liquid region

Thermal Expansion Coefficient:

$\alpha = 1.789 \times 10^{-5} + 2.398 \times 10^{-9}\,T + 3.269 \times 10^{-13}\,T^2$	solid region
$\alpha = 1.864 \times 10^{-5} + 3.917 \times 10^{-10}\,T + 2.833 \times 10^{-12}\,T^2$	liquid region

Viscosity (molten steel):

$$\log_{10}\mu = \frac{2385.2}{T} - 3.5958 \quad (\mu \text{ in kg/m·s})$$

Vapor Pressure (molten steel):

$$\log_{10}p = 11.1183 - \frac{18868}{T} \quad (p \text{ in Pa})$$

Density:

$\rho = 8084 - 0.4209\,T - 3.894 \times 10^{-5}\,T^2$ (kg/m³)	solid region
$\rho = 7433 + 0.0393\,T - 1.801 \times 10^{-4}\,T^2$ (kg/m³)	liquid region

lists the major thermophysical properties of type 316 stainless steel. The primary reasons for selecting this reference material are its excellent high temperature strength characteristics, its resistance to void swelling, its compatibility with mixed-oxide fuels and sodium coolant, and its relatively low cost.

To extend the lifetime of LMFBR cores, improved and advanced materials are being developed for ducts and cladding. They represent three major classes of alloys: solid-solution strengthened austenitic stainless steels similar to AISI 316, ferritic steels, and precipitation hardened Fe-Ni-Cr superalloys. All of these developmental alloys have demonstrated superior swelling resistance compared to AISI 316. The solid-solution austenitic steels are relative minor compositional modifications to AISI 316 and are therefore very similar to the reference material in high temperature strength properties and sodium compatibility; their improved swelling resistance will lead to longer core lifetimes when used as ducts and cladding. The ferritic steels are perhaps the most swelling resistant class of materials being considered for LMFBR's; they are typically limited in high temperature strength and therefore have primary application as duct material. If fracture toughness properties prove to be adequate following neutron irradiation, the ferritic steels have the promise of surpassing the performance of austenitic stainless steel ducts. The precipitation hardened superalloys combine excellent swelling resistance with superior high temperature stress rupture properties; they are equally well suited to duct and cladding applications but their high temperature strength make them an especially strong candidate to increase fuel pin life, compared to the other developmental alloys.

11-4 COOLANT

The choice of coolant for a fast breeder reactor probably has a greater effect on the overall physical plant layout than any other design selection. Whereas the coolant strongly influences the neutronic behavior of the core and provides a direct framework for the consideration of cladding materials, the most recognizable effect of the coolant selection is its effect on major components such as pumps and steam generators.

The liquid metal sodium has been selected as the coolant for all the major fast breeder power reactor projects underway around the world. As a consequence, the properties of sodium and its implications in design will be given particular attention. The gas coolants helium and steam will be given limited attention, with brief reference to other metal coolants, to provide a framework for weighing the pros and cons of the options that have received the most attention to date.

A. REQUIREMENTS

The principal task of an FBR coolant is to remove heat from a high power density core. However, several considerations come into play in assessing the capability of the coolant to perform this function. It is convenient to organize the discussion of these facets along the lines suggested by Wirtz;[12] namely, thermal, neutronic, hydraulics, and compatibility considerations. Having covered the principal features of each coolant candidate, from the standpoint of those four categories, an intercomparison summary will be given at the end of this section.

B. THERMAL CONSIDERATIONS

The very high power density in an FBR core ($\simeq 400$ Watts/litre compared to $\simeq 100$ Watts/litre for an LWR) places very stringent requirements on the heat transfer properties of a coolant candidate. For a given heat transfer area, A, (e.g., the outside cladding surface), the rate of heat removal, $\dot{Q}/A$, is given by Eq. (9-24):

$$q = \frac{\dot{Q}}{A} = h(T_{co} - T_b), \tag{11-5}$$

where

$q =$ heat flux $\left(\text{W}/\text{m}^2 \right)$

$h =$ convection heat transfer coefficient $\left(\text{W}/\text{m}^2 \cdot {}^\circ\text{C} \right)$

$T_{co} - T_b =$ temperature drop from cladding surface to bulk coolant $({}^\circ\text{C})$.

Obviously, a large heat transfer coefficient is desirable for optimum heat removal.

Table 11-7 provides typical numbers for the convection heat transfer coefficient, as well as a typical coolant velocity and attendant pressure loss along the length of the core for the three cooling candidates of interest. Sodium is clearly superior from the standpoint of heat transfer characteristics, although He and steam can be driven hard enough, given sufficient driving pressure, to accomplish the required task.

Table 11-8 is included to provide similar data for other liquid metals and, in particular, an intercomparison with water. The numbers given are based upon a coolant flow of 3.3 m/s in a 25 mm diameter pipe. Water has the

TABLE 11-7   Typical Heat Transfer Data for Coolant Candidates*

Property	Coolant		
	Na	He	Steam
$h(W/m^2 \cdot °C)$	85 000	2 300 (smooth surface)	11 000
		10 000 (rough surface)	
$V(m/s)$	6	115	25
$p(MPa)$ (psi)	0.7 (100)	7.0 (1 000)	15 (2 200)
$c_p(kJ/kg \cdot °C)$	1.3	5.2	2.6
$\rho(kg/m^3)$	815	4.2	41

*Augmented from page 28 of Ref. 12.

TABLE 11-8   Comparison of Heat Transfer Data for Various Fluids at High Temperatures*

Property	Coolant				
	Na	NaK	Hg	Pb	H_2O
T_{melt} (°C)	98	18	−38	328	0
T_{boil} (°C)	880	826	357	1743	100
c_p (kJ/kg·°C)	1.3	1.2	0.14	0.14	4.2
k (W/m·°C)	75	26	12	14	0.7
h (W/m²·°C)[†]	36 000	20 000	32 000	23 000	17 000
Relative Pumping Power Required ($H_2O=1$)	0.93	0.93	13.1	11.5	1.0

*Taken from page 29 of Ref. 12.
[†] For 3.3 m/s velocity in a 25 mm duct.

best heat capacity, but sodium is superior in all other categories. Water, of course, cannot be considered for FBR applications because of its major degradation of the fast neutron spectrum.

C. NEUTRONICS CONSIDERATIONS

Whereas any coolant has some moderating effect on the neutron spectrum, the degree of moderation taking place is proportional to the atomic mass and the density. Sodium (mass 23) is clearly heavier than either helium or steam, but when the density effect is taken into account, the helium-cooled reactor yields the hardest spectrum, steam yields the softest spectrum (because of the hydrogen content), and sodium provides an intermediate spectrum. Because of this spectrum difference, helium has a slight intrinsic edge regarding achievable breeding ratios.

Another neutronics consideration is the effect of coolant activation. Helium, being inert, is completely unaffected by irradiation. Steam becomes somewhat radioactive (via ^{17}O), but the impact on design is minor. Sodium activation, however, attains a high degree of short half-life activation. The reaction taking place is as follows:

$$^{23}Na + n \rightarrow {}^{24}Na \xrightarrow[15\ hr]{\beta} {}^{24}Mg$$

with an attendant release of 1.37 and 2.75 MeV gammas. This activation has led to the inclusion of an intermediate coolant loop for sodium cooled systems in order to ensure that all radiation is confined to the primary loop. Other reasons for including an intermediate loop in the LMFBR system are (1) to protect the core from possible pressure surges or positive reactivity effects due to hydrogen moderation should a steam generator leak occur and result in a sodium-water reaction, and (2) to protect the steam generator from radioactive corrosion products and fission products.

A final neutronics consideration concerns the reactivity effect of coolant loss during an accident. As noted in Chapter 6, the large density change associated with sodium loss can cause an appreciable change in the neutron spectrum and an associated positive reactivity feedback. Neither of the gas cooling media exhibits such a characteristic, although a thorough safety analysis of such systems does cover the remote possibility of water entering the core as a result of a major break or depressurization. Under such conditions, similar variations in reactivity could occur.

D. HYDRAULICS CONSIDERATIONS

In order for a reactor coolant to perform its heat removal function, a fraction of the total power output must be used to pump it through the core. Hence, it is of interest to compare the pumping requirements for the various coolant candidates.

For a given core power, $\dot{Q}$, the heat must be removed according to the relationship:

$$\dot{Q} = \dot{m} c_p \Delta T_{axial} \tag{11-6}$$

where

$\dot{Q}$ = core power (W)
c_p = specific heat (J/kg·s)
$\dot{m} = \rho V A$ = coolant mass flow rate (kg/s)
ρ = coolant density (kg/m³)
V = coolant velocity (m/s)
A = coolant flow area (m²)
ΔT_{axial} = difference between outlet and inlet coolant temperature (°C).

Hence, for a fixed flow area and axial coolant temperature increase, the coolant velocity must be

$$V = \frac{\dot{Q}}{\rho c_p A \Delta T_{\text{axial}}}.$$ (11-7)

The pressure loss over the core can be written [cf. Eq. (10-33)] as

$$\Delta p = f \frac{L}{D_e} \frac{\rho V^2}{2},$$ (11-8)

where
Δp = pressure drop over core (N/m^2)
f = friction factor
L = length of core (m)
D_e = hydraulic diameter $(m) = 4A/P_w$, where P_w is the wetted perimeter.

Since the power required to move the coolant through the core is

$$\text{Pumping power} = \Delta p \cdot A \cdot V,$$ (11-9)

combining Eqs. (11-7) through (11-9) yields

$$\text{Pumping power} = \frac{LA}{2D_e} (f \rho V^3) = \left(\frac{L\dot{Q}}{2 D_e \Delta T_{\text{axial}}} \right) \left(\frac{fV^2}{c_p} \right).$$ (11-10)

Using either expression, the strong dependence of pumping power on the coolant velocity is evident. While other parameters vary between sodium and gas (e.g., D_e and c_p are larger for helium designs than for sodium) the large velocity difference shown in Table 11-7 dominates. The numbers contained in the box below* indicate that, for typical sized plants, sodium coolant requires the least power and helium requires the most pumping power. Other liquid metals are ruled out largely because of their much larger pumping requirements, as indicated in Table 11-8.

	Sodium Cooled	He-gas Cooled	Steam Cooled
Percent pumping power of electrical output	1.5	7	3

---

*From page 28 of Ref. 12.

E. COMPATIBILITY CONSIDERATIONS

The coolant chosen for FBR application must be compatible with the cladding and should also be at least reasonably compatible with the fuel so that an in-pile breach could be accommodated without adverse fuel-coolant interactions. Helium, being inert, is compatible with any metal or fuel.

Steam, on the other hand, can be quite corrosive with the normal stainless steels. In fact, the dry-steam corrosion problem became sufficiently troublesome in the early test program[29] that a high nickel component steel, such as Incoloy®* or Inconel® was required to provide acceptable compatibility conditions. As we noted from the discussions of Section 11-3B, nickel has a deleterious neutronic effect, but even more importantly leads to considerable embrittlement from the (n, α) reaction. Steam has been dropped as an active FBR cooling candidate mainly because of the cladding corrosion problem.

Sodium is quite compatible with the preferred cladding candidates discussed above. However, because sodium has been chosen as the base coolant, much work has been directed to every facet of its in-core environment in order to head off any long term surprises. Summarized below are the main categories[1] where efforts have been focused to isolate possible problem areas.

1. *General Corrosion*

The principal metallic elements in the cladding (i.e., Fe, Cr and Ni) are very slowly dissolved in sodium from the hot core region and deposited in the cooler areas. This long-term uniform attack is called cladding *thinning* or *wastage*. Although this amounts to only tens of micrometers per year at 700°C, it must be considered in the overall load bearing capacity of the cladding, which is stressed by internal fission gas pressure. Oxygen concentrations must be kept low (<5 ppm) to minimize corrosion.

2. *Selective Leaching*

Each of the principal metallic elements mentioned above has different dissolution rates in sodium. Hence, selective leaching occurs and the composition of the steel at the outer surface is eventually altered. The effect, though small, is normally deleterious, since Cr and Ni are removed faster than Fe. Some of the more significant problems arising from the dissolution of minor constituents or activation productions from the steel in the core are discussed below.

---

*Incoloy® is a registered trademark of the International Nickel Company.

3. *Deposition*

The dissolved metal deposited in the colder region of the intermediate heat exchanger (IHX) can lead to higher heat transfer resistance (i.e., reduced heat exchanger efficiency). The deposits can become thick enough that coolant driving pressure must be increased in order to maintain the desired flow. Some long-term Si deposits have even been observed. The silicon, originating as a minor constituent in the steel, forms a compound with sodium which, if precipitated in core assembly coolant channels, can influence pumping power or local flow conditions.

4. *Transport of Radioactivity*

The major long-lived radioactive species produced in the cladding are ^{54}Mn, ^{58}Co and ^{60}Co. Though small in quantity, sodium transport and deposition of these radionuclides in the IHX and other primary system compounds can cause high enough local activity to affect routine maintenance.

5. *Carbon Transport*

Carbon, included in the component steel alloys to improve strength characteristics, is quite mobile in solid cladding and can migrate fairly readily to the outer surface. Sodium can, therefore, transport carbon, as well as all the metallic elements, to colder surfaces. Both the hot steel ("decarburization") and the cold steel ("carburization") properties can be altered by these small changes in the carbon present. This is normally not a problem with cold worked 316 stainless steel, however, because rapid carbide precipitation in the steel lowers the carbon activity sufficiently to prevent decarburization.

6. *Sodium Chemistry*

Because sodium is the transport medium for the processes described above, samples of sodium are usually extracted regularly and analyzed during normal operation. On-line instruments have been developed to monitor sodium purity for both the primary and secondary systems. In addition to determining oxygen and carbon content, it is often useful to monitor hydrogen levels, since this provides an indication of the tritium level in the cover gas (primary system) or steam generator leaks (secondary system).

7. *Sodium-Fuel Interaction*

In the event of a small cladding breach, a direct interaction of sodium and fuel could lead to the production of Na_3UO_4 or Na_3PuO_4. The concern is whether this substance could be swept into the coolant stream as a contaminant. There is also some concern about whether this material could lead to local fuel swelling aggravation and escalation of the cladding leak into a fuel rupture. If the fuel should be molten, the thermal aspect of the interaction is the principal concern. This problem will be discussed in Chapter 16.

F. SODIUM PROPERTIES

A consistent set of thermodynamic properties of sodium is reported by Fink and Leibowitz.[30] Thermodynamic properties listed in this section were obtained from that reference.

Vapor pressure of sodium as a function of temperature is plotted in Fig. 11-21. Note that the hottest sodium in a typical LMFBR system (800 K) is approximately 350°C below the boiling point at atmospheric pressure. An equation for the vapor pressure of sodium[30] developed by and consistent with the high temperature data[31,32] of Bonilla, Bhise, and Das Gupta is

$$P = \exp\left[18.832 - (13\,113/T) - 1.0948\ln T + 1.9777 \times 10^{-4}T\right],$$

where P is in atmospheres and T is in kelvins.

The critical properties, as determined by Bonilla, et al., are

$$P_c = 25.6 \text{ MPa (253 atm)}$$
$$T_c = 2509 \text{ K}$$
$$\rho_c = 214 \text{ kg/m}^3.$$

The specific heat, c_p, for sodium at typical LMFBR temperatures (e.g., from 400°C to 550°C) is 1.3 kJ/kg·K.

The density of liquid saturated sodium up to 1644 K (T in kelvins) is

$$\rho_l(\text{kg/m}^3) = 1011.8 - 0.22054T - 1.9226 \times 10^{-5}T^2 + 5.6371 \times 10^{-9}T^3.$$

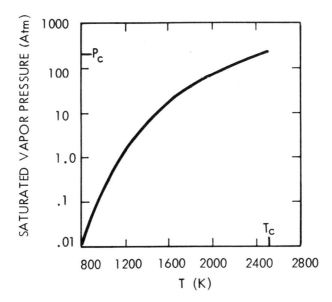

FIGURE 11-21. Saturated vapor pressure of sodium.

Expressions for ρ_l from 1644 K to the critical point, together with h_l, h_g, ρ_g, and other saturation properties, and superheat and subcooled properties, are given in Ref. 30.

Expressions for the physical properties thermal condicutivity, k,[33] and viscosity, μ,[34] as a function of temperature (kelvins) are given below:

$$k(\text{W}/\text{m}\cdot\text{K})=93.0-0.0581\ (T\text{-}273)+1.173\times10^{-5}(T\text{-}273)^2$$

$$\mu(\text{kg}/\text{m}\cdot\text{s})=A\rho^{1/3}e^{B\rho/T}(\rho\text{ in kg}/\text{m}^3)$$

where

$$A=1.235\times10^{-4}\text{ and }B=0.697\text{ for }T<773\text{ K,}$$

$$A=0.851\times10^{-4}\text{ and }B=1.040\text{ for }T>773\text{ K.}$$

G. GENERAL INTERCOMPARISON OF COOLANTS

This section summarizes the pros and cons of each coolant discussed previously, and refers to other items that bear on coolant selection. Table 11-9 lists the advantages and disadvantages of each coolant.

A principal advantage of sodium, beyond its excellent heat transfer properties, is that it does not require pressurization to prevent coolant boiling. This feature contrasts sharply with the thick-walled high-pressure systems required for He- or steam-cooled systems.

Helium is still receiving attention in the Gas Cooled Fast Reactor (GCFR) program, as a backup to the LMFBR. Steam has been essentially eliminated as a serious contender, mainly because of cladding corrosion problems.

11-5 CONTROL

Control of a large fast breeder reactor can be provided by either removing fuel or inserting a neutron absorber. Although the former method has been used (e.g., in EBR-II), the latter is by far the most widely employed technique. Neutron cross sections of all absorber materials are much smaller in the high energy range than in the thermal neutron domain, but materials widely employed for thermal reactor control purposes also tend to be good absorbers in the FBR energy range. Hence, boron carbide, consisting of either natural boron or enriched with ^{10}B, is the absorber material favored by most FBR designers. Other candidates, including tantalum and europia, have been considered because of potential gas release and swelling problems

TABLE 11-9   Intercomparison of FBR Coolants*

Coolant	Advantages	Disadvantages

Sodium	-Excellent heat transport properties
	-Low pressure system
	-Low pumping power requirements
	-Lowest fuel cladding temperature
	-Potentially high breeding ratio
	-Inherent emergency cooling of fuel
	-Extensive sodium reactor experience
	-Potential for vented fuel
		-Radioactivity (intermediate
		  sodium loop employed)
		-Unfavorable coolant reactivity
		  void coefficient
		-Chemical reactions with air and
		  water
		-Nonvisible refueling procedure
		-Solid at room temperature
		-Maintenance on primary system
		  impeded by radioactivity

Helium	-No intermediate loop
	-Coolant not activated
	-Potentially high breeding ratio
	-Visible refueling/maintenance
	-Minimal void coefficient
	-Most compatible with materials
	-Utilization of thermal GCR
	  technology
	-Potential for vented fuel
	-Potential for direct cycle
	-Flooding with H_2O tolerable
		-High pressure system
		-High pumping power requirements
		-Cladding roughening required
		-Emergency cooling provisions
		  not established
		-Unproven high power density
		  capability
		-Lack of FBR technology
		-Gas leakage difficult to control
		-High performance demands on
		  pumps and valves

Steam	-Direct cycle
	-Visible refueling/maintenance
	-Industrial capability available
	  for components
	-Minimum chemical reactions
	-Fluid at room temperature
		-High pressure system
		-High pumping power
		-Cladding corrosion
		-Lack of FBR technology
		-Emergency cooling provision
		  not established
		-Low breeding ratio
		-Fission product carryover to
		  turbine
		-Unfavorable coolant reactivity
		  coefficient

*Adapted from page 27 of Ref. 12.

with boron carbide. Silver and silver alloys have enjoyed considerable success for PWR applications but have not been seriously considered for FBR applications due to their relatively low reactivity worth and high cost.

A. REQUIREMENTS

The basic requirements of the control rod systems are (1) to compensate for built-in burnup reactivity, and (2) to provide neutronic shutdown for routine operations and safety measures. As such, enough absorber material must be provided to meet the reactivity worth requirements outlined in Chapter 6. The control system installed must be capable of long life (e.g., 3 years), which means the absorber material must be compatible with its cladding, and dimensional instability must be well characterized. Finally, overall plant economics favor absorber materials that are readily available in large quantities, i.e., an industrial product.

B. BORON

Boron, present in the form of boron carbide, B_4C, is the absorber material generally used in FBR systems. The principal advantages of boron carbide are (1) its relatively high neutron absorption cross section, (2) its availability and low cost, (3) its comparative ease of fabrication, and (4) its low radioactivity after irradiation. The ^{10}B neutron absorption mechanism is the (n, α) reaction, and the He atoms produced lead to both matrix swelling and gas release difficulties. This problem can be accommodated by operating at relatively high temperatures (cf. Fig. 11-25) and/or by venting the pins.*

Neutronics

Figure 11-22 shows the $^{10}B(n, \alpha)^7Li$ absorption cross section over the complete thermal to fast spectrum range, as well as the ^{239}Pu fission cross section from Fig. 2-7 (somewhat smoothed out) for purposes of comparison. As clearly seen from this figure, the boron-10 cross section is only of the order of one barn for the fast neutron spectrum, whereas it is of the order of 1000 barns for a thermal reactor. Furthermore, the absorption/fission cross section ratio drops with a harder spectrum, leading to a requirement for more absorber material in an FBR than an LWR. However, there is a compensating effect in that much of the neutron absorption occurs near the absorber material surface in an LWR due to self-shielding, whereas absorptions are more uniformly distributed in the absorber material for an FBR environment.

Also contained in Fig. 11-22 is the $^{10}B(n, t) 2\alpha$ cross section, which is appreciable only for the very high energy tail. The latter reaction, while favorable from the standpoint of control rod reactivity worth density, is

---

*Considerable interest exists in venting the absorber pins in order to avoid the necessity of a plenum to contain the He buildup.

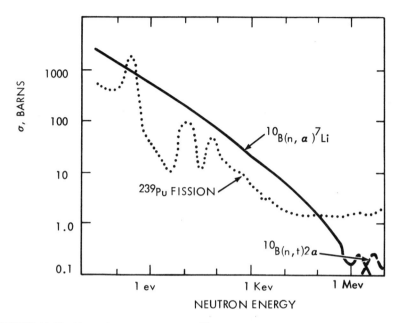

FIGURE 11-22. Absorption cross section of ^{10}B compared with the fission cross section of ^{239}Pu.

basically undesirable because the tritium therein produced must be contained to prevent leakage from the reactor systems. This reaction is responsible for about one-half of the low-level tritium radiation production in the plant. Natural boron consists of 19.8% ^{10}B, with the remainder being ^{11}B (the latter possessing a very small neutron absorption cross section). Hence, the reactivity worth of a given dilute sample of B_4C can be increased by nearly a factor of five by enriching it with the lighter ^{10}B isotope. In an FBR absorber assembly, self-shielding limits this worth increase to about 60% of the change in enrichment.

Physical Properties

Boron carbide has a rhombohedral crystalline structure. Although hot pressed pellets of 92% theoretical density are brittle, the thermophysical properties[35] of this material are generally compatible with control system requirements. Structural integrity of absorber pins is provided by cladding material, thereby allowing B_4C to be employed either in the form of powder or pellets. Below 800°C, boron carbide retains about 80% of the tritium produced. Unfortunately, cladding will not contain the tritium gas which escapes the boron carbide matrix.

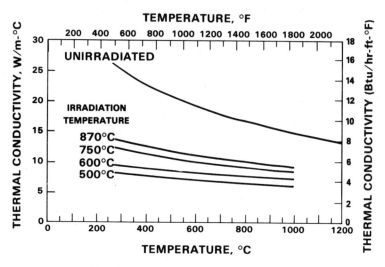

FIGURE 11-23. Effect of irradiation on thermal conductivity of boron carbide.[36]

Thermal conductivity is perhaps the most important thermophysical property of present interest to the boron carbide system. The (n, α) reaction liberates 2.78 MeV per event, and most of this energy is deposited directly within the B_4C matrix ($\simeq 75$ W/cm^3 for naturally occurring B_4C in an FBR). Hence, the thermal gradient in the absorber material is a direct function of thermal conductivity. Figure 11-23 illustrates the thermal conductivity of fully dense B_4C. It is of particular interest to note from Fig. 11-23 that the thermal conductivity drops significantly upon irradiation. The lower curves represent a large collection of irradiation data, including results for differing irradiation conditions.

Irradiation Behavior

The principal effects of irradiation on B_4C are helium generation and matrix swelling. For unenriched B_4C, a complete burnup of all the ^{10}B atoms (0.22×10^{23} captures/cm^3 of B_4C) would yield 814 cm^3 of He at standard pressure and temperature. The amount of this gas released during burnup is sensitive to both irradiation temperature and to exposure level. Figure 11-24 illustrates this effect with a three-dimensional plot. The percent gas release builds up with irradiation temperature, with a maximum occurring around 1100 K. This gas released must either be vented to the coolant or sufficient plenum volume must be provided in order to prevent excessive loading on the control cladding material. In addition to temperature, the stoichiometric composition also affects the degree of gas release.

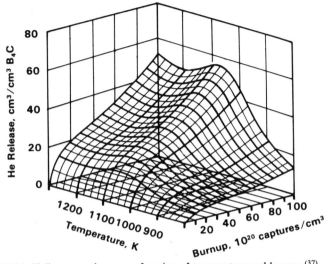

FIGURE 11-24. Helium gas release as a function of temperature and burnup.[37]

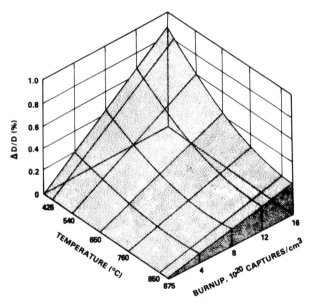

FIGURE 11-25. Boron carbide linear swelling rate ($\Delta D/D$) during fast reactor exposure.[38]

Both the He and Li atoms resulting from the neutron absorption process are larger than the original boron atoms. Hence, swelling occurs due to both the Li and the portion of He atoms remaining in the matrix. Figure 11-25 indicates that the resulting swelling is approximately linear with burnup and tends to drop as the irradiation temperature increases.

The useful life of absorber pins containing boron carbide could be enhanced by the ability of the cladding to accommodate the stresses imposed by B_4C swelling. Hence, there is considerable interest in providing cladding material for absorber pins that displays good ductility characteristics.

Compatibility

Boron carbide is quite compatible with most steel alloy cladding materials, with no appreciable metal attack below 700°C. Above that temperature, however, a layer of Fe_2B tends to build up, and the attack that occurs is about a factor of three greater in the presence of sodium. A eutectic is formed if the temperature should ever climb to 1226°C. Excess boron in the B_4C will enhance the steel attack characteristics, and the presence of lithium from the (n, α) reaction has been shown to enhance the diffusion of boron and carbon into the steel.

C. TANTALUM

Tantalum is receiving some attention as a possible substitute for boron as an FBR control material, primarily because of its favorable swelling characteristics. The fundamental reason for the improved swelling resistance is that the $^{181}Ta(n, \gamma)^{182}Ta$ reaction does not yield helium, as does the boron-10 reaction. In fact, at 530°C the material actually shrinks slightly because the tungsten atoms produced by a β-decay of ^{182}Ta to ^{182}W are smaller than the original tantalum atoms. On the other hand, since Ta is a metal, it will eventually swell to about $1\%\Delta V/V$ (by the void formation mechanisms discussed in Section 11-3B).

Additional advantages of Ta as a control material include its relative abundance (fairly low cost), ease of fabrication, moderately high absorption cross section in the fast spectrum domain, and daughter products (which are also good absorbers). The principal disadvantage is the 115-day half-life γ-decay, from ^{182}Ta to ^{182}W, which causes long-term heat removal problems. Also, Ta is soluble in Na, making cladding an absolute requirement.

D. EUROPIUM

Europium, in the sesquioxide form Eu_2O_3, has received considerable attention in recent years because of its high fast spectrum absorption cross section. Both naturally occurring isotopes (47.8% ^{151}Eu and 52.2% ^{252}Eu) have dilute sample neutron absorption cross sections over twice that of ^{10}B averaged over the LMFBR spectrum. Like tantalum, the daughter products are also good absorbers and the (n, γ) reaction avoids the need for a gas plenum.

Unfortunately, self-shielding is such that a full assembly has a reactivity worth equivalent only to that for an unenriched B_4C assembly in an FBR. The other disadvantages of europia include (1) the high level of radioactivity induced into the europium decay chain, (2) the low supply (as a rare-earth), and (3) low thermal conductivity (which requires fairly small diameter pins).

An interesting attempt to combine the attributes of both Eu and B is the europium compound EuB_6. This compound has a reactivity worth equivalent to approximately 25% ^{10}B-enriched B_4C, and it is worth about 10% more than Eu_2O_3. Furthermore, the loss of reactivity is lower than in B_4C. It is, of course, possible to enrich the boron in the EuB_6 compound to further enhance its absorption capability, but the expense goes up accordingly. Although the reactor experience is quite limited, the dimensional stability properties appear attractive. The only major problem is that the He gas release is much larger than for B_4C, so that venting is probably required.

REVIEW QUESTIONS

11-1. Make a list of the principal candidate materials for the following functions in a fast breeder reactor:
    (a) fuel
    (b) cladding and structure
    (c) coolant
    (d) control.

11-2. List the properties (a) that make mixed oxide fuel $(UO_2\text{-}PuO_2)$ attractive for LMFBR's, and (b) that are disadvantages that provide incentives for development of carbide and metal fuels.

11-3. Describe the two fundamental processes that cause fuel to swell during irradiation in a nuclear reactor. How is this swelling accommodated in the design of FBR fuel?

11-4. Suppose the cladding and duct of a fuel assembly near the center of an LMFBR core operate at 475°C in a neutron flux above 0.1 MeV of 4×10^{15} n/cm$^2 \cdot$s. If the assembly is irradiated for 2 (calendar) years at 0.7 load factor, estimate the fractional swelling $(\Delta V/V)$ of the cladding and duct.

11-5. Why is the term 0.2% offset yield strength introduced when describing austenitic steels?

11-6. What is creep rupture strength and why is it important for cladding design? How do temperature and neutron irradiation affect ductility, creep rupture, and irradiation creep?

11-7. Discuss briefly the reasons that sodium has been generally selected as the coolant for fast breeder reactors.

11-8. Why can a neutron absorber like boron be used for reactivity control when fast neutron capture cross sections are so small?

11-9. Why must B_4C control rods in an FBR be cooled?

REFERENCES

1. D. R. Olander, *Fundamental Aspects of Nuclear Reactor Fuel Elements*, TID-26711-P1, Office of Public Affairs, U.S. ERDA, 1976.
2. W. L. Lyon and W. E. Baily, "The Solid Liquid Phase Diagram for the UO_2-PuO_2 System," *J. Nuc. Mat.*, 22 (3) (1967) 332–339.
3. U. P. Nayak, A. Boltax, R. J. Skalka and A. Biancheria, "An Analytical Comparison of the Irradiation Behavior of Fast Reactor Carbide and Oxide Fuel Pins," *Proc. Topical Meeting Advanced LMFBR Fuels*, Tucson, AZ, ERDA 4455. October 10–13, 1977, 537, 539, 540.
4. E. H. Randklev, "Radial Distribution of Retained Fission Gas in Irradiated Mixed Oxide Fuels," *Trans. ANS*, 28 (June 1978) 234–236.
5. T. N. Washburn and J. L. Scott, "Performance Capability of Advanced Fuels for Fast Breeder Reactors," *Proc. Conf. on Fast Reactor Fuel Element Technology*, New Orleans, LA, April 13–15, 1971.
6. J. A. L. Robertson, *Irradiation Effects in Nuclear Fuels*, Gordon and Breach, NY, (1969) 223. (Copyright held by American Nuclear Society, LaGrange Pk, IL.)
7. R. D. Leggett, R. K. Marshall, C. R. Hann, and C. H. McGilton, "Achieving High Exposure in Metallic Uranium Elements," *Nuc. Appl. and Tech.*, 9 (1970) 673.
8. C. W. Walter, G. H. Golden, and N. J. Olson, *U-Pu-Zr Metal Alloy: A Potential Fuel for LMFBR's*, ANL 76-28, Argonne National Laboratory, November 1975.
9. A. A. Bauer, "Nitride Fuels: Properties and Potentials," *Reactor Technology*, 15, No. 2, Summer 1972.
10. V. S. Yemel'yanov and A. I. Yebstyukhin, *The Metallurgy of Nuclear Fuel—Properties and Principles of the Technology of Uranium, Thorium and Plutonium* (trans. Anne Foster), Pergamon Press, 1969.
11. J. H. Kittel, J. A. Horak, W. F. Murphy, and S. H. Paine, *Effects of Irradiation on Thorium and Thorium-Uranium Alloys*, ANL-5674, Argonne National Laboratory, 1963.
12. K. Wirtz, *Lectures on Fast Reactors*, Kernforschungszentrum, Karlsruhe, 1973.
13. J. K. Fink, M. G. Chasanov, and L. Leibowitz, *Thermophysical Properties of Thorium and Uranium Systems for Use in Reactor Safety Analysis*, ANL-CEN-RSD-77-1, Argonne National Laboratory, June 1977.
14. R. L. Fish and J. J. Holmes, "Tensile Properties of Annealed Type 316 Stainless Steel after EBR-II Irradiation," *J. Nuc. Mat.*, 4–6 (1973) 113.
15. C. W. Hunter, R. L. Fish, and J. J. Holmes, "Mechanical Properties of Unirradiated Fast Reactor Cladding During Simulated Overpower Transients," *Nuc. Tech.*, 27 (1975) 367–388.
16. E. E. Bloom and J. R. Weir, Jr., "Effect of Neutron Irradiation in the Ductility on Austenitic Stainless Steel," *Nuc. Tech.*, 16 (1972) 45–54.
17. R. L. Fish and C. W. Hunter, "Tensile Properties of Fast-Reactor Irradiated Type 304

Stainless Steel," *ASTM Symposium on the Effects of Radiation on Structural Materials*, ASTM STP 611, (1976), 119.

18. T. Lauritzen, *Stress-Rupture Behavior of Austenitic Steel Tubing: Influence of Cold Work and Effect of Surface Defects*, USAEC Report GEAP-13897, (1972).

19. H. Farrar IV, C. W. Hunter, G. D. Johnson, and E. P. Lippincott, "Helium Profiles Across Fast-Reactor Fuel Pin Cladding," *Trans. ANS*, *23* (1976).

20. C. W. Hunter and G. D. Johnson, "Fuel Adjacency Effects on Fast Reactor Cladding Mechanical Properties," *International Conference on Fast Breeder Reactor Fuel Performance*, ANS/AIME, Monterey, CA, March 1979.

21. J. F. Bates and M. K. Korenko, *MK-8 Equation for Stress-Free Swelling of 20% Cold Worked AISI 316 Stainless Steel*, HEDL-TME 80-8, Hanford Engineering Development Laboratory, May 1980.

22. F. A. Garner, E. R. Gilbert, "Stress Enhanced Swelling of Metals During Irradiation," *ASTM 10th International Symposium on Effects of Radiation on Materials*, Savannah, GA, June 3–5, 1980.

23. D. J. Mazey, J. A. Hudson, and R. S. Nelson, "The Dose Dependence of Void Swelling in AISI 316 Stainless Steel During 20 MeV C + + Irradiation at 525°C," *J. Nuc. Mat.*, *41* (1971) 257.

24. J. L. Bates and M. K. Korenko, *Updated Design Equations for Swelling of 20% CW AISI 316 Stainless Steel*, HEDL-TME 78-3, Hanford Engineering Development Laboratory, January 1978.

25. J. L. Straalsund, H. R. Brager, and J. J. Holmes, *Radiation-Induced Voids in Metals*, J. W. Corbett and L. C. Ianniello, eds. AEC Symposium Series No. 26 (CONF-710601) (1972) 142.

26. J. J. Holmes, "Irradiation-Induced Swelling in Nickel Alloys," *Trans. ANS*, *12* (1969) 117.

27. J. L. Straalsund and E. R. Gilbert, "Development of the Climb Induced Glide Concept to Describe In-Reactor Creep of FCC Materials," *J. of Nuc. Mat.*, *90* (1980) 68–74.

28. L. Leibowitz, E. C. Chang, M. G. Chasanov, R. L. Gibby, C. Kim, A. C. Millunzi, and D. Stahl, *Properties for LMFBR Safety Analysis*, ANL-CEN-RS-76-1, Argonne National Laboratory, March 1976.

29. W. Häfele, D. Faude, E. A. Fischer, and H. J. Laue, "Fast Breeder Reactors," *Annual Review of Nuc. Sci.*, Annual Reviews, Inc., Palo Alto, CA, 1970.

30. J. K. Fink and L. Leibowitz, *Thermophysical Properties of Sodium*, ANL-CEN-RSD-79-1, Argonne National Laboratory, May 1979.

31. V. S. Bhise and C. F. Bonilla, "The Experimental Pressure and Critical Point of Sodium," *Proc. International Conf. Liquid Metal Technology in Energy Production*, Seven Springs, Pa, May 1977. (Also COO-3027-21, NTIS [1976].)

32. S. Das Gupta, "Experimental High Temperature Coefficients of Compressibility and Expansivity of Liquid Sodium and Other Related Properties, D.E.S. Dissertation with C. F. Bonilla, Dept. of Chemical Engineering and Applied Chemistry, Columbia University, Xerox-University Microfilms (1977). Also COO-3027-27, NTIS (1977).

33. G. H. Golden and J. D. Tokar, *Thermophysical Properties of Sodium*, ANL-7323, Argonne National Laboratory, 1967.

34. O. J. Foust, ed., *Sodium-NaK Engineering Handbook*, Vol. 1, Gordon and Breach, New York, 1972, 23.

35. W. K. Anderson and J. S. Theilacker, eds., *Neutron Absorber Materials for Reactor Control*, Washington, D.C., U.S. Government Printing Office, 1962.

36. Neutron Absorber Technology Staff, *A Compilation of Boron Carbide Design Support Data for LMFBR Control Elements*, HEDL-TME 75-19, Hanford Engineering Development Laboratory, February 1975.

37. J. A. Basmajian and A. L. Pitner, "A Correlation for Boron Carbide Helium Release in Fast Reactors," *Trans. ANS*, *26* (1977) 174.

38. D. E. Mahagin and R. E. Dahl, *Nuclear Applications of Boron and the Borides*, HEDL-SA-713, Hanford Engineering Development Laboratory, April 1974. (See also Ref. 30.)

CHAPTER 12

REACTOR PLANT SYSTEMS

12-1 INTRODUCTION

The principal objective of the Fast Breeder Reactor is to generate electricity. This is accomplished by transferring energy from nuclear fission to a steam system to run a turbine-generator. In this chapter we describe the LMFBR systems outside the core which are needed to meet this objective. The main emphasis is on the heat transport system, focusing on the design problems unique to LMFBR's. First, the overall heat transport system is described, including the primary and secondary sodium systems and the various steam cycles in use and proposed. Discussions then follow of the main components in the sodium system—the reactor vessel and reactor tank, sodium pumps, intermediate heat exchangers, and steam generators*.

The four remaining sections address shielding problems peculiar to the LMFBR, refueling considerations, core and coolant system instrumentation, and auxiliary systems necessary for sodium-cooled plants.

A few words are in order to place the topics of this chapter in perspective relative to the other subjects covered in this book. Our treatment of plant systems is very brief; we devote only one chapter to design areas that require an extraordinary share of the development effort in order to bring the FBR to commercial status. In the long run, the successful commercialization of the FBR will depend to a large degree on the successful development of components which operate safely and reliably and are built at a reasonable capital cost.

---

*A more complete discussion of LMFBR systems appeared after the present manuscript was completed. The reader is referred to A. K. Agrawal and M. Khatib-Rahbar, "Dynamic Simulation of LMFBR Systems," *Atomic Energy Review, Vol. 18,* No. 2 (1980) IAEA, Vienna.

12-2 HEAT TRANSPORT SYSTEM

A. LMFBR SYSTEM LAYOUT

The LMFBR heat transport system consists of a primary sodium system, a secondary (or intermediate) sodium system, and a steam system. The secondary system is unique to liquid metal cooled reactors. It is employed to prevent the possibility of contact between radioactive sodium in the primary system (mainly 15 hour half-life ^{24}Na) and water in the steam generator. The main components in the sodium systems are the reactor vessel or tank, primary pumps, intermediate heat exchangers (IHX), secondary pumps, and steam generators.

Pool vs Loop Systems

Two types of primary systems are being used, as described in Chapter 2-3E, the pool and the loop system. In the pool system the entire primary system (i.e., reactor, primary pumps, and IHX's) is located in a large sodium pool in the reactor tank. In the loop system the primary pumps and IHX's are located in cells outside the reactor vessel with interconnecting piping. The heat transport systems for both the loop and pool designs are shown schematically in Fig. 12-1. The system selected for each prototype or demonstration plant design is listed in Table 12-1, as well as the main design parameters for the heat transport systems. The only test reactor to use the pool design is EBR-II; the rest use the loop system, including FFTF, BOR-60, Rapsodie, JOYO, KNK-II, Fermi, Dounreay, FBTR, and PEC.

The term loop applies to a sequential series of components in the heat transport system between the reactor and the turbine, each operating independently of the other loops in the system. In an FBR loop system, each loop consists of a single primary and secondary pump and one or more IHX's and steam generators. For the pool system, the term loop refers to the IHX's and secondary sodium system; there is no requirement that the number of primary pumps and loops be the same, though they frequently are. Most commercial size demonstration plants have four loops; most of the smaller prototype plants have three loops. In Fig. 12-1, only one loop is illustrated. The four loops in the pool reactor Super Phénix (for which there are also four primary pumps) are shown in the plan view in Fig. 12-2. This figure shows two IHX's in the pool for each loop and each primary pump.

Both pool and loop designs have particular advantages, and there are advocates of both concepts. Several *advantages of the pool system* can be

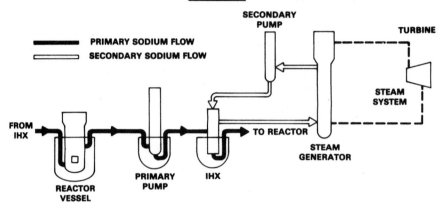

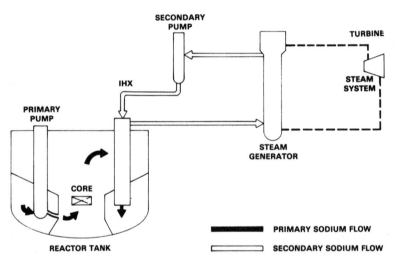

FIGURE 12-1. Heat transport systems.

identified. Leakage in the primary system components and piping does not result in leakage from the primary system and there may be a lower chance that a primary system pipe could be broken. The mass of sodium in the primary system is of the order of three times that of a loop system, thus providing three times the heat capacity. This results in a lower temperature rise in off-normal transients or a longer time to reach boiling if heat sinks are isolated. The large thermal inertia of the pool tends to dampen transient thermal effects in other parts of the system. The cover gas system can be simpler since the only free surface needed is the free surface in the tank

TABLE 12-1 Main Design Parameters for Large Fast Breeder Reactors

	BN-350 (USSR)	Phénix (France)	PFR (U.K.)	SNR-300 (Germany)	MONJU (Japan)	CRBRP (U.S.)	BN-600 (USSR)	Super Phénix (France)	CDFR (U.K.)	SNR-2 (Germany)	BN-1600 (USSR)
Electrical Rating, MW	350 (equiv)	250	250	327	280	375	600	1200	1320	1300	1600
Thermal Power, MW	1000	568	600	770	714	975	1470	3000	3230	3420	4200
System	loop	pool	pool	loop	loop	loop	pool	pool	pool	loop	pool
Number of loops	6	3	3	3	3	3	3	4	6	4	4
Primary Pump Location	cold leg	cold pool	cold pool	hot leg	cold leg	hot leg	cold pool	cold pool	cold pool	hot leg	cold pool
Number of IHX's per loop	2	2	2	3	1	1	2	2	2	2	1
IHX Temperatures:											
Reactor Outlet, °C	500	560	550	546	529	535	550	545	540	540	550
Reactor Inlet, °C	300	400	400	377	397	388	377	395	370	390	350
Secondary Outlet, °C	450	527	540	528	505	502	520	525	510	510	505
Secondary Inlet, °C				335	325	344		345	335	340	310
Steam Generator	integral; bayonet tube evaporator, U-tube superheater	modular: "S"-shaped	separate; U-tube	separate; 2-straight 1-helical	separate; helical coil	separate; hockey stick	separate; straight tube	integral; helical coil	integral; helical coil	integral; helical coil or straight tube	
Steam Cycle	recirculating	once-through (Benson)	recirculating	once-through (Sulzer)	once-through (Benson)	recirculating	once-through (Benson)	once-through (Benson)	once-through	once-through	
Number of Units per Loop:											
Integral Steam Generators	—	—	—	—	—	—	—	1	—	2	
Separate Evaporators	2	12	1	3	1	2	1	—	—	—	
Separate Superheaters	1	12	1	3	1	1	1	—	—	—	
Steam Drums	0	1	1	0	0	1	1	—	—	—	
Moisture Separators	—	—	—	1	0	—	—	—	—	—	
Reheaters	0	12	1	0	0	0	1	0	—	—	
Turbine											
Inlet Pressure, MPa	4.9	16.3	12.8	16.0	12.5	10.0	14.2	18.4	16.0	16.5	14.0
Inlet Temperature, °C	435	510	513	495	483	482	505	490	490	495	500
Number/Rating, MWe	1	1	1/250	1	1	1/434		2/600	2/660	1/1300	2/800
Type	K-100-45	Condensing Tandem Compound	Tandem Compound	Condensing	Tandem Compound	Tandem Compound	K-200-130	Condensing	Tandem Compound	Single Shaft	

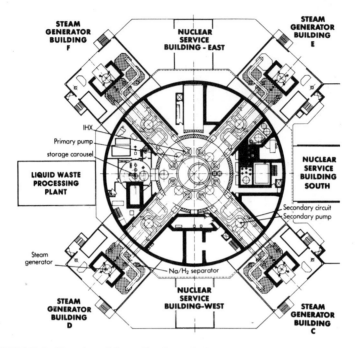

FIGURE 12-2. Plan view of Super Phénix pool, showing the 4 loops, 4 primary pumps, and 8 IHX's.

unless an actively controlled cover gas is used, as in Super Phénix, to reduce reactor vessel wall thermal stresses.

Advantages of the loop system include the following. Maintenance is simpler for the loop system since components can be isolated in cells. This feature also provides for greater flexibility in making system modifications and major maintenance during reactor operation. Less neutron shielding to prevent activation of the secondary sodium is required. The structural design of the vessel head is simpler than the large roof deck of a pool reactor. Greater difference in vertical elevation of the IHX relative to the core enhances natural circulation of a loop relative to that of a pool, and the well defined primary coolant flow path allows a more reliable prediction of natural circulation for a loop-type LMFBR. Tighter coupling (i.e., quicker response to changes) of the steam and secondary sodium systems to the primary sodium system and reactor in a loop design, due to the smaller mass of sodium in the primary system, influences the control and load-following characteristics of the overall heat transport and steam system, but it is unclear whether this involves a net advantage relative to a pool design.

Heat Exchanger Configurations

An important aspect of the heat transport system layout which is illustrated in Fig. 12-1 is the relative axial elevation of all the components. The main components of both the primary and secondary systems are arranged so that the thermal center of each component is above the thermal center of the previous one in the flow cycle. This profile provides natural circulation of the sodium in order to transfer heat from the core to an ultimate heat sink in the event of a loss of pumping capability.

Reactor inlet and outlet temperatures of LMFBR's are generally of the order of 400°C and 550°C, respectively, although these values will vary as designs become more or less conservative. These temperatures then correspond, except for small heat losses, to the temperatures across the primary side of the IHX.

Primary and secondary flows in the IHX are generally countercurrent, with log mean temperature differences between the primary and secondary sodium of the order of 30°C to 40°C. With the single exception of PFR, secondary flow is on the tube side in order to facilitate cleanup of sodium-water reaction products from the steam generator should a major leak occur there. The pressure on the secondary side is higher than on the primary side to avoid leakage of radioactive sodium to the secondary in case of tube leaks (which is another reason for putting the secondary on the tube side).

Steam generators can be categorized as *integral* or *separate*.* In an integral system, evaporation and superheating occur without separation of steam and water between the two processes. For most integral steam generators, both evaporation and superheating take place in an integral unit (i.e., within the same shell), as in Super Phénix. Other integral steam generator systems employ separate components, as in BN-350. In a separate steam generator, evaporation and superheating occur in different units, with steam separation between the two processes. Steam separation usually takes place in a steam drum (or steam separator) or moisture separator which are separate components between the evaporator and superheater. Steam separation, however, can be incorporated as an integral part of the evaporator unit. In some designs more than one evaporator feeds into a single super-

---

*The term "modular" has also been used for systems consisting of several separate modules which are interconnected to perform a single function, as in Phénix. The meaning of the terms integral and separate as applied to steam generator systems is not entirely consistent throughout the LMFBR industry. The definitions described here appear to prevail at the time of publication, but further redefinition may be expected as new variations in LMFBR plant designs evolve.

heater. Steam-water is always on the tube side with sodium on the shell side. Since steam-water pressure is higher than the sodium pressure, in the event of tube leak, steam or water will flow into the sodium instead of the reverse in order to prevent the contamination of the turbine with sodium oxide.

B. STEAM CYCLE

The high temperature of the sodium coolant allows the LMFBR to take advantage of a superheated steam cycle. This allows thermal efficiencies close to 40%, near those of modern day fossil plants and considerably higher than efficiencies of the order of 32% for light water reactors. The use of a saturated steam cycle (as used in LWR's) is also being considered for LMFBR's, perhaps in order to achieve higher reliability at the expense of thermal efficiency. All operating LMFBR's and those under construction in the early 1980s, however, use superheat cycles.

Four steam cycles being used and suggested for LMFBR's are illustrated in Fig. 12-3. All are, of course, modifications of the basic Rankine cycle.

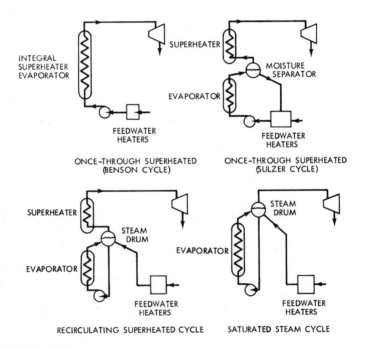

FIGURE 12-3. LMFBR steam cycles.

Two classes of superheat cycles are represented—the once-through cycle and the recirculating cycle. The once-through cycle is further subdivided into the Benson and Sulzer cycles. The Benson cycle employs an integral steam generator, whereas the Sulzer cycle uses separate evaporators and superheaters with a moisture separator in between. The exit quality from the evaporator in the Sulzer cycle is high, e.g., 95% in SNR-300 for which this cycle is used. In contrast, the exit quality from the evaporators in a recirculation superheater cycle is considerably lower, e.g., 50% in the CRBRP design. The moisture separator and the steam drums are shown in Fig. 12-3 as separate units, but they can also be incorporated as an integral part of the evaporator.

C. PLANT CONTROL

Plant control is a complex process that varies with each plant and combination of systems selected. As a consequence, we will introduce only the basic concepts in this overview book. A sense of the factors involved in the control of an LMFBR heat transport system can be derived from outlining the plant control developed for the CRBRP design, as written by members of the CRBRP staff. This description, together with Fig. 12-4, follows.

Plant control is achieved by a two-level system. It provides automatic and manual control of the reactor, heat transfer systems, turbine and auxiliary systems for normal and off-normal operation. The first level of control is *load demand* established by either an automatic signal from the grid load dispatching system or a plant-operator-determined setpoint. The supervisory controller receives the load demand signal and also input of steam temperature, steam pressure and generator output. The controller compares the steam side conditions to the load demand and, in the case of any variation, sends demands to each of the second-level controllers which adjust the reactor power level, system flows and steam supply to meet the new load demand.

Each of the second-level controllers can be operated manually. Manual control is used while the plant is shut down and during startup to bring the plant to 40% power level after which automatic control may be initiated.

The reactor controller receives a demand input of reactor power level from the supervisory controller and input of sodium temperature and flux level in the reactor. The controller compares the flux level with the power demand signal to determine the change required to meet the demand. A trim signal is provided from the steam temperature to the turbine throttle to maintain throttle conditions at a preset temperature value. At the same

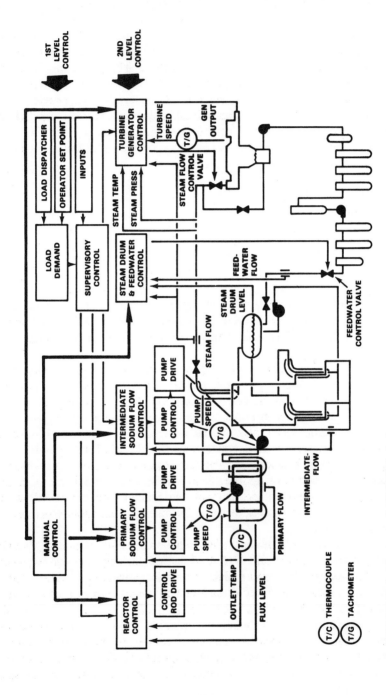

FIGURE 12-4. Plant control (CRBRP design).

time, core outlet temperature information is used to modify the controller output so that core outlet temperature is maintained within predetermined limits. The resulting signal is a demand to the control rod drive mechanism controller to adjust rod position to a point where the difference between power output and power demand is zero.

The primary and intermediate sodium flow controllers receive inputs of flow demand from the supervisory controller and sodium flow in the cold leg of each loop. The controller compares the demand with the flow signal to determine any change required. A trim signal is provided to the intermediate pump flow controller from the turbine throttle to maintain throttle pressure at the preset value. The resulting signal is a demand to the pump controllers which also receive input from tachometers on each of the pumps. Comparison of these inputs results in a demand signal from the pump controllers to the pump drive controls for any changes in pump speed.

The steam drum and feedwater controller does not receive input from the supervisory controller because steam drum level is kept constant for all power levels. The controller receives inputs of flow in the main steam line, steam drum level, and flow in the feedwater line. Comparison of these inputs results in a demand signal to the feedwater control valves which adjust flow to maintain steam drum level.

The turbine generator controller receives inputs of load demand from the supervisory controller, steam flow temperature and pressure, turbine speed and generator output. Flow, temperature and pressure signals are combined to give mass steam flow information. The generator output signal is compared to the load demand to establish any requirement for power change. Comparison of these demands results in a signal from the turbine generator controller to the turbine throttle valve which adjusts steam flow to meet the load demand. The turbine speed signal is used to trim for small variations in turbine speed.

12-3 COMPONENTS

A. REACTOR VESSEL AND REACTOR TANK

The reactor vessel of a loop system (SNR-300) and the reactor tank of a pool system (Super Phénix) are shown in Figs. 12-5 and 12-6.

Reactor Vessel (Loop System)

The reactor vessel in a loop design is a vertical, cylindrical shell with a dome shaped bottom. The vessel is hung at the top from a support ring. The

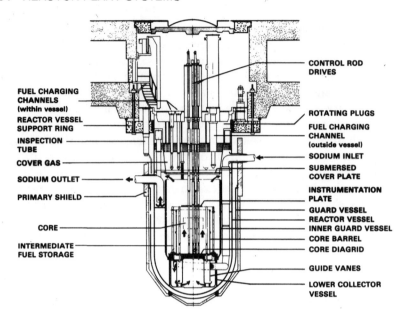

FIGURE 12-5. Reactor vessel for the loop system, SNR-300.

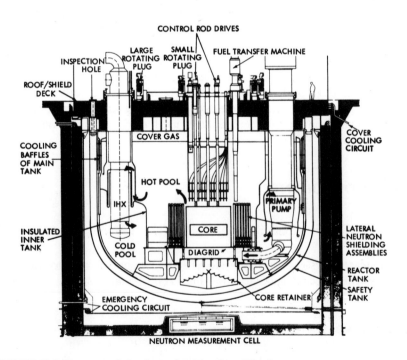

FIGURE 12-6. Reactor tank for the pool system, Super Phénix.

fuel assemblies rest on a core support structure (also called a plate, grid, or diagrid). In SNR-300 the core support structure is attached to an inner guard vessel (not present in CRBRP) which is in turn supported at the bottom by the reactor vessel. In the CRBRP design the core support structure is attached to a core support ledge which is attached to the side of the vessel. Joined to the core support structure (in both SNR-300 and CRBRP) is a core barrel or jacket which separates the sodium flowing through the core, blankets, and radial shielding from the surrounding sodium pool. An inlet flow structure guides the sodium from the inlet plenum to the fuel assemblies; an upper internals structure guides the flow from the assemblies into the outlet plenum. Containers for interim fuel storage during refueling are located outside the core barrel.

A guard vessel is located outside the reactor vessel to protect against any potential loss of sodium from the vessel. The reactor and guard vessel sit in a reactor cavity. Both inlet and outlet sodium pipes enter above the guard vessel so that any pipe rupture within the reactor cavity will not result in sodium loss to the core. In SNR-300 the inlet pipes penetrate the vessel above the outlet pipes, and the inlet pipes run downward between the inner guard vessel and the reactor vessel to the inlet plenum below the core. In the FFTF and the CRBRP design, the inlet pipes run downward between the reactor vessel and the guard vessel and enter the vessel near the bottom. In CRBRP, the inlet and outlet pipes are stainless steel with diameters of 0.6 m and 0.9 m, respectively, and wall thicknesses of 13 mm.

Several centimeters of argon gas cover the sodium (hence, the common name, cover gas), separating the sodium pool from the reactor head (also called closure head, or cover). The head provides access for both control rods and refueling ports and is usually fabricated with several rotating plugs for refueling (discussed in Section 12-5).

Reactor Tank (Pool System)

In considering a structure as large and complex as a pool reactor tank, several questions quickly come to mind. How is the tank supported? How are the hot and cold sodium separated (especially around the IHX's that must join them)? How are the IHX's and primary pumps supported? How is the large roof/shield deck above the tank constructed, supported and insulated?

Since the answers vary to some degree between every plant, a useful way to gain insight into proposed solutions is to focus on a particular design. We shall illustrate the basic design solutions adopted for Super Phénix since it is the first commercial-size pool reactor to be built.

The Super Phénix reactor tank is shown in Fig. 12-6. The tank is stainless steel with a height and diameter of 19.5 m and 21 m, respectively, and a wall thickness of ~50 mm. It is hung from the roof/shield deck that covers the tank.

Further details of the reactor tank and internals are shown in Fig. 12-7. The reactor tank is surrounded by a safety tank to contain sodium in the event of a leak in the reactor tank. Inside the reactor tank is a baffle tank to prevent the reactor tank from exceeding the sodium reactor inlet

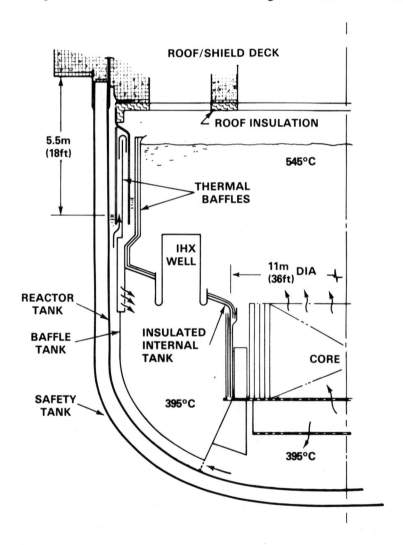

FIGURE 12-7. Primary tank and internals (Super Phénix).

temperature. This is accomplished by providing bypass sodium flow from the inlet plenum to the annular space between the baffle and reactor tanks.

Yet another internal tank is the insulated internal tank which forms both a physical and thermal barrier between the hot and cold sodium. Of particular importance in the design of this tank are the penetrations for the primary pumps and IHX's. These components are supported from the roof/shield deck. Because of the large temperature variations, from room temperature above the deck to the sodium pool temperatures below, the IHX penetrations of the insulated internal tank must accommodate 50 to 70 mm of axial expansion. A bell-jar type seal is used to accomplish this, as shown in Fig. 12-8. The pump fits loosely into a well which extends from the cold sodium to the roof/shield deck so that no seal is required. An added problem for the pumps, however, involves the pipe connection to the inlet plenum; a flexible joint must be designed to accommodate the roughly 50 mm radial and 100 mm axial thermal expansions due to differences between the deck and pump operating temperatures.

The roof/shield deck represents one of the most difficult design problems for the pool concept. The construction of the roof/shield deck is illustrated schematically in Fig. 12-9; it is composed of a steel web structure filled with concrete. This deck is supported by the concrete vault which surrounds the reactor tank, and the deck in turn supports the tank. The deck must also support the primary pumps, IHX's, control rod drives, and fuel handling

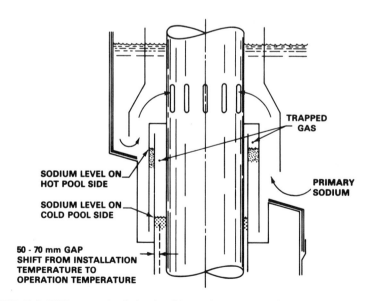

FIGURE 12-8. IHX penetration in insulated internal tank (Super Phénix).

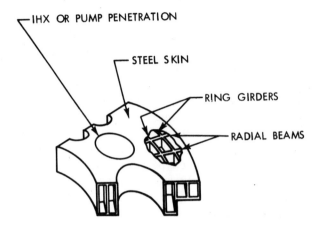

FIGURE 12-9. Section showing construction of roof/shield deck (Super Phénix).

equipment, and must in addition be designed to remain leaktight following potential deformation from a hypothetical core disruptive accident. As in the case of a loop system, the deck is separated from the sodium pool by a cover gas. In addition, the deck must be insulated from the high temperature sodium, and even with insulation some cooling of the deck is required. The Super Phénix insulation is stainless steel gauze sandwiched between stainless steel foil. (Similar insulation is also required to insulate the concrete walls surrounding the reactor tank.)

B. SODIUM PUMPS

Primary and secondary pumps in LMFBR's generally fall into the category of mechanical, vertical-shaft, single-stage, double suction impeller, free-surface centrifugal pumps. Figure 12-10 shows a diagram of the CRBRP primary pump to illustrate the nature of the pumps in general use. The sodium level is just below the thermal shield. Argon cover gas is over the sodium, and this gas is connected to the reactor vessel cover gas through a pressure equalization line.

Choices affecting pump design involve differences between primary pumps for the loop versus pool system, differences between primary and secondary pumps (often minor), and pump location. Important design choices (not discussed here) include seals and bearings and impeller and bypass flow arrangements.

Electromagnetic pumps can also be used for LMFBR's since sodium is an excellent electrical conductor. An electromagnetic pump is being used for the secondary sodium in EBR-II, and they were used in SEFOR and in the primary circuit of the Dounreay Fast Reactor. They are not used in the

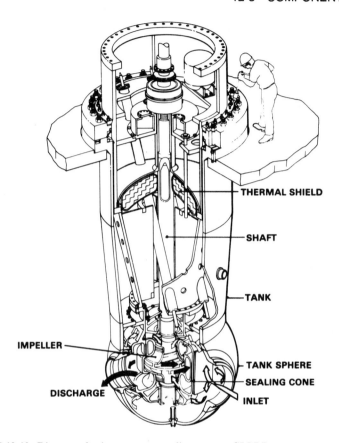

FIGURE 12-10. Diagram of primary system sodium pump (CRBRP).

main loops in large LMFBR power plants, but they are being used in some backup decay heat removal systems, e.g., in SNR-300 and Super Phénix (see Chapter 14-3C).

Only mechanical pumps are being used in large LMFBR's. Testing is underway in the U.S. on the use of a relatively small inducer-type mechanical pumping element in series with a large centrifugal pump to assist in meeting the strict suction requirements of LMFBR centrifugal pumps (discussed below on NPSH).

Pump Location

A classic selection problem for the loop design LMFBR has been the location of the pumps. For the pool design, the primary pump is always located in the cold sodium. For the loop system, however, the location of the primary pump can be either on the hot leg or the cold leg. In all secondary systems, the pump is located in the cold leg both because the advantages of hot-leg location existing for the primary loop are not present in the secondary loop and also because it is important to pressurize the secondary sodium in the IHX in order to force flow from leaking tubes in the direction from the nonradioactive secondary to the radioactive primary.

In the loop design, despite the obvious advantage (to seals and bearings, for example) of placing the primary pump in the colder sodium environment of the cold leg (as in MONJU, BN-350, and most of the early experimental loop reactors), the hot leg has been selected for several loop designs (SNR-300, SNR-2, FFTF, and CRBRP). The main reason for the hot-leg choice involves suction requirements for primary system pumps. These requirements are stringent in order to accommodate the full range of anticipated transients, a subject treated in Reference 1. The basic argument can be explained briefly, however, by examining conditions during normal steady-state operation.

The *net positive suction head* (NPSH) of a pump is the difference between the absolute pressure at the pump suction and the vapor pressure of the fluid pumped. Any pump has, inherent in its design, a minimum required NPSH at any given flow rate to prevent cavitation. The available NPSH must, therefore, always be greater than this minimum value. The available NPSH can be obtained from the relation

$$\text{NPSH} = H_p + H_z - H_l - H_v,$$

where H_p = pressure at the liquid surface of the source from which the pump takes suction (hence, the cover gas pressure for the present application),

H_z = hydrostatic head of the liquid source above the impeller,

H_l = pressure drop (losses) in the piping and equipment upstream of the pump,

H_v = vapor pressure of the fluid at the suction.

The pressure above the sodium surface in the LMFBR primary pumps is equalized with the reactor cover gas pressure, H_p, and need be only slightly above atmospheric in an LMFBR, just enough to prevent inleakage of air or inert gas. The hydrostatic head, H_z, controls the length of the pump shaft. The sodium vapor pressure, H_v, is small, on the order of only 1 kPa for the hot leg sodium.

If the pump is on the hot leg, the pressure drop H_l includes only the losses through a short distance of piping plus vessel exit and pump entrance losses. With the pump on the cold leg, however, H_l must include the pressure drop through the IHX plus additional piping. The IHX pressure drop is generally of the order of 50 to 100 kPa. The only way, therefore, to obtain a comparable NPSH in the cold leg to that in the hot leg is to increase the length of the pump shaft (hence H_z) or to pressurize the argon cover gas in the reactor (H_p)*, or a combination. Another option is to develop a more advanced pump that can operate with a lower NPSH. The difficulty of achieving these options provides the incentive for the hot leg location. The problem is further aggravated as LMFBR plant size increases since the required NPSH for a particular pump increases with increasing rated capacity.

C. INTERMEDIATE HEAT EXCHANGER

Most LMFBR intermediate heat exchangers are of similar design. Illustrations of an IHX for both a pool design (Super Phénix) and a loop design (CRBRP) are provided in Fig. 12-11.

Except for BN-350[†], the IHX's in all of the prototype and demonstration plants are vertical, counter flow, shell-and-tube heat exchangers with basically straight tubes. With the exception of PFR, the secondary sodium enters at the top and flows to the bottom through a central downcomer; flow is then reversed and returns upward through the tubes. Primary sodium generally flows downward on the shell side and exits at the bottom. Principal differences in IHX's between the loop and pool systems appear at the entrance and exit flow nozzles.

The tube bundle must be mounted to allow for differential thermal expansion between the tubes and the shell. To accomplish this, the lower tube sheet is allowed to float and, therefore, is supported by the tube bundle. The tubes are supported by the upper tube sheet. The CRBRP design shows a flexible bellows at the top of the downcomer to allow differential expansion between the downcomer and the tube bundle. An alternate method, used in FFTF, for accommodating this differential expansion is to design a bend in part of the tube bundle. A sine wave bend is used in PFR and CDFR.

Each of the eight IHX's in Super Phénix includes 5380 tubes, 14 mm outside diameter, 12 mm inside diameter, and 6.5 m length. A remotely operated machine has been developed to seal off tubes that leak.

Both 316 and 304 stainless steel are generally used in the IHX components.

---

*Pressurizing the cover gas increases the potential for leakage of radioactive gas through the cover seals.

[†]BN-350 uses U-tubes.

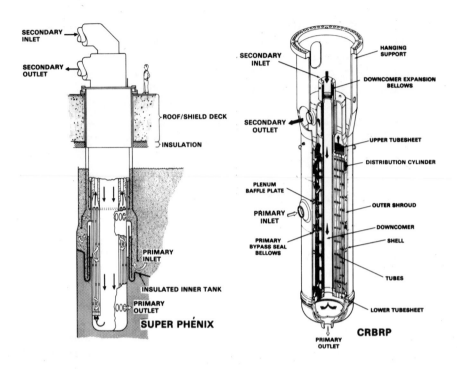

FIGURE 12-11. Intermediate heat exchanger designs for a pool system (Super Phénix) and for a loop system (CRBRP).

D. STEAM GENERATORS

Steam generators can be integral or separate, as discussed in Section 12-2A. They can use tubes that are straight, helical, U-tube, or hockey-stick shaped. Either single wall or double wall tubes can be used. Steam generator characteristics for the prototype and demonstration plants were given in Table 12-1.

The choice of integral versus separate involves the steam cycle selection, as discussed in Section 12-2B. The general shapes of the temperature distributions in an integral unit are shown in Fig. 12-12. After boiling is complete, the steam can be superheated close to the inlet sodium temperature.

In separate steam generators the subcooled heating and boiling take place in the evaporator, while the superheating occurs in the superheater. In the separate evaporator, complete evaporation does not occur. The CRBRP evaporators, as an example of a recirculating cycle, were designed for 50% exit quality. The exit quality for the SNR-300 evaporators, which operate on

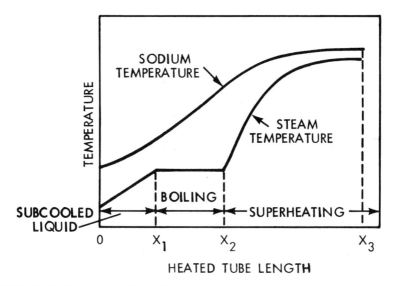

FIGURE 12-12. Temperature distributions in an integral superheater (based on Figure 7-8, Ref. 2).

the Sulzer once-through cycle, is 95%. In both cases the steam is separated from the liquid before entering the superheater. As can be observed from Table 12-1, both integral and separate systems have been widely used.

Several designs for steam generators have evolved—all having particular advantages and disadvantages. A fundamental consideration in each design is the method for accommodating thermal expansion. Figure 12-13 contains examples of the basic configurations capable of providing such accommodation: helical coil, U-tube, and hockey sticks. The straight tube configuration, also shown in Fig. 12-13, requires special provisions to accommodate thermal expansion—similar to those described for the IHX. Helical coils are used in Super Phénix (France), CDFR (U.K.), SNR-2 (Germany), and MONJU (Japan). U-tubes are used in PFR (U.K.) and in the BN-350 (USSR) superheaters. Straight tubes are employed in BN-600 (USSR) as well as in two loops of the SNR-300 (Germany). A third loop of SNR-300 incorporates a helical coil steam generator. Either helical coils or straight tubes will be used in SNR-2. In the United States, the hockey stick was selected and tested for the CRBRP design, and development of both helical coil and straight tube designs are underway. Another type of tube is used in the BN-350 evaporators called bayonet tubes. These tubes have a central tube surrounded by an annulus. Water flows down the central tube and a water-steam mixture flows back up the annulus.

Tube integrity is far more important for the LMFBR steam generator than for the LWR due to the potential chemical reaction between sodium and water. Early consideration was given to the use of double wall tubes in

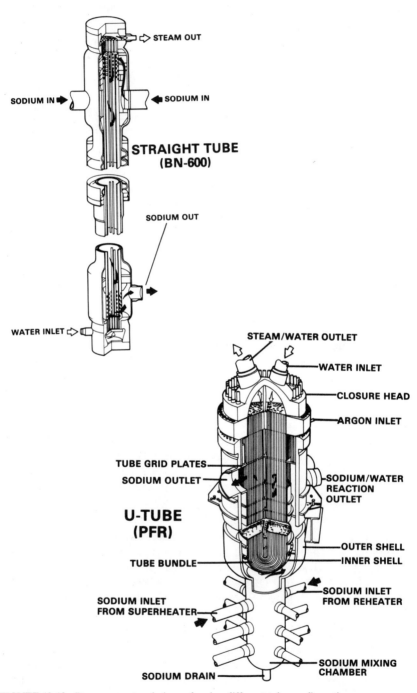

FIGURE 12-13. Steam generator designs, showing different tube configurations.

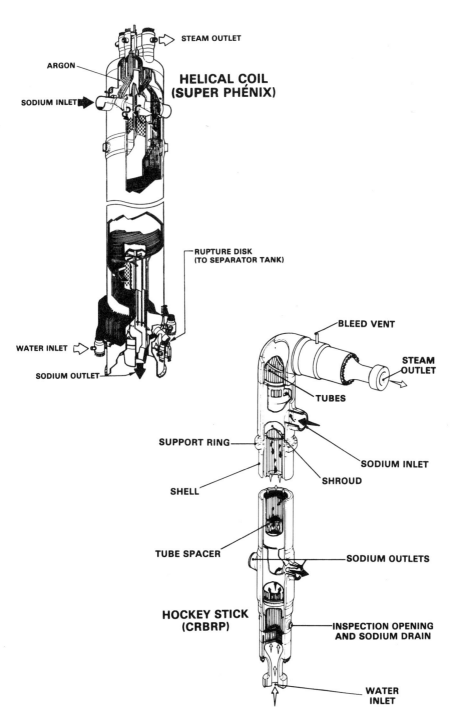

STEAM OUTLET

ARGON

SODIUM INLET

**HELICAL COIL
(SUPER PHÉNIX)**

RUPTURE DISK
(TO SEPARATOR TANK)

WATER INLET

SODIUM OUTLET

BLEED VENT

STEAM
OUTLET

TUBES

SUPPORT RING

SODIUM INLET

SHROUD

SHELL

TUBE SPACER

SODIUM OUTLETS

**HOCKEY STICK
(CRBRP)**

INSPECTION OPENING
AND SODIUM DRAIN

WATER
INLET

FIGURE 12-13 (cont'd). Steam generator designs, showing different tube configurations.

475

which leaks could be detected from gas conditions between the tube walls. Double wall tubes are used, for example, in EBR-II and Dounreay Fast Reactor steam generators. When experiments demonstrated that sodium-water reactions could be adequately contained, however, the emphasis shifted to the simpler single wall tube design, and all prototype and demonstration plants now use single wall tubes. Some interest in double wall tubes is being revived with the argument that they might lead to greater steam generator reliability which might be especially important for the early generation plants. While experience with LMFBR steam generators has been reasonably encouraging, BN-350, PFR, and the Fermi Reactor all experienced difficulties with steam generator leaks.

Pressure relief systems sealed by rupture discs are incorporated in the steam generator design to accommodate pressures resulting from potential sodium-water reactions. A sodium-water reaction outlet to a separator tank is shown for the Super Phénix and PFR steam generators in Fig. 12-13; relief lines exist on the other designs but are not shown.

Most steam generators, both tubes and shell, are made of ferritic steel containing $2\frac{1}{4}\%$ chromium and 1% molybdenum, a material selected to minimize chloride stress corrosion. In some cases this is stabilized with 1% niobium to reduce carbon loss to the sodium. Exceptions to this are the selection of Incoloy® 800 for Super Phénix and the use of austenitic stainless steel for the original PFR superheaters and reheaters.

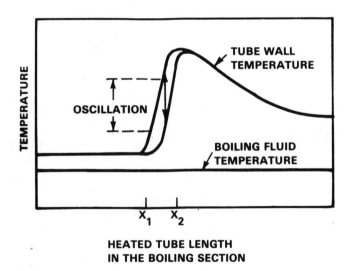

**HEATED TUBE LENGTH
IN THE BOILING SECTION**

FIGURE 12-14. Instability and rise in wall temperature at the transition from nucleate to film boiling (based on Figure 7-15, Ref. 2).

An important condition in steam generator design is the transition between nucleate boiling and film boiling, or the departure from nucleate boiling (DNB). At this point the temperature of the tube wall rises sharply and an instability in tube wall temperature occurs in the transition zone. This temperature behavior is illustrated schematically in Fig. 12-14. The transition occurs between x_1 and x_2; in this region the tube wall is intermittently in contact with water or steam, and the wall temperature oscillates rapidly. Fluctuations of this kind, if of too large a magnitude, could cause thermal fatigue of the tube or structural changes that might enhance waterside corrosion. These and other considerations regarding temperature distributions, dryout, and heat transfer correlations are discussed in Reference 2.

12-4 SHIELDING

Greater attention must be given to shielding design and analysis for fast reactors than for thermal reactors. The high energy neutron flux in an FBR is considerably higher than in an LWR, and high energy neutron leakage from the core is greater. Although the neutron production rates for an FBR and LWR of the same power level are comparable, the power density (kW/litre) is higher for the FBR. Also, the neutrons in an LWR are slowed down to thermal energies close to the fission source so that the high energy leakage source for irradiation of surrounding structures is relatively low for an LWR compared to that for an FBR.

LMFBR shielding can be discussed according to areas which require extensive shielding design and analysis. For both pool and loop designs, these areas include in-vessel radial shielding, the closure head assembly and its penetrations, and neutron flux monitors. For pool designs, special attention is required for the intermediate heat exchangers where secondary sodium can be activated by neutrons. For loop designs, other key areas include the reactor vessel support area and the primary heat transport system pipeways.

Shielding in these areas will be discussed briefly here. Other areas that require shielding, but which will not be discussed, include auxiliary piping penetrations, heating and venting system penetrations, shielding for cover gas and coolant purification systems, fuel handling equipment shielding, and biological shielding for areas that require personnel access during normal, full power operation.

Information on FFTF and CRBRP shielding is reported in References 3 and 4.

A. COOLANT AND COVER GAS RADIOACTIVITY

Before describing the shielding systems in an LMFBR, radioactivity of the coolant and cover gas will be discussed. The use of sodium as the coolant in the LMFBR introduces shielding problems different from those in the LWR as a result of neutron activation of the sodium. Sodium in nature is composed entirely of sodium-23. The (n, γ) reaction in sodium produces radioactive ^{24}Na which has a 15.0-hour half life and emits both a 1.4 and 2.8-MeV gamma with decay. An $(n, 2n)$ threshold reaction also occurs, producing ^{22}Na which has a 2.6-year half life and emits a 1.3 MeV gamma. During operation ^{24}Na is the dominant activation product, and shielding against gammas from this source in the primary sodium is one of the important shielding problems in LMFBR design. The ^{22}Na activity becomes the dominant activity in the sodium ~ 10 days after shutdown. For maintenance of primary pumps and IHX's, however, radioactivity from corrosion products (discussed in Chapter 11-4E) becomes the most important radioactive source. The calculated ^{24}Na specific activity in the CRBRP primary sodium is 30 Ci/kg (based on a primary system sodium inventory of 6.4×10^5 kg). For FFTF, the calculated ^{24}Na activity is 11 Ci/kg. For an early (1968) General Electric pool design,[5] the calculated ^{24}Na activity was 18 Ci/kg (based on a primary system sodium inventory of 1.3×10^6 kg). The calculated CRBRP ^{22}Na specific activity after 30 years of operation is 3.5 mCi/kg. The corresponding FFTF value for ^{22}Na is 1 mCi/kg.

Another radiation source in an LMFBR is the reactor cover gas. Activation of impurities in the sodium and direct activation of ^{40}A to ^{41}A contribute to activity in the cover gas. Even ^{23}Ne, appears from an n, p reaction with ^{23}Na, but its half life is short (38 s). The main design

TABLE 12-2 FFTF Design Basis Cover Gas Activity

Isotope	Activity (Ci/m³)
Xe-131 m*	5.43×10^{-1}
Xe-133 m	1.47×10^{1}
Xe-133	2.67×10^{2}
Xe-135	1.26×10^{3}
Kr-83 m	6.82×10^{1}
Kr-85 m	1.34×10^{2}
Kr-85	9.30×10^{-9}
Kr-87	1.80×10^{2}
Kr-88	2.64×10^{2}

*The m designates the metastable state.

requirement, however, is to permit reactor operation with leakage occurring in a specified fraction of the fuel pins. For FFTF, this fraction of defective pins was set at 1%. Calculated activity due to fission gas in the reactor cover gas for these design-basis conditions is given in Table 12-2. Since failed fuel will likely never approach 1%, actual activities will likely be far below these levels.

B. IN-VESSEL SHIELDING

In-vessel radial shielding is required (1) to prevent excessive radiation damage to structural materials which must remain in the reactor for the lifetime of the plant, and (2) to protect the vessel itself. Examples of in-vessel structures are the core barrel and the core restraint systems. Axial shielding below the core is needed to protect the core support structure. The shielding design must assure that permanent components have an end-of-life ductility consistent with a threshold criterion for brittle fracture. In both the FFTF and the CRBRP design, the threshold of ductility was chosen to be 10% total elongation, a level that assures a ductile mode of deformation up to failure and permits conventional structural analysis methods and criteria to be used in design.

The radial blanket serves as the first shield between the core and the radial structures.* Beyond the blanket are located removable radial shielding (RRS) assemblies. The arrangement of these assemblies for an early CRBRP homogeneous core design is shown in Fig. 12-15. A corresponding elevation view was shown in Fig. 8-21. The RRS assemblies contain rods of stainless steel or of nickel based alloys compatible with sodium, such as Inconel®, as illustrated in Fig. 8-21. Nickel has a high inelastic scattering cross section and is particularly effective in degrading the energy of fast neutrons. Iron is also effective, though less effective than nickel. Stainless steel is less expensive, however, than Inconel®. The material choice for CRBRP was narrowed to 316 stainless steel and Inconel 600®. The shielding rods in the CRBRP design extend from the bottom of the lower axial blanket to the approximate top of the upper axial blanket. The shielding is in the form of rod bundles in order to allow cooling of the shield by sodium. The main heat sources in the shield are gammas from the core and blanket and gammas generated in the shield itself from both inelastic scattering and neutron capture reactions. In FFTF, the radial reflectors are made up of

---

*FFTF has no blanket but uses removable radial reflectors to protect the fixed radial shield and the radial support structure and core barrel.

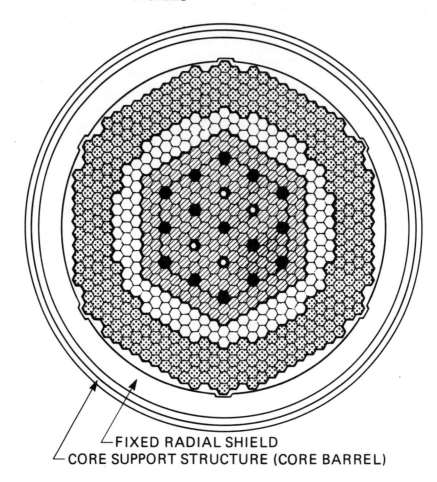

FIXED RADIAL SHIELD
CORE SUPPORT STRUCTURE (CORE BARREL)

CORE ASSEMBLY TYPE	NUMBER OF ASSEMBLIES
FUEL ASSEMBLIES	198
BLANKET ASSEMBLIES	150
RADIAL SHIELD ASSEMBLIES	324
PRIMARY CONTROL ASSEMBLIES	15
SECONDARY CONTROL ASSEMBLIES	4

FIGURE 12-15. Plan view of CRBRP reactor core layout (early homogeneous core design).

hexagonal blocks of Inconel® that are bolted together. Holes through the blocks provide cooling passages.

In the CRBRP design, the removable radial shield assemblies are surrounded by a fixed radial shield. The fixed radial shield is an annulus of 316 stainless steel 0.146 m thick, which can experience relatively high fluence since it is not a load bearing component. The FFTF fixed radial shield is formed from flat plates to create a 12-sided shield between the core and the core barrel.

In the pool design, graphite is usually incorporated into the radial shielding to moderate neutrons and allow them to be absorbed before reaching the IHX. Local B_4C shielding is used near each IHX and primary pump in order to reduce activation of the structural materials, thereby allowing maintenance as well as reducing secondary sodium activation by thermal neutrons in the IHX's.

An axial shield is located below the lower axial blanket to protect the core support structure and lower inlet modules. In CRBRP, this shielding consists of a 0.51 m long 316 stainless steel shield block in each fuel, blanket, and control assembly. In FFTF, the lower shield and inlet orifice assembly are 0.54 m long. No special upper axial shielding is required for either FFTF or CRBRP because of the shielding provided by the upper sodium pool.

The most challenging problem to be resolved in the CRBRP in-vessel shielding analysis was the prediction of neutron streaming. Pathways for streaming, that influence the analysis, include the clearance gaps required in the design of the fixed radial shield, the cooling channels of the axial shielding provided in each core assembly, the fission gas plenum of each core assembly, and interfaces between in-vessel components.

C. REACTOR ENCLOSURE SYSTEM SHIELDING

Key areas for shielding of the reactor enclosure system include component penetrations and interfaces in the closure head assembly, the reactor vessel support area, and the ex-vessel flux monitors (in the reactor cavity). These areas are illustrated in Fig. 12-16 for CRBRP, typical of a loop design. For a pool design, neutron flux levels at the tank support area and reactor cavity walls are lower than for the loop design so that the shielding problems in these areas are less severe.

We describe here the design solutions in these areas for the CRBRP design as being representative of the types of shielding problems encountered.

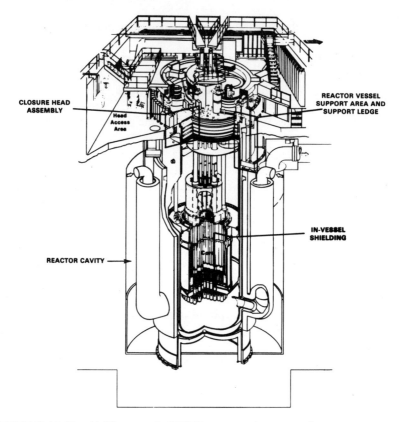

CLOSURE HEAD
ASSEMBLY

Head
Access
Area

REACTOR VESSEL
SUPPORT AREA AND
SUPPORT LEDGE

IN-VESSEL
SHIELDING

REACTOR CAVITY

FIGURE 12-16. Key shielding areas in CRBRP reactor and reactor enclosure system.

Closure Head Assembly Shielding

The CRBRP closure head assembly (CHA) is shown in Fig. 12-17. The main penetrations involve refueling components, control rod drive mechanisms, and upper internals jacking mechanisms. Radiation source terms which affect the CHA shielding design include neutron and gamma streaming up the stepped annuli of the CHA penetrations or component interfaces, the radioactive cover gas below and in the CHA penetrations, and neutron and gamma penetration through the CHA bulk shielding.

The sodium filled dip seals around the refueling plugs shown in Fig. 12-17 presents the major shield design problem for the closure head assembly. These seals form the barrier for the cover gas in the CHA rotating plug annuli. CRBRP was designed to operate with 1% failed fuel. Radioactive

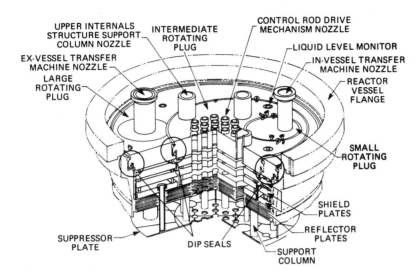

UPPER INTERNALS
STRUCTURE SUPPORT
COLUMN NOZZLE

INTERMEDIATE
ROTATING
PLUG

CONTROL ROD DRIVE
MECHANISM NOZZLE

EX-VESSEL TRANSFER
MACHINE NOZZLE

LIQUID LEVEL MONITOR

LARGE
ROTATING
PLUG

IN-VESSEL TRANSFER
MACHINE NOZZLE

REACTOR
VESSEL
FLANGE

SMALL
ROTATING
PLUG

SHIELD
PLATES

SUPPRESSOR
PLATE

DIP SEALS

REFLECTOR
PLATES

SUPPORT
COLUMN

FIGURE 12-17. Closure head assembly configuration (CRBRP).

fission product gases in the cover gas from this failed fuel requires ~ 0.3 m of steel shielding for personnel access to the head access area. Dip-seal tradeoff studies resulted in the location of the seals in the closure head as shown in Fig. 12-17. Radiation through these dip seals is the largest contributor to the dose rate in the head access area.

Reactor Vessel Support Area Shielding

Shielding the reactor vessel support area created a significant design problem for both the FFTF and CRBRP. The radiation source is predominantly due to streaming in the reactor cavity. In both the FFTF and the CRBRP design, a canned B_4C radiological shield is placed at the lower elevation of the support ledge to stop thermal neutrons, and a carbon steel collar reduces the streaming gap at the vessel flange elevation. A concrete shield ring above the support ledge reduces radiation streaming into the head access area.

Ex-Vessel Source Range Flux Monitor

The Source Range Flux Monitor (SRFM) which monitors the core during shutdown and refueling is located in the reactor cavity for CRBRP. The neutron flux from the core which reaches the flux monitors must be great enough to allow monitoring of changes in core subcriticality, while neutron fluxes from extraneous sources and gamma dose rates must be sufficiently low. These objectives are accomplished by shielding at the SRFM which consists of a graphite moderator block of 0.51 m by 0.63 m surrounded by lead and B$_4$C background shields. Neutron background, due to fuel-in-transfer or fuel storage in the fuel transfer and storage assembly, is reduced by means of B$_4$C shields in the reactor cavity. The gamma background at the SRFM is reduced to acceptable levels by surrounding the moderator block with lead to reduce gamma levels from the vessel, guard vessel, and sodium, and by the utilization of a high purity aluminum alloy as the structural material for the SRFM to minimize its neutron activation gamma background.

D. HEAT TRANSPORT SYSTEM (LOOP DESIGN)

The intermediate heat exchanger must be shielded from neutrons in order to prevent activation of the secondary sodium. In CRBRP, secondary sodium activation was to be held below 0.07 μCi/kg. Ordinary concrete structural walls of the equipment cells provide the bulk shielding. Considerable design effort was required to reduce neutron streaming through the pipeways into the cells.

Delayed neutron monitors for detecting fuel cladding failure are also located in the CRBRP heat transport piping system. Hence, the background neutron flux levels at these monitors must be minimized.

A problem encountered in the CRBRP shielding design was the photoneutron production in the concrete cell walls. Those photoneutrons are generated by the interaction of gamma rays from ^{24}Na in the primary coolant pipes with deuterium in the concrete. Since more than 80% of the neutron background at the delayed neutron monitors was attributed to photoneutrons from the concrete, non-hydrogeneous materials were specified for neutron background shielding around the monitors.

E. SHIELDING METHODS

Neutron flux distributions and radiation sources near the reactor core can be obtained from one, two, and three-dimensional diffusion theory calculations. For neutron and gamma dose rates at large distances from the core and external to the vessel, however, transport theory solutions are required. Special complex techniques, combined with transport theory, are required for streaming calculations; and it is often required that streaming calculations be verified by experiment when design margins are low.

For the CRBRP and FFTF shielding analysis, 40–60 neutron groups were used in discrete-ordinate transport theory calculations. Forward biased quadrature sets containing 100–166 angles were used to solve streaming problems, and Monte Carlo transport methods using albedo scatter data, together with experiments, were developed to verify the design calculations.

12-5 REFUELING

Fuel handling involves two phases: (1) exchanging spent fuel assemblies with fresh assemblies in the reactor, and (2) receiving new fuel at the power plant and shipping spent fuel from the plant. Both phases will be discussed in this section. Safety criteria and design features for fuel handling and storage systems are discussed in Chapter 14-6C.

The principal components and building arrangement for both pool and loop LMFBR designs are shown in Figs. 12-18 and 12-19. The pool design is for Super Phénix. The loop design is one developed by Atomics International (AI) for a demonstration size plant.[6] Refueling designs for liquid metal reactors are reviewed in Reference 7. The Phénix refueling system is described in Reference 8.

Reactor refueling (phase 1) is done with the reactor shut down. The components and cells are arranged with ex-vessel storage outside of the

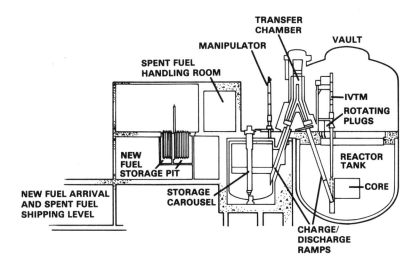

FIGURE 12-18. Refueling system for Super Phénix (pool design).

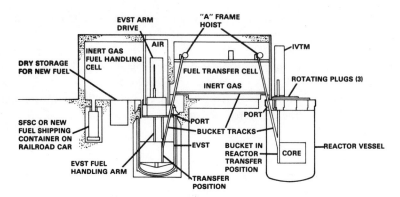

FIGURE 12-19. Refueling system for a loop design (Atomics International Design, Ref. 6).

Reactor Containment Building so that receiving and shipping (phase 2) can be done while the reactor is operating.

A. REACTOR REFUELING

Refueling normally occurs once a year and usually consists of replacing one third of the core fuel at each refueling shutdown. Refueling systems are designed to complete normal refueling in less than two weeks.

Assemblies that must be replaced periodically include core fuel, radial blanket, control, and shielding assemblies. In contrast to LWR's, refueling of LMFBR's is done without removing the head of the reactor vessel. This technique is called *under-the-plug-refueling*.

As noted in Figs. 12-18 and 12-19, there are three areas involved in the refueling process: the reactor vessel or tank, the fuel transfer cell (FTC) or transfer chamber, and the ex-vessel storage tank (EVST) or storage carousel. An In-Vessel Transfer Machine (IVTM) transfers fuel inside the reactor vessel. Fuel is transferred between the vessel and the EVST in a transfer bucket by means of an "A-frame" hoist. Transfer ports are located between the reactor vessel and the FTC and between the EVST and the FTC. A fuel handling arm, or manipulator, transfers fuel in the EVST. In the AI design, the EVST can store up to one refueling batch of new fuel plus an entire core load of spent fuel.

Fuel remains under sodium throughout the fuel transfer process. Consider the sequence of replacing a spent fuel assembly with a new assembly, starting with the new assembly in the EVST. We will illustrate this sequence with details from the AI design; different designs will vary in detail but will be similar in general.

The new assembly is lifted out of its storage position in the EVST by the EVST fuel handling arm. The assembly (under sodium) is transferred to and placed in the transfer bucket. The bucket has space for two assemblies, one new and one spent. At this point the bucket contains only the one new fuel assembly, in a vertical position. The bucket is then hoisted at an angle through a fuel transfer port and guided by tracks up into the FTC. It is then guided to the reactor vessel and down through a second fuel transfer port into the reactor vessel and placed in a vertical position in the shielding region outside the reactor core.

The IVTM is then moved to a position directly above the spent fuel assembly to be replaced and the assembly is grappled. The spent assembly is raised above the remaining assemblies and transferred through the sodium pool to the open space in the transfer bucket, into which it is then lowered. The fresh assembly is next withdrawn from the bucket by the IVTM and transferred to the position in the reactor from which the spent assembly was just removed. The spent assembly is then returned through the FTC to the EVST, and the process is ready to be repeated for the next new fuel assembly.

The FTC has an inert gas atmosphere. The inert gas is generally the same as the reactor cover gas; hence, generally argon. An "A-frame" hoist is used to guide the transfer bucket into the FTC, or transfer chamber. Note that in the pool design, the space provided by the large reactor tank allows the apex of the A-frame hoist to be low enough to be located in the transfer chamber, thus eliminating the lateral movement of the hoist shown for the loop design. The angle in the loop design, with its small diameter reactor vessel, is too steep for the apex to be in the FTC.

The bucket contains sodium during the transfer of both new and spent fuel. In the AI design, a siphon exists at the top of the bucket that lowers the sodium level to 150 mm below the top of the bucket as it is hoisted out of either sodium pool so that sodium does not spill in the FTC.

Refueling can begin about two days after reactor shutdown since a spent fuel assembly will have a decay power level of the order of 30 to 40 kW by this time. During the transfer process through the FTC, the transfer bucket is out of a sodium pool in the AI design for about 10 minutes. During this time the sodium temperature in the bucket will be increased by heat from a spent fuel assembly by about 20°C. If the bucket becomes stuck in the FTC, the decay heat will be transferred from the bucket surface to the cell walls by radiation. The bucket temperature will rise to a maximum of $\sim 500°C$ in 3 hours or more. The heat is then transferred through the cell walls to the air in the building. If the bucket gets stuck in a transfer port, forced convection is required to maintain the bucket temperature at 500°C.

The large LMFBR designs use a rotating plug concept for refueling in which several rotating plugs are located in the reactor head, and the IVTM is mounted on the smallest plug. Usually three rotating plugs are used. These plugs are illustrated in Fig. 12-17 for the CRBRP. Note that the largest plug is concentric with the reactor vessel flange but that the smaller two plugs are eccentric. Each plug can rotate independently so that the IVTM can move to a position directly above any assembly in the reactor.

In the EVST, only one rotating plug is used in the AI design, and an arm under the plug is used to transfer assemblies. In Super Phénix the assemblies are rotated on a carousel rather than having a rotating plug in the cover.

The lucid logic of the refueling designs described should not give the reader the impression that these designs were easy to develop and are the obvious or only natural way to accomplish refueling in an LMFBR. These designs are the result of long-term and still on-going evolution in design, with frequent ingenious mechanical innovations. Other refueling schemes are possible for the LMFBR and, in fact, have been used. For example, FFTF does not use three rotating plugs of varying sizes, but instead uses three small independent rotating plugs with under-the-head arms to transfer fuel, each of which can cover one third of the assemblies. In SEFOR an entirely different, and simpler, system was used. SEFOR had an inerted (argon) refueling cell over the reactor vessel. During refueling the vessel cover was removed (as in LWR refueling) and fuel was withdrawn up into the refueling cell and transferred to a storage tank. Irradiation was low enough in SEFOR that natural convection cooling by argon was sufficient, which made this simple refueling scheme particularly attractive for this case.

B. RECEIVING AND SHIPPING

Receiving and shipping is accomplished in a Fuel Handling Building which houses a new fuel arrival-spent fuel shipping area, a new fuel handling area, and a spent fuel handling area. All of these areas cannot be illustrated adequately in Figs. 12-18 and 12-19 because they are located at different azimuthal positions around the reactor.

New fuel can be received and safely stored in a dry storage tank while the reactor is operating. This fuel can then be moved to the EVST before the reactor is shut down for refueling.

Spent fuel is moved from the EVST during reactor operation into the spent fuel handling area and later placed in a shielded spent fuel shipping cask (SFSC). Low power fuel might be placed in a gas atmosphere in the SFSC for shipping. Several options are available for the higher power fuel assemblies. They could be shipped in sodium if reprocessing is to occur on site. For reprocessing away from the site, they could be stored long enough to allow shipment in a gas atmosphere SFSC, or they could be shipped in a forced convection gas-cooled SFSC.

12-6 INSTRUMENTATION

All nuclear reactor systems incorporate a high degree of instrumentation in order to provide the continued monitoring features necessary for both plant control and scientific data collection. Much of this instrumentation, such as radiation monitoring, is common to all reactor types and need not be considered in this text. However, the presence of a liquid metal poses a few instrumentation challenges which are unique to the LMFBR system. These challenges constitute the basis for the present section.

A. CORE PARAMETER MONITORING

Variables such as flux, temperature, flow, and pressure must be determined in any nuclear reactor, but the sodium environment requires measurement techniques somewhat different from those of light water systems.

Flux

It was noted in Section 12-4C that the flux monitoring for a typical LMFBR system consists of a group of neutron detectors located in the reactor cavity external to the reactor vessel. In-core or in-vessel detectors may be used for initial startup operation, but the neutron source from spontaneous fission of ^{240}Pu is normally present which is strong enough to activate such remotely located detectors. An appreciable neutron source is also possible from ^{242}Am, especially if recycle fuel is used.

Figure 12-20 shows a flux monitoring set for the CRBRP design. For this design, the highly sensitive BF_3 detectors are used for low power operation (startup, refueling), U-235 fission chambers are employed for mid-range operations, and compensated ion chambers are used for power range measurements. Three sets, identical to that shown in Fig. 12-20, provide an overlapping range of flux monitoring to continuously record neutron flux from shutdown to more than full power. Electrical signals from these detectors, which are proportional to reactor power, are used for both reactor control and the plant protection system (PPS). Such signals also feed the data logging system and provide annunciator trips in the control room for out-of-limit conditions.

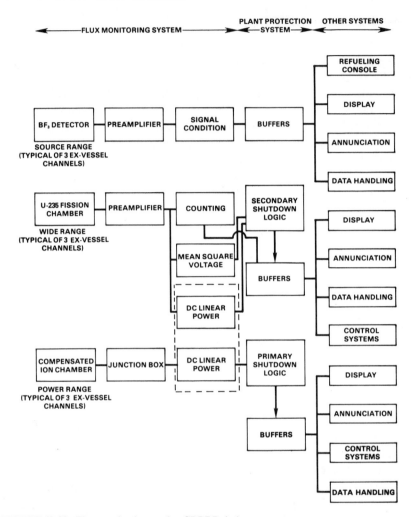

FIGURE 12-20. Flux monitoring set for CRBRP design.

Temperature

Sodium temperatures must be measured routinely throughout the primary and secondary circuits to calculate thermal power and determine loop operating conditions. Two types of detectors are in common use: Resistance Temperature Detectors (RTD's) and thermocouples. The RTD's provide a highly accurate and reliable measurement to ensure that the plant is operating within design limits. The sensor for such a device typically

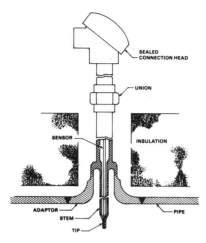

FIGURE 12-21. A typical thermowell assembly incorporating a Resistance Temperature Detector (RTD) for measuring sodium temperatures.

consists of a double element of platinum contained within a sheath which is spring loaded against the bottom of a thermowell, as illustrated in Fig. 12-21. Insertion of this sensor raises the possibility of sodium leakage from the penetration in the event of thermowell failure. Whereas well failures in such devices appear to be rare, a backup cable penetration seal (as shown by the sealed connection head in Fig. 12-21) can be provided to prevent such leakage.

Flow

Coolant flow measurements must be made to complete the thermal power calculations and loop operating characteristics. Both the standard Venturi flow meter and a magnetic flow meter are often used in liquid metal systems. The Venturi meter is highly accurate, but suffers from a response time that is often too slow for control system and PPS use. The magnetic flow meter, on the other hand, tends to be less accurate but exhibits a rapid response time. When the two are used in series, the Venturi meter can be used to provide in-place calibration* of the rapid-response magnetic flow meter.

---

*It is possible to calibrate both types of meters by activating the sodium with a pulsed neutron device and using time-of-flight recording techniques. This procedure was successfully employed in FFTF.

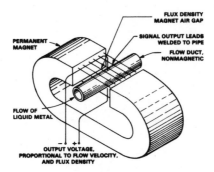

FIGURE 12-22. Schematic of permanent magnet flow meter.

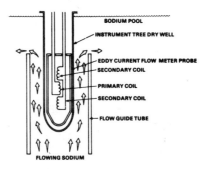

FIGURE 12-23. Simplified diagram of eddy current flow meter.

The magnetic flow meter is unique to liquid metal systems. It is a direct result of the electric properties of the liquid metal coolant. Figure 12-22 contains a schematic of a permanent magnet flow meter.

Another type of flow meter made possible by the unique properties of the liquid metal is an eddy current flow meter. Fig. 12-23 contains a simplified diagram of such a device used in measuring the rate of coolant discharge from an assembly into the upper sodium pool.

Pressure

Liquid pressure measurements are normally made by routing a small column of the high pressure liquid onto one side of a sensing diaphragm. This causes a complication when measuring sodium pressure because sodium solidifies well above room temperature. Trace heating could be provided to assure liquid sodium conditions, but this becomes unreliable for many

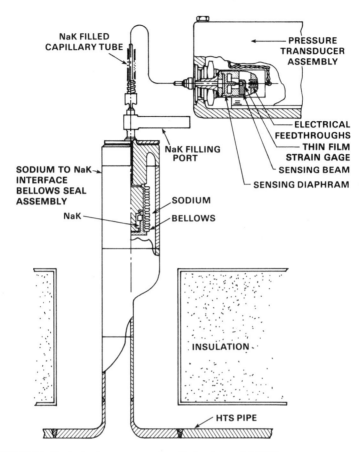

FIGURE 12-24. Typical pressure sensor installation for an LMFBR system.

applications. An alternate method often employed is to interface the sodium with NaK via a bellows system, as illustrated in Fig. 12-24, taking advantage of the fact that NaK is a liquid at room temperature in the close-tolerance pressure transducer assembly.

B. FUEL FAILURE DETECTION

The detection of a cladding breach in a fuel element can be normally accomplished by monitoring increased cover gas activity or by detecting the presence of delayed neutrons in the sodium leaving the reactor. Locating the fuel assembly containing the leaking fuel element is more difficult; the use of gas tagging provides one means to provide such identification.

Cover Gas Monitoring

A cover gas monitoring system normally exists to allow the presence of fission products which escape the fuel pin to be discovered. The most abundant fission products that escape are isotopes of the noble gases xenon and krypton. Many of the fission products emit gamma rays with relatively low energy ($\sim$100 keV) and detection of such activity in a prevailing background of high energy ^{23}Ne (440 keV) and ^{41}A (1300 keV) gammas could present a difficulty.* However, numerous Xe isotopes emit relatively high energy gammas and, as such, experience indicates that detection of cladding failure for an LMFBR system is readily attainable.

Germanium detectors, which are particularly sensitive to low level gamma detection, are often utilized in the cover gas system, along with high resolution gamma spectrometers. Detector efficiency can be improved by concentrating the xenon and krypton isotopes in the cover gas; this is accomplished by passing the gas stream through a charcoal-packed column. The time for fuel failure detection by cover gas monitoring is of the order of minutes.

Delayed Neutron Monitoring

A second system often used to discover pin cladding failure is to detect delayed neutrons which emanate from fission products circulating in the coolant stream. These neutron emissions are produced chiefly by two fission products, ^{87}Br (56 second half life) and ^{137}I (25 second half life). Both isotopes are soluble in sodium and enter the sodium via sodium contact on exposed fuel, by fuel washout through a cladding breach, or by fission gas expulsion of fuel into the coolant.

Delayed neutron detectors, usually consisting of BF_3 chambers, are typically located near primary pumps. Overall detection time is obviously a function of primary loop sodium transport time; this is typically the order of a minute.

Location Indicators

Detection of a fuel failure via the cover gas or delayed neutron detector system provides no information on which fuel assembly contains the failed

---

*The ^{23}Ne activity is considerably less in a pool type reactor, relative to a loop type system, due to the larger holdup time (which allows the 38 s half-life ^{23}Ne to decay) and less turbulence in the pool level.

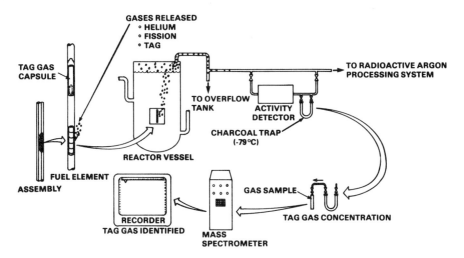

FIGURE 12-25. Fuel failure location determination by gas tagging.

fuel pin. Since a large breeder reactor may be comprised of the order of 300 fuel assemblies, it is important to have some techniques available to identify the offending assembly.

One technique developed for this identification is gas tagging. Unique blends of stable xenon and krypton gas isotopes are injected into the fission gas plenum of each pin during final fabrication; all pins within the same fuel assembly have identical blends. A three-dimensional network of $^{126}Xe/^{129}Xe$, $^{78}Kr/^{80}Kr$, and $^{82}Kr/^{80}Kr$ was used to provide over 100 unique gas tags for the fuel and absorber assemblies in the FFTF. As illustrated in Fig. 12-25, the failed assembly is identified by matching the results of the cover gas mass spectrometer analysis with previously determined analysis of all gas tags in the reactor, suitably corrected for burnup and background.

C. SODIUM LEAKS AND LEVEL MEASUREMENT

The detection of sodium leaks is important for at least three reasons: (1) sodium will burn in an air atmosphere, (2) the primary loop sodium is radioactive, and (3) loss of substantial amount of sodium could impair cooling capacity. Sodium level must be measured in all vessels containing sodium.

Leak Detection

One method used to detect the presence of sodium exploits the electrical conductivity properties of the liquid metal. Contact type sensors consists of two electrodes extended to a gap at a location where leaking sodium may be expected to collect. The presence of sodium shorts out the electrode gap and allows a signal to be delivered to the control room. Such detectors are often placed under the reactor vessel and at any low point surrounding piping systems which contain flowing sodium. The principal difficulties of such detectors are (1) oxidation of the electrodes, and (2) providing assurance that leaking sodium will actually reach the detector (a major concern in the case of small leaks).

A second type of sodium leak detector checks for the presence of sodium aerosol. Atmospheric gas samples from the area in question can be analyzed by either ionization detector or filter examination. For the ionization technique the gas stream is passed over a heated filament which produces sodium ions. Collector electrodes will then induce an ion current if sodium is present. In the case of the filter technique, a replaceable submicron filter is placed in the gas stream and the filter is then periodically removed and analyzed chemically for sodium deposits.

Sodium Level

Induction level probes can be used to measure the liquid level of sodium, as illustrated schematically in Fig. 12-26. The induction field established by

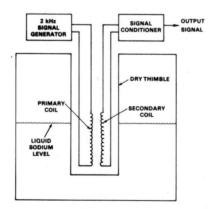

FIGURE 12-26. Induction sodium level probe schematic.

the primary coil is modified by the electrical properties of the liquid coolant, thereby leading to a secondary signal which is directly proportional to the sodium level. Though useful for maintaining inventory records in all sodium repositories, this is especially important within the reactor vessel where a fundamental safety concern is to guarantee a coolant level well above the top of the core at all times.

12-7 AUXILIARY SYSTEMS

Beyond the major systems mentioned earlier in this chapter, numerous auxiliary systems exist which are necessary to support overall plant operation. Many of these are large systems, such as heating and ventilating, but they are not unique to the LMFBR (except for the general desire to minimize the presence of water in the immediate vicinity of sodium systems). Our present purpose is to describe briefly auxiliary systems which are unique to a sodium cooled plant.

There are numerous ways to categorize such systems. We have chosen only three: (1) inert gas (which recognizes the need for an inert atmosphere surrounding a combustible coolant), (2) trace heating (necessary to keep sodium in the liquid phase for low core power levels), and (3) sodium purification (clearly unique to sodium coolant).

A. INERT GAS

An inert cover gas is a requirement for any part of an LMFBR system where a free liquid sodium surface can exist (i.e., vessel, pumps, IHX). Prudent design also normally requires an inert atmosphere in any of the cells which house sodium piping systems. Whereas the word "inert" normally implies a noble gas, the essential characteristic of the atmosphere desired is that it be chemically inert to sodium. Nitrogen satisfies this requirement and is both abundant and relatively inexpensive. Hence, it has been almost universally employed as the inert atmosphere for equipment cells. Unfortunately, it can not be used for high temperature application ($\geq 400°C$) because of nitriding problems in the steel enclosures. Consequently, argon has been selected as the cover gas *within* the vessel, the piping systems, and the refueling transfer chamber for all major LMFBR projects to date. Helium is a potential alternative.

Because of this normal split of inert gases employed, and the natural associated split in physical layout of the two subsystems, it is appropriate to discuss the key features of each separately.

Argon Cover Gas Subsystem

The argon cover gas subsystem provides an inert atmosphere and pressure control for all liquid metal-gas interfaces. Chemical purification features include both sodium vapor and oil vapor traps. Compressors and storage facilities are necessary ingredients of this subsystem as well as pressure equalization lines to keep all cover gases at the same pressure. The purging system needs to have a high enough capacity to allow complete changes of atmosphere to accommodate maintenance operations.

Because of the possibility of radioactive contamination (discussed in Section 12-4A) a key feature of this subsystem is the radioactive argon processing subsystem (RAPS) which removes krypton and xenon radioisotopes. A charcoal bed, which utilizes a cryogenic still, provides an effective way to remove the krypton and xenon isotopes from the argon stream. Surge tanks are useful to allow short-lived isotopes in the contaminated argon to decay.

Nitrogen Subsystem

A nitrogen subsystem is normally incorporated to supply and control atmospheric pressure and purity in the inerted equipment cells. This is done by means of a feed-and-bleed system to regulate pressure and fresh nitrogen purging to minimize contamination, using measurements of the oxygen or water-vapor in the cells as a control signal. Nitrogen is also normally supplied for the sodium/water reaction pressure relief system in the steam generator, for cleaning operations, and for valve actuation in inerted cells.

A key feature employed to remove radioactive contaminants is the cell atmosphere processing subsystem (CAPS). It works on the same principle as the RAPS System, complete with cryogenic features, but usually has substantially larger capacity. One reason for the large capacity requirement is to allow pressurization of equipment cells during pressure testing of plant containment.

B. TRACE HEATING

Sodium melts at $98°C$; hence, it must be heated at low reactor power levels to remain in the liquid form. The method normally employed to provide such heating is electrical trace heaters. A typical trace heating assembly, as illustrated in Fig. 12-27, consists of a nickel-chromium resis-

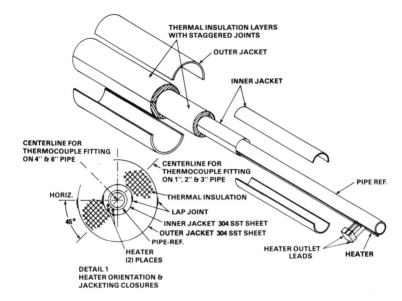

FIGURE 12-27. Trace heating and insulation for one-to-six inch pipe.

tance element insulated with magnesia, covered with a nickel-iron-chromium alloy heater sheath, and surrounded by a large thickness of thermal insulation. Such heaters provide a heat flux around 10 to 20 kW/m². For a large plant, the trace heating system may consume on the order of 10 MW during initial startup (cold core) conditions. The requirements for trace heating drop appreciably when primary and secondary pumps are activated—due to frictional heating resulting from pumping action.

C. SODIUM PURIFICATION

The principal objective of the sodium purification system is to maintain the sodium clean from chemical or radioactive particulate contaminants. As noted in Chapter 11, several trace elements from in-core structural materials dissolve into the flowing sodium coolant during normal operation. Table 12-3 provides a listing of elements which a typical sodium purification system may be designed to monitor.

The main component incorporated in such a system to remove such impurities is the cold trap. This device, which is connected to a bypass line from the main sodium loop, removes impurities by crystallization at a temperature ($\sim 150°C$) significantly below the main-stream sodium temperature. Figure 12-28 contains an illustration of a typical cold trap. Sodium

TABLE 12-3   Contaminants Monitored by a Typical
Sodium Purification System

Boron	Iron
Carbon activity	Lithium
Cesium-137	Manganese
Chlorine	Molybdenum
Chromium	Nickel
Hydrogen	Nitrogen
Iodine-131	Oxygen
Iodine-132	Plutonium
Iodine-133	Tritium
Iodine-135	Uranium

oxide crystallizes on the packing. This packing is replaced when it begins to plug.

An interesting feature of the cold trap shown is the economizer. Inlet sodium must be cooled to low temperatures prior to entering the crystallizer, but returning (purified) sodium must be reheated to nearly the bulk coolant

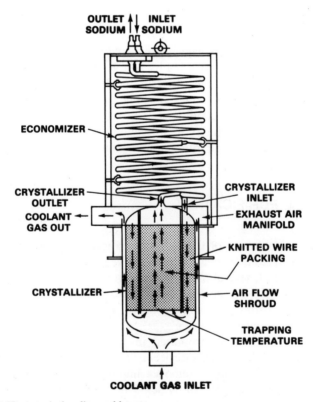

FIGURE 12-28. A typical sodium cold trap.

temperature prior to returning to the main coolant system. Both of these functions can be performed by bringing the inlet sodium through a tube which is concentrically enclosed by an outer tube containing the counterflowing purified sodium. A much smaller auxiliary cooling and heating system is then required to allow satisfactory cold trap performance than would be the case without the economizer feature.

REVIEW QUESTIONS

12-1. (a) List several advantages of both the pool and the loop heat transport systems.

    (b) From Table 12-1, discern which appears to be the more favored.

12-2. Describe the major differences between the Benson steam cycle, the Sulzer cycle, and the recirculating cycle. Why would a saturated steam cycle merit any consideration for an LMFBR?

12-3. Explain the function of the different sodium levels and the trapped gas in Figure 12-8.

12-4. (a) Draw sketches of the primary system of a loop design with the primary pump on the hot leg and on the cold leg. Identify H_p, H_z, and H_l from the equation defining NPSH on your sketches.

    (b) Recognize that the absolute pressure at the pump suction is $H_p + H_z - H_l$, and explain why $H_p + H_z$ must be larger if the pump is on the cold leg.

    (c) From Table 12-1, judge whether the LMFBR industry generally prefers one of these two locations.

12-5. (a) Intermediate Heat Exchanger (IHX):
- In the IHX, is the primary sodium generally on the shell or the tube side, and why?
- How is the relative thermal expansion between tubes and shell usually accounted for?

    (b) Steam Generator:
- Is steam or sodium on the tube side, and why?
- How is thermal expansion between tubes and shell accounted for?
- What material is generally used for the tubes and shell in LMFBR steam generators?
- Why is burnout a concern in LMFBR steam generators while, unlike an LWR, it is not a problem in the core?
- What would happen in the steam generator in response to the high pressure from an extensive sodium-water reaction?

12-6. List several reasons why the shielding design is more challenging for an LMFBR than for an LWR. Note the important differences in shielding problems between a pool and a loop design.

12-7. (a) Why is under-the-plug refueling required for an LMFBR whereas it is not needed for an LWR?

(b) List the steps involved in accomplishing refueling for the AI loop design described.

12-8. (a) If a breach occurs in the cladding of an LMFBR fuel pin, what signals indicate this to the operator?

(b) How can sodium leaks be detected?

(c) How can the sodium level in tanks and the reactor vessel be determined?

12-9. What is the function of a sodium cold trap, and how does it operate?

REFERENCES

1. J. Graham, *Fast Reactor Safety*, Academic Press, New York (1971).

2. Y. S. Tang, R. D. Coffield, Jr., and R. A. Merkley, *Thermal Analysis of Liquid Metal Fast Breeder Reactors*, American Nuclear Society, La Grange Park, IL (1978).

3. W. L. Bunch, J. L. Rathbun and L. D. Swenson, *Design Experience–FFTF Shielding*, S/A-1634, Hanford Engineering Development Laboratory, November, 1978 (presented at the US/USSR Joint Fast Breeder Reactor Shielding Seminar, Obninsk, USSR, November 1978).

4. R. K. Disney, T. C. Chen, F. G. Galle, L. R. Hedgecock, C. A. McGinnis, G. N. Wright, *Design Experience–CRBRP Radiation Shielding*, CRBRP-PMC 79-02, CRBRP Technical Review, (April 1979) 7–28.

5. A. S. Gibson, P. M. Murphy and W. R. Gee, Jr., *Conceptual Plant Design*, System Descriptions, and Costs for a 1000 MWe Sodium Cooled Fast Reactor, Task II Report, AEC Follow-On Study, GEAP-5678, General Electric Company, (December 1968).

6. J. S. McDonald (AI), C. L. Storrs (CE), R. A. Johnson (AI), and W. P. Stoker (CE), "LMFBR Development Plant Reactor Assembly and Refueling Systems," Presented at ASME Meeting, August 18–21, 1980, San Francisco, CA.

7. K. W. Foster, "Fuel Handling Experience with Liquid Metal Reactors," *Proc. Int. Symp. on Design, Construction and Operating Experience of Demonstration Liquid Metal Fast Breeder Reactors*, Bologna, Italy, (April 1978).

8. E. Benoist and C. Bouliner, "Fuel and Special Handling Facilities for Phénix," *Nucl. Eng. International*, (July 1971) 571–576.

PART IV

SAFETY

Safety analysis constitutes an integral part of a fast breeder reactor system, just as it does for any nuclear reactor or other industrial system. In covering the material for this section, it is appropriate to divide the discussion into two parts—one in which the plant protection system (PPS) operates as designed (Protected Transients), and one in which it is postulated to fail (Unprotected Transients).

In the nomenclature associated with the Environmental Impact Statement for a nuclear power plant, accidents are categorized into nine classes. Classes 1 through 8 include accidents for which the plant protection system functions as needed. Only Class 9 includes major accidents in which the plant protection system is assumed to fail. Hence, the first part of the safety analysis presented here concerns basic safety design philosophy, the plant protection system, and Class 1 through 8 accidents. These issues are addressed in Chapters 13 and 14. The second part of the safety section concerns Class 9 accidents for an LMFBR; these are considered in Chapters 15 and 16.

CHAPTER 13

GENERAL SAFETY CONSIDERATIONS

13-1 INTRODUCTION

It has long been recognized that the potential for accidental release of radioactive materials is present in a nuclear reactor system. Consequently, a major portion of the overall technical effort expended within the nuclear power industry has always been allocated to safety considerations. This safety awareness, which consciously permeates the approach all the way from conceptual design through the licensing process and long-term operation, has been implemented to a far greater extent than normally present in other areas of human endeavor.

Considerable attention has historically been given to the potential consequences of highly improbable accidents—events so remote in probability that they are correctly termed hypothetical accidents. Because the calculated consequences of such postulated events can be high, numerous investigators, notably Farmer,[1] Starr,[2] and Rasmussen,[3] have attempted to compare quantitatively the safety risks of nuclear power with other risks which we encounter in everyday life. A discussion of risks also appears in the reactor-safety text by Lewis.[4]

A perspective on the individual risk* of acute fatalities by various causes in the U.S. is given in Table 13-1.[5] The data of the Table 13-1 show that there were no fatalities due to radiation during the year given. Moreover, there have never been any fatalities among the general public as a result of the commercial operation of nuclear reactors. Nevertheless, the magnitude of individual accidents is of psychological importance to the general public. The basic concern is that a single accident which results in the loss of many lives may be viewed more harshly than several smaller accidents, even though they result in the same total number of fatalities. Hence, it has been

---

*Risk, as used here, generally denotes only the probability aspect of an occurrence. A more accurate definition, namely the product of probability and consequences, is discussed in Section 13-4.

TABLE 13-1   Individual Risk of Fatality by Various Causes[5]
(U.S. Population Average 1977)

Accident Type	Fatalities For 1977	Approximate Risk (Probability of Death Per Resident Per Year)	Relative Probability of Accidental Death By Particular Mode
Motor Vehicle	49510	2×10^{-4}	0.48
Falls	13773	6×10^{-5}	0.13
Fires and Flames	6357	3×10^{-5}	0.06
Drowning	5961	3×10^{-5}	0.06
Poison	4970	2×10^{-5}	0.05
Inhalation & Ingestion of Objects	3037	1×10^{-5}	0.03
Firearms	1982	1×10^{-5}	0.02
Air and Space Transport	1643	8×10^{-6}	0.02
Water Transport	1357	6×10^{-6}	0.01
Electrocution	1183	6×10^{-6}	0.01
Falling Objects	1096	5×10^{-6}	0.01
Railway	576	3×10^{-6}	0.006
Cataclysms (Tornadoes, Hurricanes, Floods)	202	9×10^{-7}	0.002
Lightning	116	5×10^{-7}	0.001
Bites and Stings	55	3×10^{-7}	0.0005
Radiation	0		
All others	11384	5×10^{-5}	0.11
All Accidents	103202	5×10^{-4}	1.00

important to attempt a comparison of reactor related accidents with potentially large catastrophies such as airplane crashes, dam failures, and major industrial accidents.

The most complete nuclear reactor risk-consequence study accomplished to date is the WASH-1400 Reactor Safety Study headed by Rasmussen.[3] Figure 13-1 contains some of the principal results from that study. Plotted is the estimated probability of an accident (either a natural catastrophe or an accident arising from the failure of a man-made system) causing more than N deaths per year.

An immediate difficulty in arriving at such results is the lack of a data base for accidents of such low frequency. Hence, considerable extrapolation was required. However, by using consistent extrapolation procedures, it is possible to make an estimate regarding the relative risks from nuclear power plant accidents compared to those from other potentially large accidents. Such results, based on an extensive study of potential accident paths in light water reactor nuclear power plants, though not zero, are well below those

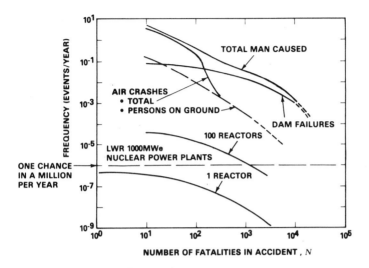

FIGURE 13-1. Estimated frequency of man-made events with fatalities greater than N. (Adapted from Refs. 3 and 4).

generally accepted by society. Despite the early stage of development of fast breeder reactors relative to light water reactors, safety analyses and preliminary risk evaluations from LMFBR designs (e.g., Reference 6) indicate that there is no reason why risks for LMFBR's cannot be held to the same low orders of magnitude as those calculated for LWR's.

The primary purpose of the present chapter is to provide an examination of how LMFBR safety analysts approach their task. A discussion of the evolution of LMFBR safety concerns and a comparison of LMFBR and LWR safety issues is important to this presentation. Our discussion will begin with the multiple barrier approach to safety, from which the logic groundwork is laid for studying both *protected* and *unprotected* accidents. In this context, the term "protected" implies successful operation of the Plant Protection System (PPS) when called upon to function, whereas "unprotected" implies failure of the PPS. Because of the considerable difference in system response, with or without proper PPS action, we have chosen to devote a separate chapter to each situation (Chapter 14, Protected Transients; Chapter 15, Unprotected Transients). It should be clearly understood, however, that the term "unprotected" only connotes failure of the PPS; it does not imply a lack of protection to plant personnel or the general public. Both intrinsic and engineered safeguards are available (e.g., containment, as discussed in Chapter 16) to provide personnel protection even for "unprotected" accidents.

We next address, in the present chapter, the principal differences between fast reactors and thermal reactors—from a safety point of view. The final section of this chapter deals with the more recent approaches employed to analyze potential accident paths, namely, the mechanistic, probabilistic, and phenomenological techniques.

13-2 MULTIPLE BARRIER APPROACH TO SAFETY

Fast breeder reactors are designed to be stable machines, and experience has demonstrated the attainability of such behavior. Though the potential for radiation release to the atmosphere exists, there are numerous natural and engineered barriers to inhibit such release. Our purpose in the present section is (1) to discuss such barriers, (2) to develop the rationale for categorizing accident consequences according to their relative probability of occurrence, and (3) to summarize one approach aimed at optimizing the fast reactor safety research and development efforts.

A. PHYSICAL BARRIERS

Three barriers exist within a reactor to physically prevent the release of fission products to the outside environment. As illustrated schematically in Fig. 13-2, these include (1) the fuel matrix itself, (2) the fuel pin cladding,* and (3) the primary coolant system.† A fourth barrier, the outer containment, represents an engineered safeguard.

The fuel matrix, normally a ceramic material, has considerable capability for retaining solid fission products, as well as gaseous fission products in the unrestructured region. Fission gas which is released during irradiation is normally contained in a fission gas plenum, in order to prevent contamination of the coolant (discussed in Chapter 8). However, reactors are designed to operate with a small fraction of failed fuel. All large reactors have a radioactive waste removal system built in as a part of the cover gas system, and small amounts of radioactive debris in the primary sodium can be removed in a routine manner via cold traps.

---

*Cladding is required in the fuel pin design to provide structural integrity and, as such, also provides a fission product barrier. Since the latter function may not be essential, some design effort has been focused on perforated cladding (vented fuel pins) to enhance fuel pin lifetime.
†In addition to the physical barriers of the primary system, sodium coolant also acts as a natural barrier to the release of significant radioactive products (discussed in Section 13-4).

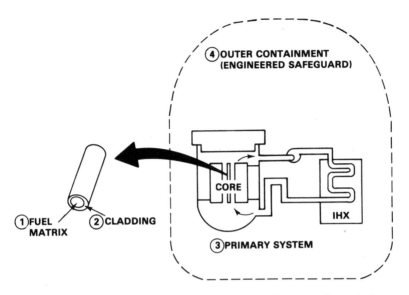

FIGURE 13-2. Physical barriers to radioactive release in an FBR system (Shown for loop-type reactor).

B. DEFENSE IN DEPTH

In addition to the physical barriers to radioactive release in an FBR system, a basic safety philosophy has long existed that appropriate engineered features be incorporated into the system to provide defense in depth.

To provide the framework for establishing such an approach, it is necessary to recognize the relationship between the anticipated frequency of occurrence for various accidents vs the accident severity levels which could be accepted for each event. Such a relationship should certainly indicate little or no damage to the reactor system for high frequency events, whereas relatively higher levels of damage could be tolerated for lower frequency events. The challenge to the reactor designer is to assure that the appropriate engineering features, such as an effective Plant Protection System (PPS), be available to ensure such a system characteristic.

Figure 13-3 illustrates such a relationship, as first proposed in 1967, by Simpson, et al.[7] Although the nomenclature adopted at that time was somewhat arbitrary, it was later selected as an RDT Standard[8] for use in the United States.*

---

*A considerable standards effort was expended during the late 1960s within the Reactor Development & Technology (RDT) division of the U.S. AEC, and the standards therein developed still bear this name.

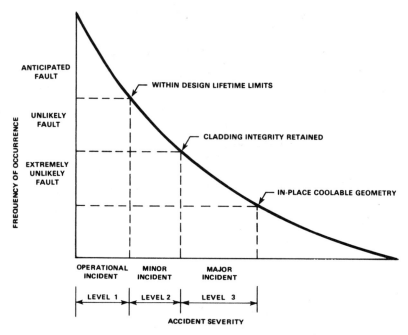

FIGURE 13-3. Defense-in-depth approach for FBR systems.

Another way to view the approach philosophically was introduced by Kintner.[9] This is the classic three-level approach, superimposed on Fig. 13-3, which is summarized as follows:

First level - Execute and build a sound, conservative, inherently safe design.

Second level - Provide protection systems designed to assure that off-normal events will be prevented, arrested, or accommodated safely.

Third level - Evaluate and provide features which add margin as additional assurance of public safety even for extremely unlikely and unforeseen circumstances.

The basic thrust in Level 1 is to place safety consciousness at the highest level in all phases of the design, construction, and operation program. It includes deliberate and detailed safety assessments of all major systems in the plant, including vigorous quality assurance and safety checks to ensure that the plant is built and operated as intended. Included in this level are operational transients, such as scrams and startup. Because of the certainty of such transients, it is important to be assured that the useful lifetime of the core is not impaired by their occurrence.

Beyond Level 1, there is the recognition that a single unlikely fault of some type (e.g., pump failure) might occur on a very infrequent basis, perhaps once or twice in the expected life of the plant. Hence, a highly reliable PPS is designed to arrest any such accidents and limit the core damage to an acceptable level (although a higher damage limit is allowed than for Level 1—consistent with the infrequency of the event).

Level 3 is addressed to that category of events which, although never expected to occur are, nevertheless, mechanistically possible. In performing analysis to assess the consequences of all accidents, the event which results in the highest damage severity level is commonly labeled the *Design Basis Accident* (DBA).

Beyond the DBA, there lies a domain of accident consequences which are less probable (normally larger in magnitude); such accidents are discussed in Section 13-3 and Chapters 15 and 16. Generally, in order for consequences of this magnitude to be calculated, at least two or more low probability failures must take place in sequence, e.g., a massive reactivity insertion event coupled with complete failure of the plant protection system. Because of the extremely low probability of such a range of accidents, they are often referred to as *incredible* or *hypothetical* events, even though they may be mechanistically possible.

C. CLASSIFICATION OF ACCIDENTS

Whereas the defense-in-depth approach, as illustrated in Fig. 13-3, provides the basic philosophy in all FBR safety studies, reactor designers and safety analysts need more specific guidance regarding (1) the treatment of specific accidents, and (2) the specification of equipment requirements in order to achieve the safety margins desired in the overall plant.

With regard to the first need, the Nuclear Regulatory Commission (NRC) in the United States has provided a categorization[10] of accident types which must be considered in the Environmental Impact Statement. Table 13-2 includes the nine classes of events which must be addressed for all nuclear power plants licensed in the U.S. The examples shown for LWR systems have undergone considerable cross examination and have certain regulatory status.[10] The LMFBR examples, however, are offered simply as possible candidates to provide a framework for many of the accidents discussed in this text; they are without any regulatory status.

It is important to recognize that an enormous effort is devoted to analyzing accidents in the 1 to 8 categories of Table 13-2. Further, it is fairly common to effect certain design changes which correct any plant deficiencies that may be exposed as a result of these analyses. Such efforts often go

TABLE 13-2 Classification of Postulated Accidents for Environmental Impact Statement

Description	LWR Examples	LMFBR Examples
1. Trivial Incidents 2. Small Releases	Small spills; small leaks outside containment Spills, leaks and pipe breaks	Single seal failures; Minor sodium leaks IHTS valve, seal leaks; Condensate storage tank valve leak; Turbine trip/steam venting
3. Radwaste System Failures	Equipment failure; Release of waste storage tank contents	RAPS/CAPS valve leaks; RAPS surge tank failure; Cover gas diversion to CAPS; Liquid tank leaks
4. Events that release radioactivity into the primary system	Fuel failures during normal operation; Transients outside expected range of variables	Fuel failures during normal operations; Transients outside expected range of variables
5. Events that release radioactivity into the secondary system	Class 4 and Heat Exchanger and SG Leak	Class 4 and Heat Exchanger Leak
6. Refueling accidents inside containment	Drop fuel element; Drop heavy object onto fuel; Mechanical malfunction or loss of cooling in transfer tube	Drop of fuel element; Crane impact on head; Inadvertent floor valve opening; Leak in fuel transfer cell/chamber
7. Accidents to spent fuel outside containment	Drop fuel element, Drop heavy object onto fuel; Drop shielding cask...loss of cooling to cask; Transportation incident *on site*	Shipping cask drop; EVST/FHC system leaks; Loss of forced cooling to EVST
8. Accident initiation events considered in design-basis evaluation in the Safety Analysis Report	Reactivity transient; Rupture of primary piping; Flow decrease, Steam-line break	SG leaks; Na/Water reaction; Fuel failure propagation; Rupture of primary piping; Pump failure or reactivity transient (with PPS operating)*
9. Hypothetical sequences of failures more severe than Class 8	Successive failures of multiple barriers normally provided and maintained	Successive failures of multiple barriers normally provided and maintained

*Such events with PPS failure have often been assessed to establish the safety margin in the containment design and the long term decay heat removal capability.

CAPS = Cell atmosphere processing system
EVST = Ex-vessel storage tank (in spent fuel)
FHC = Fuel handling cell

IHTS = Intermediate Heat Transport System
RAPS = Radioactive argon processing system (purifies contaminated core gas)
SG = Steam generator

unnoticed beyond the bounds of any given project because the analysis techniques used often lack the novelty of those characteristically developed for analyzing Class 9 accidents. Consequently, reporting of such activities at professional technical conferences or in journals is not extensive.

Another accident categorization, which focuses on the classification of safety equipment, is presented in nuclear standards developed under the sponsorship of the American Nuclear Society (ANS). These standards are developed in accordance with procedures provided by the American National Standards Institute (ANSI), which ensure a consensus of affected interests to permit their approval as American National Standards. Because of the nature of the ANSI procedures, which require industry, government and public acceptance before any proposed guidelines can attain the status of approved American National Standards, many of the LMFBR standards are yet to be finalized. Several of the proposed LMFBR standards have been published as draft standards by the American Nuclear Society for trial use and comment. Despite the incompleteness of the reactor standards work to date, heavy industry commitment has resulted in much progress. Hence, it is instructive to sketch out the classifications and indicate the general guidance available to reactor designers. Table 13-3 contains the indexing nomenclature for the LWR's and the LMFBR.

TABLE 13-3   Selected Safety Standards in Nuclear Power Reactors

Type Reactor	TOPIC	ANS No.	ANSI No.
PWR	Nuclear Safety Criteria	51.1	N18.2*
BWR	Nuclear Safety Criteria	52.1	N212*
LMFBR	General Safety Design Criteria	54.1	N214
	Safety Classification and Related Requirements	54.6	

*These standards are under revision; when they are superseded by a revision approved by ANSI, the revision shall apply.

The PWR and BWR standards were initially developed under subcommittee ANS-50, now replaced by the ANS Nuclear Power Plant Standards Committee (NUPPSCO). This program contains all the general safety criteria for reactors. The Nuclear Safety Criteria for the PWR (ANS-51.1) and the BWR (ANS-52.1) are summarized in Table 13-4 along with a preliminary version of the General Safety Design Criteria for the LMFBR (ANS-54.1). It is noted that four categories are specified for the PWR, whereas only three are used in the draft version of the proposed standard for the LMFBR. The PWR Categories II and III correspond roughly to Category II for the LMFBR.

TABLE 13-4 Accident Conditions Categorized by ANS Standards

PWR[11] and BWR[12]		LMFBR[13]	
CATEGORY	EXAMPLES	CATEGORY	EXAMPLES
I. NORMAL OPERATION (expected frequently during normal course of operation and maintenance)	Startup, shutdown, standby, power ascension from partial load to full power; cladding defects within Tech Specs. refueling	I. NORMAL OPERATION (expected frequently during normal course of operation and maintenance)	Startup, normal shutdown, standby, load following; cladding defect within Tech Specs; refueling
II. INCIDENTS OF MODERATE FREQUENCY (any one of which may occur during a *calendar year* for a particular plant)	Inadvertent control rod group withdrawal; partial loss of core cooling; moderate cool-down; loss of off-site power; single error of operator	II. ANTICIPATED OPERATIONAL OCCURRENCES (May individually occur one or more times during the *lifetime* of the plant)	Tripping of Na pumps; failure of all off-site power; tripping of Turbine-Generator set; inadvertent control rod withdrawal
III. INFREQUENT EVENTS (may occur during the *lifetime* of a particular plant)	Loss of reactor coolant (with normal coolant makeup system only); secondary pipe break; fuel assembly in violation of Tech Specs; control rod withdrawal in violation of Tech Specs; unexplained reactivity insertion; complete loss of core flow (excluding pump locked rotor)	III. POSTULATED ACCIDENTS (not expected to occur—but included in the design basis to provide additional margins for assuring no undue risk to the health and safety of the public)	Spectrum of events appropriate to a specific design considering both the probability and consequences of the events (e.g. pipe rupture, large Na fire, large Na-water reaction, rupture of radwaste system tank)
IV. LIMITING FAULTS (not expected to occur—but postulated because of potential for significant radioactive release; most drastic faults that must be designed against)	Major pipe rupture (up to and including double-ended rupture of largest pipe); fuel or structure movement due to core damage; ejection of single control rod; major secondary system pipe rupture (double-ended); coolant pump locked rotor		

TABLE 13-5 General Relationship of Safety Classes to Equipment Specifications

PWR[11] and BWR[12]		LMFBR[14]	
SAFETY CLASS (SC)	DESCRIPTION	SAFETY CLASS (SC)	DESCRIPTION
1	Components whose failure could cause a Condition III or Condition IV loss of reactor coolant	1S (STRUCTURES)	Structures that protect safety class systems and components; control post-accident release of radioactive nuclides
2 (for reactor containment components)	1. Reactor coolant pressure boundary not in SC-1 2. Necessary to remove residual heat 3. Safety Systems inside containment	1E (ELECTRICS)	Electrical equipment and systems that are essential to emergency reactor shutdown, containment isolation, core and containment heat removal
2 (other than reactor containment)	1. Components not in SC-1 or SC-2 2. Failures lead to radioactive release to environs	1M (MECHANICAL)	Mechanical equipment and systems which form a part of the reactor coolant boundary essential for safe shutdown; perform scram functions (except for routinely replaceable components such as fuel and absorber assemblies). Class 1M relates to the most critical components.
3		2M	
		3M	

Table 13-5 contains a highly condensed intercomparison of the safety class categorizations as they relate to equipment specifications. Both the light water reactor and LMFBR systems use three safety classes, but the LWR classes are organized to correspond to certain ASME code classifications, whereas the LMFBR classes are organized according to structures, electrical and mechanical equipment specifications. All of the equipment listed for the LMFBR in Table 13-5 is designated as *Seismic Category I* and should be designed to withstand the effects of the *Safe Shutdown Earthquake* (SSE) and retain its safety functions.

D. LINES OF ASSURANCE

The above categorization techniques for classifying accident conditions have proved very helpful in assuring that adequate protective measures are taken for the range of accidents reasonably expected to occur in the lifetime of a plant. Unfortunately, in practice they do not provide adequate guidance for the issues which must be addressed for the very low-probability-higher consequence type accidents (i.e., the Class 8 and Class 9 varieties listed in Table 13-2). A major deficiency of the above categorization is that it simply classifies accidents and inferred consequences without quantitative regard to the probabilities involved. The principal ingredient lacking is risk, defined to be the product of probability and consequences.

Consequently, the Lines of Assurance (LOA) concept[15] has evolved in the U.S. Department of Energy programs over the past several years to provide the framework for a risk-oriented approach to FBR safety. The basic idea in this approach is to recognize the mechanistic nature of any accident progression sequence and to evaluate the probability of any such accident path progressing through the natural restraints inherent or designed into an FBR system.

The LOA approach has been further developed within the U.S. LMFBR Safety Program as a management tool to aid in structuring, managing, and communicating a comprehensive safety development program. Program management with the LOA structure allows proper emphasis to be placed on accident prevention and accident-consequence accommodation and attenuation. The LOA approach is based on the concept that multiple restraints exist or have been designed into an LMFBR plant which protect against the release of radioactive and/or toxic materials to the environment.

The LOA's and the respective restraints are shown in Table 13-6. The areas of concern in LOA-1 correspond to those safety issues which must be addressed within the Design Basis, while LOA's 2, 3, and 4 address the capability of LMFBR's to accommodate and mitigate the consequences of accidents.

TABLE 13-6   LOA Barrier Definitions

Line of Assurance	Restraints to Accident Progression
LOA-1: Prevent Accidents	Reactor shutdown system
LOA-2: Limit Core Damage	Fuel cladding and self-actuated shutdown system
LOA-3: Maintain Containment Integrity	Primary system and reactor containment system.
LOA-4: Attenuate Radiological Consequences	Engineered systems and inherent mechanisms

Management of the LOA-structured R&D program is based on a risk-oriented end product with success criteria and, ultimately, associated goal failure probabilities defined for each element of the program. The end products are the research and development results which satisfy the success criteria. Comparison of the LOA work package against the success criteria provides insight into whether or not additional work needs to be done.

13-3   ACCIDENT ANALYSIS PERSPECTIVE

From the discussion of the preceding section, it is apparent that the approach to FBR safety considerations has paralleled and benefited from the large reservoir of experience developed from LWR systems. Indeed, for the classes of accidents envisioned to have a reasonable chance of ever actually occurring, the overall consequences for LWR and FBR systems would be generally similar.

The precise accident paths would, of course, differ—as exemplified by the accident at Three Mile Island. For an FBR system, it is conceivable that a combination of feedwater pump failure and closed valves in parallel feedwater systems could occur which would lead to overheating of a steam generator. However, the subsequent accident sequence in an LMFBR would bear little relationship to that which occurred at TMI since the sodium cooled reactor is a low pressure system; single phase primary loop flow would continue independent of the pressurization requirements evident in a LWR. As such it is highly unlikely that this common triggering event would lead to any fuel failures for an LMFBR. On the other hand, it is conceivable that other combinations of events could lead to partial core damage. The principal point is that the overall time scales and consequences of accidents in the categories implied by Fig. 13-3 are expected to be similar for LWR and FBR systems.

It is in the area where major accidents are postulated that the differences in safety characteristics between LWR and FBR systems become pronounced. Hence, we will first look at some of the principal differences in the physical characteristics of the two systems and then follow the implications of such differences through the major accident studies which have thereby evolved.

A. LMFBR VS LWR SAFETY CHARACTERISTICS

Table 13-7 contains a summary of some principal differences in physical characteristics between LWR and LMFBR systems which influence major accident evaluations. Perhaps the most significant difference, with regard to the considerations involved with an assessment of unprotected accidents, is the core configuration. In a LWR, the core components (fuel, cladding, coolant*) are arranged to maximize k_{eff}. Any rearrangement, due to fuel melting or coolant loss, tends to shut down the reactor neutronically. An FBR, on the other hand, needs no moderator and is assembled with the fuel separated only to allow sufficient heat removal to take place. Consequently, if fuel melting should occur, a core compaction process would lead to a reactivity gain (as described in Chapter 6). This possibility provided the underlying concern for the majority of the large-accident safety studies done for the early fast reactor systems. As will be noted in Chapters 15 and 16, there is now considerable evidence that the inherent forces within the core of an FBR undergoing a postulated fuel meltdown would more likely disperse the fuel, thus leading to a reactivity loss rather than a reactivity gain. Nevertheless, the potential for core compaction exists and the attendant reactivity gain associated with such an eventuality provides a fundamental departure in behavior relative to the LWR system.

Probably the principal safety disadvantage of the LWR is that the water coolant needs to be pressurized [~7 MPa (1000 psi) for a BWR and 15 MPa (2200 psi) for a PWR] in order to obtain reasonable coolant temperatures. Hence, any breach in the primary system directly leads to coolant "blowdown." This concern is absent in an LMFBR system, since the sodium is highly subcooled (by about 350°C) at normal operating conditions. The maximum system pressure [~1 MPa (150 psi)] is just high enough to insure proper coolant flow through the core. A breach in the primary system of an LMFBR would not result in coolant boiling. The primary concern would be eventual loss of coolant and any chemical reactions which might occur.

The chemical reactions of concern in an LMFBR are primarily sodium-air and sodium-water reactions. A barrier to such reactions for normal operation is provided by covering all sodium systems with an inert cover gas. Under gross accident conditions a sodium-concrete interaction also becomes

---

*The coolant often serves as a moderator in a thermal spectrum system.

TABLE 13-7 Comparative Characteristics Affecting Major Accident Evaluations

Characteristics	LWR	LMFBR
Core Assembly	Optimal reactivity geometry	Not arranged in most reactive configuration
Stored Energy in Coolant	High (pressurized system)	None (subcooled at 1 atm)
Chemical Energy Potential	None at normal temperatures	High; sodium–air/water
Loss of Coolant	Reactivity loss	Reactivity gain*
Inherent Emergency Heat Removal Capability	Low	High
Radiological Inventories: • Fission Products • Plutonium	← Equivalent → Medium	 High
Nuclear Properties • $\bar{\beta}$ • Λ	0.007 $\sim 2 \times 10^{-5}$ seconds	0.004 $\sim 4 \times 10^{-7}$ seconds
Thermal Time Constants • Fuel • Structures • Coolant	1–5 seconds Few seconds 0–5 seconds	1–2 seconds < seconds 0.1 seconds

*Strong spatial dependence (positive near core center, negative near core edge)

a concern (including water release from the heated concrete); a similar concern is present in the LWR system from the molten fuel-zircaloy reaction under major accident conditions.

As discussed in Chapter 6, loss of sodium in an LMFBR can result in a reactivity increase if the loss takes place near the central region of the core. A loss of coolant accident in a PWR or BWR results in a reactivity decrease.

For a blowdown accident in an LWR, the emergency core cooling system (ECCS) must be activated to provide sufficient water to keep the core covered in order to prevent core melting. It is inherently easier to keep the core of an LMFBR covered in an accident since the sodium is so far below the normal boiling point and natural convection cooling paths can be readily established even in the event of primary pump system failure.

With regard to the radiological inventories present, the total fission product level is essentially equivalent for comparable sized reactors. There is considerably more plutonium in an LMFBR operating on the ^{238}U–Pu cycle; for comparable size plants, an LWR operating with enriched uranium fuel contains up to ~1/6 the Pu inventory of the LMFBR due to conversion of ^{238}U (See Table 7-4).

The nuclear properties of an LWR would seem to imply easier dynamic control, relative to a Pu-fueled LMFBR, because the delayed neutron fraction is about twice the LMFBR value (due to the presence of ^{235}U rather than ^{239}Pu), and the prompt neutron generation time is about two to three orders of magnitude larger. However, these differences come into play only under major accident conditions, and concerns about the short neutron lifetime of the LMFBR, which were expressed in the early days of FBR safety studies, have been largely allayed due to inherent feedback characteristics which limit the rise of power upon attaining prompt criticality (discussed in Chapter 15).

The thermal time constants of an LMFBR tend to be shorter than those of an LWR, due mainly to the smaller fuel pin size and smaller coolant flow area per pin. Such factors are routinely taken into account in performing transient studies.

B.   ACCIDENT INITIATORS

Although there are substantial differences in the systems associated with LWR's and LMFBR's, which necessarily lead safety analysts to consider several physically different mechanisms as potential accident initiators, the neutronics response of each reactor type is very similar for the more probable accidents in which the plant protection system (PPS) works and the transient is mild. It is only in the area of very low probability accidents, caused by major system failure, that the accident scenarios differ substantially.

The major low probability-large consequence concern in the LWR is a massive rupture in the primary system, which would allow coolant flashing and subsequent blowdown. Should such an event occur, operation of the emergency core cooling system (ECCS) is required to prevent fuel melting. Hence, considerable efforts have been and continue to be expended to assure the functionability and reliability of the ECCS to provide means of cooling the core over a long enough time frame to remove the decay heat. As an upper limit to the public consequences possible from an LWR accident, a double-ended pipe break has historically served as the initiating event for LWR accident consequences evaluation.

As mentioned from the discussions associated with Table 13-7, a major pipe break in an LMFBR does not lead to the same consequences as in an LWR because rapid system depressurization would not result in coolant flashing. Hence, the accidents analyzed for public consequence considerations in an LMFBR have been traditionally associated with the core rearrangement concern, denoted as the first item in Table 13-7. In order to allow such a rearrangement to occur, fuel melting is necessary and, consequently, bounding accidents for LMFBR systems have been historically associated with a postulated power/flow imbalance—coupled with failure of the PPS to terminate the excursion neutronically. The heat generation/heat removal imbalance can come about due to either a *transient overpower* (TOP) condition, in which a reactivity insertion causes the power to rise, or a *transient undercooling* (TUC) condition, in which the power remains essentially constant, but primary coolant flow is lost.

Despite the exceedingly low probabilities for an unprotected TOP or TUC event to occur in an LMFBR, such accidents have historically provided the basis for much of the analysis discussion surrounding FBR safety and, in particular, the question of public protection. Before proceeding with the kinds of studies and results which have accrued for such an approach, however, it may be helpful to discuss two terms which are frequently used in conjunction with major accident studies; namely, the "Design Basis Accident" and the "Hypothetical Core Disruptive Accident."

Design Basis Accident (DBA)

The DBA is normally defined to be that accident that leads to the most severe consequences of all accidents considered credible. Needless to say, considerable debate has ensued on what constitutes such an accident since the limit on "credibility" does not have universal acceptance. From a logic standpoint, however, it corresponds to the upper limit of Level 3 in Fig. 13-3, or the most severe of the Class 8 accidents in Table 13-2.

Hypothetical Core Disruptive Accidents (HCDA)

Hypothetical accidents, as indicated from the Class 9 designation, are those accidents which are postulated and may be mechanistically possible, but even if physically possible are of extremely low probability. Because FBR cores are not characteristically arranged in their most reactive configuration, all of the early upper-limit studies for FBR safety evaluations were focused on exploring the consequences of core compaction. They further ignored many of the intrinsic feedback characteristics of an FBR core and assumed that ultimate neutronic shutdown could only come about by the development of fuel vapor pressure, which would physically disassemble the core. Hence, the hypothetical category of accidents has historically been strongly linked with core *disassembly* accidents, and this has led to the commonly used term "hypothetical core disruptive accident," or the "HCDA."

There is another accident path which is appropriately designated hypothetical, but does not require disruptive core pressures to effect permanent shutdown. This path involves the core melting into a subcritical array. Attention must, of course, be given to this accident both to provide assurance of adequate post-accident heat removal (to maintain permanent cooling) and to demonstrate adequate mitigation of radioactive debris which may be released.

The core disassembly scenario is often referred to as an "energetic" HCDA, and the non-energetic accident is generally called a "core melt" accident. (Analysts sometimes use the term "core melt HCDA" for the core melt accident to connote both the hypothetical nature of the accident and the disrupted geometry of the fuel—even though no high pressure core disassembly forces are involved).

C. HISTORICAL PERSPECTIVE OF FBR ACCIDENTS

Prior to launching into the various approaches under current use in analyzing FBR accidents, it is appropriate to review briefly the early approaches to FBR safety. The reason for this backward glance is that much of the present work—especially the efforts focused on major accident considerations—is strongly influenced by safety studies conducted some twenty years ago.

Early Concerns

Certainly the strongest factor which influenced the early FBR safety analysts was the question of possible core compaction. The fact that fuel densification would increase the system reactivity, in contrast to thermal reactor systems, dominated the types of accidents postulated for analysis.

In addition to the core compaction question, the small $\bar{\beta}$ (hence, small value for the dollar unit of reactivity) and short prompt neutron lifetime provided concern that a core meltdown may rapidly lead to a prompt critical condition and that the power rise beyond that point could be extremely rapid. Although fuel vaporization, an inherent shutdown mechanism, would ultimately limit the energy generated by physically moving the core apart, the concern was that the potential work energy released might be larger than practically containable.

Bethe-Tait Accident

In order to provide an order of magnitude estimate of the energy release possible for such a postulated FBR core compaction process, Bethe and Tait[16] conducted a scoping analysis using a geometry sufficiently simplified to allow a hand calculation solution. The first assumption was that the reactor in question could be represented by a core in R-Z geometry, and that the initial phases of the accident could be represented by a molten core with all sodium lost. The latter situation, for the small early generation FBR core, results in a substantially subcritical reactor. Core collapse under the force of gravity was then postulated and the reactivity insertion rate (ramp rate) at the time of attaining prompt criticality was used to drive the core disassembly phase.

As pictorially sketched in Fig. 13-4, the geometry for the core disassembly phase of the analysis was assumed to be spherical (for computational simplicity). Internal core pressures were assumed to be zero until all internal void volumes (fuel porosity plus the open coolant channels) were filled up due to fuel expansion. At that point, pressure was assumed to rise linearly with energy and this pressure buildup caused rapid core dispersal and permanent neutronic shutdown.

Because of the importance of this analysis, not so much from the standpoint of the original energy release numbers, but rather because of its impact on the activities regarding FBR safety evaluations, some of the details of the formulation will be covered later in Chapter 15. For the moment, however, the major point to be made is that this type of analytical approach—namely, assuming a gravity driven core collapse and a subsequent hydrodynamic core disassembly—was used around the world for

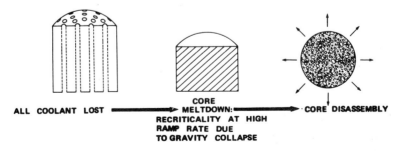

FIGURE 13-4. Pictorial overview of the Bethe-Tait accident.

many years to establish a conservative upper bound for FBR containment requirements.

FBR HCDA Studies

Since about 1960, HCDA studies have been conducted for nearly all of the fast breeder reactors in the United States and Western Europe. A major result of each study has been the calculation of a *maximum HCDA work energy*, defined as an upper limit potential for work done on the reactor environs by expanding core debris.

The original Bethe-Tait analysis would predict increasing values for maximum HCDA work energy for the larger core sizes. However, as will be discussed in more detail in Chapter 15, several major improvements have been made to such HCDA analytical techniques, most of which lead to reduced consequence predictions. Improved treatment of fuel vapor as a disassembly force, the mitigating role of Doppler feedback, and a more mechanistic manner in which to establish the reactivity insertion rate for the disassembly calculations, all represent major contributions in this regard.

As a result of such improvements, the predicted work energy for various FBR systems has tended to be reduced with time, even though the core size and steady-state power ratings have generally increased. Table 13-8 indicates this trend numerically, and Fig. 13-5 illustrates the normalized trend over the last 20 years. The normalization process used for this illustration is the steady-state power level, although the same trend would be observable if the core sizes were used (except for SEFOR, where the design power density was intentionally low).

TABLE 13-8   Maximum HCDA Work Energy Calculations for FBR Systems

Reactor	Country	Year Critical	Power MWth	Approximate Maximum HCDA Work Energy, MJ	HCDA/ Power Ratio
FERMI	USA	1963	200	2000	10
EBR-II	USA	1964	65	600	9.2
SEFOR	USA/Germany	1969	20	100	5
PFR	United Kingdom	1974	600	600–1000	1–1.7
FFTF	USA	1980	400	150–350	0.4–0.9
SNR-300	Germany	~1983	760	150–370	0.2–0.5

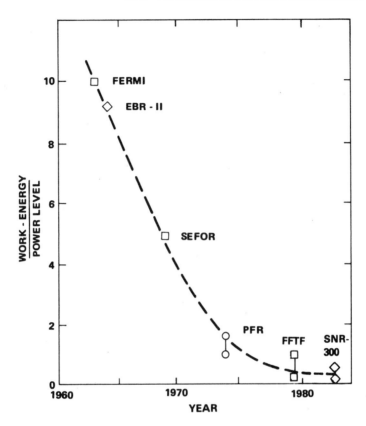

FIGURE 13-5. Historical visualization of normalized maximum HCDA work energy release.

The absolute numbers are not particularly significant; indeed, it is difficult to obtain precise numbers because of the sensitive relationship between HCDA energy release calculations and licensing considerations. However, it is clear from Table 13-8 and Fig. 13-5 that the efforts expended in better understanding the sequence and ultimate consequence of even an unprotected hypothetical core disruptive accident have made a favorable impact on containment demands.* Use of the original Bethe-Tait model for consequence prediction would lead to an increasing value for the ordinate of Fig. 13-5, rather than the sharply decreasing trend shown. Hence, it is of considerable interest to explore more carefully the basis for the marked reduction in predicted HCDA consequences. (See Chapter 15-7.)

---

*Another version of this trend, which contains more reactors and somewhat different values, is contained in Ref. 17

13-4 RISK AND ACCIDENT ANALYSIS APPROACHES

Several approaches can be employed to implement the safety studies necessary for a particular plant design. An important concept which is becoming of increasing value in providing focus to such studies is that of risk—defined as the product of accident probability and accident consequences. Methods employed to determine these two elements of risk generally fall into the categories of (1) mechanistic, (2) probabilistic, and (3) phenomenological approaches.

As to be expected, the relative emphasis given to the concept of risk, as well as to the degree to which the basic elements of risk are blended into an overall safety analysis effort, varies somewhat within the international FBR community (e.g., References 18 and 19). Despite the particular approaches used, however, it should be clear from the preceding section that substantial progress has been made over the past two decades in reducing the requirements for FBR accident containment. Much of this progress has resulted from an increased attempt to analyze postulated accident sequences mechanistically, i.e., computationally tracking through cause and effect relationships—rather than relying on upper bound calculations that contain an undetermined number of safety margin layers.

A. RISK

$R = \Sigma \Sigma P c \, D_k$

The individual risk, R_l (consequences per year), due to an accident of type l can be divided into three factors:[4]

$$R_l = P_l C_l D_l, \tag{13-1}$$

where P_l = probability per year that an accident of type l will occur (accidents per year)
C_l = amount of radioactivity released to the atmosphere as a result of an accident of type l (curies per accident)
D_l = consequences resulting from the atmospheric release of one unit of radioactive material (health consequences or dollars per curie).

The total risk, R, is simply the summation of Eq. (13-1) over all possible accident types, i.e.,

$$R = \sum_l P_l C_l D_l. \tag{13-2}$$

For thermal reactors, this expression has frequently been simplifed since one nuclide, iodine-131, has been assumed to dominate the biological effects of the radiation release—independent of the accident type. Hence D_I can be pulled out from under the summation sign (at least for scoping purposes). There is substantial evidence, however, that the presence of water significantly reduces the potential for release of elemental iodine.

In low probability FBR accidents, several radionuclides might control the biological effects, such as cesium, iodine, and plutonium. Any iodine release would probably be in the form of sodium iodide, due to the presence of sodium.

The key item to grasp from this introductory discussion is that the mechanistic approach ("consequence-oriented") is focused mainly on the determination of the C_I terms, whereas the probabilistic approach ("risk-oriented") is concerned mainly with the P_I terms (along with a less accurate estimate of the corresponding C_I values*).

B. ACCIDENT ANALYSIS APPROACHES

The discussion below summarizes the basic ingredients required to determine overall risk; namely, the mechanistic and probabilistic approaches. An overview is also provided for the phenomenological approach. As implied by the discussion of Section 13-2D above, the Lines of Assurance concept recognizes that each of these three approaches has strengths and deficiencies. A principal task of the LOA concept is, therefore, to blend these approaches in an optimal manner to allow an overall system risk assessment to be confidently attained.

Mechanistic

The mechanistic approach is conceptually straightforward. It represents an attempt, by the analyst, to track a transient from the postulated inception of the accident all the way through the various material motions and component failure sequences until the system has reached a long term

---

*The decline in accuracy of the C_I terms is not an intrinsic feature of the probabilistic approach. Rather, it is simply the normal result of an overall balance of effort.

steady-state condition. All of the governing inherent system response features are taken into account in performing this tracking procedure.

This approach is popular because it forces both the analyst and experimenter to determine the accident sequence resulting from direct cause and effect relationships. It is logically appealing because it allows one to visualize the factors of governing importance and also allows sensitivity analyses to be rationally conducted to allow for areas of greatest uncertainty. Numerous computer code systems have evolved over recent years[20] and such computational capability has been highly successful in both reducing containment requirements for major-accident calculations and providing a focus on those areas of uncertainty where both analytical and experimental effects are most urgently needed. Examples of employing the mechanistic approach are given in Chapters 15 and 16.

The principal disadvantage of this approach is that large and complex computational systems must be developed and continually improved to provide a systematic tracking of the numerous combinations of interactions that could occur—especially for analyzing whole-core transients. An additional problem is that of developing computational capability for adequate simulation beyond the point where gross core geometric changes occur.

Probabilistic

The probabilistic approach is focused primarily on (1) an evaluation of the likelihood that certain accident initiating conditions will be attained (e.g., the probability of a specified system fault along with PPS failure), and (2) the relative probability that an accident would progress down a particular path.

In the first area, the probabilistic approach takes into account a systematic evaluation of event trees, fault trees, and the appropriate mathematical techniques to evaluate overall probability for system failures. With regard to the latter area, the approach quantitatively takes into account variations in physical parameters in the reactor system and variations in the models used to describe certain complex physical phenomena (e.g., fuel pin failure models).

As inferred from this discussion, it is clear that the probabilistic approach requires some consequence evaluations to be made in order to assess the implications of probabilistic variations in various parameters in the overall analysis. The consequence calculations, however, are done with computational models that are greatly simplified from the mechanistic models discussed earlier. For example, the SAS code[21] has been used extensively to analyze unprotected FBR accident sequences mechanistically. However, it is so large and time consuming that a greatly simplified version, PARSEC,[22] has been written for use in probabilistic studies. Much of the detail of the SAS simulation capability is, of course, absent in the PARSEC code, but the global prediction capability over a fairly wide range of accident conditions can be attained in the simple PARSEC version by appropriate curve fitting and calibration techniques. By including distribution functions for reactor parameters (e.g., Doppler coefficients) into such a fast running code, accident spectra[23] useful for probabilistic predictions can be obtained.

A major difficulty in employing the probabilistic approach is that the distribution functions for the numerous parameters of uncertainty are not well known. Little failure data are available for systems that have low failure rates. Also, there are many facets of FBR accident analysis—particularly for unprotected accidents—where simplified consequence models are not available for use in probabilistic assessments.

Phenomenological

Both the mechanistic and probabilistic approaches focus to varying degrees on accident logic trees, and considerable effort is expended on training through the implications of all phases—particularly for the mechanistic approach. Experience has indicated, however, that various funnels sometimes exist within the overall logic tree structure. For example, there can be no consequences to the public unless the primary coolant system integrity is lost—no matter what the initiating cause. Hence, emphasis on phenomenological arguments which could demonstrate primary system integrity would allow considerable relaxation regarding the details of how the accident progressed to that point. This is called the phenomenological approach. A particularly elegant presentation of some of the most important phenomenological arguments has been given by Fauske.[24]

A particularly important question associated with unprotected FBR accident scenarios is that of recriticality. The concern is that a partially molten core could become sealed off from coolant flow if cladding and/or fuel should be dispersed out of the core region and freeze in the axial blanket/reflector regions. Mechanistic analyses are very difficult to perform because of the gross geometric changes which exist for such conditions. Phenomenological arguments[24, 25] however, have been used to suggest the natural dispersion tendencies of a core internally heated by radioactive decay. The success of such an approach could be very useful in relieving the safety analyst of considerable computational difficulty. Molten fuel-coolant interactions and plutonium aerosol release represent additional areas where phenomenological arguments might be useful in demonstrating intrinsic reactor system safety—thus relieving expensive and time-consuming mechanistic approaches.

The principal problem with the phenomenological approach is that specific feedback to the reactor designer is limited. Also, the approach does not address certain detail which is often important in particular safety assessments.

REVIEW QUESTIONS

13-1. (a) What is meant by defense in depth?

(b) What multiple barriers exist to protect the public against release of radioactivity in the event of an LMFBR accident?

13-2. Discuss the relationship between Lines of Assurance and the nine classes of accidents listed in Table 13-2.

13-3. Discuss and compare the principal safety problems of an LMFBR and an LWR.

13-4. What are meant by "transient overpower" and "transient undercooling" conditions?

13-5. What is the limited sense of the word "unprotected" when used in the context of an "unprotected transient" in an LMFBR? If an unprotected transient occurs, are the public and environment outside the reactor building protected?

13-6. (a) Discuss the relationship between "hypothetical core disruptive accident" and "Bethe-Tait accident."

(b) Why have predicted values for maximum HCDA work energy decreased with time, as illustrated in Fig. 13-5?

13-7. Write a brief description of each of the accident analysis approaches discussed in Section 13-4.

REFERENCES

1. F. R. Farmer, "Reactor Safety and Siting: A Proposed Risk Criterion," *Nuclear Safety, 8* (1967) 539–548.

2. C. Starr, "Benefit-Cost Studies in Sociotechnical Systems," *Proceedings Colloquium on Perspectives on Benefit-Risk Decision Making*, National Academy of Engineering, Washington, D. C., 1972.

3. N. Rasmussen, *Reactor Safety Study, An Assessment of Accident Risks in U.S. Commercial Nuclear Power Plants*, WASH-1400 (NUREG 75/014), U.S. Nuclear Regulatory Commission, 1975.

4. E. E. Lewis, *Nuclear Power Reactor Safety*, John Wiley and Sons, New York, 1977, Chapter 2.

5. *Accident Facts* (1979 Ed.), National Safety Council, Chicago, Illinois, 1979.

6. *Final Environmental Impact Statement*, NUREG-0139, Clinch River Breeder Reactor Plant, U.S. Nuclear Regulatory Commission, Document No. 50-537, February 1977.

7. D. E. Simpson, W. W. Little, and R. E. Peterson, "Selected Safety Considerations in Design of the Fast Flux Test Facility," *Proc. of the International Conference on the Safety of Fast Reactors*, Aix-en-Provence, France, September 19–22, 1967.

8. *Supplementary Criteria and Requirements for RDT Reactor Plant Protection Systems*, RDT C 16-1T, December 1969, Division of Reactor Development and Technology, U.S. Atomic Energy Commission.

9. E. E. Kintner, "Engineering of U.S. Fast Breeder Reactors for Safe and Reliable Operation," *International Conference on Engineering of Fast Reactors for Safe and Reliable Operation*, October 9–13, 1972, Karlsruhe, Germany.

10. *Preparation of Environmental Reports for Nuclear Power Stations*, NUREG-0099, Revision 2, July 1976, U.S. Nuclear Regulatory Commission.

11. *Nuclear Safety Criteria for the Design of Stationary Pressurized Water Reactor Plants*, ANS-51.1 (N18.2-1973). American Nuclear Society, La Grange Park, Illinois.

12. *Nuclear Safety Criteria for the Design of Stationary Boiling Water Reactor Plants*, ANSI/ANS-52.1-1978. American Nuclear Society, La Grange Park, Illinois.

13. *General Safety Design Criteria for an LMFBR Nuclear Power Plant*, ANS-54.1, Trial Use Draft, April 1975. American Nuclear Society, La Grange Park, Illinois.

14. *LMFBR Safety Classification and Related Requirements*, ANS-54.6, Trial Use Draft, October 1979. American Nuclear Society, La Grange Park, Illinois.

15. J. D. Griffith, R. Avery, J. Graham, P. Greebler, R. Keaten, and D. E. Simpson, "U.S. Approach to LMFBR Risk and Safety R&D Cost Benefit Assessment," *Presented at the ENS/ANS Topical Meeting on Nuclear Power Reactor Safety*, October 16–19, 1978, Brussels, Belgium.

16. H. A. Bethe and J. H. Tait, *An Estimate of the Order of Magnitude of the Explosion When the Core of a Fast Reactor Collapses*, RHM-56-113, April 1956.

17. John Graham, "Selection of Safety Design Bases for Fast Power Reactors," *Proc. Fast Reactor Safety Meeting*, Beverly Hills, California, April 2–4, 1974, CONF-740401-P3, U.S. Atomic Energy Commission (1974), 1647–1659.

18. P. Tanguy, "A French View on LMFBR's Safety Aspect," *Proc. of the Int. Mtg. on Fast Reactor Safety Technology, Vol. I*, Seattle, WA (1979) 9–16.

19. G. Kessler, "Safety Levels Satisfactory for the Commercialization of the LMFBR," *Proc. of the Int. Mtg. on Fast Reactor Safety Technology, Vol. V*, Seattle, WA (1979) 2672–2682.

20. A. E. Waltar and A. Padilla, Jr., "Mathematical and Computational Techniques Employed in the Deterministic Approach to Liquid-Metal Fast Breeder Reactor Safety," *Nuclear Sci. and Eng., 64* (1977) 418–451.

21. D. R. Ferguson, et al, "The SAS4A LMFBR Accident Analysis Code System: A Progress Report," *Proc. International Meeting Fast Reactor Safety and Related Physics*, Chicago, Illinois, October 5–8, 1976, CONF-761001, U.S. Energy Research and Development Administration (1977).

22. J. K. Vaurio and C. Mueller, "A Probabilistic/Deterministic Procedure for Analyzing LMFBR Core Disruptive Accidents," *Proc. International Meeting Fast Reactor Safety and Related Physics*, Chicago, Illinois, October 5–8, 1976, CONF-761001, U.S. Energy Research and Development Administration (1977).

23. K. O. Ott, "Probabilistic Fast Reactor Accident Analysis," *Nucl. Sci. and Eng., 64* (1977) 452–464.

24. H. K. Fauske, "The Role of Core Disruptive Accidents in Design and Licensing of LMFBR's," *Nuclear Safety, 17-5* (1976) 550–567.

25. H. K. Fauske, "Boiling Fuel-Steel Pool Characteristics in LMFBR HCDA Analysis," *Trans. ANS, 22* (1975) 386.

CHAPTER 14

PROTECTED TRANSIENTS

14-1 INTRODUCTION

A Plant Protection System (PPS) is routinely provided in an FBR, as in any nuclear reactor system, to provide a rapid and safe shutdown for abnormal situations. The prime objective of such a PPS is to assure protection of the general public, the workers, and the plant investment.

Fundamental to any transient situation which might arise is the heat generation-to-heat removal ratio. Hence, an effective PPS must be able to arrest any undesirable increase in heat generation or loss of heat removal capability. This generally becomes translated into the two major parts of the PPS: (1) a reactivity shutdown system, e.g., scrammed absorber rods, and (2) a decay, or residual, heat removal system.* Other engineered safeguards features may be included in the overall plant system, such as containment isolation, fission product removal from containment, or post-accident heat removal devices, but such systems will not be discussed in this chapter since they are not needed or activated if the PPS successfully performs its function.

It is important to recognize that successful action by the PPS to any anticipated accident condition renders the LMFBR an exceptionally safe system. This results from the low pressure operation of the system, at temperatures far below boiling, and the large heat capacity of the coolant. It is only if the PPS should fail when needed that major consequence accidents need be considered. Such a realization emphasizes the benefits to be derived by assuring an effective and reliable PPS.

The purpose of the present chapter is to focus attention on Protected Transients. In particular, this directly involves the plant protection system itself and the reliability considerations which are essential for proper PPS

---

*The decay heat removal system is not always considered an integral part of the PPS. We have chosen to include it in this manner to provide the emphasis due such an important subsystem.

action. Both localized transients resulting from local fault propagation and transients which affect the whole core are considered—assuming the PPS to function as designed. Additional accident conditions which must be considered from an overall balance-of-plant point of view are then briefly discussed.

14-2 PLANT PROTECTION SYSTEM

In designing an effective PPS, it is first necessary to determine the accident conditions for which the PPS is expected to respond. Since it is always possible to postulate accident conditions beyond the boundary for which the PPS can effectively cope, it is important to define this boundary and to demonstrate that such postulated conditions are of a physically impossible, or highly improbable, nature. The next step is to establish the core integrity limits (e.g., fuel and cladding temperature, cladding strain) beyond which positive PPS action would not be fully effective. Desired PPS response can then be specified.

A. ACCIDENT CONDITION ENVELOPE

An imbalance in the heat generation-to-heat removal ratio constitutes the primary concern for in-core accident analyses. As long as the nuclear heat generated is adequately removed, all of the intrinsic barriers to radiation release (e.g., fuel matrix, cladding) remain intact. It is only when an over-heated condition exists that a potentially serious safety question arises. Consequently, the fundamental questions which must be posed in designing an effective PPS are focused on malfunctions which could lead to overheating (i.e., reactivity insertion mechanisms) and undercooling (i.e., impairment of heat removal).

Reactivity Insertion Mechanisms

The most obvious area for scrutinizing possible reactivity insertions is the potential motion of core materials which involve significant reactivity worth such as control material, fuel, and sodium. Control rod withdrawals constitute the dominant operational mode for purposely increasing reactivity into the reactor. Hence, it is natural to ask what the core response would be if such rods were inadvertently withdrawn too fast or too far. Interlocks normally prevent more than one rod from being withdrawn at one time, and a maximum speed of withdrawal is purposely designed into the system. There are numerous barriers, of both a design and an intrinsic nature, that prevent excessive reactivity insertion from control rod withdrawal. Even given the worst combinations of errors leading to rapid control rod withdrawal, the maximum reactivity insertion rates are typically well under 1$/second.

As noted from Table 14-1, other possible reactivity insertion mechanisms include possible control assembly or fuel assembly meltdown, loss of hydraulic balance to the fuel assemblies, radial displacement of the core assemblies, fuel assembly voiding, and cold sodium insertion. The first two possibilities could conceivably lead to reactivity insertion rates of the order of a dollar per second, but only if highly pessimistic meltdown models are postulated. The concern over loss of hydraulic holddown is that some assemblies might "float" upward during steady-state operation and then suddenly fall back to their normal locations as a result of a mechanical shock to the system (e.g., earthquake). However, cores are arranged to minimize this possibility by incorporating mechanical barriers to limit axial movement of in-core assemblies. Radial core displacement normally occurs in a transient due to thermal expansion, but it is also possible that assembly "bridging" patterns could be established following fuel transients or during the core clamping process following refueling. The concern is that the resulting core porosity could be lost during a system shock and a rapid reactivity insertion might result. However, detailed analyses and full scale mockup tests indicate the maximum reactivity insertion due to such a possibility would typically be well under a dollar. Fuel assembly voiding could result in a positive reactivity insertion due to a positive sodium void coefficient, but the maximum sodium void worth is rarely more than a small fraction of a dollar for any one assembly. Cold sodium entering the core (due to an abrupt change in the heat-sink, resulting in overcooling in the IHX) could result in a positive reactivity insertion if the overall core sodium void coefficient is negative.

TABLE 14-1 Potential Reactivity Insertion Mechanisms

Mechanisms	Implications
Uncontrolled Control Rod Withdrawal	
Control Assembly Meltdown	$\leq 1\$/s$ Reactivity Insertion Rate
Fuel Assembly Meltdown	
Loss of Hydraulic Holddown	Few Dollars Maximum Reactivity Insertion
Radial Core Displacement	
Fuel Assembly Voiding	Step Reactivity Insertion Well Under One Dollar
Cold Sodium Insertion	

Impairment of Heat Removal

The loss of primary loop flow constitutes the most serious concern from the standpoint of the heat transport system, although any event which could allow the inlet coolant temperature to rise is also to be avoided. Table 14-2 contains a partial list of failures in the heat transport system which are normally investigated for required PPS response.

TABLE 14-2   Undercooling Considerations

-----PRIMARY LOOP-----

Loss of off-site electrical power
Loss of off-site and emergency diesel electric power
Loss of electric power to one primary pump
Continuous flow reduction by control equipment failure
Mechanical failure of one primary pump

-----SECONDARY LOOP-----

Loss of electrical power to one secondary pump
Mechanical failure of one secondary pump
Secondary pump control failure

-----STEAM LOOP-----

Inadvertent opening of steam generator outlet relief
    valve
Inadvertent opening of steam generator inlet dump
    valve
Steam line rupture
Feedwater line rupture
Feedwater pump failure

The loss of off-site electrical power is normally included as an anticipated event and, consequently, it is particularly important to demonstrate that the PPS system will function properly in this event. A safe and non-damaging shutdown must also be assured even if the backup electrical supply fails to come on line. A continuous flow reduction due to control equipment malfunctions tends to lead to higher cladding temperatures than the loss of power events because scram signals are not received until elevated core heating has already occurred. Individual pump failure, due either to power loss or mechanical failure (pump seizure or a breach in the shaft or impeller blades) leads to a considerable flow imbalance and the consequences must be assessed.

Failures in the secondary or steam loop systems tend to provide less severe consequences to the core since some thermal time delay is inherently present. However, the consequence of all credible events must be assessed to assure that the propagation of the initiating accident could not lead to unacceptable shutdown conditions.

Discussions of plant responses to two loss-of-flow transients, one in the primary system and one in the steam system, are provided in Section 14-5B.

B. PPS EFFECTIVENESS CONSIDERATIONS

A properly functioning PPS should arrest the type of accidents discussed above and effect a safe, cold shutdown with no appreciable core damage. However, three concerns exist regarding PPS effectiveness which warrant discussion.

First, it is conceivable that the initiating event might exceed the capability of the PPS. Second, it could be postulated that an event might progress rapidly beyond PPS capability. Finally, there is the remote possibility that the PPS fails completely when called upon to perform.

With regard to the first concern, it is possible to postulate an event taking place which could conceivably introduce reactivity into the system so rapidly that it is not practical to design an effective PPS response. Examples might include reactor vessel or core support structure ruptures, the passage of a large coherent gas bubble through the positive void reactivity portion of the core, or rapid expulsion of a control rod. It is important to demonstrate that such events are physically impossible* or that their probability of occurrence is acceptably remote.

The second concern is associated with an event which might progress undetected and unchecked until it reaches a point at which it exceeds the protective capability of the PPS. The propagation of fuel element failure is the classic concern in this regard. All evidence to date suggests little propagation potential and an even smaller likelihood that any such propagation would be very large and rapid (discussed in Section 14-4).

Complete failure of the PPS, when action is required, is a very low (but not zero) probability event. There is an extensive effort underway to gain sufficient reliability data to demonstrate that complete PPS failure is sufficiently low to be considered negligible. Key to this demonstration is the development of systems which are functionally designed to provide adequate plant protection by including redundancy to the extent that no one failure would incapacitate the entire system, employing sufficient diversity to minimize the potential for common mode failure, and incorporating

---

*It may be appropriate to introduce design changes to mitigate or obviate potential areas of concern.

system independence to the maximum extent possible (discussed in Section 14-3).

There are two other accident conditions which have caused some concern because they could possibly lead to major accident consequences even if the PPS performs precisely as designed. These include a massive, double-ended pipe rupture at a critical location for the loop-type reactor, and the loss of the ultimate heat sink. The former accident might be ruled out as hypothetical (because of the highly unlikely probability of an abrupt, catastrophic failure in a low pressure system) and it would appear that the latter can be designed against (by including an active emergency cooling system, if needed).

C. REACTOR SYSTEM INTEGRITY LIMITS

Given the spectrum of accidents listed above, it is essential to ensure that a proper response of the PPS will provide adequate protection to the reactor system. To be assured of such protection, it is necessary to determine the fuel pin and heat transport system limits which must not be exceeded during such accident conditions.

In arriving at such limits, it is common practice to relate the severity level allowed with the probability of occurrence of the accident. This approach, as outlined in Fig. 13-3, has long been used in the fast reactor industry, but the nomenclature employed for the various accident severity levels has varied. Table 14-3 is included, therefore, to provide a tie between the concept outlined in Fig. 13-3 with other descriptions commonly used. Also included in Table 14-3 are representative damage limits employed early in the FFTF project. For this reactor, it was assumed during the design that fuel pin integrity could not be assured if cladding strain exceeded 0.7% or if fuel melting exceeded 50% at the axial midpoint.

Core Damage Limits

In establishing transient design limits for reactor core structural components, the fuel pin and control rod absorber pin cladding receive particular scrutiny due to their lead role in controlling the accident sequence. The objective of such analysis is to define realistic limits, which are based upon an irradiation test program, to establish the capability of the cladding under steady-state and transient conditions. Extensive testing has, for example, been performed to establish FFTF cladding behavior characteristics and has led to the levels listed in Table 14-3. It should be noted, however, that

TABLE 14-3 Representative Damage Designations
(FFTF Design Limits)

Accident Severity Level	Fault Classi- fication*	Fuel Melting Criteria†	Cladding Temperature Criteria (°C)	Cladding Strain Criteria	Heat Transport ASME Code Condition‡
No Damage	Operational	0	650	0.2%	Normal
Operational Incident	Anticipated	0	810	0.3%	Upset
Minor Damage	Unlikely	0.1	870	0.7%	Emergency
Major Damage	Extremely Unlikely	0.5	980	-	Faulted

*The designations used for RDT Standards
†Radial fraction of fuel molten at axial midpoint of hottest pin.
‡See Reference 1. The nomenclature for these design and service limits was revised in 1980. "Normal" to "Faulted" was redesignated "Level A" to "Level D."

continued testing may reveal new limits which might be more appropriate for future plants. Indeed, the life fraction rule, as discussed in Chapter 8-3, may represent a more definitive manner in which to specify damage limits, although significant uncertainty still exists in applying the life fraction rule to transient conditions.

Fuel Melting Criteria. One of the first performance limits used in evaluating the FFTF fuel pin design[2] was the extent of fuel melting. These limits, established for various levels of accident severity, were based upon observations from early experiments that had been performed with mixed oxide fuel and upon the results of analytical investigations of the interaction between fuel and cladding during rapid overpower excursions. More recent efforts have replaced fuel melting criteria in favor of the cumulative damage techniques discussed in Chapter 8-3B.

Cladding Temperature Criteria. Fuel melting criteria can be considered applicable only in overpower transients sufficiently rapid to cause little increase in cladding temperature. For slow transients, or flow decay transients, cladding temperature is a major factor in cladding integrity. The integrity limit corresponds to the ultimate strength, which would depend upon temperature. Lower strain limits, corresponding to lower temperatures for a given design internal pressure, would apply for normal conditions and operational incidents.

Cladding temperature limits, however, are not absolute; rather, the temperature limits must be established for a particular reactor and a particular set of operating conditions. A limiting cladding temperature for one system may be overly conservative for another system (or vice versa) even if both systems use identical cladding material. The reason for this is that the loading conditions, design life, and the expected steady-state and transient operating temperatures may be significantly different for the two systems.

Cladding Strain Limits. Cladding strain limits for FFTF fuel and absorber pin design were established with the objective of defining design criteria which were independent of the reactor operating history.

An overall cladding integrity limit was established which corresponded to 0.7 percent calculated permanent cladding strain. This value was originally selected as an extrapolation of limited irradiation data on stainless steel cladding.[3] Based on the 0.7 percent integrity limit, a calculated value of 0.2 percent strain was selected to cover steady-state operation and an additional calculated 0.1 percent strain (i.e., a total of 0.3 percent) was allocated to cover anticipated faults or operational incidents. The remaining 0.4 percent strain increment was then available to allow for a single unlikely fault at the end of the fuel pin design life.[4]

The conservatism of the FFTF fuel pin design has been demonstrated by comparison of predicted steady-state design lifetime with test data from approximately 600 FFTF prototypic fuel pins (including several transient tests in capsules) with 20% cold worked 316 stainless steel cladding irradiated in EBR-II[5,6] and about 50 pins transient-tested in the Transient Reactor Test Facility (TREAT).[7]

Reactor Heat Transport System Damage Limits

The Reactor Heat Transport System (HTS) and connected systems provide the necessary functions to remove reactor heat safely under all plant operating conditions. Boundary materials and service conditions of heat transport systems should be selected and maintained to provide materials properties which will assure non-brittle behavior and avoid conditions that could lead to rapidly propagating failures.

The reactor coolant pressure boundary components are designed and analyzed in accordance with the ASME Boiler and Pressure Vessel Code, Section III, Class 1 requirements (Section III, Nuclear Power Plant Components).[1] Conformance with these requirements will assure HTS structural adequacy at operating temperatures under specified loading conditions, both normal and off-normal, including internal pressure, dead weight, seismic events, design thermal events, and design basis accidents.

Table 14-3 provides an example which shows how the relevant ASME Code conditions for the HTS can be associated with a given accident severity level or fault classification. These Code conditions, in turn, specify the maximum stress or strain allowed.

14-3 RELIABILITY CONSIDERATIONS

The need for and functions of an effective PPS have been emphasized in the preceding sections. The purpose of the present section is to discuss the role which reliability plays in assuring an effective PPS. We begin our discussion with the general elements which comprise overall reliability and then direct specific attention to the plant shutdown and heat removal paths of the overall PPS. With this background, it is then possible to deal with overall reliability goals.

A. BASIC ELEMENTS OF RELIABILITY

The five basic elements which contribute to overall system reliability include (1) functionability, (2) redundancy, (3) diversity, (4) independence, and (5) serviceability.

Functionability

By functionability, we mean the design capability of the PPS to perform its assigned task if called upon to function. It must, for example, contain sufficient negative reactivity worth to overcome any reactivity which might be inserted by the accident initiating mechanism.* Furthermore, the system must be capable of inserting such negative reactivity into the reactor fast enough to overcome the mechanism causing the accident. Fortunately, for any of the credible accident initiators which have been identified, the functionability requirements of the PPS are relatively easy to meet. For a control assembly in a commercial size plant, a reactivity worth of a few dollars (cf. Table 6-5) and an insertion time of the order of a second are quite sufficient. Spring assisted rod insertion drives are often used for certain FBR shutdown rods, but this is more a matter of system diversity

---

*It is now general practice to design the reactivity shutdown system to provide adequate reactivity shutdown with the most reactive control assembly inoperable. This has commonly become known as the "stuck rod" criterion (cf. Chapter 6-9).

than of need for rapid response. Hence, systems similar to those routinely employed in the thermal reactor industry are generally adequate for fast reactor application.

Redundancy

Redundancy implies that multiple systems exist, each fully capable of providing the required safety functions. As an example, two self-sufficient systems to provide reactivity shutdown were described in Chapter 6.

Diversity

Diversity means that different principles are employed in the multiple safety systems utilized. Diversity is important if the reactivity shutdown and heat removal systems are to be as free as possible from common cause failures. A considerable incentive exists with regard to enhancing overall PPS reliability to employ diverse systems all the way from the accident sensing devices to the activation of the power reducing or heat removal component itself.

Several examples of diversity will be given in Sections B and C, below.

Independence

Independence denotes complete separation of the multiple safety systems so that the successful operation of one system in no way is affected by a failure of another system. Elements of system independence have been shown in some of the previous examples, but there must be a conscious effort in the design and construction of an effective PPS to assure that the separate safety systems are indeed independent.

Serviceability

Serviceability refers to the capability of the PPS not to function spuriously when there exists no requirement for its operation. It would be fairly easy to design a PPS with so much redundancy, diversity, and independence that the probability of failure when called upon to function would be exceedingly low. The challenge is to design such a system which retains this attribute without spuriously initiating scrams during periods of normal operation.

Example of Reactivity Shutdown System Reliability

Several basic elements of reliability can be demonstrated by the reactivity shutdown system for FFTF, as illustrated in Fig. 14-1. Both independence and redundancy are readily identified by noting the separation and duplica-

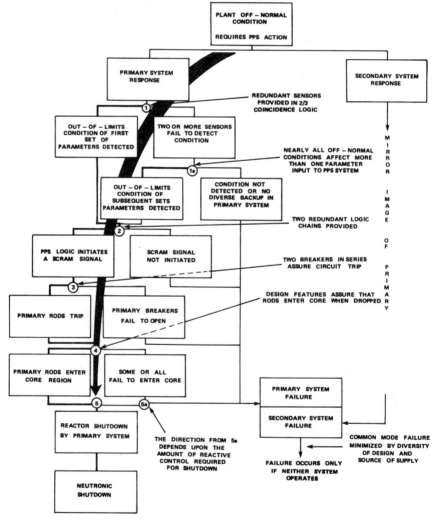

FIGURE 14-1. Schematic of the reactivity shutdown logic for primary and secondary rod system response in the FFTF.

tion exhibited by the two (primary and secondary) shutdown systems. Other features become evident by a careful study of the detail shown for the primary system.

Proper operation of the primary system is indicated by the heavy lined path along the left part of the figure. For each of the numbered junctures where a failure could occur, positive design measures are taken to reduce the probability of failure to a minimum. For example, juncture ① indicates that when the first out-of-limits condition occurs (e.g., high flux), this circumstance is detected if any two out of three sensors register in coincidence as expected, and the logic flows to juncture ②. If sensor failures on the first out-of-limits condition occur, juncture ⑴ₐ is encountered. At that point a second out-of-limits condition exists (e.g., high temperature). Positive action by two out of three sensors at this point will revert the logic back to juncture ②. (A list of potential scram signals is given later in Table 14-4.) Similar redundance continues at each step. Several such out-of-limits conditions may be available for detection, although perhaps somewhat delayed in time.

It should be noted that only partial success of the primary system is required to arrest most accident conditions. A major observation is that *both* the primary and the secondary system have to fail completely in order to render the reactivity shutdown system ineffective. Since the secondary system is a mirror image of the primary (but with different input signals*), the numerical probability of reactivity shutdown system failure is the product of the failure probability for these two independent and redundant systems. A prime consideration with regard to independence is that of providing physical barriers between the systems and/or maximizing the separation of system electrical cables.

An example of serviceability provided for this design can be noted from the two out of three logic existing for the sensors at junctures ① and ⑴ₐ. This logic allows one of the three sensors to spuriously activate without causing the system to scram.

Diversity, though not explicitly shown in Fig. 14-1, constitutes a very important part of the system. The degree to which diverse driving signals and diverse mechanical hardware is present in such a design is discussed in the following section.

B. ILLUSTRATIONS OF DIVERSITY IN THE REACTIVITY SHUTDOWN SYSTEM

A few examples of system diversity in the primary and secondary reactivity shutdown systems are included in this section to illustrate meth-

---

*See Table 14-4.

TABLE 14-4 Scram Signals for Actuating Reactivity Shutdown
Systems in the FFTF

Primary Shutdown System	Secondary Shutdown System
1. Power range nuclear high flux	1. Flux/total flow
2. Power range nuclear low setting	2. Flux-increasing delayed flux
3. High startup flux	3. Flux-decreasing delayed flux
4. Flux-decreasing delayed flux	4. Low primary loop flow
5. Flux-increasing delayed flux	5. High primary loop flow
6. $Flux^2$/loop pressure	6. Low secondary flow
7. IHX primary outlet temperature	7. Loss of off-site power
8. Reactor vessel coolant level	8. Reactor outlet plenum temperature
9. Flux/closed loop flow	9. Closed loop outlet temperature
10. Closed loop IHX primary outlet temperature	10. Experiment-associated trip functions
11. Experiment-associated trip functions	11. Flux/closed loop flow
	12. High closed loop flow

ods currently being used to minimize the potential for common cause failure.

Table 14-4 lists the scram signals which actuate the primary and the secondary shutdown systems for FFTF. An immediate observation is that there are 11 or 12 signals to initiate either shutdown system; of perhaps more importance, the input signals are different for each system.

Table 14-5 shows the diversity incorporated into the early design of the reactivity shutdown system of the CRBRP. The conscious effort to include diversity in the design wherever possible is clearly shown.

Whereas the above examples have demonstrated areas where considerable diversity can be included in the PPS design, the reactivity shutdown systems were all based on a rodded configuration for the absorber elements. It could be argued, therefore, that an accident sufficiently severe to grossly distort the control rod or safety rod assembly guide tubes might disable both systems for driving absorber material into the core. Hence, varying degrees of efforts have long existed to devise an effective shutdown system which uses a principle for core absorber insertion fundamentally different from a rod.

A concept being implemented in SNR-300 reactor[9] is to assemble the absorber material in a chain-type (articulated joint) configuration and allow a spring to thrust the absorber from its normal position below the core up into the core region upon demand. As illustrated in Fig. 14-2, the basic idea is to allocate one core assembly position for the purpose of pulling up this flexible absorber "chain," which could presumably effect neutronic shutdown even if the assembly guide tube was considerably distorted.

TABLE 14-5 Diversity in the Reactivity Shutdown System in the CRBRP design[8]

	Primary System	Secondary System
Control assembly (CA)		
Control rod	37-pin bundle	19-pin bundle
Guide geometry	Hexagonal	Cylindrical
Number of control rods	15	4
Control-rod-drive line (CRDL)		
Coupling to CA	Rigid coupling	Flexible collet latch
Connection to control-rod drive mechanism (CRDM)	CRDL lead screw to CRDM roller nuts	CRDL attached to CRDM carriage with pneumatic activation of CRDL latch through slender rod
Disconnect from CA for refueling	Manual	Automatic
Control-rod-drive mechanism (CRDM)		
Type	Collapsible roter-roller nut	Twin ball screw with translatingcarriage
Overall stroke	0.94 m	1.75 m
Scram function		
Scram release	Magnetic, release CRDM roller nuts	Pneumatic, release CRDL, latch in CA
Scram assist	Spring in CRDM	Hydraulic in CA
Scram speed vs flow rate	Increases with decreasing flow rate	Decreases with decreasing flow rate
Scram assist length	0.55 m	Full stroke
Scram deceleration	Hydraulic dashpot	Hydraulic spring
Scram motion through internals	Full stroke	6.4 m
Designer	Westinghouse	General Electric

Another concept is that of hydraulically suspended absorber balls.[10] As illustrated in Fig. 14-3, the coolant flow in this special assembly would provide sufficient force to support the absorber balls hydraulically over the top of the core during normal operation. Should a reduction in primary loop flow occur, as in the case of a TUC,* the balls would fall into the core region and provide reactivity shutdown. For the particular example shown,

---

*See Chapter 13-3B for the definition of TUC events.

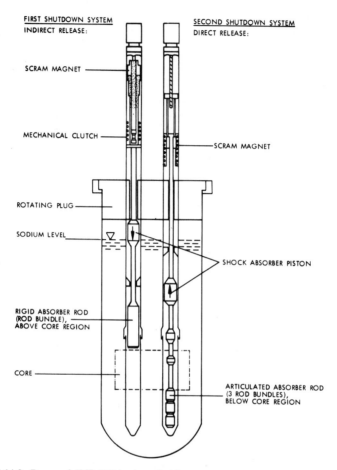

FIRST SHUTDOWN SYSTEM
INDIRECT RELEASE:

SCRAM MAGNET

MECHANICAL CLUTCH

ROTATING PLUG

SODIUM LEVEL

RIGID ABSORBER ROD
(ROD BUNDLE),
ABOVE CORE REGION

CORE

SECOND SHUTDOWN SYSTEM
DIRECT RELEASE:

SCRAM MAGNET

SHOCK ABSORBER PISTON

ARTICULATED ABSORBER ROD
(3 ROD BUNDLES),
BELOW CORE REGION

FIGURE 14-2. Proposed SNR-300 backup shutdown system.

a curie point temperature magnetic device,* located below the core in this special assembly, has been included in the design to allow the system to function even in the event of a TOP.† This is accomplished by placing a fissionable material (which would heat during a TOP event) to drive the temperature magnetic device and force a valve to close. This action would stop coolant flow, thus causing the absorber balls to drop rapidly into the core. A variation of this design is the inclusion of a cup on the assembly

---

*Such a device operates on the principle that particular materials lose their magnetic behavior once a certain temperature is attained. This concept provides an appealing basis for designing a device to automatically drop any kind of solid control material into the core in response to elevated coolant temperature.

† See Chapter 13-3B for the definition of TOP events.

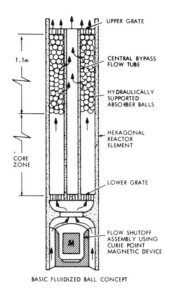

1.1m

CORE
ZONE

UPPER GRATE

CENTRAL BYPASS
FLOW TUBE

HYDRAULICALLY
SUPPORTED
ABSORBER BALLS

HEXAGONAL
REACTOR
ELEMENT

LOWER GRATE

FLOW SHUTOFF
ASSEMBLY USING
CURIE POINT
MAGNETIC DEVICE

BASIC FLUIDIZED BALL CONCEPT

FIGURE 14-3. Hydraulically suspended absorber ball concept for backup shutdown system.

outlet which would drop and seal off coolant flow to the assembly upon the receipt of an overpower signal. While a design of this type seems appealing on initial inspection, one must be cautious in the overall evaluation of such a concept. For example, if a core containing balls was disrupted, it is appropriate to ask whether one would really feel as secure when dismantling the core as would be the case if only solid rods of absorber material were in the core. Also, there is a concern of a mechanical failure allowing some balls to be suddenly washed out, i.e., the "ball popout" accident.

A third concept[11] is under development which combines the most favorable features of the two concepts mentioned above. This device employs an articulated absorber section, similar to the SNR-300 design, but uses a curie point magnetic latch feature above the core. The absorber device is released downward into the core if either (1) holding coil current is lost, or (2) the coolant temperature rises beyond the curie point.

C. DECAY HEAT REMOVAL SYSTEM*

The other major part of the PPS is the Decay Heat Removal System, also sometimes called the Residual, or Shutdown, Heat Removal System. Even if

*An extended description of decay heat removal with emphasis on natural convection was published after the present manuscript was completed. The reader is referred to A. K. Agrawal and J. G. Guppy, editors, *Decay Heat Removal and Natural Convection in Fast Breeder Reactors*, Hemisphere Publishing Corp., New York, 1981.

reactivity shutdown is successfully achieved, there can be substantial heat generation by radioactive decay (as shown in Fig. 14-4). In addition to fission product decay, some decay heat is produced by beta decay of ^{239}U and ^{239}Np plus smaller amounts from decay of activation products (e.g. steel, sodium) and higher order actinides such as ^{242}Cm (cf. Chapter 16-5A).

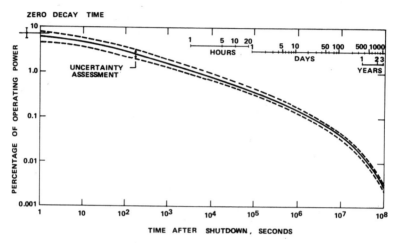

FIGURE 14-4. Fission product decay heat for FFTF.[12]

One example of reliability considerations for this system is afforded in the CRBRP design, as shown in Fig. 14-5. For this system, there are three backup systems to remove the decay heat should the normal heat sink (balance of plant steam/feedwater system) fail. The first is protected air-cooled condensers (PACC) which cool the steam drum. A second heat sink can be made available by opening the safety relief valve in the steam line between the steam drum and the turbine and venting steam to the atmosphere. A protected storage water tank is available for supplying make-up water, in addition to an alternate water supply. Two of the auxiliary feedwater pumps are electrically driven, but a steam-driven pump is also available. Finally, a completely separate Overflow Heat Removal System (OHRS) is provided to extract heat directly from the in-vessel primary loop position. The heat sink for this system is provided by an air blast heat exchanger, which represents diversity from the water system in the steam loop.

The Super Phénix decay heat removal system includes sodium-to-air heat exchangers on the secondary sodium loops plus four backup decay heat removal systems. Each backup system includes an immersion cooler in the reactor tank, a closed sodium loop, and a sodium-to-air heat exchanger. The

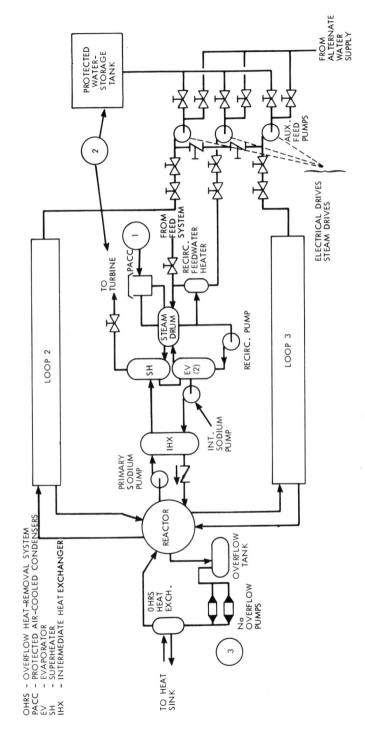

FIGURE 14-5. Decay heat removal system for CRBRP design.[8]

OHRS – OVERFLOW HEAT-REMOVAL SYSTEM
PACC – PROTECTED AIR-COOLED CONDENSERS
EV – EVAPORATOR
SH – SUPERHEATER
IHX – INTERMEDIATE HEAT EXCHANGER

SNR-300 also has two backup decay heat removal loops with immersion coolers and sodium-to-air heat exchangers.

Natural Convection Operation

In addition to the backup heat removal systems, the components in the Reactor Heat Transport System are arranged in such a way (as described in Chapter 12-2A) that decay heat can be removed by natural convection in the event of complete loss of pumping capability. For long-term cooling, this requires reactor scram and the availability of an ultimate heat sink. The backup decay heat removal systems of SNR-300 and Super Phénix are designed in such a way that core damage is avoided even with passive operation of the systems, i.e., with natural convection in the sodium-to-air heat exchangers as well as the sodium loops. It still must be demonstrated under actual LMFBR operating conditions that natural convection can, in fact, be established rapidly enough to prevent fuel damage. Extrapolation of successful early tests at both PFR and Phénix in which pumps were completely cut off at partial power indicate that natural convection can be readily established after loss of pumping capability at full power. The tests performed at FFTF to observe the transition to natural convection after shutting off all pumps at full power (including the blowers on the sodium-to-air heat exchangers) provide further confirmation of natural convection cooling capability.

D. OVERALL RELIABILITY GOALS

It is well recognized that the various aspects just discussed all contribute to the effectiveness of overall PPS reliability. A major question is the numerical level which the PPS must attain such that the consequences of PPS failure need not be considered for safety evaluation purposes.

One way to develop a framework for addressing the numerical value for sufficiency in the PPS acceptability is to observe the risks which society is apparently willing to accept for normal living patterns. Starr found that the present acceptable risk from the production of electricity is of the order of 20×10^{-6} fatalities per person per year.[13] From this, Starr later concluded [14] that a target risk level for nuclear stations of 10^{-6}, which is well below that of natural hazards, should be well within the socially accepted range.*

---

*An additional allowance may have to be made for the psychological concerns of "catastrophic" accidents, discussed in Chapter 13-1.

It should be noted that his figure denotes the risk of actual fatalities. Graham and Strawbridge[15] observe that if a goal of 10^{-6} PPS failures/reactor year for FBR's was established, then the risk of fatalities would be substantially lower. For example, if we define

10^{-x} = probability that PPS failure leads to significant radioactivity release from containment

10^{-y} = probability that released radiation will cause a public fatality

then

$$10^{-(6+x+y)} = \text{probability of fatalities to the public.}$$

This figure is considerably less than 10^{-6}, perhaps several orders of magnitude lower. In fact, the logic of the Lines of Assurance (Chapter 13) was initially based upon an attempt to determine such a number and provide the confidence necessary to gain wide acceptance of its validity.

The United States Nuclear Regulatory Commission, after much deliberation on the issue of Anticipated Transients Without Scram (ATWS) for present power reactors,[16] concluded the following: For a population of one thousand nuclear plants in the U.S., the safety objective will require that there be no greater than one chance in one million per year for an individual plant of an accident with potential consequences greater than 10 CFR Part 100 guidelines.

Although no firm goals have yet been established for FBR industry acceptance, a figure of 10^{-6} PPS failures per reactor year has been widely discussed and, from the above background discussion, would appear to represent a conservative goal, consistent with the ATWS guidance.

14-4 LOCAL FAULT PROPAGATION

The evaluation of the potential for fuel element failure propagation (FEFP) in liquid-metal-cooled fast breeder reactors has been the subject of considerable research over a number of years in most countries involved in the development of the LMFBR.[17] The study of FEFP is concerned with the occurrence of initially small local faults in the core that could possibly grow into more widespread serious faults.

Pin-to-pin failure propagation is defined as a failure in one pin causing failure in an adjacent pin (Fig. 14-6). In concept, such propagation could be either self-limiting, with damage confined to a region of the pin bundle, or progressive, with the potential for involving the whole assembly. Assembly-to-assembly propagation is defined as the condition in which an assembly with an initial failure causes damage in a neighboring assembly.

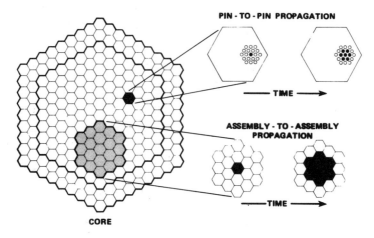

FIGURE 14-6. Local fault propagation concerns.

One original concern was that pin-to-pin failure propagation might begin slowly, but then proceed more rapidly and possibly involve even assembly-to-assembly failure propagation on a time scale that would be difficult to arrest with the Plant Protection System. Hence, in assessing such possibilities, the two overriding questions were as follows: (1) Can such failure propagation readily occur, and (2) if so, is individual assembly instrumentation necessary for PPS initiation? There now appears to be overwhelming evidence that such propagation for the breeder reactors of current design (mixed uranium-plutonium oxide pellets, stainless steel cladding) is very improbable. The basis for this determination is outlined below.

A. PIN-TO-PIN FAILURE PROPAGATION

Potential Initiating Mechanisms

There are three general categories of phenomena which could initiate a pin-to-pin failure sequence. Random failures due to manufacturing defects could cause fission gas release as a mechanism to induce failure propagation. Local blockage formed by particulate matter could lead to higher local coolant temperatures (possibly including limited local boiling). Finally, fuel pin loading enrichment errors could cause local melting and the release of molten fuel. Other subcategories include insufficient fuel pin heat transfer and stochastic fuel pin failure.

Pin Failure Detectability

Before investigating the potential of each of these mechanisms to cause propagation, it is of interest to touch briefly on the issue of pin failure detectability. At least four possible methods exist for detecting fuel pin failure. First, an increased assembly outlet coolant temperature could be detected, but a substantial fraction of the pins would have to be failed before a sufficient assembly flow perturbation could be caused to affect the outlet temperature. The second method is the use of delayed neutron detectors, in which delayed neutrons from fission product delayed neutron precursors from failed fuel are detected in the coolant. A third possibility is that of monitoring fission product release. As a subset of this method, a fourth procedure is to tag the fission gas with a special blend of isotopic mixtures such that the failed pin (or at least the assembly in which it resides) can be uniquely identified. In general, the detection of a single pin failure via any of these techniques is fairly slow. Hence, it is important to assure that the propagating mechanisms identified above do not lead to rapid failure propagation.

Fission Gas Release

The principal concern associated with fission gas release is that of local coolant starvation. Figure 14-7 indicates, however, that adiabatic pin heating (i.e., total coolant removal from the cladding) would be required for approximately 250 ms in order for the hottest cladding in a typical LMFBR to reach its potential failure temperature. If a large cladding rupture followed by rapid fission gas release should occur, the fission gas would be essentially vented in only about 100 ms. If a slow leak were to occur (in the order of hours or days), there would be no appreciable flow perturbation. Hence, there appears little or no chance of pin failure due to fission gas blanketing.

Furthermore, if gas-jet impingement did lead to failure of the neighboring fuel pin, the area of flow starvation on the adjacent pin would be directly across the coolant channel from the initially failed pin. Hence, any subsequent fission gas jet would tend to come right back toward the initial failure. This results in a self-limiting propagation process.

Finally, it has been postulated that transitory mechanical loads associated with fission gas release might be high enough to fail neighboring fuel pins. However, fission gas release does not appear to occur fast enough to cause such high loading patterns.

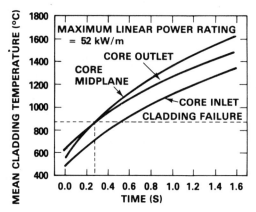

FIGURE 14-7. Typical LMFBR cladding heatup assuming complete thermal blanketing.[17]

Local Coolant Blockage

In order for local blockages to take place, foreign materials in the coolant stream must get to the site of the postulated blockage.* However, it is now standard practice to include strainers in the coolant inlet ports to prevent particles above a certain size from entering the core. If fuel particles from a failed pin were to enter the coolant stream, they would be few in number prior to being detected by the fission gas release monitoring system. Blockage could conceivably occur due to swelling of failed pins, but this is a very slow process.

For any planar blockage which is impervious to flow, nearly 50% blockage of the total flow area would need to occur in order to reduce the overall assembly flow by 5%. Hence, a local blockage would have to be very large to cause an appreciable amount of damage. Further, most blockages arising from the accumulation of particulate matter would be at least partially porous.

Molten Fuel Release

The primary protection against fuel enrichment errors is the strict quality assurance program that normally requires individual pin nondestructive

---

*It is possible that fuel pin swelling could cause local blockage patterns. This provides one of the incentives to minimize swelling potential (discussed in Chapter 8-3D).

assay tests prior to assembly into a duct array. Even if local enrichment were to occur, however, early scoping in-pile tests indicate little potential concern.

A larger concern would be the potential thermal interaction between molten fuel and sodium coolant. This could be a problem for some fuel/coolant systems, but it has been largely allayed for the mixed oxide-sodium system because of the spontaneous nucleation arguments (discussed in Chapter 16-3B).

In summary, pin-to-pin failure propagation is viewed to be highly unlikely. Even if such propagation should occur, it is sufficiently slow to allow detection before serious safety conditions arise.

B. ASSEMBLY-TO-ASSEMBLY PROPAGATION

Principal Propagation Concerns

There are four general categories of concerns regarding assembly-to-assembly failure propagation potential. These include (1) pin-to-pin failures which lead to full assembly involvement, (2) full assembly flow blockage, (3) pressure generation from a molten fuel-coolant interaction, and (4) the reactivity consequences of assembly meltdown. The potential for pin-to-pin failure propagation was discussed above. A brief synopsis of the other concerns is contained below.

Flow Blockage

Fuel assembly failure propagation could conceivably result from a large flow blockage. The potential for flow blockage formation and its detectability are related to design and operating practices. Proper assembly designs (e.g. multiple coolant inlets, etc.), in addition to proper construction and operating practices prior to assembly installment (e.g., gas flow testing), make a sudden and complete assembly blockage unlikely. The formation of a severe flow blockage, if it should occur at all, is therefore likely to occur on a time scale which is slow compared with the time needed for detection. An inlet blockage needs to become very large ($\sim 90\%$ planar blockage for typical LMFBR designs) before the onset of exit coolant boiling. Furthermore, recent progress in understanding fluid mechanics and heat transfer in the wake behind blockages [18] rules out the concern for developing dangerous internal blockages during operation prior to the occurrence of limited fuel pin failures that can be detected by fission product detectors (both solid and gaseous products) or delayed neutron detectors. Acoustic detection could be applied against more rapid propagation but does not appear necessary on the basis of failure propagation technology.

Pressure Generation

The concern over molten fuel/coolant interaction (MFCI) pressure gener-* ation is that such loadings might cause sufficient core distortion to inhibit control insertion and, thereby jeopardize the ability to effect reactor shutdown. Much of the pressure generation concern can be ruled out, however, because any initial MFCI which did occur would tend to cause a two-phase dispersal state, and the vapor generated would prevent the liquid-liquid contact necessary to sustain high pressure vapor explosion conditions. Furthermore, for the UO_2-Na systems, there is strong evidence to suggest that spontaneous nucleation, a condition which may be needed for a high pressure vapor explosion, cannot occur (discussed in Chapter 16-3B).

Fuel Compaction

The final concern arises since a complete meltdown and compaction of only 6 to 7 assemblies could theoretically cause a sufficient reactivity insertion to lead to a whole-core meltdown, if the PPS were to fail. However, the in-pile evidence to date suggests that a dispersive—rather than a compactive—flow path would likely ensue. Fuel dispersion would, of course, lead to negative (rather than positive) reactivity consequences. Indeed, such a dispersal process could cause a positive reactivity component due to dispersed sodium and steel, but the reactivity worth of the fuel would override the former effects. Finally, axial propagation of failure (within the affected assemblies) would be faster than radial failure propagation to other assemblies.

14-5 WHOLE-CORE TRANSIENTS

Whole core transients are those events for which a reactivity or flow imbalance affects the entire reactor. Examples of possible whole core accident initiators were discussed in Section 14-2. For any particular reactor design, extensive studies are normally conducted to assure that none of these transients violate the system integrity limits, given positive PPS action.

Outlines are contained below for three sample calculations, one for a reactivity transient and the other two for transient undercooling events. Much of the detail will necessarily be omitted, but a brief sketching of the salient features should be instructive to help visualize the types of analyses which must be done to assure effective PPS performance.

A. REACTIVITY TRANSIENTS

The sample overpower transient chosen is a 3.4¢/s reactivity insertion rate into the FFTF at full power.[19] This reactivity insertion rate corresponds to the normal speed of control rod withdrawal (0.25 m/min), but assumes that the rod is continuously withdrawn and can only be arrested by PPS action. The analytical tool used was the MELT-III Code.[20] Hot-channel factors were used to assure a 99.9% confidence level that the calculated fuel pin temperatures would not be exceeded.

Figure 14-8 contains the peak core cladding temperatures assuming primary PPS action, as well as assuming only secondary PPS action. The peak temperatures (801°C for primary action and 831°C for secondary action) are well below the 870°C limit listed in Table 14-3. Hence, this calculation indicates there is considerable margin-to-failure (using the

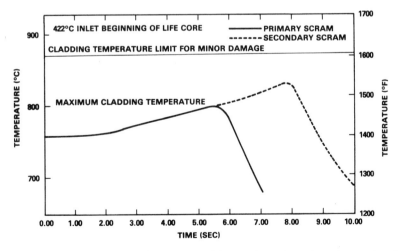

FIGURE 14-8. Sample protected whole core reactivity transient. (3.4¢/s insertion rate into full power FFTF).

cladding temperature limit*)—even if the primary PPS fails to perform.

B. TRANSIENT UNDERCOOLING EVENTS

Loss of Primary System Flow

An unlikely transient undercooling event is exemplified by the loss of all off-site electric power and on-site standby electric power. The sample case is taken for the FFTF where again the accident is assumed to be initiated from full power conditions. The IANUS Code [21] was used to perform the calculation and the same hot-channel factors were employed as in the case of the reactivity transient. Complete loss of all forced convection in the primary and secondary sodium systems and in the sodium-to-air heat exchangers was assumed.

Figure 14-9 contains the key calculated results. Shown in this illustration is the inside wall temperature of the average core cladding plus the comparable value for the hottest cladding in the core (including the hot-channel factors). The cladding temperature drops immediately following the power loss because the primary loop pump coastdown characteristics retain the flow well above the power curve (which drops rapidly due to PPS action). After about 50 seconds, however, the normalized power exceeds the normalized flow. This unfavorable power/flow imbalance causes the cladding temperature to rise beyond its steady-state value. Temperature increases continue until natural convection is established, after which cladding temperatures recede. Even in this extreme case as shown in Fig. 14-9, the maximum hot channel cladding temperature just approaches the 870°C limit. More realistic hot channel factors, noted in Ref. 22, reduce the peak cladding temperature by over 80°C from that shown. Hence, even for this unlikely event, cladding integrity is concluded to be maintained.

Flow Failure in the Steam System

As a second example of a transient undercooling event, we consider a transient initiated by a flow failure in the CRBRP steam system. The specific failure is a breach in the saturated steam line connecting the

---

*Cladding strain is often considered to be a more appropriate failure criterion than cladding temperature for transient overpower conditions, For the present case, the calculated cladding strain is well within acceptable limits for anticipated and unlikely events (as given in Table 14-3).

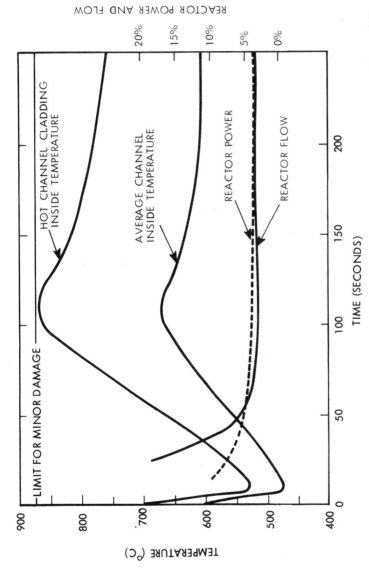

Figure 14-9. Sample loss of all electric power accident calculated for FFTF. NOTE: Actual FFTF test results (obtained from a full power scram to complete natural convection with all electrical power off) yielded maximum cladding temperatures significantly below the full power steady-state values. It should be noted that the calculations plotted in this figure include all the saftey hot channel factors and, therefore, are considerably higher than expected actual values for such an event. The actual test results, which are in close agreement with best-estimate pretest predictions, emphasize the safety margin imbedded in the hot channel analyses.

evaporator to the superheater. For the CRBRP design, the transient would cause an immediate cessation of superheater steam flow in that loop and initiation of a reactor trip. The superheater would then rapidly become isothermal at the sodium inlet temperature due to the loss of cooling. Sodium leaving the evaporators of the affected loop would initially drop in temperature due to overcooling as the water flow increased and flashed to atmospheric pressure through the steam drum. Then, as the loop blows dry through the rupture, the evaporator sodium temperature would rapidly increase to the superheater inlet temperature. The transient would then propagate through the secondary cold leg piping and result in similar severe transients on the secondary pump, the IHX secondary inlet, the IHX primary outlet, the primary cold leg piping and check valve, and the reactor vessel inlet nozzle. Subsequently, these components would experience decreasing temperature transients as a result of the reactor scram.

14-6 OTHER ACCIDENT CONSIDERATIONS

In addition to the power/flow imbalance transients, which directly affect core integrity conditions, there are other accidents for which protection must be applied. These include steam generator faults, sodium leaks and spills, fuel handling and storage faults, and external events (both natural catastrophic and man-made events). A brief summary of each class is contained below.[23]

A. STEAM GENERATOR FAULTS

In the design of steam generators for an LMFBR system, the potential for water-to-sodium leaks is the major concern. Small faults in steam generators can, if undetected, lead to wastage of surrounding tubes and ultimately major damage to the steam generator. The possibility of large failures of tubes is also considered in the design of steam generator systems. Assuring the high integrity of the boundary between sodium and water/steam is, therefore, an area of high priority in steam generator design. Designing steam generator systems to the ASME Code, Section III, Class 1, and RDT Standards, as well as requiring all tube welds to be inspectable and crevice-free are among the measures taken towards minimizing the potential for tube leaks.

Provisions must be made to accommodate major leaks to avoid damaging the steam generator beyond repair. In systems using more than one steam generator in the same loop, the other steam generators in the loop could be severely damaged by the effects of a sodium-water reaction in one of the steam generator units, if adequate relief for the sodium-water products is not provided. Also, the intermediate heat exchanger (IHX) could be damaged, which could potentially open a direct path to the environment for radioactive primary loop sodium. To design such a relief system, the current U.S. practice is to assume a hypothetical failure of steam generator tubes. The basis for design involves a guillotine failure of a tube followed by the secondary failure of six adjacent tubes. It is assumed that when the initiating failure occurs, the flow in the ruptured tube reaches a choked condition instantaneously. Flows in the six secondary tube failures are assumed to reach their choked condition in approximately 0.1 second following the initial failure. This event is classified as an Extremely Unlikely Event; replacement of the damaged steam generator is an acceptable design basis, but loss of all the steam generators in the affected loop is not an acceptable consequence. Therefore, ASME Code rules for Faulted Conditions are used in designing for this accident in the faulted unit, while the remaining steam generators in the same loop are designed to meet the ASME Code rules for Emergency Conditions.

Excessively high design pressures are prevented by providing a relief system which allows safe rapid expulsion of sodium and reaction products from the secondary sodium system. The relief system is also designed to protect the IHX against major leaks in the steam generator. The present approach in the U.S. is to design the IHX for loads from a major leak using ASME Code rules for Emergency Conditions.

B. SODIUM SPILLS

Large sodium spills are considered unlikely in the LMFBR; however, the possibility for such spills is recognized in the design of the plant. The consequences of radioactive sodium spills are mitigated by use of elevated piping (low static pressure) enclosed with metal-clad insulation, enclosure of major equipment within secondary or "guard" vessels, and reduced oxygen (~ 1 volume percent) in primary sodium equipment cells. As a secondary precaution, the major primary sodium equipment cells are located in the sub-grade region of the reactor containment building. Cells are enclosed with a gas-tight steel membrane or "liner" to contain the lower-oxygen atmosphere. The liner also protects structural concrete from reaction with sodium. Where the possibility and potential consequences of a sodium spill are considered significant, the cell liners are installed with provisions for thermal expansion, and the space between the cell liners and concrete is vented to limit buildup of steam pressure due to water release from the heated concrete behind the liners.

Secondary sodium systems are enclosed in air-filled cells since there is no radiological hazard; structural concrete for these cells is protected from sodium attack by lining the cell floor and lower side walls with a basin or "catch pan" of steel plate.

Even though no major sodium spills are expected in an LMFBR, safety analyses consider limiting accidents such as large leaks and loss of system inventory. These accidents would result in off-normal heat releases accompanied by a pressure and temperature rise within the cells. Releases from postulated sodium pool and spray fires are evaluated and the resulting pressures and temperatures are compared to limiting structural values for the cells.

Experimental work with sodium pool and spray fires has been utilized to verify computer codes such as SOFIRE,[24] SPRAY,[25] and CACECO[26] which are used for such safety calculations. Experiments with large scale sodium fires (300–1000 kg) have been utilized to test steel liners and catch pans and to test nitrogen flooding as a fire extinguishing method in air-filled cells. Further discussions of sodium fires appears in Chapter 16-6.

C. FUEL HANDLING AND STORAGE FAULTS

The safety criteria and definition of faults for design of the fuel handling and storage systems consider five safety functions:

(1) remove residual heat from spent fuel
(2) maintain subcriticality
(3) provide containment of radioactive material
(4) provide biological shielding
(5) prevent loss of safety functions of other systems.

Once the safety functions have been identified, generic types of faults are postulated (for example, loss of cooling of spent fuel in each location where it is handled or stored). The worst accident of each generic class is then selected for analysis of consequences. In some cases, several accidents of a class have a possibility of exceeding the accident severity criteria, and all require analysis. The accidents are then analyzed and compared to the allowable consequences. If the guideline exposure limits are approached, redesign to eliminate the fault or limit its consequences is implemented.

As a result of the above type of analysis, a number of specific requirements and design features have been developed for spent fuel storage facilities and equipment which transports spent fuel. These include in-vessel and ex-vessel transfer machines, grapples in fuel handling hot cells, and spent fuel shipping casks. At least two means of cooling are provided, i.e., both a primary and a backup system. All active and standby components required for maintenance of these two systems must be accessible for periodic inspection and/or checkout.

Guard jackets are provided for sodium vessels used for storage of spent fuel, and the liquid sodium level in the vessel may not be less than the minimum safe level (that is, the level below which the fuel temperature would exceed the allowable criteria) following a vessel or cooling system leak. Siphon breakers are included in any vessel inlet or outlet lines where the possibility of siphoning exists. Either there are no permanently connected drain lines, or any such lines contain double valves locked closed. Unless the backup cooling system is inherent, the primary or backup means of cooling, or both, are connected to an emergency power supply. At least one system adequate for cooling spent fuel is designed to Seismic Category I, which means that the cooling function will be maintained following a Safe Shutdown Earthquake.

One significant difference between LMFBR designs and U.S. light water reactors in handling of spent fuel is that LMFBR individual spent fuel assemblies are lifted out of the coolant during transfer from the reactor vessel. Considerable design and development effort has gone into the design of reliable and inherent cooling for the ex-vessel transfer machine which performs this operation. The primary means of cooling employs redundant blowers, and the backup means relies on inherent natural convection.

Experience with sodium reactors has indicated that two of the most common means for release of radioactivity to containment were leakage through gas seals and spread of contamination through sodium drippage from the movable fuel transfer machine. To minimize the amount of radioactive gas leakage, double seals are used on fuel handling and storage equipment, with pressurized buffer gas in between if the contained level of radioactivity is high. Design features are also incorporated to minimize the amount of sodium drippage and to collect all such drippage in specified locations where it will not be released to the containment. Measures are also included to minimize the surface exposed to the contaminated interior of the transfer machine which could subsequently be exposed to the containment.

D. EXTERNAL EVENTS

Natural phenomena and external, man-made conditions to be accommodated in the design of LMFBR's are the same as those for light water reactors; the guidelines and criteria for consideration of such events are established by regulatory authorities. A summary of events considered is given below.

Natural Phenomena

The plant must be designed so that it can be safely shut down and maintained in a safe condition following any of the extremely unlikely natural phenomena, including the most severe that have been reported or can be foreseen for the site and its surroundings. The natural phenomena that provide design bases for the plant include:

Probable Maximum Flood. This is determined by evaluating potential flooding conditions resulting from severe storms, ice conditions, and potential failures of upstream dams.

Minimum Possible Low Water Level. Depending on the design features, this may affect the source of cooling water and/or the ability of the ultimate heat sink to perform its function. This low water level is determined from the most severe drought conditions or dam failure conditions considered possible at the site.

Tornado. A limiting tornado appropriate to the site must be factored into the design, including provisions to protect against associated tornado missiles.

Safe Shutdown Earthquake. The earthquake which could cause the maximum vibratory ground motion at the site must be considered.

Man-Made External Conditions

The plant must be designed so that it can be safely shut down and maintained in a safe condition for man-made external conditions that are considered to represent a significant risk to the plant. These conditions are highly site dependent and can be discussed here only in general terms.

Missiles which could be generated within the plant must be considered if they could jeopardize the functioning of safety systems. For example, turbine missiles must either be included in the plant design bases or shown to be sufficiently improbable to be excluded. A specific orientation of the turbine shaft relative to the containment building is frequently used to minimize the probability of damage from turbine missiles.

Examples of other potential conditions considered are an airplane crash (if the plant is located in the vicinity of substantial air traffic), external explosions (from nearby industrial installations or transportation on nearby waterways, railways or highways), toxic chemical or radioactive releases (from nearby industrial or nuclear plants), and fires.

REVIEW QUESTIONS

14-1. What functions are included in the Plant Protection System (PPS)?

14-2. List the potential mechanisms for (a) accidental reactivity insertion, and (b) undercooling.

14-3. Write a definition for each of the five basic elements of reliability.

14-4. Describe the processes which might occur at juncture ② on Fig. 14-1 and explain the protection path still existing if a scram signal at juncture ② is not initiated.

14-5. List the three backup systems in the CRBRP design for removal of decay heat and explain the principle of each.

14-6. Why is it improbable that failure of a single fuel pin can propagate to a larger accident?

14-7. What are the maximum cladding temperatures predicted for FFTF (a) for a protected overpower transient caused by the continuous withdrawal of a single control assembly, and (b) for a protected transient undercooling event caused by a loss of all off-site plus on-site standby electric power?

14-8. List the principal areas outside the reactor core that must be safeguarded against accidents in an LMFBR plant, together with an example in each case of a method to mitigate such an accident.

REFERENCES

1. American Society of Mechanical Engineers, *Boiler and Pressure Vessel Code* Section III, Division 1, 1975 edition.

2. D. E. Simpson, W. W. Little, and R. E. Peterson, "Selected Safety Considerations in Design of the Fast Flux Test Facility," *Proc. of the International Conference on the Safety of Fast Reactors*, Aix-en-Provence, France, September 1967.

3. P. Murray, "Problems in the Irradiation Behavior of LMFBR Core Materials," *Int. Conf. Proc. on Fast Reactor Irradiation Testing*, TRG-Report-1911, Thurso, Scotland, April 1969, 402–410.

4. R.˙Simm and A. Veca, *FFTF Pin Final Design Support Document*, FCF-213, Westinghouse Electric Corp., Advanced Reactor Division, December 1971.

5. R. J. Jackson, *Evaluation of FFTF Fuel Pin Design Procedures Vis-Vis Steady State Irradiation Performance in EBR-II*, Addendum to HEDL-TME 75–48, Hanford Engineering Development Laboratory, October 1975.

6. R. E. Baars, *Evaluation of FFTF Fuel Pin Transient Design Procedure*, HEDL-TME 75–40, Hanford Engineering Development Laboratory, September 1975.

7. R. E. Baars, T. Hikido, and J. E. Hanson, "Fast Reactor Fuel Pin Performance Requirements for Off-Normal Events," *Proc. Int. Conf. Fast Breeder Fuel Performance*, Monterey, CA, 1979, 155–164.

8. J. Graham, "Nuclear Safety Design of the Clinch River Breeder Reactor Plant," *Nuclear Safety*, *16*, 5, September–October, 1975.

9. F. H. Morgenstern, J. Bucholz, H. Kruger, and H. Rohrs, "Diverse Shutdown Systems for the KNK-1, KNK-2, and SNR-300 Reactors," CONF-740401, *Proc. of the Fast Reactor Safety Meeting*, Beverly Hills, CA, April 1974.

10. E. R. Specht, R. K. Paschall, M. Marquette, and A. Jackola, "Hydraulically Supported Absorber Balls Shutdown System for Inherently Safe LMFBR's" CONF-761001, *Proc. of the Int. Meeting on Fast Reactor Safety and Related Physics*, Chicago, IL, October 1976. Vol III, p. 683.

11. R. B. Tupper, M. H. Cooper, and C. E. Swenson, compilers, "Self-actuated Shutdown System Development," *Nuclear Safety and Reliability Engineering Annual Progress Report, Period Ending September 30, 1979*, W-ARD-SR-940004, Westinghouse Electric Corp., 1979.

12. D. R. Marr and W. L. Bunch, *FTR Fission Product Decay Heat*, HEDL-TME 71-72, Hanford Engineering Development Laboratory, February 1971.

13. C. Starr, "Social Benefit Versus Technological Risk," *Science*, *165* (1969) 1232-1238.

14. U.S. Congress, Joint Committee on Atomic Energy, *Possible Modification or Extension of the Price-Anderson Insurance and Indemnity Act, Hearings Before the Joint Committee on Atomic Energy on Phase II: Legislative Proposals*, H. R. 14408, S. 3452, and S. 3254, 93d Congress, 2nd Session, Pt. 2, Testimony of C. Starr, May 16, 1974, 617.

15. J. Graham and L. E. Strawbridge, "Design Margin Approach in LMFBR Nuclear Safety," *Proc. European Nucl, Conf. Vol. 5.*, April, 1975, p 368-386.

16. *Anticipated Transients Without SCRAM for Water-Cooled Power Reactors*, WASH-1270, September 1973.

17. J. B. van Erp, T. C. Chawla, R. E. Wilson and H. K. Fauske, "Pin-to-Pin Failure Propagation in Liquid-Metal-Cooled Fast Breeder Reactor Fuel Subassemblies," *Nuclear Safety, 16 3*, May-June 1975, 391-407.

18. C. V. Gregory and D. J. Lord, "The Study of Local Blockages in Fast Reactor Subassemblies," *J. of the British Nucl. Energy Soc., Vol. 16, No. 3* (1977) 251-260.

19. E. D. Schrull, *Protected Reactivity Insertion Events in the FTR*, HEDL-TI-76027, Hanford Engineering Development Laboratory, August 1976.

20. A. E. Waltar, W. L. Partain, D. C. Kolesar, L. D. O'Dell, A. Padilla, Jr., J. C. Sonnichsen, N. P. Wilburn, and H. J. Willenberg (HEDL) and R. J. Shields, (CSC), *Melt-III, A Neutronics Thermal-Hydraulic Computer Program for Fast Reactor Safety*, HEDL-TME 74-47, Hanford Engineering Development Laboratory, December 1974.

21. S. L. Additon, T. B. McCall, and C. F. Wolfe, *Simulation of the Overall FFTF Plant Performance (IANUS—Westinghouse Proprietary Code)*, HEDL-TC 556, Hanford Engineering Development Laboratory, December 1975.

22. *FFTF Final Safety Analysis Report*, HEDL-TI-75001, Hanford Engineering Development Laboratory. December 1975.

23. D. E. Simpson, A. Alker, H. K. Fauske, K. Hikido, R. W. Keaten, M. G. Stevenson, L. Strawbridge, "*Accident Considerations in LMFBR Design*," ERDA-76-103, Presented at the US/USSR Seminar on LMFBR Reactor Safety, Argonne National Laboratory, Argonne, IL, January 12-15, 1976. Paper #7.

24. P. Beiriger, J. Hopenfeld, M. Silberberg, R. P. Johnson, L. Baurmash, R. L. Koontz, "*SOFIRE-II User Report*," AI-AEC013055, Atomic International, March 1973.

25. P. R. Shire, "Reactor Sodium Coolant Hypothetical Spray Release for Containment Accident Analysis Comparison of Theory with Experiment," *Proc. of the Fast Reactor Safety Meeting*, April 2-4, 1974, Beverly Hills, CA, 473-487.

26. R. D. Peak and D. D. Stepnewski, "Computational Features of the CACECO Containment Analysis Code," *Trans. ANS, 21* (1975) 274-275.

CHAPTER 15

UNPROTECTED TRANSIENTS

15-1 INTRODUCTION

As indicated in the two previous chapters, there are several features which combine to make an LMFBR system safe and reliable.* Only when a major off-normal condition is encountered, combined with a postulated failure of the Plant Protective System (PPS), can serious accident consequences be predicted. Although the occurrence of these circumstances is highly improbable, such hypothesized unprotected transients[†] have been analyzed extensively and with increasing degrees of precision for the past two decades. These studies provide considerable insight regarding the *intrinsic* response of an LMFBR should plant protection equipment fail when needed.

This chapter and Chapter 16 provide an introduction to the types of analyses performed to study unprotected fast breeder reactor behavior. It is the authors' experience that considerable time is required for a reader new to the field of fast reactor safety to organize, assimilate, and put into perspective the large number of complex phenomena involved in unprotected transients. Indeed, the terminology developed to describe these phenomena is a living, constantly changing language, expanding as accident analysis techniques become more sophisticated.

Our emphasis, therefore, will be on the principal physical processes and phenomenological interactions which tend to be most influential in the unprotected accident scenarios. It will be necessary to leave out many of the

---

*All of the examples for this chapter, plus Chapter 16, apply to the LMFBR, though many phases of the analysis are quite similar for the Gas Cooled Fast Breeder Reactor. Analyses and concerns unique to GCFR safety are discussed in Chapter 18.

[†]As noted from the discussion in Chapter 13, the term "unprotected" is used in the present context to connote failure of the PPS when called upon to function. Several barriers exist to provide protection to plant personnel and the general public even in the event of an "unprotected" transient.

important and interesting details and mathematical models associated with mechanical analyses in an overview book of this type. Fortunately, many of the analytical models involve extensions of the steady-state heat transfer, fluid flow, and fuel pin models presented in detail in Chapters 8 through 10; hence, the groundwork for these models has already been laid. Given the perspective of the material presented in this chapter, the serious reader should have a usable road map from which to depart into new areas of research.

Although our presentation is necessarily restricted, we have selected a few areas in which to present analytical methods in some detail. For example, we will present the basic conservation equations because they serve as the framework for so many of the safety models. We will also describe the disassembly equations because they are classical and have played an especially important role in containment design. Limited treatment is given to fuel and sodium expansions in Chapter 16, as well as the bases for the major theories regarding molten fuel/coolant interaction (MFCI). Finally, some of the computational systems commonly used to investigate the various accident phases will be discussed briefly.

Unprotected transients are generically categorized as transient overpower (TOP) and transient undercooling (TUC) accidents.* Given a steady-state condition in which the rate of heat generation (nuclear fission) is just balanced by the rate of heat removal (coolant flow), a damaging condition could arise due to an increase in the heat generation term (with the heat removal term fixed), or due to a loss of heat removal capability (with the heat generation fixed). Hence, the TOP accident refers to an accident in which a reactivity insertion occurs (e.g., unlimited control rod withdrawal), leading to a power rise, but where normal coolant flow is maintained. Conversely, the TUC accident refers to a class of accidents in which coolant flow is interrupted (e.g., loss of power to the pumps), but where fission heating is maintained.

This chapter provides focus to the salient features of unprotected transients during the nuclear transient, i.e., from an intact geometry through a transition phase or core disassembly. Chapter 16 provides focus to the subsequent events; namely, mechanical consequences, post-accident heat removal (PAHR), containment, and radiological consequences.

Some of the basic phenomenological processes are relevant to more than one phase of the accident sequence. The molten fuel/coolant interaction (MFCI) process is perhaps the prime example. This particular topic will be

---

*The transient undercooling accident has also been referred to as the loss-of-flow (LOF) accident. These terms were used interchangeably in the early literature.

treated in some detail in Chapter 16 since it is in the area of mechanical consequences that the MFCI is so important. However, frequent references to the MFCI will be noted throughout Chapter 15.

15-2 PHYSICAL PHASES OF ACCIDENT SEQUENCES

Before entering into a discussion of the various processes governing fast reactor behavior under postulated unprotected accident conditions, it is instructive to introduce an overview of the principal physical phases where accident analysts provide special attention in establishing accident scenarios. Figures 15-1 through 15-4 depict the various phases in which the analytical output from one phase is usually processed and evaluated before proceeding to the next phase.

Included in this section is a brief description of the salient features (definition) of each phase. Such discussion is only intended for orientation and perspective; more detailed discussion of each phase constitutes the remainder of this chapter and Chapter 16.

Whereas the computational capability for making assessments of each of the phases depicted is becoming increasingly sophisticated,[1] it must be

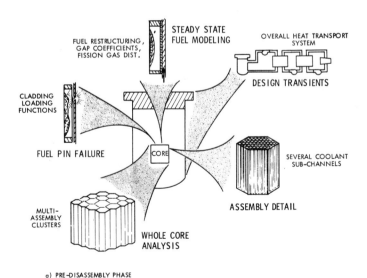

a) PRE-DISASSEMBLY PHASE

FIGURE 15-1. Essentially intact geometry.

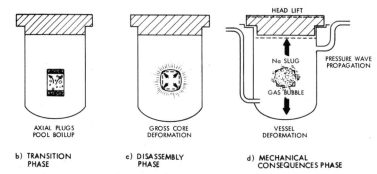

FIGURE 15-2. Considerable geometric distortion.

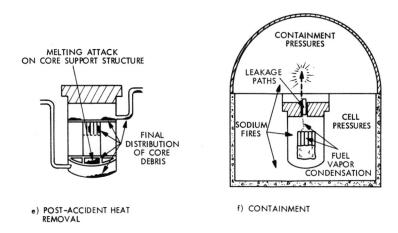

FIGURE 15-3. Longer-term distribution problem.

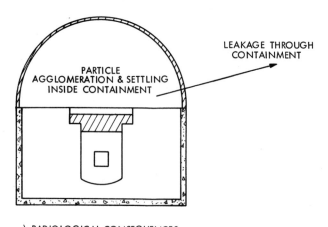

FIGURE 15-4. Ultimate radiological dose considerations.

emphasized that a strong experimental data base* is essential to guide model development and provide calibration requirements.

A. PRE-DISASSEMBLY PHASE

The *pre-disassembly phase* (Fig. 15-1) deals with the transient sequence from accident initiation to the point of (a) neutronic shutdown with essentially intact geometry, (b) a gradual core meltdown where the computational models needed for tracking the accident further must be significantly modified, or (c) the onset of conditions for hydrodynamic core disassembly due to generation of high internal pressures from vaporized core materials.

Important topics include accident initial conditions defined by steady-state fuel and thermal-hydraulic performance, transient response of the overall heat transport system for design and accident conditions, and core behavior under HCDA conditions.[†] The latter situation includes sodium boiling, fuel pin failure mechanisms, and motion of molten cladding and fuel as driven by fission gases, sodium liquid, and/or sodium vapor. Whole-core accident computational models are required to treat the coupled neutronics and thermal-hydraulic core response, which includes determining pin failures and material motion in several channels (each representing an assembly or group of assemblies).

B. TRANSITION PHASE

Should the above accident path lead to a permanent neutronic shutdown with essentially intact geometry, the accident progression terminates. However, if there are insufficient intrinsic negative feedbacks to terminate the postulated excursion prior to coolant loss and subsequent core melting, it is conceivable that large segments of the core (fuel and cladding) could melt into a pool of molten core debris and eventually boil.

---

*A particularly important aspect of obtaining meaningful results when interfacing several computational systems is employing a consistent data set for materials properties. The demand for consistent data tends to be more difficult to meet in the safety analysis area than in the neutronics systems area, mainly because of the wide range of temperatures and pressures potentially involved, as well as the sheer breadth of areas of interest (e.g., neutronics constants, multi-phase materials behavior, and radiological dispersion patterns, to cite a few). Efforts are underway to obtain such data in a readily usable format.[2]

[†] See Chapter 13-3B for a discussion of the term "HCDA."

This slowly developing boiling process has been termed the *transition phase,* since it is sandwiched between the classical pre-disassembly and disassembly phases. The principal question in this phase is whether the core will boil until a permanently cooled noncritical configuration could be established, or whether criticality conditions could reoccur.

C. DISASSEMBLY

As mentioned in Chapter 13, the early work in LMFBR safety analysis was based on an assumed core compaction that yielded a high rate of reactivity gain and led to physical *disassembly* of the core due to buildup of internal core pressures. The calculation of energy generation during disassembly, originally based on the classic Bethe-Tait[3] analysis,* has long prevailed as the basic guide to containment design for FBR systems.

D. MECHANICAL CONSEQUENCES

If a postulated excursion is found to lead to a core disassembly, the question after nuclear shutdown becomes that of determining the mechanical consequences to the reactor environs. A major consideration is that of determining the appropriate working fluid (fuel, steel, or sodium) for conversion of thermal energy to expansion work on the surrounding sodium, vessel internals, vessel head, vessel walls, and associated piping. This phase of analysis is usually terminated when the high-pressure core region (generated during the nuclear excursion) expands or condenses to the point where no more work can be accomplished.

E. POST-ACCIDENT HEAT REMOVAL

Having generated an excursion wherein at least some molten materials are formed, a question arises with respect to long-term cooling capability. Considerable efforts are expended at this stage to assure that long-term cooling of core debris can be established.

---

*Current disassembly analyses are conducted using extensively modified and more sophisticated models.

F. CONTAINMENT

Once mechanical damage calculations are completed, an important link in providing the driving force for ultimate containment studies is to define the *radiological source term*, i.e., the concentration of fuel and sodium aersols and fission products that may be present in the outer containment. It is important to establish potential pathways for plutonium and fission products from the core region up through the sodium and structural surroundings to the outer containment. Likewise, it is essential to predict the effects of sodium fires or sodium/core debris reactions with structural materials which might result in containment pressurization.

G. RADIOLOGICAL CONSEQUENCES

The final step in the accident analysis sequence is to determine the type of *containment* or *containment/confinement system* necessary to assure adequate public protection against any radiation released from the accident. This determination requires the prediction of particle agglomeration and settling, leakage through containment, and downwind plume formation and transport capabilities.

15-3 CONSERVATION EQUATIONS

Many of the events in unprotected transients involve extended motion of molten single-phase and two-phase core materials. Analyses of these phenomena invariably utilize some form of the basic fluid mechanics conservation equations, together with appropriate equations of state. These equations are generally written in differential form and solved with computer models by finite difference techniques.

Because of the frequency with which the conservation equations are encountered in fast reactor safety analysis, it is useful to present them in an organized and consistent manner before further discussing the accident events for which they are needed. The equations will not be derived here; derivations can be found in textbooks on fluid mechanics, e.g., Reference 4. In studying transient hydrodynamics codes, it is often useful to compare the differential equations reported with a standard set, as presented here, in order to understand the assumptions being made.

Many transient hydrodynamics computer models are based on a Lagrangian frame of reference in which mass does not transfer across boundaries of the finite difference nodes; i.e., the nodes move with the flow and consequently change in geometry. Other codes are based on an Eulerian coordinate system in which the mass is allowed to move across node boundaries while the node geometry remains fixed. Lagrangian coordinates are easier to program in finite difference form and are more useful when material motion is either limited in extent or constrained to one or two dimensions. A Lagrangian coordinate system, therefore, is quite useful for the relatively small displacements involved in disassembly. Eulerian coordinates are more useful when material motion is extensive, as in the transition phase and in the mechanical consequences phase following disassembly. The relative ease of programming the Lagrangian equations led to an early limited success with their application to the mechanical consequences phase, but the earlier Lagrangian codes are being replaced by Eulerian codes for this analysis. The VENUS and REXCO codes (discussed in Section 15-7C and Chapter 16-4B) were written in the Lagrangian frame of reference. The SIMMER code (Section 15-6E) utilizes the Eulerian approach.

The equations provided here include the general terms usually encountered in fast reactor safety codes. Special terms do appear in particular models, however, and these will appear only as X_1, X_2, and X_3 in the present equations. For example, momentum transfer due to friction and heat transfer by convection to surfaces are not specifically included in the equations presented here, but sometimes need to be included. Most safety codes neglect viscous effects in both the momentum and the energy equations. Where they are important, they would appear in the X_2 and X_3 terms as would friction and surface convection heat transfer. Multi-phase codes with separate equations for each phase have mass, momentum, and energy transfer terms between phases that must be included under X_1, X_2, and X_3, as will be noted during the discussion of the SIMMER code.

The equations that follow include the three conservation equations in both the Eulerian and Lagrangian frames of reference. The principal differences between the two systems of equations are the inclusion of a convective term on the left-hand side of each Eulerian equation that does not appear in the Lagrangian equations, and the *substantial* derivative, D/Dt, in the Lagrangian equations.

The substantial derivative is a time rate of change that applies to a fluid mass moving with the flow. In a Lagrangian code in finite difference form, each node is allowed to move with the flow. In the continuity and energy equations, the substantial derivative is the time rate of change of the density and energy of the mass in the node. In the momentum equation the substantial derivative of the velocity is the acceleration of the node itself.

This is in direct contrast to a node in an Eulerian system in which the node does not move, but material flows into and out of the node.

Continuity equations are first presented in vector notation and are then expanded in Cartesian and cylindrical coordinates. Azimuthal symmetry is assumed in cylindrical coordinates, following the practice in most fast reactor safety codes. The body force per unit mass, represented by $\vec{g}$, is limited to gravity and hence operates only in the z direction, such that $g_z = -g$.

A. EULERIAN COORDINATE SYSTEM

The independent variables in the Eulerian frame of reference are x, y, z, t in Cartesian coordinates and r, θ, z, t in cylindrical coordinates.

Continuity

$$\frac{\partial \rho}{\partial t} + \nabla \cdot \rho \vec{v} = X_1, \tag{15-1}$$

where X_1 accounts for mass transfers between phases when a separate equation is written for each phase; X_1 is zero otherwise.

In Cartesian coordinates, $\vec{v} = \hat{i}v_x + \hat{j}v_y + \hat{k}v_z$.

In cylindrical coordinates, $\vec{v} = \hat{i}_r v_r + \hat{i}_\theta v_\theta + \hat{i}_z v_z$.

Cartesian coordinates:

$$\frac{\partial \rho}{\partial t} + \frac{\partial}{\partial x}(\rho v_x) + \frac{\partial}{\partial y}(\rho v_y) + \frac{\partial}{\partial z}(\rho v_z) = X_1. \tag{15-1a}$$

Cylindrical coordinates (with azimuthal symmetry):

$$\frac{\partial \rho}{\partial t} + \frac{1}{r}\frac{\partial}{\partial r}(r\rho v_r) + \frac{\partial}{\partial z}(\rho v_z) = X_1. \tag{15-1b}$$

Momentum

$$\frac{\partial}{\partial t}(\rho\vec{v})+\nabla\cdot\rho\vec{v}\vec{v}=-\nabla p+\rho\vec{g}+\vec{X_2}. \tag{15-2}$$

$$\vec{X_2}=0 \text{ (Euler equation).}$$

$$\vec{X_2}=\mu\nabla^2\vec{v} \text{ (Navier Stokes equation).}$$

$$\vec{X_2}=\nabla\cdot\tau \text{ (General single-phase equation}$$
where τ is the stress tensor as
defined in Ref. 4).

For multi-phase flow with a separate equation for each phase, X_2 can account for momentum transfer between phases.

Equation (15-2) can be reduced to another form frequently quoted by expanding the terms on the left hand side as follows:

$$\frac{\partial}{\partial t}(\rho\vec{v})=\rho\frac{\partial\vec{v}}{\partial t}+\vec{v}\frac{\partial\rho}{\partial t}.$$

$$\nabla\cdot\rho\vec{v}\vec{v}=\rho\vec{v}\cdot\nabla\vec{v}+\vec{v}\nabla\cdot\rho\vec{v}.$$

Note that the sum of two of these terms ($\vec{v}\,\partial\rho/\partial t+\vec{v}\nabla\cdot\rho\vec{v}$) can be replaced by the continuity equation [Eq. (15-1)], so that Eq. (15-2) can be rewritten as

$$\rho\frac{\partial\vec{v}}{\partial t}+\rho\vec{v}\cdot\nabla\vec{v}+X_1\vec{v}=-\nabla p+\rho\vec{g}+X_2. \tag{15-2a}$$

Cartesian coordinates (3 scalar equations):

$$\frac{\partial}{\partial t}(\rho v_x)+\frac{\partial}{\partial x}(\rho v_x v_x)+\frac{\partial}{\partial y}(\rho v_x v_y)+\frac{\partial}{\partial z}(\rho v_x v_z)=-\frac{\partial p}{\partial x}+X_{2x}, \tag{15-2b}$$

$$\frac{\partial}{\partial t}(\rho v_y)+\frac{\partial}{\partial x}(\rho v_y v_x)+\frac{\partial}{\partial y}(\rho v_y v_y)+\frac{\partial}{\partial z}(\rho v_y v_z)=-\frac{\partial p}{\partial y}+X_{2y}, \tag{15-2c}$$

$$\frac{\partial}{\partial t}(\rho v_z)+\frac{\partial}{\partial x}(\rho v_z v_x)+\frac{\partial}{\partial y}(\rho v_z v_y)+\frac{\partial}{\partial z}(\rho v_z v_z)=-\frac{\partial p}{\partial z}+\rho g_z+X_{2z}. \tag{15-2d}$$

Combining these equations with the continuity equation, together with $X_1=0$ and $X_2=0$, gives the following more familiar form of the momentum equation in the Eulerian frame of reference:

$$\rho\left(\frac{\partial v_x}{\partial t}+v_x\frac{\partial v_x}{\partial x}+v_y\frac{\partial v_x}{\partial y}+v_z\frac{\partial v_x}{\partial z}\right)=-\frac{\partial p}{\partial x}, \tag{15-2e}$$

$$\rho\left(\frac{\partial v_y}{\partial t}+v_x\frac{\partial v_y}{\partial x}+v_y\frac{\partial v_y}{\partial y}+v_z\frac{\partial v_y}{\partial z}\right)=-\frac{\partial p}{\partial y}, \tag{15-2f}$$

$$\rho\left(\frac{\partial v_z}{\partial t}+v_x\frac{\partial v_z}{\partial x}+v_y\frac{\partial v_z}{\partial y}+v_z\frac{\partial v_z}{\partial z}\right)=-\frac{\partial p}{\partial z}+\rho g_z. \tag{15-2g}$$

Cylindrical coordinates (with azimuthal symmetry):

$$\frac{\partial}{\partial t}(\rho v_r)+\frac{1}{r}\frac{\partial}{\partial r}(r\rho v_r v_r)+\frac{\partial}{\partial z}(\rho v_r v_z)=-\frac{\partial p}{\partial r}+X_{2r}, \tag{15-2h}$$

$$\frac{\partial}{\partial t}(\rho v_z)+\frac{1}{r}\frac{\partial}{\partial r}(r\rho v_z v_r)+\frac{\partial}{\partial z}(\rho v_z v_z)=-\frac{\partial p}{\partial z}+\rho g_z+X_{2z}, \tag{15-2i}$$

or, combined with continuity and with $X_1=0$ and $X_2=0$,

$$\rho\left(\frac{\partial v_r}{\partial t}+v_r\frac{\partial v_r}{\partial r}+v_z\frac{\partial v_r}{\partial z}\right)=-\frac{\partial p}{\partial r}, \tag{15-2j}$$

$$\rho\left(\frac{\partial v_z}{\partial t}+v_r\frac{\partial v_z}{\partial r}+v_z\frac{\partial v_z}{\partial z}\right)=-\frac{\partial p}{\partial z}+\rho g_z. \tag{15-2k}$$

Energy

$$\frac{\partial}{\partial t}\left[\rho\left(u+\frac{v^2}{2}\right)\right]+\nabla\cdot\rho\vec{v}\left(u+\frac{v^2}{2}\right)=-\nabla\cdot p\vec{v}+\nabla\cdot k\nabla T+Q+\rho\vec{v}\cdot\vec{g}+X_3. \tag{15-3}$$

The $-\nabla\cdot p\vec{v}$ term accounts for pdV work (i.e. flow work in the Eulerian system), $\nabla\cdot k\nabla T$ accounts for conduction heat transfer, Q is the volumetric heat source, and $\rho\vec{v}\cdot\vec{g}$ accounts for potential energy.

Multiplying the flow work term, $-\nabla\cdot p\vec{v}$, by ρ/ρ and recognizing that enthalpy, h, is equivalent to $(u+p/\rho)$, we can combine the flow work term with the second term on the left hand side of Eq. (15-3) to give the following form of the energy equation:

$$\frac{\partial}{\partial t}\left[\rho\left(u+\frac{v^2}{2}\right)\right]+\nabla\cdot\rho\vec{v}\left(h+\frac{v^2}{2}\right)=\nabla\cdot k\nabla T+Q+\rho\vec{v}\cdot\vec{g}+X_3.$$

$$(15\text{-}3a)$$

Cartesian coordinates:

$$\frac{\partial}{\partial t}\left[\rho\left(u+\frac{v^2}{2}\right)\right]+\frac{\partial}{\partial x}\left[\rho v_x\left(h+\frac{v^2}{2}\right)\right]$$

$$+\frac{\partial}{\partial y}\left[\rho v_y\left(h+\frac{v^2}{2}\right)\right]+\frac{\partial}{\partial z}\left[\rho v_z\left(h+\frac{v^2}{2}\right)\right]$$

$$=\frac{\partial}{\partial x}\left(k\frac{\partial T}{\partial x}\right)+\frac{\partial}{\partial y}\left(k\frac{\partial T}{\partial y}\right)+\frac{\partial}{\partial z}\left(k\frac{\partial T}{\partial z}\right)+Q+\rho v_z g_z+X_3,$$

$$(15\text{-}3b)$$

where $\qquad v^2=v_x^2+v_y^2+v_z^2.$

Cylindrical coordinates (with azimuthal symmetry):

$$\frac{\partial}{\partial t}\left[\rho\left(u+\frac{v^2}{2}\right)\right]+\frac{1}{r}\frac{\partial}{\partial r}\left[r\rho v_r\left(h+\frac{v^2}{2}\right)\right]+\frac{\partial}{\partial z}\left[\rho v_z\left(h+\frac{v^2}{2}\right)\right]$$

$$=\frac{1}{r}\frac{\partial}{\partial r}\left(rk\frac{\partial T}{\partial r}\right)+\frac{\partial}{\partial z}k\frac{\partial T}{\partial z}+Q+\rho v_z g_z+X_3, \qquad (15\text{-}3c)$$

where $\qquad v^2=v_r^2+v_z^2.$

B. LAGRANGIAN COORDINATE SYSTEM

The independent variables in a Lagrangian coordinate system are time, t, plus the initial position coordinates of each fluid mass (or each node in a finite difference approximation). The current position coordinates of each fluid mass are dependent variables given by:

Cartesian coordinates:

$$\frac{dx}{dt} = v_x; \; \frac{dy}{dt} = v_y; \; \frac{dz}{dt} = v_z. \tag{15-4a}$$

Cylindrical coordinates:

$$\frac{dr}{dt} = v_r; \; r\frac{d\theta}{dt} = v_\theta; \; \frac{dz}{dt} = v_z. \tag{15-4b}$$

Continuity

$$\frac{D\rho}{Dt} + \rho(\nabla \cdot \vec{v}) = X_1. \tag{15-5}$$

(The quantities X_1, X_2, and X_3 in the Lagrangian equations account for the same phenomena as those described in the presentation of the Eulerian equations.)

Cartesian coordinates:

$$\frac{D\rho}{Dt} + \rho\left(\frac{\partial v_x}{\partial x} + \frac{\partial v_y}{\partial y} + \frac{\partial v_z}{\partial z}\right) = X_1. \tag{15-5a}$$

Cylindrical coordinates:

$$\frac{D\rho}{Dt} + \rho\left[\frac{1}{r}\frac{\partial}{\partial r}(rv_r) + \frac{\partial v_z}{\partial z}\right] = X_1. \tag{15-5b}$$

Momentum

$$\rho\frac{D\vec{v}}{Dt} + X_1\vec{v} = -\nabla p + \rho\vec{g} + X_2. \tag{15-6}$$

Cartesian coordinates (3 scalar equations):

$$\rho\frac{Dv_x}{Dt}+X_1v_x=-\frac{\partial p}{\partial x}+X_{2x}, \tag{15-6a}$$

$$\rho\frac{Dv_y}{Dt}+X_1v_y=-\frac{\partial p}{\partial y}+X_{2y}, \tag{15-6b}$$

$$\rho\frac{Dv_z}{Dt}+X_1v_z=-\frac{\partial p}{\partial z}+\rho g_z+X_{2z}. \tag{15-6c}$$

Cylindrical coordinates (with azimuthal symmetry):

$$\rho\frac{Dv_r}{Dt}+X_1v_r=-\frac{\partial p}{\partial r}+X_{2r}, \tag{15-6d}$$

$$\rho\frac{Dv_z}{Dt}+X_1v_z=-\frac{\partial p}{\partial z}+\rho g_z+X_{2z}. \tag{15-6e}$$

Energy

$$\rho\frac{D}{Dt}\left(u+\frac{v^2}{2}\right)+X_1\left(u+\frac{v^2}{2}\right)=-\nabla\cdot p\vec{v}+\nabla\cdot k\nabla T+Q+\rho\vec{v}\cdot\vec{g}+X_3. \tag{15-7}$$

Cartesian coordinates:

$$\rho\frac{D}{Dt}\left(u+\frac{v^2}{2}\right)+X_1\left(u+\frac{v^2}{2}\right)=-\left[\frac{\partial}{\partial x}(pv_x)+\frac{\partial}{\partial y}(pv_y)+\frac{\partial}{\partial z}(pv_z)\right]$$

$$+\frac{\partial}{\partial x}\left(k\frac{\partial T}{\partial x}\right)+\frac{\partial}{\partial y}\left(k\frac{\partial T}{\partial y}\right)+\frac{\partial}{\partial z}\left(k\frac{\partial T}{\partial z}\right)+Q$$

$$+\rho v_z g_z+X_3. \tag{15-7a}$$

Cylindrical coordinates (with azimuthal symmetry):

$$\rho\frac{D}{Dt}\left(u+\frac{v^2}{2}\right)+X_1\left(u+\frac{v^2}{2}\right)=-\left[\frac{1}{r}\frac{\partial}{\partial r}(rpv_r)+\frac{\partial}{\partial z}(pv_z)\right]$$

$$+\frac{1}{r}\frac{\partial}{\partial r}\left(rk\frac{\partial T}{\partial r}\right)+\frac{\partial}{\partial z}k\frac{\partial T}{\partial z}+Q$$

$$+\rho v_z g_z+X_3. \qquad (15\text{-}7\mathrm{b})$$

It is useful to note that the Lagrangian and Eulerian equations are equivalent since the time derivative for the Lagrangian system is related to the Eulerian derivatives by the following operator relationship:

$$\frac{D}{Dt}=\frac{\partial}{\partial t}+\vec{v}\cdot\nabla.$$

This means that each of the equations in the Lagrangian system can be derived directly from the equations in the Eulerian system by utilizing the continuity equation together with the above relationship for D/Dt.

C. PSEUDO-VISCOUS PRESSURE

A comment about the treatment of steep pressure gradients in fast reactor hydrodynamics codes is appropriate before concluding the discussion of the conservation equations. Except possibly for localized propagation of thermal interactions between molten materials and sodium, core disruptive events in fast power reactors such as disassembly, if they can occur at all, would proceed on a time scale too slow to cause significant structural damage from shock waves. This topic, together with a discussion of relative time scales between potential nuclear excursions and chemical detonations, is discussed further in Chapter 16-2. Nevertheless, steep pressure gradients may exist during compression of fluids in HCDA's, and these gradients are difficult to treat with finite difference approximations to the momentum equation.

Nearly all fast reactor hydrodynamics codes utilize a technique developed by von Neumann and Richtmyer[5] that adequately solves this problem through the use of a *pseudo-viscous pressure*, q (also called *artificial viscosity* even though the units are those of pressure). The concept is difficult and beyond the scope of this book even though the equations for the calculation of q are relatively simple. We only point out here that whenever a fluid is being compressed, q is calculated and added to the actual pressure p in the momentum and the energy equations. The effect of adding this pseudo-viscous pressure is to spread out the pressure gradient over several mesh cells in order to allow an acceptable solution by the finite difference method. Comparisons with more accurate shock wave techniques have demonstrated the adequacy of this approach for fast reactor safety analysis.

15-4 TRANSIENT OVERPOWER ACCIDENT (TOP)

A transient overpower accident (TOP) in an FBR refers to an off-normal condition in which a reactivity insertion occurs. This reactivity excursion could be initiated from any steady-state power condition (e.g., low flow shutdown, full-flow/full-power) or even during the approach to power. As noted from the discussion in Chapter 14, a plant protection system (PPS) is designed to provide reactor protection against such an occurrence. Hence, it is only in the case of PPS failure that a TOP incident becomes a concern. Because no core damage results unless PPS failure occurs, the only transient overpower incident that is of safety concern is the postulated unprotected TOP case. As a consequence, all further references in this chapter to the TOP event include the assumption of total PPS failure.

The TOP accident is discussed below by introducing the salient features of this generic accident type and then presenting an illustrative case. This is followed by a discussion of some of the key phenomenology which control the ultimate consequences of such an accident.

A. TOP SYNOPSIS

Given the assumption of a reactivity insertion occurring in an operating FBR, combined with the hypothesis of complete failure of the PPS, the problem is to determine the intrinsic dynamic response of the reactor. In essence, the fuel, cladding and coolant heat up, Doppler feedback and core expansion mitigate the power rise, the hottest fuel pins eventually rupture, molten fuel is ejected into the coolant channel, and the integrated result of this material movement is either to augment the accident or to effect a rapid reactivity shutdown. The actual accident path depends strongly on several factors, including the magnitude of the reactivity insertion ramp rate ($\$/s$), the amount of total reactivity inserted, the magnitude of the reactivity feedback mechanisms resulting from core heating, the type and axial location of pin failure, the post-pin failure molten fuel motion, and the final disposition of fuel, cladding and coolant.

Computer Modeling Techniques

Several computational systems have been developed to perform the coupled neutronics, thermal-hydraulics calculations necessary to simulate the whole-core response to unprotected transient overpower conditions. One key feature of such systems consists of a model to solve the point kinetics equations* (cf. Eqs. 6-5 and 6-6) along with spatially dependent reactivity feedback terms such as Doppler broadening, sodium voiding, core structural variations, and molten material movement.

A second required feature is the capability of the system to model the transient behavior of several representative fuel pins within the reactor core to allow for *incoherency* in the accident sequence. The term incoherency refers to the fact that pin failures and/or other accident phenomena would not occur simultaneously throughout the core. Rather, there would exist strong variations in the time and spatial location of these phenomena due to localized differences in power/flow ratios, fuel burnup, fuel power density, and cladding history. The transient model must include the ability to predict the time and location of pin failure, and the subsequent events such as molten fuel motion, molten fuel-coolant interactions, coolant ejection dynamics, and debris freezing patterns. Detailed heat transfer and hydraulic calculations are made only for selected fuel pins, each of which is normally assumed to be representative of portions or all pins within an assembly or cluster of assemblies. Each of these groupings is called a *channel*. The evolving family of multichannel codes from the SAS[6] and MELT[7] series have served as baseline tools for such calculations in the United States. The CAPRI[8] and FRAX2[9] codes are representative of such code systems employed in Europe. The advent of such multichannel codes has allowed the treatment of incoherency, and these more realistic treatments have led to

---

*The point kinetics equations can be used in a fast breeder reactor system for those portions of the transient over which variations from the original core geometry are small—so long as the reactivity feedback terms adequately reflect the spatial variations in temperature and material density change. This simplification, relative to that of a thermal reactor, is allowed because the relatively large neutron mean free path in an FBR allows the spatial shape of the flux to be rapidly established after a system reactivity perturbation is introduced. The point kinetics approximation may be less accurate for a heterogeneous core, relative to a homogeneous core, because of the reduced core coupling in the heterogeneous design.

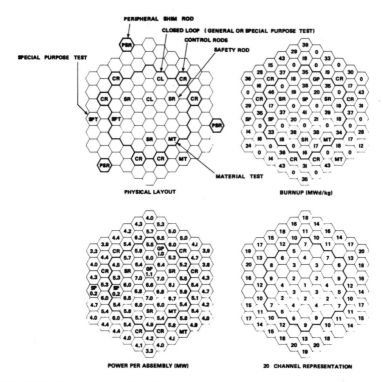

FIGURE 15-5. Beginning of cycle 4 (BOC-4) 20-channel representation for the FFTF.

predictions of reduced accident severity relative to calculations by earlier coherent core analysis methods.

A better appreciation of how such code systems work, and the manner in which the controlling phenomena interact during a TOP event, is afforded by the sample calculation depicted in Figs. 15-5 and 15-6.

Core Modeling

Figure 15-5 contains a plan view of the FFTF as well as the burnup and power per assembly for a Beginning of Cycle 4 core configuration.* One way to conduct a transient analysis would be to model a representative pin from each of the assemblies. However, for this case some 76 thermal-

---

*This Beginning of Cycle 4 configuration designates the status of the core at the inception of the fourth burn cycle. It is very similar to an equilibrium cycle (see Chapter 7-2).

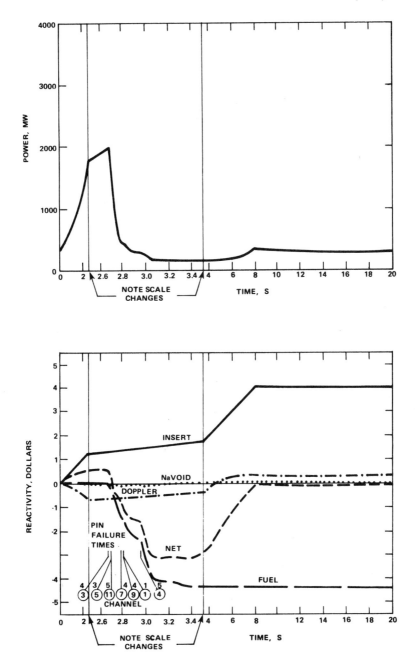

FIGURE 15-6. Typical TOP response characterization for 20-channel representation.[11] The encircled numbers denote the channel number for the sequence of failure; the numbers above the circles indicate the number of assemblies per channel.

hydraulics calculations would thereby be required to model the driver fuel assemblies at each time step during the transient. Whereas the capability for such analysis exists,[10] it is generally possible to attain an acceptable degree of accuracy by computationally lumping the assemblies into similar clusters (channels). For the example shown, this task was accomplished by grouping fuel pins with similar burnup and power patterns (with care to avoid merging assemblies with different flow orificing). The 20-channel structure resulting is shown at the bottom right of the figure.

Sample Transient

Figure 15-6 contains key reactor response indices for a transient initiated from full power conditions by a reactivity insertion of 0.5$/s (50¢/s). The net reactivity initially follows the inserted reactivity, causing the power to rise, but rising fuel temperatures soon yield a negative Doppler feedback which tends to offset the reactivity inserted by the postulated accident condition. Fuel temperatures continue to increase, however, and eventually fuel pin failures occur. The designation near the bottom of the figure indicates the channel failure sequence (numbers encircled) and the number of assemblies represented by each channel. Most of these failures occurred at an axial position well above the axial midplane of the core. As a consequence, molten fuel moved axially inside the central void region within the pins up to the cladding rupture location and into the coolant channel. This resulted in a negative fuel-motion reactivity effect, which dropped the net reactivity well into the subcritical region. The corresponding drop in reactor power is shown in the upper part of the figure.

For this particular example, the molten fuel—upon entering the coolant channel—was assumed to transfer sufficient heat to the flowing sodium that almost instantaneous coolant flashing took place. This caused a momentary flow reversal; but as coolant vapor pressures receded, flow was reestablished and much of the fragmented fuel was swept hydraulically upward and out the reactor, a process called *hydraulic sweepout*.

Since the identification of the mechanism for the reactivity insertion is often absent in such a postulated accident, it is not possible to logically define the magnitude of either the insertion rate or the total reactivity insertion. Hence, it is usually necessary to conduct sensitivity calculations by varying this arbitary input parameter over a wide range of values. For this example, an upper bound of $4 total insertion was employed. It was this continued reactivity insertion that raised the net reactivity back up from a negative value to approximately zero.

Flow Diagram

From this illustrative calculation, several questions arise. The actual accident sequence could go in several directions, depending upon the types of system response at several key junctures.

Figure 15-7 indicates the principal branch points which require detailed consideration. The first major concern is the axial location for fuel pin failure. The hollow arrow denotes that the bulk of both analytical and experimental evidence indicates a failure location well above the core midplane. The major subsequent question is whether the molten fuel entering the coolant channel is washed out of the core or whether local blockages would take place. A partial blockage may be an acceptable consequence, since only a 10% flow reduction would occur for an impervious radial blockage of up to 90% of the total flow area. Hence, early benign accident termination would occur and in-place cooling could be established. However, if a total flow blockage were to occur, coolant flow would be lost and

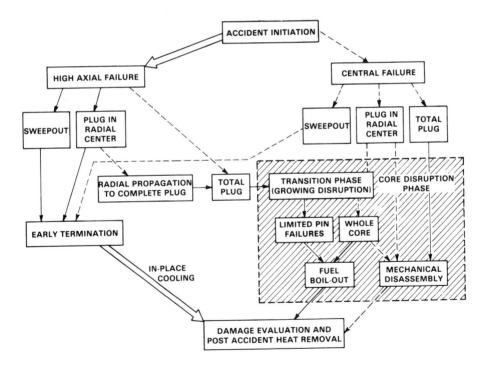

FIGURE 15-7. Potential accident sequence paths for the unprotected transient overpower accident.

the uncooled core would gradually tend toward the transition phase (cf. Fig. 15-2).

The right hand side of Fig. 15-7 denotes the possible sequence of events if the pins were to fail in large statistical numbers at or near the axial core midplane. Such a failure location would allow molten fuel to flow within the cladding jacket toward the core center—thus leading to a positive reactivity insertion. Whereas hydraulic sweepout could occur to terminate the transient, the time scales associated with the power rise would become very short and any such sweepout might be ineffective. Further, the rapid melting rates might cause a sufficiently high rate of molten fuel ejection that channel blockage would be difficult to avoid. Hence, hydrodynamic disassembly of the core might become the mode of permanent neutronic shutdown.

The principal parameters governing the ultimate sequence of the unprotected transient overpower accident are discussed in more detail below.

B. FUEL FAILURE DYNAMICS

From the above discussion, a parameter which heavily influences the ultimate course of a TOP event is the time and axial position of fuel pin failure. The timing is important because that determines the amount of molten fuel which can move, and the axial position of failure basically determines whether the initial response to cladding rupture is a positive or a negative reactivity contribution. Relative pin failure locations within an assembly are also important for determining potential flow blockage patterns.

Transient Considerations

Perhaps the most important parameter in determining transient fuel pin behavior is the interrelationship between steady-state power level and burnup. Fresh (unirradiated) pins tend to resist failure up to high energy depositions because the cladding is ductile and there is no fission gas available to provide a cladding loading force besides thermal expansion. For irradiated cases, the low power pins have little if any restructuring and, therefore, retain most of the fission gas generated during burnup within the unrestructured fuel matrix. Hence, upon heating, this fission gas is readily released and this loading function, combined with thermal expansion, tends to cause such pins to fail relatively early. Further, the failure location could be near the axial center since that is where the cladding loading function is maximized. However, early failure of such pins may be of minimal significance to the subsequent course of the accident because the amount of molten fuel available at the time of failure is relatively small. The principal concern is that an early failure location near the core axial midplane might serve as the site for molten fuel to enter the coolant channel later in the transient.

It is the failure of the high power, irradiated pins that are expected to most affect the accident sequence. These pins tend to have considerable restructuring, and most of the fission gas generated in the columnar and equiaxed regions is released to the fission gas plenum during steady-state irradiation (cf. Chapter 8-2). As a consequence, substantial fuel melting can be accommodated prior to pin failure. In addition, the cladding tends to fail near the top of the restructured region—where the internal cladding loading patterns and high cladding temperatures combine to produce the location most susceptible to failure.

Empirical Approach

The qualitative arguments given above must be quantified to conduct safety analysis. Effective modeling of the fuel pin behavior under transient conditions is difficult. Consequently, the basis for predicting pin failure patterns for the analyses required for reactors such as FFTF was provided by direct experimental data, gathered over the transient domain of maximum interest. Though the in-pile data base was clearly recognized to reflect an environment for failure which is different in several respects from the environment expected in actual whole-core reactor conditions, useful empirical relations were derived.[12]

Mechanistic Approach

Several mechanistic (or deterministic) modeling approaches have been pursued to place the important area of fuel pin failure prediction on a firmer basis. Computer code systems such as BEHAVE,[13] LAFM,[14] DSTRESS,[15] and FPIN[16] reflect encouraging efforts in this pursuit. Each of these codes uses a life fraction rule based on the Larson-Miller parameter for the pin failure criterion in a manner similar to that described for steady-state pin failure in Chapter 8. Despite some success in correlating pin failures in transient tests, such as those in the TREAT reactor,[17] reliable prediction of pin failures during transients is still under development, and efforts are being made to incorporate strain cladding criteria into life fraction calculations.

C. OTHER CONSIDERATIONS

In addition to the above issues of axial failure location and the quantity of molten fuel within the pin at the time of failure, there are other key areas of uncertainty in the TOP accident as discussed below.

Fuel Sweepout vs Partial Blockage

There is experimental evidence[18] that substantial amounts of the molten fuel entering a full flowing stream of sodium coolant will be fragmented and axially transported away from the cladding rupture site. However, there are other data[19] which suggest significant local blockages forming from the once-molten fuel which becomes rapidly solidified. Much of the ambiguity in the results is due to the difficulty in attaining test conditions which appropriately simulate the accident conditions of interest. Computational systems such as PLUTO[20] and EPIC[21] have been developed to address the problem and determine the controlling parameters. Presumably, the potential for efficient hydraulic axial transport of fuel debris is considerably higher for wire wrapped pins than for assemblies which employ grid spacers. The wire wrap allows an unobstructed axial flow path and the swirling flow field intrinsically encourages the dislodging of localized blockages. However, it is conceivable that the wire could contribute to the formation of certain types of blockage patterns.[22]

Long-Term Coolability

Once nuclear shutdown of the TOP accident has occurred, it is necessary to demonstrate that long term coolability of the core debris can be established and maintained. In this regard, it is important to recognize that all of the fuel pins within any given assembly do not behave identically (even though they are normally assumed to be identical in the clustering schemes employed in the SAS and MELT-type codes). As will be recalled from Figs. 10-9 and 10-10, there exist strong power-to-flow variations among the pins within the assembly duct. This difference becomes even more pronounced during a thermal upset condition. As a consequence, detailed studies[23] of pin behavior within assemblies during a TOP indicate a substantial time and spatial variation in the pin failure patterns. Such behavior suggests that even if large blockages should result from failure of the initial pins, flow paths would still exist in the unfailed portion of the assembly[24] to provide long-term cooling.

Figure 15-8 is a sketch which qualitatively illustrates some of the above considerations. Contained in this sketch are four progressive time frames, each including a side view of a single pin, a top view of a small group of pins, and a perspective of an entire assembly. Though not corresponding to any particular calculation, the sketch indicates the sequence which must be considered by the analyst. Initial cladding failure results in molten fuel escaping through the rupture site and subsequently fragmenting and/or impinging on neighboring pins. The competition between the tendencies for hydraulic sweepout and local plate-out (blockage) is illustrated. The final time frame illustrates the possibility of a localized blockage near the upper

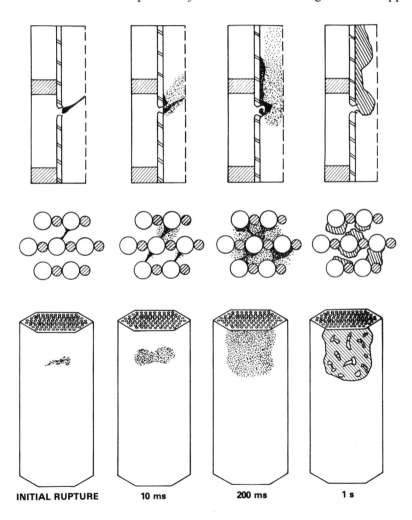

INITIAL RUPTURE   10 ms   200 ms   1 s

FIGURE 15-8. Possible geometric conception of the accident path down the left side of Fig. 15-7 as a function of time (not drawn to scale).

central region of the assembly while preserving intact fuel pins in the periphery.

Pin Failure Incoherencies

In addition to the variations in pin failure time and location due to power-to-flow differences between fuel assemblies (as accounted for in multichannel codes), it is likely that statistical factors could cause an additional degree of pin failure incoherencies in the failure patterns *within* an assembly. Differences in fuel-cladding gap widths, cladding imperfections, localized flow perturbation, etc. could all lead to such variations. The net result of this broader degree of failure incoherency would be to reduce the concentration of fuel debris in the coolant (relative to coherent pin failure)—thereby tending to enhance the prospects for establishing and maintaining long-term cooling.

15-5 TRANSIENT UNDERCOOLING ACCIDENT (TUC)

Transient undercooling, in the context of FBR safety studies, denotes a class of accidents in which the coolant system is postulated to fail. The most prominent of the particular accident sequences analyzed in this category is the flow coastdown incident, where the primary pumps are presumed to become inoperable (e.g., loss of power to the drive motors). Other transient undercooling accidents postulated include the loss of piping integrity (major reactor coolant pipe break) and the loss of the ultimate heat sink.

For the flow coastdown accident, a normal response from the PPS would effect a plant shutdown with no resulting damage. Hence, it is only in the event of complete PPS failure that the flow coastdown could lead to serious consequences. For the other two events, however, normal PPS response is assumed. The major question for these postulated incidents is whether the reactor decay heat can be adequately removed under the most extreme conditions of coolant starvation.

Since a loss of forced sodium flow is a fundamental assumption in each of the above accident postulates, the issue of sodium boiling becomes a common denominator in the accident analysis procedure. Hence, the first major topic treated in this section is the phenomenon of sodium boiling and its implications to heat removal capability. Attention will then be focused on the flow coastdown event, along with a sample calculation to allow the reader to visualize the time scales involved and develop an appreciation for the competing phenomena involved. An overview of the remaining transient undercooling events will then be given to complete the section.

A. SODIUM BOILING

An understanding of the sodium boiling process is crucial in predicting accident sequences involving flow starvation. The principal issues concern the two-phase flow regimes involved, degree of superheat, liquid film stability, entrainment, and the characteristics of coolant re-entry and re-wetting. Due to extensive investigations of sodium boiling (References 25 through 31) the basic phenomena are reasonably well understood.

Flow Regimes

Figure 15-9 illustrates the various flow regimes for boiling fluids. For liquid-metal boiling near atmospheric pressure (typical of LMFBR conditions), the *slug and annular flow* regimes tend to prevail—as opposed to the *bubbly flow* regime typical of high pressure light water systems. Prior to the inception of bulk boiling, the development of a single bubble which grows and then collapses represents typical behavior. With increased coolant heating, a multi-slug (building and collapsing) pattern begins to develop which can readily evolve into an annular flow regime. If flow reversal occurs, a bidirectional annular flow pattern (with high void fraction) rapidly

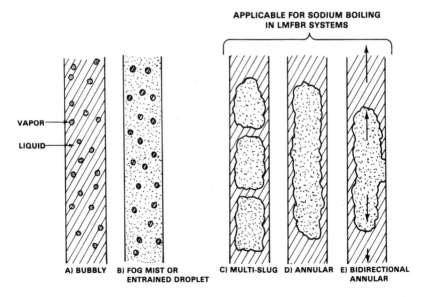

FIGURE 15-9. Typical flow regime patterns in vertical flow. Annular and mist flows are forms of dispersed flow. Adapted from Ref. 32.

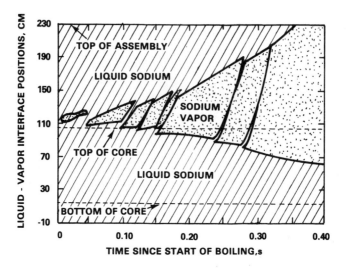

FIGURE 15-10. Multi-slug boiling patterns predicted for the initial phase of a flow coastdown accident.

develops. The annular and *mist* flows are forms of *dispersed flow*, in which the vapor phase is continuous (as opposed to bubbly flow in which the liquid phase is continuous).

The SAS code,[6] which is widely used for the prediction of coolant boiling patterns in postulated LMFBR transient undercooling events, employs a multi-slug flow model. Figure 15-10 illustrates a typical coolant boiling pattern predicted for a flow coastdown accident. Boiling initiates at the top of the core, where the coolant is hottest. The initial bubble is axially transported upward into a cooler region where it collapses in about 50 ms. Another bubble is formed, which lasts longer (due to increasing system temperatures, i.e., hotter sodium vapor) but also ultimately collapses. This process is repeated, with a general tendency for larger and more sustained bubbles as the expanding sodium vapor retards and finally reverses the upward flow of liquid sodium. It should be noted that for the low pressure operation of an LMFBR, the density of sodium vapor is about 2000 times lower than for the liquid phase. Hence, the vaporization of only a small mass of sodium results in a large volume of vapor—which has a substantial influence on the momentum relationship which governs the coolant flow.

Superheat

Early laboratory studies of sodium boiling revealed several instances in which a very high degree of superheat was measured prior to bulk boiling. Data representative of this era was reviewed by Fauske.[27] It was noted that superheat values up to 900°C had been attained under carefully controlled laboratory conditions. Such data caused considerable apprehension among safety analysts because of (1) the uncertainty such superheat levels would introduce into the predictions for the onset of boiling, and (2) the large pressure which could theoretically result when flashing did occur.

The processes necessary to attain such levels of superheat are now well understood and there is essentially universal agreement that such conditions cannot exist in an operating LMFBR system. Of the various mechanisms identified, the presence of small quantities of inert gas appears to provide sufficient nucleation sites to prevent significant superheat. An examination of previous data claiming strong velocity and heating effects indicates that these globally observed effects were likely caused by inert gas present in active cavities in the heating substance and/or gas bubbles in the liquid.[28] Forced convection experiments[29] and observations[30] of long-term LMFBR behavior indicate that the superheat at the time of boiling inception following a system transient is expected to be very close to zero.

Film Dryout

In all of the flow regimes observed for the initial stages of boiling in an LMFBR system, a film of liquid remains on the cladding wall. Vapor from this heated film contributes significantly to the driving force for bubble expansion, particularly after flow reversal when the surface area of the film becomes large. Also, the presence of this film provides effective cooling to the cladding and, thus, delays cladding failure and the impending escape of fission gas into the coolant channel. Because of the influence of the film on core cooling and boiling dynamics, it is important to be able to predict how long the film can be maintained, i.e., the time to *film dryout*.

Once boiling starts, the liquid volume fraction remaining in the two-phase region quickly drops to a value between 0.15 and 0.20. By way of perspective, for a 217-pin wire-wrapped assembly typical of FFTF dimensions, a liquid fraction of 0.2 would correspond to a liquid film thickness of about 0.1 mm if uniformly distributed. There are two principal mechanisms for reducing the size of this film thickness: evaporation and entrainment. The evaporation rate can be calculated in a fairly straight-forward manner, but entrainment of the film into the streaming vapor field is more complicated.

The onset of *entrainment* is preceded by a phenomenon called *flooding*. For an upward vapor velocity above a critical value (which is a function of the density of both the vapor and liquid film, the viscosity and surface tension of the liquid, and the relative velocity between the two phases), the liquid film develops large surface waves—sometimes substantially larger than the average thickness of the film. When the upward viscous drag caused by such surface roughness becomes sufficiently strong to offset the gravitational draining force, net vertical motion in the film ceases. This constitutes the flooding threshold. For yet higher upward vapor velocities, portions of the liquid film are torn off and become entrained in the vapor field.

A correlation of the physical properties vs Reynolds number can be constructed, based on available data, which indicates the minimum vapor velocity for the entrainment of the sodium film to be approximately 37 m/s.[31] By way of comparison, a minimum sodium vapor velocity of greater than 120 m/s is required to entrain molten cladding (due principally to the much larger density of stainless steel).

For an LMFBR accident involving boiling at full power conditions, a typical film dryout time due to evaporation only is about 0.2 to 0.3 seconds. However, sodium vapor velocities greater than that required for entrainment can be readily established very early in the boiling process and, as such, film thickness reduction due to axial mass transport typically proceeds at a rate approximately ten times faster than that due to evaporation alone. Consequently, a calculation for the local dryout time based upon a stationary film might overestimate this time by as much as an order of magnitude.[31]

Liquid Reentry and Rewetting

The phenomenon of *rewetting* is of particular importance in assessing the problem of liquid coolant *reentry* as it relates both to the question of extended coolability and a possible strong thermal interaction between molten cladding-fuel and coolant. Reentry will not take place as long as the rate of evaporation exceeds the rate of condensation, which is generally the case in whole-core loss-of-flow accidents or single assembly events if the generated power is not severely reduced as a result of the early voiding event. However, in the case of local boiling caused by a local blockage within an assembly, condensation is likely to override evaporation; and sufficient cooling can be maintained to prevent overheating of the cladding.[33] Reentry can occur if the fuel pins become sufficiently hot to prevent rewetting, i.e., if film boiling can take place. Preliminary considerations of contact temperatures and nucleation temperatures indicate that the LMFBR cladding material must be extremely hot (approximately at the melting temperature) before film boiling dominates the boiling process.[34] This fact minimizes the possibility for any hard reentry,* since nucleate

---

*"Hard" reentry refers to coolant return via a liquid slug as opposed to reentry via a stochastic vapor/liquid mixture.

boiling will lead to pressurization and therefore prevent the coolant from penetrating very deeply into the core.

B.   UNPROTECTED FLOW COASTDOWN EVENT

The unprotected flow coastdown TUC event results from a postulated sudden loss of power to the primary coolant pumps, causing the coolant flow to coast down at a rate controlled by the rotational inertia of the pump and motor.* Several seconds are normally available to accomplish effective reactor shutdown prior to reaching coolant boiling conditions. However, the analyses for this event are based on the assumption that *all* such protection systems are inoperable. The subsequent consequences of coolant boiling, cladding melting, and fuel melting are analyzed on this basis.

Computational Modeling Techniques

As for the case of modeling the unprotected TOP event, large multichannel neutronics, thermal-hydraulics computer codes are necessary to track the numerous interacting phenomena which govern the course of an unprotected flow coastdown accident. The same tools referenced for the TOP analysis also have varying degrees of capability for analyzing the flow coastdown TUC accident, but the SAS code family has been the tool most extensively tested and employed for such use. The most advanced version of this family of codes contains a sodium boiling module (including the capability for modeling film dryout and liquid coolant reentry)[35] and modules to estimate the movement of molten cladding[36] and molten fuel.[37] Predictions from the coolant boiling module have been compared to experimental data and reasonable agreement achieved. Less confidence is possible for those models describing later consequences in the accident sequences, both due to the lack of appropriate experimental data and the complexity of modeling core configurations in which considerable geometric distortion is predicted to occur.

Sample Transient

Figure 15-11 displays the overall transient calculation[38] for the Clinch River Breeder Reactor Plant (CRBRP) subjected to an unprotected flow

---

*Primary loop sodium inertia also contributes to the flow coastdown characteristics.

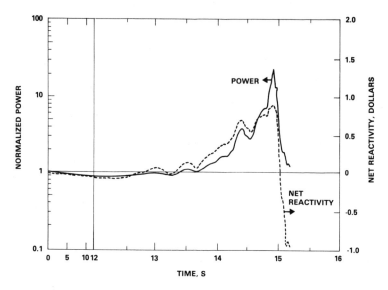

FIGURE 15-11. Typical power and reactivity response for an unprotected flow coastdown TUC event.

coastdown event.* As the coolant initially heats up, the effect of fuel pin thermal expansion, Doppler feedback (from the gradually heating fuel), and change in coolant density combine to yield a small net negative reactivity. This causes a very slow drop in core power. However, coolant boiling begins in the hottest assemblies some 12 seconds into the transient. As boiling spreads to more assemblies, the reactor power starts climbing due to the positive voiding reactivity. Eventual cladding melting and relocation adds some additional positive reactivity. At approximately 14.4 seconds into the transient, fuel in the first assemblies to void is predicted to disrupt and to begin dispersing away from the core centerline, being driven by a combination of fission gas and sodium vapor pressures. Continuing sodium voiding and slow collapse of some of the previously-dispersed fuel lead to a second mild power burst, which melts fuel throughout the rest of the core. Sufficient fuel dispersal then occurs to terminate this second burst.

Figure 15-12 contains a map of the hydraulic and heat transfer processes occurring in the hottest assembly for the sample transient.[38] Flow reversal follows boiling inception by about 0.5 seconds. Intermittent reentry of liquid sodium at the ends of the assembly, followed by increased vapor generation as the sodium contacts hot cladding, causes the chugging action.

---

*This calculation was for the end-of-equilibrium cycle homogeneous core configuration. The hottest channel consisted of 12 assemblies.

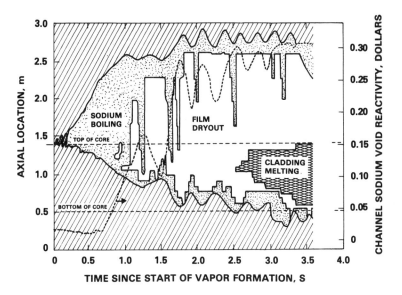

FIGURE 15-12. Hydraulic response in hottest channel for the transient of Fig. 15-11. (The dashed line shows the sodium voiding reactivity for this channel only.)

Film dryout follows boiling inception at the top of the core in less than one second, and cladding melting commences soon afterward. Near the middle of the core, note that dryout occurs at some points in about 0.1 s after boiling inception, which is reasonably consistent with the discussion of film dryout in Chapter 15-5A.

From this particular calculation, it was not possible to provide a clear conclusion regarding the long-term cooling of the core. Conditions typical of hydrodynamic core disassembly were not attained, but the potential for coolant channel blockage (from freezing cladding and fuel debris) could not be ruled out. Hence, conditions for the transition phase were established.

Flow Diagram

It should be clear, from the preceding example, that the analysis for a postulated unprotected flow coastdown becomes very difficult, and there are several branch points in the accident sequence which need careful investigation. Figure 15-13 includes an accident flow diagram which denotes the general direction which such an accident sequence could take, with the hollow arrows indicating the most likely sequence. Whereas early termination with in-place cooling established appears quite possible for the unpro-

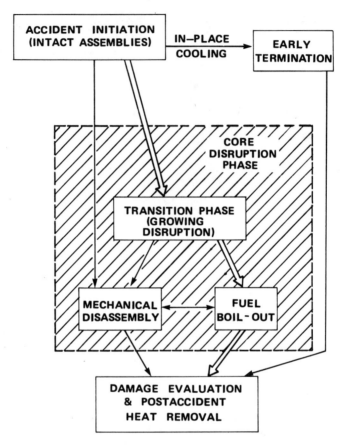

FIGURE 15-13. Potential accident sequence paths for the unprotected transient undercooling accident.

tected TOP event, this eventuality is less likely for the unprotected flow coastdown accident. Once substantial coolant boiling begins, it is difficult to prevent the accident from engulfing the bulk of the core and leading to significant fuel pin melting.* The question is whether the reactivity effect of coolant voiding, and/or cladding and fuel motion, would lead to core disassembly conditions or a transition phase involving fuel and cladding boiling. Several of these key phenomena and branch points are discussed below.

---

*With regard to the onset of sodium boiling, one of the interesting features of Super Phénix is the flowered core design. Analysis indicates that core heating during a postulated TUC event would cause sufficient negative reactivity feedback to prevent the onset of coolant boiling.

ILLUSTRATION OF NONCOHERENCE EFFECTS AND CLADDING RELOCATION

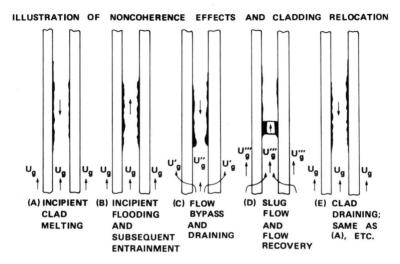

(A) INCIPIENT CLAD MELTING

(B) INCIPIENT FLOODING AND SUBSEQUENT ENTRAINMENT

(C) FLOW BYPASS AND DRAINING

(D) SLUG FLOW AND FLOW RECOVERY

(E) CLAD DRAINING; SAME AS (A), ETC.

FIGURE 15-14. Illustration of noncoherence effects and cladding relocation during a transient undercooling event. (Figure constructed from logic developed in Ref. 39.)

Cladding Motion

Aside from the magnitude of the sodium void coefficient, the questions which provide the key bearing on the ultimate consequences of the unprotected flow coastdown accident concern molten cladding and fuel movement.

A major question involves both the transient behavior and ultimate position of the cladding. Observation of Fig. 6-10 reveals that movement of the cladding out of the core results in a positive reactivity feedback; also any movement could result in refreezing in a configuration which could prevent coolant flow. Figure 15-14 indicates one possible visualization of molten cladding movement.[39] The diagram at the left depicts the initial cladding melting condition. Assuming sufficiently high sodium vapor velocities, cladding flooding and then entrainment is shown to occur. However, this flooding process greatly increases the local axial pressure drop and, as a result, the sodium vapor would tend to bypass the molten channel and allow cladding draining to occur. Flow recovery would then occur and the cycle be repeated.* The principal argument here is that bulk cladding motion may

---

*This model is based on the assumption that several adjacent pins are behaving in a similar manner.

not occur, even though considerable local sloshing would be present. The validity of this argument, as opposed to bulk transport of the molten cladding to the axial extremities of the core (where refreezing could occur), has substantial influence on the subsequent course of events. In-pile experimental evidence suggests this mechanism may indeed be influential in retarding axial cladding relocation, though it may not be effective in completely preventing the formation of porous cladding blockages in the upper core region.

Fuel Motion

Fuel motion represents the other principal concern. Early work in this field was focused around the potential for molten fuel slumping under the force of gravity. Later research[40] suggested that fission gas release from irradiated fuel might cause fuel pellets to crack and disperse in a powdery fashion prior to melting. Evidence from more recent experimental and analytical investigations[41] points toward irradiated fuel generally dispersing as a mixture of molten liquid and irregular chunks once the fuel has become significantly molten, so long as the breakup is occurring at power levels well above nominal power.

Transient Undercooling—Transient Overpower Accident (TUCOP)

For small-sized LMFBRs up to and including FFTF, the sodium void reactivity effect is small enough that direct reactivity effects attributed to sodium voiding during a TUC event contribute fairly small perturbations to its overall reactivity balance. For the CRBRP transient illustrated above, the sodium void effect was sufficiently strong to accelerate cladding and fuel melting patterns, but early fuel dispersal in the hottest assemblies tended to offset subsequent positive reactivity insertions from sodium voiding.

However, for larger LMFBR cores, where the neutron leakage component is smaller and the sodium void coefficient thereby larger, it is possible that the reactivity consequences due to coolant voiding could be sufficiently large and positive to cause a whole-core power excursion typical of the TOP event. Such a condition is possible, even for the example calculations shown, if an assumption was made that fuel dispersal in the hot assemblies did not occur. This ensuing accident is called the transient undercooling-transient overpower, TUCOP, event.* The concern for this type of accident is how the fuel pins would fail—particularly for those regions of the core still containing sodium. The axial failure location could be close to the core midplane because the low flow rates force the peak cladding temperature towards the axial center of the core (more in concert with the axial power profile). Also, depending upon the magnitude of the sodium void coefficient, the reactivity insertion rate could be larger than the rates normally considered in the TOP event.

C. LOSS OF PIPING INTEGRITY

As noted from the discussion of Chapter 13-3, a postulated double ended rupture of a primary coolant pipe constitutes the limiting accident analyzed in a light water reactor. Such an accident is considered even less probable in an LMFBR because of the much lower system pressure. Also, ductile steel is used for the primary piping and even if a defect or flaw should develop, it would lead to a small penetration in the piping wall such that leakage could be detected prior to rupture. Nonetheless, attention has been given to the remote possibility of such ruptures to assess the inherent system response. Because of the extremely rapid flow reduction occurring for such a condition, coolant boiling might occur even if the plant protection system performed as designed.

Initiating Event

The worst possible location for a postulated rupture of the primary piping in a loop-type LMFBR appears to be at the inlet nozzle to the lower coolant plenum. This is because such a position represents the lowest point in the primary system piping, the fluid column between the reactor inlet plenum and the rupture location is minimal (resulting in very little fluid inertia), and the associated frictional pressure drop over this length is very small. Even with a rupture at this location, the core remains covered by sodium due to the presence of a guard vessel around the inlet downcomer piping.

Should a rupture occur, this condition typically would be detected by instrumentation which monitors both the coolant pressure and the neutron flux. A ratio of the square root of the pressure to the neutron flux is often used to detect off-normal conditions; this is roughly equivalent to a core

---

*This accident is also sometimes referred to as the loss-of-flow driven TOP event (LOF 'd' TOP).

flow-to-power ratio. A scram signal should be effective within about one-half second from the time of rupture.

Transient Conditions

By the time scram is completed, coolant temperatures would likely be climbing due to the combination of a large heat capacity within the fuel pins and low coolant flow. Coolant boiling is a distinct possibility.* However, due to the large coolant temperature gradient between assembly flats, boiling would initially occur within the central regions of the assembly—with condensation occurring at some downstream recombination point. The principal question is whether such localized boiling could be prevented from engulfing the full assembly cross section prior to removing sufficient heat from the hottest pin bundles to allow restoration of liquid phase flow.

D. LOSS OF ULTIMATE HEAT SINK

Initiating Event

The final transient undercooling event considered in this section is the loss of the ultimate heat sink. During normal scrammed conditions, an active heat removal system is employed to remove the decay heat. It is conceivable that the loss of all AC power and failure of backup diesel generators would result in the loss of forced cooling. Even if this should occur, the heat transport system is designed to assure adequate heat removal from the core via natural circulation. Testing of whole-core systems provides considerable confidence that natural convection will, indeed, adequately remove decay heat for an LMFBR system (discussed in Chapter 12-2A and 14-3C).

If, however, it is further postulated that the ultimate heat sink were lost (i.e., no communication to secondary or tertiary heat transport systems and loss of all emergency decay heat removal systems), natural circulation could eventually be interrupted; and the subsequent sequence of events could lead to a loss of in-place coolable geometry.

---

*The likelihood of boiling is very small for LMFBR systems employing four or more primary loops.

Boiling Natural Circulation Flow and Heat Transfer

Although single-phase cooling is the normal mode for natural convection conditions, there may be sufficient heat removal to prevent gross cladding melting even if saturation temperatures are attained. Both analysis[42] and experiments[43] suggest coolant boiling may be stable for decay power heat fluxes. Vaporization rates for assembly power levels less than $\sim 5\%$ of full power appear to be insufficient to cause flow starvation. Flow starvation refers to the condition when liquid is flowing over a surface at a rate too low for boiling heat transfer to remove heat at the required rate.

The relatively cold upper assembly structure provides a large condensation potential which initially far exceeds any possible vapor generation rate. However, as the vapor condenses in the upper assembly, or heat is absorbed from the single-phase coolant, the temperature profile in the upper assembly would elevate, increasing the average sodium temperature in the assembly and thereby the natural circulation head. Furthermore, the increased buoyancy resulting from the boiling would cause an increase in coolant flow which might restore single-phase cooling temporarily, thereby resulting in the possibility of alternating cycles of single-phase and two-phase flow.

If a loss of heat sink accident should proceed to the point of appreciable coolant boiling, the subsequent consequences become difficult to predict. Depending on the burnup of the assembly, pin failures could occur even with coolant temperatures below the saturation point. The release of fission gas would complicate the analysis, but the long-term implications of such release do not appear to be major. Any melting of the fuel would not likely result in recriticality, because the reactor is already highly subcritical. However, the possibility of regaining a critical configuration cannot, in general, be ruled out and such a core configuration could provide the starting point for a transition phase analysis.

15-6 TRANSITION PHASE

The transition phase[44] of a postulated unprotected accident sequence represents the time domain in which core melting patterns lead to gross geometric changes, but the net reactivity is low and the time scale is far longer than characteristic of the core hydrodynamic disassembly phase. It refers to that phase of an accident where neither the traditional predisassembly computational tools referred to in Sections 15-4 and 15-5 (which are based on essentially as-fabricated geometry) nor the disassembly tools to be discussed in Section 15-7 (which are based on all fluidized geometry) are applicable. During this phase the molten core debris may boil, and hence form a *boiling pool*.* The basic question to be answered is whether a core in such a condition of geometric disarray can regain criticality and, if so, at what rate. If the void fraction of a boiling pool can be maintained sufficiently high (i.e., if the pool is *boiled up*), then criticality is unlikely. *Collapse* of the pool (i.e., change to low void fraction) could lead to recriticality.

A key issue in determining whether boil-up exists, so that recriticality is avoided, is whether a clear pathway exists between the boiling core debris and the upper sodium pool (*open system*), or whether such pathways are all blocked by frozen steel and/or fuel (*bottled system*). Consequently, the first topic to be summarized in this section is that of axial blockage potential. The second major topic concerns the boiling pool flow regime. The reason for the importance of the pool flow regime is that pool dynamics (which directly determine the potential for recriticality) are governed by the dominate flow regime. The heat loss mechanisms are also important because such heat transfer directly determines the potential for boil-up within the pool.

Given this background, attention is then addressed to issues affecting open system and bottled system core pool dynamics. Finally, brief mention is made of representative computational systems under development for analyzing transition phase accident sequences.

A. AXIAL BLOCKAGE POTENTIAL

As noted from the previous section, once an accident proceeds to the point of cladding melting there exists the potential for a combination of upward sodium vapor and downward gravitational forces to relocate this molten cladding to the cooler axial extremities of the core where it can freeze. Both experimental and analytical studies provide a strong indication that sufficient quantities of molten steel (and possibly some fuel) will eventually relocate in the lower core region to freeze and block the coolant inlet region.

The question of blocking the upper core outlet region is less well defined. The in-pile data relating to such questions are presented in a review article by Dickerman, et al.[19] Whereas substantial upper cladding plugs were

---

*It is normally assumed, once this stage of an unprotected accident is reached, that sufficient molten cladding draining has occurred to freeze and plug the core inlet region. Hence, lack of coolant reentry from the top leads to eventual core debris boiling due to the generation of decay heat.

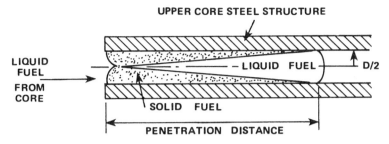

FIGURE 15-15. Conduction limited freezing model.

found, it was recognized that the single pin and seven pin experiments may be misleading, particularly with regard to the "cladding sloshing" (cf. Fig. 15-14) arising from bypass flow considerations which are difficult to simulate in tests with a small fuel-pin cluster. Out-of-pile experiments[45] using 28-pin clusters with simulant materials provided encouraging backup for Fauske's prediction[39] that the large axial relocation of molten cladding observed in the in-pile experiments may not occur during whole-core accident conditions.

However, there are at least two phases of the unprotected accident sequence where the freezing of molten cladding or fuel in the upper core structure is of interest. The first is in conjunction with cladding motion immediately after initial melting, i.e., the conditions mentioned above. The second is the disposition of a molten steel/fuel mixture ejected into the upper structure due to the dynamics of a boiling pool. In the attempt to define the potential for such material freezing in the upper blanket and/or reflector or gas plenum region, three different models have been proposed: the conduction limited, bulk-freezing, and ablation-induced freezing models.

The *conduction limited* model, as denoted* in Fig. 15-15, is based on a classical conduction freezing mode; i.e., freezing is governed by the transient heat conducted through a growing frozen fuel layer at the channel wall. Once the growing frozen shell closes at the radial center of the channel, flow ceases and the remaining molten fuel already inside the channel freezes in situ. Cheung and Baker[46] were among the first to determine experimentally the penetration distances of simulant fluids freezing in tubes. Empirical

---

*For simplicity of argument, the two freezing models illustrated in Figs. 15-15 and 15-16 depict pure fuel as the molten material ejected upward into the cooler upper structure. The same basic concept would apply if this were a molten fuel/steel mixture. Also, it should be noted that the pins in these figures are drawn in a horizontal position, rather than in the actual vertical configuration.

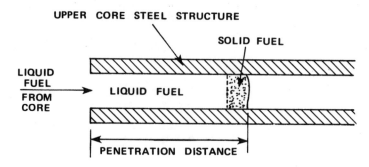

FIGURE 15-16. Bulk freezing model.

correlations of their data were in good qualitative agreement with such a conduction freezing model.

Ostensen and Jackson[47] earlier argued that the surface of the upper channel wall would melt upon contact with the flowing molten fuel and, as such, would not provide an underlying surface to which any frozen fuel could attach. Hence, they proposed a turbulent mode of heat transfer to the steel wall and suggested that the leading edge of this molten fuel slug would become a "slush" which would freeze in a bulk-type fashion once the heat of fusion was removed. Early experiments[48] supported this *bulk-freezing* model concept. This model is illustrated in Fig. 15-16.

Out-of-pile experiments using a thermite reaction* were performed by Epstein, et al.[49] to measure penetration depths of molten UO_2 into a dry upper plenum region simulation. The results generally supported the bulk-freezing model concept, but better agreements were obtained by Epstein, et al.[50] after modifying the bulk freezing model to allow for steel *ablation* into the flowing fuel mixture. In this context, ablation refers to the process of molten steel directly mixing with the fuel, such that the flowing fluid reflects the properties of a mixture of fuel and steel, rather than pure fuel (as in the bulk-freezing model).

In comparing both the available data and computational models related to molten fuel/cladding axial relocation and freezing, Epstein[51] concluded in his review article that the probability for maintaining an open path from the boiling pool to the upper sodium plenum is quite high.† Consequently, most of the developmental work has been addressed to this area. Neverthe-

---

*The thermite reaction commonly used for creating molten UO_2 out-of-pile is to chemically react uranium powder with Mo_2O_3.

†More recent investigations suggest less optimism for this conclusion. Long axial penetration distances are still expected, but eventual blockage cannot be precluded with present designs.

less, there still remains sufficient uncertainty in this area that bottled up core boiling dynamics remains a subject of considerable interest.

B. FLOW REGIMES

The potential two-phase flow regimes for boiling sodium were pictorially represented in Fig. 15-9. It was noted in the discussion that bubbly flow does not occur for a low pressure sodium boiling situation. Rather, a slug or annular type flow regime prevails.

Such is not the case if internally heated fuel is the boiling medium and the geometry is of assembly or full core dimensions. *Bubbly* flow tends to prevail at low power levels, and for higher power levels a transformation to *churn-turbulent* flow appears to occur. At even higher heating levels, *foam* flow has been observed. A transformation to *entrained droplet* (or *fully dispersed* flow) has yet to be experimentally observed for materials of interest to the core boil-up question. *Slug* or *annular* flow is never expected to occur for core boil-up situations principally due to the larger geometries involved. Surface tension and viscous effects stabilize the slug flow during sodium boiling in the normal closely-packed pin geometry, but no pin structure remains for conditions associated with the transition phase.

With regard to the pictorial hierarchy depicted in Fig. 15-9, both churn-turbulent and foam flow regimes occur between bubbly and entrained droplet regimes. Churn-turbulent, as observed from the microwave heating experiments of Ginsberg, et al,[52] tends to exhibit a void fraction in excess of 50% and represents a highly dynamic two-phase mixing process. Foam flow (which can be likened to shaving cream from a pressure can dispenser) exhibits even higher void fractions and has also been experimentally observed[52] for heating rates similar to those which produced churn-turbulent flow. A prediction of which of these two flow regimes will occur is difficult because seemingly small factors, such as the presence of mixture impurities, can cause a controlling effect.

For the purpose of evaluating boiling flow regimes, Fauske has noted[53] that it is possible to estimate the approximate boundaries separating the prevailing flow regimes for a given geometry and set of materials by neglecting viscous effects. By starting with the Kutateladze[54] stability criterion, which relates drag (or dynamic pressure) forces to buoyancy forces, the stability parameter, k, takes on the form

$$k = \frac{u^* \sqrt{\rho_c}}{[g\sigma(\rho_H - \rho_L)]^{\frac{1}{4}}}, \tag{15-8}$$

where $\qquad$ u^* = critical superficial velocity of the lighter phase
ρ_c = density of the continuous phase
σ = surface tension of the heavy fluid
ρ_H = density of the heavy fluid
ρ_L = density of the light fluid
g = gravitational constant.

The velocity, u^*, can be determined for an open volume-heated boiling pool according to the relationship

$$u^* = \frac{Q(1-\bar{\alpha})H}{\rho_L h_{fg}},$$ (15-9)

where $\qquad$ Q = volumetric heat generation rate
$\bar{\alpha}$ = average void fraction below height H
H = axial position within the pool
h_{fg} = latent heat of vaporization.

By eliminating u^* between these two equations, and using stability parameters appropriate to different flow regimes, Fauske[53] developed flow regime maps for a boiling pool, as shown in Fig. 15-17. The variables chosen are relative power level vs axial position in the pool. The major point made from Fauske's work is that a dispersed flow regime may exist over a large segment of an LMFBR boiling pool even down to low power levels (characteristic of decay heat). However, these plots were developed for an open system (free communication to an upper low pressure region) and are not necessarily valid for the bottled core. Also, it should be noted that the dispersed flow regime predicted in Fig. 15-17 has yet to be validated experimentally.

C. OPEN SYSTEM

Substantial interest has developed in the boiling pool behavior for an open system. The basis for this is twofold. First, if an unprotected accident should ever reach conditions which typify the transition phase, there is some evidence[51] (as discussed in 15-6B above) that a pathway may exist between the boiling core and the upper low pressure coolant plenum. Second, given this situation, Fauske has developed arguments which suggest that pool

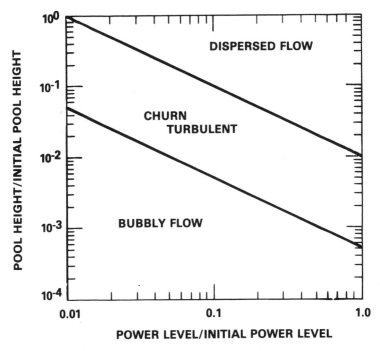

FIGURE 15-17. Flow regime map for boiling fuel.[53]

boil-up might be assured even for power levels down to the order of 1% nominal power. Therefore, since a boiled-up core would be subcritical, the potential exists for an LMFBR system to be intrinsically subcritical even if all PPS action is lost and the core develops into a fully molten state.

The basis for this far reaching postulate is shown in Fig. 15-18. Here the potential for criticality has been folded into the curves of Fig. 15-17. This figure also provides estimates for the dimensions of non-boiling layers based on bottom heat losses. Percent nominal power, along with the time required to reach the low levels depicted, is shown as the abscissa. The vertical scale denotes the fuel column thickness which can exist from a stability stand-point. For example, the thickness of a non-boiling (fully dense) layer of liquid fuel at 3% nominal power (the decay heat level approximately one minute after shutdown) is only about 1 cm. Above that layer, which is at the bottom of the pool, boiling would be incurred with a high associated void fraction. Above approximately 20 cm in the pool, dispersed flow (i.e., the region above the fluidization line) would exist. Such a physical situation would be highly subcritical. From the data shown in the figure, the pool

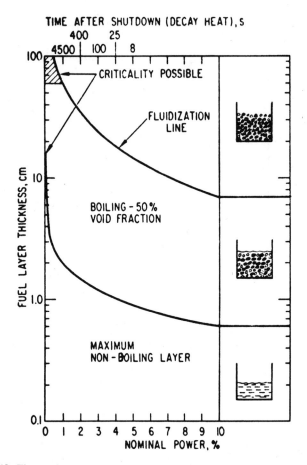

FIGURE 15-18. Flow regime map indicating required conditions for fuel criticality.[51]

thickness would approach conditions characteristic of a critical configuration only for a power level of the order of 1% or less.

It should be noted, however, that this figure is based on rather minimal heat loss from the boiling pool, and the prediction of a dispersed droplet regime is tentative. Nevertheless, the implications of such an intrinsic core dispersal possibility provide considerable incentive for confirmation.

Because rapid heat transfer from the boiling pool could negate the above boil-up arguments by allowing pool collapse, considerable attention has been given to the modes of heat transfer to the surrounding structure. For an open pool, the two heat transfer directions of interest are horizontal (side walls) and downward.

The rate of downward heat flow appears to depend heavily upon whether a solid fuel crust could form to inhibit heat transfer. Dhir, et al.[55] have argued that a molten steel layer would develop between an upper molten fuel pool and a lower steel container. If so, buoyant forces would cause this thin molten steel layer to rise up periodically through the molten fuel in the form of large steel droplets (Taylor instability), and this agitation could prevent the formation of a solid fuel crust. Experimental support for this model was offered by covering dry ice with water and observing that no ice formation took place at the horizontal interface. Epstein,[56] however, in reviewing their model, noted that the volatility of dry ice would naturally break up the formation of any ice. He suggested that a fuel crust would form between the upper molten fuel and lower steel container for materials present in an LMFBR system since the contact temperature is well below the fuel melting point. His experimental results, which appear to support his position, suggest that sufficient fuel crusting would form to minimize the downward flow of heat from a molten pool.

Several experiments have been performed[57,58] using simulant electrolyte materials to provide guidance on the radial heat loss from a boiling pool. In general, the experiments indicate that a standard boundary layer approach is appropriate in estimating horizontal heat losses. Baker[59] correlated these data by replacing the pool depth dimension in the Nusselt number by a ratio of the horizontal heat flux to the volumetric heat-generation rate. By this maneuver he was able to demonstrate that the radial heat fluxes should be insensitive to the pool depth. Other experimenters[60,61] found that natural circulation effects adjacent to the vertical wall could influence and, in some cases, dominate the Nusselt number.

Whereas a frozen fuel crust appears to provide a considerable barrier to downward heat flux, the structural stability of any such crust formation along the vertical walls is much more questionable. Such formation could provide considerable impedance to heat flow. However, even if crust break-up should occur, conditions exist for rapid initiation of a new crust formation. Thus, the fuel melting point, as opposed to the steel melting point, should be used to establish the temperature difference between the pool and the vertical wall.

D. BOTTLED SYSTEM

From the analytical and experimental data base presently available, it would appear unlikely that an upper core blockage could form so completely as to totally seal off all communication between the core and the upper plenum region. However, lacking conclusive evidence for such a statement, several studies have been focused on the pool dynamics of a bottled-up system.

The principal concern[62] for a bottled-up system is whether sufficient condensation on the above pool structure can occur (1) to keep the system boiled up, and (2) to prevent overpressurization. Without sufficient condensation, pool collapse (potential recriticality) could occur, and without sufficient overall heat loss it is possible that high internal pressure (potential mechanical disassembly) could occur. Hence, the modes of heat transfer become significantly more important for the bottled core. Work in this area is focused mainly on the potential for maintaining pool boil-up until blockages are melted out and open system blowdown can occur. If this cannot be shown to be the case, the subsequent question is whether recriticality could occur and, if so, the rate at which recompaction could take place.

Fauske[62] argued that for such a sealed boiling configuration, the energy exchange at the steel boundaries would be due principally to droplet vapor impingement which should allow enough heat transfer to prevent pool collapse and overpressurization. Analytical studies by Ostensen, et al.,[63] which were based on a churn-turbulent flow regime for a whole-core bottled up geometry, indicated a strong potential for maintaining pool boil-up until permanent leakage paths could be established. A second study[64] focused more on the earlier phase of the accident, where complete axial blockages may be formed only in isolated assemblies prior to involvement of the whole core (the so-called "transition-to-transition phase"). Again, it did not appear likely from that study that assembly pool collapse could result with sufficient severity to adversely affect a benign accident termination. However, the importance of correctly modeling the neutronics aspects of a grossly disconfigured core was clearly noted.

At least two physical models of the hydrodynamics of a bottled up boiling pool have been presented. Condiff, et al.[65] visualize the flow patterns as sketched in Fig. 15-19, where a fluidized flow of molten fuel/steel drops rise during the boiling process. They condense at the top boundary, causing the growth of fog droplets. These large size condensed droplets then descend downward through the fluidized pool like rain in a fog. The other model, proposed independently by Martin[66] and by Greene, et al.,[61] is based on a refluxing of the condensed fuel/steel mixture on the sidewalls (see Fig. 15-20). A growing boundary layer of molten steel joins with this condensed fuel and steel and reenters the lower region of the pool as a liquid. It is important to determine the appropriate physical model for bottled core boil-up since the refluxing model tends to represent a larger potential for criticality than other suggested visualizations.

As a matter of perspective, it should be noted that at least two other mechanisms have been identified as potential paths to remove fuel from a bottled up core, in addition to top plug melt-out. The first is the potential

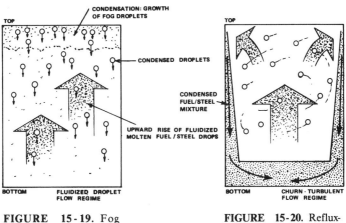

FIGURE 15-19. Fog model for bottled core boil-up.

FIGURE 15-20. Refluxing model for bottled core boil-up.

for molten fuel to be ejected between assembly duct walls early in the transition phase sequence,[67] and the second is the potential for molten fuel draining through control rod assembly chambers.[68] If either of these two mechanisms could be verified, more confidence could be established that recriticality could be prevented—even in the event of total core blockage.

E. COMPUTATIONAL APPROACHES

As might be expected, computational modeling techniques for simulating the transition phase are in an early stage of development and considerably less confidence is afforded their predictions than for the pre-disassembly phase tools. The basic reason for this situation arises from the sheer complexity of attempts to model a core which has neither the intact geometry of the earlier accident phase nor the completely fluid configuration of the disassembly phase. Though major portions of the core may be molten, or even boiling, the time scales are such that heat transfer is very important.

Nevertheless, important progress has been made to model at least portions of the transition phase in a pseudo-mechanistic fashion. The FUMO-T Code[69] has been developed to model the hydrodynamics of a bottled boiling pool in 1-D geometry, and the TRANSIT Code[70] has been designed to allow studies of the incoherencies of the progression of an uncooled core into the transition phase.

The SIMMER Code

The most ambitious attempt to model a core that loses its original geometry is the SIMMER Code[71] developed under the direction of J. E. Boudreau of Los Alamos and C. N. Kelber of the USNRC. Although the size of this model may make routine parametric studies economically prohibitive to perform (due to the computer running time involved), the growing capability of the model is expected to allow considerable insight into problems involving disrupted geometries and extended material motion. Hence, the essence of this model will be presented.

SIMMER stands for S_n, Implicit, Multifield, Multicomponent, Eulerian, Recriticality. Neutronics and fluid dynamics behavior during a core disruptive accident are coupled in SIMMER in two-dimensional cylindrical (r, z) coordinates. Either the time-dependent multigroup diffusion equation or the neutron transport equation is solved by a quasi-static method to predict neutron behavior. By comparison with critical experiments, it has been shown that the enhanced accuracy of transport theory calculations are important when voided regions develop in a disrupted core, as anticipated for the transition phase.[72] The major new contribution made by SIMMER lies in its treatment of hydrodynamics by techniques developed by A. Amsden and F. Harlow[73] at Los Alamos.

Although the SIMMER code is described in the present section on Transition Phase, it is also useful for disassembly phase analysis, and it may provide the most realistic analysis yet developed for the processes during expansion of the fluid following disassembly (cf. Fig. 15-2).

Hydrodynamics behavior is calculated in SIMMER by defining three fields—a structure field, a liquid field, and a vapor field. Core materials are referred to as components, and two types of components are defined—density components and energy components. Density components are used to follow material motion, and energy components are used to predict material temperatures.

SIMMER-II,[71] an updated version of the SIMMER code, has six basic components: fertile fuel, fissile fuel, steel, sodium, control material, and fission gas. Each field, however, keeps track of more components than this; for example, the structure field can distinguish between as-fabricated fuel and fuel that has melted and refrozen, and it can distinguish between steel cladding and steel in the duct walls.

The structure field consists of solid fuel, cladding and structure, and unreleased fission gas. Also, when a material is frozen, it is removed from the liquid field and added to the structure field. The liquid field consists of a uniform mixture of all molten materials. In addition, the liquid field contains solid particles of fuel and steel which may be flowing with the liquid. The vapor field contains a uniform mixture of all vapors and inert gases.

The conservation equations are solved for each field in Eulerian coordinates, together with the appropriate equations of state. A single velocity is calculated for each fluid field; within a fluid field all components move with the velocity of the field. The structure field remains stationary and thus has no momentum equation; the field does, however, influence the motion of the fluid fields. For the structure and liquid fields, separate energy equations are solved for each component. Only one energy equation is used for the vapor field since all gases are assumed to be mixed and at the same temperature.

A complete list of all of the density and energy components in each field in SIMMER-II is given in Table 15-1. The difference between density components and energy components results from the need to keep track of

TABLE 15-1   Density and Energy Components in Each Field in Simmer-II

Field	Density Components	Energy Components
Structure	Fabricated fertile fuel	Fabricated fuel
	Fabricated fissle fuel	
	Frozen fertile fuel	Frozen fuel
	Frozen fissile fuel	
	Cladding	Cladding
	Duct wall	Duct wall
	Control	Control
	Intragranular fission gas	
	Intergranular fission gas	
Liquid Field	Liquid fertile fuel	Liquid fuel
	Liquid fissile fuel	
	Liquid steel	Liquid steel
	Liquid sodium	Liquid sodium
	Liquid control	Liquid control
	Solid fertile fuel particles	Solid fuel particles
	Solid fissile fuel particles	
	Solid steel particles	Solid steel particles
Vapor Field	Fertile fuel vapor	Fuel vapor
	Fissile fuel vapor	
	Steel vapor	Steel vapor
	Sodium vapor	Sodium vapor
	Control vapor	Control vapor
	Fission gas	Fission gas

fertile and fissile material separately in the mass flow equations. This need arises because fuel of different fissile fractions (i.e., with different burnups and core versus blanket fuel) must be mixed. The fertile and fissile fuel components are assumed to be intimately mixed in each field, however, so that they are always at the same temperature; therefore, only a single fuel component is needed for the energy equation.

Coupling between the fields and components is taken into account in the conservation equations through the terms X_1, X_2, and X_3 in Eqs. (15-1), (15-2), and (15-3).

Phase transitions are modeled with considerable complexity in SIMMER. Mass transfer between fields is included in X_1 in Eq. (15-1). Melting and freezing are calculated by comparison of material energies with solidus and liquidus energies. Vaporization and condensation are calculated using a nonequilibrium conduction-limited phase transition model in which the phase transition rates are proportional to the difference between saturation and actual temperatures of the components. A multicomponent vaporization/condensation model is used to treat the effects of noncondensibles and multicomponent interference.

Momentum coupling is accounted for by drag force terms and by momentum exchange with mass transfer between fields, all incorporated into X_2 in Eq. (15-2). The drag correlations are in a simple form applicable for two-phase dispersed flow (cf. Fig. 15-9) which is satisfactory for a wide range of events postulated for HCDA's, but treatment of other flow regimes will be required for boiling-pool transition-phase phenomena. Rapid compression of fluids involving steep pressure gradients (i.e., mild shock-wave phenomena) are handled in SIMMER by the pseudo-viscous pressure method described in Section 15-3C.

Energy coupling is accounted for by three terms in X_3 of Eq. (15-3). The first term is energy transfer by convection and radiation between fields or between components within a field. The second term is energy exchange with mass transfer. The third term is energy transfer due to drag heating. In SIMMER all of the energy from $p\,dV$ work (expansion or compression) appears in the vapor field energy equation.

15-7 CORE DISASSEMBLY

As noted in Chapter 13-3C, calculations of the energy release from a fast reactor subjected to a large unprotected reactivity insertion have become a standard part of the overall FBR safety analysis. The original analysis, done by Bethe and Tait in 1956,[3] was intended as an order of magnitude study of the maximum energy which could be generated if the only excursion

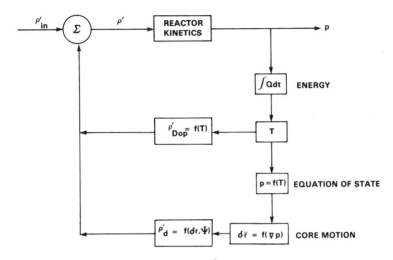

FIGURE 15-21. Flow diagram for disassembly calculation.

shutdown mechanism was the development of internal core pressures which could physically disassemble the reactor. Though simplistic in the modeling assumptions employed, that analysis has provided the basis for an extensive body of subsequent work. We will begin the presentation here with the techniques currently used to calculate disassembly. We will then summarize the original Bethe-Tait calculation and describe some of the results which were obtained during the evolution of the model to the present form.

The principal ingredients of models used to simulate the dynamic behavior of an FBR core disassembly are illustrated in the block diagram of Fig. 15-21. Given a large positive reactivity* insertion, ρ'_{in}, the power level rises and the time integrated energy buildup causes fuel temperatures to increase. The rising fuel temperature will cause a negative reactivity feedback due to the Doppler effect which will mitigate the power rise. At some point, the elevated fuel temperature will lead to high internal core pressures (e.g., vapor pressure formation) and the resulting pressure gradient will physically begin to push the core apart. This effect is called disassembly. This core motion reduces material density, causing a negative reactivity feedback; only a small expansion of the core is required to produce a large disassembly reactivity. The reactor rapidly becomes sufficiently subcritical that any continued external reactivity insertion mechanism has no appreciable bearing on the ultimate consequences. This marks the conclusion of the neutronic excursion and the disassembly phase of the accident.

---

*We define reactivity in this chapter as ρ' in order to distinguish it from density, ρ.

Analyses of several tests have shown that the disassembly process can be adequately modeled. The SEFOR program provided experimental proof that Doppler feedback kinetics were modeled correctly for a fast oxide-fueled reactor on a prompt critical transient (cf. Chapter 6-6A). The KIWI-TNT test, in which a small graphite-moderated reactor was purposely caused to disassemble on a prompt critical transient in the Nevada desert, has been analyzed by both Stratton[74] and Jackson.[75] Both analyses indicate that the disassembly model is basically reliable.

A. DISASSEMBLY EQUATIONS

The equations corresponding to each block on Fig. 15-21 are presented in this section. The hydrodynamics equations are based on the assumption that the core is a homogeneous single-component fluid, with void space taken into account through the equation of state. Lattice structure that might still exist offers inertial resistance but no structural resistance to flow.

We present the equations in cylindrical geometry, as normally encountered in disassembly codes. The hydrodynamics equations are written here in Lagrangian coordinates since all of the early disassembly codes were written in this system although an Eulerian system can also be used, as is done in SIMMER. These equations are converted to finite difference form in computer solutions and are integrated over time and space to obtain total reactivities and energy generation.

Net Reactivity

The reactivity to be used in the reactor kinetics equation is a sum of four quantities: the reactivity that describes the condition of the core at the initiation of the disassembly phase (i.e., ρ'_0), the input reactivity ramp (α, in units of ρ'/second), and the Doppler and disassembly feedback reactivities (ρ'_{Dop} and ρ'_d).

The disassembly calculation is usually started when the reactor is near or has reached prompt critical; hence $\rho'_0 \sim \bar{\beta}$. The input reactivity usually results from sodium expulsion and/or from fuel (and cladding) motion that is slow relative to disassembly core motion. Since the calculated time scale of the disassembly phase is generally in the 10 to 100 millisecond range, the disassembly proceeds too rapidly for the ramp rates from the input-reactivity phenomena to change appreciably during disassembly. Hence, the input ramp rate is normally held constant throughout the disassembly phase.

The net reactivity in the reactor kinetics equations can then be written as

$$\rho' = \rho'_{in} + \rho'_{Dop} + \rho'_d = \rho'_0 + \alpha t + \rho'_{Dop} + \rho'_d. \tag{15-10}$$

Reactor Kinetics

Point kinetics are used in the form given by Eqs. (6-5) and (6-6), where the symbol Q is used in place of p to avoid confusion with pressure. The use of point kinetics implies that the power shape does not change during the excursion. The kinetics equations are

$$\frac{dQ}{dt} = \frac{\rho' - \bar{\beta}}{\Lambda} Q + \sum_{i=1}^{6} \lambda_i y_i \tag{15-11}$$

$$\frac{dy_i}{dt} = \frac{\bar{\beta}_i}{\Lambda} Q - \lambda_i y_i \quad (i = 1 \text{ to } 6), \tag{15-12}$$

where Q is used here for the average power density, which is also the volumetric heat source in the fuel since the fluid is assumed to be homogeneous.

The spatial distribution of the power density is needed for the energy equation; it is obtained by multiplying Q by the normalized spatial shape of the power distribution, $N(r, z)$, such that

$$\frac{1}{V} \int N(r, z) \, dV = 1,$$

where V signifies volume.

Energy Equation

The energy equation is given by Eq. (15-7). Conduction heat transfer is neglected due to the rapidity of the disassembly, changes in kinetic and potential energy are negligible, and all effects which might be included in X_3 are negligible. For these conditions, the rate of change of specific internal energy, u, is

$$\frac{\partial u(r, z)}{\partial t} = -\frac{1}{\rho} \left[\frac{1}{r} \frac{\partial}{\partial r} (r \rho v_r) + \frac{\partial}{\partial z} (\rho v_z) \right] + \frac{QN(r, z)}{\rho}. \tag{15-13}$$

Fuel Temperature

Equation-of-state data are used to obtain the fuel temperature as a function of internal energy. Neglecting the influence of the small amount of vaporization occurring during disassembly, the fuel temperature T_f, can be approximated by

$$T_f = T_{f0} + \int \frac{du}{c_p}, \qquad (15\text{-}14)$$

where T_{f0} is the fuel temperature before the excursion, c_p is the specific heat at constant pressure (J/kg·K), and the integration is over the duration of the transient.

Doppler Reactivity

The Doppler feedback reactivity is calculated by the methods presented in Chapter 6. This reactivity can be estimated from Eq. (6-28) as

$$\rho'_{\text{Dop}} = \sum_i \frac{K_{Di}}{V_i} \int_{V_i} W(r, z) \ln \frac{T_f(r, z)}{T_{f0}(r, z)} dV. \qquad (15\text{-}15)$$

Pressure

The pressure of the core material is calculated from the equation of state, assuming equilibrium thermodynamics. Pressure equation-of-state data are described in Part B of this section. For the present we simply indicate pressure as

$$p = f(T_{\text{fuel}}), \text{ or } p = f(u), \qquad (15\text{-}16)$$

where u is specific internal energy in the fuel.

Motion

The fluid velocity is obtained from the momentum equation, Eq (15-6), with $X_1 = X_2 = 0$. Gravity can be neglected during disassembly. The equations for the r and z components of the velocity are

$$\frac{\partial v_r}{\partial t} = -\frac{1}{\rho}\frac{\partial p}{\partial r}.$$

(15-17a)

$$\frac{\partial v_z}{\partial t} = -\frac{1}{\rho}\frac{\partial p}{\partial z}.$$

(15-17b)

Displacement

Displacement of volume elements leads to the disassembly reactivity. The radial and axial components of the displacement, δr and δz, are obtained simply from integration of the velocity, i.e.,

$$\frac{\partial}{\partial t}(\delta r) = v_r.$$

(15-18a)

$$\frac{\partial}{\partial t}(\delta z) = v_z.$$

(15-18b)

Disassembly Reactivity

A reactivity worth distribution of core material, $\psi(r, z)$, is obtained from perturbation theory in order to calculate reactivity due to displacement of core materials:

$$\psi(r, z) = \text{reactivity per m}^3 \text{ of core material at } r, z.$$

Examples of reactivity worth distributions in units of reactivity per kg were shown in Figs. 6-8 through 6-10.

It is then assumed (for most disassembly calculations) that $\psi(r, z)$ does not change during disassembly since displacements are so small during this phase of an HCDA. The gradient $\nabla \psi(r, z)$ is also obtained from the worth distribution. Finally the differential equation for the disassembly reactivity,

ρ'_d is given by*

$$\frac{d\rho'_d}{dt} = \int \left(\frac{\partial \psi}{\partial r} \frac{\partial(\partial r)}{\partial t} + \frac{\partial \psi}{\partial z} \frac{\partial(\delta z)}{\partial t} \right) dV,$$

$$= \int \left(\frac{\partial \psi}{\partial r} v_r + \frac{\partial \psi}{\partial z} v_z \right) dV. \tag{15-19}$$

The basic disassembly model, in summary, consists of six differential equations (considering the six delayed neutron equations as one and the momentum and displacement equations each as one), together with an input reactivity function, equations of state, and data for Doppler reactivity and material worth distributions.

B. EQUATION OF STATE

The pressure in the momentum equation, Eq. (15-17), is the vapor pressure of the core materials whenever space is available for vapor to exist. The vapor pressure of most materials can be approximated as a function of temperature by an equation of the form:

$$p = A \exp\left[-\frac{B}{T} + C \right]. \tag{15-20}$$

This form is consistent with the Clausius-Clapeyron equation which, approximating the saturated vapor as a perfect gas with negligible saturated liquid specific volume relative to saturated vapor specific volume, is

$$\frac{dp}{p} = \frac{h_{fg}\,dT}{RT^2}, \tag{15-21}$$

where $\qquad h_{fg} =$ molal latent heat of vaporization (J/kg)

$\qquad\quad R =$ gas constant for material $(J/kg\cdot K)$.

---

*Often Eqs. (15-17) through (15-19) are combined to give

$$\frac{d^2\rho'_d}{dt^2} = -\int \frac{1}{\rho} \nabla \psi \cdot \nabla p \, dV$$

which shows a direct proportionality between the second derivative of the disassembly reactivity and the pressure gradient.

Integrating from an initial pressure and temperature p_0 and T_0, and approximating h_{fg} by a constant gives

$$p = p_0 \exp\left[\frac{h_{fg}/R}{T} + \frac{h_{fg}/R}{T_0}\right], \qquad (15\text{-}22)$$

which is the same form as Eq. (15-20). This equation is a straight line on a plot of $\ln p$ versus $1/T$, as illustrated in Fig. 15-22 for several of the correlations commonly employed for UO_2 fuel vapor. A small deviation from linearity is provided by an additional term that is often used for UO_2. For the FFTF safety analysis, the following equation of state was used for

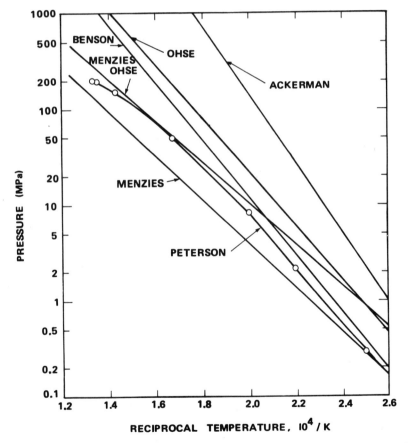

FIGURE 15-22. Various correlations for UO_2 fuel vapor equations of state: R. J. Ackerman;[76] D. A. Benson;[77] D. C. Menzies;[78] R. W. Ohse;[79] D. M. Peterson (taken from Ref. 80).

oxide fuel:

$$p = \exp\left(-\frac{78847}{T} + 53.152 - 4.208 \ln T\right), \qquad (15\text{-}23)$$

where p is in MPa.

As long as volume is available for vapor, the fuel will remain in two phases. However, the liquid specific volume increases with increasing temperature until eventually liquid may fill all of the available volume. At this point the fuel again becomes entirely liquid, hence, single phase. Further heating of the fuel will produce very large pressures which will tend to disassemble the core even faster than the lower two-phase pressures. A two-phase system is commonly referred to as a "soft system" (due to the "softness" of the gas pressure) while an all-liquid system is called a "hard system." If the core does not disassemble while it is still a soft system, the high liquid pressures of the hard system will disassemble it very quickly.

The equation of state illustrating the high liquid pressures of the hard system is shown by Fig. 15-23. Here the reduced pressure, P_r, is plotted as a function of the reduced temperature, T_r, where reduced values of a property are the ratio of the property to its value at the critical point.

The lower envelope of the family of straight lines represents the saturated vapor pressure curve. The straight lines represent lines of constant reduced specific volume in the liquid region.

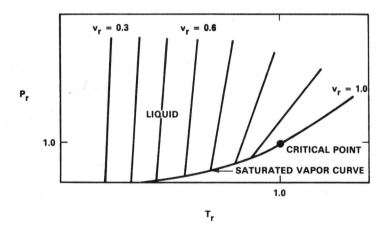

FIGURE 15-23. Density dependent form of the equation of state.

Other Fluids Causing Disassembly

Three other fluids besides fuel may be the source of the high pressures that could cause disassembly: sodium, fission gas, and steel.

If sodium is still in the core, there will be little or no space available for fuel vapor. Hence, the system will be hard. In this case the liquid sodium will be compressed by the liquid fuel and the sodium pressure will cause a rapid disassembly.[81]

It has been argued that fission gas released from melting fuel will have a higher vapor pressure than fuel and will cause disassembly before the fuel is heated enough for fuel vapor to cause disassembly. This would be a beneficial effect, but it has to be demonstrated that fission gas pressure can be relied upon to cause an early disassembly.

Steel vaporizes at a significantly lower temperature than fuel, and the steel cladding may be intimately mixed with the fuel. If heat transfer to the steel occurs fast enough, steel vapor may disassemble the core faster than fuel vapor. Again, however, this has yet to be demonstrated.

C. DISASSEMBLY CODES AND EXAMPLE RESULTS

The early core disassembly codes were one-dimensional [e.g., WEAK,[82] MAX,[83] PAD[84]]. Two-dimensional computer programs [e.g., MARS,[85] VENUS,[86,87] POOL,[88] KADIS[89]] provided considerably more capability for assessing the cases of most interest. All these programs employ point kinetics and perturbation theory, except for PAD (which utilizes a time-dependent neutron transport technique that allows reliable computations for larger time scales after reactor shutdown). One version[90] of VENUS is interfaced with a two-dimensional space-time kinetics program to allow a more appropriate treatment of the neutronics. The SIMMER code series[71] provides two-dimensional kinetics for use during disassembly and for a variety of disrupted geometrics of interest.

VENUS has developed into the workhorse code for core disassembly parametric calculations. Figures 15-24 through 15-27 illustrate a typical VENUS calculation for a "soft" (sodium-out) core of about 1000 liters in size, for an input ramp reactivity rate of 50$/s. Considerable geometric distortion in the Lagrangian grid can be observed at the time of nuclear shutdown, but such motion is still within the general domain of applicability of perturbation theory. Another interesting feature is the time of the negative feedback reactivities. The Doppler reactivity begins contributing

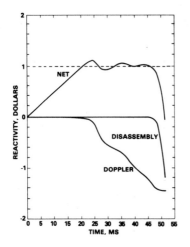

FIGURE 15-24. Reactivity plot for 50$/s sodium-out "soft-system" case.

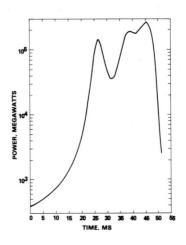

FIGURE 15-25. Power trace for 50$/s sodium-out "soft-system" case.

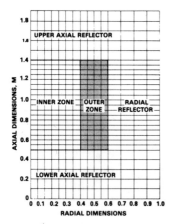

FIGURE 15-26. Initial reactor core grid.

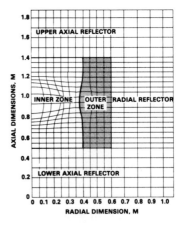

FIGURE 15-27. Core grid distortion for 50$/s sodium-out case at time of nuclear shutdown.

negative reactivity simultaneously with the energy generation from the rising power. In contrast, the disassembly reactivity is delayed since it must await core motion; but once the disassembly process starts, the negative disassembly feedback is extremely rapid.

D. BETHE-TAIT MODEL

Because the original Bethe-Tait analysis[3] was intended only as an order-of magnitude estimate of potential energy release from a worst-case disassembly accident, many approximations were made to allow a hand calculated solution. The principal assumptions included (1) use of the point kinetics formulation, (2) use of perturbation theory (small core movement), (3) no delayed neutrons, (4) no Doppler feedback, (5) a threshold equation of state, (6) spherical geometry, (7) a completely homogeneous core, and (8) constant fuel density. The third approximation was justified in the original model by analyzing the core dynamic response only during the prompt critical domain, and Doppler feedback was neglected because the early cores in the United States were metal fueled (hence, little Doppler effect). Spherical geometry was selected simply to make the solution tractable without resorting to large scale computers.

The essential aspects of the original Bethe-Tait analysis are presented in books by Thompson and Beckerley,[91] Wirtz,[92] and Lewis[93] and will not be repeated here. It is of interest, however, to review the final result of the analysis in order to provide insight into the important parameters and to understand why the introduction of the Doppler effect, which is large in modern-day FBR's, so greatly reduces the energy releases predicted by the original analysis.

In the Bethe-Tait model, reactivity is added at a constant ramp rate until a threshold energy density is reached. Beyond this time, internal pressures quickly rise and disassembly occurs. The neutronic transient terminates when the magnitude of the negative disassembly reactivity equals the input reactivity in excess of prompt critical. The final expression for the energy density generated (J/m^3) during the transient, as given by Lewis,[93] has the following two forms:

$$e \propto \frac{(\rho'_*)^3 R^4}{\Lambda^2} \tag{15-24}$$

$$\propto \frac{(\alpha)^{3/2} R^4}{\Lambda^{1/2}} \ , \tag{15-25}$$

where $\rho'_* =$ reactivity in excess of one dollar, R is the core radius, α is the input reactivity ramp rate, and Λ is the neutron generation time.

This proportionality suggests that both the reactivity ramp rate and the reactivity inserted before reaching the disassembly threshold are important. The expressions also show that the energy density would be expected to increase rapidly as the size of the reactor increases, a result that would pose a serious problem for the FBR concept were it not for other mitigating effects, particularly the Doppler effect. Refinements to Eq. (15-25), e.g., Reference 94, suggest a weaker dependence on α and Λ for milder transients, but the above expressions provided the principal motivation for the major advances in modeling that followed.

Equation (15-25) indicates that the energy release increases with decreasing Λ for an excursion controlled by disassembly reactivity feedback. This is in contrast to the effect of Λ for a transient controlled by prompt feedback, where prompt feedback refers to a negative reactivity that occurs instantaneously with energy generation. For such a transient with a ramp input reactivity, the energy density is proportional to $(\alpha\Lambda)^{1/2}$, as explained, for example, in Reference 92. The reason for this difference in response to Λ is that the disassembly feedback is delayed relative to a prompt feedback since disassembly feedback is inertially controlled. The inertial nature of ρ'_d can be seen most clearly from the footnote to Eq. (15-19); while the second derivative of ρ'_d is directly proportional to the pressure gradient, which in turn is proportional to the energy density, the integrated ρ'_d is delayed relative to the energy generation.* For a disassembly controlled transient, therefore, a short neutron generation time Λ allows a large energy generation before ρ'_d terminates the transient. For a prompt feedback, on the other hand, the short generation time is favorable in that the feedback, acting simultaneously with the energy generation, stops the excursion quickly. The Doppler reactivity is just such a prompt reactivity feedback.

Improvements to the Bethe-Tait Model. Whereas the original Bethe-Tait model was instructive regarding physical insight into the core disassembly process, as well as in helping to identify the parameters of key importance, Eq. (15-25) indicates that the energy density release predicted by such a model would increase rapidly with increasing core volume. For the small reactors existing at the time the model was developed, namely EBR-II and FERMI, the bounding energy release values predicted (even for very high reactivity insertion rates) could be accommodated by mechanical design. However, there was a serious question whether the energy releases predicted

---

*This delay is clearly illustrated in Figs. 15-24 and 15-25.

by such a model for larger cores were physically reasonable, or whether such predictions were a result of model simplicity. Investigators began to refine the model to include physical processes neglected in the original work. Moreover, improvements to the Bethe-Tait model forced abandonment of any attempt at a closed solution. Instead, the calculations are currently done by numerically integrating the equations in Section 15-7A and accounting for all the relevant feedback terms.

Major improvements included a more realistic equation of state and Doppler feedback. Since dynamic response of a core disassembly accident is inertially controlled, any internal core pressure developing early in the excursion is very effective in mitigating the overall energy release since this causes some core acceleration early in time. Nicholson[95] pioneered the computational efforts to include this effect into the Bethe-Tait model. Equations of state now used were discussed in Section 15-7B.

Meyer and Wolfe[96] first clearly illustrated the large reduction in energy release for large cores that could be achieved by including the Doppler effect. A typical curve of energy* vs Doppler constant (K_D or $T dk/dT$) is shown in Fig. 15-28 for two constant ramp rate conditions. As noted from the logarithmic scale used, a "knee" appears in the curve somewhere in the vicinity of $T dk/dT = -0.002$ for the example given; i.e., for a Doppler constant smaller than this magnitude, the total energy generated increases very rapidly. Hence, the argument for a strong Doppler coefficient.

An interesting feature of both curves in Fig. 15-28 is their oscillatory nature. Whereas the reason for this behavior represents a rather fine point of the calculations, an explanation is in order because several features of the disassembly process are revealed by a discussion of this point. We shall consider the 100\$/s ramp case and focus attention on the two Doppler constant values of -0.0045 and -0.005. The total energy generation is seen to be larger for the latter value. Figure 15-29 contains the power traces for the two cases. As noted from the figure, the power in the first peak is larger for the case of -0.0045. This, of course, is a direct result of less Doppler feedback. In Fig. 15-30, however, the fuel temperature rise is noted to be fairly small (the core initially having been assumed to be shut down at approximately room temperature) and fuel pressures are well below the point of being significant for core disassembly. Since reactivity is assumed to be added constantly by the external 100\$/s ramp, the net reactivity (as illustrated in Fig. 15-31) exceeds prompt critical, and a second power pulse is generated. Again, the power peak is larger for the -0.0045 case and the resulting fuel temperature rise is larger. This process repeats until fuel

---

*Here the ordinate represents the total energy generated during the excursion from initial steady-state conditions.

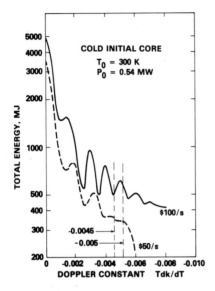

FIGURE 15-28. Typical energy release versus Doppler constant for an FBR core disassembly excursion.

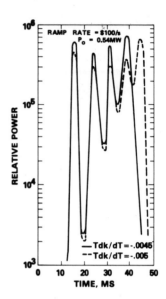

FIGURE 15-29. Power traces for Doppler constants of −0.0045 and −0.005.

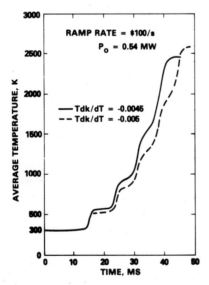

FIGURE 15-30. Temperature buildup for Doppler constants of −0.0045 and −0.005.

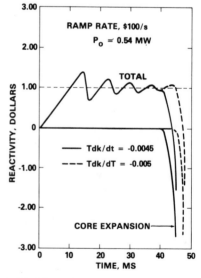

FIGURE 15-31. Net reactivity traces for Doppler constants of −0.0045 and −0.005.

temperatures are elevated to the point where significant disassembly pressures are generated. For the case of -0.0045, this point is reached at the fourth power pulse. For the case of -0.005, however, another power pulse is required to elevate temperatures to the core disassembly point. It is the energy generated in the fifth pulse that is responsible for the larger energy generation associated with the larger (more negative) Doppler coefficient. For a Doppler constant slightly higher than -0.005, however, sufficiently high fuel temperatures are generated in five pulses to disassemble the core; hence, the decrease in energy generation versus Doppler coefficient up to the point the processes explained above again cause a situation where $N+1$ pulses are required.

Despite the sensitivity of energy generation versus Doppler to the fine details of the calculation, an envelope can be drawn (to bound the results typified in Fig. 15-28) that clearly has a knee shape, and the argument for a large (highly negative) Doppler coefficient to reduce the magnitude of a core disassembly accident remains valid.

PROBLEMS

15-1. Provide a one-sentence description of each of the seven phases of an unprotected LMFBR accident described in Section 15-2.

15-2. Select one of the conservation equations in the Eulerian system and show that the corresponding equation in the Lagrangian system can be obtained from the relationship:

$$\frac{D}{Dt} = \frac{\partial}{\partial t} + \vec{v} \cdot \nabla.$$

15-3 Discuss the physical process described by each term (a) in Eq. (15-2a), i.e., the momentum equation, and (b) in Eq. (15-3a), i.e., the energy equation.

15-4. Does fuel melting begin before sodium boiling starts (a) in the unprotected TOP transient, (b) in the unprotected TUC transient?

15-5. (a) In modeling the unprotected TOP transient, why is it important to include many channels in the analysis?

    (b) At what axial position is fuel failure most likely to occur in an unprotected TOP accident?

    (c) What is sweepout?

15-6. (a) In an unprotected TUC transient, at what axial location does sodium boiling begin?

    (b) Where is sodium vapor present a few tenths of a second after boiling begins?

  (c) What is meant by film dryout, entrainment, and flooding?
15-7. (a) What is the transition phase of an HCDA?
  (b) How can an axial blockage occur during the transition phase?
  (c) What is the difference between the conduction limited freezing model and the bulk freezing model?
  (d) What are bottled systems and open systems as applied to the core in the transition phase?
  (e) Why is the boiling pattern in the core during the transition phase of interest?
15-8. (a) If core disassembly were to occur, what would be the two principal negative feedback mechanisms operating?
  (b) What might cause positive input reactivity that drives the disassembly?
  (c) The net reactivity in Fig. 15-24 hovers around 1 dollar for much of the transient, which is typical for disassembly calculations. Explain this behavior.
  (d) The sodium void reactivity insertion rate can usually be considered constant during disassembly. Why?
15-9. (a) Why does perturbation theory represent a reasonable approximation for calculating disassembly reactivity?
  (b) What are the potential sources of pressure tending to drive the core apart during disassembly?
  (c) What is the difference between a hard and a soft system with regard to the disassembly process?

REFERENCES

1. A. E. Waltar and A. Padilla, Jr., "Mathematical and Computational Techniques Employed in the Deterministic Approach to Liquid Metal Fast Breeder Reactor Safety," *Nucl. Sci. & Eng.*, *64* (1977) 418–451.
2. H. Alter, G. F. Flanagan, and N. M. Greene, "Central Computerized Data Base for Liquid Metal Fast Breeder Reactor Safety Codes," *Proc. of the Intl Mtg on Fast Reactor Safety and Related Physics*, CONF-761001, Chicago, IL, October 5-8, 1976, 1057–1065.
3. H. A. Bethe and J. H. Tait, *An Estimate of the Order of Magnitude of the Explosion When the Core of Fast Reactor Collapses*, RHM (56)/113, U. K. Atomic Energy Research Establishment, Harwell, 1956.
4. R. B. Bird, W. E. Stewart, and E. N. Lightfoot, *Transport Phenomena*, John Wiley & Sons, Inc, New York, 1960.
5. J. von Neumann and R. D. Richtmyer, "A Method for the Numerical Calculation of Hydrodynamic Shocks," *J. Appl. Phys. 21* (3) (1950) 232.
6. D. R. Ferguson, W. R. Bohl, C. H. Bowers, J. E. Cahalan, F. E. Dunn, T. J. Heames, J. M. Kyser, W. L. Wang, and H. U. Wider, "The SAS4A LMFBR Accident Analysis Code System: A Progress Report," *Proc. of the Int. Meeting on Fast Reactor Safety and Related Physics*, CONF-761001, Chicago, IL, October 1976, 1225–1235.

7. C. H. Lewis and N. P. Wilburn, *MELT-IIIA: An Improved Neutronics, Thermal-Hydraulics Modeling Code for Fast Reactor Safety Analysis*, HEDL-TME 76-73, Hanford Engineering Development Laboratory, 1976. (See also Ref. 20, Chapter 14.)

8. D. Struwe et al., "CAPRI-A Computer Code for the Analysis of Hypothetical Core Disruptive Accidents in the Predissassembly Phase," *Proc. of the Intl Mtg Fast Reactor Safety and Related Physics*, CONF-740401-P1, April 2-4, 1974, Beverly Hills, CA.

9. T. P. Moorhead, "Clad Strain and Melt-Through Failure Mode Analysis for Fast Running Application," *Specialist Workshop on Prediction Analysis of Material Dynamics in LMRBR Safety Experiments*, LA-7938-C, Los Alamos, NM, March 13-15, 1979, 85–98.

10. N. P. Wilburn, "Intersubassembly Incoherencies and Grouping Techniques in LMFBR Hypothetical Overpower Transients," *Trans. ANS, 27* (1977) 538.

11. A. E. Waltar, N. P. Wilburn, D. C. Kolesar, L. D. O'Dell, A. Padilla, Jr., L. N. Stewart (HEDL) and W. L. Partain (NUS), *An Analysis of the Unprotected Transient Overpower Accident in the FTR*, HEDL-TME 75-50, Hanford Engineering Development Laboratory, June 1975.

12. R. E. Baars, J. H. Scott, and G. E. Culley, "Failure Threshold Correlation for Fast Reactor Transient Overpower Accident Analysis," *Trans. ANS, 21* (1975) 303.

13. R. R. Sherry and D. B. Atcheson, *User's Manual for the BEHAVE-SST (Steady-State and Transient) Fuel Rod Mechanics Program*, GEFR-0001, General Electric Company, 1977.

14. P. K. Mast, *The Los Alamos Failure Model (LAFM): A Code for the Prediction of LMFBR Fuel Pin Failure*, LA-7161-MS, Los Alamos Scientific Laboratory, March 1978. (See also P. K. Mast, J. H. Scott, and J. J. Dorning, "A Single Computational Model for the Prediction of Fuel Pin Failure in the Transient Overpower Accident," *Trans. ANS, 28* (1978) 474.)

15. G. L. Fox, *DSTRESS User's Manual*, HEDL-TME 76-95 Addendum, Hanford Engineering Development Laboratory, May 1977.

16. A. E. Klickman, T. H. Bauer, T. H. Hughes, J. M. Kramer, R. K. Lo, C. C. Meek, R. J. Page and L. W. Person, "Methods of Predicting Fuel Failure—Development and Evaluation," *Proc. of the Int. Meeting of Fast Reactor Safety*, Seattle, WA, August 1979, 944–952.

17. R. D. Burns III and J. H. Scott, "Statistical Analysis of 7 TREAT Experiments," *Trans. ANS, 30* (1978) 463.

18. R. N. Koopman, et al., "TREAT Transient Overpower Experiment R12," *Trans. ANS, 28*, (1978) 482.

19. C. E. Dickerman et al., "Status and Summary of TREAT In-Pile Experiments on LMFBR Response to Hypothetical Core Disruptive Accidents," Vol 2, Liquid Metal Fast Breeder Reactors, *Symposium on the Thermal Hydraulic Aspects of Nuclear Reactor Safety, ASME Winter Meeting*, Atlanta, GA, November 27-December 2, 1977.

20. H. V. Wilder, "PLUTO 2: A Computer Code for the Analysis of Overpower Transients in LMFBRs," *Trans. ANS, 27*, (1977) 533.

21. P. A. Pizzica and P. B. Abramson, "EPIC: A Computer Program for Fuel-Coolant Interaction," *Proc. of the Int. Meeting on Fast Reactor Safety and Related Physics*, CONF-761001, Chicago, IL, October 5-8, 1976.

22. V. K. Dhir, K. Wong, and W. E. Kastenberg, *A Mechanistic Study of Fuel Freezing and Channel Plugging During Fast Reactor Overpower Excursions*, UCLA-ENG-7679, University of California, Los Angeles, August 1976.

23. S. C. Yung and N. P. Wilburn, "Failure Pattern Within an FTR Subassembly Under a TOP Accident," *Trans. ANS, 27* (1977) 536.

24. N. P. Wilburn, E. D. Smith, D. D. Atcheson, R. E. Baars, and B. W. Spencer, "Fuel Pins and Core Response Under LMFBR TOP Accident Conditions," *Presented at the ENS/ANS Topical Meeting on Nuclear Power Reactor Safety*, Brussels, October 16-19, 1978.

25. J. Costa, "Contribution to the Study of Sodium Boiling During Slow Pump Coastdown in LMFBR Subassembly," O. C. Jones and S. G. Bankoff, eds., *Liquid Metal Fast Breeder Reactor*, Vol 2, The American Society of Mechanical Engineers, New York (1977) 155–170.

26. W. Peppler, "Sodium Boiling in Fast Breeders: A State of the Art Review," O. C. Jones and S. G. Bankoff, eds., *Liquid Metal Fast Breeder Reactor*, 2, American Society of Mechanical Engineers, New York (1977) 123–154.

27. H. K. Fauske, "Superheating of Liquid Metals in Relation to Fast Reactor Safety," *J. of Reactor and Fuel-Processing Tech.*, *II*, No. 2 (1968).

28. H. K. Fauske, "Nucleation of Liquid Sodium in Fast Reactors," *J. of Reactor Technology*, *15* No. 4, 1972–73.

29. R. E. Henry, et al., "Incipient Superheat in a Convection Sodium System," *Proc. Fifth Int. Heat Transfer Conference*, Tokyo, September 3-7, 1974.

30. D. France, et al., *Sodium Superheat Experiments using LMFBR Simulation Parameters*, ANL-CT-73-25, Argonne National Laboratory, December 1973.

31. M. Ishii and M. A. Grolmes, "Prediction of Onset of Entrainment for Liquid Metals," *Trans. ANS, 21* (June 1975) 325.

32. R. T. Lahey, Jr., and F. J. Moody, *The Thermal Hydraulics of a Boiling Water Nuclear Reactor*, American Nuclear Society, La Grange Pk, IL, 1976, 201.

33. H. K. Fauske, "Some Aspects of Pin-to-Pin Failure Propagation in LMFBR's," *Nuc. Sci and Eng.*, *54* (1974) 10–17.

34. R. E. Henry, "A Correlation for the Minimum Wall Superheat in Film Boiling," *Trans. ANS*, *15*, 1 (June 1972) 420.

35. G. Hoppner and F. E. Dunn, "Sodium Film Motion Model in SAS2A Voiding," *Trans. ANS*, *17* (1973) 244.

36. W. R. Bohl and T. J. Heames, "Cladding Motion Model for LMFBR Loss-of-Flow Accident Analysis," *Trans. ANS*, *17* (1973) 358.

37. W. R. Bohl and M. G. Stevenson, "A Fuel Motion Model for LMFBR Unprotected Loss-of-Flow Accident Analysis," *Proc. Conference Mathematical Models and Computational Techniques for Analysis of Nuclear Systems*, CONF-730414, Ann Arbor, MI, (1974).

38. *Hypothetical Core Disruptive Accident Considerations in CRBRP*—Vol. 1: Energetic and Structural Margin Beyond the Design Base, CRBRP-3, CRBR Project Office, Oak Ridge, TN, November 1978, 4–37.

39. H. K. Fauske, "Some Comments on Cladding and Early Fuel Relocation in LMFBR Core Disruptive Accidents," *Trans. ANS*, *21* (1975) 322–323.

40. C. A. Hinman and O. D. Slagle, *Ex-Reactor Transient Fission Gas Release Studies, Fuel Pin PNL 2–4*, HEDL-TME 77–83 Hanford Engineering Development Laboratory, (May, 1978).

41. L. W. Deitrich, "An Assessment of Early Fuel Dispersal in the Hypothetical Loss-of-Flow Accident," *Proc. of the Int. Meeting on Fast Reactor Safety*, Seattle, WA, August 1979, 615–623.

42. F. E. Dunn, "Fuel Pin Coolability in Low Power Voiding," *Trans. ANS*, *28* June 1978, 430.

43. A. Kaiser, W. Peppler, and M. Straka, "Decay Heat Removal From a Pin Bundle," *Proc of the Intl. Mtg. on Fast Reactor Safety and Related Physics*, Vol. IV, Chicago, IL, October 1976, 1578.

44. J. F. Jackson, M. G. Stevenson, J. F. Marchaterre, R. H. Sevy, R. Avery, and K. O. Ott, "Trends in LMFBR Hypothetical Accident Analysis," *Proc. of the Int. Meeting on Fast Reactor Safety and Related Physics*, CONF-74041-P3, Beverly Hills, CA, April 1979, 1241–1264.

45. R. E. Henry, W. C. Jeans, D. J. Quinn, and E. A. Spleha, "Cladding Relocation Experiments," *Proc. of the Intl Mtg on Fast Reactor Safety and Related Physics*, Chicago, IL, October 5–8, 1976.

46. F. B. Cheung and L. Baker, Jr., "Transient Freezing of Liquids in Tube Flow," *Nucl Sci Eng*, *60* (1976) 1–9.

47. R. W. Ostensen and J. F. Jackson, "Extended Fuel Motion Study," in *Reactor Development Program Progress Report*, ANL-RDP-18, Argonne National Laboratory, July 1973, 7.4–7.7.

48. R. W. Ostensen et al., "Fuel Flow and Freezing in the Upper Subassembly Structure Following an LMFBR Disassembly," *Trans. ANS, 18*, (1974) 214–215.

49. M. Epstein, R. E. Henry, M. A. Grolmes, H. K. Fauske, G. T. Goldfuss, D. J. Quinn, and R. L. Roth, "Analytical and Experimental Studies of Transient Fuel Freezing," *Proc. of the Int. Meeting on Fast Reactor Safety and Related Physics*, Chicago, IL, October 1976.

50. M. Epstein, M. A. Grolmes, R. E. Henry, and H. K. Fauske, "Transient Freezing of a Flowing Ceramic Fuel in a Steel Channel, *Nucl Sci Eng*, *61* (1976) 310–323.

51. M. Epstein, "Melting, Boiling and Freezing: The 'Transition Phase' in Fast Reactor Safety Analysis," *Liquid Metal Fast Breeder Reactor, Symposium on the Thermal and Hydraulics Aspects of Nuclear Reactor Safety*, ASME Winter Meeting, Atlanta, GA, November 27–December 2, 1977.

52. T. Ginsberg, O. C. Jones, Jr., and J. C. Chen, "Volume-Heated Boiling Pool Flow Behavior and Application to Transition Phase Accident Conditions, *Presented at the ENS/ANS Topical Meeting on Nuclear Power Reactor Safety*, Brussels, October 16–19, 1978.

53. H. K. Fauske, "Boiling Flow Regime Maps in LMFBR HDCA Analysis," *Trans. ANS, 22* (1975) 385.

54. S. S. Kutateladze, "Elements of the Hydrodynamics of Gas-Liquid Systems," *Fluid Mechanics-Soviet Res, 1*, 4, 29 (1972).

55. V. K. Dhir, J. S. Castle, I. Catton, W. E. Kastenberg, and J. B. Doshi, "Role of Wall Heat Transfer and Other System Variables on Fuel Compaction and Recriticality," *Proc of the Intl Mtg on Fast Reactor Safety and Related Physics*, Chicago, IL, October 5–8, 1976.

56. M. Epstein, "Stability of a Submerged Frozen Crust," *ASME Winter Meeting*, New York, December 1976, Paper No. 76-WA/HT-31. (See also M. Epstein and G. A. Lambert, *Reactor Development Program Progress Report*, ANL-RDP-59, Argonne National Laboratory, March 1977, 6.55–6.57.)

57. R. P. Stein, J. C. Hesson, and W. H. Gunther, "Studies of Heat Removal from Heat Generating Boiling Pools," *Proc of the Intl Mtg on Fast Reactor Safety and Related Physics*, CONF-740401-P2, 1974, 865–880.

58. J. D. Gabor, L. Baker, Jr., J. C. Cassulo, and G. A. Mansoori, "Heat Transfer From Heat-Generating Boiling Pools," Presented at *AMSE-AICHE Natl Heat Transfer Conf*, St. Louis, MO, August 1976.

59. L. Baker, Jr., "Analysis of Heat Transfer from Molten-Fuel Layers," *Reactor Development Progress Report*, ANL-RDP-49, Argonne National Laboratory, March 1976, 9.8–9.13.

60. W. R. Gustavson, M. S. Kazimi, and J. C. Chen, "Heat Transfer and Fluid Mechanics in a Volume-Heated Boiling Pool," *Proc of the Int Mtg on Fast Reactor Safety and Related Physics*, Chicago, IL, October 5–8, 1976.

61. G. A. Greene, O. C. Jones, Jr, and C. E. Schwartz, *Thermo-Fluid Mechanisms of Volume-Heated Boiling Pools*, BNL-NUREG-50759, Brookhaven National Laboratory, Upton, NY (1977).

62. H. K. Fauske, "Boiling Fuel-Steel Pool Characteristics in LMFBR HCDA Analysis," *Trans. ANS, 22* (1975) 386.

63. R. W. Ostensen, R. J. Henninger, and J. F. Jackson, "The Transition Phase in LMFBR Hypothetical Accidents," *Proc of the Intl Mtg on Fast Reactor Safety and Related Physics*, CONF-7611001, Chicago, IL, October 5–8, 1976.

64. A. E. Waltar, J. Muraoka, and F. J. Martin, "A Computational Model for Analyzing Postulated LMFBR Accident Fuel Boil-Up Conditions," *Proc of the Int Mtg on Fast Reactor Safety and Related Physics* CONF-7611001, Chicago, IL, October 5–8, 1976.

65. D. W. Condiff, M. Epstein, and M. A. Grolmes, "Transient Behavior of Boiling Fuel Steel Mixtures," *Reactor Development Program Progress Report*, ANL-RDP-53, Argonne National Laboratory, September 1977 6.11–6.16.

66. F. J. Martin, *Bottled Transition Phase Analysis—Preliminary Report*, HEDL-TME 78–64, Hanford Engineering Development Laboratory, April 1980.

67. B. W. Spencer, et al., "Fuel-Freezing Tests," *Reactor Development Progress Report*, R. G. Sachs, J. A. Kyger, compilers, ANL-RDP-74, Argonne National Laboratory, April 1978.

68. S. J. Hakim and J. M. Kennedy, "Development of Transition-Phase Code," *Reactor Development Program Progress Report*, ANL-RDP-70, Argonne National Laboratory, August 1978.

69. F. J. Martin, "FUSS: The Integrated FUMO, SAS3A, Neutronics Update System," *Trans. ANS*, 27 (1977) 539.

70. S. J. Hakim and R. J. Henniger, "TRANSIT-A Code for Analysis of the Transition Phase in an FBR Accident," *Trans. ANS*, 14 (1976) 261.

71. L. L. Smith, J. E. Boudreau, C. R. Bell, P. B. Bleiweis, J. F. Barnes, and J. R. Travis, "SIMMER-I, An LMFBR Disrupted Core Analysis Code," *Proc. of Int. Meeting on Fast Reactor Safety and Related Physics*, Chicago, IL, CONF-871001, October 1976, 1195–1202. (See also L. L. Smith et al, *SIMMER-II: A Computer Program for LMFBR Disrupted Core Analysis*, LA-7615-MS, NUREG/CR-9453, Los Alamos Scientific Laboratory 1978.)

72. R. Curtis, C. N. Kelber, E. Gelbard, L. Lesage, L. Luck, L. R. Smith, and D. Wade, "The Use of Benchmark Criticals in Fast Reactor Code Validation," *Int. Symposium on Fast Reactor Physics*, IAEA and OECD-NEA, IAM-SM-244, Aix-en-Provence, France, September 1979.

73. A. A. Amsden and F. H. Harlow, *KACHINA: An Eulerian Computer Program for Multifield Fluid Flows*, LA-5680, Los Alamos Scientific Laboratory (1974).

74. W. R. Stratton, D. M. Peterson, P. M. Altomare, "Analysis of the KIWI-TNT Experiment," *Trans. ANS*, 8, No. 1 (1965) 126–127.

75. T. F. Bott, J. F. Jackson, "An Analysis of the KIWI-TNT Experiment with the Venus-II Disassembly Code," *Trans. ANS*, 23 (1976) 323–324.

76. R. J. Ackerman, et al., "High Temperature Thermodynamic Properties of Uranium Dioxide," *J. Chem. Phys.*, 25 (1956) 1089.

77. D. A. Benson, *Application of Pulsed Electron Beam Vaporization to Studies of UO_2*, SAND77-0429 (1977).

78. D. C. Menzies, *The Equation of State of Uranium Dioxide at High Temperatures and Pressures*, UKAEA TRG-1119 (D) (1966).

79. R. W. Ohse, et al., *Contribution of Uranium Plutonium Oxide Pressure, and Influence of Radial Fission Product Distributions on Pin Failure of Fast Breeder Fuels Under Accident Conditions*, IAEA-SM-190/8 (1974).

80. J. T. Larkins, T. P. McLaughlin, and W. R. Stratton, *Estimates of the Upper Limit of the Destruction Energy Created Subsequent to Postulated Fast Reactor Power Excursions*, LA-UR-79-834, Los Alamos Scientific Laboratory, 1979.

81. L. D. O'Dell and A. E. Waltar, "Effect of Distributed Voids in LMFBR Core Disassembly Calculations," *Trans. ANS*, 15 (1972) 358.

82. J. W. Stephenson, Jr. and R. B. Nicholson, *Weak Explosion Program*, ASTRA-417-6.0, Astra, Inc., Raleigh, NC, 1961.

83. W. W. Little, Jr., *Max-A One-Dimensional Maximum Hypothetical Accident Code in FORTRAN-IV*, BNWL-612, Pacific Northwest Laboratory, Richland, WA, March 1968.

84. D. M. Peterson, W. R. Stratton and T. P. McLaughlin, *PAD: A One-Dimensional Coupled Neutronic-Thermodynamic-Hydrodynamic Computer Code*, LA-6540-MS, Los Alamos Scientific Laboratory, December 1976.

85. H. Hirakama, *MARS: A Two-Dimensional Coupled Neutronics-Hydrodynamics Computer*

Program for Fast Reactor Power Excursions, ANL-7701, Argonne National Laboratory, October 1970.

86. W. T. Sha and T. H. Hughes, *VENUS: A Two-Dimensional Coupled Neutronics-Hydrodynamics Computer Program for Fast Reactor Power Excursions*, ANL-7701, Argonne National Laboratory, October 1970.

87. J. F. Jackson and R. B. Nicholson, *VENUS-II: An LMFBR Disassembly Program*, ANL-7951, Argonne National Laboratory, 1972.

88. P. B. Abramson, "Core Disruptive Accident and Recriticality Analysis with FX2-POOL," *Proc. of the Int. Meeting on Fast Reactor Safety and Related Physics III*, CONF-761001, Chicago, IL, October 1976, 1253.

89. P. Schmuck, G. Jacobs, G. Arnecke, *KADIS: A Program to Analyze the Disassembly Phase of Hypothetical Accidents in (LMFBRs)*. DOE-tr-113. (Trans. from KFK 2497, Inst. of Neutronenphysik und Reaktortechnik Karlsruhe, West Germany, Nov. 1977).

90. D. P. Weber and B. D. Ganapol, "Multi-Field Hydrodynamics Disassembly Calculations with Space-Time Kinetics: FX2/VENUS-III," *Proc of the Int Mtg on Fast Reactor Safety and Related Physics*, CONF-761001, Vol III, Chicago, IL, October 5–8, 1976, 1244.

91. T. J. Thompson and J. G. Beckerley, eds., *The Technology of Nuclear Reactor Safety*, MIT Press, Vol. 1, Cambridge, MA (1964). See particularly Chapter 10, by W. J. McCarthy, Jr., and D. Okrent.

92. K. Wirtz, *Lectures on Fast Reactors*, Kernforschungszentrum, Karlsruhe, 1973, Chapter 4.

93. E. E. Lewis, *Nuclear Power Reactor Safety*, John Wiley and Sons, New York, 1977, Chapter 5.

94. V. Z. Jankus, *A Theoretical Study of Destructive Nuclear Power Bursts in Fast Power Reactors*, ANL-6512, Argonne National Laboratory, 1962.

95. R. B. Nicholson, *Methods for Determining the Energy Release in Hypothetical Reactor Meltdown Accidents*, APDA-150, Atomic Power Development Associates, Inc., December 1962.

96. R. A. Meyer, B. Wolfe, and N. F. Friedman. "A Parameter Study of Large Fast Reactor Meltdown Accidents," *Proc of the Conf on Safety, Fuels, and Core Design in Large Fast Power Reactors*, ANL-7120, Argonne National Laboratory, October 11–14, 1965.

CHAPTER 16

CONTAINMENT CONSIDERATIONS

16-1 INTRODUCTION

The last major discipline of fast reactor safety analysis is that of containment. Once a complete spectrum of accidents has been considered and analyzed through the point of permanent nuclear shutdown, the remaining questions focus on mechanical consequences, long-term heat removal capability, potential chemical interactions of high temperature materials, and the ultimate issue of radiological doses that might result from the escape of radionuclides from the containment or confinement system.

This chapter addresses such issues. In order to describe the phenomena and methods involved in the analysis of potential radiological doses from an unprotected core disruptive accident, we begin with mechanical consequences within the reactor vessel that might lead to release of materials from the primary system. We then follow the ensuing potential events in the general sequence that they might occur.

In order to provide a perspective of where events such as interactions of core material and sodium with structures outside the primary system can occur, Fig. 16-1 shows the principal features of the reactor building for typical loop and pool designs. This figure illustrates the relative positions of the primary system, reactor cavity, equipment cells, refueling areas, and the large containment or confinement area in the reactor containment building.

Mechanical damage to the primary system might result from a two-phase expansion of either fuel or sodium. Fuel expansion could directly cause mechanical deformation of the primary system structures; or heat might first be transferred from fuel to sodium in a molten fuel/coolant interaction (MFCI) and the resulting sodium vaporization and expansion could lead to mechanical damage. The process of changing the form of energy generated during disassembly (i.e., thermal, or heat energy) to a form that can cause damage to primary system structures is called thermal/work energy conversion.

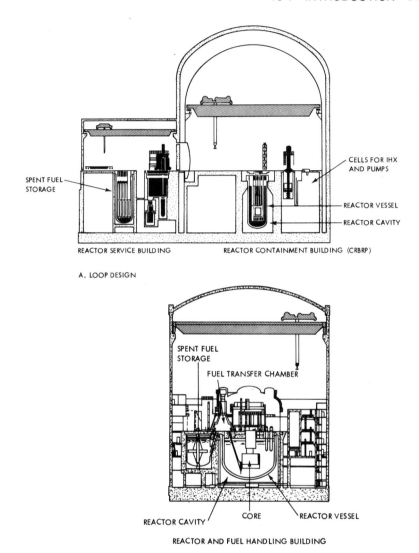

FIGURE 16-1. Typical reactor building layout for LMFBR loop and pool designs.

The generation of thermal energy by disassembly was described in Chapter 15-7. Mechanical work done during the expansion of the resulting high temperature, high pressure fuel is described in Section 16-2. Molten fuel/coolant interaction (MFCI) is the subject of Section 16-3. The maximum mechanical work from the fuel expansion or the MFCI is the maximum HCDA work energy discussed in Chapter 13-3B. Section 16-4 is concerned with *energy partitioning*, i.e., the potential mechanical consequences from the work done by expanding fuel or sodium.

Attention is then focused on post-accident heat removal aspects, where the principal question is how decay heat can be effectively removed for various distributions of core debris. The accident can be terminated at this point if heat removal is sufficient to allow the reactor vessel and associated piping and components in the primary system to remain intact.

If primary system integrity is lost, however, several other issues become important. Sodium fires, for example, become possible if a system rupture should occur in the vicinity of an atmosphere containing a sufficient oxygen concentration to support combustion. Hence, the conditions for ignition and sustainability of sodium fires requires attention. Furthermore, sodium and/or core debris outside of the primary system could encounter concrete, causing water release due to concrete heating and producing direct chemical interactions. Such possibilities, resulting from primary system failure, lead us to consider overall ex-vessel containment transient analysis. This requires a description of the various systems for containment and/or confinement that are being considered for fast breeder reactors in order to assure that radiological doses to the public are within acceptable levels. These topics are discussed in the concluding section.

Before launching into this final safety chapter, the reader should be made aware of the preliminary nature of the material presented. Each topic discussed is under intensive research throughout the international fast reactor community, and definitive answers will evolve over a relatively long time. There are many scientists long associated with fast reactor safety development who expect that it will eventually be possible to demonstrate a negligible probability (and, indeed, impossibility for some phenomena) for many of the accident scenarios described in this chapter. Based on experience to date, it is clear that the nature of the problems described here will be modified and better defined with continued research and development.

Space does not permit the presentation of quantitative models for the many phenomena involved in containment design. Our objective is rather to provide a qualitative description of the major problems in containment design toward which LMFBR safety research is being directed, together with an introduction to the selected analytical approaches for some of these problems.

16-2 FUEL EXPANSION

Two approaches have been used to determine the mechanical work that could be done by a partially vaporized core. The first is a bounding approach in which the fraction of the energy generated during the nuclear excursion which is theoretically available to do work on the environs is determined, and the second is a mechanistic approach in which the expansion is analyzed mechanistically in order to determine the extent to which this work potential can be dissipated within the structural surroundings.

A. BOUNDING APPROACH

One way to estimate the mechanical work is to compute the PdV work due to expansion of the two-phase fuel down to some post-accident system pressure. A final pressure of one atmosphere has often been used to provide an upper bound to this potential, although more realistic assessments are based on the recognition that pressures inside the reactor vessel would likely be considerably higher at the time the bubble expansion process is complete.

A useful way to determine the maximum theoretical work energy possible for an expanding mixture of two-phase (liquid-vapor) fuel is to follow the accident sequence on a temperature-entropy (T-s) diagram for the fuel, as illustrated in Fig. 16-2. If the core disassembly calculation is initiated for a core operating at normal steady-state conditions, the prompt burst begins at

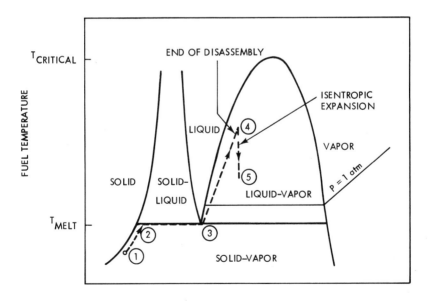

FIGURE 16-2. Sequence of core disruptive accident on a fuel T-s diagram. The melting temperature shown on this diagram is about 3040 K for UO_2-PuO_2; the saturation temperature at 1 atmosphere is about 3500 K, and the critical point is of the order of 8000 K.

State 1 with the fuel in solid form and in equilibrium with fuel vapor at that low temperature.

The reactivity input causing the power excursion leads to an increase in fuel temperature. The fuel melts between States 2 and 3. Subsequent heating raises the fuel temperature while partially vaporizing the fuel, causing the pressure rise that ultimately disassembles the core. At the end of the nuclear excursion, the fuel would either remain under the dome on the *T-s* diagram (though near the saturation line, since expansion volumes and qualities are small), or would be in the liquid region if the liquid expands to fill all the space and produces a "hard system," as described in Chapter 15-7B. The former path is the case most often encountered in calculated transients. Hence, in our example in Fig. 16-2, we illustrate the fuel state at the end of disassembly by State 4.

It is then assumed that the two-phase fuel expands isentropically (i.e., adiabatically and reversibly) to a lower pressure State 5. This expansion process constitutes the work energy phase. This expansion of the "HCDA bubble" could damage the vessel internals, strain the vessel and pipes, and accelerate the sodium above the core upward—where it might impact the closure head, thus damaging the head and further straining the vessel. These processes are discussed further in Section 16-4. The expansion would cease after the sodium impacted the head and the bubble filled all the space left by the rising sodium and strained vessel, a state represented by State 5 on Fig. 16-2.

To determine the work done by the fuel in the expansion process, we can write the first law of thermodynamics for the control mass of two-phase fuel:

$$dQ + dU + dW = 0 \ , \tag{16-1}$$

where dQ represents the heat transferred from the fuel during the expansion, dU is the change in internal energy, and dW is the work done on the surroundings (e.g., on the sodium, and ultimately, the vessel). If we assume the expansion process to be isentropic, then $dQ = 0$, and the work calculated will be the maximum possible from the expansion of the fuel alone. Making this assumption,

$$W_{max} = U_4 - U_5. \tag{16-2}$$

In an accident calculation, some thermodynamics property at State 5 is generally given—either P_5 (pressure), or v_5 (specific volume) from the available expansion volume. The fuel properties (hence, U_4 and the specific entropy s_4) at the end of disassembly are calculated. From s_4 (which equals s_5) and either P_5 or v_5, together with the fuel equation of state, the final internal energy U_5 can be calculated. Hence, this work can be obtained.

A simple expression for the work can be derived from Eq. (16-2) in terms of quantities more easily obtained than U_4 and U_5 by making the following approximations: the vapor acts as a perfect gas; h_{fg} is constant; v_l is negligible relative to v_g; and c_p for liquid fuel is constant. Thus (cf. Problem 16-1),

$$W_{\max} = M_f \big[c_p(T_4 - T_5) - h_{fg}(x_5 - x_4) + R(T_5 - T_4) \big], \qquad (16\text{-}3)$$

where

$M_f =$ fuel mass expanded (kg)
$c_p =$ specific heat for liquid fuel (J/kg·K)
$h_{fg} =$ fuel latent heat of vaporization (J/kg)
$x =$ quality
$R =$ gas constant for fuel, $8317/270$ J/kg·K for UO_2-PuO_2.

The quality at State 4 is determined from the expression

$$x_4 = \frac{\dfrac{V_4}{M_f} - v_{l4}}{v_{g4} - v_{l4}}, \qquad (16\text{-}4)$$

where V_4 is the appropriate core volume that contains M_f at the end of disassembly, and the v's are saturated vapor and liquid specific volumes at temperature T_4.

The quality at State 5 can be determined from the relation (cf. Problem 16-2)

$$x_5 = x_4 \frac{T_5}{T_4} + \frac{c_p T_5}{h_{fg}} \ln \frac{T_4}{T_5}. \qquad (16\text{-}5)$$

If P_5 is known, T_5 can be found from the vapor pressure equation of state, e.g., Eq. (15-23). Then x_5 and, finally, $W_{\max}$ can be calculated.

If the available volume, or V_5, is the final known condition instead of P_5, the problem is more complicated. Neglecting v_l and assuming $v_{g5} = RT_5/P_5$ (perfect gas),

$$v_5 = \frac{V_5}{M_f} = x_5 v_{g5} = x_5 \frac{RT_5}{P_5}.$$

Substituting this expression for x_5 into Eq. (16-5) gives

$$\frac{v_5 P_5(T_5)}{RT_5} = x_4 \frac{T_5}{T_4} + \frac{c_p T_5}{h_{fg}} \ln \frac{T_4}{T_5}, \qquad (16\text{-}6)$$

where $P_5(T_5)$ is obtained from Eq. (15-23). This equation must be solved for T_5 in order to find x_5 and, finally, W_{max}.

The actual expansion problem has an additional degree of complexity not yet addressed. We have presented Eqs. (16-2) through (16-6) as if the entire core could be represented by a single temperature at State 4 and another at State 5. The actual condition at State 4, however, is a distribution of temperatures, with the hottest temperature at the center. If one were to expand each fuel volume (e.g., each mesh cell in an actual disassembly calculation) independently to the final pressure, the expansion work would be significantly greater than if the core were first thermally mixed to obtain a uniform (average) temperature before expansion occurred. Some examples of the magnitude of this effect are provided in Reference 1. The reason for the large difference is that thermal mixing of the fuel is a highly irreversible process. The effect of this irreversibility is to decrease the potential of the fuel to do expansion work. This mixing prior to expansion (or during the expansion process) is called *self-mixing*. If significant self-mixing does not take place before (or during) expansion, and if average fuel temperatures are used in Eq. (16-3), the work calculated from this procedure would be too low.

B. MECHANISTIC APPROACH

Several phenomena would be present in an actual expansion process in an LMFBR HCDA that would significantly reduce the deliverable work below the theoretically available work. Heat and frictional losses to the surrounding structures would lower the potential for mechanical damage. Self-mixing in the core (referred to above) and pressure gradients in the expanding fluid would lower the efficiency of the conversion of thermal energy to work energy. Abrupt area changes resulting from structural geometry variations in the expansion path would lead to a further reduction. These reductions are particularly large if the internal structure above the core (including axial blanket and fission gas plenum in the fuel pins) remains in place. If above-core structures are forced out of position by high core pressures, the reduction would be less. Analysis of work done with these phenomena taken into account is an example of the mechanistic approach (cf. Chapter 13-4B).

Early calculations of fuel expansion and two-dimensional coolant hydrodynamics were made with the REXCO and ICECO computer codes,* and some examples of REXCO results are described in Section 16-4. In addition,

---

*Computer codes such as REXCO and ICECO are described in Section 16-4B; SIMMER was described in Chapter 15-6E.

experimental investigations of bubble expansion with simulant materials have been carried out at a number of facilities both in the United States and Europe, and with actual fuel at the TREAT (ANL-Idaho Falls), ACPR (Sandia-Albuquerque), and CABRI (CENC-Cadarache) reactors.

The most sophisticated computational tool for analysis of mechanistic effects during fuel expansion is the SIMMER code, which was described briefly in Chapter 15-6E. Early calculations with the SIMMER code indicated that the phenomena described above might reduce the actual work potential of an expanding fuel bubble by an order of magnitude. Figure 16-3 shows the result of a SIMMER calculation for an LMFBR HCDA in which mechanistic results of damage potential (work) are compared with isentropic fuel expansion results.[2] Damage potential is plotted as a function of average fuel temperature at the end of the disassembly phase. The lower curves (model uncertainties) show the range of values calculated by SIMMER to account for the phenomena described above for reducing

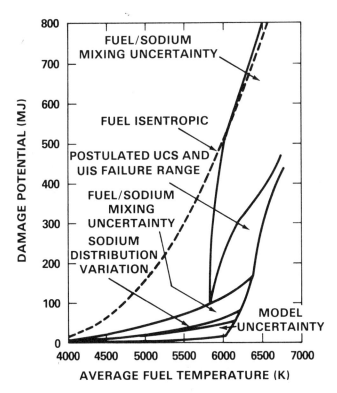

FIGURE 16-3. Overall influence of the mechanistic treatment and uncertainties on damage potential.[2] UCS and UIS signify upper core structure and upper internal structure.

damage potential. On top of these uncertainties are added the uncertainties resulting from fuel-sodium mixing and the subsequent vaporization of sodium (MFCI, as discussed in the next section). The calculations indicate that the work potential remains low relative to the results for isentropic fuel expansion as long as the average fuel temperature does not approach 6000 K, a value that can probably be ruled out by core design.

16-3 MOLTEN FUEL/COOLANT INTERACTION

The possibility of molten fuel interacting with sodium during an LMFBR accident sequence has long been a concern because of the potential for sufficiently rapid heat transfer to trigger a sodium vapor explosion. As readily evident from the preceding discussions, there are several phases of an unprotected accident sequence where a molten fuel/coolant interaction (MFCI) could significantly influence the course of events. However, it is in the mechanical consequences phase that attention was first focused on the role of sodium during a power excursion.

A molten fuel/coolant interaction (MFCI) is said to occur if fuel mixes with sodium in such a manner that the rate of heat transfer from the fuel to the sodium is much faster than the rate which occurs during normal boiling. The consequence of this interaction is a rapid rise in pressure due to sodium vaporization, followed by expansion of high pressure two-phase sodium. Since the boiling temperature of sodium is well below the melting temperature of mixed oxide fuel, such a process would enable the sodium to be a more efficient working fluid than expanding fuel. Hence, for a given energy deposition at the end of the nuclear excursion, more damage might be imparted to the core surroundings if sodium rather than fuel were the expanding medium.

Frequently MFCI's are subdivided into two categories—energetic MFCI's (or vapor explosions) and low energy MFCI's (or vigorous boiling). In a vapor explosion the time scale for heat transfer between the fuel and the sodium is small compared with that for the expansion of the fuel/coolant mixture. In this case, a significant fraction of the thermal energy in the fuel can be converted into mechanical work on the surroundings. In low energy MFCI's, heat transfer rates are lower and mechanical work from sodium expansion may be quite low. If the MFCI zone is constrained, however, by the large pool of sodium above it so that expansion cannot occur rapidly, enough sodium might be vaporized even in a low energy MFCI to generate more mechanical work than would have resulted from fuel expansion alone.

Efforts to determine the damage potential of MFCI's in an HCDA have taken two directions. First, estimates have been made of the work that can be done by sodium expansion in LMFBR geometries based on various assumptions or approximations with regard to specific phenomena in the interaction. Second, a large research effort has been directed toward a more complete understanding of the physical phenomena that occur in a vapor explosion. We shall describe several sodium expansion models and then discuss briefly the more important aspects of the physics of vapor explosions.

A. SODIUM EXPANSION MODELS

Hicks-Menzies

Hicks and Menzies[3] were the first to provide a quantitative analysis of MFCI's in an LMFBR. They offered a bounding calculation in which they determined the thermodynamically maximum possible work that could be done by the sodium for a given amount of energy generated in the fuel. Figure 16-4 contains a plot of the Hicks-Menzies results for an initial fuel temperature of 3450 K and an initial sodium temperature of 1150 K. This plot shows the expansion work as a function of both the sodium-to-fuel mass ratio and the final pressure.* As expected, the work potential increases as the final pressure is lowered. The nonlinear dependence upon the sodium-to-fuel mass ratio can be explained as follows. For a small sodium mass there is not enough working fluid to cause much PdV work to be done, whereas for a high sodium mass the heat transferred from the fuel is absorbed by so much sodium that the peak sodium temperature remains low. The maximum work occurs for a sodium-to-fuel mass ratio slightly less than 0.1, which also corresponds roughly to the mass ratio in a typical undisturbed LMFBR core.

The principal approximations made by Hicks and Menzies were (1) thermal equilibrium, (2) temperature-independent material properties, (3) neglect of the heat of fusion for fuel, and (4) perfect gas behavior for the sodium. The first approximation is by far the most important. It implies that perfect heat transfer between fuel and sodium exists throughout the sodium expansion process. Hence, the calculation is purely thermodynamical; no potentially limiting rate processes are considered.

---

*An updated calculation was reported by Judd[4] which used an improved (Himpan) equation of state for sodium instead of perfect gas behavior assumed by Hicks and Menzies. Maximum work values were increased by about 30%.

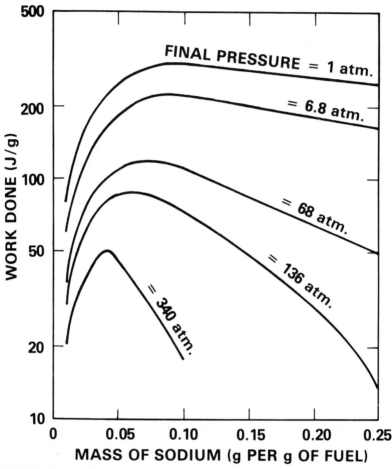

FIGURE 16-4. Hicks-Menzies curve for the work done by sodium expansion for an initial fuel temperature of 3450 K and an initial sodium temperature of 1150 K.[3]

The Hicks-Menzies model is a two-step process. First, fuel and sodium are mixed and energy is instantaneously transferred as heat from molten fuel to liquid sodium until thermal equilibrium is reached, i.e., until the temperatures of the two fluids are equalized. In the second step, the sodium vaporizes and expands doing $P\,dV$ work on the surroundings. Throughout the expansion process, as the sodium temperature is reduced by the expansion, heat transfer from the liquid fuel to the sodium is allowed to continue so that the two fluids remain in thermal equilibrium. (Physically this would require an infinitely rapid heat transfer rate across an infinitesimal temperature difference—which is the reason that this model represents a bounding calculation.)

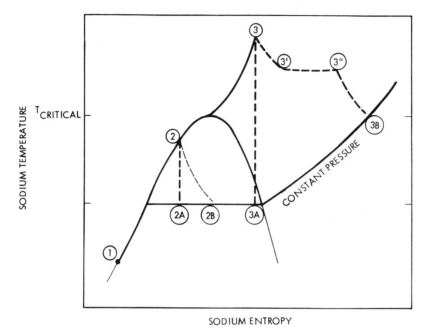

SODIUM ENTROPY

FIGURE 16-5. Schematic HCDA sequences on a sodium *T-s* diagram.

As was done for the fuel expansion, the sodium heating and expansion processes can be illustrated on a *T-s* diagram for sodium, shown in Fig. 16-5. Saturated liquid sodium initially at State 1 is heated by the molten fuel to State 2, or possibly up through the critical point ($T_c = 2509$ K) to State 3.* An expansion process accompanied by heat transfer from the fuel to the sodium (as in the Hicks-Menzies model) is indicated by paths $2 \rightarrow 2B$ and $3 \rightarrow 3B$. On the other hand, after the sodium is heated, the expansion might occur so rapidly that it could follow an isentropic path, as indicated by $2 \rightarrow 2A$ and $3 \rightarrow 3A$, in which case the work done would be significantly reduced below the Hicks-Menzies result. The path $3 \rightarrow 3'$ represents the expansion while heat is transferred from molten fuel; $3' \rightarrow 3''$ represents a constant temperature expansion as the fuel solidifies; and $3'' \rightarrow 3B$ represents heat transfer from solidified fuel.

Assuming no phase changes during the initial mixing process and constant values for specific heats, the equilibrium sodium temperature prior to the expansion is simply

$$T_{eq} = \frac{c_f T_f + m c_{Na} T_{Na}}{c_f + m c_{Na}}, \qquad (16\text{-}7)$$

---

*The saturated vapor pressure equation for sodium was given in Chapter 11-4F.

where the subscripts f and Na refer to the initial fuel and sodium conditions. The quantity m represents the sodium-to-fuel mass ratio. During the expansion, the sodium is the fluid that does work on the surroundings. The fuel transfers its heat to the sodium and is always at the same temperature as the sodium. Based on a control mass consisting of fuel plus sodium, the expansion is adiabatic. The first law for the expansion of the combined fuel and sodium control mass is

$$dW + dU = 0, \tag{16-8}$$

where $dW = P\,dV$ of the expanding sodium, and $dU = dU_{\text{Na}} + dU_{\text{fuel}}$. This is the calculation that was performed by Hicks and Menzies. The calculation of W depends strongly on the equilibrium temperature, T_{eq}, at the beginning of the expansion.

It is unlikely that the sodium temperature will ever rise to the sodium critical temperature, T_c. If a vapor explosion occurs, the sodium temperature will likely be limited to the spontaneous nucleation temperature (discussed in Section 16-3B) which is about $0.9\ T_c$. For low energy MFCI's there are several barriers to heat transfer from the fuel to the sodium that tend to limit the sodium temperature and prevent the instantaneous heat transfer assumed in the Hicks-Menzies model, and hence, to limit the work potential of the sodium. These barriers include low thermal conductivity of the fuel, vapor blanketing by sodium vaporization, and blanketing by fission gas.

Time-Dependent Models

Both the heat transfer from the fuel to the sodium and the motion of the expanding sodium are time-dependent processes. Accounting for these time-dependent effects reduces the work potential of the sodium below that obtained from the Hicks-Menzies model. Some of the essential features of time-dependent models are described below.

The rate of heat-transfer from fuel to sodium is reduced first by the fuel conductivity of the fuel droplets. A temperature gradient exists in the fuel that prohibits the fuel and sodium from being at identical temperatures. Resistance to heat transfer may also exist at the fuel-sodium interface, particularly after sodium vaporization begins or fission gas is released. In addition, the time required for fuel to fragment and mix with the sodium would influence the time-dependence of the expansion process.

The motion of the expanding sodium in the MFCI zone is constrained by the material surrounding the zone, and particularly by the colder sodium pool above the core. The constraint of the MFCI zone is modeled in two stages, an early *acoustic constraint* followed later by an *inertial constraint*.

During the first part of the expansion which is controlled by the acoustic constraint, the constraining sodium above the MFCI zone is compressed so that pressure from the MFCI zone is transmitted through the sodium pool by an acoustic wave. For this acoustic constraint time domain, the MFCI pressure, P, is related to the velocity, V, of the upper surface of the expanding zone by the relation

$$P - P_0 = \rho_0 c_0 V \quad, \tag{16-9}$$

where P_0, ρ_0, and c_0 are initial values of pressure, density, and sonic velocity in the constraining sodium.

Later, the sodium above the MFCI zone moves upward as an incompressible inertial mass; the constraint is then said to be inertial. The relation between the MFCI pressure and the expansion velocity of the zone is then obtained directly from $F = ma$, or

$$P - P_0 = \rho H \left(\frac{dV}{dt} + g \right) \quad, \tag{16-10}$$

where H is the height of the sodium being accelerated above the MFCI zone, P_0 is the pressure in the cover gas over the sodium, and g is the gravitational acceleration.

SOCOOL Model

Padilla[5] pioneered the first effort to calculate the reduction in work energy of expanding sodium due to the time-dependent effects of heat transfer and constraint. The SOCOOL model assumed an acoustic constraint on the MFCI zone until the pressure wave traversed the distance to the top of the sodium pool and back to the MFCI zone—a time called the *unloading time*. Uninhibited heat flow was allowed from the droplet surface to the liquid sodium prior to the unloading time, but the low thermal conductivity of the oxide fuel was taken into account by considering the time dependence of the heat transfer process within the spherical droplet. After unloading, heat transfer from the fuel to the sodium was stopped entirely based on the assumption that the droplet would then be vapor blanketed, and a thermodynamic, non-time-dependent expansion was then calculated to obtain the work done. In this kind of model, the first law would be written separately for the sodium in terms of time derivatives as

$$\dot{Q} = \dot{U} + \dot{W} \quad, \tag{16-11}$$

where $\dot{Q}$ represents the rate of heat transfer from the fuel to the sodium.

The rate of heat transfer and the expansion work calculated by SOCOOL were strongly affected by the fuel droplet size. The thermal equilibrium case (as in Hicks-Menzies) corresponds to a droplet size of zero. Experimental data on fragmentation size of UO_2 fuel in sodium suggests mean particle diameters of the order of 100 to 1000 micrometres (cf. Figure 16-9, at the end of this section) for which calculated work energies are significantly lower than for the thermal equilibrium case. Smaller sizes cannot be ruled out, however, if a detonation wave were possible as discussed later for the Board-Hall model.

ANL Parametric Model

Cho and Wright[6] expanded and generalized Padilla's model. First, they allowed a more general constraint by using both an acoustic constraint followed by an inertial constraint on the MFCI zone. They also added further heat transfer models and included a time constant to account for the time delay associated with fuel fragmentation and mixing with sodium. Inclusion of these effects led to further reductions in the calculated work energy potential of expanding sodium. Additional modifications were made in parametric models elsewhere.

For some time, a problem with the parametric transient models was the unavailability of appropriate experiments with which to compare the models. Some measure of success was achieved in the late 1970s by Jacobs with the Karlsruhe (Germany) version of a parametric-type code, MURTI.[7] Jacobs, together with French and American colleagues, has been reasonably successful in calculating MFCI's between molten UO_2 and sodium in experiments at both the Centre d'Études Nucléaires de Grenoble[8] and the Sandia Laboratories.[9]

B. VAPOR EXPLOSIONS

The primary concern with regard to an MFCI in LMFBR safety analysis is whether a large energetic vapor explosion could ever occur. Considerable progress has been made toward an understanding of vapor explosions, and the major results are reviewed in References 10-12.

Many experiments have been conducted in which molten UO_2 fuel and sodium have been brought into contact in order to observe under what conditions, if any, a vapor explosion could occur between these materials. In nearly every case, no vapor explosion occurred. In a few cases involving small quantities of sodium, energetic small-scale explosions have taken place.[13] In several experiments,[8,9] significant pressure pulses occurred that may be more accurately described as low energy or small-scale MFCI's rather than energetic or large-scale vapor explosions. This extensive experience indicates that a large-scale vapor explosion between oxide fuel and sodium may be either impossible or extremely unlikely in an LMFBR.

Even more experiments on vapor explosions have been conducted with materials other than UO_2 and sodium. These tests have contributed to an understanding of specific mechanisms involved in vapor explosions. Moreover, this experience has led to substantial agreement about many, though not all, aspects of vapor explosions.

Stages in a Vapor Explosion

It is generally agreed that a large-scale vapor explosion involves several necessary stages, usually described as (1) coarse premixing, (2) triggering, (3) escalation, and (4) propagation. Each of these stages is described below. This discussion is followed by an outline of the two theories around which most of the interest in vapor explosions is centered, namely: 1) spontaneous nucleation theory, proposed by Fauske,[14] and 2) detonation theory, proposed by Board and Hall.[15]

Coarse Premixing. The first stage in a vapor explosion is the coarse mixing of the hot liquid throughout the cold liquid. Coarse mixing refers to a mixture of relatively large particles of molten fuel as opposed to the tiny droplets later needed for the very rapid heat transfer required for a vapor explosion. Film boiling appears to be necessary to allow coarse mixing, and it prevents large-scale contact of molten fuel and sodium for a sufficient time to allow significant premixing to occur.

The severity of a vapor explosion is often associated with whether the mixing and the resulting interaction are *coherent*. In a coherent interaction, rapid premixing of hot and cold liquids occurs throughout a large volume so that the ensuing explosion can involve a large fraction of the two liquids almost instantaneously. In an LMFBR, coherent mixing of large quantities of fuel and sodium would be required for there to be any chance of a sufficiently energetic interaction to breach the vessel.

Triggering. Another requirement for a vapor explosion is liquid-liquid contact between the hot and cold liquids. A triggering mechanism must exist to cause this contact. Liquid-liquid contact would occur upon collapse of the vapor blankets present during film boiling. Mechanisms that could cause this collapse are a pressure pulse (shock wave) and cooling of the hot liquid.

Escalation. The coarse fuel droplets must break up into smaller droplets in order to provide the large heat transfer surface required for the rapid heat transfer rates in a vapor explosion. Some questions related to droplet fragmentation mechanisms that might lead to escalation of a MFCI are still unresolved. Local pressurization on contact between the hot and cold liquids may provide mechanisms for droplet breakup. Fauske[14] has presented the theory that the initial contact temperature between the fuel and the sodium must be at least as high as the spontaneous nucleation temperature of the sodium in order for sufficient local pressurization to take place to escalate the interaction into a vapor explosion. Further discussions of both spontaneous nucleation theory and fragmentation are provided below.

Propagation. The final stage in a vapor explosion is the propagation of the interaction through the coarse mixture of fuel and coolant. Early theories assumed that when a vapor explosion occurred, collapse of the vapor blankets and subsequent fragmentation occurred essentially simultaneously throughout the mixture. Colgate[16] suggested the first time-dependent propagation mechanism. Board and Hall[15] then proposed that the mechanism responsible for propagation is a detonation wave. Their theory is described briefly below, following the discussion of spontaneous nucleation theory.

Spontaneous Nucleation Theory

Fauske[14] proposed that the following conditions are required for a large-scale vapor explosion:

(1) coarse premixing, allowed by film boiling;
(2) liquid-liquid contact;
(3) initial contact temperature equal to or greater than the spontaneous nucleation temperature of the coolant; and
(4) adequate constraints.

Above a temperature called the *spontaneous nucleation temperature** embryonic vapor bubbles with a critical radius sufficient to maintain stability

---

*In a pure uniform liquid, the spontaneous nucleation temperature is called the *homogeneous nucleation temperature*. The value of the homogeneous nucleation temperature is roughly 0.9 of the absolute critical temperature. (For sodium, $T_c = 2509\,\text{K}$). If nucleation sites (e.g., impurities or surfaces) are present in the liquid, bubbles may form spontaneously on these sites at temperatures somewhat lower than the homogeneous nucleation temperature.

nucleate at an extremely rapid rate, and very rapid phase transformation follows. An impressive number of experiments has been performed, particularly with water and organic liquids, to show that a vapor explosion occurs only when the contact temperature is at least as high as the spontaneous nucleation temperature of the cold liquid.

When two fluids are brought into contact, the interface, or contact, temperature, T_i, is given by

$$\frac{T_{h0} - T_i}{T_i - T_{c0}} = \sqrt{\frac{\left(\rho c_p k\right)_c}{\left(\rho c_p k\right)_h}} \quad , \tag{16-12}$$

where the subscripts h and c refer to the hot and cold fluids and subscripts 0 refer to initial values. The quantities ρ, c_p, and k are density, specific heat, and thermal conductivity, respectively.

Figure 16-6 compares the spontaneous nucleation temperature, T_{SN}, for sodium with the contact temperature, T_i, between oxide (UO_2) fuel and sodium for a particular set of initial fuel and sodium temperatures. The differences between T_{SN} or T_i and the sodium saturation temperature, T_{sat}, at the system pressure (i.e., the *degrees of superheat*) are plotted against system pressure. Note that the interface temperature is well below the

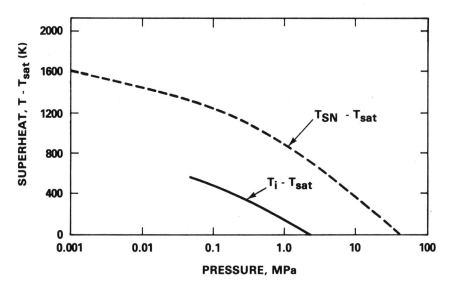

FIGURE 16-6. Comparison between spontaneous nucleation temperature for sodium and sudden contact temperature between UO_2-PuO_2 fuel initially at 3470 K and sodium initially at 1070 K[14].

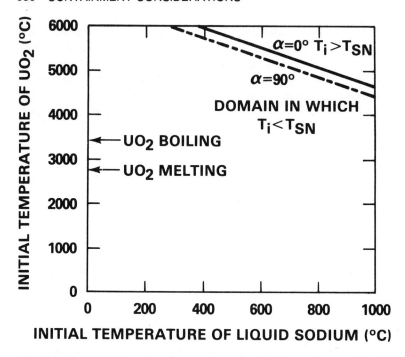

FIGURE 16-7. Illustration that the spontaneous nucleation temperature is unlikely to be reached on contact between molten oxide fuel and liquid sodium for temperatures of interest.[17] $T_i < T_{SN}$ for all initial fuel and sodium temperatures below the diagonal lines. α is the angle between a sodium embryonic bubble and a nucleation site. $\alpha = 0$ corresponds to homogeneous nucleation.

spontaneous nucleation temperature. Hence, according to Fauske's hypothesis, no large-scale vapor explosion would occur when oxide fuel is mixed with sodium at these temperatures.* A second figure, Fig. 16-7, shows the domain of initial temperatures for which $T_i < T_{SN}$ for mixtures of UO_2 and sodium, which includes virtually all temperatures that might be encountered in an HCDA. The contact angle α refers to the angle for wetting between sodium and impurities or surfaces which might serve as nucleation sites; an angle of zero corresponds to homogeneous nucleation.

Even if spontaneous nucleation theory rules out a large-scale vapor explosion between oxide fuel and sodium, this is not necessarily the case for carbide fuel. Equation (16-12) yields a higher interface temperature between carbide fuel and sodium due to the higher thermal conductivity of carbide

---

*A small-scale, delayed explosion might occur if sodium were entrapped in fuel and heated to the spontaneous nucleation temperature. Fauske suggests that this is what happened in the small vapor explosions reported in Reference 13.

fuel. Even in the oxide fuel case, contact temperatures between high-temperature molten steel cladding and sodium can be above the spontaneous nucleation temperature of sodium. In these systems, however, the other conditions listed above must also be satisfied in order for a large-scale vapor explosion to occur.

Detonation Theory

Coherence for efficient large-scale MFCI's implies the existence of a propagation mechanism which couples the regions of explosive energy release to the adjacent unexploded regions. Colgate[16] proposed that explosive expansion of the cold fluid would cause further mixing between hot and cold fluids and, hence, propagation of the interaction. Board and Hall[15] expanded on Colgate's ideas by developing a treatment of the explosion dynamics for the case of a steadily propagating one-dimensional interaction. Three stages are postulated in the model. For the first stage, it is assumed that fuel and coolant become coarsely intermixed. In the second stage, an unidentified trigger mechanism is assumed to result in a shock wave. In the third stage, the shock wave traveling through the coarse mixture causes fine fragmentation of the fuel and further mixing, which in turn produces the rapid heat transfer necessary for sustaining the wave.

These conditions are illustrated in Fig. 16-8, in which the coarse premixture is shown, followed by the fragmentation due to passage of the shock wave. As in the case of spontaneous nucleation theory, experimental verification has been obtained for various aspects of this model. Details of the

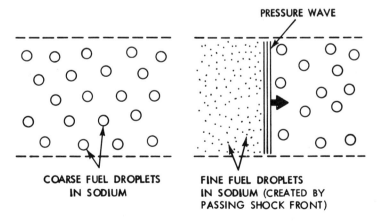

PRESSURE WAVE

COARSE FUEL DROPLETS
IN SODIUM

FINE FUEL DROPLETS
IN SODIUM (CREATED BY
PASSING SHOCK FRONT)

FIGURE 16-8. Premixing, fragmentation, and propagation in the Board-Hall Model.

fuel fragmentation mechanisms behind the shock front are not resolved, however.

C. FUEL FRAGMENTATION

For any of the MFCI models, a prerequisite to the rapid heat transfer rate is a large heat transfer surface and, hence, extensive fragmentation of the molten fuel. Numerous fragmentation models have been suggested, and different ones may apply to different types of MFCI's.

One important category of breakup process is hydrodynamic fragmentation caused by high relative velocities between the fuel and the coolant. This mechanism may account for the high-velocity breakup of the coarse premixture by a shock wave as postulated in the Board-Hall model. Hydrodynamic fragmentation mechanisms are categorized according to a dimensionless parameter called the Weber number:

$$\text{We} = \frac{\rho D V^2}{\sigma},\qquad(16\text{-}13)$$

where ρ = density of the liquid being fragmented,
D = droplet diameter,
V = relative velocity,
σ = surface tension of the liquid being fragmented.

Although breakup can occur with We as low as ~ 12, the processes that can produce the small particles needed for the Board-Hall model in a short enough time period are boundary layer stripping and catastrophic fragmentation from Taylor instabilities, both of which occur at higher We and, hence, very high relative velocities.

Weber number correlations for these fragmentation processes, together with breakup time correlations as functions of Bond number (Bo) and drag coefficients (C_D) (where $\text{Bo} = \frac{3}{8} C_D \text{We}$), have been developed for gas-liquid systems. Experimental data[18] indicate that the correlations hold also for liquid-liquid systems, though the entire area of liquid-liquid hydrodynamic fragmentation is one for which further experimental and theoretical development is needed.

Another class of fragmentation mechanisms is broadly identified by Bankoff[11] under the category of boiling mechanisms. These include many suggested mechanisms variously called violent boiling, compression waves, bubble collapse, jet penetration, coolant entrapment, and vapor blanket collapse. Some of these could be cyclic mechanisms, e.g., coolant could be trapped with fuel, followed by explosive vaporization, collapse of the resulting two-phase bubble, jet penetration of fuel by the coolant, and repeat of the entrapment cycle. Much research has been addressed to the relation between collapse of vapor blankets and the strength of the subsequent interaction. If the liquid-liquid contact temperature on collapse of the vapor blanket is greater than the spontaneous nucleation temperature, experiments indicate that the interaction is strong, an observation that lends support to the spontaneous nucleation theory of Fauske.

Other fragmentation theories have also been proposed that generally lead to slower and less violent breakup. Thermal stress fragmentation is one such mechanism.[19] This model is based on rapid freezing of the outer shell of the fuel droplet which causes high thermal stresses and causes the fuel surface to shatter. This mechanism could explain much of the fragmentation data obtained from the relatively quiescent injection of fuel into sodium. However, the time required to solidify fuel (~ 50 ms) is too long to allow this mechanism to contribute to a vapor explosion. Another mechanism suggested is the rapid release of gas (such as fission product gas) from within a molten fuel droplet. This mechanism is likely not very important for oxide fuel due to the low solubility of inert gases in UO_2.

Many experiments have been performed, both in-pile and out-of-pile, in which oxide fuel has been quenched by sodium. Fragmentation always occurs, and the resulting particle sizes vary over a wide range. Example particle size distributions are shown in Fig. 16-9. The in-pile tests were run in the TREAT reactor in which a power transient causes the ejection of molten fuel from fuel pin test samples into sodium. These particles are generally larger than the sizes needed for propagation according to the Board-Hall model. Further work on particle breakup in shock tubes is needed to study the potential for smaller size particles.

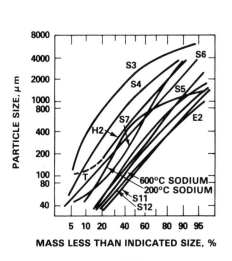

a) TREAT AND SMALL-SCALE TESTS

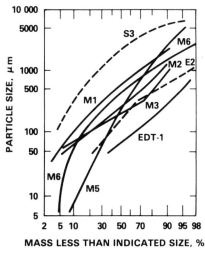

b) LARGE SCALE OUT-OF-PILE EXPERIMENTS

FIGURE 16-9. Oxide fuel particle size distributions after fragmentation in sodium.[20]

16-4 ENERGY PARTITIONING AND MECHANICAL CONSEQUENCES

Once the energy deposition source term of the HCDA has been defined, the problem becomes that of determining the consequences to the in-vessel components and to the vessel and its associated piping. Figure 16-10 is a sketch showing the key factors that must be considered for a loop-type system.

An expanding high temperature-pressure bubble would be felt on the reactor vessel internals and would be transmitted through the fluid to the boundaries of the system. The downward force against the core support structure might be transmitted through the fluid to the lower head of the reactor vessel and, in any case, would be transmitted via the vessel walls to the reactor vessel support system. The radially propagating pressure wave would be felt at the wall of the reactor vessel. The pressure, if sufficiently high, could cause deformation of the reactor vessel wall radially outward at approximately the same elevation as the maximum pressure.

In the upward direction, the sodium slug (i.e., the pool above the high temperature-pressure bubble) would be accelerated upward although the

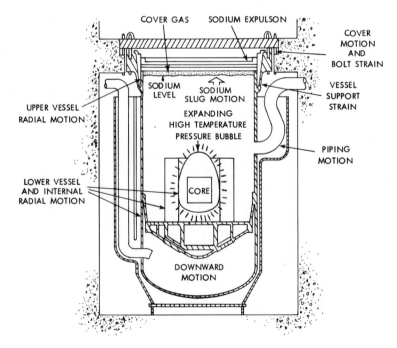

FIGURE 16-10. Reactor geometry and mechanics of potential damage for a loop-type system.

upper core and upper internal structures would provide significant resistance to the bubble expansion. The acceleration of the sodium slug would impart a kinetic energy to the slug. Upon impact of this slug with the reactor vessel head, the momentum of the slug would be transferred to the head, imparting some upward velocity. This impact would be inelastic, and the excess energy would be dissipated radially by deformation of the upper vessel wall. The upward motion of the vessel head would generally be resisted by its supporting structure and holddown system.

Effects of an energetic accident on the remainder of the heat transport system would result from displacement of the reactor vessel, with corresponding nozzle displacement that could result in a deflection of piping, and from pressure waves which could be propagated throughout the primary heat transport system through the outlet and inlet sodium nozzles.

The section below begins with a historical perspective in early scoping methods for correlating mechanical damage. This is followed by a summary of experimental support, including both the SL-1 accident and subsequent simulant experiments. Computational methods are then reviewed, together with a few results selected from a vast amount of literature on the subject to illustrate the nature of the problem involved.

A. EXPERIMENTAL DATA

Early Studies

For the early FBR's the fraction of energy generated during disassembly that was assumed to be converted to mechanical work was based on extremely conservative assumptions. Later research succeeded in reducing this conservatism to some extent, and there is research still underway to establish the degree to which additional physical phenomena may reduce this fraction still further.

Early assessments of damage potential due to such energy release were based on equating this potentially available work energy to an equivalent charge of TNT, i.e.,

$$2 \text{ MJ} \simeq 1 \text{ lb TNT}.$$

Hence, every MJ of work potential was assumed to represent the explosive potential of about $1/2$ lb TNT. The principal reason for using this equivalence was that substantial experience existed, particularly within the U.S. Naval Ordnance,[21] regarding the damage potential for detonating TNT.

Any attempt to predict the actual mechanical response of an FBR system to an HCDA excursion using the TNT energy equivalence model is ambiguous because the pressure-time characteristics of a TNT detonation are considerably different from those of a nuclear excursion. Mechanical damage from an explosion or pressure transient can be caused by both a shock wave, which is transmitted rapidly to a structure, and the more slowly expanding bubble of reaction products or vaporized material. Pressures in a TNT detonation build up on a microsecond time scale and reach the order of 5000 MPa. Hypothesized energy releases from an HCDA excursion, on the other hand, build up over a millisecond time scale, and peak pressures are orders of magnitude lower. As a consequence, much of the damage potential of a TNT detonation to surrounding structures comes from shock wave effects, whereas longer-term bubble expansion would be the predominant damage mode for the slower time-scale pressure buildup of an FBR excursion.

SL-1 Accident

The SL-1 accident,[22] involving a small military thermal reactor at the Idaho Falls test site in 1961, provides an interesting data base upon which to estimate the potential mechanical damage from a major power burst. The energy source term for SL-1 is substantially different (aluminum-clad metal fuel in water coolant) from that of an LMFBR; consequently, very little insight is possible with regard to the magnitude or time scale regarding a postulated LMFBR HCDA event. However, given the total work energy potential, some of the resulting mechanical deformation effects are quite instructive.

The total energy released from the nuclear excursion in SL-1 was estimated[22] to be 130 MJ. Of this total, some 50 to 60 MJ was believed to be generated in the central fuel elements and transferred to water in less than 30 ms in an MFCI between the aluminum and water. It was this prompt release that caused the damage to the core, thermal shields, and reactor vessel.

Proctor[23] attempted to reconstruct the sequence of major events, given the 50 to 60 MJ prompt energy generated, by observing the permanent vessel strain and the vessel jump. Figure 16-11 is a sketch of the post-accident vessel (exaggerated for emphasis), along with the actual vessel strains measured. Proctor deduced that about 2.5 MJ of energy was deposited in the lower vessel, core structure and thermal shield as a result of the initial pressure burst in the core. Another 3.6 MJ of energy was attributed to the water slug, which caused the vessel to lift about 3.3 m, thereby damaging the

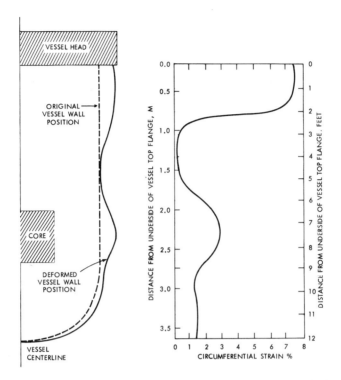

FIGURE 16-11. Vessel strains following the SL-1 accident. (Adapted from Ref. 23)

vessel head and causing permanent strain near the top of the vessel. This total work energy, 6.1 MJ, represented about 12% of the prompt nuclear release (50 MJ) or about 5% of the total nuclear release (130 MJ).

The major goal of Proctor in conducting this analysis was to see whether the earlier naval studies on containment of explosives could be employed for nuclear reactor HCDA mechanical consequences assessment. Much of this earlier work relating to the damaging effects of underwater explosions had been documented in a classic book by Cole, entitled *Underwater Explosions*.[24] This work was augmented in the late 1950s and early 1960s by U.S. Naval Ordnance Laboratory experiments conducted with models of the Fermi reactor to determine the effects of a hypothetical explosion in that system. Additional experiments, using a variety of test vessel sizes and configurations, resulted in a set of empirical explosion "containment laws."[21]

A key observation in these tests was that the radial vessel strain at the point of vessel failure was always greater than one-third the ultimate elongation of failure in an ordinary tensile test for the reactor vessel material. In these explosion containment laws, the observed vessel wall strain was correlated to the high energy release for characteristic explosions similar to chemical high explosives (i.e., TNT).

In his SL-1 evaluations,[23] Proctor showed that the vessel strain correlated reasonably well with the reactor explosion containment laws. The reactor vessel wall strain was considerably less than the predicted failure strain, and no wall failure was observed. In the SL-1, the liquid surface was well below the reactor vessel head at the time of the accident. When the water slug struck the reactor vessel head, the impact on the lower face of the head lifted the entire reactor vessel from its original position and also radially deformed the upper wall of the vessel.

Simulant Experiments

Experimental programs to assess mechanical damage in an LMFBR HCDA have been conducted with simulant materials both in the United States[25,26,27] (e.g., at SRI International and Purdue University) and in Europe[28,29] (e.g., at Foulness, Winfrith, and Ispra). Since expansion times would be much slower (e.g., in the millisecond range) in an FBR HCDA than in a chemical explosion, and since bubble expansion would cause more damage than shock waves in an HCDA, a considerable effort was expended to develop simulant explosives with a burning rate more compatible with the millisecond time frame.[30] In addition to the vessel damage experiments referenced above, tests with simulant explosives under sodium have been conducted at Cadarache to study breach of the cover and sodium transport through the cover in an HCDA.[31]

Small scale in-pile tests involving fuel and/or sodium expansion and energy partitioning have been carried out at several transient test reactors, e.g., TREAT, ACPR, and CABRI.

B. COMPUTATIONAL PROCEDURES

Several computational systems have been developed to assess the structural consequences of an HCDA. Of these systems, the REXCO series[32] has been the most widely used, presumably because it was developed specifically for reactor accident analysis.* In order to more fully analyze systems that encounter large geometric distortions, Eulerian codes, such as ICECO,[33] have also been developed.

---

*REXCO, as well as other Lagrangian codes, grew out of common ancestry, i.e., the HEMP code at the Lawrence Livermore Laboratory or F-MAGEE at LASL.

Figure 16-12 represents a typical model set up for a REXCO calculation of a loop-type reactor system, and Fig. 16-13 illustrates the time sequence of events. An expanding high temperature, high pressure HCDA gas bubble in the core region is assumed to become manifest at time zero. The expansion of this high pressure region is clearly seen in the first 15.9 ms, and by 38.3 ms the sodium slug has impacted the vessel head. In this REXCO calculation the sodium slug impacts the vessel head uniformly. Some analyses with the SIMMER code, e.g., Reference 2, indicate that the gas bubble tends to jet upward so that the slug strikes the head in a nonuniform manner, impacting first at the center of the cover. This results in a decreased upward force on the head. The vessel plenum length/diameter ratio is an important factor in determining the extent of nonuniform slug impact.

Following impact of the sodium slug on the reactor vessel head, REXCO can then be used to calculate the further partition of energy. The results of an early REXCO calculation,[34] shown in Fig. 16-14, illustrate the concept. Time zero in this figure corresponds to the time when the sodium slug strikes the head. An initial triangular shaped force on the head is followed

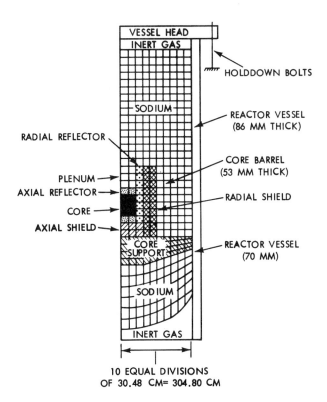

FIGURE 16-12. REXCO model of LMFBR loop system.

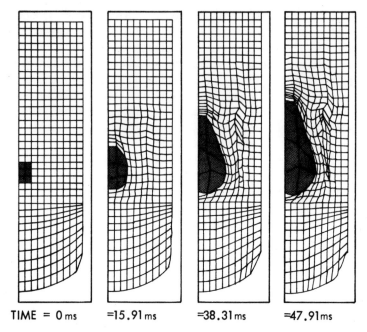

TIME = 0 ms   =15.91 ms   =38.31 ms   =47.91 ms

FIGURE 16-13. Example of reactor configurations predicted by the REXCO code at various times during the mechanical consequence phase of an HCDA.

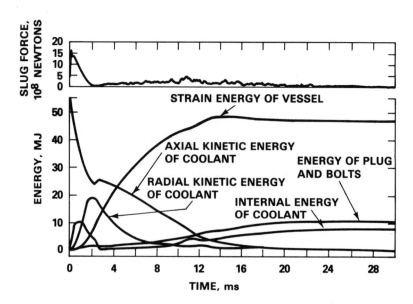

FIGURE 16-14. Example calculation of slug force and energy partition after sodium slug impacts the vessel head in an HCDA.[34]

by a longer and smaller residual force. The axial kinetic energy of the sodium is first transferred into internal energy and radial kinetic energy of the sodium. Some of the original energy is converted to strain energy in the head and hold-down bolts, but the bulk eventually becomes manifest as strain energy in the vessel. Some energy remains as internal energy of the sodium. The initial conditions for the calculations (i.e., at head impact) were 30 m/s velocity, 54 MJ slug kinetic energy (1.2×10^5 kg sodium slug), 86 mm vessel-wall thickness, 7×10^5 kg head, 0.32 m^2 total area of holddown bolts, and 1.4 MPa core vapor pressure.

Both REXCO and ICECO calculations are compared[35] to radial strains measured in the SNR-300 simulation experiments at Ispra[29] in Fig. 16-15. The similarity with the SL-1 results in Fig. 16-11 is evident.

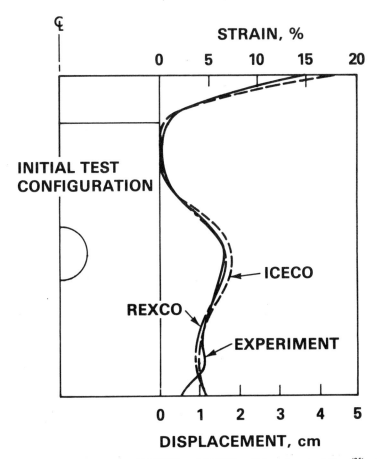

FIGURE 16-15. Comparison of REXCO and ICECO radial strain calculations[35] with SNR-300 simulant experiments.[29]

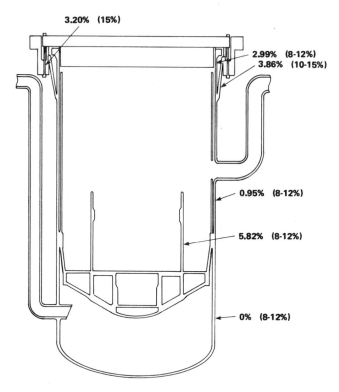

FIGURE 16-16. Prototype vessel strains predicted for upper-limit HCDA at the end of the mechanical consequences phase in FFTF. Values in parenthesis are failure strains.

Figure 16-16 gives examples of typical permanent strains calculated as compared to the strain levels anticipated for failure (in parenthesis). In addition to the vessel and vessel support features, it is also of interest to consider the effects of pressure wave propagation around the primary system piping. Elaborate computational systems have also been developed to treat these problems. An important result of such analyses is the realization that substantial pressure peak attenuation may occur by means of plastic deformations in the piping.

16-5 POST-ACCIDENT HEAT REMOVAL

One of the key questions that must be addressed in studying accidents which involve significant core damage is post-accident heat removal. The principal issues are outlined in this section; the topic is reviewed further in References 20 and 36. It is appropriate first to define the heat generation

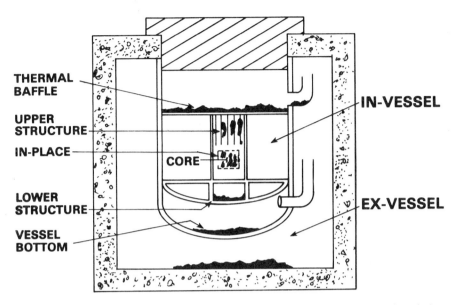

THERMAL
BAFFLE

UPPER
STRUCTURE

IN-PLACE

CORE

LOWER
STRUCTURE

VESSEL
BOTTOM

IN-VESSEL

EX-VESSEL

FIGURE 16-17. Potential locations of fuel debris for post-accident heat removal analysis (loop-type system).

sources and then to focus on the principal geometric locations where heat transfer and cooling occur. These locations can be conveniently divided into *in-vessel* and *ex-vessel*. As illustrated in Fig. 16-17, in-vessel cooling refers either to in-place cooling (within the core) or to cooling debris scattered within the reactor vessel pressure boundary, but removed from the original core location. Ex-vessel cooling focuses on cooling debris beds within structures outside the reactor vessel.

A. POST-ACCIDENT HEAT SOURCES

The first step in assessing post-accident coolability is defining the distribution of heat sources following reactor shutdown. This involves determining the location of the fuel inventory and the distribution of heat sources.

Fuel Relocation

Knowledge of the fuel location following a whole-core accident is necessary since the fuel matrix comprises most of the nonvolatile fission fragments. These radionuclides make up the bulk of the decay heat generation source.

Based on the discussion in Chapter 15 regarding the transient overpower (TOP) and transient undercooling (TUC) accidents, most of the fuel would remain essentially in-place for the TOP scenario. Some axial fuel relocation would occur in scattered assemblies and some of these assemblies may contain partial blockages; most of the fuel would remain at or near its original location.

It is in the unprotected TUC event that substantial fuel relocation might take place. Based on the discussion in Chapter 15-5, a considerable inventory of fuel could be ejected out of the top of the core into the upper core structure region; some fuel could penetrate into the upper sodium pool region and come to rest in the thermal baffle area or even reach the outlet piping (cf. Fig. 16-17).

With regard to downward penetration, large heat sinks below the reactor core would likely cause fairly thick steel blockages to form as a result of molten cladding relocation. In-vessel retention devices might be used in some LMFBR designs in order to prevent penetration of the vessel by molten fuel. Such a device, consisting of a series of trays that could be cooled by the sodium pool, was illustrated for Super Phénix in Fig. 12-6. Without in-vessel retention capability, molten fuel, if present in large amounts, could melt through the bottom of the vessel and fall, together with the sodium still in the vessel, into the reactor cavity.

Recriticality Potential. One of the concerns associated with loss of original core geometry in an FBR is the potential for recriticality of core material. Long-term post-accident heat removal can be successful only if assurance is provided that recriticality cannot occur. Hence, the geometry of potential locations for particulate beds or molten pools must be considered. Although a discussion of critical fuel geometries is beyond the scope of this book, the topic is reviewed in References 20 and 36.

Designs for structures below the core have been developed for the specific purpose of distributing molten fuel or core debris formed in an HCDA into non-critical configurations. An example of an in-vessel design is the Super Phénix core retainer (Fig. 12-6), in which each tray in the retainer is designed to catch a quantity of core material too small to achieve criticality. An ex-vessel design is illustrated in Fig. 16-20 in Section 16-5C on ex-vessel phenomena. This figure shows a distribution device which spreads out core debris into a subcritical geometry on the Core Retention System.

Heat Sources

The principal heat source in a damaged subcritical core is decay heat from fission products (see Fig. 14-4). Beta decay of ^{239}U (23.5 minute half life) is important for several hours and beta decay of ^{239}Np (2.35 day half life) provides a significant heat source for several days. Smaller sources include decay heat from activated steel and sodium and higher order actinides such as ^{242}Cm. Heat from the fission process itself decays rapidly after neutronic shutdown, following the longest ($\sim$80 second) decay period of the delayed neutrons.

Decay heat from the fission products (including the noble gases) follows the curve contained in Fig. 14-4. Computer codes such as the U.S. codes ORIGEN,[37] CINDER,[38] AND RIBD[39] can be used to obtain the contributions of individual radionuclides to the decay heat.

Fission products have been categorized in various ways for convenience. The four groups listed in Reference 20 are given in Table 16-1, together with the elements in each group and the principal form in which they are present in the oxide fuel.

TABLE 16-1   Fission Product Categories

Group	Category	Family	Principal Form	Element
I	Noble Gases	Noble Gas	Elemental	Xe, Kr
II	Halogens	Halogen	Elemental	I, Br
III	Volatile Solids	Alkali Metal	Metal	Cs, Rb
		Transition	Metal	Ag, Cd
		—	Metal	As, Se, In, Sn, Sb, Te
IV	Low Volatility (or "Nonvolatile") Solids	Transition	Metal	Tc
		Noble Metal	Metal	Ru, Rh, Pd
		Alkali Earth	Oxide	Sr, Ba
		Transition	Oxide	Mo, Y, Zr, Nb
		Rare Earth (Lanthanides)	Oxide	La, Ce, Pr, Nd, Pm, Sm, Eu, Gd

The low volatility fission products are sometimes subdivided further according to whether oxides are formed in oxide fuel. This distinction is made since the oxides tend to remain with the fuel, whereas the metals tend to concentrate with the steel. The noble gases and halogens and some of the volatile fission products are normally assumed to separate from the fuel upon loss of fuel pin integrity, but the Group II and III elements do not necessarily escape from the primary sodium because the halogens may react with sodium and the Group III elements are soluble in sodium.

Fission yields for ^{239}Pu and ^{235}U are shown in Fig. 16-18. Differences are noted around $A = 90$ (e.g., Sr) and between 105 and 110, but the effects of

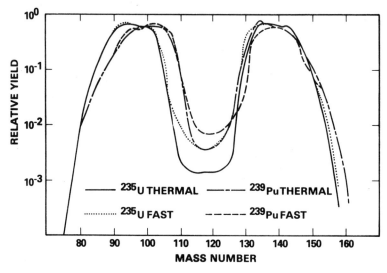

FIGURE 16-18. Fission product yields for ^{235}U and ^{239}Pu.

the differences are not large. The fission product decay heat level as a fraction of operating power in LMFBR fuel is similar to that of LWR fuel. However, the LMFBR decay heat density is substantially higher due to the higher operating power density.

B. IN-VESSEL COOLING

Sodium is a most desirable coolant—and this applies also for cooling a damaged core or core debris. It maintains a single phase liquid condition up to 880°C at atmospheric pressure and has an adequate heat capacity. Furthermore, as noted in Section 16-3, there is ample evidence that molten ceramic fuel (and also molten steel) will fragment upon contact with sodium, thereby exposing large heat transfer areas for removal of decay heat.

Such properties can be used to advantage in an LMFBR system to provide long-term cooling of a damaged core or of core debris within the reactor vessel. Topics of importance include in-place cooling (i.e., where the damaged core is basically intact) and particulate bed cooling (where debris has relocated to other locations within the vessel). The ability of the sodium to cool the debris, however, is predicated on open flow paths to the heat generating sources. If large flow blockages occur, or if sodium boils dry from the debris bed, then a molten core-debris pool could form. Hence, cooling of each of these configurations must be considered.

In-Place Cooling

Four conditions have been proposed for cooling fuel material remaining in a substantially damaged core region, each entailing a different cooling configuration and associated heat transfer mechanism: (1) cooling the outer rows of pins within each assembly, even though the central region may be blocked, (2) cooling voided assemblies (possibly to the point of flow recovery) by radial heat transfer to neighboring intact assemblies, (3) cooling particulate debris in blocked assemblies via sodium leakage through the debris, and (4) cooling molten debris in the lower shielding section via radial heat transfer to neighboring assemblies.

Studies indicate that Condition 1 can be satisfied with natural convection flow for heat loads below 10 to 20% of full power, and that Condition 2 can be satisfied if power in the local voided assembly is below 3% of operating power and flow in neighboring assemblies is maintained by pony motor flow. Natural convection cooling of debris Conditions 3 and 4 is possible only at low power levels (perhaps 0.5% full power). Hence, as would intuitively be expected, in-place cooling is readily achieved for essentially original geometry, even under natural convection conditions alone, but cooling capability is increasingly jeopardized as the degree of fuel concentration increases.

Particulate Bed Cooling

Since molten fuel and steel will fragment upon contact with liquid sodium, there is considerable interest in determining the cooling characteristics of packed beds of particulate core debris. Potential locations of particulate beds are shown in Fig. 16-17. The first step in assessing the cooling of such a bed is to determine its depth and composition. Given a horizontal surface upon which such a bed can form, the depth is determined by knowing the total debris mass available, the particular surface area and the debris porosity (typically about 50%).

Debris beds that form on in-vessel surfaces will likely be immersed in sodium. Experiments have shown that the predominant direction of heat flow from coolable beds will be upward to the overlying sodium. For shallow beds the heat flow will be by conduction and convection; for deeper beds, sodium boiling occurs within the bed, forming vapor channels that vent from the interior of the bed to the overlying sodium. As the bed depth exceeds coolable limits, *particulate bed dryout* occurs, meaning that sodium cannot penetrate into the bed to provide a high heat flow upward from the bed. Coolability beyond this point will depend on the rate of heat removal by conduction in both the upward and downward directions.

Molten Pool Heat Transfer

If a particulate bed of fuel debris is not sufficiently cooled, its temperature rises until a bed of molten fuel and steel is obtained. For this situation, the principal questions concern the direction of heat flow—whether upward, downward, or sideward.

Natural convection occurs in internally heated molten pools, with hot fluid rising and colder fluid moving downward. The onset of motion in a fluid layer occurs when the buoyancy forces caused by temperature gradients exceed the viscous forces. Inertial forces also affect the transport process after flow develops. The Grashof number (Gr) combines inertial, viscous, and buoyancy forces. Since natural convection is a combined momentum and energy exchange process, the Prandtl number (Pr) also influences the heat transfer. Hence, natural convection is correlated with the product of Gr and Pr, a product known as the Rayleigh number (Ra). For an *internally heated fluid* between two horizontal plates at equal temperatures and separated by a distance L, the Rayleigh Number, labeled Ra_I, is*

$$Ra_I = \frac{g\beta Q L^5}{\nu\alpha k} \ , \qquad (16\text{-}14)$$

where

β = coefficient of thermal expansion
g = gravitational acceleration
Q = volumetric heat generation rate
ν = kinematic viscosity
α = thermal diffusivity
k = thermal conductivity
L = distance between plates.

Figure 16-19 contains data and correlations from several sources showing the fraction of heat from a molten pool flowing downward (for equal boundary temperatures).[36] The abscissa is $Ra_I/64$ where Ra_I is defined by Eq. 16-14. Values of Ra_I of interest for PAHR are generally in the higher range of values on Fig. 16-19 for which only 10 to 30% of the heat generated would be transferred downward.

---

*The Rayleigh number for natural convection driven by a temperature difference ΔT instead of an internal heat source is given by $g\beta\Delta T L^3/\nu\alpha$. To obtain Ra_I, the ΔT is replaced by QL^2/k.

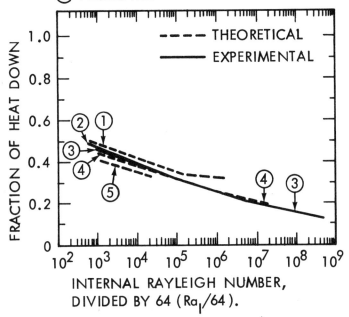

① SUO·ANTILLA AND CATTON
② KULACKI AND GOLDSTEIN
③ JAHN AND REINEKE
④ JAHN AND REINEKE
⑤ PECKOVER

FIGURE 16-19. Comparison of heat transfer correlations (fraction of heat flowing downward) for internally heated horizontal pools with equal boundary temperatures[36]. Numbers encircled indicate experimenters as found in Reference 36.

Sideward heat transfer correlations are less well defined. Work on sideward as well as other conditions for upward and downward heat transfer from molten pools is an area occupying increasing attention for both LWR and FBR safety analysis.

Heat transfer correlations for upward, downward, and sideward heat transfer by natural convection from molten pools with internal heat sources are generally of the form:

$$Nu = C \cdot Ra_I^m. \tag{16-15}$$

The use of a correlation of this form is illustrated in Chapter 18-5B.

C. EX-VESSEL PHENOMENA

If the core debris melts through the vessel, heat transfer and interaction of the debris in the reactor cavity must be examined. Two topics are of principal interest—first, techniques that might be employed to reduce the consequences of melt-through such as engineered core retention concepts (i.e., devices specifically designed to stop the flow of molten fuel debris), and, second, the interaction of sodium and core debris with structural material below the reactor vessel.

Engineered Ex-Vessel Core and Sodium Retention Concepts

The engineered concepts so far devised to mitigate the consequences of vessel melt-through include *steel liners*, *crucibles*, and various types of *sacrifical beds*.

An important objective of reactor cavity design is to prevent contact of sodium with concrete since water released from concrete could interact with sodium to form hydrogen. The concrete reactor cavity walls are lined with a stainless *steel liner* to separate sodium from the concrete. This liner is sufficient to contain sodium but not core debris after particulate bed dryout.

Conceptually, a *crucible* retention system is designed to contain core debris indefinitely. It consists of a container or structure in which a sacrificial material is located. Cooling is generally provided by a system completely independent of the reactor coolant system. Figure 16-20 shows such a system, as proposed for the SNR-300 reactor. The distribution device is designed to distribute the core debris over the crucible surface both to prevent recriticality and to spread out the heat source. The SNR-300 crucible structure is cooled by NaK.

Three types of sacrificial beds have been proposed as ex-vessel core retention systems according to the extent of cooling designed. The *completely passive* sacrificial bed represents a device with no cooling system, constructed of a high thermal capacity/low thermal conductivity material and arranged such that hot core debris can penetrate only at a slow rate. Materials considered for such a bed include ThO_2, depleted UO_2, magnesium oxide (MgO), and graphite. Penetration is especially slow in MgO due to its large specific heat and heat of fusion, and almost no gas or aerosols evolve from MgO beds during penetration.

Passively cooled sacrificial beds have been proposed where the outer surface of the bed is cooled by natural convection. This cooling would

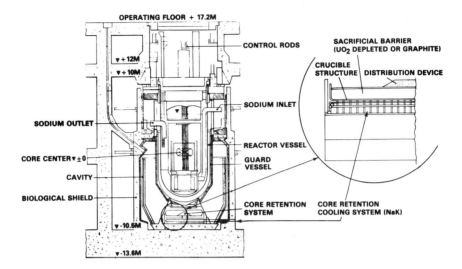

FIGURE 16-20. SNR-300 ex-vessel core retention system.

provide improved potential for retaining core debris and also enhanced thermal protection of the concrete cavity walls. An obvious deviation from the above concept is an *actively cooled* sacrificial bed.

Table 16-2 summarizes the debris accommodation (core retention) provisions employed in recent LMFBR systems.

Interaction of Sodium and Core Debris with Concrete

Concrete Thermal Response. Although it is common practice to provide steel liners on the inner surface of all concrete cells where sodium could spill (in order to prevent direct contact between sodium and concrete), hot sodium on a steel-lined slab of concrete could still cause appreciable heating of the concrete.*

Substantial amounts of water can be liberated from structural concrete if it is heated, as evidenced from the water release data in Fig. 16-21, for any kind of structural concrete.[40] Note that the free water (or capillary water) is readily released at fairly low temperatures. At about 450°C, chemically bound water is released. There are some data to suggest that at even higher

---

*Some designs include insulating material, such as firebrick, between the concrete and the steel liners in order to minimize concrete heating in the event of a sodium spill. Such liners are often referred to as "hot" liners.

TABLE 16-2 Debris Accommodation (Core Retention) Provisions in LMFBR's (Adapted from Ref. 36)

Reactor	Provisions	
United States		
• EBR-I	None	
• EBR-II	None	
• FERMI-I	In-vessel:	Zirconium meltdown pan
	Ex-vessel:	Graphite crucible below vessel with primary shield tank
• SEFOR	Ex-vessel:	Sodium catch tank 45 ft below vessel. Fuel dispersion cones below vessel and in catch tank
• FFTF	In-vessel:	Debris cooling in core support structure dome and above thermal baffle, but no special provision
	Ex-vessel:	Steel lined cavity
• CRBRP	In-vessel:	Some small amount of debris cooling on reactor internal structures, but not special design provisions
	Ex-vessel:	Lined and insulated reactor cavity, vented pipe chase and ex-containment cooling-venting purge system
United Kingdom		
• DFR	Fuel dispersion cone and melt tubes to bedrock	
• PFR	Single layer of trays within tank; capable of retaining seven assemblies	
• CDFR (proposed)	Three layers of trays within tank; retention capacity for entire core	
France		
• Rapsodie	None	
• Phénix	Externally cooled outer vessel	
• Super-Phénix	In-vessel catch trays, external cooling of safety vessel	
Federal Republic of Germany/ Netherlands/Belgium		
• SNR-300	In-vessel:	Catch trays in lower plenum
	Ex-vessel:	High-temperature crucible with NaK cooling

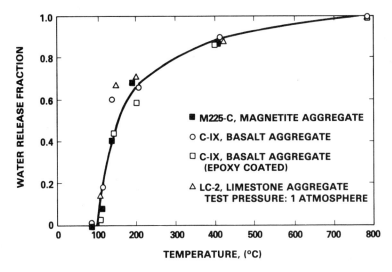

FIGURE 16-21. Water release from concrete.[40]

temperatures, additional release may occur due to complete decomposition and dehydration of the aggregate material. The important point, however, is that a large quantity of water can be released from concrete at an elevated temperature, and some accommodation (such as vents) must be provided to relieve the resulting pressure.

In addition to water release, substantial amounts of CO_2 gas could be released from heated concrete via the dissociation of $CaCO_3$ if limestone aggregates are present:

$$CaCO_3 + heat \rightarrow CaO + CO_2.$$

Another concern for unlined concrete is the breaking away of large pieces of concrete by cracking or through a process known as *spalling*. The thermal shock caused by the direct contact of molten core debris would almost certainly cause some surface spalling. There has been some concern that large concrete sections could be penetrated via fissures without the need to expend the energy required to melt through the section.

Sodium Concrete Reactions. Direct interaction of hot sodium on bare concrete produces chemical reactions that liberate hydrogen gas. Most of this hydrogen is generated by exothermic sodium-water reactions:

$$2Na + H_2O \rightarrow Na_2O + H_2,$$

$$Na + H_2O \rightarrow NaOH + \tfrac{1}{2}H_2.$$

In a series of sodium-concrete reaction tests,[41] hydrogen was initially produced at a rate of approximately 5 kg/m^2·h. Unless the hydrogen is vented or oxygen is provided to remove the hydrogen by recombination,* a sustained sodium-water reaction could lead to severe overpressurization of the containment.

Data from such tests indicate sodium initially heats the concrete, driving off water, and then the chemical reaction of sodium with water and aggregate material releases both hydrogen gas and energy. The released energy can accelerate sodium-concrete reactions since it causes increased sodium and concrete heating, further liberation of water, and further chemical reaction with sodium.

On the other hand, the initial sodium attack rates on concrete can subside in a few hours due to buildup of heavy, viscous reaction products that settle on any horizontal surfaces and impede the reaction. A similar saturation effect is noted for vertical surfaces, but the attack rate is greater (presumably since the reaction products tend to gravitate away, leaving a better interaction geometry). This saturation effect could be negated if sufficient concrete cracking could occur to expose fresh concrete surfaces. Quantitative rates of sodium reaction with concrete are not fully resolved; research is underway, especially at the Sandia National Laboratory and the Hanford Engineering Development Laboratory, to define these processes further.

Two preliminary expressions have been derived[41] to fit the sodium penetration depth for horizontal surfaces of magnetite and limestone concrete:

$$\text{Magnetite:} \quad d = 17.5 \, (1 - \exp^{-0.2t})$$
$$\text{Limestone:} \quad d = 10.4 \, (1 - \exp^{-0.4t})$$

where
$$d = \text{penetration depth (mm)}$$
$$t = \text{time (hours).}$$

Figure 16-22 compares these curves with actual penetration data for horizontal concrete surfaces. Similar equations were also fit to a more limited data base for sodium penetration into vertical surfaces, and this fit is also included in the figure. A penetration rate for a horizontal surface of 13 mm/h for four hours is often used for containment transient calculations.

Other important exothermic sodium reactions can occur during sodium interaction with concrete (in addition to sodium burning in oxygen—to be

---

*Devices provided to perform such a function are called hydrogen recombiners.

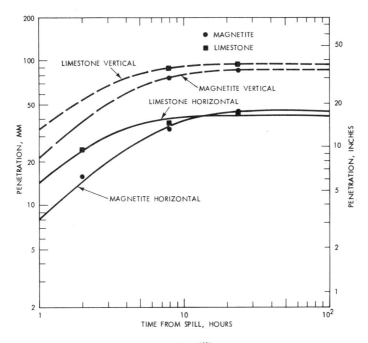

FIGURE 16-22. Concrete penetration by sodium.[41]

discussed in Section 16-6). These are listed in Fig. 16-27 in the final section of this chapter.

Core Debris-Concrete Interactions. In addition to potential chemical attack by sodium, the possibility of direct concrete contact with molten steel and molten fuel must be considered for an ex-vessel accident. The H_2O and CO_2 releases from concrete due to heating are capable of oxidizing the steel content of the penetrating core debris. Such oxidation could lead to the generation of H_2 and CO to add to the hydrogen liberated from a sodium/concrete interaction.

Prediction of the rate of concrete penetration by a core melt is a difficult problem that has not yet been fully resolved, though preliminary analytical methods are available (e.g., the GROWS[42] and the USINT[43] codes in the U.S.). The problem is complicated by the addition of molten products from the concrete to the pool as the melt penetrates. This addition of low-density material reduces the density of the molten pool to the extent that it might eventually float above molten steel, thus reversing the layering of material in the pool during the penetration process. In addition, gas rising from the concrete will remove heat and may sparge some fission products from the pool.

16-6 SODIUM FIRES

If any boundary in the primary coolant system has been ruptured, the possibility of a sodium *pool* or *spray fire* exists. A pool fire could occur if sodium is spilled in an equipment cell or on the floor of the Reactor Containment Building. A spray fire could result from sodium being ejected from an equipment or pipe leak or from the vessel cover in an HCDA. It is important to understand the environmental conditions that allow sodium fires to exist so that design measures can be incorporated to reduce the likelihood of fires. A large amount of research has been done to determine the rate of sodium combustion, the pressure buildup from combustion gases, and the physical properties of aerosols produced in sodium fires.

Sodium fires occur according to the following exothermic reactions:

$$\text{Sodium Oxide:} \quad 4\text{Na} + \text{O}_2 \rightarrow 2\text{Na}_2\text{O}$$
$$\text{Sodium Peroxide:} \quad 2\text{Na} + \text{O}_2 \rightarrow \text{Na}_2\text{O}_2$$

where the first reaction is the predominant mode. The fire produces low flames and is characterized by a dense white oxide smoke.

The best design measure for reducing the possibility of sodium fires is to provide inerted cells wherever there is potential for a sodium fire. A typical oxygen content in such a cell is less than 2%.

A. POOL FIRES

A sodium pool fire will not normally ignite below about 250°C (although it could ignite as low as 200°C if the pool is disturbed). Since the fire occurs only at the pool surface, it is meaningful to characterize the combustion rate on an area basis. A fairly standard rate to use for a pool fire in air is 25 $\text{kg/m}^2 \cdot \text{h}$.

Figure 16-23 illustrates a typical temperature profile for the sodium liquid below the surface and the atmosphere immediately above. Figure 16-24 typifies the pressure responses one would expect for a sodium pool fire in a given cell. The pressure would initially rise as the cell temperature increases, but would then diminish as the cell oxygen becomes depleted. For a given sodium inventory, the pressure would be lower for larger cell sizes. SOFIRE[46] is the principal U.S. code used for computing the pressure and temperature response of a sodium pool fire.

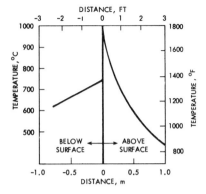

FIGURE 16-23. Temperature in the pool and atmosphere near the pool surface in a sodium pool fire.[44]

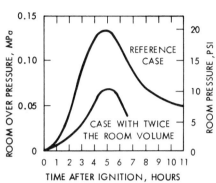

FIGURE 16-24. Typical cell pressure for a sodium pool fire.[45]

B. SPRAY FIRES

The ignition temperature for a sodium spray fire is substantially lower than for a pool fire. If the sodium is in large droplet form, ignition may occur around 120°C, but ignition at lower temperatures could occur for fine mist conditions.

Peak cell pressures depend somewhat on the heat transfer conditions for the cell walls, but pressures in the 1 MPa range are theoretically possible. Experimentally measured pressures, however, are consistently lower than theoretical upper limit pressures. Figure 16-25 contains a plot of theoretical and measured pressures as a function of sodium-to-oxygen ratio for a sodium spray fire in air. (The pressure is independent of cell size since the basis is the ratio of Na to O_2.) The pressure rises rapidly for the oxygen rich domain (where sodium peroxide could be formed) and continues to rise as sodium oxide is formed. However, heat loss effects begin to reduce the pressure as oxygen becomes depleted and the sodium itself becomes a considerable heat sink.

Experimental data of the type depicted in Fig. 16-25 are generally gathered from two types of sodium sprays. In one case, the spray is created as the result of an explosive ejection, whereas in the other case the sodium results from a high pressure spray discharge. Since the former situation results in smaller droplet sizes, the burn rate is faster and the resulting pressure is higher (as shown in Fig. 16-25).

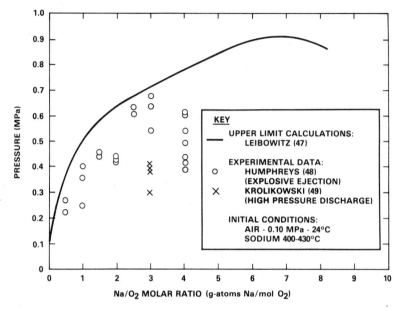

FIGURE 16-25. Cell pressure for sodium spray fires in air as a function of Na/O_2 molar ratio.[46]

SPRAY[50] and SOMIX[51] are examples of U.S. codes used for determining pressures and temperatures resulting from sodium spray fire conditions.

16-7 CONTAINMENT AND CONFINEMENT DESIGN AND TRANSIENT ANALYSIS

A. CONTAINMENT SYSTEMS

A key decision that must be made early in the design of an FBR is the type of containment, confinement, or combination containment/confinement system to be used. Table 16-3 itemizes the types of containment systems selected for several breeder reactor plants, together with a brief description of the salient features of each system.[52] Three of the systems are depicted schematically in Fig. 16-26.

Selection of the particular scheme to use involves many trade-offs. Clearly, economics and the effectiveness of each system in reducing offsite doses in the event of various accidents are important factors. Some of the

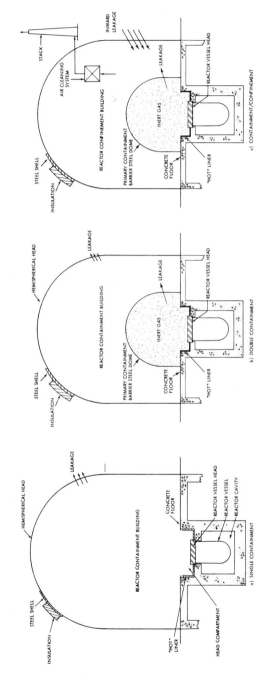

FIGURE 16-26. LMFBR containment systems (multiple, with pumpback, not shown).

687

TABLE 16-3 Types of Containment Systems Used in Current Breeder Reactor Plants (Adapted from Ref. 52)

Containment Type	Description	Reactors
Single Containment	Open head compartment and low-leakage outer containment building	FFTF, EBR-II, JOYO
Double Containment	Sealed, inert high pressure inner containment barrier, surrounded by a low-leakage outer containment building*	FERMI, SEFOR
Containment/Confinement	Sealed, low-leakage inner containment barrier, surrounded by a ventilated low pressure outer confinement building with discharge to stack via an air cleaning system	PFR, CRBRP,† SUPER PHÉNIX, BN-350, BN-600
Multiple Containment with Pumpback	Sealed, high pressure inner containment barrier, surrounded by one or more outer barriers.* A negative pressure zone is maintained in the outermost space by pumping back leakage to the inner containment space. Eventual venting to a stack via the air cleaning system is provided.	SNR-300

*The Reactor Containment Buildings (RCB's) in these cases are designed to withstand 0.17 MPa (10 psig) or greater differential pressures.

†The CRBRP design incorporates a concrete structure outside of the RCB. Its principal purposes are for shielding and missile protection. During certain postulated core disruptive accident situations, the RCB is designed to vent through a cleanup system into the gap between the building and the concrete structure. At other times, the gap is maintained at a negative pressure and any flow from the gap is filtered before release.

principal phenomena involved in the analysis of potential offsite doses are described below.

B. CONTAINMENT TRANSIENT ANALYSIS

It should be apparent from the earlier sections of this chapter that determining the consequences of failure of the primary system requires a systems treatment of many interacting phenomena. The purpose of this part of Section 16-7 is to indicate how such an analysis might be conducted, and to illustrate some key results for a typical analysis.

Cell Conditions and Leakage Rates

The first large computational system used in the United States to estimate cell temperatures, pressures, constituent composition, leak rates, etc., during a postulated FBR core melt-through and sodium spill was the CACECO code.[53] Figure 16-27 summarizes the principal interacting modes treated in CACECO. The chemical reactions considered by the code include those sodium reactions listed in Sections 16-5 and 16-6, together with the exothermic hydrogen recombination reaction:

$$2H_2 + O_2 \rightarrow 2H_2O.$$

As indicated in Fig. 16-27, temperatures and pressures in interconnecting cells (e.g., reactor cavity, heat transport cells, outer containment) are determined by the chemical reactions, heat transfer, and cell-to-cell mass transfer which might occur. The principal objectives of such a computational package are to determine sodium leakage rates from cell to cell and to estimate the time dependent pressure, temperature and material inventory in the outer containment. Updates of this computational system include the CONTAIN system at Sandia and the CONACS system at the Hanford Engineering Development Laboratory.

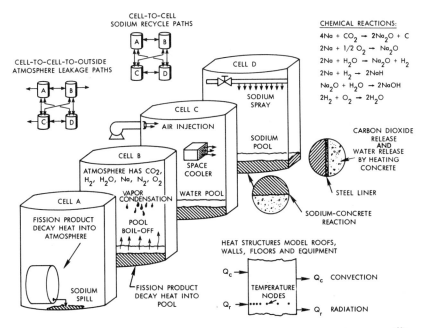

FIGURE 16-27. Principal features of the CACECO cell transient computation system.[53]

Radiological Source Term

In order to estimate radiological leak rates to the environment and potential doses to the public, it is necessary to obtain a *radiological source term*, i.e., a concentration of radioactive materials in the reactor containment building (RCB). This source term includes principally fission products, plutonium, and sodium oxide.

A determination of how core debris might be transported to the RCB following an HCDA is a difficult undertaking. One phase of the problem that is well resolved, however, is the fission product and actinide sources in the core itself. Computer codes [e.g., the U.S. codes ORIGEN,[37] CINDER,[38] and RIBD[39]] are available to calculate detailed inventories of all fission product isotopes from the various fuel isotopes. Actinide inventories are calculated according to the methods in Chapter 7.

The COMRADEX[54] code is an example of the type of model that can be used to estimate the source term. This code follows the radiological inventory from an HCDA through a series of four *chambers* within the containment. The initial chamber monitored is normally the reactor vessel or reactor cavity. This chamber then leaks to the second (e.g., the equipment cells) and/or to the outer containment. For each isotope in each chamber at each time step, the following differential equation is solved:

$$\frac{dN_i^k}{dt} = \lambda_L^{(k-1)} N_i^{(k-1)} + \lambda_R^{(i-1)} N_{(i-1)}^k + \left[\lambda_R^{(i-1)} M_{(i-1)}^k\right]^* \qquad (16\text{-}16)$$
$$- \lambda_R^i N_i^k - \lambda_C^k N_i^k - \lambda_L^k N_i^k,$$

where

$i=$ isotope index (i is this isotope; $i-1$ is the preceding member of this decay chain)

$k=$ chamber index (k is this chamber; $k-1$ is the previous chamber)

$N=$ number of atoms (N_i^k is the number of atoms of this isotope in this chamber)

$\lambda_L=$ leak rate from the chamber indicated and is calculated by a CACECO-type code

$\lambda_R=$ radioactive decay constant for the isotope indicated

$\lambda_C=$ cleanup or fallout rate[†] for the chamber indicated and calculated by an aerosol code such as an updated version of HAA-3 (cf. below)

---

*This term is for the reintroduction of noble gases (e.g., Xe, Kr); thus, it is not considered if the isotope is not a noble gas.

[†]Fallout is not applied to noble gases. Sodium is normally sufficiently abundant that there are no free halogens; consequently, the sodium halides (e.g., NaI, NaBr) are usually treated as particulates with respect to fallout and filtration.

M = number of atoms that have been captured (removed from the chamber by fallout or from the exhaust from the previous chamber by filtration) with decay of previously captured precursors adding to this value.

Although fission product actinide and radioactive sodium inventories at the start of a postulated accident can be determined accurately, an accurate calculation of the initial radiological atom densities, N, in the COMRADEX code also requires a quantitative understanding of processes and mechanisms involved in the release and transport of radioactive material from the fuel and sodium into the chambers. The noble gases (Kr and Xe) are usually assumed to be released from both the fuel and the sodium. Release of fuel aerosols and remaining fission products are more difficult to assess. There are several paths by which some of these materials can become airborne and, hence, contribute to the radioactive aerosol source term. First, in a disassembly accident, they might move upward to the cover gas in a large two-phase bubble of fuel or of fuel and sodium produced during or after core disassembly, and then escape from the reactor vessel through openings in the head resulting from mechanical damage caused by the accident. The halogens (principally iodine) and some of the volatile fission products could be released by this path, and some of the fuel and solid fission products could be suspended as aerosols. Alternately, if fuel melting occurs without disassembly, some fuel and fission products will escape from the resulting particulate bed or molten pool into the sodium. A tracing of the sodium through a computational model such as CACECO, combined with COMRADEX, could provide an estimate of the percentage of these elements that escape to the RCB. These source term release and transport processes (along with research programs directed toward their understanding) are reviewed in Reference 55.

In a disassembly accident a small fraction of the core can be vaporized. Condensation of vaporized fuel and fission products might produce small primary aerosol particles in a two-phase bubble, although the particles would rapidly agglomerate into larger clusters of particles. Some molten fuel may fragment into relatively small particles by hydrodynamic processes such as flashing or by two-phase flow through openings in the reactor head. Whether a large two-phase bubble can rise to the surface carrying aerosols with it or collapse and break up before reaching the cover gas is not yet clear. Several experimental programs are directed toward this phase of a potential accident; one example is the study of partial vaporization of UO_2 pellets in a pool of sodium by capacitor discharge at Oak Ridge National Laboratory.[56]

Since experimentally verified mechanistic models are still being developed to evaluate the release of solids from the reactor vessel and to follow them through the various barriers to the RCB, early FBR analyses were generally based on an arbitrary source term in the RCB, such as 1% of the solids initially in the core.

Agglomeration and In-Containment Settling

Substantial amounts of suspended particulate matter (e.g., sodium oxide or UO_2-PuO_2) can agglomerate and fall-out or plate-out on the floor and walls of the various cells through which such material must pass before escaping into the environment. Extensive experimental studies have provided good understanding of the behavior of nuclear aerosols.

The computer code HAA,[57] together with its successors [e.g., HAARM-2[58]] was developed primarily for the purpose of calculating aerosol behavior during an HCDA. A similar model, PARDISEKO,[59] was developed at Karlsruhe, and PARDISEKO-III compares well with HAARM-2. Other models include AEROSIM (U.K.) and ABC (Japan). These models, which are based on the assumption of a log-normal particle size distribution, will allow combinations of Brownian agglomeration, gravitational agglomeration, settling, wall plating, and leakage from the cell in question. The codes are used extensively to determine aerosol depletion rates in the reactor cavity, equipment cells, and the RCB. Useful reviews of aerosol modeling for LMFBR safety analysis appear in References 60 and 61.

Dose Calculations

Calculations of the dose that might be received by a person outside the RCB is essentially the same for fast breeder reactors as for thermal reactors, with the possible exception that the effect of plutonium must be more carefully assessed for the fast reactor. As in thermal reactors, the dose can come initially from three sources: direct dose received by direct gamma radiation from sources within the RCB, external dose received from gamma radiation in clouds fed by leakage from the RCB, and internal dose received from inhalation of isotopes leaking from the RCB. An additional longer-term source is through various food and water chains.

Once the contaminated atmosphere leaks through or is exhausted from the outer containment (according to some determined leak or exhaust rate), local meteorological data are used to establish the dispersion characteristics of the airborne radioactivity in the same way as for thermal reactors. The COMRADEX code can also perform this calculation, and it integrates the three sources of dose listed above at any distance downwind from the RCB.

Whereas a code such as COMRADEX will determine doses, it is some-times of interest to determine integrated biological consequences to a population mass downwind from the reactor site. The CRAC code, origi-nally developed for use in the WASH-1400 Reactor Safety Study,[62] is often used for this purpose. Features of both the COMRADEX and CRAC codes have been incorporated into the CRACOME computational system.[63]

Sample Cell Transient Calculation

An effective way to depict the principal parameters of interest during a vessel melt-through containment transient is to present results from a particular cell transient analysis. The sample case is taken from a study[52] of various containment systems with a single containment configuration (Fig. 16-26, concept a), i.e., an inert reactor cavity sealed off during normal operation from a 6.4×10^4 m³ air-filled steel reactor containment building (RCB), which has a 0.1% leak rate per day [at an absolute pressure of 0.17 MPa (10 psig)]. The postulated containment transient is a core melt-through of the vessel (without disassembly), accompanied by a spill of 10^6 kg of primary loop sodium into the reactor cavity.

Figures 16-28 through 16-33 illustrate the principal features of the calcu-lated transient. Figure 16-28 shows the heat source in the reactor cavity, which is principally the result of fission product decay heat in the core debris (cf. Section 16-5A). Increasing sodium temperature in the reactor cavity leads to an increase in the sodium vapor pressure. This pressure buildup eventually causes seal failure and allows sodium to flow into the RCB where it can react with oxygen. For this case, the water released from

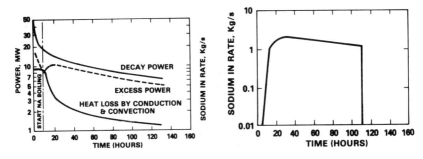

FIGURE 16-28. Energy balance in cavity pool. **FIGURE 16-29.** Sodium entry rate to reactor containment building (RCB).

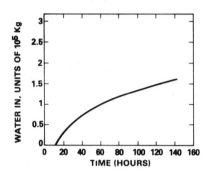

FIGURE 16-30. Total amount of water entering RCB.

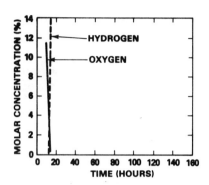

FIGURE 16-31. RCB hydrogen and oxygen concentration.

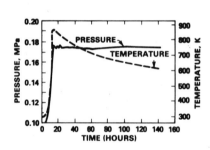

FIGURE 16-32. RCB pressure and temperature.

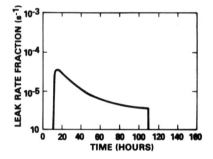

FIGURE 16-33. Leak rate from the RCB.

the heated concrete is assumed to be vented to a condensing room and the steel liners do not fail. During the first eight hours sodium vapor begins to leak to the RCB, and increasing temperatures cause equipment cells to pressurize and nitrogen to leak to the RCB; thus the RCB begins to pressurize.

At about eight hours the sodium in the reactor cavity begins to boil, thus further pressurizing both the cavity and the RCB. This process continues for about 100 hours, until all the sodium boils out. As noted from Fig. 16-29, the rate of sodium entry into the RCB becomes fairly high around ten hours, and this sodium reacts with the oxygen in the RCB, causing a rapid depletion of oxygen, as shown in Fig. 16-31. Heating of the concrete in the RCB releases water vapor (Fig. 16-30). This water vapor, together with hydrogen formed by the chemical reaction between water and sodium, further pressurizes the RCB. The rapid increase of hydrogen starts at about 14 hours, as shown in Fig. 16-31, wherein hydrogen recombination was not taken into account.

At about 11 hours the RCB pressure reaches 0.17 MPa (10 psig). For this example calculation, a pressure relief valve between the RCB and the outer atmosphere is activated at 0.17 MPa, thus holding the pressure constant at this value as shown in Fig. 16-32. The corresponding RCB leak rate is shown in Fig. 16-33.

Other scenarios are possible if different assumptions regarding system failures are made. For example, the reactor cavity could be linked to surrounding inert cells with rupture seals so that cavity pressures could be relieved by venting into these cells. Such a system has the advantage of retaining the bulk of the radioactive debris within inert areas with a minimized probability of venting to the RCB. Sample calculations for such a system[53] indicate substantially reduced dynamic pressures and temperatures in the RCB.

It should be noted that several mechanisms exist to cause a rapid pressure rise in the RCB (given the postulate of a massive sodium spill). Hence, it may be preferable to use a containment/confinement system—in preference to either a single or double containment system—to allow controlled and filtered release.

C. COMPARISON OF CONTAINMENT AND CONTAINMENT/CONFINEMENT SYSTEMS

Although many combinations of failure modes must be considered before any absolute statements can be made regarding the validity of the dose calculations attained for any particular containment system, a comparison of the *relative* doses calculated for the three systems shown in Fig. 16-26 is instructive. For the calculated doses given in Reference 52, the single containment system corresponds to the case considered in Figs. 16-28 through 16-33. The double containment considered in the study has an inner low-leak containment over the reactor head area, whereas the containment/confinement system is characterized by a small inner low leakage containment surrounded by a high leakage outer confinement system. For the latter system, the atmosphere between the inner containment and the outer confinement (RCB) is at a slight negative pressure (inward leakage from the environment), and this atmosphere is pumped through a high exhaust stack.

Regulatory guidelines in the United States provide bases for maximum allowable radiological doses from a design basis nuclear accident. These

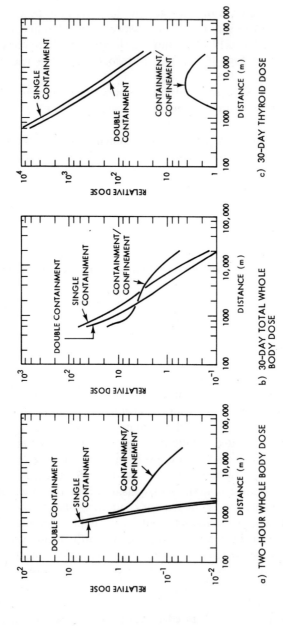

FIGURE 16-34. Dose comparisons for three containment systems (relative to a 2-hour whole body dose of 1 at 1000 m for the containment/confinement system).

guidelines define dose limits for whole-body exposure and for thyroid exposure from iodine for a two-hour period in an exclusion area and for a longer period in a low population zone.

Relative doses as a function of distance from the reactor are shown for three containment systems in Fig. 16-34, taken from the analysis in Reference 52. All doses are normalized to the two-hour whole-body dose at 1000m for the containment/confinement system. Comparison of these doses indicates the relative effectiveness of each containment system in reducing offsite doses.

Figure 16-34 (a) shows relative two-hour whole-body doses vs distance for the three systems. Note that the double containment provides only a marginal improvement over the single containment system, whereas the containment/confinement system is slightly inferior—especially at large distances. The containment systems appear somewhat better for the two-hour dose because venting is not required in these cases for about eight hours, whereas some filtered exhaust is released to the outer atmosphere in the containment/confinement system.

For the 30-day whole-body and thyroid doses depicted in Figs. 16-34 (b) and (c), the relative improvement of the double vs single containment configuration is still marginal. However, the containment/confinement system appears to represent a considerable improvement, particularly for the 30-day thyroid dose. The "S-shape" of the whole-body dose vs distance curve for the containment/confinement system derives from the continual discharge of low-level radionuclides, mostly noble gases. The much better thyroid characteristics of this system occur because of the filtration and cleanup of all discharged iodine particulates, plus high stack discharge. This marked improvement in 30-day thyroid dose is of considerable interest because this dose represents the limiting dose for low population distances in the majority of cases.

REVIEW QUESTIONS

16-1. In Table 13-8, maximum HCDA work energy values were listed for several reactors. What is the relationship of these energies to the calculations described in Chapter 16?

16-2. An energetic HCDA leading to core disassembly would result in expansion work by the fuel. Explain the meaning of this work, i.e., on what would the work be done, and what physical damage might result from this work?

16-3. What is calculated (i.e., the end result) in the Hicks-Menzies model?

16-4. If the spontaneous nucleation model for molten fuel/coolant interaction is valid, why is it unlikely that a large energetic interaction between UO_2-PuO_2 fuel and sodium can occur in a core disruptive accident in an LMFBR?

16-5. In a time-dependent expansion calculation, what is a "constraint?" Explain the difference between an acoustic constraint and an inertial constraint.

16-6. Energy released during disassembly is eventually partitioned into several different forms of energy. List these forms of energy.

16-7. What are the sources of heat that must be removed after an HCDA, and where might these sources be located?

16-8. List the ex-vessel phenomena, structures and devices that are considered in assessing post-accident heat removal in an LMFBR.

16-9. What mechanism is responsible for releasing water from concrete? What gases would be generated if sodium and core debris heated and chemically reacted with the concrete walls of the reactor cavity?

16-10. What problems can result from sodium fires in an LMFBR accident? List places where sodium fires are possible, and whether you would expect the fire to be a spray or a pool fire.

16-11. In an LWR, a large containment building is required to contain the pressure caused by the potential flashing of the high pressure water coolant in the event of a pipe break. A rise in pressure in the Reactor Containment Building (RCB) of an LMFBR might occur following an HCDA even though the coolant is subcooled during normal operation. List several potential phenomena that might cause a rise in pressure in the RCB in an LMFBR accident.

16-12. Explain the similarities and differences between the four types of containment described in Section 16-7.

16-13. (a) Explain Figs. 16-30, 16-31, and 16-32.
(b) Discuss your views on double containment, single containment, and containment/confinement based on Fig. 16-34 together with any other information that you may wish to consider.

16-14. What are nuclear aerosols and what is their significance?

16-15. What do the following computer codes calculate?

CACECO	PARDISEKO
COMRADEX	REXCO
ORIGEN	SOMIX

16-16. Make a list, in the sequence in which they might occur, of all the potential phenomena described in Chapter 16 for an energetic HCDA, starting after core disassembly.

PROBLEMS

16-1. Derive Eq. (16-3), using the approximations that the vapor acts as a perfect gas, h_{fg} is constant, v_l is negligible relative to v_g, and c_p for saturated liquid fuel is constant.

16-2. Derive Eq. (16-5), using the fact that $s_4 = s_5$ can be used as a starting point, together with the relationship, $s_{lg} = h_{lg}/T$. To evaluate $s_{l4} - s_{l5}$, integrate along the saturated liquid curve using the Gibbs equation

$$T\,ds = du + P\,dv = dh - v\,dP,$$

and recognize that, for a saturated liquid, $P\,dv \ll du$, or $v\,dP \ll dh$.

REFERENCES

1. M. Kirbiyik, P. L. Garner, J. G. Refling, and A. B. Reynolds, "Hydrodynamics of Post-Disassembly Fuel Expansion," *Nucl. Eng. and Des.*, 35, (1975) 441–460.
2. C. R. Bell, J. E. Boudreau, J. H. Scott, and L. L. Smith, "Advances in the Mechanistic Assessment of Postdisassembly Energetics," *Proc. of the Int. Mtg. on Fast Reactor Safety Technology, Vol. I*, Seattle, WA (1979), 207–218.
3. E. P. Hicks and D. C. Menzies, "Theoretical Studies on the Fast Reactor Maximum Accident," *Proc. of the Conf. on Safety, Fuels, and Core Design in Large Fast Power Reactors*, ANL-7120, (1965) 654–670.
4. A. M. Judd, "Calculation of the Thermodynamic Efficiency of Molten-Fuel-Coolant Interactions," *Trans. Am. Nucl. Soc.*, 13, (1970) 369.
5. A. Padilla, Jr., "Analysis of Mechanical Work Energy for LMFBR Maximum Accidents," *Nucl. Tech.*, 12, (1971) 348–355.
6. D. H. Cho, R. O. Ivins, and R. W. Wright, "Pressure Generation by Molten Fuel-Coolant Interactions Under LMFBR Accident Conditions," *Proc. of the Conf. on New Developments in Reactor Mathematics and Applications*, Idaho Falls, ID, CONF-710302, I (1971) 25–49.
7. H. Jacobs, "Computational Analysis of Fuel-Sodium Interactions with an Improved Method," *Proc. of the Int. Conf. on Fast Reactor Safety and Related Physics*, Chicago, IL, CONF-761001 III, (1976) 926–935.
8. M. Amblard and H. Jacobs, "Fuel Coolant Interaction; The CORRECT II UO_2-Na Experiment," *Proc. of the Int. Meeting on Fast Reactor Safety Technology, III* Seattle, WA (1979) 1512–1519.
9. H. Jacobs, M. F. Young, and K. O. Reil, "Fuel-Coolant Interaction Phenomena Under Prompt Burst Conditions," *Proc. of the Int. Meeting on Fast Reactor Safety Technology, III*, Seattle, WA, (1979) 1520–1528.
10. S. J. Board and L. Caldarola, "Fuel Coolant Interactions in Fast Reactors," *ASME Symposium on the Thermal and Hydraulic Aspects of Nuclear Reactor Safety*, ASME Winter Meeting, New York, NY, 1979.
11. S. G. Bankoff, "Vapor Explosions: A Critical Review," *Sixth International Heat Transfer Conference*, Toronto, Canada, 1979.

12. A. J. Briggs, T. F. Fishlock, and G. J. Vaughan, "A Review of Progress with Assessment of MFCI Phenomenon in Fast Reactors Following the CSNI Specialist Meeting in Bournemouth, April 1979," *Proc. of the Int. Meeting on Fast Reactor Safety Technology, III*, Seattle, WA (1979) 1502–1511.

13. D. R. Armstrong, F. J. Testa, and D. Raridon, Jr., "Molten UO_2-Sodium Dropping Experiments," *Trans. Am. Nucl. Soc., 13*, (1970) 660.

14. H. K. Fauske, "The Role of Nucleation in Vapor Explosions," *Trans. Am. Nucl. Soc., 15*, (1972) 813.

15. S. J. Board, R. W. Hall and R. S. Hall, "Detonation of Fuel Coolant Explosions," *Nature, 254*, (1975) 319–321. See also M. Baines, S. J. Board, N. E. Buttery, and R. W. Hall, "The Hydrodynamics of Large Scale Fuel Coolant Interactions," *Nuc. Tech., 49*, June 1980, 27–39.

16. S. A. Colgate and T. Sigurgeirsson, "Dynamic Mixing of Water and Lava," *Nature, 244*, (1973) 552–555.

17. H. K. Fauske, "The Role of Core-Disruptive Accidents in Design and Licensing of LMFBR's," *Nuclear Safety, 17*, (1976) 550–567.

18. M. Baines and N. E. Buttery, "Differential Velocity Fragmentation in Liquid-Liquid Systems," RD/B/N4643, Berkely Nuclear Laboratories, September 1979.

19. A. W. Cronenberg and M. A. Grolmes, "Fragmentation Modeling Relative to the Breakup of Molten UO_2 in Sodium," *J. Nucl. Safety, 16* (6), 1975.

20. E. L. Gluekler and L. Baker Jr., "Post Accident Heat Removal in LMFBR's," O. C. Jones and S. G. Bankoff, eds., *Liquid Metal Fast Breeder Reactors*, Vol. 2, The American Society of Mechanical Engineers, NY (1977) 287–325.

21. W. R. Wise and J. F. Proctor, "Explosion Containment Laws for Nuclear Reactor Vessels," NOLTR-63-140, U.S. Naval Ordnance Laboratory, 1965.

22. J. F. Kunze (ed.), "Additional Analysis of the SL-1 Excursion," USAEC Report IDO-19313 (TM-62-11-707), General Electric Company, 1962, and "Final Report of SL-1 Recovery Operation," USAEC Report IDO-19311, General Electric Company, Idaho Test Station, July 27, 1962.

23. J. F. Proctor, "Adequacy of Explosion-Response Data in Estimating Reactor-Vessel Damage," *Nucl. Safety, 8* (6), (1967) 565–572.

24. R. H. Cole, *Underwater Explosions*, Princeton University Press, 1948.

25. A. L. Florence, G. R. Abrahamson, and D. J. Cagliostro, "Hypothetical Core Disruptive Accident Experiments on Single Fast Test Reactor Model," *Nucl. Eng. and Des., 38*, (1976) 95.

26. D. D. Stepnewski, G. L. Fox, Jr., D. E. Simpson, and R. D. Peak, "FFTF Scale Model Structural Dynamics Tests," *Proc. of the Fast Reactor Safety Meeting*, CONF-740401-P2, Beverly Hills, CA (1974).

27. M. Saito and T. G. Theofanous, "The Termination Phase of Core Disruptive Accidents in LMFBR's," *Proc. of the Int. Meeting on Fast Reactor Safety Technology, III*, Seattle, WA (1979) 1425–1434.

28. H. Holtbecker, N. E. Hoskin, N. J. M. Rees, R. B. Tattersall, and G. Verzeletti, "An Experimental Programme to Validate Wave Propagation and Fluid Flow Codes for Explosion Containment Analysis," *Proc. of the Int. Mtg. on Fast Reactor Safety and Related Physics*, CONF-761001, Vol. III, Chicago, IL (1976) 1304–1313.

29. M. Egleme, N. Brahy, J. B. Fabry, H. Lamotte, M. Stievenart, H. Holtbecker, and P. Actis-Dato, "Simulation of Hypothetical Core Disruptive Accidents in Vessel Models," *Proc. of the Int. Mtg. on Fast Reactor Safety and Related Physics*, CONF-761001, Vol. I, (1976) 1314–1323.

30. D. J. Cagliostio, A. L. Florence, G. R. Abrahamson, and G. Nagumo, "Characterization of an Energy Source for Modeling HCDA's in Nuclear Reactors," *Nucl. Eng. and Des., 27*, (1974) 94.

31. J. P. Breton, A. Lapicote, A. Porrachia, M. Natta, M. Amblard, and G. Berthoud, "Expansion of a Vapor Bubble and Aerosols Transfer," *Proc. of the Int. Mtg. on Fast Reactor Safety Technology, III*, Seattle, WA (1979) 1445–1454.

32. Y. W. Chang and J. Gvildys, *REXCO-HEP: A Two-Dimensional Computer Code for Calculating the Primary System Response in Fast Reactors*, ANL-75-19, Argonne National Laboratory, 1975.

33. Chung-yi Wang, *ICECO-An Implicit Eulerian Method for Calculating Fluid Transients in Fast-Reactor Containment*, ANL-75-81, Argonne National Laboratory (1975).

34. Y. W. Chang, J. Gvildys, and S. H. Fistedis, "A New Approach for Determining Coolant Slug Impact in a Fast Reactor Accident," *Trans. Am. Nucl. Soc., 14*, (1971) 291.

35. A. H. Marchertas, C. Y. Wang, and S. H. Fistedis, "A Comparison of ANL Containment Codes with SNR-300 Simulation Experiments," *Int. Mtg. on Fast Reactor Safety and Related Physics*, CONF-761001, Vol. III, (1976) 1324–1333.

36. M. S. Kazimi and J. C. Chen, "A Condensed Review of the Technology of Post-Accident Heat Removal for the Liquid-Metal Fast Breeder Reactor," *Nucl. Tech., 38* (1978) 339–366.

37. M. J. Bell, *ORIGEN-The ORNL Isotope Generation and Depletion Code*, ORNL-4628, Oak Ridge National Laboratory (1973).

38. T. R. England, R. Wilezynski, and N. L. Whittemore, *CINDER-7: An Interim Report for Users*, LA-5885-MS, Los Alamos Scientific Laboratory, (1975).

39. R. O. Gumprecht, *Mathematical Basis of Computer Code RIBD*, DUN-4136, Douglas United Nuclear, Inc., 1968, and D. R. Marr, *A User's Manual for Computer Code RIBD-II, A Fission Product Inventory Code*, HEDL-TME 75-26, Hanford Engineering Development Laboratory, January 1975.

40. J. D. McCormack, A. K. Postma, and J. A. Schur, *Water Evolution from Heated Concrete*, HEDL-TME 78-8, Hanford Engineering Development Laboratory, February 1979.

41. J. A. Hassberger, *Intermediate Scale Sodium-Concrete Reaction Tests*, HEDL TME 77–99, Hanford Engineering Development Laboratory, March 1978. See also J. A. Hassberger, *Intermediate Scale Sodium-Concrete Reaction Tests with Basalt and Limestone Concrete*, HEDL-TME 79-55, Hanford Engineering Development Laboratory, September 1980; R. P. Colburn, et al., "Sodium Concrete Reactions, *Proc. of the International Meeting on Fast Reactor Safety Technology*, IV, Seattle, WA, August 1979, 2093.

42. L. Baker, F. B. Chenug, R. Farhadieh, R. P. Stein, J. D. Gabor, and J. D. Bingle, "Thermal Interaction of a Molten Core Debris Pool with Surrounding Structural Materials," *Proc. of the Int. Conf. on Fast Reactor Safety Technology*, I, Seattle, WA (1979) 389–399.

43. R. L. Knight and J. V. Beck, "Model and Computer Code for Energy and Mass Transport in Decomposing Concrete and Related Materials," *Proc. of the Int. Conf. on Fast Reactor Safety Technology*, IV, Seattle, WA (1979) 2113–2121.

44. *Liquid Metals Handbook*, NAVEXOS P-733 (Rev), Richard N. Lyon, ed., AEC, (1952).

45. J. Graham, *Fast Reactor Safety*, Academic Press, New York, 1971 (Chapter 4).

46. P. Beiriger, J. Hopenfeld, M. Silberberg, R. P. Johnson, L. Baurmash, and R. L. Koontz, *SOFIRE-II User Report*, AI-AEC-13055, Atomic International (1973).

47. L. Leibowitz, "Thermodynamic Equilibria in Sodium-Air Systems," *J. Nucl. Materials, 23* (1967) 233–235.

48. J. R. Humphreys Jr., "Sodium-Air Reactions as they Pertain to Reactor Safety and Containment," *Proc. of Second International Conference on Peaceful Uses of Atomic Energy*, Geneva, Vol. 22, (1958) 177.

49. T. S. Krolikowski, L. Leibowitz, R. E. Wilson, J. C. Cassulo, and S. K. Stynes, "The Reaction of a Molten Sodium Spray with Air in an Enclosed Volume Part I. Experimental Investigations," *Nuc. Sci. Eng., 38* (1969) 156–160. See also T. S. Krolikowski, L. Leibowitz, R. O. Ivins, and S. K. Stynes, same publication, "Part II. Theoretical Model," 161–166.

50. P. R. Shire, *Spray Code Users Report*, HEDL-TME 76–94, Hanford Engineering Development Laboratory (1977).

51. M. P. Heisler and K. Mori, *SOMIX-I Users Manual for the LBL-CDC 7600 Computer*, N707TI130045, Atomic International (1976).

52. S. E. Seeman and G. R. Armstrong, *Comparison of Containment Systems for Large Sodium-Cooled Breeder Reactors*, HEDL-TME 78-35, Hanford Engineering Development Laboratory, April 1978.

53. R. D. Peak, *User's Guide to CACECO Containment Analysis Code*, HEDL-TME 79-22, Hanford Engineering Development Laboratory, June 1979.

54. G. W. Spangler, M. Boling, W. A. Rhoades, C. A. Willis, *Description of the COMRADEX Code*, Atomic International, AI-67-TDR-108, 1967.

55. A. B. Reynolds and T. S. Kress, "Aerosol Source Considerations for LMFBR Core Disruptive Accidents," *Proc. of CSNI Spec. Meeting on Nuclear Aerosols in Reactor Safety*, Gatlinburg, TN, April 1980.

56. A. L. Wright, T. S. Kress, and A. M. Smith, "ORNL Experiments to Characterize Fuel Release from the Reactor Primary Containment in Severe LMFBR Accidents," *Proc. on CSNI Spec. Meeting on Nuclear Aerosols in Reactor Safety*, Gatlinburg, TN, April 1980.

57. R. S. Hubner, E. U. Vaughan, and L. Baurmash, *HAA-3 User Report*, AI-AEC-13038, Atomic International, 1973.

58. L. D. Reed and J. A. Gieseke, *HAARM-2 User's Manual*, BMI-X-665, Battelle Columbus Laboratory, 1975.

59. H. Jordan and C. Sack, *PARDISEKO-III, A Computer Code for Determining the Behavior of Contained Nuclear Accidents*, KFK-2151, Karlsruhe Nuclear Center (1975).

60. W. O. Schikarski, "On the State of the Art in Aerosol Modeling for LMFBR Safety Analysis," *Proc. of the Int. Conf. on Fast Reactor Safety and Related Physics*, CONF-761001, Chicago, IL, Vol. IV, (1976) 1907–1914.

61. M. Silberberg, Chairman, *Nuclear Aerosols in Reactor Safety*, A State-of-the-Art Report by a Group of Experts of the OECD NEA Committee on the Safety of Nuclear Installations (June 1979).

62. N. Rasmussen, Director, *Reactor Safety Study, An Assessment of Accident Risks in U.S. Commercial Nuclear Power Plants*, WASH-1400 (NUREG 75/014), U.S. Nuclear Regulatory Commission, 1975.

63. M. G. Piepho, *CRACOME Description and Users Guide*, HEDL-TME 80-56, Hanford Engineering Development Laboratory (1981).

PART V

GAS COOLED
FAST REACTORS

The Gas Cooled Fast Reactor (GCFR) represents a possible alternative to the Liquid Metal Fast Breeder Reactor (LMFBR). Whereas many earlier parts of the book are equally relevant to both the GCFR and the LMFBR (e.g., neutronic techniques and material properties), the use of a gaseous coolant in the GCFR results in certain design and safety considerations which are fundamentally different from the LMFBR. This concluding section of the book is addressed to such differences. Chapter 17 is focused on GCFR design features and Chapter 18 provides a review of relevant GCFR safety considerations.

CHAPTER 17

GAS COOLED FAST REACTOR DESIGN

17-1 INTRODUCTION

The Gas Cooled Fast Reactor (GCFR) is under development in the United States and Europe as an alternative to the Liquid Metal Fast Breeder Reactor (LMFBR). Gas cooling leads to a harder neutron spectrum than sodium cooling, thus allowing a higher breeding ratio and a lower doubling time relative to the LMFBR. The GCFR also has the potential for lower capital costs. Since helium is a chemically and neutronically inert single-phase gas which does not become radioactive, the GCFR does not require an intermediate heat transport loop, special cell liners or fire protection, or special cleanup and decontamination systems. A more extensive listing of the advantages and disadvantages of the GCFR system is given in Section 17-2 below.

Initial conceptual design work on the GCFR was originated in the United States at General Atomic (1962). Since that time, studies have included preliminary designs of a 300-MWe demonstration plant and a 1000-MWe commercial power plant. Essentially all of the studies prior to 1978 were based on downward coolant flow through the core. In 1979 the reference design was modified to provide upward coolant flow through the core in order to incorporate natural circulation cooling capability in the event of loss of all forced coolant circulation.

Since the late 1960s the program has proceeded with considerable international cooperation from the German national laboratories at Karlsruhe (GfK) and Jülich (KfA), the German industrial firm Kraftwerks Union (KWU), and the Swiss National Laboratory at Wuerenlingen (EIR). An independent GCFR study program of a commercial size GCFR plant was begun in the late 1960s under the auspices of the European Association for Gas Cooled Breeder Reactors (GBRA).

Several potential gas coolants, including steam, carbon dioxide and helium, were examined prior to the selection of helium. Advantages and disadvantages were noted for each coolant. Steam was rejected due to cladding compatibility problems, positive coolant reactivity effects, and reduced breeding performance. Several factors led to the rejection of carbon dioxide, including the higher pressure drop and associated forces across components, increased acoustic loadings, and the economic penalty assessed to providing the increased primary coolant pumping power required. A novel concept based on N_2O_4 coolant is being pursued in the Soviet Union. This cycle involves partial dissociation of N_2O_4 into NO_2 during heating, and then recombination during the turbine expansion phase. It is possible that N_2O_4 could be liquified in the cold-leg return, thus reducing pumping power requirements.

Because of the predominance of helium as the base GCFR coolant, the remainder of this chapter will be devoted to a description of the helium-cooled GCFR. Since it is assumed at this stage of the text that the reader is reasonably familiar with LMFBR systems, the GCFR will be presented from the perspective of how it *differs* from the LMFBR. The many features which are similar will not be emphasized.

Our discussion will begin with an overview listing of advantages and disadvantages of helium as a coolant. Given this background, the broad considerations of overall system aspects are introduced and then attention is focused ever increasingly on the fine structure aspects of core design. Safety considerations unique to the GCFR are covered in Chapter 18.

17-2 FEATURES OF GAS COOLING

Of all the basic design choices to be made in developing a fast breeder reactor system, selection of the coolant probably has the largest influence on the resulting plant configuration and performance. This is particularly evident when comparing the implications of helium vs sodium cooling systems. Table 17-1 contains a condensed set of advantages and disadvantages of employing helium rather than sodium as the cooling medium. The table is constructed to emphasize design related factors first, followed by safety related considerations, although substantial overlap is readily apparent.

A. ADVANTAGES OF HELIUM COOLING

A principal advantage of helium is that it results in less neutron absorption and moderation than sodium or other coolants. The lower absorption increases the fraction of neutrons available for breeding, while the harder

TABLE 17-1   Relative Advantages of Helium vs Sodium Coolant

	Advantages	Disadvantages
Design Related	• High breeding ratio, low doubling time: Good fuel conservation; small burnup reactivity swing, low control requirements	• High neutron leakage: Commercial core size required for good economics
	• Nonradioactive coolant: No intermediate loop (lower capital cost); maintenance and inspection access	• Relatively poor heat transfer properties: Cladding surface roughening required; increased pumping power required
Safety Related	• Low coolant neutronic interaction: More room to accommodate metal swelling; small coolant coefficient of reactivity	• High pressure required for economic power generation (8-10 MPa); Low coolant heat capacity; high pumping power required for gas
	• Inert and transparent coolant: No corrosion; nonflammable coolant; remote visual inspection	• Reduced heat removal capability upon depressurization: Natural circulation inadequate for depressurized decay heat removal; reliable forced circulation systems required
	• Single phase coolant: No abrupt loss of cooling potential; no potential for mechanical damage from flashing coolant; no interruption of natural circulation possible	• Lower negative Doppler constant: Slower response to mitigate reactivity transients

spectrum significantly increases the total number of neutrons produced per fission. The advantage of the harder spectrum is most pronounced for the Pu/U fuel cycle since the neutrons per fission in Pu and fast fission rate in ^{238}U are quite dependent on the incident neutron energy. This neutronic advantage is less obvious in early plant designs with lower circulator power and low fuel volume fractions since the overall breeding ratio decreases with decreasing fuel volume fraction. Advanced designs with higher circulator power and fuel volume fractions have higher breeding ratios and shorter doubling times. For FBR core designs with approximately the same fuel volume fraction a gas-cooled core will have a higher breeding ratio, require a larger fissile inventory (due to increased leakage) and have a shorter doubling time than a sodium cooled fast reactor. This has obvious advantages regarding fuel resource utilization.

A second major advantage is that helium does not become radioactive. Consequently, the intermediate coolant loop normally present in the LMFBR (to isolate radioactive primary coolant from the steam generator) can be omitted. The principal advantage of this simplification is lower capital cost.

Helium has a very small neutron absorption cross-section, and at the atom densities present, the neutron scattering cross-section of helium is likewise small. This feature of helium neutronic insensitivity allows the core lattice to be more open than in an LMFBR. Such a configuration can be used to advantage in accommodating metal swelling. An additional consequence of helium neutronic insensitivity is that the He-cooled GCFR does not exhibit the strong positive reactivity coefficient typified by sodium (discussed in Chapter 6).

Since helium is inert, incompatibility problems between coolant and structural or fuel materials are essentially absent. This allows the potential for higher exit coolant temperature. The transparent helium gas also allows visual observation of components, presumably resulting in more flexibility and ease in maintenance. A safety feature of the helium-inertness property is that the coolant does not react chemically with either water or air, a problem inherent in the LMFBR system.

Finally, since helium is always employed in the gas phase, there is no possibility of an abrupt change in cooling characteristics, as is the case for sodium flashing from the liquid to the gaseous phase. The molten fuel/coolant interaction (MFCI), was noted in Chapter 16 to provide concern for the LMFBR during postulated unprotected accident conditions. There is no mechanism for a rapid thermal energy transfer from molten fuel to coolant to cause a phase change in a GCFR. Consequently, there is no potential for the mechanical damage resulting from a postulated major accident condition to be augmented by flashing coolant.

Two other advantages of helium cooling, not indicated in the table, deserve brief mention. Because helium is always utilized in the gaseous phase, there is the design potential of employing a direct cycle, i.e., ducting the primary helium to the turbine without resorting to the intermediate step of generating steam. Though clearly attractive from a capital cost point of view, one fundamental problem of the direct cycle is incorporating the turbine as an integral part of the high pressure primary system.

The other advantage is that fission gas escaping through a cladding defect or rupture site would not appreciably alter the heat transport capabilities for a helium-cooled system. A similar circumstance in an LMFBR could result in displacement of liquid coolant, possibly partially starving fuel pins of adequate coolant in the immediate vicinity of the rupture site. In perspective, however, it is noted in Chapter 14 that this problem appears to have minimal implications even for an LMFBR system.

B. DISADVANTAGES OF HELIUM COOLING

We now touch on the principal disadvantages of using helium coolant relative to sodium. Whereas the lack of neutronic interaction of helium was previously noted to have advantageous features, this property also has a negative design feature for small systems. A low power GCFR (of the prototype range of 300 MWe or less), with its relatively small core, has substantially higher core neutron leakage than an LMFBR of the same power level. Hence, favorable core economics can only be obtained for a high power (~ 1000 MWe or more) GCFR for which the core is relatively large and the neutron leakage is reduced.

Because of the inherently poorer heat transfer properties of helium, relative to sodium, it is necessary to induce turbulence by roughening the surface on the fueled portion of the fuel pin cladding exterior walls. The result of such required roughening is to increase the coolant axial pressure drop over the core and to complicate the thermal-hydraulic design. Another direct result of the poor heat transfer properties of helium is the high pumping power required for adequate circulation (cf. Chapter 11-4D).

Probably the biggest disadvantage of the GCFR system is that helium must be pressurized up to the 7-10 MPa (1050 to 1500 psi) range in order to provide sufficient cooling for the high power density cores of interest for economic power generation. High system pressures are required, in part, to compensate for the low heat capacity of helium. Sodium systems can operate at atmospheric conditions (the only pressurization necessary being that required in the inlet plenum to get adequate flow). The direct consequence of the need for high pressure helium is to require assurance that depressurization cannot occur rapidly. Indeed, it is the concern about the consequences of depressurization—despite the remote probability of the occurrence—which has provided the major safety concern for the GCFR concept. This requirement of reliable system pressurization is the principal driving force for employing a prestressed concrete reactor vessel (PCRV), as described in Section 17-3.

Due to the potential for accidental depressurization of the GCFR, provisions must be made to cool the core at the reduced pressures which result. In addition, reliable core cooling must also be provided for planned depressurizations such as equipment maintenance and refueling. At the low pressures expected for both operational and accidental depressurizations, natural circulation cannot provide adequate heat removal. Therefore, reliable forced convection systems must be provided to accommodate depressurized operation. Natural circulation is effective in LMFBR's for all accident and operational modes.

Finally, the Doppler coefficient, which is clearly a parameter of considerable safety importance, has a smaller negative value for the GCFR relative to the LMFBR. There are two factors leading to this result. First, the neutron spectrum for a GCFR is harder than for a LMFBR (cf. Fig. 17-1), thus leaving fewer neutrons in the Doppler resonance region. Second, the GCFR has a larger fissile mass fraction which is required in order to achieve criticality in the more loosely coupled core. (The advantage for the LMFBR in the Doppler coefficient is lost, however, when the sodium-out Doppler coefficient is relevant.)

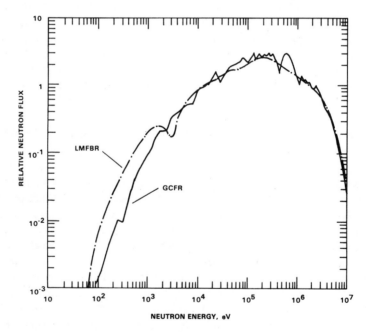

FIGURE 17-1. Comparison of flux spectra of a GCFR and an LMFBR. (GCFR spectrum supplied by C. J. Hamilton of General Atomic Co. July, 1980; LMFBR spectrum taken from Fig. 4-10.)

Whereas the above discussions provide no clear-cut choice between the GCFR and LMFBR systems, significant advantages of the GCFR are evident. Perhaps the strongest barrier to widespread adoption of the GCFR concept is that there is very little actual experience with such systems. The heavy international commitment to the LMFBR has uncovered several problems, although none appears to be without a solution. If a similar commitment were to be given to the GCFR, it is likewise expected that problems not yet evident would surface. Nonetheless, there are enough attractive features inherent to the GCFR concept to warrant careful study of such systems.

17-3 SYSTEM DESIGN

The overall features of a GCFR plant are presented below for a demonstration plant design delivering 360 MWe. A plant layout is presented in Fig. 17-2. The plan view and side view are presented in parts a and b, respectively.

Figure 17-3 illustrates the nuclear steam supply system of a proposed demonstration GCFR. It is readily apparent from this illustration that the primary coolant system (core, main steam generator and auxiliary heat exchanger, main and auxiliary circulators, etc.) is located entirely within the massive prestressed concrete reactor vessel (PCRV) structure.

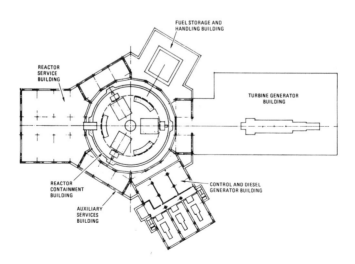

FIGURE 17-2a. Plan view GCFR plant layout.

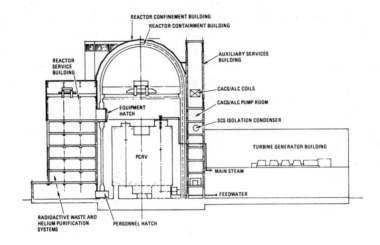

FIGURE 17-2b. Side view GCFR plant layout.

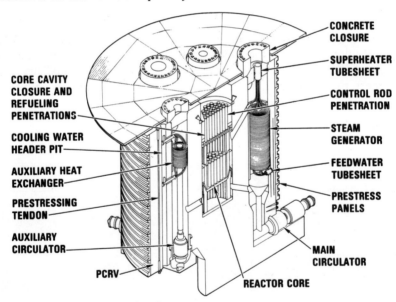

FIGURE 17-3. GCFR nuclear steam supply system (300 MWe).

A. REACTOR VESSEL

In the GCFR, high primary system helium pressures are required to provide the operating conditions required for economical electric power generation. Typical GCFR primary system pressures are in the range of 7 to

10 MPa (1050 to 1500 psi), which is comparable to that of LWR's. In order to keep the possibility of sudden system depressurization to an acceptably low value, the prestressed concrete reactor vessel (PCRV) concept has been adopted. This system, originally developed for the High Temperature Gas Reactor (HTGR), consists of a massive block of concrete which is prestressed by longitudinal tensions and by circumferential cable wrapping so as to maintain the PCRV in compression at all times. The PCRV for the demonstration GCFR is approximately 32 m high by 33 m in diameter.

The reactor core and its associated structural support and shielding components are located in the central PCRV cavity. The three main cooling loops, each of which includes a steam generator and a helium circulator, are located in three peripheral cavities in the vessel walls. Three auxiliary loops are contained in smaller cavities located in the PCRV walls between the steam generator cavities. Each of these cavities has a concrete closure plug at the top which can be opened for initial installation of components and for their repair and replacement if required. The inlet to the auxiliary heat exchanger is via a circumferential duct from the adjacent steam generator inlet plenum. The purpose for this design, as opposed to directly ducting hot gas from the core outlet through the PCRV, is to preserve the temperature profile when transferring from forced circulation in the main loop to natural circulation in the auxiliary loop.

All cavities and ducts within the PCRV are lined with a 13 mm thick steel liner to provide leak-tightness. This liner is protected from the high helium temperatures by a specially engineered insulation system on the coolant side, and by cooling water circulating in embedded piping on the concrete side. Liner stresses are compressive and they are relatively low because the effect of helium pressure is more than offset by the effect of the prestressing system.

Design criteria typically require the PCRV to accommodate a steady-state pressure twice the maximum operating cavity pressure. Penetration seals are sized so that the primary make-up system is capable of providing sufficient helium to replace that lost through leaks. In particular, even during a low probability seal failure event, primary system depressurization via failed seals is expected to be sufficiently slow to assure successful operation of the decay heat removal system.

B. HEAT TRANSPORT SYSTEMS

The schematic shown in Fig. 17-4 illustrates the GCFR heat transport systems. The normal operating or power generation system consists of the primary coolant system using high pressure helium, the steam and feedwater system and the plant heat rejection system. The primary coolant system uses

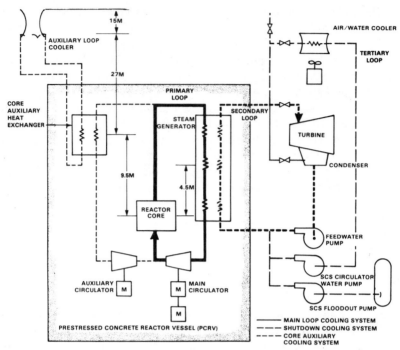

FIGURE 17-4. GCFR heat transport systems.

motor-driven centrifugal compressors to circulate helium upward through the core where it receives nuclear heat and downward through the steam generators where it transfers heat to the entering feedwater to produce superheated steam for use in the main turbine generator. The primary coolant system and all of its associated components are housed within the PCRV.

The steam and feedwater system starts with condensate supplied from the main condenser mounted below the main steam turbine. The feedwater delivered by the steam turbine driven boiler feedpumps then flows through a series of feedwater heaters which utilize steam extracted from the main turbine at various bleed points to heat the feedwater before entering the three steam generators. The steam generators are of the integral once-through type with helical coils. Water flows upward, boils and becomes superheated steam before exhausting at the top enroute to the main steam turbine-generator.

The plant heat rejection system includes a circulating water system (not shown in Fig. 17-4), an evaporative cooling tower, a water makeup system and a cooling tower blowdown system. The principal difference between the power generation systems for the GCFR and the LMFBR (cf. Fig. 12-1) is the lack of an intermediate loop. This intermediate loop is not required for the GCFR because the primary loop helium does not become radioactive.

There are three separate systems for residual heat removal in the GCFR:

(1) Steam bypass to the condenser using the normal power conversion heat transfer system components,

(2) Operation of the three Shutdown Cooling System (SCS) loops which make use of the steam generators and main helium circulators (driven by pony motors) with safety-class feed and heat dump components, and

(3) Operation of the three Core Auxiliary Cooling System (CACS) loops which employ separate safety-class circulators, heat exchangers and heat dump circuits. The CACS system is also designed to transfer residual heat from the core by natural convection to the ultimate heat sink, ambient air.

Steam bypass is used for residual heat removal when the normal heat transfer system components are operable. It is accomplished by opening the turbine bypass valve and closing the turbine stop valve, diverting steam through desuperheaters to the main condenser. Helium coolant circulation is continued by the main circulators and the steam generators are fed by the main feedpumps. For longterm cooldown, necessary steam supplies are provided from auxiliary boilers.

If the normal heat transfer system components are unavailable, plant cooldown is automatically initiated in the safety-class Shutdown Cooling System, as shown in Fig. 17-4. In the SCS mode of operation, circulation in each of the three main helium loops is provided by the main helium circulators driven by variable speed pony motors. Heat is removed by each steam generator to separate closed loop water-to-air cooling systems. The steam generators are flooded out by the SCS floodout pumps, and long term residual heat removal continues via water recirculation. The condensate produced is returned to the steam generator inlet by the SCS circulating water pumps. Water in the tank, which is the ultimate heat sink, boils and is vented to the atmosphere. The tank holds enough water for about 20 minutes operation before makeup is required.

Should both the above systems be unavailable, plant cooldown is accomplished using the Core Auxiliary Cooling System, (CACS), as shown in Fig. 17-4. The CACS is comprised of three independent loops which are also independent of the SCS. Each loop consists of a Core Auxiliary Heat Exchanger (CAHE), an auxiliary circulator driven by a variable speed electric motor, and a core auxiliary cooling water loop. The water loop consists of a pressurized surge tank, circulating water pumps, and an auxiliary loop cooler which is an air-blast type heat exchanger. The CACS is designed to provide abundant core cooling following all design basis events, including the design basis depressurization of the primary cooling system.

Additionally, if power to all active components is unavailable, the heat transfer components within the CACS are located at sufficient elevation differences such that natural circulation will transfer heat from the core all the way to the ultimate heat sink (atmospheric air), providing the PCRV is pressurized.

C. MAJOR COMPONENTS

The major components in the power conversion system for the GCFR include the main circulators and the steam generators. Figure 17-5 contains a cutaway of a demonstration plant main circulator design. Each of the three main helium circulators is rigidly mounted with closures at the bottom of the prestressed concrete reactor vessel, below the steam generator cavities. They are driven by electric motors horizontally located outside the PCRV. Helium leaving the steam generator enters the circulator and is accelerated by a centrifugal impeller, passes into a series of pipe diffusers, and then discharges into a plenum surrounding the circulator. Helium from this plenum then flows through the cold duct to the core inlet plenum. A check valve that acts as an isolation valve is located in the circulator inlet duct.

Figure 17-6 illustrates the integral steam generator employed for a GCFR demonstration plant. The steam generators are located in cavities in the prestressed concrete reactor vessel. Hot helium enters at the top of the steam generator, flows downward on the shell side of the tube bundle, gives up heat to water and steam, and exits at the bottom. The water enters through a feedwater tubesheet located at the bottom, flows upward inside the tubes and exits at the top through a superheater tubesheet.

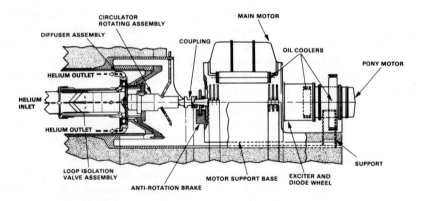

FIGURE 17-5. GCFR main circulator.

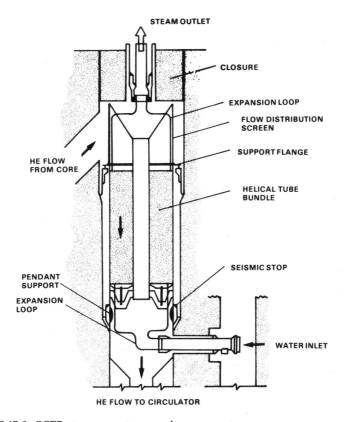

FIGURE 17-6. GCFR steam generator general arrangement.

The steam generator is supported at the top by a flanged connection to the PCRV liner and is provided with a redundant support structure and a lateral seismic restraint at the bottom. A flow distribution device at the inlet ensures uniform helium flow. Helical coiled tubes, which make up the heat transfer section of the steam generator, are threaded through perforated tube support plates which transmit dead weight and seismic loads out to the supporting shrouds. A steam generator of approximately 350 MW thermal capacity may be approximately four metres in diameter and eight metres long.

17-4   REACTOR CORE DESIGN

The reactor core design of an upflow GCFR is similar in many ways to that of an LMFBR. For a comparable power output the GCFR core tends to be slightly larger since a more open lattice is required for cooling. The similarity of the GCFR core design to that of an LMFBR is not by coincidence, but rather by intent. The GCFR design is based upon utilizing the fuel technology developed in the LMFBR programs to minimize fuel development costs. As a consequence, the fuel design utilizes the same fuel and structural materials operating under similar temperature and irradiation conditions typical of LMFBR fuel designs. Mixed oxide fuel is employed for the active core, although either carbide or possibly metal fuel* could be used. Depleted UO_2 is used in the axial and radial blankets, but ThO_2 could also be used as the fertile material. Twenty-percent cold-worked type 316 austenitic stainless steel is being specified for the cladding, duct, and associated structural components based upon the extensive experience with this material. Alternative structural materials are also being evaluated. These include the ferritics, other austenitic stainless steels, and high nickel alloys.

A principal geometric difference between a GCFR and LMFBR core is the spacing between fuel pins. The fuel pin spacing in a GCFR is determined by a combination of factors, the most significant being a trade-off between pumping power and core neutronic performance. Table 17-2 includes typical volume fractions for the two types of cores. Whereas the GCFR requires a much more open lattice for heat transfer considerations, this turns out to be an advantage in accommodating void swelling for the core pin cladding and duct materials.

TABLE 17-2 Typical Core Volume Fractions

	GCFR	LMFBR
Fuel	0.30	0.45
Coolant	0.55	0.40
Cladding	0.05	0.15
Structure	0.10	

Figure 17-7 contains a core schematic for a GCFR design of approximately 360 MWe. The general arrangement of the core configuration is very much like that of a typical homogeneous LMFBR. The central core region is surrounded by a radial blanket region of fertile material. Fertile blanket material is also included in the fuel assemblies, above and below the active core. Reflector/shield assemblies are located beyond the radial blanket

---

*As pointed out in Chapter 2, metal fuel may impose heat fluxes too high for adequate cooling by a gas. However, metal fuel could probably be used in the GCFR if the linear heat rate was reduced from that characteristic of a metal fueled LMFBR.

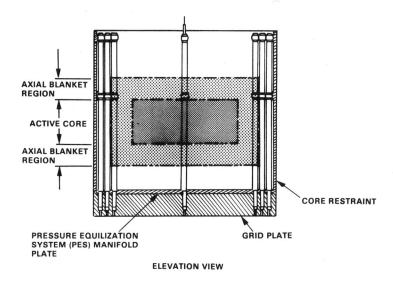

AXIAL BLANKET
REGION

ACTIVE CORE

AXIAL BLANKET
REGION

CORE RESTRAINT

PRESSURE EQUILIZATION
SYSTEM (PES) MANIFOLD
PLATE

GRID PLATE

ELEVATION VIEW

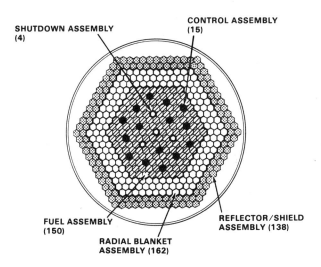

SHUTDOWN ASSEMBLY
(4)

CONTROL ASSEMBLY
(15)

FUEL ASSEMBLY
(150)

REFLECTOR/SHIELD
ASSEMBLY (138)

RADIAL BLANKET
ASSEMBLY (162)

PLAN VIEW

FIGURE 17-7. GCFR general core configuration. (NOTE: Numbers in parenthesis indicate the number of assemblies.)

TABLE 17-3   Typical 367 MWe GCFR Design Parameters

System Parameters (See also Table 17-4 and 17-5.)

Reactor thermal power, MWth	1088	
Net electrical power, MWe	367	
PCRV operating pressure, MPa	10.5	(1523 psia)
Core pressure drop, kPa	183	(26.5 psi)
Core inlet temperature, °C	298	(568°F)
Core outlet temperature, °C	530	(986°F)
Maximum cladding temperature, °C	750	(1380°F)
Coolant flow rate, kg/s	949	(2093 lb/s)

Core Design

Number of fuel assemblies	150
Number of control assemblies	15
Number of shutdown assemblies	4
Number of radial blanket assemblies	162
Number of reflector/shield assemblies	138
Total number of core assemblies	469
Assembly cross-section	hexagonal
Assembly length, mm	4900
Assembly across flats OD, mm	201
Interassembly gap, mm	7.0
Assembly pitch, mm	211.0
Load pad gap, mm	0.8
Duct wall thickness, mm	4.0

Fuel Assembly

Active core length, mm	1200
Axial blanket length, mm	600 each end
Number of fuel pins/assembly	265
Fuel pin spacer system	spacer grid
Number of spacer grids	12
Fuel pin diameter, mm	8
Fuel pin pitch, mm	11.5
Pin-to-duct gap, mm	2.48

Blanket Assembly

Blanket length, mm	2100
Number of blanket rods/assembly	61
Blanket rod spacer system	wire wrap
Blanket rod diameter, mm	21.7
Blanket rod pitch, mm	23.25

Control and Shutdown Assemblies

Absorber rod length, mm	1200
Number of absorber rods/assembly	61
Absorber rod spacer system	spacer grid
Absorber rod diameter, mm	16.8
Absorber rod pitch, mm	20.9

Shield Assembly

Shield length, mm	3200
Number of shield rods/assembly	19
Shield rod spacer system	wire wrap
Shield rod diameter, mm	40.1
Shield rod pitch, mm	41.35

assemblies to protect the core restraint structures. Shielding material is also provided in the core assemblies below the lower axial blanket to protect the core support grid plate and above the upper axial blanket to protect the control rod guide structures. More shielding is required in a GCFR than in an LMFBR due to the lower coolant density.

A cross-section view of each of the core assemblies is shown in Fig. 17-8 and design configuration parameters are given in Table 17-3. The pin sizes used in each of the assembly types (fuel, blanket, control, and shield) are comparable to those used in LMFBR designs. Figure 17-8 also illustrates the load pads (stand-off ribs) used to provide core restraint at two axial levels above the core support structure (cf. Chapter 8-6C).

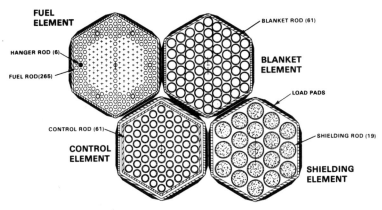

FIGURE 17-8. Fuel, control, blanket, and reflector shield assemblies. (NOTE: Numbers in parenthesis indicate the number of pins in the respective assemblies.)

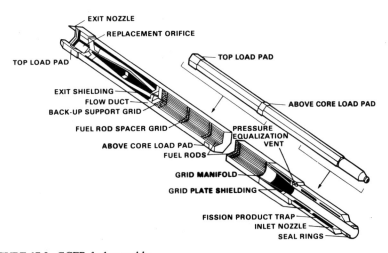

FIGURE 17-9. GCFR fuel assembly.

Figure 17-9 contains a cutaway of a GCFR fuel assembly. A fixed area replaceable orifice is included in the exit nozzle to maintain uniform outlet temperatures from each assembly. The lateral spacing of the fuel pins is maintained by means of spacer grids. Spacer grids are used rather than a wire wrap system due to the relatively large pitch-to-diameter ratio of the fuel pins. The use of spacer grids also provides greater design freedom than in a wire wrap system in the sizing of the edge gap between the outer pins and the flow duct. In a wire wrap system the conventional practice is to provide the same gap between the pins and the duct as used between adjacent pins (100% edge gaps). Since the edge flow channels are different from the interior channels, a somewhat smaller edge gap is more optimal. In the GCFR, representative edge gaps are in the 50 to 80% range.

17-5 FUEL PIN DESIGN

The fuel pins of the GCFR are similar to the fuel pins of the LMFBR, except for the cladding surface roughening and the pressure equalization system. Figure 17-10 illustrates a typical GCFR fuel pin and the associated axial temperature distribution.

The cladding surface of the GCFR fuel pins is roughened over the entire core length to enhance the heat transfer. Whereas it was recognized early in the development of the GCFR that the thermal performance of the GCFR could be improved by surface roughening, such a design measure directly increases the cooling system pressure drop. Hence, it has become useful to define a performance index, I, which relates these two processes as

$$I = St^3/f ,$$

where

$$St = Stanton \ number \ (cf. \ Eq. \ 17\text{-}1)$$
$$f = friction \ factor.$$

The performance index is proportional to the ratio of the thermal power to the pumping power. Hence, it is desirable to maximize I.

For a typical two-dimensional roughness,[1] the Stanton number is about twice that for a smooth surface and the friction factor is about four times that for a smooth surface. Hence, the performance index for a rough surface is about twice that of a smooth surface. A three-dimensional roughening with a performance index of about 4 times that of a smooth surface has also been proposed for the GCFR.[2]

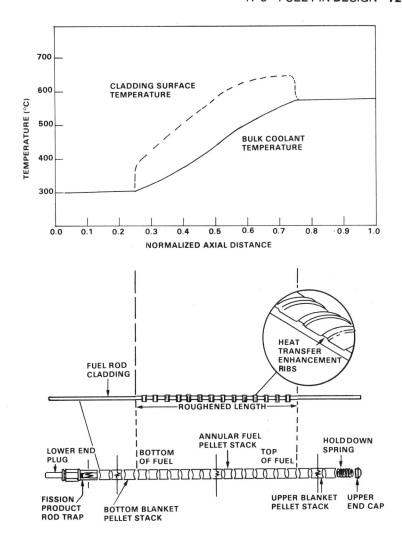

FIGURE 17-10. A typical GCFR fuel pin and associated axial temperature distribution.

In addition to improving the thermal performance of the GCFR, surface roughening reduces the temperature difference between the grid spacers and the fuel pins, thus mitigating the problems associated with radiation induced differential metal swelling.

Several designs for cladding surface roughening have been developed, with the basic goal of keeping the cladding hot spot below an acceptable limit while minimizing the core pressure drop. The design most seriously considered is a ribbed design as illustrated in the inset of Fig. 17-10. The design is characterized by a rib height, the rib width, and the rib pitch, i.e.,

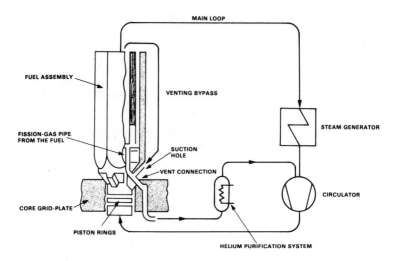

FIGURE 17-11. Schematic diagram of GCFR Pressure Equalization System (PES).

distance between ribs. The heat transfer correlations for such designs are discussed in the following section.

Pressure inside the fuel pins is equalized to that of the reactor coolant by collective venting of fission gas through the fission product trap and through a vent manifold to the helium purification system, as shown in Fig. 17-11. This pressure equalization relieves the cladding from mechanical stresses due to external gas coolant pressure and internal fission product gas pressure. This feature also limits the release of activity from breached pins into the reactor coolant and provides a means for detecting and locating fuel elements with failed cladding. This is accomplished by using radioactivity monitors on the vent lines from separate groups of elements. The pressure equalization feature also eliminates the need for a large fission gas plenum in the fuel pin, such as is necessary in the LMFBR fuel pin design.* The use of pressure equalized and vented fuel pins in the GCFR eliminates one of the primary failure mechanisms in LMFBR fuel pins, namely, creep rupture of the cladding due to internal pressure.

17-6 CORE THERMAL HYDRAULICS

Considerable attention must be given the thermal hydraulic design of a GCFR. This aspect of the design is relatively more important than for the

---

*As noted in Chapter 11-2B, vented pin designs are also being considered for the LMFBR.

LMFBR for at least two reasons:

(1) .The temperature drop from the cladding outer surface to the coolant is much larger for the GCFR (since sodium has considerably superior heat transport properties relative to helium). Hence, there is a strong incentive to minimize this temperature drop.

(2) Surface roughening on the cladding for a GCFR leads to turbulent flow, but laminar flow prevails for low flow conditions and must likewise be understood. Furthermore, the thermal hydraulic conditions may be considerably different for the smooth vs roughened portions of the cladding.

The Peclet number is frequently used for liquid metal heat transfer correlations (Chapter 9-3). Much of the thermal hydraulic work for gas-cooled systems is correlated according to the Stanton number, defined as

$$St = \frac{Nu}{Re\,Pr}. \tag{17-1}$$

Given the Reynolds number, Re, and the Prandtl number, Pr, the Nusselt number, Nu, can then be used directly to obtain the heat transfer coefficient, h, via Eq. (9-28).

Because of the relative importance of determining accurate heat transfer to the coolant in a GCFR system, considerable detail will be given below. However, the reader should be cautioned that correlations tend to change as new data become available, particularly for turbulent flow conditions. Also, the correlations given are only for interior channels (cf. Fig. 9-13). Refinements are necessary for edge (wall) channels and corner channels, as well as for the effects of grid spacers.[2]

In addition to the complications mentioned above, hot channel factors must also be included. These procedures, however, are very similar to those used in an LMFBR system (cf. Chapter 10-4).

The discussion below is divided into laminar and turbulent flow considerations, followed by a brief section to guide the analyst in choosing the appropriate correlation schemes. The final section illustrates thermal hydraulic results for a typical GCFR assembly.

A. LAMINAR FLOW

The pressure drop across a channel can be expressed according to Eq. (10-1), i.e.,

$$\Delta p = f\frac{L}{D_e}\frac{\rho V^2}{2},$$

where f is the friction factor. For laminar flow in a gas system, the friction factor can be expressed as

$$f_l = \frac{K}{Re}\left(\frac{T_w}{T_b}\right),\tag{17-2}$$

where K, T_w and T_b represent the channel shape factor and the wall and bulk coolant temperatures, respectively. Figure 17-12 contains a plot of K vs the pitch-to-diameter ratio.[3] For the principal range of interest for the GCFR (P/D = 1.3 to 1.5), the channel shape factor can be expressed[2] as

$$K = -128 + 260\left(\frac{P}{D}\right) - 60\left(\frac{P}{D}\right)^2.\tag{17-3}$$

For fully developed laminar flow, the Nusselt number for helium can likewise be determined as a function of only the pitch-to-diameter ratio. The results[4] are also included in Fig. 17-12. For the P/D range of 1.3 to 1.5, the appropriate correlation[2] can be written as

$$Nu_\infty = -13.7 + 24.1\left(\frac{P}{D}\right) - 5\left(\frac{P}{D}\right)^2.\tag{17-4}$$

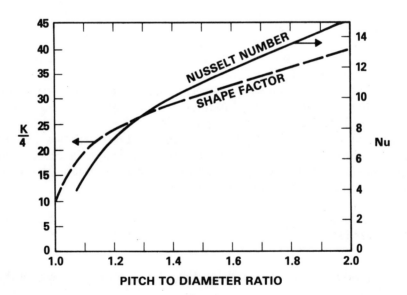

FIGURE 17-12. Shape factor and Nusselt number as a function of pitch-to-diameter ratio for laminar helium flow.

It should be noted that the above expression is for *fully developed* laminar flow. For a typical GCFR assembly, entrance effects may be significant nearly halfway through the assembly. These entrance effects are described in terms of the Graetz number, which is defined as

$$Gz = \frac{Re\,Pr}{x/D_e}, \tag{17-5}$$

where x and D_e represent the thermal entrance length and hydraulic diameter, respectively. The actual Stanton number at any given axial position, x, can then be determined by a curve fit to the numerical solution of Reference 5, namely,

$$St_x = St_\infty \left(1 + \frac{1.99}{Gz^{-0.506}\exp(-0.041\,Gz)}\right). \tag{17-6}$$

The local heat transfer coefficient is then determined by combining Eqs. (17-6), (17-5), (17-4), (17-1), and (9-28).

B. TURBULENT FLOW

Heat transfer correlations for turbulent flow are of interest for both the smooth and the roughened positions of the fuel pin (cf. Fig. 17-10). Hence, separate attention is given in the discussions which follow.

Smooth Tubes

The recommended friction factor for turbulent flow in smooth tubes is

$$f'_t = 1.04(0.0056 + 0.5\,Re^{-0.32}), \tag{17-7}$$

where the multiplier (1.04) is a bundle enhancement factor[6] and the quantity in parentheses represents a correlation[7] for circular tubes. This correlation is valid for $Re \geqslant 10\,000$.

The Nusselt number is given as[6,8]

$$Nu = Nu_{cir}[1 - \phi(Pr)]\left[\frac{D_{eq}}{D}\right]^{n(Pr)}\left(\frac{T_w}{T_{in}}\right)^{-0.2}, \tag{17-8}$$

where

$$\text{Nu}_{cir} = \frac{(f'/8)\,\text{Re}\,\text{Pr}}{\overline{K} + 12.7\sqrt{f'/8}\,(\text{Pr}^{2/3} - 1)}$$

$$f' = (1.82\,\ln\text{Re} - 1.64)^{-2}$$

$$\overline{K} = 1.07 + \frac{900}{\text{Re}} - \frac{0.63}{1 + 10\,\text{Pr}}$$

$$\phi(\text{Pr}) = \frac{0.45}{2.4 + \text{Pr}}$$

$$D_{eq} = \sqrt{\frac{2\sqrt{3}}{\pi}}\;P$$

$$P = \text{pitch}$$

$$D = \text{rod diameter}$$

$$n(\text{Pr}) = 0.16\,\text{Pr}^{-0.15}$$

$$T_w = \text{wall temperature}$$

$$T_{in} = \text{inlet coolant temperature.}$$

The Reynolds number, Re, in the above equation is based on the hydraulic diameter of the equivalent annulus ($D_{eq} - D$). This correlation is valid for

$$4 \times 10^3 \leqslant \text{Re} \leqslant 6 \times 10^5.$$

Rough Tubes

Heat transfer correlations for turbulent flow over a roughened surface are intrinsically complex. The general approach in determining the friction factor is to start with the basic principles of turbulent-boundary-layer heat transfer theory by integrating the universal velocity profile,[9] the non-dimensional coolant velocity, as

$$U^+ = 2.5\ln\frac{y^+}{h^+} + C(h^+), \tag{17-9}$$

where

$$U^+ = \frac{\text{actual flow velocity}}{\text{shear velocity}} = \frac{\overline{U}}{u^*}$$

$\overline{U} = $ average flow velocity

$u^* = $ shear velocity $= \overline{U}\sqrt{f_t''/8}$

$f_t'' = $ friction factor in the interior channel [cf. Eq. (17-11)]

$C(h^+) = $ roughness parameter [Eq. (17-10)]

$y^+ = $ non-dimensional distance $= yu^*/\nu$

$\nu = $ kinematic viscosity at bulk coolant temperature

$y = $ distance from wall

$h^+ = $ roughness Reynolds number $= \varepsilon \, \mathrm{Re}\sqrt{f_t''/8}$

$\varepsilon = $ relative roughness $= h_r/D_e$

$h_r = $ rib height

$$D_e = \text{hydraulic diameter} = \frac{R}{2\left(R_0^2 - R^2\right)}$$

$R = $ rod radius

$$R_0 = \text{radius of zero shear} = P\sqrt{\frac{\sqrt{3}}{2\pi}}.$$

For a GCFR selected roughened surface, the roughness parameter, $C(h^+)$, has been determined as[10]*

$$C(h^+) = C_0 + 0.4\ln\left[\frac{h_r}{0.01(R_0 - R)}\right] + \frac{5}{\sqrt{h_w^+}}\left[\frac{T_w}{T_b} - 1\right]^2, \quad (17\text{-}10)$$

where

$$C_0 = 4.0 + \frac{2.75}{(h^+)^{0.216}},$$

and $C(h^+)$ is within the range

$$C(h^+) \leqslant 5.5 + 2.5\ln(h_b^+).$$

The subscripts w and b refer to wall and bulk, respectively. For surface heat transfer correlations, all flow parameters are evaluated at the wall.

---

*Eq. (17-10) is applicable only for the GCFR geometry. A general expression for rough surfaces with two-dimensional ribs is found in Reference 1.

The relationship between the roughness parameter, $C(h^+)$ and the friction factor in the interior channel, f_t'', is[1]

$$\sqrt{\frac{8}{f_t''}} = C(h^+) + 2.5 \ln\left[\frac{R_0 - R}{h_r}\right] - B, \qquad (17\text{-}11)$$

where

$$B = \frac{3.75 + 1.25 R_0 / R}{1 + R_0 / R}.$$

Equation (17-11) is solved iteratively to find the friction factor, f_t''.

The heat transfer coefficient is obtained via the Stanton number. To make this determination, the approach is to integrate the logarithmic temperature profile

$$T^+ = 2.5 \ln\frac{y^+}{h^+} + G(h^+), \qquad (17\text{-}12)$$

where the correlation for the parameter $G(h^+)$ is given as[10]

$$G(h^+) = g_0 Pr^{0.44}\left(\frac{T_w}{T_b}\right)^{0.5}\left[\frac{h_r}{0.01(R_0 - R)}\right]^{0.053}. \qquad (17\text{-}13)$$

In this expression

$$g_0 = 4.95(h^+)^{0.24} + \frac{10.3}{(h^+)^{0.7}},$$

and

$$R = \text{radius of the equivalent annulus.}$$

The Stanton number is then determined[1] from

$$St = \frac{f_t''/8}{1 + \sqrt{f_t''/8}\,[G(h^+) - C(h^+)]}. \qquad (17\text{-}14)$$

The actual Stanton number to be used in Eq. (17-1) is slightly smaller[11] due to the large variation in the local heat transfer coefficient which can occur over a rib pitch.

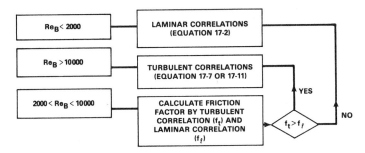

FIGURE 17-13. Procedure for determining friction factors and heat transfer coefficients for the transition flow regime in a GCFR.

C.   TRANSITION FLOW

For the transition flow range $(2000 < \mathrm{Re}_B < 10000)$, the general procedure is to calculate the friction factor by using both laminar and turbulent correlations. The larger of the friction factors will decide the range of all the correlations to be used for the analysis. This method is shown schematically in Fig. 17-13.

D.   TYPICAL FUEL ASSEMBLY CONDITIONS

Tables 17-4 and 17-5 contain typical calculated operating parameters for a GCFR fuel assembly. For this design, the total pressure drop over the

TABLE 17-4   GCFR Fuel Assembly Normal Operating Conditions

Assembly pressure drop (MPa)	0.29
Helium temperature at reactor inlet (°C)	298
Helium temperature at reactor outlet (°C)	530
Maximum (hot spot) cladding temperature, midwall (°C)	750
Maximum linear power at rated power (kW/m)	41.0
Radial maximum-to-average power ratio	1.25
Axial maximum-to-average power ratio	1.21
Average power per fuel assembly (MW)	5.34
Average flow per fuel assembly (kg/s)	5.00
Average Reynolds number	10^5
Velocity of flow (m/s)	70
Peak fluence (n/cm^2) ($E > 0.1$ MeV)	2.3×10^{23}
Maximum burnup (atom %)	10

TABLE 17-5 Pressure Drop in Various Parts of the GCFR Fuel Assembly

Region	Δp (%)
Inlet	8.3
Lower axial blanket	4.8
Rough core region	44.0
Upper axial blanket	7.0
Spacers	22.6
Acceleration	2.1
Outlet	11.2

assembly is only 0.29 MPa (43 psi), which is less than a typical LMFBR fuel assembly. As noted from Table 17-5, about half of this pressure drop is over the roughened portion of the pins.

Whereas the assembly pressure drop is lower for the GCFR, the total pumping power is over four times that required for the LMFBR (cf. Chapter 11-4D).

One particularly interesting aspect of the GCFR blanket assemblies is that the relatively low power output of such pins allows the use of smooth cladding throughout. Furthermore, the P/D ratio can be small enough to allow the use of wire-wrapped pin spacers. Consequently, during operation the Reynolds number is lower in the blanket assemblies than in the fuel assemblies. At decay heat levels, relatively more power is generated in the radial blanket pins since their power comes principally from gamma heating. The net effect of these conditions is that the flow requirement for a GCFR core during low power operation is dictated by the radial blanket rather than the fuel assembly.

17-7 NEUTRONICS CONSIDERATIONS

The GCFR has several unique neutronic characteristics which result from the neutronic transparency of the helium coolant and the high void fraction in the lattice. The low material density results in a hard neutron spectrum (cf. Fig. 17-1) and a high leakage of neutrons to the axial and radial blanket. The hard spectrum reduces the importance of absorption in the resonance energy range, therefore reducing the impact of resonance broadening with temperature. As a result, the Doppler coefficient is only about one-half as large as for the softer LMFBR spectrum.

Calculational methods for nuclear analysis of the GCFR and the LMFBR are very similar. In fact, many of the same computer code systems are employed. The most notable difference is the need to account for neutron streaming in the GCFR when diffusion theory methods are employed. In some calculations, e.g., feed or criticality searches, a simple reactivity bias is adequate. For reactivity coefficient determinations, however, directional diffusion modifiers are required. More detailed calculations require transport theory or Monte Carlo techniques to properly account for neutron streaming in the low density regions (coolant channels).

Present GCFR designs are based on fuel loading schemes similar to the LMFBR; namely, a three-batch loading with a refueling interval of approximately one year. Up to four radial enrichment zones have been proposed, which could result in a fairly flat radial peak-to-average power (as low as 1.20).

As noted from the discussion of Section 17-4, variable orificing capability is designed with each fuel assembly. This allows the potential to maintain a constant coolant temperature rise over the core as the power shifts during each burn cycle.

REVIEW QUESTIONS

17-1. What significant advantages make the GCFR an attractive reactor concept?

17-2. (a) At what pressure is the GCFR operated?

(b) Why must a GCFR be pressurized while an LMFBR need not be?

(c) What is the advantage of using a PCRV to accommodate the high pressure of the GCFR?

17-3. (a) Discuss the significant differences between the heat transport systems of the GCFR and the LMFBR.

(b) How do these differences influence the thermal efficiency of the plant?

17-4. (a) Describe the backup heat removal systems for a GCFR.

(b) Under what conditions can natural convection accomplish decay heat removal?

17-5. Discuss the reasons for and ramifications of the more open lattice of the GCFR relative to the LMFBR.

17-6. (a) Why and how are GCFR fuel pins roughened?

(b) Explain the relevance of the performance index I.

17-7. Values of Reynolds number, flow velocity, and maximum linear power are given for a GCFR in Table 17-4. Compare these with typical values for an LMFBR.

17-8. Why is the Stanton number used for heat transfer correlations for the GCFR whereas the Peclet number is sufficient for the LMFBR?

17-9. Why is the neutron spectrum harder for the GCFR than for the LMFBR?

REFERENCES

1. M. Dalle Donne and L. Meyer, "Turbulent Convective Heat Transfer from Rough Surfaces with Two-Dimensional Rectangular Ribs," *Int. J. Heat Mass Transfer*, 20 (1977) 581–620.
2. C. B. Baxi and M. Dalle Donne, *Fluid Flow and Heat Transfer in the Gas Cooled Fast Breeder Reactor*, GA-A 15941, General Atomic Report, July 1980.
3. E. M. Sparrow and A. L. Loeffler, Jr., "Longitudinal Laminar Flow Between Cylinders Arranged in Regular Array," *AIChE Journal*, 5 (1959) 325–329.
4. E. M. Sparrow, A. L. Loeffler, Jr., and H. A. Hubbard, "Heat Transfer to Longitudinal Laminar Flow Between Cylinders," *Trans. of ASME, J. of Heat Transfer*, 83 (1961) 415–422.
5. U. Grigull and H. Tratz, "Thermischer Einlauf in Ausgebildeter Laminarer Rohrströmung," *Int. J. Heat Mass Transfer*, 8 (1965) 669–678.
6. A. Martelli, *Thermo-und fluiddynamische Analyse von gasgekuhlten Brennelementbundeln*, KfK 2436 (EUR 5508d), KfK Karlsruhe, 1977.
7. T. B. Drew, E. C. Koo, and W. H. McAdams, "The Friction Factor for Clean Round Pipes," *Trans. Am. Int. Chem. Engr.*, 28 (1932) 56-72.
8. B. S. Petukhov and L. I. Roizen, "Generalized Dependence for Heat Transfer in Tubes of Annular Cross-Section," *High Temp.*, 12 (1975) 935.
9. J. P. Holman, *Heat Transfer*, McGraw-Hill Book Co., New York, 4th Ed., 1976, 179.
10. M. Dalle Donne, et al., *EIR, KfK Joint Heat Transfer Experiment on a Single Rod, Roughened with Trapezoidal Ribs and Cooled by Various Gases*, Kernforschungszentrum, Karlsruhe, W. Germany, KfK 2674, October 1978.
11. M. Hudina and S. Yanar, *"The Influence of Heat Conduction of the Heat Transfer Performance of Some Ribbed Surfaces Tested in ROHAN Experiment,"* Swiss Federal Institute for Reactor Research, Würelingen, Switzerland, TM-IN-572, 1974.

CHAPTER 18

GAS COOLED FAST REACTOR SAFETY

By A. Torri and B. Boyack*

18-1 INTRODUCTION

The general features of the Gas Cooled Fast Reactor (GCFR) were outlined in Chapter 17. This concluding chapter is devoted to safety considerations which are unique to the GCFR.

It is assumed that the reader has studied both Chapter 17 and the safety chapters for the LMFBR (Chapter 13 through 16). Much of the FBR safety material presented earlier is directly relevant to the GCFR. Hence, the present chapter is condensed considerably from that which would be required for a stand-alone discussion of GCFR safety considerations.

The structure of this chapter follows the format of the four preceding safety chapters; namely, general safety considerations, protected transients, unprotected transients, and finally accident containment. Whereas these discussions will be directed principally to the differences in GCFR and LMFBR safety considerations which result from employing different cooling systems, some mention will also be given to the variations in nomenclature and approach which have arisen in the parallel development programs.

18-2 GENERAL SAFETY CONSIDERATIONS

General safety considerations for fast breeder reactors were presented in Chapter 13. Specific objectives of GCFR safety studies are (1) to provide analyses and evaluations necessary to assess the risk to public health and safety resulting from operation of the facility, and (2) to determine the margins of safety during all stages of plant operation as well as the adequacy of safety-related structures, systems, and components. This assess-

*General Atomic Company

TABLE 18-1 LOP Definitions and Success Criteria

LOP Barrier	Function	Probability	Plant Consequence	Public Consequence
1 Operating Systems	Shutdown/Cooldown Core following anticipated operational occurrence	$<10^{-2}$	Reoperable without extensive repair	Plant contributions less than 1% to background exposure (10CFR50*App. I)
2 Dedicated Safety Systems	Shutdown/Cooldown Core in the event the operating systems in LOP-1 fail	$<10^{-2}$	No lifetime reduction to permanent components; No expected fuel lifetime loss	Annual background exposure not exceeded (10CFR20*)
3 Inherent Features	Shutdown/Cooldown Core in the event the active systems in LOP-2 fail	$<10^{-2}$	No loss of core geometry	Annual radiation worker exposure limit not exceeded (10CFR20*)
4 Reactor Vessel	Contain debris/energy release following core meltdown from failure of first 3 LOP's	$<10^{-1}$	No loss of linear or penetration integrity of vessel which could consequentially cause loss of containment integrity	No immediate health effects (10CFR100*); no significant long-term effects
5 Containment	Delay/Control the release of activity from LOP-4 failure	$<10^{-1}$	No substantial loss of containment leaktight integrity	No immediate fatalities
6 Natural Attenuation	Attenuate radiological consequences resulting from LOP-5 failure	$<10^{-1}$	No criteria for plant; possible site criteria	Maximum LWR consequences not exceeded

*These designations refer to Title 10 of the U.S. Code of Federal Regulations, Parts 20, 50, or 100.

ment of risk has traditionally been made within the context of "multiple levels of safety design" on the basis of deterministic evaluations of conservative plant conditions ranging from anticipated operating modes all the way to accident conditions of exceedingly low probability.

The GCFR approach to ensuring that the design meets all applicable safety requirements is via the implementation of a comprehensive safety program plan. This plan defines Lines of Protection (LOP) which are conceptually similar to the Lines of Assurance (LOA) described in Chapter 13-2. The barrier, function, probability target, and success criteria expressed in terms of plant and public consequence for each LOP are shown in Table 18-1. Each LOP provides a sequential and quantified barrier between the public and the radiological hazards associated with postulated GCFR accidents.

It should be noted that LOP's 1 and 2 deal with design features provided in the normal course of considering those safety issues which must be addressed within the design basis, while LOP's 3 through 6 address the capability of the GCFR to accommodate and mitigate accidents traditionally considered beyond the design basis. The LOP approach therefore extends the traditional defense in-depth concept to consider the accommodation of postulated accidents much more severe than the design basis. Additionally, it may be noted that it is the function of LOP's 1 through 3 to render an extremely low probability to any accident which could potentially lead to significant releases of radioactivity to the environment and of LOP's 4 through 6 to mitigate the consequences of these low probability accidents in the unlikely event that they should occur.

18-3   PROTECTED TRANSIENTS

The GCFR is designed to respond to and accommodate a complete spectrum of accident initiators. Each of the events is selected for one or more of the following reasons: (1) it is a design basis event for a system or component, included for completeness to demonstrate that it is bounded by a design basis event; (2) it is required by the relevant licensing agency to be included in safety analysis reports; or (3) it is required by industry code or standard. Accident initiators are grouped into the following categories:

(1) Breaks in feedwater system piping
(2) Increase/decrease in core heat removal
(3) Reactivity and power distribution anomalies
(4) Moisture ingress into primary coolant system
(5) Decrease in reactor coolant inventory

(6) Radioactivity release from a subsystem or component
(7) Failure of normally operating auxiliary systems
(8) Events external to the plant.

The complete list of accident sequences to be evaluated is extensive since several combinations of accident initiators and event sequences must be considered in each category. Previous risk assessments such as the Reactor Safety Study (WASH-1400) have shown that the more frequent transients may represent the more dominant sequences that could ultimately lead to loss of core cooling. To combat this situation, reliability assessments of safety performance are included as an integral part of GCFR safety activities. At the conceptual design stage, such activities assist in the selection of basic system configurations. As the design advances to progressively more detail, control and instrumentation systems are included. A "later stage" includes an evaluation of test, maintenance, and operating procedures.

Since the list of potential accident sequences is large, only a limited number of events of particular interest for the GCFR will be discussed. The first category of events are reactivity anomalies. The first accident event to be discussed is a steam tube leak. This is of interest because the steam can flow directly through the core and induce a reactivity effect. Extensive calculations have determined that the maximum reactivity effect is small (a few cents at most) and is quite tolerable. Second, a helium leak would also lead to a reactivity insertion though less severe than the loss of sodium in an LMFBR. Here again, calculations reveal that the net effect would be small (the order of $+3$ cents). The third accident selected for mention is a control rod ejection. This accident is reduced by design features to an extremely low probability, which removes the event from the design basis envelope. One of the design features is the use of cover plates to back up the primary closures for the control rod drive penetrations.

The second category of interest is the decrease in reactor coolant inventory. The *design basis depressurization accident* (DBDA) is of primary importance since this accident, as evaluated using standard licensing practices for equipment availability and analysis conservatisms, leads to the highest expected fuel cladding temperatures of any event in the GCFR design duty cycle. The DBDA in the GCFR is based on a non-mechanistic, assumed 484-cm^2 free-flow leak area open to the reactor containment building. Under DBDA conditions, the reactor coolant pressure would decrease exponentially with time and would become equal to the containment pressure of about 200 kPa in about $1\frac{1}{2}$ minutes. Since the DBDA is an extremely low probability event, one residual heat removal system is considered to be adequate. The Core Auxiliary Cooling System (CACS) was chosen to perform the residual heat removal function under DBDA conditions in the GCFR (cf. Fig. 17-4).

A typical sequence of events during DBDA is as follows. Soon after initiation of a DBDA, a reactor trip is actuated. The plant protection system identifies this condition, and the CACS is phased in to take over the residual heat removal function. Following the reactor trip, the power from emergency diesel generators is used to bring the CACS equipment into a "standby" condition. A depressurization accident is detected by a signal resulting from the primary coolant pressure decrease, and the auxiliary circulators are accelerated to their full design speed. As the auxiliary circulators accelerate, the auxiliary circulator pressure rise forces the auxiliary loop isolation valves open and the main loop isolation valves closed, completing the loop transfer.

The flow is expected to be laminar through the core when the core becomes depressurized. The heat transfer and friction characteristics of laminar flow (cf. Chapter 17-6) are accounted for in the DBDA transient analyses conducted to investigate the consequences of such an event.

Figure 18-1 illustrates the hot spot cladding temperature response assuming transfer to two of the three available CACS loops at one minute after reactor trip and incorporating the conservative model for the system parameters. The maximum hot spot cladding temperature of a maximum-powered fuel rod is about 100°C lower than the cladding faulted limit of 1260°C and 210°C lower than the cladding melting point of 1370°C. This allows an adequate margin to accommodate local anomalies, such as the fuel assembly edge channel effect, and a margin for potential fuel rod damage below the cladding melting temperature. Higher cladding temperatures are permissible

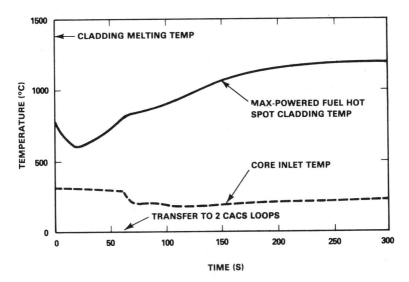

FIGURE 18-1. Core inlet and maximum-powered fuel hot spot cladding temperature following a DBDA.

than in the LMFBR because of the absence of coolant boiling and because of the vented pin design. Lower temperatures could be obtained with larger CACS motors; the counter balancing issue is the increased cost of such motors for this low probability event.

As noted in Table 18-1, inherent GCFR features are relied upon in the event that the required operating systems and dedicated safety systems are unavailable. A significant inherent safety feature (assuming GCFR primary helium pressure greater than 2 MPa) is the availability of natural circulation to cool the reactor core. Design requirements are placed on the CACS to ensure the availability of adequate natural circulation and the highest degree of passive operation possible. The elevations required to provide the required natural circulation in the primary coolant, secondary, and tertiary (air) systems are shown in Fig. 17-4.

The effectiveness of natural circulation cooling of the core is illustrated in Fig. 18-2 for the simultaneous loss of electric power to all main circulators. Upon loss of electrical power, the three main circulators coast down for 90 seconds during which time it is conservatively assumed that the pony motors for the Shutdown Cooling System (SCS) are unavailable to maintain flow in the main loops. In addition, common mode failure of all three CACS circulators to start is assumed. At 90 seconds after loss of the main circulator power and reactor trip, the main loop isolation valves close by gravity and the auxiliary loop isolation valves open by gravity. The natural

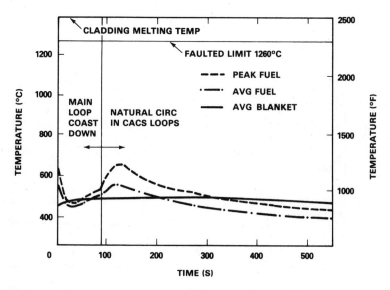

FIGURE 18-2. Hot spot cladding temperature of fuel rod during natural circulation core cooling in an upflow GCFR.

circulation flow through the core increases from near zero to 3% of its normal design flow within three seconds of valve action. This natural circulation flow is found to maintain the hot spot cladding temperatures below 650°C (1200°F).

18-4 UNPROTECTED TRANSIENTS

In the GCFR, as in the LMFBR, accidents can be postulated which hypothesize the total failure of either all reactor shutdown systems or all heat removal systems. These accidents are very unlikely because they require that an abnormal event be compounded by a multitude of common mode failures in two or three diverse systems. However, since by definition these postulated accidents lead to core damage, the consequences to the environment may be considerably larger than the consequences of accidents where core integrity is maintained. These accidents are termed *Beyond Design Basis Accidents** because a multitude of engineered safety systems specifically provided to prevent core damage have been postulated to fail. Nevertheless, such accidents are studied to determine the potential magnitude of public consequences due to these very low probability accidents. Such studies also help determine the extent to which the reactor vessel and the containment are effective in mitigating the consequences. Knowledge of the principal phenomena important in the progression of these accidents permits considerations of containment design options which further mitigate the public consequences of beyond design basis accidents.

A. CLASSES OF UNPROTECTED ACCIDENTS

The Plant Protection System (PPS) is designed to bring into service the engineered safety systems if in the sequence of an abnormal occurrence the normal operating systems are not capable of maintaining the appropriate core design limits either because the initiating event is too severe or because additional failures occur in the operating systems. The principal systems controlled by the PPS are the primary and the secondary reactor shutdown systems, dedicated to attaining core subcriticality, and the Shutdown Cooling System (SCS) and Core Auxiliary Cooling System (CACS), which are

---

*This term is used by the GCFR safety community rather than the HCDA terminology generally used by LMFBR analysts (cf. Chapter 13-3B).

dedicated to maintaining decay heat removal following reactor shutdown. Any sequence of events which causes the total inability to accomplish either reactor shutdown or decay heat removal following shutdown from power operation is termed an Unprotected Accident. Three classes of unprotected accidents are discussed separately because they differ in important aspects of accident initiation and progression. The three classes are

(1) *Loss of Shutdown Cooling* (LOSC) *Accidents*: LOSC accidents postulate the total loss of core cooling for an extended period of time following reactor shutdown.
(2) *Transient Undercooling* (TUC) *Accidents*: TUC accidents postulate the failure of the normal core heat removal system during power operation combined with the total failure of both the primary and the secondary reactor shutdown systems. These accidents have in the past frequently been referred to as Loss of Flow (LOF) accidents.
(3) *Transient Overpower* (TOP) *Accidents*: TOP accidents postulate the occurrence of a positive reactivity insertion while operating at power, combined with the total failure of both the primary and the secondary reactor shutdown systems.

Each of these three classes of unprotected accidents has a particular set of physical phenomena important to the determination of consequences that makes it necessary to consider them separately. In the following, each accident class is discussed briefly with particular emphasis on features and phenomena unique to GCFRs. Most of the discussion of LMFBR safety issues in Chapters 15 and 16 is at least qualitatively or phenomenologically applicable to the GCFR—with one notable exception. All effects related specifically to the sodium coolant in LMFBR accidents are distinctly different in GCFR's. The two phenomena which bear most directly on this difference are the sodium heat capacity and the effects of sodium phase change. It is assumed in the following discussion that the reader is familiar with the basic content of Chapters 15 and 16.

B. LOSS OF SHUTDOWN COOLING (LOSC) ACCIDENT

The LOSC accident for the GCFR will be discussed in somewhat more detail than the other accident types for two reasons. First, this accident is recognized as having a probability of occurrence comparable to the other unprotected accidents, and it tends to be a bounding accident in the GCFR. Second, the TOP and TUC accidents in the GCFR and LMFBR bear sufficient similarity that only a brief discussion of key differences is appropriate.

Core cooling system failures during decay heat removal will proceed to core damage only if a total failure of all core cooling systems is postulated. The operation of any single cooling loop at rated conditions will prevent cladding overheating. Thus, the set of initial conditions for a LOSC accident require postulating the total failure of all of the following:

- Main Cooling System (MCS)
- Shutdown Cooling System (SCS)
- Core Auxiliary Cooling System (CACS)
- Natural Circulation Cooling by CACS.

Detailed reliability analyses of both the probability of initiating events requiring the decay heat removal and the probability of failure for the decay heat removal systems have provided adequate assurance that the LOSC type accidents are of sufficiently low probability to be properly classified as Beyond Design Basis Events. Common mode failures of redundant components within a system must be explicitly considered since such occurrences have been observed with sufficient frequency[1] to dominate the probability of rare events.

Given this set of initiating conditions, the physical/phenomenological sequence of events for an LOSC accident is shown in Fig. 18-3. Following reactor shutdown, the inertia coastdown of the main circulators will provide a period of core cooling prior to core heatup. Axial temperature profiles in the core and between the core inlet and outlet plenum are preserved during this coastdown phase such that following cessation of coastdown induced flow, a transition to natural convection controlled core heatup occurs. While in the GCFR the coolant heat capacity is not as large as for sodium, the coolant inertia effects are likewise smaller and the development of natural circulation flow profiles occurs rapidly. Even in a configuration where the core cavity is completely isolated from the peripheral heat exchanger cavities, there is analytical and experimental evidence[2] that significant axial heat transport to the axial blanket, the assembly shielding structure, and the internal structures in the core cavity can be expected.

Two basic flow patterns are possible according to whether flow recirculation occurs within each assembly or not. With recirculation, helium rises in the central portion of each assembly where the fuel rods are hotter than in the peripheral region. The peripheral fuel rod temperature rise lags behind the central rods because of heat losses to the duct wall. Hot helium leaving the top of the core region loses heat to the upper blanket and shield structure within the assembly and/or the upper cavity structures. Cooled

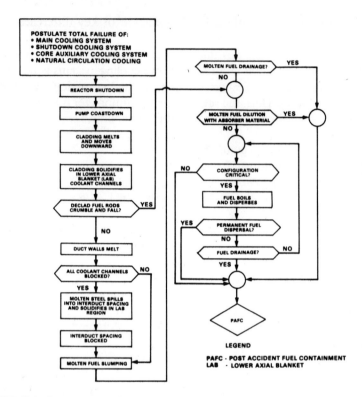

FIGURE 18-3. Event sequence for GCFR loss of shutdown cooling (LOSC) accident.

helium recirculates down along the inside of the duct wall. An alternative flow pattern, without assembly recirculation, can be established by upflow convection in the high powered core assemblies and downflow in the lower power radial blanket assemblies. The actual flow distribution is expected to be a combination of these two effects resulting in significant axial heat transport to cooler structures outside the core fuel region, causing at least a substantial delay in the time of cladding melting.

The onset of cladding melting marks the time when the core geometry is beginning to change. Molten cladding will drain downward and refreeze in the lower core or lower axial blanket region[3] and form substantial blockages. Fuel rods will declad from the top of the blockage to the upper axial blanket. The stability of the declad portion of the fuel rods and the location of the blockage are important in considering the potential for reactivity addition by fuel crumbling. While breakage of declad fuel rods is expected, large scale crumbling tends to be prevented by fuel pellet bonding during normal operation and during the slow heatup to cladding melting with only very small temperature gradients. This heatup process also may permit fuel melting to reduce the potential for fuel crumbling.

The completeness of blockages across an assembly is an important issue for two reasons. First, an incomplete blockage would allow continued natural circulation heat transport axially through the assembly after cladding melting, thus delaying the time of fuel melting. Second, an incomplete blockage can provide drainage paths for molten fuel and prevent the buildup of a molten fuel pool above the blockage. Early experimental evidence indicates that such flow paths may exist in the outer fuel rod rows next to the duct wall, at least in the early stages of blockage formation. The tendency for such partial blockages is enhanced by natural convection flow effects between the core inlet and outlet plenum during the steel relocation phase.

Fuel melting is eventually expected and if molten fuel cannot drain past incomplete blockages, molten fuel will accumulate on top of the blockage. Two mechanisms may prevent the accumulation of a critical configuration at this stage. First, molten fuel may drain sideward through the control rod assemblies which may have a less restricted geometry for molten fuel flow than the blocked fuel assemblies. The principal consideration is the timely melting of the control assembly duct wall relative to the accumulation of a critical pool depth. Once a clear drainage path for molten fuel is established, the gravity driven drainage process is expected to be governed by the conduction limited freezing process without being complicated by the presence of a liquid coolant level in the drainage path. Second, the radial and axial melting progression may allow sufficient neutron absorber material to mix with the molten fuel to maintain subcriticality either until drainage is effected or until all fuel is melted.

In Chapter 15-6, fuel boil-up was considered an important phenomenon in the TUC and TOP accidents in an LMFBR. For LOSC accidents fuel boil-up is unlikely to occur because either the system pressure is too high for fuel boiling (GCFR, 10 MPa) at decay heat or the time scale is too long and the decay heat has dropped below the 1 to 2% threshold for boil-up (LMFBR, GCFR depressurized).

If none of the several phenomena for averting the accumulation of a molten fuel pool of critical height in the core region is operative, criticality may occur. The characterization of this criticality differs from the circumstances normally considered in accidents without shutdown. In LOSC cases, the fuel to be dispersed is essentially a liquid pool, and the delayed neutron precursors have decayed. Furthermore, the neutron spectrum is very hard because the fuel is separated from the steel and because of the inserted absorber rods. While these effects alter the details of the fuel dispersal process, the fundamental process still is the axial separation of fuel driven by the phase change of fuel in the hottest region. The phenomena of nuclear subcriticality depend on the extent to which fuel is dispersed out of the core region for a time period sufficient to establish the weakest link drainage path. At this point sufficient fuel is permanently removed from the core region that subcriticality can be controlled by geometric means and the emphasis then shifts to molten fuel containment in a subcritical configuration.

C. TRANSIENT UNDERCOOLING (TUC) ACCIDENTS

Transient Undercooling Accidents encompass those postulated events which are initiated by a coolant mass flow reduction while operating at power in combination with a total failure of all available reactor shutdown mechanisms. A wide range of initiators which may cause a coolant mass flow reduction is considered, including those low probability events which are not expected to occur in the lifetime of an individual plant but which may occur when considering a number of plant lifetimes. Typical initiators include loss of power to all main circulators, loss of feedwater, or accelerated coastdown of a circulator due to loss of bearing function. While the occurrence frequency for very slow depressurization accidents may be within the frequency of initiators considered for TUC accidents, the time scale on which reactor shutdown is required is sufficiently slow that dependence on automatic reactor shutdown is not necessary.

The GCFR employs two reactor shutdown systems which are required to be independent and diverse from each other. Each system employs a multitude of individual absorber assemblies with a combined total of about 20 absorbers. Each system by itself can adequately respond* to any initiator and bring the reactor to a safe shutdown condition. Furthermore, the insertion of any one or two absorbers from the fully withdrawn position is adequate to shut down the core to a hot standby condition. Therefore, a TUC accident can occur only if all the absorber assemblies in each shutdown system are postulated to fail entirely.

The phenomenology of GCFR TUC accidents is in many respects similar to the LMFBR. This has justified the adoption of the LMFBR code SAS-IIID for the GCFR. This code is known as SASGAS. The principal differences from the LMFBR TUC phenomenology are (1) the absence of sodium related rapid reactivity effects and the attendant coupling of TUC phenomena in high power channels with TOP phenomena in low power

---

*Such positive response includes provision for the stuck-rod criterion.

channels, and (2) the elevated boiling point of fuel at the GCFR system pressure makes the draining of fuel from low power assemblies possible and also suppresses fuel boilup at decay heat level.

The event sequence for GCFR TUC Accidents is displayed in Fig. 18-4. The solid lines depict the expected dominant path—this expectation being based mostly on analysis. Alternate paths which cannot be fully ruled out are shown by the dashed lines. In TUC sequences the fuel is undercooled relative to the full flow condition and results in extensive cladding melting prior to fuel melting. Residual flow effects at the time of cladding melting may be sufficient to cause upward levitation of molten cladding to the upper axial blanket, with subsequent freezing and plugging of flow paths also a possibility. Subsequently, molten cladding is expected to start draining towards the lower axial blanket. The positive reactivity effect of cladding melting and draining increases reactor power to approximately 30 times nominal. The negative reactivity feedback from the Doppler effect

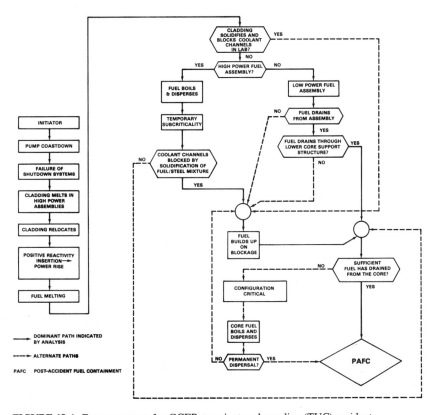

FIGURE 18-4. Event sequence for GCFR transient undercooling (TUC) accident.

maintains reactivity a few cents below prompt critical until fuel melting occurs. If fuel slumping initiates close to prompt critical, the prompt critical condition is only slightly exceeded. Hence, a large ramp rate cannot develop and only a mild power burst would occur.

A significant difference between LMFBR and GCFR in predicted fuel behavior during an LOF arises from the difference in coolant pressure. At the 10 MPa coolant pressure in the GCFR, fuel boiling occurs at substantially higher temperatures, i.e., 4650 K vs 3500 K at atmospheric pressure. During the power excursion, the maximum power density fuel would be heated to above the boiling point at 10 MPa and the subsequent vaporization of fuel at the core center would force molten fuel to relocate towards the axial blankets and cause a neutronic shutdown. A peak fuel vapor pressure of 16.6 MPa is typical of such calculations. However, since the fuel can expand only to the coolant pressure (10 MPa) and since only a small fraction of the fuel is at a temperature higher than the saturation temperature at 10 MPa, only a small volume of fuel would vaporize and the mechanical work potential would remain small.

As fuel penetrates into the blankets, a frozen fuel crust formation is expected to begin on the blanket rods. At the penetration velocities anticipated for the high power channels, the fuel crust may not be stable. The penetration process is probably controlled by the bulk freezing or ablation model discussed in Chapter 15, and plugging in these channels may occur. As a consequence of the large temperature difference between fuel melting (3040 K) and fuel boiling at 10 MPa (4650 K), the low power density regions of the core would also have melted when the first fuel vaporizes. The low power regions of the core would not experience fuel vaporization and, therefore, major portions of the core would exist as liquid fuel columns when neutronic shutdown occurs. These regions would begin draining towards the lower axial blanket. Molten cladding draining and molten fuel draining would occur in such short succession that there would be insufficient time for cladding to refreeze in the lower axial blanket. Furthermore, a stable fuel crust is predicted to develop on the lower axial blanket rods when molten fuel drains under gravity forces alone. This fuel crust thermally insulates the molten fuel and is expected to prevent blockages from developing in the lower axial blanket regions of low power assemblies. Therefore, the drainage of molten steel and fuel from low power assemblies has been identified as a mechanism in the GCFR that may assure sufficient fuel removal following initial neutronic shutdown to prevent recriticality. This mechanism for accident termination is important in the GCFR because, for the GCFR system pressure, fuel boil-up would not be expected to occur at decay heat levels. Out-of-pile experiments using the thermite reaction technique may be useful for initial investigation of the characteristics of molten fuel draining through a rod bundle.

D. TRANSIENT OVERPOWER (TOP) ACCIDENT

Transient Overpower Accidents encompass those postulated events which include an inadvertent and unterminated reactivity insertion while the reactor is operating at power in combination with the total failure of all available reactor shutdown mechanisms. A wide range of reactivity insertion mechanisms in the GCFR has been studied[4] with the conclusion that in a GCFR there are no mechanisms which have the potential for rapid insertion of reactivity. Reactivity effects due to postulated coolant density changes which could result from primary system depressurization are both small and slow to develop in a single phase coolant. The forced withdrawal of a control rod with the pressure difference between the primary system and the containment is prevented by design, i.e., either by providing for each control rod penetration two structurally independent closures or by providing a positive clutch on each control rod drive which is only disengaged during the controlled motion of a control rod for reactivity control purposes. Other mechanisms by which reactivity insertions in a GCFR can be postulated include steam and water ingress, seismic effects, temperature distribution effects, core distortions, inadvertent control rod withdrawals, and core pressure differential effects. The reactivity effects from all these sources can be bounded by reactivity insertion rates of less than 0.10 $/s or equivalent step changes in reactivity of less than about 0.10 dollars.

The comments with respect to the capability of the GCFR shutdown systems under the discussion of TUC accidents apply to the TOP accidents as well. It is, however, noteworthy to mention that an unmitigated forced withdrawal of an inserted control rod at power, which is driven by the primary coolant to containment air pressure difference, is considered for the purpose of establishing the minimum response time for the reactor shutdown systems.

The possible event sequences for a TOP accident in the GCFR are shown in Fig. 18-5. The expected sequence of events, on the basis of analysis, is shown in the solid line paths. The dashed paths represent potential alternative paths which on the basis of current knowledge cannot be completely ruled out. The most commonly considered initiating event for a TOP condition is the continued inadvertent withdrawal of a control rod with a concurrent failure of reactor shutdown. Current models predict the following sequence events. Since full cooling is maintained, the fuel temperatures increase more rapidly while cladding temperature increases lag. Upon fuel melting, the pressure inside the cladding increases due to the fuel volumetric expansion and accelerated gas release. Cladding failure by mechanical rupture in the outlet half of the core region is the predominant failure mode for higher ramp rate initiation, while cladding melting in the upper core half

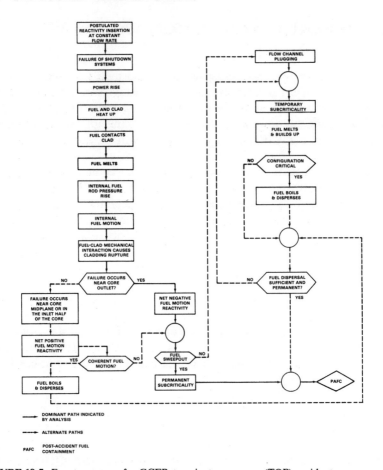

FIGURE 18-5. Event sequence for GCFR transient overpower (TOP) accident.

may be the failure mode for low ramp rate initiation. In either case, molten fuel is ejected into the high velocity coolant stream. Very rapid fragmentation and freezing of molten fuel into particles with a consistency of fine sand is predicted to occur very close to the ejection location. The high velocity coolant stream is predicted to continuously sweep the fuel particles out of the fuel assemblies. This fuel removal mechanism induces a nuclear shutdown while maintaining a coolable core geometry for long term residual heat removal.

18-5 ACCIDENT CONTAINMENT

Two principal barriers are provided in the GCFR, as in any reactor concept, for accident containment. These are the primary system pressure vessel, which in the GCFR consists of a massive Prestressed Concrete

Reactor Vessel (PCRV), and the containment, which in the GCFR consists of an inner low leakage containment and an outer massive concrete confinement. A cutaway of the PCRV and the components located in the PCRV was shown in Fig. 17-3. These two barriers are designed to certain criteria to accommodate design basis accidents, but they also offer inherent capability to mitigate the consequences of accidents beyond the design basis. Both topics are introduced below, with the bulk of the discussion focused on the mitigating features applicable to the less probable accidents.

Design Basis Accidents. Both the PCRV and the containment are required to meet a number of criteria for the accommodation of the design basis accidents. Typical design basis requirements for the PCRV are (1) to assure that the maximum credible breach of the primary coolant boundary does not exceed the design basis depressurization area, and (2) to provide overpressure protection to assure that the maximum cavity pressure is not exceeded. For the containment, typical design basis requirements are (1) to provide a minimum containment backpressure for core cooling following depressurization accidents, (2) to maintain containment leakage rates within design limits for the maximum containment pressure following a design basis depressurization accident or a maximum credible steam pipe rupture in the containment, and (3) to maintain radiological dose consequences within 10CFR100 limits for a postulated release of activity into the containment, defined as the *Site Suitability Source Term*.

Beyond Design Basis Accidents. For accidents which are less probable but also more severe than the plant design basis accidents, both the PCRV and the containment fulfill important functions in the mitigation of public consequences. Given the postulated occurrence of a sequence of events which leads to core melting or core disruption, accident containment is concerned with three specific considerations: (1) containment of energy releases, (2) containment of core debris, (3) containment of noble gas activity and fuel/fission product aerosols.

Each of these aspects is discussed in the following for each containment barrier.

A. CONTAINMENT OF ENERGY RELEASES

The ability of the GCFR to contain large energy releases is an inherent feature of the massive prestressed concrete reactor vessel (PCRV). To appreciate this design feature it is necessary to place the range of potential energy releases from unprotected GCFR accidents into perspective with the PCRV energy release containment capability. Mechanical damage to the primary coolant system boundary as a result of processes occurring during fuel melting and relocation is related to the potential for generating mechanical work and the potential for transmitting work to the primary coolant boundary. Therefore, the mechanical work potential for each accident category discussed in the previous section must be assessed.

While only preliminary analyses of these effects have been completed using the isentropic expansion model, they have been sufficient to identify the important phenomena. For TOP accidents, no mechanical work potential is anticipated due to the predicted fuel sweepout characteristics of a GCFR design. TUC accidents are anticipated to be terminated by fractional fuel vaporization in the high power channels combined with fuel drainage from the low power channels. Boiling in the high power channels occurs when the fuel temperature exceeds the saturation temperature corresponding to the system pressure. At the GCFR system pressure of 10 MPa, the pressure ratio between the fuel vapor pressure and the ambient pressure always remains small, typically less than 2, and as a result the predicted mechanical work potential is small—typically a few tens of MJ. Furthermore, the mechanism for transmitting mechanical work from the core to the primary coolant boundary through an ideally compressible gas (helium) is very inefficient, and most of the mechanical work transmitted through the coolant would be absorbed by the massive internal shielding structures. Therefore, the primary coolant boundary is expected to experience mainly the loading which results from the small static pressure increase without any significant shock wave or water hammer type effects.

One phenomenon has been identified for TUC transients which may have the potential for increasing the energy release in the core. Radial homogenization of the fuel into the coolant channels can reduce the neutron streaming from the core to the radial blanket and thus introduce a positive reactivity effect at the time when other fuel motion mechanisms would begin to turn the power transient around. In order for this mechanism to be effective, two conditions must be satisfied.[5,6] First, the radial fuel homogenization time constant must be very short, i.e., substantially less than 0.1 s, and, second, radial fuel homogenization must occur more than 500°C below the fuel bulk boiling temperature. Neither of these parameters is well defined, and the need for experimental resolution of these parameters has been identified. It should be recognized that this particular phenomenon is generic to fast reactors and applies to GCFR's as well as voided LMFBR's.

The potential for Loss of Shutdown Cooling (LOSC) accidents to generate mechanical energy releases has only recently received attention and is related to changing the design to a bottom supported upflow core configuration. The pivotal question in this regard is the potential for recriticality due to slow fuel melting and relocation in a shutdown core with the control rods inserted. Recriticality can only occur if a substantial fraction of the core fuel melts and is prevented from draining out of the core region. A mechanism for molten fuel accumulation in the core has been identified in the form of steel blockages forming in the lower axial blanket as a result of cladding melting, draining and refreezing. Although other mechanisms for preventing recriticality exist, such as molten fuel draining through the control assemblies or neutron absorber introduction, the potential for recriticality cannot at this time be ruled out. Only preliminary scoping analyses of the consequences of a recriticality during an LOSC accident have been made[7] and range in the 100 to several hundred megajoules. Due to the absence of a liquid coolant, most of this energy would be dissipated by core deformation rather than being transmitted to the primay coolant boundary.

The energy containment capability of the massive PCRV is much larger than the energy releases conservatively predicted from unprotected GCFR accidents. Connor[8] estimated a minimum energy absorption capability of a GCFR PCRV of 2000 MJ. More recent analyses at ANL[9] have investigated the PCRV energy absorption capability for potential LMFBR applications. Even with the more severe loading conditions resulting from sodium slug acceleration, ANL concluded that PCRVs can absorb without failure an energy release of 10 000 MJ. Therefore, the PCRV energy containment capability is on the order of 100 times the energy release which could be transmitted to the primary coolant boundary, and gross failure of the PCRV is not considered possible with respect to generating missiles which could impact containment integrity. It should be noted that the opening of a PCRV relief valve as a result of overpressure is considered possible but has not yet been investigated. The opening of a relief valve could represent a mechanism for the release of fission products and fuel aerosols from the primary system to the containment. However, a complete depressurization would require that (a) the relief valve fails to reseat, and (b) the operator fails or is unable to close the block valve located downstream of the relief valve.

B. CONTAINMENT OF CORE DEBRIS

The potential capability of reactor systems to contain molten fuel has been studied in the past for the purpose of understanding the key considerations. In the GCFR program, the capability for molten fuel containment both inside the primary coolant boundary (in-vessel) and outside the primary vessel (ex-vessel) has been studied. Relatively more emphasis has been placed on in-vessel molten fuel containment because of certain unique design features normally provided in the GCFR. The prestressed reactor vessel (PCRV) shown in Fig. 17-3 includes a massive concrete base of about 5 m thickness without any penetration into the central core cavity. Further-

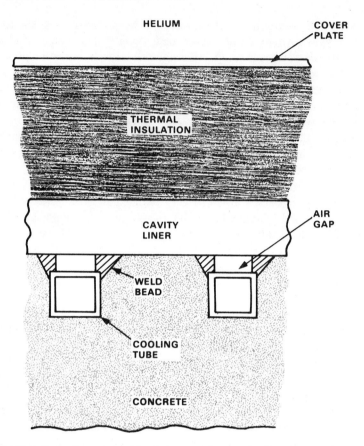

FIGURE 18-6. Cavity liner, thermal insulation, and liner cooling detail in the GCFR.

more, each cavity and duct is lined with a steel liner that provides the primary coolant barrier. This steel liner is thermally insulated on the inside and cooled by two independent liner cooling systems on the concrete side of the liner as shown in Fig. 18-6. The liner thus represents a normally cooled barrier which is not available in reactor designs that employ a steel pressure vessel. Much of the emphasis in GCFR post-accident fuel containment has been directed to examining the extent to which this cooled barrier can be exploited for containment of molten fuel in the central cavity below the core. The following conditions are required to be satisfied for successful in-vessel post-accident fuel containment:

(1) The core melting and fuel relocation processes do not cause damage to the liner cooling system,

(2) The liner temperature does not exceed the failure temperature due to the heat load imposed by molten fuel collected in the bottom of the central cavity,

(3) Liner cooling flow rate and heat rejection capability is sufficient to keep the liner cooling water from boiling,

(4) Spillover of molten fuel from the central cavity to the side cavities through the bottom crossover ducts is prevented.

Based on the discussion in Section 18-A, it is expected that Condition 1 will be met since the liner cooling system is located on the outside of the PCRV liner, which is protected by the thermal barrier and shielding structures from direct impact of potential debris generated in the core. Conditions 2 to 4 will be discussed below.

The objectives in evaluating in-vessel Molten Fuel Containment System (MFCS) concepts are

(1) to evaluate the technical feasibility of maintaining the PCRV liner integrity following core melting,

(2) to identify a lead concept which would be selected in the event that in-vessel molten fuel containment should become a design objective, and

(3) to define the functional requirements which such a system would be required to meet.

The first step in evaluating potential MFCS concepts is to establish a generic set of initial conditions which reasonably represents the range of conditions for the various accident categories. The volume of debris to be accommodated by an MFCS must consider the range of core material which may melt, both in the initial transient as well as in the longer term as a result of conditions created by the initial transient. TUC and LOSC transients are more likely to generate potentially large quantities of molten fuel. TUC tranients are unlikely to cause melting in the radial blanket assemblies because the transient is dominated by fission power heat generation and the decay heat removal systems remain functional. The more slowly developing LOSC accidents which are controlled by decay heat levels may lead to cladding melting in the first few rows of radial blanket rods where the cladding heat fluxes at decay heat are comparable to the corresponding heat fluxes in the core. The coolability of the first row of radial blanket assemblies may, therefore, be affected by the potential for forming flow blockages as a result of cladding draining and refreezing.

The axial extent of core melting and particularly the extent to which the core support plate melts is uncertain. Molten fuel may drain through the assembly inlet nozzle either by melting through the lower axial blanket or by draining laterally into control assembly channels. Unless blockages are formed in the region of the core support plate which allows molten fuel to accumulate, it is unlikely that the core support plate would melt as a result of fuel relocation.

Following the relocation of molten fuel into the molten fuel container, some portion of the sensible heat and decay heat in the debris is radiated upward to the core support plate. Failure of the core support plate, while increasing the debris volume, also increases the amount of heat that can be absorbed in the debris and thereby delays the time at which the peak heat fluxes are experienced by the liner cooling system. Initial MFCS sizing considerations have included the melting and relocation into the MFCS of all the material between the bottom of the grid plate and the top of the assemblies over a radial dimension which includes the core and the first row of radial blankets.

Molten fuel containment concepts for GCFR's have been developed in Germany[10] and the United States.[11] Among the many concepts, the ceramic crucible, the borax bath, the uranium metal bath, and the steel bath concepts have been studied. The analytical methods used for the evaluation of alternate concepts include Baker's[12] empirical model for two-dimensional heat transfer in internally heated pools, conduction heat transfer from the side and bottom structures, and convective and radiative heat transfer from the pool surface to the PCRV internal structures.

The *ceramic crucible concept* (cf. Chapter 16-5C) utilizes refractory materials which form a crucible inside the liner to contain the molten core debris without melting or chemical attack of the crucible surface. At the same time it provides the required shielding for normal operation. The concept can be applied to the GCFR design with minor modifications. The thick crucible wall provides a stored heat capacity that can last 30 h after core meltdown. The peak heat flux which eventually reaches the cavity liner is sufficiently low that only a moderately enhanced liner cooling capacity will remove the entire downward-flowing heat. However, because of this thick crucible wall, the debris pool temperature reaches 3000°C, and the margin for fuel boiling under depressurized conditions is small. The major disadvantage of this concept is that inherently a large fraction (60–80%) of the debris decay heat is directed from the pool surface upward into the PCRV cavity. Removal of this upward heat flux without overheating internal structures requires dependence on the normal core cooling systems, which is undesirable since loss of these systems may have caused the accident.

The *Borax bath concept* for the GCFR was proposed by Dalle Donne.[13] Steel boxes filled with borax ($Na_2B_4O_7$) are installed in the lower reactor cavity. Following a core meltdown, the oxide fuel is expected to dissolve in the liquid borax to form a compound solution pool. The dissolving process is controlled by steel box melting, so that the liquid borax is already at the melting point of steel, where a fast dissolving rate may be achieved. The low boiling point of borax (1700°C) may cause a borax vapor blanket to form at the fuel-borax interface, so that the fuel and steel may sink through the borax bed without dissolving the fuel. In addition, the borax pool may become separated from the fuel by an intermediate steel layer, interrupting the dissolving process. Small-scale simulation tests[13] indicate that UO_2 dissolution can be accomplished in the presence of steel. Only 20% to 30% of the decay heat flows upward because of the low pool temperature. Sidewall and downward heat fluxes are increased, but the peak heat flux does not occur until about 10 h. For normal operation, the individual steel boxes must be designed to accommodate the maximum design pressure or be vented. A vented box design has disadvantages for water ingress accidents and borax vapor transport.

The *heavy metal bath concept* uses a large mass of high-density low-melting-point uranium metal alloy installed inside the lower reactor cavity. Following a core meltdown, a low-temperature pool of uranium alloy is expected to form with solid fuel fragments in suspension. The molten pool is contained by the unmelted solid edge of the heavy metal. The principal advantage of this concept is its self-sealing feature. Air gaps between structural alloy blocks become filled by the uranium alloy, which has a higher density than UO_2, thereby preventing penetration of molten UO_2 into structural gaps and cracks, which can locally increase the heat flux to the cavity liner. A heat transfer study shows that heat removal from the heavy metal bath is feasible with a wide range of suitable pool temperatures. The disadvantages of this concept include the high cost of uranium materials, the potential for metal-water reactions if the liner is breached, and the possibility of crusting on top of the heavy metal which could suspend a significant fraction of the UO_2 above the pool. Uranium alloys also have a low heat capacity, requiring a 2-meter thick layer for a 4-hour heat capacity. For normal operation, the uranium alloy must be properly shielded for neutron fluxes and protected from oxidation by moisture ingress.

The *steel bath concept* employs a large mass of steel plates which melts following a core meltdown to form a "light metal bath." This concept is similar to the uranium bath except that the core debris is heavier than the pool material and is collected at the bottom of the steel pool. A refractory layer placed between the steel and the cavity liner is thus needed to protect the liner from potential hot spot effects. A steel core retention system has a greater stored heat effect than the uranium system and, therefore, the liner heat flux and temperatures are lower. This concept, similar to the ceramic crucible concept, can be accommodated without a large cost or significant design changes.

Table 18-2 compares the four concepts and lists the important parameters. The ceramic crucible is the simplest concept, but it is highly dependent on upward heat removal; the borax bath and uranium bath concepts offer better performance, but require major design changes and experimental development. The steel bath concept appears to be an interesting compromise. A combination of the essential features of two concepts, i.e., a heavy metal base with an overlaying steel bath, may offer the best combination of desirable features.

TABLE 18-2 Comparison of Molten Fuel Containment Concepts

| Parameter | Molten Core Retention Concepts | | | |
	Ceramic Crucible	Borax Bath	Uranium Bath	Steel Bath
Pool temperature, °C	High (>3000)	Low (1427)	Low (>1200)	Medium (>1500)
Cavity liner temperature, °C	Low (150–200)	High (300–400)	High (280–350)	Medium (250–300)
Time of maximum liner heat flux, hr	Long (20–40)	Medium (6–10)	Short (3–4)	Medium (6–10)
Maximum liner heat flux, kW/m^2	Low (50–100)	High (200–300)	High (200–300)	Medium (150–250)
Fraction of upward heat removal	High (0.6–0.8)	Low (0.2–0.3)	Medium (0.3–0.4)	Low (0.1–0.3)
Central cavity height required	Small	Large	Large	Large
Design changes needed	Minor	Major	Major	Minor
Needs for experimental work	Low	High	High	Medium
Recriticality potential	Low	Low	Future Work	Future Work
Pool manageability	Medium	Low	High	High
Fuel penetration and material flotation	Yes	Yes	No	Yes
Scalability	High	Low	Medium	Medium
Cost	Low	Medium	High	Low
Disadvantages for normal operation	No	Yes	Yes	No

Such a combination concept is shown in Fig. 18-7. The MFCS structure consists of three basic material stratifications: crucible, refractory material, and sacrificial steel layers. Typical materials and their thicknesses are given in Table 18-3 based on functional requirements and thermal considerations. The quantity of sacrificial steel is such that a 10 hour stored heat capacity for the decay heat (excluding noble gases and volatiles) is provided without exceeding a quasi steady-state temperature of 2000°C in the stainless steel pool. The recommended heavy metal is depleted uranium alloy which is sized to fill all the gaps and open porosities of the crucible blocks upon melting and to prevent both fuel penetration into the crucible and flotation

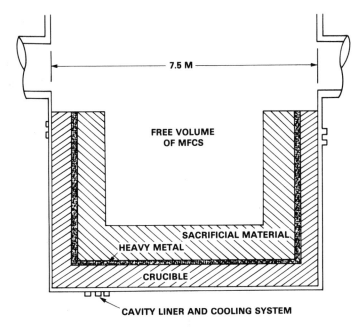

FIGURE 18-7. Combined concept Molten Fuel Containment System (MFCS) configuration.

of the lighter crucible blocks. Magnesia or a mixture of magnesia and boron carbide would be recommended for the crucible material. Since graphite is a good conductor, a layer of 0.05 m thick fused silica would be placed between the graphite and the cavity liner.

The fuel debris configuration in the MFCS at 10 hours, when a quasi steady-state condition is expected to be established, is shown in Fig. 18-8. The permanent crucible contains both the core debris and materials melted

TABLE 18-3 MFCS Candidate Materials and Thicknesses

Region	Materials	Bottom Wall Thickness, m	Side Wall Thickness, m
Sacrificial material	Stainless steel	0.33	0.33
Heavy metal	Depleted uranium alloy	0.03	0.05
Crucible	1. $MgO + B_4C$ or	0.25–0.3	0.15–0.2
	2. $Graphite + B_4C$	0.3–0.4	0.15–0.2
	Fused silica (beneath graphite)	0.05	0.05

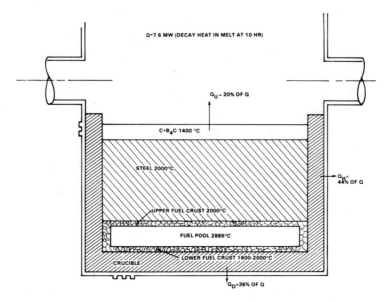

FIGURE 18-8. Quasi steady-state MFCS configuration at 10 hours.

from the MFCS wall. In the heat transfer analysis, a model of separated fuel and steel pools was assumed. The correlations of G. Fieg[14] were used to determine the directional heat transfer in the fuel and steel pools. For pools in the form of right circular cylinders with a height-to-diameter ratio, called the *aspect ratio*, of 0.25, Fieg's correlations give

$$\left.\begin{aligned}
\overline{Nu}_U &= 0.414 \, \mathrm{Ra}_I^{0.216} \\
\overline{Nu}_H &= 0.163 \, \mathrm{Ra}_I^{0.244} \\
\overline{Nu}_D &= 1.12 \, \mathrm{Ra}_I^{0.103}
\end{aligned}\right\} \, 10^7 \leqslant \mathrm{Ra}_I \leqslant 4 \cdot 10^9 \qquad (18.1)$$

where $\overline{Nu}_U$, $\overline{Nu}_H$ and $\overline{Nu}_D$ are average upward, sideward (horizontal), and downward Nusselt numbers, respectively. The internal Rayleigh number, Ra_I, is defined by Eq. (16-14).

Properties of molten fuel and molten steel were taken from the "Properties for LMFBR Safety Analysis."[15] Data and results of the analyses are shown in Table 18-4.

Upward heat removal was limited to 20% by the selected MFCS configuration. The calculated sideward and downward heat removal was 44% and 36%, respectively, of the total 7.6 MW decay heat in the debris. The heat flux at the cavity liner level is reasonably uniform and has a maximum of 95 kW/m² at the sideward location. The total peak heat for the liner cooling system in the lower portion of the central PCRV cavity which removes the

TABLE 18-4 Data and Results of MFCS Heat Transfer Analysis

Expected time of quasi steady-state	10 hrs (after shutdown)
Decay heat in melt at 10 hrs	7.6 MW
Steel pool temperature at 10 hrs	2000°C
Upward heat removal from melt	1.52 MW
Water temperature in cavity liner cooling system	80°C
Effective heat transfer coefficient from water to liner	795 W/m·°C
Fuel pool temperature	2865°C
Upper fuel crust thickness	0.02 m
Lower fuel crust thickness	0.06 m
Sideward heat flux at cavity liner	95 kW/m
Downward heat flux at cavity liner	61 kW/m
Sideward heat removal from fuel and steel pools	3.34 MW
Downward heat removal from fuel and steel pools	2.75 MW
Sideward cavity liner temperature	250°C
Lower cavity liner temperature	190°C
Average temperature of shield materials	1400°C
Maximum temperature of MgO crucible	2000°C

sideward and downward heat flux from the MFCS was 6 MW. This heat load is approximately 10 to 15 times the design heat load for normal operating conditions. A substantially increased water heat-up span can be tolerated for molten fuel containment conditions because design temperature limits for concrete are not applicable in this case. Nevertheless, a several fold increase in water flow rate would be necessary to accommodate this heat load. Based on assumed cavity liner cooling conditions given in Table 18-4, the maximum cavity liner temperature is 250°C at the maximum heat flux location. These liner cooling conditions are believed to be achievable and tolerable within the technology developed for gas-cooled reactor liner cooling system designs.

The heat load in the upper cavity should be removed by mechanisms already provided by the design. A detailed two-dimensional transient thermal analysis of the upper PCRV cavity with a geometry shown in Fig. 18-9 indicated that a heat removal capacity of 2 MW is required at 10 hours after accident initiation in order to maintain all structural components in the upper cavity below 800°C. This heat load is made up of two components; the noble gases and volatile fission products which have left the debris pool, and the upward flowing heat from the top of the debris pool. Peak structural temperatures in excess of 800°C would raise concern of high temperature creep failures, which could add additional debris mass to the molten fuel container. While this in principle would be acceptable, it would have to be demonstrated that no substantial fuel debris spillover into the side cavities could occur through the bottom crossover duct. Five potential mechanisms for upward heat removal include (1) heat removal to the PCRV

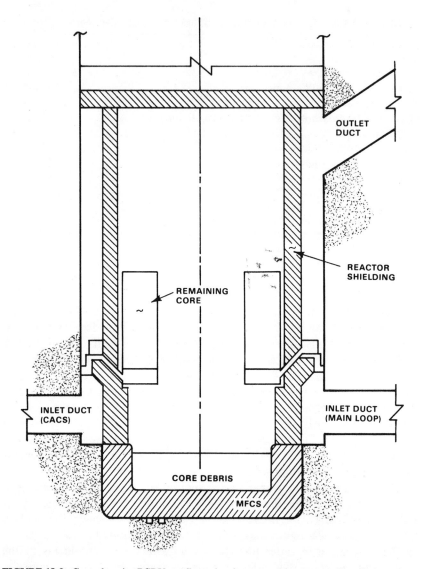

FIGURE 18-9. Central cavity PCRV configuration for upward heat removal analysis.

liner cooling system in the upper central cavity alone, (2) heat removal to the liner cooling system in the main steam generator cavities, (3) heat removal to the liner cooling system in the CACS cavities, (4) natural circulation to the ultimate heat sinks in the CACS loops with three loops depressurized, and (5) natural circulation to the ultimate heat sink in a single CACS loop pressurized.

Material compatibility considerations are most important in establishing the feasibility of molten fuel containment. While larger uncertainties remain in this respect, experimental studies at Sandia[16] have indicated that UO_2 and MgO are compatible up to temperatures exceeding 2000°C. No material compatibility assessments between UO_2, MgO and the uranium-iron alloys to date have been completed. The need for the uranium-iron alloy may be eliminated by a mechanical form of holddown for the MgO blocks to prevent their flotation into the denser UO_2 pool. The provision of several hours time delay before requiring any active cooling of the core debris is nearly as important a consideration as material compatibility. While time delays of several hours are feasible, all in-vessel molten fuel containment concepts require the restoration of some form of heat removal.

In-vessel molten fuel containment is not the only possible approach to contain molten fuel. The massive PCRV lower head provides a massive heat sink which can delay the penetration of molten fuel for many days and results in a very dilute but large debris pool with a very low power density. However, the decomposition and dissolution of concrete by molten fuel releases byproducts which must be considered in the response of the containment. First, the liberation of free and chemically bound water from the concrete is likely to react with the overlaying steel to form hydrogen. Flammable concentrations of hydrogen could accumulate after a few days of concrete decomposition and would require consideration of continuous hydrogen recombination or containment venting before a flammable mixture accumulates. Second, the most commonly used concrete aggregate, namely limestone, contains a large fraction of $CaCO_3$ which decomposes into CaO and CO_2. The latter, being a noncondensable gas, constitutes a containment pressurization source which ultimately would also require containment venting. However, the CO_2 problem can largely be circumvented by use of less common aggregates.

C. CONTAINMENT OF FISSION PRODUCTS AND FUEL AEROSOLS

The GCFR program relies largely on the technology developed for the LMFBR program for the containment of fission products and fuel aerosols. While a detailed secondary containment design for the GCFR has not been developed, the containment-confinement configuration for LMFBR's has been adopted in principle. As long as containment integrity is maintained, this type of containment can accommodate extremely large releases of fission products and fuel aerosols before the dose levels in the environment would be expected to exceed the dose limits stipulated in 10CFR100. This is true in the U.S. because the Nuclear Regulatory Commission (NRC) guidelines for the radiological design of containments are based on a site suitability source term which can only result from core melt accidents.

Permissible containment design leakage rates are established on the basis of this source term such that, with very conservative analysis assumptions, the calculated doses for the limiting organs (bone and lung) are only a small fraction of the accident dose limits currently defined by NRC for exposure to plutonium isotopes. Therefore, safety considerations with respect to the containment of fission product and fuel aerosols essentially focus upon mechanisms which could potentially lead to a loss of the containment integrity. In this respect, there are a number of important differences between a GCFR and an LMFBR:

(1) The massive strength of the PCRV, which was discussed in Section 18-A, above, renders the possibility of generating missiles in the containment as a result of energy releases inside the PCRV extremely remote.

(2) The location of potential fission product and fuel aerosol release from the PCRV is well defined, namely, through the PCRV overpressure relief valve. If necessary, non-gaseous fission products and fuel aerosols could be trapped at the relief valve discharge and prevented from entering the containment directly.

(3) The primary coolant which is released, together with a release of fission products and fuel aerosols in the GCFR, is helium and, therefore, no flammability, heat source, or health hazard is associated with primary coolant releases.

(4) Since the primary coolant is a noncondensable gas, any coolant releases will increase the containment pressure which, after reaching an equilibrium, will persist until the containment gas can be purged or transferred to other containers. However, the containment building is designed for a full primary system depressurization. Furthermore, the presence of any helium acts as a diluent which increases the quantity of hydrogen which could be accommodated without reaching a flammable concentration.

(5) Sodium in an LMFBR has the beneficial effect of retaining a large portion of the iodine and fuel vapor aerosols which mitigates potential releases of these activities into the containment. This effect is absent in the GCFR, where only plateout and settling mechanisms mitigate the releases to the containment. However, since the dominant release in the GCFR is occurring through the PCRV relief valve, either filtration of the valve discharge or closing of the downstream block valve can be utilized to mitigate activity releases to the containment.

GCFR containment analysis methods utilize the aerosol analysis codes developed under the LMFBR programs. However, the physical responses of the containment such as pressure and temperature effects from noncondensable gases and heat loads from sensible heat, decay heat, and hydrogen recombination utilize the analysis methods developed under the High Temperature Gas Cooled Reactor (HTGR) program.

REVIEW QUESTIONS

18-1. Based on the design probabilities for each LOP given in Table 18-1, what is the overall probability objective that the consequence of a GCFR accident not exceed the maximum consequences of an LWR accident?

18-2. Describe the GCFR Design Basis Depressurization Accident (DBDA).

18-3. (a) What is the peak cladding temperature in a GCFR when the only cooling available is by natural circulation in the CACS loops?

(b) What helium pressure must be maintained in order for natural circulation in the CACS loops to be effective?

18-4. (a) Why is the LOSC accident given more attention in the GCFR than in the LMFBR?

(b) What is the status of the reactivity shutdown system and the decay heat removal systems during an LOSC?

18-5. Why is molten fuel/coolant interaction of no significance for the GCFR but is important for the LMFBR?

18-6. List the design basis requirements for (a) the PCRV and (b) the containment.

18-7. List the principal GCFR components designed for the following functions in a Beyond Design Basis Accident:

(a) Containment of energy release

(b) Continment of core debris

(c) Containment of noble gas activity and fuel/fission product aerosols.

18-8. What are the important differences in containment phenomena between the GCFR and the LMFBR?

REFERENCES

1. K. N. Fleming and P. H. Raabe, "A Comparison of Three Methods for the Quantitative Analysis of Common Cause Failures," *Trans. Probabilistic Analysis of Nuclear Reactor Safety, May 8–10, 1978*, Los Angeles, CA, ISBN: 0-89448-101-0, Vol. 3 (1978).

2. M. Croft, J. V. Del Bene and A. Torri, "Natural Convection Effects in a GCFR Subassembly During Loss of Flow with Scram Conditions," *Proc. of the International*

Meeting on Fast Reactor Safety Technology, III, Seattle, WA, (1979) 675.

3. D. L. Hanson and A. J. Giger, "PLOF Accident Preview: Results of a Test Employing 34 Simulated GCFR Fuel Rods," *Proc. of the Int. Meeting on Fast Reactor Safety Technology*, Seattle, WA, (1979) 1868.

4. A. Torri and M. J. Driscoll, *Reactivity Insertion Mechanisms in the GCFR*, General Atomic Co., GA-A12934, April 1974.

5. T. Y. C. Wei, R. H. Sevy and E. M. Gelbard, "Neutron Streaming Effects During an Unprotected Loss of Flow Accident in a GCFR," *Proc. of the International Meeting on Fast Reactor Safety Technology*, Seattle, WA, (1979), 171.

6. T. Y. C. Wei and R. H. Sevy, "Neutron Streaming Effects During a GCFR HCDA—Mitigation Possibilities," *Trans. ANS, 34* (1980) 528.

7. E. E. Morris and R. H. Sevy, "Energetics Following Loss of Cooling in a Shutdown GCFR," *Trans. ANS, 34* (1980), 529.

8. J. G. Connor, Jr. and M. E. Giltrud, *Evaluation of GCFR PCRV Containment Capability*, NSWC/NOL/TR 77-42, Naval Surface Weapons Center (1977).

9. R. W. Seidensticker and A. H. Marchertas, "The Energy Absorption Potential of PCRV in Primary LMFBR Containment," *Trans. ANS, 34* (1980), 530.

10. M. Dalle Donne, S. Dorner and K. Schretzman, "Post-Accident Heat Removal Considerations for Gas-Cooled Fast Breeder Reactors," *Presented at the NEA Coordinating Group on GCFR Development Specialist Meeting on the Design, Safety and Development of the GCFR*, Tokyo, Japan, March 1976.

11. C. S. Kang and A. Torri, "Evaluation of Molten Fuel Containment Concepts for Gas Cooled Fast Breeder Reactors," *Proc. of the Int. Meeting on Fast Reactor Safety Technology*, Seattle, WA (1979), 412.

12. L. Baker, Jr., M. G. Chasanov, J. D. Gabor, W. H. Gunther, J. C. Cassulo, J. D. Bingle, G. A. Mansoori, "Heat Removal from Molten Fuel Pools," *Proc. of the International Meeting on Fast Reactor Safety and Related Physics*, (CONF-761001), Chicago, IL, October 5–8, 1976, 2056.

13. M. Dalle Donne, S. Dorner and G. Schumacher, "Development Work for a Borax Internal Core-Catcher for a Gas-Cooled Fast Reactor," *Nuc. Tech. 39* (1978) 138–154.

14. C. Fieg, "Heat Transfer Measurements of Internally Heated Liquids in Cylindrical Convection Cells," *Proceedings of the 4th PAHR Information Exchange Meeting*, Varese, Italy, October 10–12, 1978, 144.

15. L. Leibowitz, E. C. Chang, M. G. Chasanov, R. L. Gibby, C. Kim, A. C. Millunzi, D. Stahl, *Properties for LMFBR Safety Analysis*, ANL-CEN-RSD-76-1, Argonne National Laboratory, March 1976.

16. D. W. Varela, "High Temperature Magnesium Oxide Interactions with UO_2," *Trans. ANS, 34* (1980) 546.

APPENDICES

APPENDIX A

FAST REACTOR DATA

Compiled by LaVern D. Zeller*

Page

INDIVIDUAL FBR SCIENTISTS AND ENGINEERS WHO ASSISTED IN COMPILING AND CHECKING DATA

1. France — J. Petit and A. P. Schmitt
2. Germany — J. Gilles and R. Fröhlich
3. Italy — F. Granito
4. Japan — K. Aizawa, Y. Kani, and A. Watanabe
5. United Kingdom — A. M. Judd, A. G. Edwards, M. R. Hayns, and H. Teague
6. United States — E. L. Fuller, J. I. Sackett, J. M. Keeton, A. E. Klickman, and R. B. Rothrock
7. Soviet Union — M. F. Troyanov and Y. Bagdasarov

PRINCIPAL GENERAL DATA SOURCES

1. E. R. Appleby, *Compilation of Data and Descriptions for United States and Foreign Liquid Metal Fast Breeder Reactors*, HEDL-TME 75-12, August 1975.
2. International Working Group on Fast Reactors, *LMFBR Plant Parameters*, IWGFR/14/Rev. 1, March 1979.
3. John G. Yevick, *Fast Reactor Technology: Plant Design*, The MIT Press, 1966.

---

*Westinghouse Hanford Company

ABBREVIATIONS USED IN TABLES

Fuel Pellet Form	A–Annular
	S–Solid
	R–Rods
Fission Gas Plenum Location	T–Top
	B–Bottom
Reflector/Blanket Material Form	P–Pellet
	R–Rods
Steam Generator:	
Type	I–Integral
	M–Modular
	S–Separate
Tube Configuration	B–Bayonet
	U–U-tube
	S–Straight
	H–Helical
	HCY–Hockey stick
Pump Type	E–Electromagnetic
	C–Centrifugal
Pump Position	Cold–Cold Leg
	Hot–Hot Leg

——————— Containment Structures ———————

Configuration:	S = Single Containment	
	D = Double Containment	
	CC = Containment/Confinement	
	M = Multiple Containment	
	(with pump back)	
	I = Industrial Grade Building	
Materials:	1 = steel	
	2 = concrete	
	3 = steel-lined concrete	
	(*i*-steel inside; *o*-steel outside)	

TABLE A-1 EARLY EXPERIMENTAL REACTORS

REACTOR	CLEMENTINE	EBR-I	BR-1	BR-2	BR-5	LAMPRE	SEFOR	BR-10
Country	USA	USA	USSR	USSR	USSR	USA	USA*	USSR
GENERAL								
Date Critical	1946	1951	1955	1956	1958	1961	1969	1973
Date Full Power	1949	1951			1959	1961	1971	
Electrical Power, MW	—	0.2					—	—
Thermal Power, MW	0.025	1.2	(30W)	0.1	5	1	20	10
CORE PARAMETERS								
Core Volume, l	2.5	5.9			17.2	3.1	566	
Core Height, m	0.14	0.22			0.28	0.15	0.93	
Core Diameter, m	0.15	0.18	0.13		0.28	0.16	0.88	
Core Volume Fractions								
%Fuel		48				52	43	
%Sodium		15**				34	30	
%Other		37				14	27	
Flux, 10^{15} n/cm²·s	0.004	0.11		0.1	1.0		0.6	2.0
Power Density, kW/l		170			500		35	
Λ, μs		0.04					0.6	
βeff		0.0068	0.002				0.0032	
Doppler Constant, $-Tdk/dT$							0.008	

*Joint Venture with Germany and Euratom
**NaK

FUEL								
Fuel Type	Pu	Enr. U	Pu	Pu	PuO_2, UC	(a)	UO_2 – PuO_2	PuO_2
Fuel Pellet Forms	R	S	R	R	S	S	S	
REFLECTOR/BLANKET								
Axial Material		U	U	U	U-9%Mo	U-9%Mo	Ni	U
Radial Material		U	U	U	U	SS	Ni	
CONTROL								
Material	^{10}B+U	U			Ni	Ni+SS	Ni	
No. Assemblies	4(b)	12			2+2(c)	4(d)	10	
REACTOR VESSEL								
Height, m	1.2	0.7					4.1	
Diameter, m	0.2	0.4			0.4		1.8	
Thickness, m	0.002	0.01						
CONTAINMENT STRUCTURES								
Configuration	I	I	I	I	I		D	I
Materials	—	2	—	—	—			—
PRIMARY HEAT TRANSPORT								
Coolant	Hg	NaK	None	Hg	Na	Na	Na	Na
Cover Gas		Ar			Ar		Ar	
Type	Loop	Loop			Loop	Loop	Loop	Loop
No. Loops	1	1			2	2	1(e)	2
Pump Type	E	E			C	E	E	E
Total Flow, kg/s	2	36			60	7		
Inlet Core Temp., °C	38	230			430	450	370	
Outlet Core Temp., °C	121	322			500	560	430	
Max. Fuel Temp., °C	135	477		60	1300	870		

(a) Molten plutonium alloy
(b) Radial uranium reflector also used.
(c) 2 Control rods + 2 cylindrical reflectors
(d) 4 Control rods + annular reflector
(e) Auxiliary loop also used.

TABLE A-2 EARLY POWER AND TEST REACTORS

REACTOR Country	DFR UK	FERMI USA	EBR-II* USA	RAPSODIE** France	BOR-60 USSR	KNK-2† Germany	JOYO†† Japan	FFTF USA	PEC Italy
GENERAL									
Date Critical	1959	1963	1963	1967/1970	1969	1977	1977/81	1980	1985
Date Full Power	1963	1970	1965	1967/1970	1973	1979	1979/82	1980	
Electrical Power, MW	15	65	20	—	12	21	—	—	
Thermal Rating, MW	60	200	62.5	24/40	60	58	75/100	400	118
CORE PARAMETERS									
Core Volume, l	120	400	73	49/42	60	250/70	304/238	1040	325
Core Height, m	0.53	0.77	0.36	0.34/0.32	0.40	0.60	0.60/0.55	0.91	0.65
Core Diameter, m	0.53	0.82	0.51	0.43/0.41	0.43	0.82	0.79/0.76	1.21	0.83
No. Enrichment Zones	1	1	1	1	1	2	1	2	1
Core Volume Fractions									
%Fuel	40	29	32	40/37	40	32	33/35	35	35
%Sodium	40	50	49	34/36	28	50	41/41	41	38
%Other	20	21	19	26/27	32	18	26/24	24	27
Peak Flux, 10^{15} n/cm²·s	2.5	4.5	2.5	2.0/3.2	3.3	1.8/2.2	3.0/5.0	7.2	4.1
Ave. Flux, 10^{15} n/cm²·s		2.6	1.5	1.2/2.3		1.0	2.0/3.0	4.5	2.7
Peak Linear Power, kW/m	37	28	27	39/43	56	35/43	32/40	42	36
Ave. Linear Power, kW/m	35	17	23	26/31	35	16/31		24	25
Peak Power Density, kW/l	900	774	1002	650/1080	1180	150/440	480/701	730	512
Ave. Power Density, kW/l		458	860	430/770	750		279/391	460	350
Λ, μs		0.14	0.1	0.11/0.24		0.38	0.23/0.30	0.5	
βeff		0.007	0.0068	0.005		0.006	0.005/0.004	0.0032	0.0036
Doppler Constant, $-Tdk/dT$							0.002	0.005	0.0037

*MK-IA/MK-II Cores
**Rapsodie/Rapsodie Fortissimo
†Driver/Test Zone
††MK-I/MK-II Cores

FUEL ASSEMBLY

No. Driver Assemblies	342(a)	102	127/77	64/54	90	22/7	82/67	76	78
Fuel Type	U-7%Mo Alloy	(b)	(c)	UO_2-PuO_2	Enr. UO_2	UO_2/UO_2-PuO_2	UO_2-PuO_2	UO_2-PuO_2	UO_2-PuO_2
Fuel Pellet Form	(d)	(e)	(e)	S	A	S	S	S	A
Fuel Pellet Density, %TD		100		96/92	93.5	90/86	94/93	90	95
Smear Density, %TD		85/75		89/84	73.5	83/80	87	86	88
Bond Material	Na	Na	(f)	He/He	He	He+Ar	He	He	He
Fuel Pellet Dia., mm	16.5	3.7/3.3	3.76	5.6/4.2	5.0	7.0/5.1	5.4/4.6	4.9	5.6
Pin Diameter, mm	20.0	4.4	4.01	6.7/5.1	6.0	8.2/6.0	6.3/5.5	5.8	6.7
Cladding Thickness, mm		0.23/0.30	0.13	0.45/0.37	0.3	0.5/0.38	0.35	0.38	0.42
Cladding Material	Nb	Zr	(g)	316SS/316SS	SS	SS	316SS	316SS	316SS
Pin Spacers		Grid	wire	Fins/Wire	Wire	Grid	Wire	Wire	Wire
Pin Pitch, mm		5.1	5.7	7.1/5.9	6.7	10.1/7.9	7.6/6.5	7.3	7.9
Pin Pitch/Diameter		1.26	1.29	1.06/1.16	1.12	1.23/1.32	1.21/1.18	1.24	0.18
F.G. Plenum Location	T	None	T	T/T&B	B	T	T	T	B
F.G. Plenum Length, m	0.06	NA	0.08/0.24	0.1/(h)	0.40	1.36/1.55	0.41/0.55	0.9	0.50
Pin Length, m		0.83	0.46/0.61	0.48/0.53	1.1		1.91/1.53	2.38	1.65
No. Pins per Assembly	1	140	91	37/61	37	(i)	91/127	217	91
Duct Flat-to-Flat, mm		(i)	58.2	49.8/49.8	44	124	78.5	116	82.6
Duct Thickness, mm		2.4	1.0	1.0/1.0	1	2.6	1.9	3	2.4
Duct Pitch, mm		68.4	58.9	50.8/50.8	45	129	81.5	120	86
Duct Material		SS	SS	SS/SS	SS	SS	SS	SS	SS
Assembly Length, m		2.45	2.33	1.66/1.66	1.58	2.09/2.30	2.97	3.7	3.0
REFUELING									
Refueling Interval, days	60	14	3/5	35/90		365	60	100	60
Max. Fuel Burnup, MWd/kg		10	80	50/90	127	80	50/60	80	33
Ave. Fuel Burnup, MWd/kg	30	8	68	40/40	80	50	42/50	45	40
Storage Positions		35	75	40/40	None	—	20	(k)	76

(a) Number of Rods
(b) Enriched U-10%Mo Alloy
(c) Uranium-Fissium Alloy
(d) Annular Fuel Rods
(e) Fuel Pins
(f) Metallurgical Bond
(g) 304LSS/316 SS
(h) 0.1 Top & 0.06 Bottom
(i) 102-121/211-169
(j) 67.2 mm Square Duct
(k) 3 Modules of 19 positions each within Reactor Vessel

TABLE A-2 EARLY EXPERIMENTAL REACTORS (cont'd.)

REACTOR / Country	DFR UK	FERMI USA	EBR-II* USA	RAPSODIE** France	BOR-60 USSR	KNK-2† Germany	JOYO†† Japan	FFTF USA	PEC Italy
REFLECTOR/BLANKET									
Material	(a)	(b)	(g)	UO₂/(c)	UO₂-axial SS-radial	UO₂	UO₂/SS	Inconel®	UO₂
Axial:									
Material Form		Pins	(h)		P	P	P	R	Pins
Material Dia., mm		10.0	N.A.	13.3	5.0	5.1	5.4/4.6	4.8	5.4
Pin Diameter, mm		11.2	N.A.	14.5	6.0	6.0	6.3/5.5	5.8	5.8
Top Length, m		0.43	0.35/0.39	0.27/—	0.10	0.20	0.40/0.30	0.14	0.18
Bottom Length, m		0.43	0.61/0.57	0.27/0.24	0.10	0.20	0.40/0.35	0.14	0.25
No. Pins per Assembly		16	N.A.	7+7(d)/7	37	217/166	91/127	217	91
Radial:									
No. Assemblies	(e)	537	162/366	500/300	168	5	176/191	99	199
Material Form	R	R	(i)			P	P/(k)	(f)	(f)
Material Dia., mm	34	10.0	N.A.	15.4	14.5	8.0	13.6/—	—	—
Pin Diameter, mm		11.2	N.A.	16.5		9.2	15.0/—	—	—
Assembly Length	2.33	2.45	2.33	1.12		2.14	2.97	3.7	3.0
No. Pins per Assembly	1	25	6/19	7	7	121	19/—	—	—
CONTROL									
Material	(p)	Enr. B₄C	(j)	Enr. B₄C	B₄C and Eu₂O₃	Enr. B₄C	Enr. B₄C	B₄C	Enr. B₄C
No. Assemblies	12(q)	10	10	6	7	8	6	9	11
REACTOR VESSEL									
Height, m	6.3	11.1	2.28	9	6.2	10.2	9.9	13.3	10.7
Diameter, m	3.2	4.4	2.31	2.3	1.4	1.9	3.6	6.3	3.2
Thickness, m	0.012	0.05	0.019	0.015	0.018	0.012–0.016	0.025	0.019	0.030
CONTAINMENT STRUCTURES									
Configuration	S	S	S	S	I	CC	S	S	D
Materials	1	1	3.0	1	—	2/1	1	1	1/1
Des. Press., MPa, gauge	0.137	0.22	0.17	0.24	—	0.25	0.13	0.067	0.149
Des. Leak Rate, Vol %/Day	0.075	0.1	0.2	10(1)		1.0	3.0(m)	0.1	0.5

(a) Natural Uranium, Ni, & SS
(b) Depleted U-2.75 Mo Alloy
(c) UO₂ Below & SS Above; UO₂ Radial
(d) Top + Bottom
(e) 1872 Rods
(f) Hexagonal Inconel® blocks
(g) SS & Depleted Uranium
(h) Tri-fluted block SS/plug w/Flow holes SS
(i) SS Hex Blocks/Pins
(j) Enr. U w/B₄C follower
(k) 3 types (rod cluster, plate layer, block)
(l) at 0.025 MPa
(m) at 0.13 MPa, 360°C

HEAT TRANSPORT AND STEAM SYSTEMS

Coolant	NaK	Na	Na	Na	Na	Na	Na	Na	Na
Cover Gas	Ar	Ar	Ar	Ar/He	Ar	Ar	Ar	Ar	Ar
Primary:									
Type	Loop	Loop	Pool	Loop	Loop	Loop	Loop	Loop	Loop
No. Loops	24	3	1	2	2	2	2	3	2
Pump Type	E	C	C	C	C	C	C	C	C
Pump Position	Cold	Cold	Cold	Cold	Cold	Hot	Cold	Hot	Cold
Total Flow, kg/s	450	1120	481	210/230	260	280	600	2200	630
Inlet Reactor Temp., °C	200	279	371	405/400	340	360	370	360*	400
Outlet Reactor Temp., °C	350	418	473	495/510	520	525	468/500	503*	550
Max. Fuel Temp., °C	650	602	688	2000/2180	700	2055	2330/2500	2250*	2340
Max. Clad. Temp., °C	500	552	599	585/650		~685	620/650	670*	650
Secondary:									
No. Loops	12	3	1	2	2	2	2	3	2
Pump Type	E	C	E	C	C	C	C	C	C
Pump Position	Cold	Cold	Cold	Hot	Cold	Cold	Cold	Cold	Cold
Total Flow, kg/s	900	1120	302	210/200	260	250	600	2200	624
Inter. Heat Exchanger:									
Number	24	3	1	2	4	2	2	3	2
Primary Side	Shell	Shell	Shell	Shell	Shell		Shell	Shell	Shell
Steam Generator:									
Number	12	3	8	—	—	2	—	—	—
Type	(a)	M	M	—	I/M	—	—	—	—
Tube configuration	(a)	H	S	—	—	—	—	—	—
Superheater	Yes	Yes	2	—	Yes	No	—	—	—
Reheater	No	Yes	No	—	No	No	—	—	—
Turbine									
Number	1	1	1	—	—	1	—	—	—
Inlet Pressure, MPa	1.0	3.97	8.62	—	8.82	7.85	—	—	—
Inlet Temp., °C	270	404	435	—	460	485	—	—	—
Dump Heat Exchanger Number	12	0	1	2	—	—	4	12	—

(a) Total separation in copper matrix
* Initial operating temperature

TABLE A-3   PROTOTYPE REACTORS

REACTOR	BN-350	PHÉNIX	PFR	BN-600	SNR-300	MONJU	CRBRP
Country	USSR	France	UK	USSR	Germany	Japan	USA
GENERAL							
Date Critical	1972	1973	1974	1980	1984	1987	1987
Date Full Power	1973	1974	1977	1980	1985	1988	1988
Electrical Rating, MW	150(a)	250	250	600	327	280	375
Thermal Rating, MW	1000	568	600	1470	762	714	975
CORE PARAMETERS							
Core Volume, l	1870	1227	1500	2500	2230	2340	2900
Core Height, m	1.06	0.85	0.91	0.75	0.95	0.93	0.91
Core Diameter, m	1.58	1.39	1.47	2.06	1.78	1.79	2.02
No. Enrichment Zones	2	2	2	2	2	2	1(b)
Core Volume Fractions							
%Fuel	46	36	36	45		32	32
%Sodium	32	36	42	33		41	42
%Other	22	28	22	22		27	24
Peak Flux, 10^{15} n/cm$^2\cdot$s	8.0	7.2	8.9	10.0	6.4	6.0	5.5
Ave. Flux, 10^{15} n/cm$^2\cdot$s			4.5		3.2	3.0	3.4
Peak Linear Power, kW/m	44	45	48	53	36	36	42
Ave. Linear Power, kW/m	21	27	27	36	23	20	27
Peak Power Density, kW/l	730	646	770	840	520		740
Ave. Power Density, kW/l	430	406	380	550	330	295	380
Λ, μs		0.33	0.49		0.4	0.44	0.41
βeff			0.0034		0.0035	0.0036	0.0034
Doppler Constant, $-Tdk/dT$					0.006	0.006	0.008
Breeding Ratio	1.0	1.16	1.0	1.3	1.0	1.2	1.24(c)
Doubling Time, years		40		10			30(d)

(a) Plus 5000 Mg desalted water/hr
(b) Heterogeneous Core
(c) 1.29 if LWR recycled fuel used
(d) 24 if LWR recycled fuel used

FUEL ASSEMBLY							
No. Driver Assemblies	226	103	78	371	205	198	156(c)
Fuel Type	Enr. UO$_2$	UO$_2$-PuO$_2$	UO$_2$-PuO$_2$	UO$_2$-PuO$_2$	UO$_2$-PuO$_2$	UO$_2$-PuO$_2$	UO$_2$-PuO$_2$
Fuel Pellet Form	A	S	A	A	S	S	S
Fuel Pellet Den., %TD	95		80	95	87	85	91
Smear Density, %TD	73.5			77	80	80	85
Bond Material	He	He	Gas	He	He+Ar	He	He
Fuel Pellet Dia., mm	5.2		5.8	5.9	5.1	5.4	5.1
Pin Diameter, mm	6.1	6.6		6.9	6.0	6.5	5.8
Cladding Thickness, mm	0.35	0.45	0.38	0.4	0.38	0.47	0.38
Cladding Material	SS	316SS	316SS	SS	(a)	316SS	316SS
Pin Spacers	Wire	Wire	Grid	Wire	Grid	Wire	Wire
Pin Pitch, mm	7.0	7.7	7.4	8.0	7.9	7.9	7.3
Pin Pitch/Diameter	1.15	1.18	1.26	1.17	1.32	1.22	1.26
F.G. Plenum Location	T	T&B	B	B	B	T	T
F.G. Plenum Length, m	0.02		0.71	0.80	0.65	1.15	1.22
Pin Length, m	1.14	1.80	2.3	2.4	2.48	2.8	2.9
No. Pins per Assembly	169	217	325	127	166	169	217
Duct Flat-to-Flat, mm	96	124	142	96	110	111	114
Duct Thickness, mm	2.0		2.9	2.0	2.6	3	3.0
Duct Pitch, mm	98	127	145	98	115	116	121
Duct Material	SS	316SS	SS&(b)	SS	(a)	316SS	316SS
Assembly Length, m	3.5	4.3	3.8	3.5	3.70	4.2	4.57
REFUELING							
Refueling Interval, days	60	65	45	150	150	180	275
Max. Fuel Burnup, MWd/kg	50	72	65	100	83	100	80
Ave. Fuel Burnup, MWd/kg		40	75		57	80	50
Storage Positions	41	41	20	124	—	—	—

(a) Austenitic Stainless Steel

(b) Nimonic—DS® Stainless Steel*

(c) Plus six additional alternate fuel/blanket assemblies

*Mond Nickel Co., London, England

†Initially loaded with UO$_2$

TABLE A-3 PROTOTYPE REACTORS (cont'd.)

REACTOR	BN-350	PHÉNIX	PFR	BN-600	SNR-300	MONJU	CRBRP
Country	USSR	France	UK	USSR	Germany	Japan	USA
BLANKET							
Material	UO_2	UO_2	UO_2	UO_2	UO_2	UO_2	UO_2
Axial:							
Material Form	P	P	P	P	P	P	P
Material Dia., mm	12	(a)	5.08	5.9	5.1	5.4	5.1
Pin Diameter, mm			5.8	6.9	6.0	6.5	5.8
Top Length, m	0.60	0.26	0.45	0.40	0.40	0.30	0.36
Bottom Length, m	0.60	0.30	0.45	0.40	0.40	0.35	0.36
No. Pins per Assembly	37	(b)	325		166	169	217
Radial:							
No. Assemblies	412	90	51	380	96	172	214(c)
Material Form	P	P	P		P	P	P
Material Dia., mm	13.2		12.5		10.3	10.4	12.1
Pin Diameter, mm	14.2	13.4	13.5	14.2	11.6	11.6	12.8
Assembly Length, m	3.5	4.3	3.81	3.5	3.70	4.2	4.57
No. Pins per Assembly	37	61	85	37	61	61	61
CONTROL							
Material	B_4C/UO_2	Enr.B_4C	Enr.B_4C	B_4C	B_4C	B_4C	B_4C
No. Assemblies	12	6	11	27	12	19	15
REACTOR VESSEL							
Height, m	11.9	12.0	12.8	12.6	15.0	17.8	18.2
Diameter, m	6.0	11.8	12.2	12.8	6.7	7.1	6.2
Thickness, m	0.03	0.015	0.016	0.03–0.05	0.04–0.06	0.04	0.06
CONTAINMENT STRUCTURES							
Configuration	I	I	CC	I	M	CC	CC
Materials	—	2	2/2	—	3,1/2*	1/2	1/2
Design Pressure, MPa gauge	—	0.004	0.005	—	0.13/0.025	0.049	0.067
Design Leak Rate, Vol %/Day					3.2**	1.0†	

(a) 13.0 Top & 6.6 Bottom
(b) 37 Top & 217 Bottom
(c) Less six alternate fuel/blanket assemblies
* The outer concrete containment contains steel liner.
** At $\Delta P=0.0005$ MPa
† At design pressure, room temperature

HEAT TRANSPORT AND STEAM SYSTEMS

Coolant	Na	Na	Na	Na	Na	Na	Na
Cover Gas	Ar	Na	Ar	Ar	Ar	Ar	Ar
Primary:							
Type	Loop	Pool	Pool	Pool	Loop	Loop	Loop
No. Loops	6	3	3	3	3	3	3
Pump Type	C	C	C	C	C	C	C
Pump Position	Cold	Cold	Cold	Cold	Hot	Cold	Hot
Total Flow, kg/s	4460	2760	3090	6050	3500	4260	5200
Inlet Reactor Temp., °C	300	400	400	377	377	397	388
Outlet Reactor Temp., °C	500	560	550	550	546	529	535
Max. Fuel Temp., °C	1800	2300		2500	1850	2200	2350
Max. Clad. Temp., °C	680	700	700	710	685	675	657
Secondary:							
No. Loops	6	3	3	3	3	3	3
Pump Type		C		C	C	C	C
Pump Position	Cold	Cold	Cold	Cold	Cold	Cold	Cold
Total Flow, kg/s		2280	2920	5310	3250	3120	4860
Inter. Heat Exchanger:							
Number	6	6	6	6	9	3	3
Primary Side	Shell	Shell	Tube	Shell	Shell	Shell	Shell
Steam Generator:							
Type	I	M	S	S	S	S	S
Tube Configuration	B&U	"S"-shaped	U	S	S&H	H	HCY
Separate Evaporators	12	36	3	3	9	3	6
Separate Superheaters	6	36	3	3	9	3	3
Reheaters	0	36	3	3	0	0	0
Turbine:							
Number	1	1	1	3	1	1	1
Inlet Pressure, MPa	4.9	16.3	12.8	13.7	16.0	12.5	10.0
Inlet Temp., °C	435	510	513	505	495	483	482

TABLE A-4   DEMONSTRATION REACTORS

REACTOR Country	SUPER PHÉNIX France	CDFR UK	SNR-2 Germany	BN-1600 USSR
GENERAL				
Date Critical	1983	~1990	~1990	~1990
Date Full Power	1984			
Electrical Rating, MW	1200	1320	1300	1600
Thermal Rating, MW	3000	3300	3420	4200
CORE PARAMETERS				
Core volume, l	10766	6600	12000	8800
Core Height, m	1.00	1.00	0.95	1.00
Core Diameter, m	3.7	2.9	4.14	3.35
No. Enrichment Zones	2	2	2	2–3
Core Volume Fractions				
%Fuel	36.6	35	37	
%Sodium	33.8	45	41	
%Other	29.6	20	22	
Peak Flux, 10^{15} n/cm²·s	6.2	9.4	5.5	10.0
Ave. Flux, 10^{15} n/cm²·s	3.6	5.8	3.0	
Peak Linear Power, kW/m	47	42	42	
Ave. Linear Power, kW/m	28	30		
Peak Power Density, kW/l	440	620	460	710
Ave. Power Density, kW/l	275	450	270	470
Λ, μs	0.42			
βeff	0.004	0.003	0.004	
Doppler Constant, $-Tdk/dT$	0.0086	0.008	0.009	
Breeding Ratio	1.25	1.22	1.15	1.3–1.4
Doubling Time, years	23	27		
FUEL ASSEMBLY				
No. Driver Assemblies	364	350	492	
Fuel Type	UO_2-PuO_2	UO_2-PuO_2	UO_2-PuO_2	UO_2-PuO_2
Fuel Pellet Form				
Fuel Pellet Den., %TD		92		
Smear Density, %TD		80	85	
Bond Material	He	He	He+Ar	
Fuel Pellet Dia., mm		5.08	6.4	
Pin Diameter, mm	8.5	5.84	7.6	
Cladding Thickness, mm		0.35	0.50	
Cladding Material	316SS	316SS	SS	
Pin Spacers	Wire	Grid	Grid	
Pin Pitch, mm	9.8	7.35	9.1	
Pin Pitch/Diameter	1.15	1.25	1.20	
F.G. Plenum Location	T/B	B	B	
F.G. Plenum Length, m	0.15/0.85		0.65	
Pin Length, m	2.7	2.29	2.71	

TABLE A-4   DEMONSTRATION REACTORS (cont'd.)

REACTOR Country	SUPER PHÉNIX France	CDFR UK	SNR-2 Germany	BN-1600 USSR
FUEL ASSEMBLY (cont'd.)				
No. Pins per Assembly	271	325	271	
Duct Flat-to-Flat, mm	173	135	153	
Duct Thickness, mm		2.6	4.5	
Duct Pitch, mm	179	147.8	170	
Duct Material	SS		SS	
Assembly Length, m	5.4	4.3	4.0	
REFUELING				
Refuel. Interval, days	320	124	274	120–180
Max. Fuel Burnup, MWd/kg	70	90	70	100
Ave. Fuel Burnup, MWd/kg	44	50	50	70
Storage Positions	12		—	
BLANKET				
Material	UO_2	UO_2	UO_2	UO_2
Axial:				
Material Form	P	P	P	
Material Dia., mm		5.08	6.4	
Pin Diameter, mm	8.5	6.7	7.6	
Top Length, m	0.30	0.40	0.50	
Bottom Length, m	0.30	0.40	0.50	
No. Pins per Assembly	271	325	271	
Radial:				
No. Assemblies	233	202	270	
Material Form	P	P	P	
Material Dia., mm		12.0	10.3	
Pin Diameter, mm	15.8	13.5	11.6	
Assembly Length, m	5.4		4.95	
No. Pins per Assembly	91	91	127	
CONTROL				
Material	B_4C	B_4C	B_4C	†
No. Assemblies	24	29	55	
REACTOR VESSEL				
Height, m	19.5	22.8	26	18
Diameter, m	21	23.3	15	18.3
Thickness, m	0.06	0.025		
CONTAINMENT STRUCTURES				
Configuration	CC	P		
Materials	1/2	1/2		
Design Pressure, MPa	0.003*			

*Inside containment designed for 0.294 MPa
†Depleted UO_2 used for reactivity compensation

TABLE A-4 DEMONSTRATION REACTORS (cont'd.)

REACTOR Country	SUPER PHÉNIX France	CDFR UK	SNR-2 Germany	BN-1600 USSR
HEAT TRANSPORT AND STEAM SYSTEMS				
Coolant	Na	Na	Na	Na
Cover Gas	Ar	Ar	Ar	Ar
Primary:				
Type	Pool	Pool	Loop	Pool
No. Loops	4	4	4	4
Pump Type	C	C	C	C
Pump Position	Cold	Cold	Hot	Cold
Total Flow, kg/s	16900	15000	18000	16600
Inlet Reactor Temp., °C	395	370	390	350
Outlet Reactor Temp., °C	545	540	540	550
Max. Fuel Temp., °C			2200	
Max. Clad. Temp., °C	620	616	650	
Secondary:				
No. Loops	4	4	4	4
Pump Type	C	C	C	
Pump Position	Cold	Cold	Cold	Cold
Total Flow, kg/s	13200	14400	16000	
Inter. Heat Exchanger:				
Number	8	8	8	4
Primary Side	Shell	Tube	Shell	
Steam Generator:				
Type	I		I	
Tube Configuration	H		H or S	
Integral Steam Generators	4		8	
Separate Evaporators	—	—	—	
Separate Superheaters	—	—	—	
Reheaters	0	—		
Turbine:				
Number	2	2	1	2
Inlet Pressure, MPa	18.4	16.0	16.5	14.0
Inlet Temp., °C	490	486	495	490–510

APPENDIX B

COMPARISON OF HOMOGENEOUS AND HETEROGENEOUS CORE DESIGNS FOR THE CRBRP*

As mentioned in Chapter 1, a major change occurred in the CRBRP design in the late 1970s; namely, converting from a homogeneous to a heterogeneous core. Because of the evolving international interest in the heterogeneous design for the FBR, we have included in this appendix some of the principal differences in core parameters resulting from this change. A direct comparison of such core parameters for homogeneous and heterogeneous cores optimized for the same reactor system provides insight which is not readily available from scoping studies.

The early homogeneous CRBRP core layout shown in Fig. B-1 met all project design requirements except for a reduced breeding ratio relative to the design goal of 1.2. This breeding ratio deficit could have been corrected by increasing the fuel pin diameter, assuming the availability of light water reactor-grade plutonium. However, evaluations of core layouts having blanket assemblies dispersed among the fuel assemblies showed that the resulting benefits far outweigh those that could be achieved by increasing the fuel pin diameter. This new configuration, commonly identified as a heterogeneous core, incorporates annular rings of blanket and fuel assemblies where previously there were only fuel assemblies. Using the original CRBRP fuel and blanket assembly designs, this new configuration results in breeding ratios well in excess of the design goal of 1.2.

A second benefit of the change is that safety margins are more easily demonstrated by this concept. A third major benefit is a cost savings due to

---

*Taken essentially intact from R. A. Doncals, N. C. Paik, R. M. Vijuk, and C. L. Wilson, "Clinch River Breeder Reactor Plant Core Flexibility," *CRBRP Technical Review*, CRBRP-PMC 80-02, Spring 1980, pp. 8–12. (Availability: U.S. DOE Technical Information Center)

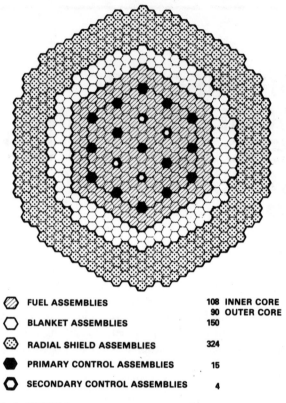

⬡ FUEL ASSEMBLIES		108	INNER CORE
		90	OUTER CORE
⬡ BLANKET ASSEMBLIES		150	
⬡ RADIAL SHIELD ASSEMBLIES		324	
⬡ PRIMARY CONTROL ASSEMBLIES		15	
⬡ SECONDARY CONTROL ASSEMBLIES		4	

FIGURE B-1. Early CRBRP homogeneous core.

the replacement of more expensive fuel assemblies by blanket assemblies. The heterogeneous core concept is now under serious evaluation by design groups in Europe and Japan. The U.S. Department of Energy has recently selected the heterogeneous core concept for incorporation in the first large U.S. LMFBR.

While the feasibility of increasing the breeding ratio of CRBRP was being examined, the design, fabrication and procurement of hardware proceeded as scheduled. Therefore, it was necessary to increase the breeding ratio without significantly changing the CRBRP hardware. Within these ground rules, the CRBRP core design evolved into a new heterogeneous configuration[1] which has blanket assemblies interspersed with the fuel assemblies. This new design has been adopted as the reference core and is shown in Fig. B-2. The core is made primarily of alternating fuel and blanket assemblies. The initial core includes 156 fuel assemblies, 82 inner blanket assemblies and 132 radial blanket assemblies. There are nine fully enriched (92% ^{10}B) primary control assemblies and six fully enriched secondary control assemblies. The core contains a single fuel enrichment zone.

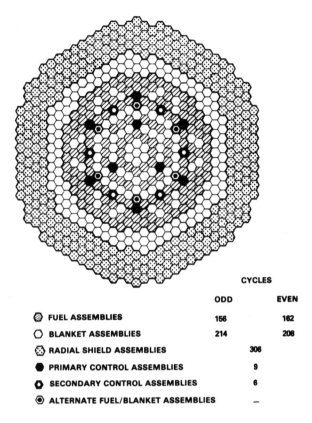

	CYCLES	
	ODD	**EVEN**
⊘ FUEL ASSEMBLIES	156	162
○ BLANKET ASSEMBLIES	214	208
⊙ RADIAL SHIELD ASSEMBLIES	306	
● PRIMARY CONTROL ASSEMBLIES	9	
◐ SECONDARY CONTROL ASSEMBLIES	6	
◉ ALTERNATE FUEL/BLANKET ASSEMBLIES	—	

FIGURE B-2. CRBRP heterogeneous core.

Unlike the scatter-refueling scheme used in the homogeneous core design, the fuel and blanket management for this heterogeneous core design is based on a batch refueling concept. At the mid-cycle of a two-year operation, six inner blanket assemblies are replaced by six fresh fuel assemblies in order to gain the system reactivity to operate for the second half of the cycle. At the end of a full two year cycle, all fuel assemblies, inner blanket assemblies and some radial blanket assemblies are discharged from the reactor. There are significantly fewer fuel assemblies in this design than in the (previous) homogeneous core design, which had 198 fuel assemblies, and, as a consequence, there is a significant cost saving for the Project.[2]

The breeding ratios of the homogeneous and heterogeneous core designs are compared in Table B-1. The low equilibrium breeding ratio of 1.08 for the homogeneous core with the FFTF fuel assembly design was increased to an acceptable value of 1.24 (above the design goal of 1.20) for the heterogeneous core design. In fact, when LWR plutonium becomes available, the heterogeneous design can provide, even with the present fuel design, an equilibrium breeding ratio of 1.29. This breeding ratio of 1.29 is higher than current prototype LMFBR's, and results in a doubling time of 26 years. The breeding ratio of the heterogeneous design with low ^{240}Pu fuel is also 1.29 at the start of the initial cycle, a significant improvement relative to the 1.15 value for the homogeneous design.

TABLE B-1 Breeding Ratio Comparison for Homogeneous and Heterogeneous Cores

	Breeding Ratio with Low ^{240}Pu Fuel	
	Initial Cycle	Equilibrium Cycle
Homogeneous Core	1.15	1.08
Heterogeneous Core	1.29	1.24

A comparison of linear powers in the fuel pins, peak fast flux, and peak fuel burnups is shown in Table B-2. Based on the two year fuel lifetime, the maximum linear power is increased by 10%, and the peak fuel burnup is increased by only 6%. The number of fuel assemblies was reduced, however, by 20% in the heterogeneous core. This increased peak fuel linear power is within the allowable fuel linear power rating. Because of the significantly reduced flux as shown in Table B-2, the equilibrium fuel lifetime of two years for the heterogeneous core was found to be achievable. No significant changes were noted in the power behavior of the blanket assembly rods for the heterogeneous core design.

TABLE B-2 Power, Flux and Fuel Burnup Comparison

Measure of Performance Parameters	Homogeneous Core	Heterogeneous Core
Linear Power (kW/m)		
Peak Linear Power	36	42
Maximum Linear Power ($3\sigma + 115\%$ Overpower)	47	52
Peak Fast Flux (n/cm$^2\cdot$s)	4.2×10^{15}	3.4×10^{15}
Peak Fuel Burnup (MWd/kg)	104	110

In January, 1977, it was established that the improved breeding performance of the heterogeneous core was within the desired ground rules of the study. Consequently, the Clinch River Breeder Reactor Plant Project requested Argonne National Laboratory (ANL) to perform a safety analysis of the new core design using assumptions consistent with those used for the

CRBRP Preliminary Safety Analysis Report (PSAR). The ANL safety analysis indicated[3] that the calculated lower fuel sodium void reactivity worth (Table B-3) and improved incoherency in the voiding phenomena resulted in improved safety margins with respect to hypothetical core disruptive accident energetics. Specifically, ANL investigated undercooled flow transients without reactor scrams in the heterogeneous core under the various levels of conservative assumptions of fuel-coolant and fuel-clad-coolant interactions. In all cases studied, the reactor transient behavior was sluggish and energetically benign due to the reduced sodium void reactivity and increased voiding incoherence. It is also shown in Table B-3 that the fast-acting fuel Doppler feedback is reduced in the heterogeneous core since there is a time lag in heating the in-core blankets. Although the fuel Doppler is reduced, studies of a wide variety of reactor and plant transients did not result in any non-allowable or unstable condition. In fact, the slower acting inner blanket assembly Doppler was found to be desirable in the transition phase of the undercooled transients without reactor scram.

TABLE B-3   Comparison of Key Safety Parameters

Measure of Safety Related Parameters	Homogeneous Core	Heterogeneous Core
Maximum Positive Sodium Void Reactivity (Fuel Assemblies, $)	4.00	2.31
Doppler Constant ($-Tdk/dT$)		
Fuel	0.005 59	0.002 58
Inner Blankets	—	0.004 40
Radial Blankets	0.000 70	0.001 18
Axial Blankets	0.000 44	0.000 26

Further detailed investigations of CRBRP have shown that there are several additional generic advantages of the heterogeneous design. These are (1) improved power doubling time for certain power growth scenarios, (2) less performance sensitivity to fuel pin diameter, (3) improved core restraint response, (4) flexibility to use alternate fuel cycles, and (5) lower fluence-to-burnup ratio. The last advantage affords the opportunity for increased fuel life and extended time between refuelings or decreased allowance for swelling and creep which permits a greater fuel volume fraction.

One of the early concerns with the heterogeneous design was the possibility that existing fast reactor physics analysis methodology might not be applicable to the new design because the methodology was developed and tested against a series of homogeneous critical assemblies. Therefore, the CRBRP heterogeneous core design was simulated in the Zero Power Plutonium Reactor (ZPPR), located within the Idaho National Engineering

Laboratory. To achieve the objectives of the physics simulation tests, the physical characteristics of CRBRP were modeled as closely as possible within the framework of the ZPPR matrix. Measurements performed included criticality, sodium voiding worths, reaction rate distributions, and control rod and fuel/blanket interchange worths. Analyses of these experiments showed that the conventional methodology is quite applicable to the designs of heterogeneous cores. In addition, the high breeding ratio and the reduced sodium void worth in the CRBRP heterogeneous core were experimentally verified.

REFERENCES

1. P. W. Dickson, "Engineering Aspects of Heterogeneous and Homogeneous Reactors," *Optimization of Sodium-Cooled Fast Reactors*, British Nuclear Energy Society, London, 1978, 269–278.
2. C. L. Wilson, "Alternate Core Designs and Fuels for the CRBRP," *CRBRP Technical Review*, CRBRP-PMC-79-01, Winter 1979, 7–46 (Availability: U.S. DOE Technical Information Center).
3. R. J. Henninger, C. H. Bowers, J. E. Cahalan, B. D. Ganapol, D. R. Ferguson, W. R. Bohl, W. L. Wang and S. J. Hakim, "An Analysis of Selected Transient Undercooling Accidents without Scram for the Clinch River Breeder Reactor with a Parfait Core," ANL/RAS 77-16, May 6, 1977 (Availability: U.S. DOE Technical Information Center).

APPENDIX C

4-GROUP AND 8-GROUP CROSS SECTIONS

The cross sections in this appendix are characteristic of a typical neutron spectrum and material composition for the core of an oxide-fueled LMFBR. They were obtained from 4- and 8-group cross sections supplied by D. R. Haffner, R. W. Hardie, and R. P. Omberg of the Hanford Engineering Development Laboratory (1978). These cross section sets were collapsed from a 42-group cross section set developed from ENDF/B using the shielding-factor method. Doppler cross sections for ^{238}U appearing at the end of the appendix were estimated from earlier calculations. The cross sections in this appendix are useful for educational purposes and for rough estimates and comparisons; they are not intended, however, for any actual design applications.

All cross sections are in units of barns. "Fission products" refers to fission product pairs, such that one fission product pair is produced per fission. The cross sections are presented in Sections C-1, 2, and 3 as follows:

C-1 4-GROUP CROSS SECTIONS

4-Group Structure

Group	Δu	Lower Energy of Group	χ
1	2.5	820 keV	0.76
2	2.0	110 keV	0.22
3	2.0	15 keV	0.02
4	–	0	0

4-Group Cross Sections

Material	Group	σ_{tr}	σ_c	σ_f	$\sigma_{er}+\sigma_{ir}$	ν
Boron	1	1.6	0.06		0.45	
(natural)	2	3.3	0.2		0.33	
	3	2.8	0.6		0.15	
	4	3.9	2.0		–	
Boron-10	1	1.8	0.3		0.45	
	2	4.2	1.1		0.33	
	3	5.2	3.0		0.15	
	4	12.2	10.3		–	
Carbon	1	1.8	0.001		0.39	
	2	3.4	0		0.38	
	3	4.3	0		0.27	
	4	4.4	0		–	
Oxygen	1	2.2	0.007		0.40	
	2	3.8	0		0.26	
	3	3.6	0		0.16	
	4	3.6	0		–	
Sodium	1	2.0	0.002		0.51	
	2	3.6	0.001		0.17	
	3	4.0	0.001		0.13	
	4	7.0	0.009		–	
Chromium	1	2.4	0.006		0.34	
	2	3.6	0.005		0.13	
	3	4.5	0.02		0.05	
	4	11.1	0.07		–	
Iron	1	2.2	0.007		0.40	
	2	2.8	0.005		0.08	
	3	5.1	0.010		0.03	
	4	7.2	0.03		–	
Nickel	1	2.3	0.073		0.31	
	2	4.4	0.010		0.11	
	3	12.7	0.03		0.06	
	4	21.1	0.05		–	
Molybdenum	1	3.6	0.02		0.75	
	2	7.0	0.06		0.11	
	3	8.2	0.13		0.06	
	4	9.5	0.7		–	
Thorium-232	1	4.5	0.08	0.07	1.20	2.34
	2	7.4	0.19	0	0.21	–
	3	11.8	0.41	0	0.05	–
	4	15.3	1.48	0	–	–

4-Group Cross Sections, cont'd

Material	Group	σ_{tr}	σ_c	σ_f	$\sigma_{er}+\sigma_{ir}$	ν
Uranium-233	1	4.4	0.03	1.81	0.93	2.69
	2	6.4	0.17	2.05	0.12	2.52
	3	11.5	0.30	2.74	0.04	2.50
	4	17.9	0.74	6.49	–	2.50
Uranium-235	1	4.6	0.06	1.2	0.79	2.69
	2	7.4	0.3	1.3	0.20	2.46
	3	12.5	0.6	1.9	0.04	2.43
	4	18.7	2.0	5.0	–	2.42
Uranium-238	1	4.6	0.06	0.32	1.39	2.77
	2	7.8	0.13	0	0.22	–
	3	12.1	0.35	0	0.05	–
	4	12.9	0.9	0	–	–
Plutonium-239	1	4.9	0.02	1.83	0.83	3.17
	2	7.5	0.16	1.55	0.13	2.92
	3	12.0	0.45	1.63	0.07	2.88
	4	17.4	2.4	3.25	–	2.87
Plutonium-240	1	4.8	0.07	1.59	0.74	3.18
	2	7.4	0.18	0.27	0.22	2.95
	3	12.0	0.5	0.07	0.05	2.88
	4	17.2	2.1	0.13	–	2.87
Plutonium-241	1	4.8	0.08	1.65	0.82	3.23
	2	8.0	0.20	1.72	0.31	2.98
	3	12.6	0.48	2.48	0.05	2.94
	4	19.8	1.74	6.32	–	2.93
Plutonium-242	1	4.5	0.04	1.46	0.65	3.12
	2	7.1	0.12	0.17	0.18	2.89
	3	12.6	0.33	0.04	0.05	2.81
	4	22.0	1.54	0.02	–	2.81
Fission Products	1	7.8	0.05		1.83	
(pairs)	2	11.4	0.17		0.20	
	3	14.7	0.50		0.09	
	4	19.1	1.88		–	

4-Group Scattering Matrices (Elastic & Inelastic), $\sigma_{g' \to g}$

Material	$\sigma_{1 \to 2}$	$\sigma_{1 \to 3}$	$\sigma_{2 \to 3}$	$\sigma_{3 \to 4}$
Boron (and ^{10}B)	0.45	0	0.33	0.15
Carbon	0.39	0	0.38	0.27
Oxygen	0.40	0	0.26	0.16
Sodium	0.51	0	0.17	0.13
Chromium	0.32	0.02	0.13	0.05
Iron	0.37	0.03	0.08	0.03
Nickel	0.29	0.02	0.11	0.06
Molybdenum	0.71	0.04	0.11	0.06
Thorium-232	1.15	0.05	0.21	0.05
Uranium-233	0.87	0.06	0.12	0.04
Uranium-235	0.77	0.02	0.20	0.04
Uranium-238	1.32	0.07	0.22	0.05
Plutonium-239	0.79	0.04	0.13	0.07
Plutonium-240	0.72	0.02	0.22	0.05
Plutonium-241	0.78	0.04	0.31	0.05
Plutonium-242	0.59	0.06	0.18	0.05
Fission Products (Pairs)	1.75	0.08	0.20	0.09

C-2 8-GROUP CROSS SECTIONS

8-Group Structure

Group	Δu	Lower Energy of Group	χ
1	1.5	2.2 MeV	0.365
2	1.0	820 keV	0.396
3	1.0	300 keV	0.173
4	1.0	110 keV	0.050
5	1.0	40 keV	0.012
6	1.0	15 keV	0.003
7	3.0	750 eV	0.001
8	–	0	0

8-Group Cross Sections

Material	Group	σ_{tr}	σ_c	σ_f	$\sigma_{er}+\sigma_{ir}{}^{*}$	ν
Boron-10	1	1.4	0.3		0.4401	
	2	2.0	0.3		0.6502	
	3	3.7	0.7		0.79	
	4	4.6	1.5		0.57	
	5	4.7	2.3		0.39	
	6	5.8	3.8		0.34	
	7	10.9	9.1		0.06	
	8	31.5	29.8		−	
Carbon	1	1.6	0.003		0.4760	
	2	1.9	0		0.57	
	3	3.0	0		0.65	
	4	3.8	0		0.67	
	5	4.2	0		0.39	
	6	4.4	0		0.61	
	7	4.4	0		0.10	
	8	4.5	0		−	
Oxygen	1	1.2	0.02		0.3023	
	2	2.8	0		0.58	
	3	3.7	0		0.69	
	4	3.8	0		0.46	
	5	3.6	0		0.39	
	6	3.6	0		0.37	
	7	3.6	0		0.06	
	8	3.6	0		−	
Sodium	1	1.5	0.005		0.623	
	2	2.2	0.0002		0.6908	
	3	3.6	0.0004		0.4458	
	4	3.5	0.001		0.29	
	5	4.0	0.001		0.35	
	6	3.9	0.001		0.30	
	7	7.3	0.009		0.04	
	8	3.2	0.008		−	
Chromium	1	2.3	0.006		0.9998	
	2	2.5	0.006		0.40	
	3	2.6	0.005		0.1201	
	4	4.6	0.005		0.22	
	5	5.5	0.012		0.28	
	6	3.1	0.033		0.12	

*Values for removal cross sections are accurate to two digits only. Digits beyond two are provided in order to maintain consistency with the scattering matrices so that a neutron balance can be achieved.

8-Group Cross Sections, cont'd

Material	Group	σ_{tr}	σ_c	σ_f	$\sigma_{er} + \sigma_{ir}^*$	ν
Chromium (cont'd)	7	11.5	0.069		0.02	
	8	4.5	0.027		–	
Iron	1	2.2	0.02		1.0108	
	2	2.1	0.003		0.46	
	3	2.4	0.005		0.12	
	4	3.1	0.006		0.14	
	5	4.5	0.008		0.28	
	6	6.1	0.012		0.07	
	7	6.9	0.032		0.04	
	8	10.4	0.020		–	
Nickel	1	2.2	0.02		0.994	
	2	2.4	0.016		0.304	
	3	3.2	0.008		0.19	
	4	5.5	0.012		0.20	
	5	6.9	0.019		0.26	
	6	21.3	0.049		0.13	
	7	21.4	0.053		0.07	
	8	17.0	0.037		–	
Molybdenum	1	2.9	0.005		1.5708	
	2	4.1	0.024		0.878	
	3	6.1	0.046		0.273	
	4	7.9	0.063		0.17	
	5	8.1	0.088		0.14	
	6	8.2	0.20		0.13	
	7	9.3	0.57		0.03	
	8				–	
Thorium-232	1	4.2	0.13	0.14	2.627	2.47
	2	4.6	0.11	0.04	1.105	2.13
	3	5.7	0.18	0	0.3704	–
	4	9.0	0.20	0	0.3634	–
	5	11.6	0.33	0	0.25	–
	6	12.1	0.51	0	0.08	–
	7	14.5	1.15	0	0.02	–
	8	27.7	6.59	0	–	–
Uranium-233	1	4.6	0.02	1.72	1.667	2.91
	2	4.4	0.05	1.85	0.88	2.59
	3	5.2	0.13	1.91	0.2883	2.53
	4	7.6	0.21	2.16	0.202	2.51
	5	10.4	0.25	2.39	0.15	2.50
	6	13.3	0.36	3.19	0.07	2.50
	7	17.1	0.65	5.89	0.01	2.50
	8	28.5	2.04	15.82	–	2.50

8-Group Cross Sections, cont'd

Material	Group	σ_{tr}	σ_c	σ_f	$\sigma_{er}+\sigma_{ir}^*$	ν
Uranium-235	1	4.2	0.04	1.23	0.1394	2.90
	2	4.8	0.09	1.24	0.853	2.59
	3	6.2	0.18	1.18	0.4746	2.48
	4	8.7	0.32	1.40	0.312	2.44
	5	11.7	0.53	1.74	0.15	2.43
	6	13.9	0.79	2.16	0.08	2.42
	7	17.7	1.71	4.36	0.01	2.42
	8	33.0	5.76	15.06	–	2.42
Uranium-238	1	4.3	0.01	0.58	2.293	2.91
	2	4.8	0.09	0.20	1.49	2.58
	3	6.3	0.11	0	0.3759	–
	4	9.3	0.15	0	0.2935	–
	5	11.7	0.26	0	0.20	–
	6	12.7	0.47	0	0.09	–
	7	13.1	0.84	0	0.01	–
	8	11.0	1.47	0	–	–
Plutonium-239	1	4.5	0.01	1.85	1.495	3.40
	2	5.1	0.03	1.82	0.826	3.07
	3	6.3	0.11	1.60	0.3709	2.95
	4	8.6	0.20	1.51	0.1905	2.90
	5	11.3	0.35	1.60	0.15	2.88
	6	13.1	0.59	1.67	0.09	2.88
	7	16.5	1.98	2.78	0.01	2.87
	8	31.8	8.54	10.63	–	2.87
Plutonium-240	1	4.3	0.02	1.61	1.534	3.40
	2	5.1	0.09	1.58	0.723	3.07
	3	6.1	0.15	0.51	0.3713	2.96
	4	8.5	0.20	0.09	0.2929	2.90
	5	11.0	0.34	0.06	0.22	2.88
	6	13.6	0.77	0.08	0.09	2.87
	7	17.1	1.85	0.13	0.02	2.87
	8	19.7	5.92	0.16	–	2.87
Plutonium-241	1	4.5	0.05	1.61	1.838	3.46
	2	5.0	0.11	1.67	0.642	3.13
	3	6.6	0.15	1.53	0.8246	3.01
	4	9.2	0.25	1.87	0.504	2.96
	5	11.9	0.40	2.31	0.22	2.94
	6	13.7	0.58	2.70	0.07	2.94
	7	18.6	1.47	5.50	0.01	2.93
	8	37.7	5.88	19.23	–	2.93
Plutonium-242	1	4.3	0.01	1.67	1.033	3.34
	2	4.5	0.05	1.36	0.68	3.00

8-Group Cross Sections, cont'd

Material	Group	σ_{tr}	σ_c	σ_f	$\sigma_{er}+\sigma_{ir}^*$	ν
Plutonium-242	3	5.8	0.11	0.34	0.2846	2.89
(cont'd)	4	8.2	0.13	0.04	0.3002	2.84
	5	11.2	0.24	0.03	0.23	2.82
	6	14.6	0.45	0.05	0.10	2.81
	7	21.0	1.25	0.02	0.03	2.81
	8	36.9	6.10	0	–	–
Fission Products	1	6.5	0.02		3.767	
(pairs)	2	8.6	0.07		1.908	
	3	10.4	0.11		0.499	
	4	12.3	0.21		0.343	
	5	13.9	0.38		0.20	
	6	15.8	0.65		0.21	
	7	18.5	1.58		0.04	
	8	28.3	6.49		–	

8-Group Scattering Matrices (Elastic & Inelastic)

$$\sigma_{g' \to g}$$

Boron-10

g' \ g	2	3	4	5	6	7	8
1	0.43	0.008	0.002	0.0001	0	0	0
2		0.65	0.0002	0	0	0	0
3			0.79	0	0	0	0
4				0.57	0	0	0
5					0.39	0	0
6						0.34	0
7							0.06

Carbon

g' \ g	2	3	4	5	6	7	8
1	0.47	0.005	0.0009	0.0001	0	0	0
2		0.57	0	0	0	0	0
3			0.65	0	0	0	0
4				0.67	0	0	0
5					0.39	0	0
6						0.61	0
7							0.10

8-Group Scattering Matrices (Elastic & Inelastic), cont'd

$$\sigma_{g' \to g}$$

Oxygen

g' \ g	2	3	4	5	6	7	8
1	0.30	0.002	0.0003	0	0	0	0
2		0.58	0	0	0	0	0
3			0.69	0	0	0	0
4				0.46	0	0	0
5					0.39	0	0
6						0.37	0
7							0.06

Sodium

g' \ g	2	3	4	5	6	7	8
1	0.52	0.09	0.003	0.009	0.001	0	0
2		0.69	0	0.0004	0.0004	0	0
3			0.44	0.005	0.0008	0	0
4				0.29	0	0	0
5					0.35	0	0
6						0.30	0
7							0.04

Chromium

g' \ g	2	3	4	5	6	7	8
1	0.79	0.16	0.04	0.009	0.0008	0	0
2		0.31	0.06	0.02	0.01	0	0
3			0.12	0	0	0.0001	0
4				0.22	0	0	0
5					0.28	0	0
6						0.12	0
7							0.02

8-Group Scattering Matrices (Elastic & Inelastic), cont'd

$$\sigma_{g' \to g}$$

Iron

g' \ g	2	3	4	5	6	7	8
1	0.75	0.20	0.05	0.01	0.0008	0	0
2		0.33	0.10	0.02	0.01	0	0
3			0.12	0	0	0	0
4				0.14	0	0	0
5					0.28	0	0
6						0.07	0
7							0.04

Nickel

g' \ g	2	3	4	5	6	7	8
1	0.67	0.22	0.08	0.02	0.004	0	0
2		0.25	0.04	0.01	0.004	0	0
3			0.19	0	0	0	0
4				0.20	0	0	0
5					0.26	0	0
6						0.13	0
7							0.07

Molybdenum

g' \ g	2	3	4	5	6	7	8
1	1.09	0.39	0.08	0.01	0.0008	0	0
2		0.62	0.20	0.05	0.008	0	0
3			0.23	0.04	0.003	0	0
4				0.17	0	0	0
5					0.14	0	0
6						0.13	0
7							0.03

8-Group Scattering Matrices (Elastic & Inelastic), cont'd

$$\sigma_{g' \to g}$$

Thorium-232

g' \ g	2	3	4	5	6	7	8
1	1.20	1.01	0.34	0.07	0.007	0	0
2		0.86	0.20	0.04	0.005	0	0
3			0.36	0.008	0.002	0.0004	0
4				0.36	0.003	0.0004	0
5					0.22	0.03	0
6						0.08	0
7							0.02

Uranium-233

g' \ g	2	3	4	5	6	7	8
1	0.61	0.73	0.26	0.06	0.007	0	0
2		0.58	0.24	0.05	0.01	0	0
3			0.28	0.008	0.0003	0	0
4				0.20	0.002	0	0
5					0.14	0.01	0
6						0.07	0
7							0.01

Uranium-235

g' \ g	2	3	4	5	6	7	8
1	0.72	0.48	0.16	0.03	0.004	0	0
2		0.72	0.12	0.01	0.003	0	0
3			0.43	0.04	0.004	0.0006	0
4				0.29	0.02	0.002	0
5					0.14	0.01	0
6						0.08	0
7							0.01

8-Group Scattering Matrices (Elastic & Inelastic), cont'd

$$\sigma_{g' \to g}$$

Uranium-238

g' / $g \to$	2	3	4	5	6	7	8
1	1.28	0.78	0.20	0.03	0.003	0	0
2		1.05	0.42	0.01	0.01	0	0
3			0.33	0.04	0.005	0.0009	0
4				0.29	0.003	0.0005	0
5					0.18	0.02	0
6						0.09	0
7							0.01

Plutonium-239

g' / $g \to$	2	3	4	5	6	7	8
1	0.66	0.60	0.19	0.04	0.005	0	0
2		0.64	0.15	0.03	0.006	0	0
3			0.31	0.05	0.01	0.0009	0
4				0.18	0.01	0.0005	0
5					0.13	0.02	0
6						0.09	0
7							0.01

Plutonium-240

g' / $g \to$	2	3	4	5	6	7	8
1	0.75	0.58	0.17	0.03	0.004	0	0
2		0.60	0.11	0.01	0.003	0	0
3			0.33	0.04	0.001	0.0003	0
4				0.29	0.002	0.0009	0
5					0.21	0.01	0
6						0.09	0
7							0.02

8-Group Scattering Matrices (Elastic & Inelastic), cont'd

$$\sigma_{g' \to g}$$

Plutonium-241

g' \ g	2	3	4	5	6	7	8
1	0.62	0.77	0.34	0.10	0.008	0	0
2		0.57	0.06	0.01	0.002	0	0
3			0.76	0.06	0.004	0.0006	0
4				0.45	0.05	0.004	0
5					0.19	0.03	0
6						0.07	0
7							0.01

Plutonium-242

g' \ g	2	3	4	5	6	7	8
1	0.47	0.39	0.14	0.03	0.003	0	0
2		0.40	0.21	0.05	0.02	0	0
3			0.26	0.02	0.004	0.0006	0
4				0.30	0.0002	0	0
5					0.21	0.02	0
6						0.10	0
7							0.03

Fission Products (pairs)

g' \ g	2	3	4	5	6	7	8
1	2.12	1.26	0.32	0.06	0.007	0	0
2		1.40	0.42	0.08	0.008	0	0
3			0.48	0.01	0.007	0.002	0
4				0.32	0.02	0.003	0
5					0.20	0	0
6						0.21	0
7							0.04

C-3 URANIUM-238 DOPPLER CROSS SECTIONS

Tabulated below are approximate increases in ^{238}U effective capture cross sections due to the Doppler effect resulting from an increase in average fuel temperature in an LMFBR core from 700 K to 1400 K.

4-Group		8-Group	
Group	$\Delta\sigma_c$ (barns)	Group	$\Delta\sigma_c$ (barns)
1	0	1	0
2	0	2	0
3	0.004	3	0
4	0.05	4	0.0001
		5	0.0008
		6	0.004
		7	0.045
		8	0.20

LIST OF SYMBOLS*

A = area (cm^2 or m^2)
A = atomic weight of scattering or target isotope
A_r = fractional fuel area restructured
A_{ri} = projected area for one wire wrap in channel i over one lead (m^2)
A_T = sum of flow areas of all the channels in a fuel assembly (m^2)
A_u = fractional fuel area unrestructured
A_i' = flow area in channel i without the wire (m^2)
B = fuel burnup (MWd/kg or atom %)
Bo = Bond number
BR = breeding ratio
B^2 = geometric buckling (cm^{-2})
c = sonic velocity (m/s)
c_p = specific heat at constant pressure (J/kg·K)
C = delayed neutron precursor concentration (precursors/cm^3)
C = cycles per year (y^{-1})
C = electrical capacity (GWe)
C = empirical constant for closed gap conductance (m$^{-1/2}$)
$C(h^+)$ = roughness parameter
CR = conversion ratio
C_D = drag coefficient
C_l = amount of radioactivity released to the atmosphere as a result of an accident of type l (curies/accident)
CDF = cumulative damage function
CSDT = compound system doubling time (y)
d = penetration depth (mm)

$D =$ average spacing (eV)

$D =$ damage fraction

$D =$ diameter (m)

$D =$ diffusion coefficient (cm)

$D =$ droplet diameter (m)

$D =$ number of direct hot channel factors

$D_e =$ effective hydraulic diameter (m)

$D_l =$ consequences resulting from the atmospheric release of one unit of radioactive material in accident type l (health consequences or \$/curie)

$e =$ fissile mass fraction in the fuel

$e =$ energy density (J/m^3)

$E =$ atom fraction of Pu in the fuel

$E =$ neutron energy (eV)

$E =$ Young's modulus (Pa)

$EF =$ out-of-reactor (or ex-reactor) factor

$E_c =$ neutron energy in center-of-mass system (eV)

$E_g =$ lower energy of group g (eV)

$E_0 =$ resonance energy (eV)

$f =$ friction factor

$f =$ load factor (fraction of time at full power)

$f =$ shielding factor

$f_l =$ laminar friction factor

$f_t' =$ turbulent friction factor for smooth tube

$f_t'' =$ turbulent friction factor for rough tube

$F =$ displacements per atom (dpa)

$F =$ fission source (n/s·cm)

$F =$ fraction of fission gas released

$F =$ volume fraction

$F =$ hot channel factor

$FAE =$ fuel adjacency effect

$FBOC =$ fissile material in core and blankets at the beginning of the fuel cycle (kg)

$FD =$ fissile material destroyed per cycle (kg/cycle)

$FDIS =$ fissile material in the fuel discharged from the core and blankets at the end of the cycle (kg)

$FE =$ fissile material external to the reactor per cycle (kg/cycle)

$FEOC =$ fissile material in the core and blankets at the end of the fuel cycle (kg)

$FG =$ fissile material gained per cycle (kg/cycle)

$FL =$ fissile material lost external to the reactor per cycle (kg/cycle)

$FLOAD =$ fissile material loaded into the core in the fresh fuel at the beginning of the fuel cycle (kg)

FP = fissile material produced per cycle (kg/cycle)

FPL = fissile material lost in processing per cycle (kg/cycle)

$F(E)$ = collision density (reactions /cm$^3 \cdot$ s)

F_b = hot channel factor for coolant enthalpy and temperature rise

F_c = hot channel factor for temperature rise across cladding

F_d = hot channel factor based on direct uncertainties

F_s = hot channel factor based on random uncertainties

F_{film} (cladding) = hot channel factor for temperature rise across film at hot spot beneath the wire wrap, to be used in obtaining peak cladding temperatures

F_{film} (fuel) = hot channel factor for temperature rise across film to be used in obtaining peak fuel temperature

F_{fuel} = hot channel factor for fuel conductivity

F_{gap} = hot channel factor for gap conductance

F_i = shear force per unit length (N/m)

F_q = hot channel factor for heat flux

F_r = gas release fraction for restructured fuel

F_u = gas release fraction for unrestructured fuel

g = jump distance at gap (m)

g = gravitational acceleration (m/s^2)

g_z = axial component of gravitational acceleration, $g_z = -g$(m/s^2)

$\vec{g}$ = body force per unit mass (m/s^2)

g_J = statistical spin factor

G = breeding gain

G = number of energy groups

G = thickness of gap between fuel and cladding (m or mm)

G_c = residual fuel-cladding gap thickness (mm)

G_0 = fabricated fuel-cladding gap thickness (mm)

Gr = Grashof number

Gz = Graetz number

h = convective heat transfer coefficient (W/m$^2 \cdot$ K)

h = specific enthalpy (J/kg)

h^+ = roughness Reynolds number

h_{fg} = latent heat of vaporization (J/kg)

h_g = gap conductance (W/m$^2 \cdot$ K)

h_r = rib height (m)

H = core height (m)

H = height of sodium column accelerated (m)

H = axial position within the pool (m)

H = lead (axial length of one complete wire wrap spiral) (m)

$H =$ Meyer hardness (Pa)

$H_l =$ pressure drop (losses) in the piping and equipment upstream of the pump (Pa)

$H_p =$ pressure at the liquid surface where the pump takes suction (Pa)

$H_v =$ vapor pressure of the fluid at the suction (Pa)

$H_z =$ hydrostatic head of the liquid in the pump above the impeller (Pa)

$I =$ performance index, GCFR fuel pin

$I =$ spin quantum number of the target nucleus

$I_i =$ fraction of fuel atoms that are isotope i

$J =$ spin quantum number of the compound nucleus

$k =$ criticality factor

$k_\infty =$ criticality factor for an infinite medium

$k =$ stability parameter

$k =$ thermal conductivity (W/m·K)

$k =$ neutron wave number (m^{-1})

$k =$ Boltzmann's constant

$k_b =$ criticality factor at the beginning of the cycle, with control rods withdrawn

$k_e =$ criticality factor at the end of the cycle, with control rods withdrawn

$k_{eff} =$ effective criticality factor

$K =$ linear electrical growth rate (GWe/y)

$K =$ form loss coefficient

$K' =$ form loss coefficient per unit length (m^{-1})

$K =$ channel shape factor, GCFR

$K_D =$ Doppler constant (T dk/dT)

$k_m =$ thermal conductivity of the gas mixture in the gap (W/m·K)

$k_p =$ thermal conductivity of fuel with porosity p

$k_s =$ effective conductivity of surface materials (W/m·k)

$l =$ angular momentum quantum number

$l =$ neutron lifetime (s)

$l =$ characteristic distance between coolant channels (m)

$L =$ neutrons lost unproductively

$L =$ length or distance (m)

LMP = Larson-Miller Parameter (K)

$L_f =$ active fuel length (m)

$L_p =$ plenum length (m)

$m =$ exponent in friction factor correlation

$m =$ mass (kg)

$\dot{m} =$ coolant mass flow rate (kg/s)

$\dot{m}_T$ = total mass flow rate in a fuel assembly (kg/s)

M = friction factor multiplier

M = mass of target nucleus

M = migration length (cm^2)

M = minimum Cumulative Damage Fraction (CDF) at failure $\simeq 2 \times 10^{-3}$

M = number of atoms that have been removed by fallout or filtration

M = mass (kg)

M = molecular weight (g/g-mol or kg/kg-mol)

M^2 = migration area (cm^2)

$\dot{M}_g$ = fissile material gained during one year (kg/y)

M_0 = initial fissile inventory in a reactor (kg)

M_0/P = specific fissile inventory (kg/MW)

n = neutron density (n/cm^3)

n = kg-moles fission gas produced per m^3 of fuel

N = atom density (atoms/cm^3 or atoms/barn·cm)

$\bar{N}$ = average atom density over fuel cycle (atoms/cm^3)

N = number of atoms

N = number of 1-GWe LWR plants operating and decommissioned

N = number of mesh intervals

N = number of cycles

$N(r, z)$ = normalized spatial shape of the power distribution

N_r = number of cycles to rupture

N_A = Avogardo's number (6.023 × 10^{23} atoms/g-mol)

$\bar{N}_m$ = average atom density during the cycle (averaged over all batches *and* over the duration of the cycle)

$N_{m,b}$ = average atom density (averaged over all batches) at the beginning of the cycle

$N_{m,e}$ = average atom density (averaged over all batches) at the end of the cycle

$N_m(q)$ = atom density of material in a batch after q cycles if all of the core were composed of this batch

N_{other} = atom density of other materials present for the homogenized cell (atoms/barn·cm)

NR_J = number of resonances

$\overline{Nu}$ = Nusselt number

$\overline{Nu}_D$ = Nusselt number for downward heat transfer

$\overline{Nu}_H$ = Nusselt number for sideward heat transfer

$\overline{Nu}_U$ = Nusselt number for upward heat transfer

$NPSH$ = net positive suction head (Pa)

p = pressure, fluid mechanics applications (Pa)

$p =$ power density (W/cm^3)

$p =$ porosity

$P =$ probability that an arriving neutron at energy E suffers its next collision in region 1 of heterogeneous geometry

$P =$ rated power (MW)

$P =$ pitch (radial distance between pin centerlines) (m)

$P/D =$ pitch-to-diameter ratio

$P =$ pressure, thermodynamics and stress applications (Pa)

Pe $=$ Peclet number

Pe$_c =$ critical Peclet number

PLF $=$ processing loss fraction

Pr $=$ Prandtl number

$^{241}PuBOC = {}^{241}$Pu in the core and blankets at the beginning of the cycle (kg)

$^{241}PuEOC = {}^{241}$Pu in the core and blankets at the end of the cycle (kg)

$^{241}PuED =$ out-of-reactor decay of ^{241}Pu per cycle (kg/cycle)

$P_{fc} =$ fuel-cladding interface pressure (Pa)

$P_l =$ probability per year that an accident of type l will occur

$P_{NL} =$ nonleakage probability

$P_p =$ gas pressure in the fission gas plenum (Pa)

$P_r =$ reduced pressure

$P(V_z) =$ Maxwellian distribution of the z component of the target velocities

$P_{wi} =$ wetted perimeter in channel i (m)

$q =$ heat flux (W/m^2)

$q =$ pseudo-viscous pressure (Pa)

$q_{OP} =$ product of the nominal hot-channel heat flux and the overpower factor

$Q =$ number of batches and number of cycles for discharge fuel

$Q =$ activation energy (J/kg·mol)

$Q =$ volumetric heat source; power density (W/m^3)

$Q =$ heat transferred (J)

$\dot{Q} =$ rate of heat transfer (W)

$r =$ annual U$_3$O$_8$ requirement for a 1-GWe LWR plant (kg/GWe·y)

$R =$ annual U$_3$O$_8$ requirement for all LWR plants (kg/y)

$R =$ gas constant for a particular material (J/kg·K)

$R =$ potential scattering radius (cm)

$R =$ radius (m)

$R =$ swelling rate parameter in units of % per 10^{22} n/cm^2 (E>0.1 MeV)

$R =$ total risk (health consequences or \$/year)

$\mathbf{R} =$ universal gas constant, 8317 J/kg-mol·K

$\text{Ra} =$ Rayleigh number

$\text{Ra}_I =$ Rayleigh number for internally heated fuel

$\text{RDT} =$ reactor doubling time (y)

$\text{Re} =$ Reynolds number

$\text{RF} =$ refueling fraction

$R_c =$ hot cladding radius

$R_c =$ mid-wall cladding radius

$R_f =$ fuel pellet radius (m)

$R_f =$ hot fuel radius

$R_l =$ risk from accident of type l (health consequences or $/y)

$R_0 =$ core radius

$R_0 =$ radius of zero shear (m)

$s =$ coolant channel gap thickness (m)

$s =$ specific entropy (J/kg·K)

$s =$ wire wrap diameter (m)

$S =$ neutron source term (n/cm^3·s)

$S =$ weighted source term in perturbation theory expression

$S =$ number of statistical hot channel factors

$S =$ surface between cell regions (cm^2)

$\text{SDT} =$ system doubling time (y)

$\text{St} =$ Stanton number

$t =$ time (s, hr, or y)

$t_c =$ refueling interval (s or y)

$t_r =$ fuel pin time-to-rupture (hours)

$T =$ temperature (K)

$\text{TD} =$ theoretical density (kg/m^3)

$\dot{T} =$ heating rate (K/s)

$T_{b,0}$ (inlet) $=$ nominal value of the inlet bulk coolant temperature (K)

$T_{ci,m} =$ maximum value of the inner cladding temperature (K)

$T_{co,m} =$ maximum value of the outer cladding temperature (K)

$T_{\text{col}} =$ minimum temperature for columnar fuel (K)

$T_{\text{eq}} =$ minimum temperature for equiaxed fuel (K)

$T_{\text{eq}} =$ equilibrium sodium temperature (K)

$T_i =$ contact temperature (K)

$T_m =$ characteristic nuclear temperature for fissile isotope m (K)

$T_r =$ reduced temperature

$T_{\text{sat}} =$ sodium saturation temperature (K)

$T_{\text{SN}} =$ spontaneous nucleation temperature (K)

$T^+ =$ logarithmic temperature profile

$u =$ lethargy

$u =$ U$_3$O$_8$ commitment over the lifetime of a 1-GWe LWR plant (kg/GWe)

$u =$ velocity (m/s)

$u =$ axial velocity component (m/s)

$u =$ elastic deformation or displacement required to accommodate a negative G (m)

$u^* =$ critical superficial velocity of the lighter phase (m/s)

$u^* =$ shear velocity (m/s)

$u =$ radial displacement (m)

$U = U_3O_8$ commitment over the lifetime of all LWR plants (kg)

$U =$ internal energy (J)

$U^+ =$ universal velocity profile

$\bar{U} =$ average flow velocity (m/s)

$(U_I)_z =$ axial velocity in the central region (m/s)

$(U_{II})_z =$ axial velocity in the outer region (m/s)

$(U)_s =$ circumferential velocity in the outer region (m/s)

$v =$ neutron velocity (cm/s)

$v =$ flow velocity (m/s)

$v =$ transverse velocity component (m/s)

$v =$ volume per unit height (m³/m)

$V =$ coolant velocity (m/s)

$V =$ relative velocity (m/s)

$V =$ volume (cm³ or m³)

$\bar{V} =$ average axial coolant velocity (m/s)

$V_{cell} =$ total heterogeneous cell volume (cm³)

$V_f =$ final specimen volume (m³)

$V_p =$ velocity component perpendicular to the wire (m/s)

$V_z =$ z-component of target velocities (cm/s)

$V_z =$ axial velocity component (m/s)

$V_1 =$ fuel volume in heterogeneous cell (cm³)

$V_2 =$ cladding-coolant volume in heterogeneous cell (cm³)

$w =$ thickness of thin-walled vessel (m)

$w_{ij} =$ net diversion crossflow per unit height (kg/s·m)

$w'_{ij} =$ turbulent cross flow rate per unit length (kg/s·m)

$W = U_3O_8$ consumed by all LWR plants (kg)

$W =$ work done by expanding fluid (J)

$W(r, z) =$ normalized spatial variation of the Doppler effect

$W(E' \rightarrow E) =$ inelastic scattering kernel

We $=$ Weber number

$x =$ quality

$x =$ thermal entrance length, GCFR

$x =$ dimensionless energy variable (Eq. 5-33)

$x_i =$ triangular mesh boundary (m)

$X =$ neutron fluence in units of 10^{22} n/cm² >0.1 MeV

$X_i =$ flow distribution factor (or flow-split parameter) for channel i

$y =$ variable proportional to delayed neutron precursor concentration (W/cm^3)

$y =$ distance from wall (m)

$y =$ dimensionless energy variable (Eq. 5-53)

$y^+ =$ non-dimensional distance

$y/r_0 =$ normalized distance from the tube wall

$y_i =$ triangular mesh boundary (m)

$z =$ fluence (n/cm^2)

GREEK SYMBOLS

$\alpha =$ capture-to-fission ratio (σ_c/σ_f)

$\alpha =$ fraction such that $(1-\alpha)$ is the maximum fraction of energy that can be lost in an elastic scattering collision.

$\alpha =$ contact angle (for coolant boiling)

$\alpha =$ curvature parameter (for swelling) in units of $(10^{22}$ $n/cm^2)^{-1}$

$\alpha =$ input reactivity ramp rate $(\$/s$ or $s^{-1})$

$\alpha =$ mean linear thermal expansion coefficient (K^{-1})

$\alpha =$ thermal diffusivity $(k/\rho c_p)$ (m^2/s)

$\bar{\alpha} =$ average void fraction below height H

$\alpha_0 =$ volume of fission gas released per cubic meter of fuel at STP

$\beta =$ coefficient of thermal expansion (K^{-1})

$\beta =$ delayed neutron fraction

$\beta =$ over-relaxation factor

$\bar{\beta} =$ effective delayed neutron fraction

$\Gamma =$ total resonance width (eV)

$\Gamma_f =$ fission resonance width (eV)

$\Gamma_\gamma =$ radiative capture resonance width (eV)

$\Gamma_n =$ neutron width of the resonance level (eV)

$\Gamma_n^0 =$ reduced neutron resonance width $(eV^{1/2})$

$\delta =$ extrapolated distance (cm)

$\delta =$ roughness

$\delta r =$ radial component of displacement (m)

$\delta r =$ radial mesh spacing, cylindrical geometry (m)

$\delta z =$ axial component of displacement (m)

$\delta z =$ axial mesh spacing, slab geometry (m)

$\Delta =$ Doppler width (eV)

$\Delta p =$ pressure drop over core (Pa)

$\Delta p_r =$ form drag losses from the component of velocity in the direction perpendicular to the wire

Δp_s = skin friction losses

ΔT_{axial} = difference between outlet and inlet coolant temperature (K)

$\overline{\Delta T_b}$ = average temperature rise of bulk coolant (K)

$\Delta T_{b,\text{OP}}$ = product of nominal hot-channel coolant temperature rise and overpower factor (K)

$\Delta V/V$ = volumetric swelling

$\Delta D/D$ = linear swelling

Δy = mesh height, triangular mesh (m)

Δx = mesh width, triangular mesh (m)

ε = convergence criterion

ε = eddy diffusivity (m^2/s)

ε = relative roughness

ε = strain

ε_θ = hoop strain

ε_r = radial strain

ε_z = axial strain

ε^c = creep strain

ε^s = swelling strain

ε^T = total strain

$\varepsilon_{\text{ir creep}}$ = irradiation creep strain

$\varepsilon_{\text{th creep}}$ = thermal creep strain

$\dot{\varepsilon}$ = thermal creep rate (s^{-1})

ζ = thermal conductivity multiplier to account for winding sodium path

η = number of neutrons emitted in fission per neutron absorbed

θ = ratio of total resonance width to Doppler width

θ = Dorn parameter (hr)

λ = DeBroglie wavelength (m)

λ = delayed neutron decay constant (s^{-1})

λ = exponential constant for electrical growth rate (y^{-1})

λ_C = aerosol cleanup or fallout rate for a chamber (s^{-1})

λ_L = leak rate from the chamber (s^{-1})

$\lambda^{\nu+1}$ = ratio of the fission source of the $(\nu+1)$th iteration to that of the previous iteration

λ_R = radioactive decay constant for the isotope indicated (s^{-1})

λ_{tr} = transport mean free path (cm)

Λ = neutron generation time (s)

μ = reduced mass

μ = viscosity ($kg/m \cdot s$)

$\bar{\mu}$ = average cosine for elastic scattering

$\nu=$ kinematic viscosity (μ/ρ) (m^2/s)
$\nu=$ number of neutrons produced per fission
$\nu=$ Poisson's ratio
$\xi=$ average logarithmic energy decrement
$\rho=$ density (kg/m^3)
$\rho=$ reactivity (absolute or \$)
$\rho'=$ reactivity (Safety chapters) (absolute or \$)
$\rho_c=$ density of the continuous phase (kg/m^3)
$\rho'_d=$ disassembly feedback reactivity
$\rho'_{Dop}=$ Doppler feedback reactivity
$\rho_f=$ fuel density in the pellet (kg/m^3)
$\rho'_*=$ reactivity in excess of one dollar
$\rho'_{in}=$ positive reactivity insertion
$\rho_H=$ density of the heavy fluid (kg/m^3)
$\rho_L=$ density of the light fluid (kg/m^3)
$\rho'_0=$ reactivity at initiation of disassembly
$\sigma=$ stress (Pa or MPa)
$\sigma=$ surface tension (N/m)
$\sigma=$ microscopic cross section $(cm^2$ or barns)
$\sigma_0=$ background cross section $(cm^2$ or barns)
$\sigma_{ohetm}=$ background scattering cross section for fuel isotope m modified for heterogeneity $(cm^2$ or barns)
$\sigma_p=$ potential scattering cross section $(cm^2$ or barns)
$\sigma_r=$ radial stress (Pa)
$\sigma_\theta=$ hoop stress (Pa)
$\sigma_{x,DOP}=$ Doppler broadened cross section $(cm^2$ or barns)
$\sigma_Y=$ yield stress (Pa)
$\sigma_z=$ axial stress (Pa)
$\sigma(\infty)=$ infinitely dilute cross section $(cm^2$ or barns)
$\sigma^*=$ reference cladding hoop stress (930 MPa)
$\bar{\sigma}=$ effective stress (Pa)
$\sigma^2=$ variance
$\Sigma=$ macroscopic cross section (cm^{-1})
$\Sigma_{c,nf}=$ macroscopic capture cross section for the non-fuel material (cm^{-1})
$\Sigma_t=$ effective macroscopic total cross section for the homogenized cell (cm^{-1})
$\tau=$ incubation parameter (for swelling) in units of 10^{22} n/cm^2 $(E>0.1$ MeV)
$\tau=$ plant lifetime (y)
$\tau=$ reactor period (s)
$\phi=$ neutron flux $(n/cm^2\cdot s)$

$\phi(E) =$ flux per unit energy (n/cm$^2 \cdot$s$\cdot$eV)

$\phi(u) =$ flux per unit lethargy (n/cm$^2 \cdot$s)

$\phi_0(E) =$ asymptotic flux per unit energy (n/cm$^2 \cdot$s$\cdot$eV)

$\phi t =$ neutron fluence (n/cm^2)

$\phi^* =$ adjoint flux (n/cm$^2 \cdot$s)

$\chi =$ fission spectrum fraction

$\chi =$ linear power (kW/m)

$\chi_g =$ fraction of neutrons produced in fission that appear in group g

$\chi_m =$ linear power-to-melting (kW/m)

$\chi_{OP} =$ product of nominal hot channel linear power and overpower factor (kW/m)

$\psi =$ Doppler broadened line shape

$\psi =$ reactivity worth distribution of core material (reactivity/m^3)

SUBSCRIPTS

$a =$ absorption

$b =$ bulk coolant

$c =$ capture

$c =$ cladding

$c =$ core

$c =$ cold

$CL =$ centerline

$Dop =$ Doppler

$e =$ elastic scattering

$er =$ elastic removal

$f =$ fission

$f =$ fuel

$f_p =$ fission product pairs

$F =$ failure

$g =$ energy group index

$g' \to g =$ scattering from group g' to group g

$G =$ energy coarse group index

$h =$ hot

$i =$ delayed neutron group index

$i =$ coolant channel index

$i =$ inelastic scattering

$i =$ inside
$i =$ heavy metal isotope index
$i =$ region index
$i =$ mesh index, triangular mesh
$i =$ interface
$in =$ inlet
$ir =$ inelastic removal
$j =$ mesh index, triangular mesh
$j =$ coolant channel index
$k =$ hot channel effect
$k =$ chamber index
$m =$ material index
$m =$ maximum values
$n =$ incident neutron
$n, \gamma =$ capture
$N =$ nuclear
$o =$ outside
$o =$ background
$0 =$ center
$0 =$ initial value
$0 =$ standard temperature and pressure conditions
$0 =$ nominal value
$0 =$ cover gas
other $=$ all materials other than resonance material m
$p =$ fission gas plenum
$r =$ reflector
$r =$ rupture
$r =$ r component
$s =$ scattering
$s =$ surface
$t =$ total
$tr =$ transport
$T =$ total
$w =$ wall
$x =$ reaction type
$x =$ x component
$y =$ y component
$y =$ yield
$z =$ z component
$\infty =$ free stream properties
$\infty =$ fully developed flow, GCFR
$\theta = \theta$ component

SUPERSCRIPTS

k = spatial mesh index
q = batch index
Q = discharge batch
ν = iteration index
* = property in channel from which crossflow originates

COPYRIGHT ACKNOWLEDGMENTS

Acknowledgment is hereby expressed to the following publishers who have graciously granted us copyright permission to reprint their work. Direct references for the copyright material used are included in parentheses following each citation, e.g., Fig. 5-10 is taken from Reference 2 as listed in Chapter 5.

American Nuclear Society: Figures 5-10(2), 6-2(5), 8-9(13), 8-22(26), 8-24(26), 8-25(25), 8-26(25), 8-27(25), 10-15(28), 11-5(6), 11-6(12), 11-12(16), 11-15(20), 11-20(26), 11-24(31), 12-12(2), 12-14(2), 14-7(17), 15-9(32), 15-14(39), 15-17(53), 16-3(2), 16-6(14), 16-14(34), 16-15(29), 16-19(36), 16-25(46), Tables 4-5(9), 6-1(3), 7-3(9), 11-1(3), 11-2(3), 11-3(13), 11-4(12), 11-5(12), 11-6(13), 11-7(12), 11-8(12), 11-9(12), 13-4(11, 12, 13), 13-5(11, 12, 14), 16-2(36)

Addison-Wesley Publishing Company: Figure 8-4(5) and Table 1-3(1)

Academic Press: Figure 16-24(45)

American Institute of Physics: Figure 15-22(76)

Annual Reviews, Inc.: Figure 4-10(8)

Argonne National Laboratory: Figure 16-4(3) and Tables 7-6(11), 11-3(12, 13) and 11-6(13)

American Society of Mechanical Engineers: Figures 9-16(11), 12-19(6), 15-18(51), 16-9(20) and Table 10-2(27)

Ballinger Publishing Company: Table 1-6(3)

Brookhaven National Laboratory: Figures 2-7(3), 2-8(3), 5-4(14), 5-5(14), 5-6(14), 5-7(14)

Centre d'Études Nucléaires de Cadarache: Figures 8-2(2,3) and 10-10(18)

W. H. Freeman and Company: Chapter 3 quotation from Edward Teller "Energy from Heaven and Earth"

Harcourt, Brace and Jovanovich, Inc.: Chapter 3 quotation from John Maynard Keynes "The General Theory of Employment, Interest and Money"

International Atomic Energy Authority, Vienna: Figure 15-22(79)

Los Alamos National Laboratory: Figure 15-22(80)

North-Holland Publishing Company: Figures 8-5(6), 10-11(9), 10-12(1), 10-13(1), 11-4(2), 11-8(14), 11-9(14), 16-25(46)

Gesellschaft fur Kernforschung mbH, Karlsruhe: Figure 8-3(4)

National Safety Council: Table 13-1(5)

Prager Publishers, Inc.: Chapter 3 quotation from Kurt Mendelssohn "The Secret of Western Domination"

Random House, Inc.: Chapter 3 quotation from Adam Smith, "An Inquiry into the Nature and Causes of the Wealth of Nations"

Sandia National Laboratory: Figure 15-22(77)

United Kingdom Atomic Energy Authority: Figures 9-2(2) and 15-22(78)

Westinghouse Electric Corporation: Figures 9-9(6), 9-20(6) and Tables 2-2(2), and 6-3(9)

John Wiley and Sons: Figure 13-1(4)

INDEX

ABOUT THE AUTHORS

Alan E. Waltar is Manager of Fast Reactor Safety Development at the Hanford Engineering Development Laboratory, Westinghouse Hanford Company, Richland, Washington. He received his B.S. (Electrical Engineering) from the University of Washington, an M.S. (Nuclear Engineering) from the Massachusetts Institute of Technology, and his Ph.D. (Engineering Science) from the University of California at Berkeley. He developed large neutronics, thermal-hydraulics computational systems for fast breeder reactors at Battelle-Northwest (1966–1970), served as Manager of Reactor Dynamics at the Westinghouse Hanford Company from 1970 to 1976, and accepted a one-year appointment as Visiting Associate Professor of Nuclear Engineering at the University of Virginia for the 1976–1977 school year. Dr. Waltar teaches part-time at the Joint Center for Graduate Studies (an extension of Oregon State University, Washington State University, and the University of Washington) in Richland, Washington. He is active in the American Nuclear Society, having served as program chairman and executive committee member of the Nuclear Reactor Safety Division, as well as associate technical program chairman of the ANS/ENS 1979 International Meeting on Fast Reactor Safety Technology. He is also a member of the American Association for the Advancement of Science.

Albert B. Reynolds is a Professor of Nuclear Engineering in the School of Engineering and Applied Science at the University of Virginia in Charlottesville, Virginia. He received his S.B. (Physics), S.M. (Nuclear Engineering) and Sc.D. (Chemical Engineering) degrees from the Massachusetts Institute of Technology. Dr. Reynolds was with the General Electric Company in San Jose and Sunnyvale, California from 1959 to 1968 in the fast breeder reactor program where he was Manager of the SEFOR and Reactor Plant Physics units. He joined the faculty of the University of Virginia in 1968 where he introduced courses in fast reactor technology and developed a research program in thermal hydraulics aspects of fast reactor safety. He has been a frequent consultant to the staff of the U.S. Nuclear Regulatory Commission on LMFBR safety and, in 1979, was a consultant to the technical staff of the President's Commission on the Accident at Three Mile Island. He spent 1977 on fast reactor safety research at the Centre d'Etudes Nucléaires de Grenoble in France. He is a member of the Executive Committee of the Nuclear Reactor Safety Division of the American Nuclear Society and is a member of the American Society of Mechanical Engineers.